Evaluating Contract Claims

Evaluating Contract Claims

John Mullen BSc(Hons), MSc, FRICS, FInstCES, FCIArb
and
R. Peter Davison BA, MSc, DipArb

Third Edition

WILEY Blackwell

This edition first published 2020
© 2020 John Wiley & Sons Ltd

Edition History
2nd Edition: Blackwell Publishing 2009
1st Edition: Blackwell Publishing 2003

The right of John Mullen and R. Peter Davison to be identified as the authors of this work has been asserted in accordance with law.

Registered Offices
John Wiley & Sons, Inc., 111 River Street, Hoboken, NJ 07030, USA
John Wiley & Sons Ltd, The Atrium, Southern Gate, Chichester, West Sussex, PO19 8SQ, UK

Editorial Office
9600 Garsington Road, Oxford, OX4 2DQ, UK

For details of our global editorial offices, customer services, and more information about Wiley products visit us at www.wiley.com.

Wiley also publishes its books in a variety of electronic formats and by print-on-demand. Some content that appears in standard print versions of this book may not be available in other formats.

Library of Congress Cataloging-in-Publication Data

Names: Mullen, John, 1959- author.| Davison, R. Peter, author.
Title: Evaluating contract claims / John Mullen, BSc(Hons), MSc, FRICS,
    FInstCES, FCIArb and R. Peter Davison, BA, MSc, DipArb.
Description: Third edition. | Hoboken, NJ, USA : Wiley-Blackwell, 2020. |
    Revised edition of: Evaluating contract claims / R. Peter Davison and John
    Mullen. 2009. | Includes bibliographical references and index. |
Identifiers: LCCN 2019017935 (print) | LCCN 2019018685 (ebook) | ISBN
    9781118918135 (Adobe PDF) | ISBN 9781118917800 (ePub) | ISBN 9781118918142
    (hardback)
Subjects: LCSH: Buildings–Maintenance–Costs. | Construction contracts. |
    Construction industry–Law and legislation. | Remedies (Law) | Claims.
Classification: LCC TH425 (ebook) | LCC TH425 .D38 2020 (print) | DDC
    692/.8–dc23
LC record available at https://lccn.loc.gov/2019017935

Cover Design: Wiley

Set in 10/12pt WarnockPro by SPi Global, Chennai, India

10 9 8 7 6 5 4 3 2 1

To Anne and Melanie

For their patience and forbearance during our long absences on projects locally and internationally.

Contents

Reviews

The previous editions have helped numerous construction professionals to apply a consistent approach to valuing claims based on good practice and founded on sound legal principles.

The third edition includes many necessary updates to deal with new standard forms, recent court decisions in the UK and overseas, the changes to the SCL Protocol, changes to the Rules of Measurement and the introduction of International Construction Measurement Standards. Above all, the third edition has now taken on an increased international tone.

The book includes detailed assistance in evaluating every type of claim which practitioners will encounter. These vary from evaluating direct or time consequences of claims to evaluating termination claims or post completion claims. The third edition also refers to claims arising from calls on bonds. No matter what type of claim, whether straightforward valuations of variations or problems of complex delay and disruption claims, the third edition of the book contains a comprehensive guide to the current state of knowledge and experience in evaluating those claims.

Whilst the previous editions were more focussed on the UK domestic market, they were used overseas by practitioners who needed to have access to specialist knowledge on evaluation of claims. This has necessarily led the authors to provide more assistance to those dealing with claims overseas. The third edition should therefore find a place, worldwide, on the shelves of all those involved in evaluating claims whether construction professional or construction law practitioners. John Mullen and Peter Davison are to be commended for providing such a practical and informative book which will assist in claims being properly formulated and established under construction contracts.

Sir Vivian Ramsey QC
Former England and Wales High Court Judge, and International Judge, Singapore
Joint editor, Keating on Construction Contracts

The authors' extensive experience is obvious from the book which contains a meticulous analysis of all (or, if not all, the vast majority of) the potential heads of claims which arise in construction disputes. The book provides an excellent route map through the issues which arise when quantifying a construction related claim and should be the starting point if you are advising on, evaluating or making a claim, whether that be a 'typical' final account claim involving claims for variations and prolongation costs or a more unusual claim involving a call on a bond or incomplete or defective work.

Rachel Ansell QC
Barrister, 4 Pump Court
Construction Silk of the Year, Chambers UK Bar Awards 2017

The main purpose of a review, is to inform the potential reader whether the book is worth his or her time and money. The answer for this magnificent book is undoubtedly 'yes'. It is not a 'how to' guide for preparing claims. Rather, it provides carefully thought-out explanations of the various types of valuation issues which arise on construction projects. The book provides in-depth analysis of many of the most conceptually difficult subjects, such as disruption, acceleration and global claims, by reference to the available legal authorities, the authors' considerable experience, and practical worked examples. There will be few people involved with the construction industry who would not benefit from careful study of this comprehensive work.

Richard Harding QC
Barrister, Keating Chambers
Founding and current chairman of the Society of Construction Law (Gulf)

As with previous editions, the particular value of this work is the perspective which the authors bring to the discussion, particularly their long experience as practitioners, consultants, experts, adjudicators and arbitrators in the field. They address an impressively wide array of topics. Their method is to bring together for each topic not only a clear exposition of the issues which arise but also a comprehensive review of the latest authorities. The authors do not hold back in discussing the difficulties which face the practitioner. The result is a valuable work of reference for anyone concerned in the pursuit of such claims.

John Marrin QC
Barrister and Chartered Arbitrator, Keating Chambers

This is the most wide ranging and detailed reference source on the perennial issue of quantifying construction contract entitlements. It is surely a 'must have' for all those involved in what the authors call 'the claims industry'.

This edition has been expanded to cover such 'hot topics' as acceleration, termination and duplication. It now covers new editions and forms of contract and of QS 'standards' and computer modelled evaluation methodologies. Perhaps most valuable of all are the practical insights from these experienced global experts – for example illustrations from their experience of the different approaches of international arbitration tribunals to that of an English Court Judge.

Neal Morris
Partner and Head of Construction Advisory & Disputes, Pinsent Masons LLP

Previous versions of Evaluating Contract Claims have been an indispensable resource for my team and I. I am happy to see that while there have been substantial updates in the 3rd edition, notably with respect to adding a more international dimension, the authors have retained their focus on the fundamentals of claim quantum evaluation which makes this book so useful to anyone working with building or engineering contracts.

This book successfully integrates the evolution of claims evaluation while reflecting the depth and breadth of the authors' experience domestically and internationally. A particularly welcome addition is the chapter on termination claims given the complexities involved in such claims.

Erin Miller Rankin
Global Partner, Freshfields Bruckhaus Deringer LLC

John Mullen's and Peter Davison's Third Edition of 'Evaluating Contract Claims' is a must have for those in the construction industry, and is also an ideal reference for those lawyers looking to be involved with construction disputes. This weighty and comprehensive reference book authored by two well-known and respected experts would appear to be primarily directed at those involved with disputes and claims. It is, but there is a lot that is equally relevant and helpful to those that prepare contract documentation either on the ultimate client's behalf or those further down the construction chain.

Paul Barry FRICS
Founding partner, GBsqd LLP

After an almost apologetic introduction, this book turns out to be impressively comprehensive. The logical progression in structure is retained from previous editions – starting with the need to understand the effects of change and disruption, before moving on to the basis for evaluating those effects and the standards of substantiation needed not only to succeed in achieving compensation, but also to guide those on the receiving end as to what they can and should expect before granting or recommending that compensation. The book concludes with an up-to-date analysis of the movement towards partnering and the forging of better relationships within the construction industry under the auspices of novel contractual arrangements.

Christopher Ennis MSc FRICS FCIArb
Past chairman of the Society of Construction Law
Director, Time | Quantum Expert Forensics Ltd

Preface

The idea for the first edition of this book originated in spending a number of years in countless meetings on construction and engineering sites and in the offices of contractors, client organisations, consultants, and solicitors. Many of these meetings were centred around the need to define the amount of compensation due as a result of claims made for additional payment under particular forms of contract. A recurring theme was the lack of understanding among parties to such contracts and their administrators as to what could, and could not, be included in such claims. The causes of the claims varied, although certain themes recurred, and they were generally well documented. However, the calculation of the quantum was often vague and lacking in substantiation, often without any reference to the contract terms and conditions, and with an equal lack of regard for any underlying principles of quantification.

Since the first edition, and indeed the second, much has changed; however, much has remained the same. As one travels the world, the same problems of quantification appear but exacerbated by the lack of local guidance in many jurisdictions on the principles to be adopted, whether through the local courts or through authoritative publications. Much of the guidance in some parts of the world is taken from the UK because it is more mature and developed in relation to construction and engineering contract law. Furthermore, this often means that, in addition to English being adopted as the language of the contract (as an international language for an international project), English law is adopted as the substantive law of the contract. As a result, projects can be found in all parts of the world where UK practitioners are employed on commercial and claims management. In many cases, they bring good knowledge and practice. In this regard, the growth of the Society of Construction Law and the increasing international reference to its Delay and Disruption Protocol are particularly beneficial forces. However, this internationalisation can also bring some bad habits, for example, where outdated quantification methods are imported or English legal principles are applied that have no place in the context of the applicable law.

This book quotes widely from the decisions of the UK courts, and to a lesser extent some Commonwealth countries. It is recognised that these may not be relevant in a particular overseas jurisdiction. As a retired South African High Court judge once said to two English programming experts who were in the 'hot tub' explaining to his arbitral tribunal how delay analysis should be carried out by reference to the UK judgments: 'Gentlemen can we just for a moment suspend your reality and pretend that we are actually here in South Africa?' Notwithstanding this humorous interjection, that arbitration saw extensive reference to the SCL Protocol and UK court judgments on a number of the more difficult quantification issues before it, both in relation to time and money.

Subject to the overriding consideration of the applicable law and express terms of the contract, in this book, we have attempted to set out some of the principles and methods that we believe should be reflected in the evaluation of claim quantum and the standard of substantiation that may ultimately be required. The views are our own and we naturally accept that other views are also possible on some issues and that in an international context applicable laws will be paramount. We have tried to indicate where those alternative views may lie. We also accept that some of our views may not be received too happily by some others and have tried to explain the reasoning for our views, in the hope that they will, at least, promote some constructive thoughts and considerations.

There is also a large degree to which in the quantification of construction and engineering claims, 'one size does not fit all'. Variables between different claims will include the applicable law, the express contract provisions, the facts of the case, the records that are available, the relationship between the parties, and the need for proportionality of approach. As a result, in many areas, this book sets out a number of alternative approaches, any of which could be applicable in the right circumstances, depending on such variables.

There are many volumes available on the subject of construction and engineering contract claims, but most of them concentrate on the establishment of legal liability or are concerned with the requirements of a particular standard form of contract. In this book, we have tried to concentrate on the quantification of claims after legal liability has been established, by setting out matters of principle that may apply regardless of the form of contract used. This said, all claims are prepared within a contractual framework, and therefore, some standard forms of contract are used as illustrations.

It is almost ten years since the second edition of this book, and since then, much new has arisen on its subjects. This includes seismic economic developments, extensive new and revised standard forms of contract, voluminous new measurement rules, and the continuing attention of the commercial courts to construction and engineering disputes, particularly in the UK.

The worldwide economic crash of 2007 and 2008 saw dramatic changes in the business climate of many markets. Construction and engineering projects were suspended or terminated, mostly temporarily, but some permanently. This gave rise to a spike in claims related to suspension and termination. Accordingly, this third edition adds a new chapter devoted to the quantification of termination claims (Chapter 7) and includes a new section on suspension claims (in Chapter 8).

After lengthy consultation and debate, in February 2017 the Society of Construction Law published the long awaited second edition of its *Delay and Disruption Protocol*. This continues the good work of the first edition in providing much needed practitioner consensus as to good practice on many aspects of the evaluation of delay and disruption claims. The fact that the SCL Protocol provides useful guidance on resolving issues in relation to these topics is reflected regularly in international arbitrations and can be seen in court decisions in both the UK and abroad over recent years. That said, it must be considered as nothing more than guidance. One of the ironies of its increasing currency is how when many UK practitioners had long concluded that the Protocol's first edition's preference for 'time impact analysis' was misplaced, that method was being increasingly relied on by parties overseas by reference to its first edition's preference for it.

There is still no magic formula in relation to the quantification of disruption to the productivity of labour, plant, or equipment and this remains a thorny issue. The last ten years have seen some rise in the use of computer models and reliance on, for example, Cumulative Impact and System Dynamics Modelling as the basis of disruption claims. The extent to which these have helped or hindered those seeking to reach agreement of the quantification of disruption is questionable. They are the subject of a new section in Chapter 6.

Global claims, particularly for disruption, continue to be the subject of heated debate in the resolution of contract claims worldwide. Many employers and their advisers seem to believe that attaching that term as a pejorative label kills a claim at birth. Contractors respond with arguments such as that the claim is not global, global claims are allowed, the claim has been particularised as far as practicable, or its global nature is the result of the extent of the failures by the employer and its consultants. However, such excuses sometimes hide other motives. In the UK, the continuing difficulties on this topic are notwithstanding significant judgments addressing global claims in the UK courts over the past few years.

As a result of the above, Chapter 6 has been significantly re drafted and expanded to cover difficult heads of claims for the time consequences of change, such as disruption, acceleration, and overheads and profit. This chapter also continues our previous historical analysis of the topic of global claims, which is brought up to date and extended to cover related terms such as 'total loss' and 'total costs'.

One area of the quantification of contractual claims that is hardly ever written about is the need to avoid duplication between the quantification of different heads of claim, particularly on a major project with large numbers of claim heads. Chapter 6 therefore includes a new section on some of the common areas of duplication and how suitable adjustment might be made.

In terms of the standard forms of contract, changes have arisen apace since our second edition. The NEC suite of contracts, first published in 1993, continues to gain attention both in the UK and internationally. They are now the only forms of contract fully endorsed by the UK Office of Government Commerce, which recommends them for all public sector construction projects in the UK. In 2009, they became the contracts of choice of the Institution of Civil Engineers, which withdrew from the ICE Conditions of Contract. In 2017, the NEC launched its new 'complete family of contracts', NEC4. Worldwide, the NEC contracts are currently in use in more than 20 countries. The proactive and prospective approaches of NEC have many benefits in the avoidance of disputes in evaluating contract claims, but they also give rise to problems of their own if they are not properly followed. Furthermore, the degree to which, in practice, these approaches suit the prevalent commercial cultures of some international markets is a matter of some debate.

Since the Institution of Civil Engineers withdrew its support for its conditions of contract, they were passed to the Association for Consultancy and Engineering and the Engineering Contractors Association, who published their Infrastructure Conditions of Contract in 2011. Since then, they have continued to publish further forms, with a significant revamp in 2014. In this book, the Infrastructure Conditions of Contract,

Measurement Version 2011, is referred to as an example of a remeasurable form of contract in use in the UK.

December 2017 witnessed the launch of the new FIDIC suite of contracts as updates to its 1999 'Rainbow Suite'. However, the 1999 suite is still in some use and forms the basis of many bespoke contracts around the world, particularly for government projects. In this book, the 1999 edition of the FIDIC Red Book is therefore referred to as an example of a remeasurable form of contract, which has been used in more than 160 countries.

The Joint Contract Tribunal has also been busy since the second edition of this book. A new suite of standard contracts was published by the JCT in 2011. In 2016, an updated set was published. In this book, the 2011 edition of the Standard Building Contract with Quantities (SBC/Q) is referred to as an example of a lump sum form of contract in use in the UK.

Given the rise in popularity of the NEC suite of contracts, NEC4-ECC is also referred to in relation to a number of types of claims, by a particular reference to its Option A: Priced contract with activity schedule. Features that are particularly covered are its prospective valuation of what NEC refers to as 'compensation events' by reference to the actual or forecast costs, rather than contract rates and prices, and time assessment by impacting the contractor's current accepted programme.

Other relevant market changes since our second edition are in relation to the measurement of construction work in February 2009 when the Royal Institution of Chartered Surveyors published its New Rules of Measurement. These new rules comprise more than 1000 pages of rules for measurement and the preparation of cost estimates and elemental cost plans. July 2017 witnessed the publication of the International Construction Measurement Standards by a worldwide coalition of bodies with the aim of creating a global common standard for reporting and costing construction works. The extent to which this improves things such as cost prediction, management, accounting, and benchmarking with a positive impact in relation to the incidence of contract claims and their evaluation is yet to be seen.

We have also taken the opportunity to bring the text up to date and reflect some pertinent decisions of the UK commercial court. These courts have continued to be busy considering issues arising out of construction and engineering contracts. Much of this has addressed subjects outside the remit of this book, such as the UK's security of payment legislation and the enforcement of adjudication decisions, but there have been a number of decisions that are significant to the evaluation of contract claims. Particular issues addressed here include 'penalties', the valuation of omissions, global claims, claims for wasted management time, and head office overheads. In addition to UK judgments, this book also considers a few interesting overseas decisions.

We also set out some examples of our experience of how arbitration tribunals in international arbitration have resolved some issues of claim quantification. This is not because those awards are in any way even persuasive to future tribunals but because they may illustrate the potential different approaches to some heads of claim that practitioners could adopt and how other tribunals might address similar issues.

As part of the increased international tone of this third edition, Chapter 8 addresses some heads of claim that were not in the second edition because they are rare in the UK but common overseas. These include, for example, claims in relation to the extension or calling of various types of contractual bonds and for the contractor's additional costs after handover of the works.

In terms of statutes, 14 countries have now passed security of payment legislation similar to the UK's Housing Grants, Construction and Regeneration Act 1996. In fact, much of this is even more 'on point' in relation to payments because it focusses on security of payment and does not also address the UK Act's other topics of grants and repairs. These statutes can have significant impacts in relation to rates of interest and also in the consequences of the adoption of adjudication as a means of temporarily resolving disputes. Even where there is no statutory legislation, recent years have seen increasing adoption of ad hoc adjudication procedures internationally. As explained in this book, this can have significance to the quality of claim submissions, particularly from subcontractors.

Information technology has played an increasing role in relation to construction and engineering contract claims, particularly in the form of Building Information Modelling, including the use of multidimensional modelling. Although the extent of use of such systems for creating, modelling, and managing information on construction and engineering projects might best be described as 'embryonic' in some countries, if it lives up to its billing, then it may have profound implications for the extent to which change occurs on engineering and construction projects, particularly unplanned change, and also the methods adopted to evaluate the consequences of such change where it does occur.

Although this book is largely written with reference to claims between contractors and employers, many of the discussions and methods are equally applicable in relation to claims between contractors and subcontractors and even suppliers.

John Mullen
R. Peter Davison

Acknowledgements

We would like to acknowledge all the help, advice, assistance and experiences we have received or gained from all those we have worked with down the years. In particular, our colleagues in Driver Group plc and the dedicated expert witness service, Diales, for their constructive comments and encouragement in producing this edition.

While not detracting in any way from the acknowledgements above, we should make clear that the opinions expressed and contents are the responsibility of the authors, as are any errors or omissions.

The authors and publisher would like to express their thanks to the following organisations for permission to reproduce extracts from their publications.

The Society of Construction Law (www.scl.org.uk), for permission to reproduce material from the second edition of the Delay and Disruption Protocol.

NEC Contracts (www.neccontract.com), for permission to reproduce material from the NEC4 Engineering and Construction Contract (ECC).

The International Federation of Consulting Engineers (FIDIC, www.fidic.org) for permission to reproduce material from the FIDIC Construction Contract (1999 Red Book).

The Association for Consultancy and Engineering (www.acenet.co.uk) for permission to reproduce material from the Infrastructure Conditions of Contract Measurement Version (August 2011).

1

Introduction

It may be thought that there is enough literature on claims in the construction industry, although the continuing incidence of disputes arising from such claims suggests that recent developments in the means of addressing such problems have not eliminated contentious claims. That such disputes feed commercial courts and arbitration centres of many countries with a ready supply of cases to hear at great expense further emphasises the point. This book aims to examine the quantification of contract claims on the basis that many disputes arise from disagreement of the financial consequences of events, even where the liability for those events may not be contested.

The objective of this text is to examine various aspects of evaluating claims for additional reimbursement arising from contracts for construction projects. There is no intention to produce a legal treatise or to address the issues of establishing liability for additional reimbursement. That said, the operation of any contractual machinery relies on the express terms of that contract and the legal background. It is of course necessary to have a basis for considering how remuneration should be properly established. Whilst a few overseas judgments are also considered, this text considers the issues assuming English law applies and is therefore referred to, where appropriate, to establish relevant authorities.

Before commencing any evaluation it is preferable if the person undertaking the task understands how change and disruption to a contract can arise in a manner that requires evaluation of the financial consequences on behalf of one party or another. This chapter briefly considers aspects of the process that provide the basis for evaluation.

Succeeding chapters then go on to consider how the base from which evaluation of additional payments may be established, the effect of changes on the programme of work, the sources of information for evaluation of additional payments, the evaluation of the direct consequences of change in terms of the impact on unit rates, etc., and the evaluation of the time consequences of change in terms of such as prolongation, disruption, acceleration, etc. Some other sources of claims (such as suspension and termination) and the means of minimising the impact of claims are also considered.

The approach taken is to attempt to demonstrate the process, principles and standard of analysis that will be required to produce acceptable claims for additional payment, not to produce a guide to calculating payments under any specific form of contract. The approach does, however, provide those on the receiving end of such claims with guidance on what they should expect to receive in the form of a properly detailed claim and also how to respond to claims that are not properly supported. We have also set out

Evaluating Contract Claims, Third Edition. John Mullen and R. Peter Davison.
© 2020 John Wiley & Sons Ltd. Published 2020 by John Wiley & Sons Ltd.

alternative approaches to many of the claims considered, including some unusual and hopefully thought provoking methodologies.

1.1   The Legal Basis

This is not a legal textbook and it goes without saying that proper advice on the applicable law should always be sought before taking any contractual position based on a legal premise. There are, however, many references in the text to the decisions of the courts in relation to a number of matters, with relevant extracts from the judgments. These extracts and quotations are included to illustrate the various principles under discussion and to underline the standard of analysis and substantiation that is required for claims taken before a formal tribunal. There is no better source for this purpose than published judgments, and the standard required by the courts is the standard by which all evaluations can be judged. However, whilst English legal principles and precedents may have significance in Commonwealth countries and be of interest elsewhere, the difference particularly with codified civil jurisdictions can be marked. Local legal advice must always be sought, particularly from lawyers with experience of construction contracts and disputes. This latter criteria can, however, sometimes be difficult to satisfy in some parts of the world.

The case references and extracts are not intended to be exhaustive but are intended to provide a basis for the reader to conduct further research if he or she so wishes. Full details of cases can be obtained through the internet from sites such as the British and Irish Legal Information Institute (BAILII) (www.bailii.org). Other sites are available both free of charge and commercially.

1.1.1   Forms of Contract

The number and range of published standard forms of contract for construction works are extensive. Not only does this text not address all of the many published forms, it is not a guide to any one of the more commonly used forms. The intention of this book is to provide guidance on matters of principle that will have to be addressed under most, if not all, construction contracts under English law. That said, it is obviously useful to apply the provisions to be found in different types of contract to illustrate various points. References are therefore made in the text to the following contracts, using the abbreviations shown below, to show the way in which they deal with specific issues:

'Infrastructure Conditions'	Infrastructure Conditions of Contract published in August 2011, Measurement Version, published by the Association for Consultancy and Engineering (ACE) and Civil Engineering Contractors Association (CECA).
'FIDIC Red Book'	The Fédération International des Ingénieurs-Conseil's (FIDIC) Conditions of Contract for Construction, for Building and Engineering Works designed by the Employer, First Edition 1999.
'SBC/Q'	The Joint Contract Tribunal's Standard Building Contract with Quantities (SBC/Q), published in 2011 by Sweett and Maxwell. The successor to the previously published 'JCT' Standard Form of Contract.
'NEC4-ECC'	The Engineering and Construction Contract, Fourth Edition 2017, published by Thomas Telford Ltd. In this book references to 'NEC4-ECC' are generally to its Option A: Priced contract with activity schedule.

The FIDIC Red Book and the Infrastructure Conditions are used to illustrate how contracts that contemplate complete remeasurement of the works address certain evaluation issues, internationally and in the UK domestic market respectively. SBC/Q illustrates the approach of lump sum contracts subject to adjustment under stated circumstances. The NEC Form of Contract is used to examine some of the concepts that have gained this contract increasing popularity over recent years, particularly among critics of what are considered by some to be the more adversarial traditional forms of construction contracts. For example, NEC's proactive and prospective approach to the evaluation of change and its use of actual costs, or forecast actual costs, rather than contract rates and prices, as a basis of remuneration for change.

There are, of course, many different forms of contract (both standard and bespoke) that can be adopted by the parties to a construction project depending upon, among other matters, the nature of the enterprises concerned and the nature and size of the project. To consider the detailed requirements of every standard form of contract would need a considerably larger volume than this. It would also require detailed consideration of the applicable law. It is therefore necessary to restrict the consideration to matters of principle, using the requirements of the various contracts considered in this book to illustrate particular points. That is not to say that the principles examined will not relate to other standard forms of contract, or to ad hoc contracts agreed between parties, but that the discussions herein will need to be considered in the light of specific requirements in particular contracts as well as the underlying applicable law. The prime source of information for any evaluation has to be the contract between the parties and its requirements. There is no substitute for reading the contract and any incorporated relevant documents. Regrettably, this is often a starting point more often honoured in the breach than in observance in practice. Domestically, the early use of the NEC contracts approach saw many practitioners applying the traditional approaches of such as the JCT and ICE contracts that they had grown up with, without reading what NEC actually requires them to do and when. As the NEC contracts have gained popularity internationally, many of the problems arising have been the result of practitioners applying approaches that were accustomed to conform to more traditional international forms such as the FIDIC Red Book. These included the very approaches that NEC sought to change. Internationally, many practitioners also fail to consider the applicable law and assume that the law and approaches of their home country will apply. Where such naivety and/or laziness leads to otherwise avoidable disputes is unforgivable.

The parties to a contract can of course agree additional reimbursement in any manner they wish, and can also waive the requirements of their contract if that is expedient and acceptable to both parties. This is often the case in commercial negotiations of additional reimbursement, where the parties may not wish to insist on the detailed substantiation of every component of the evaluation.

This text, however, assumes that the evaluation needs to be substantiated in detail to the standard required in formal dispute resolution procedures under English law. These are also the methodologies that the authors have found are regularly applied internationally. A theme of this book is the benefit that can be obtained by good substantiation in avoiding unnecessary disputes. Such a standard is not only necessary in the event of some form of dispute procedure but is of course the standard of substantiation required by the contract itself. This raises the question of defining the standard required in a formal dispute resolution process.

1.2   The Standard of Substantiation

While there may be many facets to the standard required, there are two general principles that should always be borne in mind:

- The first principle is that he or she who asserts must prove, i.e. the party claiming an item of damage, cost expense, loss or value, will have to support it with evidence.
- The second principle is the general standard of proof in many jurisdictions, as it is put in English law, that matters need to be established as being correct 'on the balance of probability'. This contrasts with the standard required in criminal matters where 'beyond reasonable doubt' is the test under English law. This second principle might, however, be subject in practice to a 'sliding scale', i.e. major and central parts of the issues need to be fully substantiated while ancillary or subsidiary parts may be subject to a lesser degree of substantiation. Those minor parts may perhaps be assessed by reference to the results of the more rigorous analysis of the major parts, perhaps on a pro rata basis. If a sufficient and representative sample of preliminaries and general item costs in a prolongation claim have been agreed following detailed checking at $x\%$ of their claimed values, then it might be concluded that the unsampled costs could also be agreed at that $x\%$.

The apparently lower standard of proof in civil matters does not imply that assertions need not be fully evidenced where it is reasonable to expect such evidence. Thus, for instance, a matter of evaluation that involves establishing the cost of materials bought specifically for a contract will require production of invoices and possibly other procurement documents (such as a matching purchase order and delivery receipt note) if relevant. Where such project-specific support for a claimed item of cost is not possible, for instance in establishing off-site overhead charges in a prolongation evaluation, it will still be necessary to produce evidence of the claimant's wider company overhead costs incurred, such as audited company accounts, with a reasoned analysis of the amount considered to be relevant to the claim.

This introduces the two tiers of evaluation common to most change items: its direct consequences in terms of the value of the physical work done and its time consequences in terms such as delay, disruption and acceleration. In many instances the evaluation may require only one or the other but in many cases both tiers will be necessary.

The level of substantiation for the evaluation may vary depending upon the particular instance and circumstances. As noted elsewhere, the express terms of the contract should always be a first point of reference. Local law and other authorities may also be relevant. Internationally there is often a lack of express provision in the contract and also guidance by way of interpretation from the local courts and other authorities. One of the great benefits of English law is that these do exist, as highlighted throughout this book.

For example, in *C.J. Sims Ltd v Shaftesbury PLC* (1991) 60 BLR 94, deciding what was meant by the expression, 'such costs to include loss of profit and contributions to overheads, all of which must be substantiated in full to the reasonable satisfaction of our quantity surveyor', Judge John Newey stated:

> Its words are peremptory – 'all … must be substantiated in full' and the substantiation is to be 'to the … satisfaction of (the defendants') quantity surveyor'. The only qualification is that the quantity surveyor cannot require more than is 'reasonable', which I think means that he cannot require more than the ordinary competent quantity surveyor would.

In this provision, the extent of substantiation to be produced, in the absence of specific requirements, is therefore that required by the ordinary competent quantity surveyor, and it is that substantiation that is the subject of this book.

Having considered the standard to which substantiation is required for such evaluations, the matter arises of the extent of support or analysis deemed necessary to establish that any particular sum would satisfy the principle. Thankfully, the courts have also had to consider such support and analysis by experts on a regular basis and have given useful guidance to those seeking to present reasoned evaluation of claims for additional payment.

For example, in *McAlpine Humberoak Ltd v McDermott International Inc.* (1992) 58 BLR 1, during the course of considering a decision by an Official Referee relating to the analysis of time, delay and disruption in a contract for the fabrication of steel sections of deck for an offshore drilling platform, the Court of Appeal made the following comment on the evidence given by one party's expert and the judge's treatment of that evidence:

> The judge dismissed the defendant's approach to the case as being 'a retrospective and dissectional reconstruction by expert evidence of events almost day by day, drawing by drawing, TQ by TQ [technical query] and weld procedure by weld procedure, designed to show that the spate of additional drawings which descended on McAlpine virtually from the start of the work really had little retarding or disruptive effect on its progress'. In our view the defendant's approach is just what the case required.

While these comments relate to the examination of time and the analysis of delay and disruption, there is no reason to believe that similar comments would not have been made in respect of the calculation of additional payment. However, the *McAlpine Humberoak* case was decided before the introduction of the Civil Procedure Rules 1998 (CPR), following the review of the litigation system by Lord Woolf and the concept of proportionality as an overriding objective in civil litigation, i.e. that the amount of analysis and evidence should be proportionate to the issues in question. Rule 1.1(c) of the CPR requires cases to be dealt with in ways that are proportionate to the amount of money involved, the importance of the case, the complexity of the issues and the financial position of the parties. It is therefore possible that a lesser standard may be satisfactory in some circumstances but that is unlikely to mean that the level of analysis and evidence will be materially reduced, or that evidence that should be available and would be expected by the ordinary competent quantity surveyor, e.g. invoices, receipts, etc., will not be required. A sampling approach might particularly be appropriate. For large sums of money it may be 'proportionate' to expect full substantiation, but lesser sums may be addressed by an abbreviated method. For instance, if the cost of additional visits to site by engineers has been established as being a necessary part of the claim and the costs of the engineer's time has been fully substantiated, it may be quite reasonable to simply present the travel expenses as a schedule without producing every receipt and invoice. Such costs are generally known, and any exceptional differences should be recognisable without production of a full 'audit trail'.

1.2.1 SCL Delay and Disruption Protocol

The Society of Construction Law (SCL) is a body of lawyers, surveyors, engineers, architects and others with an interest in the subject of law as applied to construction projects.

Whilst it is UK based, the SCL has various branches around the world and its work is referred to regularly by practitioners as a source of some authority in relation to the evaluation of construction contract claims. Its international reach and influence continues to grow.

In October 2002 the SCL published the First Edition of its Delay and Disruption Protocol ('the Protocol') which deals with the analysis of those matters and the compensation that may be due when they occur. The First Edition was not without its critics and was regarded, by some commentators and practitioners at least in some respects, as controversial. However, it represented a body of thought and opinion from a respected group of specialists, only reached after a long and extensive consultation process with interested parties in the industry. The Second Edition was published in February 2017, following further lengthy consultation, and this has addressed some of the criticisms of the First Edition. It remains to be seen what critics make of the changes that have been made. Reference is made in this book to some of the conclusions of the Second Edition where they are relevant to the discussion of aspects of the quantification of claims.

For these authors, one of the ironies of the increasing reference to the First Edition of the SCL Protocol internationally over recent years was the knowledge that whilst, for example, its description of time impact analysis as the recommended method of delay analysis was being relied on by practitioners and referred to in arbitrations in many jurisdictions, the SCL committee was busy rather watering down that recommendation.

Judicial comment has been limited. There are a number of examples of it being referred to in support of a party's approach. However, in *Adyard Abu Dhabi v SD Marine Services* [2011] EWHC 848 (Comm), [2011] BLR 384, Mr Justice Hamblin noted that:

> … the SCL Protocol is not in general use in contracts in the construction industry and nor has it been approved in any reported case. There was no evidence that the parties were aware of it or that they contracted with it in mind. Further, the SCL Protocol itself says that 'it is not intended to be a contractual document. Nor does it purport to take precedence over the express terms of a contract or be a statement of law.

He concluded that:

> In such circumstances the SCL Protocol can be of little assistance in relation to the legal causation issues which arise in this case.

In relation to approaches to analysis of delays and extensions of time, it has been quoted a number of times in reported cases by parties seeking support for their particular approach: for example, in the English courts in *Great Eastern Hotel Company Limited v John Laing Construction Ltd and Another* [2005] EWHC 181 (TCC); *Balfour Beatty Construction Ltd v The Mayor of the London Borough of Lambeth* [2002] EWHC 597; and *Mirant Asia-Pacific Construction (Hong Kong) Ltd v Ove Arup and Partners International Ltd &Anor* [2007] EWHC 918 (TCC) (20 April 2007); and similarly in the Hong Kong case *Leighton Contractors (Asia) Ltd v Stelux Holdings Ltd* HCCT 29/2004. In Australia, the Supreme Court noted in *Alstom Limited v Yokogawa Australia Pty Ltd* (No. 7) [2012] SASC 49 that the SCL Protocol supported several different approaches to analysis and considered that one of these was to be preferred to a method proposed by the claimant's expert that had no support in any current literature. More recently, in

Santos v Fluor Pty Ltd [2-17] QSC 153, the Supreme Court of Queensland took guidance from the most recent edition of the SCL Protocol in relation to the use of a measured mile analysis in relation to a claim for delay and disruption.

The Protocol is not intended to be a contract document, i.e. it is not framed with the intention that it should itself form part of the construction contract. This is made very clear in its Introduction. However, its First Edition did contain model clauses for possible incorporation in contracts, addressing the requirements of a contractor's programme and record keeping. Internationally, the authors have seen those model clauses incorporated into construction contracts, usually with some amendments, as the basis for ad hoc programming and record keeping clauses. The intention of the Protocol is that it should provide a scheme of guidance for the analysis of delay and disruption in construction contracts and the matters that should be addressed in the drafting and negotiation of the construction contract. It contains a thoughtful and well-researched set of guidelines for the methods that can be adopted to resolve the issues of delay in construction contracts, bearing in mind that many of the issues do not have finite, or absolute, answers and the Protocol can only offer a set of balanced and considered views. As ever, it must be remembered that this is subject to the express terms of the contract and the underlying applicable law.

It should, however, be borne in mind that any analysis of the delay and financial consequences of events on a construction contract will only be as sound as the facts on which it is based. There is no substitute for properly recorded factual information as the basis of any analysis. Many citations of the SCL Protocol are often misplaced, as parties seek to legitimise a claim or position without reference to the actual facts of the case.

1.2.2 Direct and Time Consequences

Many changes or events requiring evaluation for additional payment will have valuation rules set out in the contract. An obvious example is the rules for valuation of variations, discussed in Chapter 5, contained in standard form contracts such as the SBC/Q (section 5) and FIDIC forms (clause 12). For other matters, such as the evaluation of payments for prolongation of the contract period, or disruption to the progress of the works, there will usually be little or no detailed guidance in the contract for evaluation purposes beyond the principle that 'loss and expense' or defined 'cost' can be recovered, and perhaps that profit can be added in some circumstances.

The evaluation of a variation will usually be subject to the rules expressly set out in the contract, but for items that cause prolongation and disruption there may be two tiers of evaluation required. Firstly, the direct consequences of the event or change will be required to be valued, usually in the form of an analysis of the effect on the site contractor's resources and working methods. Secondly, any indirect consequences, such as off-site increased overhead or financial charges, will be necessary. The guidance for supporting the valuation of such time consequences is generally the same as that set down by the courts for the evaluation of damages for breaches of contract, albeit that many instances will not actually be breaches of contract, but events contemplated by the parties in the contract. The guiding principle, when considering breaches of contract under English law for example, is that if the plaintiff has suffered damage that is not too remote, it must, so far as money can achieve it, be restored to the position it would have been in had that particular damage not occurred.

This does not, however, mean that the claimant can as of right recover every item of cost arising from a breach. The recoverable damages will be restricted to those which could reasonably be foreseen as arising from the breach, and not necessarily all damage. This principle is the rule stated as long ago as 1854 in the case of *Hadley v Baxendale* (1854) 9 Ex. 341 as:

> Where two parties have made a contract which one of them has broken, the damages which the other party ought to receive in respect of such breach of contract should be such as may fairly and reasonably be considered arising naturally, i.e. according to the usual course of things from such breach of contract itself, or such as may reasonably be supposed to have been in the contemplation of both parties at the time they made the contract, as the probable result of the breach of it.

This introduces the doctrine of remoteness of damage, by limiting recovery to costs incurred under the two 'branches' or 'limbs' of *Hadley v Baxendale*. That is, an injured party may not necessarily recover every item of damage resulting from a breach but may be limited to matters considered to arise naturally from such a breach, i.e. 'according to the usual course of things', the first 'limb'. They may also recover for matters considered to have been 'in the contemplation of both parties at the time they made the contract', the second 'limb'.

This of course raises the issue as to what might be considered to arise naturally, or be within the parties' contemplation, and the distinction between consequential losses that may be considered to be recoverable and those which are not. Consequential costs such as loss of profit or finance charges can be recoverable providing they can be shown to fall within the principles of *Hadley v Baxendale*. However, the loss of exceptional profits available from another contract but lost by the late completion of a project were excluded from recovery under the second limb of *Hadley v Baxendale* in *Victoria Laundry (Windsor) v Newman Industries Ltd* [1949] 2 KB 528, which developed the principles applicable to the recovery of damages under the second limb of *Hadley v Baxendale*.

The *Victoria Laundry* case set down three tests for the recovery of damages under the second limb as:

> In cases of breach of contract the aggrieved party is only entitled to recover such part of the loss actually resulting as was at the time of the contract reasonably foreseeable as liable to result from the breach …
>
> What was at the time reasonably so foreseeable depends on the knowledge then possessed by the parties or, at all events, by the party who later commits the breach …
>
> For this purpose, knowledge 'possessed' is of two kinds; one imputed, the other actual. Everyone as a reasonable person, is taken to know the 'ordinary course of things' and consequently what loss is liable to result from a breach of contract in that ordinary course … But to this knowledge, which a contract breaker is assumed to possess whether he actually possesses it or not, there may be added in a particular case knowledge which he actually possesses, or special circumstances outside the ordinary course of 'things', of such a kind that breach in those special circumstances would be liable to cause more loss.

In construction terms this means that an experienced contractor, professional design and consultant team, project manager and developer would all know at the outset of a contract to construct a new office building that it may be acquired as a long-term investment for profit by an investor and used for a short-term trading profit by an occupier. The losses that may stem from this type of knowledge are usually brought within the express terms of the contract by the inclusion of express provisions for an agreed rate of delay damages to be applied in the event of contractor culpable delays to the contract completion date. Where such losses are not brought within the contract they will be capable of being pursued as a result of breaches on an actual damages basis.

It is possible for construction, and other, contracts to expressly exclude the recovery of 'consequential losses'. However, such an exclusion may not, of itself, exclude all costs arising under the second limb of *Hadley v Baxendale*, as a consequential loss may be within the first limb of the rule. For instance, in a contract for the supply of concrete masonry blocks the vendors included a clause excluding their liability for consequential loss or damage as a result of late delivery. However, this clause was held not to exclude claims against the purchasers, pursued by their blockwork subcontractors, for damages incurred as a result of delays in the subcontract works caused by the late delivery (*Croudace Construction Ltd v Cawoods Concrete Products Ltd* [1978] 2 Lloyds Rep 55). The loss was considered to be in the normal course of matters and did not need to have been in the contemplation of the parties at the time of the contract.

However, exclusion clauses can in some circumstances exclude the recovery of consequential losses. In *British Sugar PLC v NEI Power Projects Ltd* (1998) 87 BLR 42 an exclusion clause was inserted in the contract limiting the recovery of consequential losses to the value of the contract. This was subsequently held to include the consequential loss of profit as a head of claim.

In summary, the principle for damages evaluations under English law is that the offended party should be put, as far as money can achieve it, in the same position as it would have been but for the intervening event. This is providing the nature of the damage can be demonstrated to be a natural consequence or within the contemplation of the parties when they entered into the contract, and is not restricted by express agreements or exclusions in that contract.

1.2.3   Duty to Mitigate

This does not mean that the party suffering a breach of contract by another can treat the breach as a 'blank cheque'. There is a duty on the party incurring additional cost as a result of a breach of contract to mitigate that cost. The principle 'imposes on a plaintiff the duty of taking all reasonable steps to mitigate the loss consequent on the breach, and debars it from claiming any part of the damage which is due to his neglect to take such steps' (*British Westinghouse Electric & Manufacturing Co Ltd v Underground Electric Railway Co of London Ltd* [1912] AC 673). This case laid down a number of principles in relation to the duty to mitigate and the extent to which this might limit a claimant's recoveries.

In *'The Solholt' (Sotiros Shipping Inc) and Another v Shmeiet Solholt* (1983) Com LR 114 the judge stated in respect of mitigation:

A plaintiff is under no duty to mitigate his loss, despite the habitual use by lawyers of the phrase 'duty to mitigate'. He is completely free to act as he judges to be in his best interests. On the other hand, a defendant is not liable for all losses suffered in consequence of his so acting. A defendant is only liable for such part of the plaintiff's loss as is properly caused by the defendant's breach of duty.

This does not, however, allow the party in breach to sit back and criticise the steps taken by the party suffering the breach; it only requires that the steps taken shall be reasonable in the circumstances. Lord McMillan succinctly summarised the position in *Banco de Portugal v Waterlow & Sons Ltd* [1932] AC 452 506:

It is often easy after an emergency has passed to criticise the steps which have been taken to meet it, but such criticism does not come well from those who themselves created the emergency. The law is satisfied if the party placed in a difficult situation by reason of the breach of a duty owed to him has acted reasonably in the adoption of remedial measures and he will not be held disentitled to recover the cost of such measures merely because the party in breach can suggest that other measures less burdensome to him might have been taken.

In summary of these cases, a number of principles can be said to apply:
- There is a duty on a party suffering from a breach by another party to take reasonable measures to mitigate its loss resulting from that breach.
- That duty does not limit how the injured party acts; it is free to act as it judges to be in its best interests, but the defendant is only liable for such part of the loss as is properly caused by the breach.
- That duty does not extend to carrying out measures that a prudent person would not normally take in the course of its business.
- The injured party cannot recover that part of its losses that was the consequence of its failure to comply with this duty.
- The costs of reasonable measures that are taken to mitigate loss are recoverable.
- The cost of such measures will not be disallowed merely because the party in breach can suggest that other measures less burdensome to it might have been taken by the claimant.
- The onus of proving that reasonable measures were not taken is on the party defending the claim.
- Whether the measures were reasonable is to be judged at the time and in the knowledge and circumstances under which they were decided upon, not with the benefit of hindsight.
- A party cannot recover any loss that was avoided by its actions, even where those mitigation steps went beyond what is reasonable.

In a construction and engineering context, care therefore needs to be taken when considering, for instance, criticisms that a contractor has not taken the most economical means to overcome the effects on its programme of delays that are the responsibility and liability of the employer or its consultants. This might mean that, where a contractor has introduced an additional work shift at considerable expense to expedite works delayed by instructed variations increasing the work scope, it might still recover those costs even if later analysis suggests (with the benefit of hindsight) that the works might have been suitably expedited by the original workforce working overtime at less expense. Only if it can be shown that the contractor has been incompetent or has made decisions that no reasonable contractor would take (judged at the time and with the knowledge available when the decision was taken) should the reasonable costs of its measures be discounted in such circumstances.

These principles can have particular relevance in the context of a claim for costs of such as increased overtime working or the introduction of additional resources to a project. Such claims are considered further in Chapter 6 of this book.

By way of illustration of this, and in relation to context of a different type of claim, reference is made to *Maersk Oil UK Ltd (formerly Kee-McGee Oil (UK) PLC) v Dresser-Rand (UK) Ltd* [2007] EWHC 752 (TCC), a case that neatly illustrates a common complaint of defendants, i.e. that costs have been needlessly or recklessly incurred.

The brief facts of the case were that Maersk had purchased a compression facility from Dresser-Rand for use on *Janice A*, a semi-submersible vessel used on the Janice Field in the North Sea. Under the contract, Dresser-Rand undertook to provide a complete compression process and mechanical design package for over £3 million. Maersk contended that the equipment supplied by Dresser-Rand was dangerous and had caused excessive vibration, resulting in fatigue, gas escapes and the production of damaging liquids in the compressor trains.

The costs of investigating the problems and carrying out rectification works were all expenses that flowed naturally from the breach and were recoverable under the first limb of *Hadley v Baxendale*. However, the right to recover such expenses was not unqualified.

Dresser-Rand alleged that Maersk had failed to mitigate the costs incurred in that they had not used Dresser-Rand's knowledge and expertise but had engaged third parties to undertake unnecessary and unreasonable work. Dresser-Rand alleged that Maersk had simply 'thrown' resources at the alleged problems without giving Dresser-Rand the opportunity to carry out rectification works and had not adopted a reasonably cost-effective approach. This is a common complaint of defendants of defects claims in construction contracts where the claimant has had corrective, replacement or remedial works carried out and that they could have solved the problems more easily and economically themselves.

The test the court applied was whether the advice from the third parties used by Maersk had been such a completely inappropriate resource so as to break the chain of causation. The judge applied the test in *Webb v Barclays Bank PLC* [2000] PIQR 8 in which it was held:

> Whether a tortfeasor can avoid liability for subsequent injury tortiously inflicted by a second tortfeasor depends on whether the subsequent tort and its consequences are themselves foreseeable consequences of the first tortfeasor's negligence ... Where any injury is exacerbated by medical treatment, the exacerbation may easily be regarded as a foreseeable consequence, for which the first tortfeasor is liable. If the plaintiff acts reasonably in seeking or accepting the treatment, negligence in the administration of the treatment need not be regarded as [an intervening event], which relieves the first tortfeasor of liability for the plaintiff's subsequent condition. The original injury can be regarded as carrying some risk that medical treatment might be negligently given.

In the *Maersk* case, the costs incurred by Maersk through the third parties were clearly recoverable. The work had been undertaken sensibly and by competent persons and while the outcome of some investigations had been spurious, the lines of enquiry followed had been reasonable in the circumstances at the time.

As will be explained in the relevant chapters of this book, the duty to mitigate has particular relevance to certain types of claim arising on engineering and construction contracts. It most commonly arises in relation to employer's claims for the remedying or completion of defective and incomplete works that have been left by a contractor. In this respect, see Chapter 8 and the judgments in cases, for example, such as *Pearce and High v Baxter and Baxter* (1999) EWCA Civ 789 (where the employer could not recover the full costs of remedying defects where there was a defects liability period in the contract but the employer had not given the contractor an opportunity to remedy them even though it was willing and able to remedy them). However, it also has particular relevance to claims for extensions of time arising from delays caused by an employer and the recovery of costs associated by such employer breaches.

1.3   Risks

The undertaking of construction projects of any substantial size will involve risk to the parties involved. As a general rule, the larger the project, the greater the risks being assumed by the parties. External circumstances such as changes in legislation, tax regimes, climatic conditions or the general economic climate may, among many other factors, impact upon the progress and/or costs of the works. Practically all projects will also be subject to some degree of internal risk, for instance from unpredictable site conditions or the need to complete some element of design after commencement of the works. The contract should address how all these risks are apportioned between the parties and their respective rights, responsibilities, obligations and liabilities arising in the event that any particular risk materialises.

While this text is concerned with the evaluation of additional payments in construction contracts, it is important that anyone undertaking such evaluation understands the reason why many such additional payments are required. It is not necessarily the result of incompetence or failure by one of the parties or its agents that such payments may be required. The need may arise simply as a consequence of the allocation agreed in the contract of known risks that were recognised and planned for by the parties.

That is not to suggest that risk should be regarded as something unavoidable or unmanageable. Risk analysis and management is a subject on its own but the consequences of failure to properly address the risk management process should be recognised by anyone involved in the management and evaluation of costs and payments on a construction project.

1.3.1   Design Risks

It is also important to understand the apportionment of risk inherent in the design of a project and how that is catered for in the contract.

Construction projects, by their very nature, usually contain some risk related to the design of the structures or buildings. That risk may be small where the design is relatively straightforward and the project conditions well known and documented, or may be substantial when difficult or innovative design is involved or where the precise nature of the project conditions cannot be ascertained at the design stage.

If the design is undertaken by the client, either in-house or by retained consultants, and the construction to that design is executed by the contractor, then the demarcation between design and construction risk will usually be quite easy to determine. However, if there are elements of specialist design with which the contractor or its subcontractors are involved, the contract needs to make clear the elements for which the contractor has responsibility. This is often achieved by means such as the 'Contractor Designed Portion' agreement used with JCT contracts.

However, the delineation of design responsibility can become more blurred when the contractor is responsible for design and construction. It may be that whilst the employer engaged design consultants to prepare and issue drawings and specifications to the contractor, the contractor in turn is responsible for the preparation of detailed drawings or shop drawings. Alternatively, the client may have the design progressed to a particular stage and then novate the design, and usually the contract with the designer, to the contractor. The obligation on the contractor is to complete the design and construct the scheme to the completed design.

It is also the case in some jurisdictions that even where the contract places no express design duty on the contractor, the applicable law implies a duty not to construct to a design that poses a wider public risk. This can come as something of a surprise to international contractors working in another legal jurisdiction for the first time. In such instances issues may arise as to where the respective responsibilities are delineated and where any failure actually occurred.

Where there is novation, a difficulty that can arise is that if the design progressed by the client and novated to the contractor proves to have substantive defects. At what stage was the error made? Has the contractor accepted responsibility for those defects in accepting the novated design? Some in the industry interpreted the decision in *Co-operative Insurance Society v Henry Boot Scotland Ltd* (2002) as extending the contractor's express responsibility to complete the design under the JCT With Contractor's Design form of contract (JCT WCD) to include a duty to verify the design handed over by the client.

This problem was clarified in 2005 when the JCT WCD form expressly stated that the contractor is not responsible for any inadequacy in the design provided by the client as the contractor is not required to check the design so provided. However, the contractor is still responsible for checking that any design by the client complies with statutory requirements, unless the contract information expressly warrants that the design is compliant. It will be necessary with other forms of contract, or bespoke contracts, and other jurisdictions, to check the inclusion or extent of the contractor's responsibility for checking the client's design.

It is therefore quite conceivable that claims for payment may arise from problems arising in the project design and allocated in the contract. The incidence and cost of such claims will need to be analysed and managed in much the same manner as construction risks.

1.3.2 Design Review

The JCT WCD contains a Contractor's Design Submission Procedure, based on that contained in the JCT Major Projects Construction Contract, providing a timetable for submission of the design by the contractor and approval by the client. This has been incorporated into the JCT's 2016 Design and Build Contract. Observance of such a procedure and the approval scheme will of course impact on potential liability for design defects and subsequent liability for the costs of any defects.

1.3.3 Professional Indemnity Insurance

The requirement to obtain, and maintain, insurance cover for design matters may be a requirement of the contract. The JCT WCD 2005 included provision for the contractor to obtain suitable professional indemnity insurance, although previously this was a common requirement by amendment of the previous version of the contract. In SBC/Q clause 6.12 requires the contractor to take out such insurance in relation to any work that is the subject of a Contractor's Design Portion.

By contrast, the NEC4-ECC does not specifically require professional indemnity insurance, unless such a requirement is expressly included by the employer in the Contract Data on the basis that it does not want the risk of the contractor carrying out defective design and being unable to bear the costs of the consequences. Whether the additional costs outweigh the benefit to it of reduced risk will be for the employer to decide.

If a claim is to be made under such insurance the assessment of quantum will usually be on the basis of restitution, i.e. to put the claimant under the policy in the position it would have been in but for the insured event and the contract valuation rules may not be appropriate. Such a claim will usually be on a 'damages' basis.

1.3.4 Risk Analysis and Management

If it is recognised that risks are inherent in a construction project, and they are recognised in the contractual arrangements, it is obviously necessary that an objective and thorough appraisal of the risks for a particular project is carried out with the aim of eliminating risks where possible and considering mitigation strategies for those risks that cannot be eliminated should they arise.

Risk analysis can be undertaken by any party to a construction project for that part of the project with which they are concerned. There is, however, only one party that can undertake a comprehensive analysis of the risks and instigate a reasoned allocation and mitigation strategy, and that is the employer or project sponsor. Many contractors, particularly on large projects, will undertake a risk assessment as part of the tender process, but this will be limited to consideration of the risks within their own scope of work and contractual obligations. Often this will be limited to an analysis of 'what can go wrong?' It will rarely consider any of the employer's risks or those of other contractors or subcontractors employed on the project where they do not impact on the contractor.

This, at best haphazard, approach to risk analysis has often been exacerbated by the scant attention paid in the past to any rigorous approach to risk management as the project proceeds. Whilst risk management has become a much more widely recognised requirement and discipline in many markets, in some parts of the world it is still almost ignored.

A discussion of risk management techniques is beyond the scope of this text, but it needs to be recognised that management and mitigation of risks will lead to mitigation of the need to make additional payments in the construction phase. The prime mover in this process has to be the employer, and the focus of the process should be the whole scope and lifetime of the project encompassing all aspects including finance, environmental issues, construction, operation and ultimately redundancy or decommissioning. There is a great volume of literature on risk analysis and management and there are published schemes of implementation, which can be applied to the lifetime of a project.

Any proper system of risk analysis will need to consider all aspects of the project, but from a commercial perspective it is important that, when deciding on risk transfer, the following matters are taken into account:

- Which party is likely to be best able to control the events leading to, and consequences of, a risk if it occurs?
- Should the client retain some involvement in controlling the risk?
- Which party should carry the risk if it cannot be controlled?
- Is any premium charged for the transfer of risk likely to be acceptable to the transferee?
- Does the premium to be paid exceed the likely benefit of transferring the risk, particularly where the transfer is to a party that is less able to manage the risk economically?
- Will the risk, if it occurs, result in other risks arising? If so, those risks need to be considered as above.

As we are concerned here with the construction phase and evaluation of additional payments, the consequences of the failure to conduct proper risk analysis and mitigation can be simply illustrated using one of the constant sources of dispute in construction contracts, the provision of information relating to physical conditions of the site.

Consider a contract for capital dredging and reclamation works to be undertaken in commercial docks as the preliminary work to the construction of a new dock and quay facility. Part of the tender information issued to the bidding contractors included a marine survey described in the tender documents as 'indicating the depths of water in the dock with the lock gates closed', i.e. the lowest level of water to be anticipated in the docks. The successful contractor deployed a substantial range of marine equipment and vessels to undertake the reclamation phase of the project only to discover its vessels grounding on the dock bottom at times when the dock gates were open, and the minimum level of water was therefore being exceeded! It was apparent that the depths of water shown on the tender survey information did not exist. Not unnaturally the contractor was somewhat perturbed and ultimately had to reorganise its programme and methods at some considerable cost, for which it looked to the client for reimbursement.

In the ensuing protracted discussions it emerged that: the survey provided as part of the tender information was six years old at the time of tender; the docks were subject to silting as they were located adjacent to an estuary; and no maintenance dredging had been undertaken in the intervening period in the area of the works. This in turn led to further protracted discussions as to what an experienced marine contractor could, or should, know, including comment by the engineer that the reference to the survey 'indicating the depths of water in the dock with the lock gates closed' should have been disregarded by tenderers as it was included as part of a preamble inserted by the engineer's quantity surveyor!

Whatever the outcome of such a scenario, and it had to be expensive for one or more of the parties involved, one simple truth emerged. The whole costly episode could have been avoided if the employer, or engineer, had recognised that the survey information was out of date and taken one of two possible courses, either obtaining an up-to-date survey and issuing it with the tender documents, or not issuing any survey information but making the facility available to the tenderers to undertake their own surveys. The former would be preferable, requiring only one survey and avoiding any difficulties of access for multiple surveys by different companies. The real lesson is that a proper risk analysis should have identified the need for a minor expenditure on survey information, which could have prevented considerable expense and delay and consequent dispute, if only the risk had been identified and mitigated.

With the development of risk analysis and management techniques, and their greater acceptance and adoption, it may not be too fanciful to presume that failure to implement such a scheme could in the future be taken into account in the assessment of costs arising from the incidence of a risk. This could at some point result in a defence against a claim failing on the allegation that a proper risk assessment process would have identified the problem. Perhaps at some point the rule in *Hadley v Baxendale* will need to be considered in the light of, or absence of, a risk analysis and management process.

In any event, whatever the legal implications, it has to be good commercial sense for all parties to a contract to identify and consider all potential risks before commencing on the project. 'A gramme of prevention is worth a tonne of cure' should be a guiding principle for everyone involved in the construction process.

1.3.5 Risk Registers

Many large projects now include a risk register agreed and maintained jointly by the client and contractor. The NEC3-ECC contract included in its clause 16 requirements for such a register and the conduct of risk reduction meetings (formerly early warning meetings) between the project manager and contractor (together with others such as subcontractors, suppliers and the employer as desirable) to review the risks, update the register and track progress on identified risk issues. In NEC4-ECC the equivalent provisions are now in clause 15. Reference is now to an 'Early Warning Register' and the form has returned to the reference to 'early warning meetings'. It is interesting to consider that this ongoing review and updating might impact on the risk allocation in the project and therefore will need to be carefully monitored to ensure that there is no significant transfer of risk from one party to the other without proper sanction.

This emphasis on managing and providing early identification and warning of risks is carried through to the valuation of compensation events in NEC4-ECC as clause 63.7 requires that in the event that the contractor should have given early warning but did not, the compensation is to be assessed as if it had given such a warning. This could result in the contractor suffering a reduced level of cost recovery or reduction in time allowed for an event. The project manager is likely to treat any absence of early warning as a lost opportunity to take alternative action or issue other appropriate instructions in relation to the event. Similar provisions are not uncommon in bespoke engineering contract, although there are no such requirements in the Infrastructure Conditions, SBC/Q or the FIDIC Red Book. When assessing quantum in such circumstances it is therefore necessary to ensure that any alternative view of the actions and consequences is properly taken into account.

Incidentally, many international arbitrations see arguments regarding the quantification of claims preceded by arguments regarding notice and whether the lack of notice is a condition precedent under the terms of the contract and the applicable law. This is usually put as an all or nothing defence, without considering it as a question of potential mitigation. That notices were not given may not preclude the claims, but what about their effects on quantum? Had the employer and its representatives known at the time of the claimed effects of an event or decision, would they have acted differently, issued an instruction or similar, in order to mitigate the effects? It is striking how often this issue is not raised by employers as a secondary issue relating to notices. A potential position may be that, even if the lack of notice does not entirely defeat the claim on the basis of a condition precedent, it still might reduce its quantum. This is presumed on the basis either that there was a failure of a duty to mitigate or that the lack of notice was a breach by the contractor for which the resulting and unnecessary costs in the claim are damages incurred by the employer. Whether this second suggestion is possible may particularly depend on the applicable law.

1.3.6 Risks and Records

It has been stated on many occasions that the root cause of many claims and disputes on construction and engineering projects is the failure to place risk plainly on one party or the other. This is sadly true, but often overlooked in the preparation of many contracts, and any failure to identify clearly the party carrying a particular risk is likely to result in disputed claims for additional payment if the risk occurs. In addition, further disputes as to the valuation of change and disruption also occur where the risk is placed plainly on one party and where the contract allows the party carrying the risk to claim additional payment in defined circumstances. The risk allocation may be plain, but the financial consequences of its occurrence may be open to a range of opinion and argument.

There may also be circumstances where the risk apportionment is defined and the entitlement to claims for additional payment is clear, but the value of the additional payment cannot be calculated from the contract provisions. It is in such instances that a further risk element may be thought to enter the contract, the risk carried by a party failing to keep adequate or proper records of the events and consequences of such a risk occurrence.

It is often said that a party to a construction contract dispute will soon realise that unless it has good records, its case will be at least diminished or, at worst, lost. This is no exaggeration and the need for careful substantiation of claims for additional payment is a theme of the following chapters. Any party to a construction contract who does not understand, or fails to implement, the requirement to keep proper records runs a very real risk of seeing subsequent claims for additional payment reduced or even negated.

What constitutes a proper, or reasonable, record may vary with the nature of the works and the terms of the contract. Many contracts, for instance those with a large mechanical engineering content with weld examination and approval procedures, may require extensive records for quality control, which will also be useful and reliable for evaluation of additional payments. However, these records may not, of themselves, be sufficient in particular instances and there should be a mechanism for the parties to record the following events with reasonable accuracy:

1. The progress of the works with reference to physical milestones and significant events. The means of achieving such records will vary but they may not be restricted to paper records such as revised programmes or programmes marked up with progress. Devices such as photographs and video recording can be equally useful, and in some instances preferable alternatives.
2. The deployment of resources, both labour and plant, in a manner which not only identifies the scale of the resources but also allows identification of the activities undertaken in the recorded period.
3. Deliveries of critical materials and items for incorporation in the works, such as equipment packages for mechanical installations. Procurement documentation should normally be available to establish the sequence and timing of pre-delivery activities.
4. Deliveries of other prerequisites to construction such as design drawings and details or access to work areas.

It is important to consider such issues as: who should keep these records; when it is appropriate for the record keeping to be implemented; the extent to which they should be agreed; and if they should be taken jointly. Many disputes arise in the evaluation of claims for additional payment as a result of inadequate records, but probably just as many arise from disagreements as to the veracity and accuracy of records submitted by one party to the other.

While it might not be appropriate for both parties to verify all records throughout the course of a large project there is certainly an argument that such an approach might significantly reduce the risks of disputes generated by disagreement of the consequences of a specific event. At the very least it is both necessary and reasonable for both, or all, parties to agree records if notice has been given that a claim for additional payment may arise, as is required by some contract terms. This is not always the case and neglect of this aspect can exacerbate an existing dispute as to evaluation or create one where a dispute should not have been necessary.

Many standard forms have express provisions for record keeping in the event of claim situations. For instance, the Infrastructure Conditions contain provisions for record keeping when dealing with additional payments in clause 53 paragraphs (2) and (3), in the same terms as the previous ICE Conditions. The FIDIC Red Book clause 20.1 requires the contractor to keep contemporary records to substantiate any claim, and allows the engineer to monitor those records and instruct that further records are kept. These provisions allow the engineer to require any specific records that it may believe will be material, but there is, in any event, a requirement that the contractor shall keep any contemporary records that may be necessary to support a subsequent claim. Even where a standard form does not set out such requirements, it would be open to the parties to incorporate those of such as the model records provision of the First Edition of the SCL Protocol or similar. These provide no more than the express statement of something that should be implied in any event. That is, if the contractor intends to submit a claim for additional payment as a result of events on site, it should keep such contemporary records of the events and their consequences that are required to establish the claim 'on the balance of probabilities' or whatever other standard of proof applies.

The record keeping requirements set out in these conditions of the FIDIC Red Book and the Infrastructure Conditions are plain and reasonable, but it should be noted that, while the contractor is under a duty to keep records once notice has been given, there is no compulsion on the engineer to verify the records, keep copies or issue instructions as to any further records it may require. In most instances this may not be a problem and the contractor's records when inspected may be sufficient and acceptable. It would seem prudent, however, to sound a note of caution where adjudication provisions required by such as the UK's Housing Grants, Construction and Regeneration Act of 1996 ('the HGCRA') apply, as updated by the Local Democracy Economic Development and Construction Act of 2009. The implications of these Acts are outside the remit of this book, but the timing issues of the Adjudication process introduced in 1996 have to be considered in the context of record keeping.

The adjudication provisions contained in the Infrastructure Conditions at clause 66B mirror the timing aspects of the Scheme issued under the HGCRA, and give an adjudicator 28 days from referral to reach a decision, extendable by 14 days only by consent of the referring party. If the engineer has to verify extensive records and consider their implications at the same time as submitting the employer's case for consideration by the adjudicator it could be under some difficulty. The effect of the adjudication scheme in respect of records is to reinforce the advisability of joint contemporary records, and careful consideration by both parties of the nature of the records required. It is possible that, in the light of adjudication provisions and timetables, the day is not too far away when an employer's supervisory consultant will be held to be negligent in not ensuring they were in a position to answer allegations in an adjudication based on records of the works. A further requirement of the speed of adjudication may be that records are required instantly and in a suitable format without time to forensically create them or put them into suitable form.

The increasing adoption of fast-track adjudication schemes similar to those of the UK in other jurisdictions means that this need for quick access to readily useable records could apply anywhere. A problem with the introduction of what is sometimes referred to as a 'quick and dirty' third party dispute resolution process in some jurisdictions is that they do not have a matching discipline for keeping such records.

Many contracts do not contain specific clauses setting out record keeping requirements in the manner of the FIDIC Red Book. It is not unusual in international engineering 'ad hoc' forms to find provisions for a contractor to give notice that it is undertaking works that it considers are a change from, or additional to, the contract requirements but for which the employer has declined to issue a required 'change order' or other necessary instructions. Such provisions are often termed 'disputed change orders', and there may be a further process in the contract for acknowledged change orders to have disputed consequences, where the employer acknowledges the validity of a change order but not the claimed consequences. The purpose of such clauses is to enable the parties to proceed with the works without the delay generated by arguments as to whether or not a particular matter is, or is not, a change or addition to the contract requirements. Such clauses therefore typically contain requirements for notification of the potential dispute, but equally typically are less specific on the requirements for the contractor to keep records or the employer's right to require particular records to be kept and made available. The sanction for ultimately inadequate records in such instances is often stated to be the right of the employer to refuse payment.

The reasoning behind the often vague record keeping requirements of such clauses is often that large international engineering contracts, such as those for the oil and gas industries, have substantial requirements for the administration of the contract written into their specifications, together with extensive quality control and safety records, which may be considered to provide the basic sources of information required. It does, however, seem that some more specific record keeping for particular notified instances of possible dispute, such as that contained in the FIDIC Red Book, would have the potential of further reducing the potential for disputes.

In this context the provisions of the General Conditions of Contract issued by the CRINE Network (Cost Reduction in the New Era) in June 1997, superseded by the LOGIC Marine Construction contract in Edition 2 in 2004, for the offshore oil and gas industry, contain in clause 14.7 similar provisions to those previously in the ICE contracts. It is suggested that these represent best practice. Although such projects are outside the HGCRA's remit and its adjudication scheme does not apply, the CRINE/LOGIC contracts contain their own resolution of disputes provision at clause 37. This clause contemplates a more internal approach to dispute resolution, but the requirement to keep records of disputed variation works is bound to increase the chances of success in such an internal process.

1.3.7   Reimbursable Risks

Not all risks entitle the party carrying them to additional payment if they arise. There are reimbursable risks and non-reimbursable risks.

The principle often adopted in contract drafting is that the party best able to control the risk is the party who should be responsible for that risk in the contract. Consider, for example, the treatment of unforeseen physical conditions in the Infrastructure Conditions and other standard forms of construction and engineering contracts based on the old ICE approaches.

The FIDIC Red Book is no different to the ICE Conditions that were its basis, in that clause 4.12 allows the contractor to claim additional payment if unforeseen physical conditions are encountered. This is often referred to mistakenly as the 'unforeseen ground conditions' clause, but it covers any physical condition that could not be foreseen. In *Humber Oil Terminal Trustees v Harbour & General* (1991) 59 BLR 1 the clause was held by an arbitrator to include foreseeable ground conditions that acted in an unpredictable manner when subjected to particular forces. It is good sense for the contract to make the contractor responsible for notifying such conditions if they are encountered and, in conjunction with the engineer, devising remedial or other measures. The contractor is the party that can be expected to become aware of such problems before others and therefore the contract makes it responsible for proper notification and action.

From a payment perspective there are two points to note. Firstly, that clause 4.12 of the FIDIC Red Book is inextricably linked to the preceding clauses 4.10 and 4.11, which set out the contractor's responsibility for interpreting information provided and making its own investigation of the site and its surroundings. Secondly, that the interpretation of information, inspections and examinations contemplated by clause 4.11 is to be incorporated in the contractor's tender. The principle is that the contractor cannot escape responsibility for errors or deficiencies in its tender and must be considered to have included for all predictable circumstances as would be anticipated by a reasonable and competent contractor.

This means that when assessing entitlement to additional payments it is not the tender computation that is the starting point but the provisions that should have been made in the light of available information and reasonable enquiry. This starting point is not unique to these clauses of the FIDIC Red Book or any other clauses or conditions, including those also based on the old ICE Conditions. It is a general principle that is sometimes forgotten when trying to establish a starting point for the evaluation of additional payment in relation to reimbursable risks.

1.3.8 Non-reimbursable Risks

In contrast to the risks that can generate entitlement to additional payment obligations, there are risks that are identified as being clearly the responsibility of one party in any event, but without additional payment to the other party if they occur. However, that does not mean they will not impact on other parties to the contract.

For instance, most construction contracts place the risk of obtaining the quantity and quality of labour resources plainly on the contractor, but if the contractor cannot provide the required resources, the employer, or owner, will almost certainly be affected because completion of the project is likely to be delayed. While the contractor will not be able to recover additional costs in respect of any prolongation of the contract period, and related resource costs, through the construction contract, the employer will usually suffer costs of its own from the delayed completion.

Most contracts expressly provide some recovery for the employer, usually in the form of delay damages, discussed in Chapter 6 of this book, but these will often not provide full recompense for all costs incurred. It is not unusual in large engineering projects for such damages to be restricted by a cap expressed as a percentage of the contract price, resulting in major delays being even more harmful to the employer, where its actual costs and losses exceed its capped recoveries from the contractor. Indeed it is not unusual for the rate of liquidated damages stated in many building and civil engineering contracts to be less than the employer's anticipated loss as the insertion of realistic losses as damages would deter or prevent contractors from undertaking the works. Employers do, however, have the option of not including a rate of delay damages in the contract and relying on their actual damages arising from the contractor's failure to complete on time. Dangers with this alternative approach are that the employer will have to prove its actual damage and that bidding contractors may be deterred by a lack of limit on their potential liability. These issues are discussed in detail in Chapter 6.

A further alternative for the employer, where the contractor is in breach and its rate of damages would be insufficient to reimburse its true losses, might be to determine the employment of the contractor. For example, clause 8.4 of SBC/Q empowers the employer to terminate the contractor's employment for 'failure to proceed regularly and diligently with the Works'. Effectively, the employer might here be said to be trying to limit its losses. It was particularly seen as an approach on some commercial projects in the Middle East after the crash of 2007/2008, where the preceding efforts of contractors were hampered by boom and then bust, and developers saw an opportunity to both avoid making a claim against an insolvent party and to retender the works at significantly lower rates and prices.

Delays in completing a construction project can have significant knock-on effects on related parts of the employer's business and can cause substantial disruption and cost to even large, well-managed and well-financed organisations. It is therefore in the interests of all parties to a construction project that the risks inherent in its undertaking are properly understood, analysed, allocated and managed throughout the lifetime of the project. Many disputes arise simply from an imperfect understanding of risks at the outset and an often remarkable failure to attempt to manage and mitigate the risks as the project proceeds, particularly where the result is late completion and a substantial cost overrun.

1.4   Sources of Change

Many disputes over additional payments arise from the failure to record and detail the consequences of risks when they do arise. The most successful route to minimising disputes and their effects is first to ensure that the commonly arising risk events are understood and either anticipated and prevented or controlled, and then to ensure that any consequences that arise are properly recorded and the claims for associated additional time and money are properly presented. A full and well-supported presentation of a problem will usually be the first requirement to ensuring that the cause and effect are understood by all and are capable of rational analysis and resolution without the need for them to evolve into dispute with the potential for protracted, contentious and expensive formal proceedings.

A review of such as the lists of 'Relevant Events' in SBC/Q's clause 2.29 or the 'compensation events' in NEC4-ECC's clause 60.1 will identify most of the causes of change under construction and engineering contracts. An in-depth analysis of all of the potential causes of claims related to change is beyond the scope of this book, but it is useful to consider very briefly the most common causes of such issues.

1.4.1   The Process of Analysis

The previous discussion of risk management and the need to maintain records where the occurrence of a risk event involves a departure from the anticipated path of the project works makes clear that, while there should be a clear and defined risk register and risk management plan for any major project, the effects of risks cannot always be totally avoided or mitigated. On projects other than those of a very minor size, there will be instances where a risk event occurs, and its effects are alleged to give rise to an entitlement to additional time and/or payment. In such circumstances all the previous comments about records and substantiation will apply and if the required information is presented in a logical and comprehensible format the chances of avoiding unnecessary disputes will be increased.

In breach of contract cases lawyers often refer to the required chain of analysis as being:

Duty – Breach – Cause – Effect – Damage

This chain of analysis anticipates that the claiming party will fulfil each of the parts set out in it. That is, it will establish: what duty was owed to it by the other party; the cause of the alleged breach of duty; the facts of the breach itself and the effects of the breach; and lastly the damage that results from the breach and is required as compensation.

While this book is primarily engaged with the last step in this chain, that of substantiating the damage or value of an additional entitlement, there is no doubt that the task of establishing the quantum of a claim is much simplified and stands a greater chance of success if the claim itself has been fully analysed and established from the root causes as anticipated by the chain of analysis. An example of this is in relation to quantifying disruption claims. Analysis that goes back part way through the chain of causation to better detail of the effects of disruption can facilitate a more detailed approach to quantification.

The legal analysis of a claim will be concerned with liability, i.e. establishing whether or not there has been a breach that gives rise to an entitlement, but the same analysis can be applied to quantum. Using this chain of analysis, and considering it with the evaluation of quantum in mind, the various steps might involve the following matters:

Duty. This first step in the chain of analysis is primarily concerned with establishing the obligations and responsibilities of the parties to the contract. It is important in doing so to also confirm what the financial duties and responsibilities are. Among the issues to be addressed will be: What is the contract sum? What is its basis? How is it to be measured and adjusted? What notice, if any, has to be provided of financial impacts?

Breach. In applying the financial provisions of the contract have any of the obligations and responsibilities established under the 'Duty' analysis been breached? If so, in what respect? Is the breach material or is it of little consequence?

Cause. What is the cause of any breach in the obligations and responsibilities? Is the cause relevant to any analysis of the financial impacts?

Effect. This is where the quantum analysis often begins to assume equal importance with the liability analysis. Along with the effects of the liability analysis it will be necessary to analyse the financial effects of the liability breaches, and the impact, if any, of the first three stages of the financial analysis. What are the financial impacts of the liability breaches? Are there any breaches of the financial obligations and responsibilities that affect the financial impacts?

Damage. Where a damages claim is being considered it will be necessary to consider aspects of sustainability and proof of the financial effects. Are any of the financial impacts too remote to be claimed? Are the records of events and costs sufficient to support the claim being made? Has the legal burden of proof been satisfied? Should the financial impact have been mitigated from that being claimed? Does the claim overlap with other claims or additional recoveries elsewhere?

All these, and usually many other questions (such as the legal issue of contractual notice), will arise in the course of analysing a claim for additional payment. It is not possible to produce a comprehensive listing of possible questions as many will stem from the type and terms of the contract, the circumstances of the claim being made and the financial impacts being claimed. What is important is that a rigorous and logical analysis is employed and that the results are incorporated into the claim evaluation.

If this process is followed it should help provide properly established and supported claims for payment. This is, in itself, one of the most important steps to avoiding disputes. Sadly, it is also one of the most often ignored steps.

1.4.2 Inadequate Pre-contract Design and Documentation

One of the perennial causes of claims for additional payment under construction and engineering contracts is error or omission in the design and specification documentation issued for a project at the outset.

This problem should not be confused with contracts where design information is issued at the outset in an incomplete form for known and planned reasons. The most common incidence of contracts commencing with incomplete design is that of major projects where the length of time required to complete the design before commencement of work on site would be unacceptable to the employer. There is nothing intrinsically wrong with such an approach and it is often necessary for large schemes to begin the early phases of construction before the design of later stages is completed. However, that reality must be recognised by all from the outset of the project and its contract terms drafted accordingly.

It is, however, essential in such circumstances that the contract anticipates the completion of design during the construction phase and that procedures are implemented to monitor and control both the required processes and the effects. In particular, from a quantum viewpoint, the payment terms and provisions need to be set up to allow the payment for the later stages to be calculated during the contract period.

Such a situation is radically different from that where the design is intended to be complete, but is not, or where there are omissions in the documentation and/or contradictions between contract documents. Such problems often result from inadequate or badly managed pre-contract phases where the preparation of design and contract documentation is compressed into too short a time frame. The pressures in pre-contract periods are understandable; the client almost invariably views the construction process as a means to its end and is anxious not to spend either more time or expense on it than it absolutely has to. The client wants its plant, factory, office, etc., as soon as possible so that its end is achieved without delay and too much expense. There has to be an education process to ensure that any client, particularly one not regularly engaged in construction projects, understands that apparent savings in time and expenditure in the pre-contract phase can result in more substantial delays and expenditure in the construction phase. The client may still have reasons to pursue early commencement of the project, but in such circumstances it should be acknowledged that it is a commencement with incomplete information, and adapt accordingly.

One of the devices sometimes adopted in such circumstances is to include extensive provisional sums for work that has not been fully, or sometimes not even partly, designed or defined. While this may be acceptable in some circumstances, it is suggested that it should be kept to a limited extent, where 'needs must' and too many employers and their consultants misuse this as an excuse for not getting the design properly firmed up at contract stage. See in this regarding the judgment of Mr Justice Akenhead's judgment in the High Court's Technology and Construction Court in *Walter Lilly & Company Limited v (1) Giles Patrick Cyril Mackay and (2) DMW Developments Limited* [2012] EWHC 1773 (TCC) (12 July 2012). In that case the contract was based on the JCT Standard Form of Building Contract 1998 Edition Private Without Quantities, but the design was so undeveloped that all of the building works were the subject of provisional sums. The judge described the project as 'a disaster waiting to happen'. In the event costs overran several-fold.

Particular problems may arise if the contract makes a distinction between 'defined' and 'undefined' provisional sums. As an example, if it states that the bills of quantities have been prepared in accordance with the measurement rules in NRM2 (which are the same in this respect as its predecessor SMM7, incorporated into the definitions of SBC/Q). NRM2 rule 2.9.1.3 states that the contractor is deemed to have made provision in its programme of works for work covered by 'defined' provisional sums, but not those covered by 'undefined' provisional sums. Without definition for significant portions of the work, the programme can become nothing better than guesswork and the seedbed of future dispute. Issues can also arise as to whether a provisional sum was actually 'defined' or 'undefined' and hence covered by the programme or not.

A useful description of a provisional sum was provided by Lord Justice May in the judgement of the Court of Appeal in *Midland Expressway Ltd v Carillion Construction Ltd (No. 1)* in 2006, thus:

> [a provisional sum is] used in pricing construction contracts to refer either to work which is truly provisional, in the sense that it may or may not be carried out at all, or to work whose content is undefined, so that the parties decide not to try to price it accurately when they enter into their contract. A provisional sum is usually included as a round figure guess. It is included mathematically in the original contract price but the parties do not expect the initial round figure to be paid without adjustment. The contract usually provides expressly how it is to be dealt with. A common clause in substance provides for the provisional sum to be omitted and an appropriate valuation of the work actually carried out to be substituted for it. In this general sense, the term 'provisional sum' is close to a term of art but its precise meaning and effect depends on the terms of the individual contract.

For defined provisional sums to be incorporated in a contract using NRM2, rule 2.9.1.2 requires it to provide: the nature and construction of the work that is the subject of the sum; the construction of the work; a statement where and how it is to be fixed; a statement of what is to be fixed to it; and quantities to indicate the extent and scope of the work and any specific limitations. If such a sum is included in the contract then the contractor is deemed to have included in its programme of works and pricing of the contract preliminaries and general items for the work covered by the provisional sum. In practice, the information is sometimes very vague and on occasion completely absent, notwithstanding it is still described as a defined provisional sum. Such abuses can only lead to later disagreement. Furthermore, there may be disagreement regarding the information and whether it does actually satisfy the requirements of NRM2 rule 2.9.1.2, for example where a statement as to where and how provisional sum work is to be fixed is vague or incomplete.

Furthermore, where the requirements of NRM2 rule 2.9.1.2 are met and the information required for a 'defined' provisional sum is available, it does often beg the question as to why the works are included in that provisional sum rather than being measured in the bill of quantities, perhaps with the rider that they are only 'provisional' quantities.

It is useful to note that the NEC4-ECC form of contract does not provide for the use of provisional sums. That approach is based on the position that if an employer cannot clearly define part of the works at the time the contract was formed that work should not be included in the contract because it is not reasonable to expect the contractor to include work without a clear indication of its cost and programme implications. Under NEC4-ECC the preferred approach is to deal with such work by use of the early warning system and register with the work being valued as a compensation event in accordance with the contract terms. Employers may wrongly believe that reduces certainty of price in relation to the subject work, but the real effect may be the advantage of removing an area that is particularly prone to disagreement.

1.4.3 Design Development and Approval

In many contracts, especially those with large process facilities, it is often the case that the design will continue to develop after the construction contract has been let, for sound technical reasons. This could be because technology has moved on or legislative requirements in relation to the end product of the process have changed. The same reservations as stated elsewhere apply, in that the contract must be set up to cater for such developments. If substantial portions of the works either cannot or will not be fully defined or designed, or the design is likely to alter in significant respects, then the payment and planning provisions should be structured to cater for the anticipated design development. The use of provisional sums in this regard has been discussed elsewhere.

Where the design is particularly incomplete, and intended to be completed later, one approach to adopt is for the contractor to be obliged to produce a detailed programme for the first period of the works, perhaps six months, with the adoption of milestone dates to be achieved for completion of the works and detailed programmes produced when the works are fully designed. This may also require the adoption of agreed sums for the designed work, where full definition is possible at the outset, and the agreement of a means of pricing the works yet to be designed, whether by schedules of rates agreed in the contract or by agreement of payment on some form of reimbursement of cost plus agreed overhead and profit additions.

If the agreements at the outset do not realistically reflect the intended manner by which the works will be procured then future problems are almost guaranteed. However, even where the contract terms reflect the reality of the approach, the position can be subject of further abuse, giving rise to different sorts of problems.

Even where the employer appoints architectural and engineering design consultants, it is common for contractors to be required to develop the design of the works to an extent set out in the contract. This is recognised, for example, in clause 4.1 of the FIDIC Red Book. The specifications will then usually set out the details of the extent to which the contractor will design as well as the procedures for gaining the approval of the employer's architect or engineer of that design before the related work can be carried out. Typically the contract documents will set out an approval procedure involving alternative codes from the employer's representative, such as:

Code 1	'Work may proceed'	The contractor can commence construction against that design.
Code 2	'Revise and resubmit. Work may proceed subject to incorporation of comments'	The contractor can commence construction so long as it complied with the comments.
Code 3	'Revise and resubmit. Work may not proceed'	The contractor cannot commence construction and must resubmit the design, addressing the comments.
Code 4	'Review not required. Work may proceed'	The contractor can commence construction against that design.

Failure by the employer's representative in relation to such approval processes should entitle the contractor to an extension of time. In the case of the FIDIC Red Book, this is covered by clause 1.9 'Delayed Drawings or Instructions', where it will also entitle the contractor to any resulting costs incurred and reasonable profit thereon. However, such procedures for contractor design and approval are often abused. For example:

1. Where it is used to require the contractor to develop the design through stages that ought to have been carried out by the employer's designers. Thus, the contractor is not just required to produce 'shop drawings' but to complete the designers' incomplete or inadequate work.
2. Where the evaluation of the contractor's drawings unfairly rejects the design or requires revision and resubmission on spurious grounds. Thus, for example, a drawing that ought to have been given Code 1 is designated Code 3.
3. Where the approver takes an inordinate time to evaluate or respond to the design, be it longer than the approval period stated in the specifications or a reasonable time if no specific period is stated.
4. Where the employer's designer uses the approval process to require what should be covered by variation instructions under the contract. The contract documents should clarify that the review of shop drawings should be limited to checking that they comply with information given and the design provided to the contractor. This does not mean that the employer can instruct variations to that design under the pretext of a Code 2 or Code 3 requirement to revise and resubmit.

1.4.4   Access or Possession

It is obvious that a fundamental duty of an employer is to provide the contractor with access to, or possession of, the site of the works. For example, the obligation is expressly spelt out in clause 2.1 of the FIDIC Red Book, clause 2.4 of SBC/Q and clause 33.1 of NEC4-ECC. This may include sectional access dates. Particularly on linear transport projects, it may be agreed via a schedule of access dates that the contractor will be given possession in parcels as it proceeds down the chainage of the railway or highway.

In practice, particularly on linear transportation projects such as roads and railways, where land for the route is publicly purchased from many private owners, access or possession is often given at dates later than expected and/or in a manner different to the contract requirements. This may include failure of an obligation to provide 'unhindered access' to the lineage of, for example, a new railway line, including to each side of the line to allow access and working where possession is given but with restrictions such as the owner remaining in occupation.

A feature of many infrastructure projects internationally is a failure of the employer to procure the land, or access to it, at dates and in the sequence that it then undertakes to provide to the contractor under the construction contract. A common retort when the inevitable claims follow is for the employer to assert that the contractor was in delays of its own and did not need the land when promised! This can evolve into a 'chicken and egg' argument in which the contractor says its progress was delayed by land being handed over late against the contract schedule, but the employer says it procured the land to suit the contractor's actual progress.

Particularly damaging for the contractor's costs is where parcels of land are not only handed over late but in a different order to that in the contract and with no foreseeable date as to when outstanding parcels will be released.

1.4.5 Early Taking over or Beneficial Use

Contracts may provide for the completion of the works in contractual sections, with procedures for extensions of time to be addressed in relation to each, as well as completion of the works as a whole and delay damages agreed in relation to each contractual due date.

In the absence of a set of sectional completion dates agreed in the contract, contracts may also provide for the employer to elect to take possession of the parts of the works as they are completed. The FIDIC Red Book clause 10.2 allows the employer this facility. The clause prohibits the employer from making use of any part of the works other than as a temporary measure specified in the contract or agreed by the contractor, unless the engineer has issued a Taking-Over Certificate for that part. If the employer does use part of the works without such a certificate, then that part is deemed to have been taken over by the employer. The clause entitles the contractor to its costs and reasonable profit arising out of the employer taking over or using a part of the work other than as specified in the contract or agreed by the contractor.

Whilst the FIDIC Red Book constrains the employer from premature use of part of the works, the Infrastructure Conditions provide for it at clause 48(3). It provides that where a substantial part of the works has been used or occupied by the employer, other than as provided by the contract, the contractor may request, and the engineer shall issue, a Certificate of Substantial Completion for that part. In clause 48(4) the engineer can issue a Certificate of Substantial Completion for any part that is considered to have reached that state.

SBC/Q clause 2.6 also provides for early use or occupation of parts of the site or works by the employer. This is dependent on the contractor's consent. Clauses 2.33 to 2.37 also provide for partial possession by the employer, again with the contractor's consent.

Whilst such express contractual provisions provide the facility for an employer to take early possession or use parts of the works, they are often abused and sometimes ignored where the employer wishes to take the benefit without the disadvantages of releasing the contractor for liability for those parts, such as liability for delay damages and retention monies. Chapter 8 particularly considers claims in relation to post-handover costs that contractor's incur as a result of employers taking possession before the full desired scope of works are completed.

1.4.6 Changes in Employer Requirements

Just as design may change for technical or legislation related reasons there will be instances where the client's requirements may change, often for unanticipated reasons. Most lump sum and remeasurement contracts, such as the SBC/Q Form and FIDIC Red Book, where the design is undertaken by a team of consultants on behalf of the employer, provide for such changes, within limits, as variations to the contract and contain detailed provisions for quantification of such variations.

Problems can, however, arise with design and build contracts, or other variations on this theme such as EPIC (Engineer Procure Install and Commission) contracts in the oil and gas industries, if the contract does not include sufficient detailed information to establish the chain of analysis discussed elsewhere. If the contract information does not allow proper definition of what the baseline was and hence what has changed and to what extent, together with the financial consequences, then disputes may seem to follow as an inevitability.

A common example of the problems that can be encountered in this respect is the provision of large elements of the work as performance-specified equipment or packages. If, for instance, a contract includes the provision of a large piece of mechanical equipment costing, say, $2.2 million, but the definition of the package is by specification of its required input and output performance (perhaps with some physical constraints also specified) and a change is required to one or more of the input or output requirements, the analysis of the financial impact of that change becomes very difficult without recourse to information or details from outside the contract, if such information is available.

If the employer, and often also the prime contractor, are not to be left entirely at the mercy of the supplier in such instances, the need for an analysis of the purchase price and potential rates for possible future adjustments should be incorporated wherever possible in the procurement procedure.

Further problems can arise where the degree of change is high and their timing is late against the contractor's progress. The *Walter Lilly* case has been mentioned elsewhere as an example of where works were under-designed at the contract stage and all were subject to provisional sums, such that the employer could decide what it wanted in terms of specification and detail as the works progressed. This is a feature of many high-end housing projects. It is also a feature of public procurement in some markets where the end-users are allowed involvement during construction that lets them chose and change what is required as it is constructed. It is suggested that this approach is not what lump sum designed works contracts such as SBC/Q are intended for. Apart from discouraging such levels and timing of change and also getting greater involvement of those same end-users during the pre-contract design phase, where such a flexible approach is desired, the procurement route and contract should be designed to address that approach. All too often they are not, and the result is a protracted and expensive dispute, as was the consequence in *Walter Lilly*.

It may not be easy to determine whether a change has actually occurred and, even if it has, what exactly the scope of the change is. This may relate to problems with the contract documents, where the contract design or documents are not sufficiently detailed to enable precise comparison with the work as it is actually instructed and built.

1.4.7   Contract Documents

The preparation of contract documents, the express terms of those contracts and the effects of implied terms from the underlying law of the relevant jurisdiction are all outside of the remit of this book. However, shortcomings in such documents are a regular cause of claims under construction contracts.

When the contract documents for construction of a tram system to a UK city can span 8506 pages, excluding drawings, it is hardly surprising that problems can arise regarding the interpretation of those documents. Voluminous documents might be assumed to ensure a comprehensive agreement between the parties. In practice, those preparing documents often seem to believe that just including every document since the initial enquiry will ensure this. All too often what it actually results in is contradiction and confusion. It also often seems to lure the drafter into assuming that because every document has been included and they span thousands of pages, then all issues must be covered, when in fact important matters have been overlooked.

Ad hoc contracts or tailored amendments to standard forms are usually drafted to suit the best interest of the employer commissioning them. However, they are often just the cause of confusion as to what the amendments mean and how they sit with unamended provisions. In jurisdictions applying a 'contra proferentem' or similar interpretation of bespoke provisions, they can often work directly against the party they were intended to benefit. This is a particular problem in international jurisdictions where the adviser drafting the amendments is insufficiently aware of the applicable law and how the changes will fit with them.

One area of drafting of contract documents that quantity surveyors will be particularly involved with is the preparation of bills of quantities. Errors in bills are all too often the basis of contractual claims. Standard methods of measurement such as NRM, the Civil Engineering Standard Method of Measurement ('CESMM'), Fourth Edition, or the Principles of Measurement International ('POMI') are intended to provide instruction to those drafting bills of quantities and certainty for contractors pricing them as to what work is to be allowed for in each item. In practice, failures to follow the principles laid out in such rules causes dispute as to whether work was covered or should be treated as a variation. In lump sum contracts with quantities, such as SBC/Q, there may be express terms to the effect that any departure, error or omission in the measured quantities compared to the requirements of the adopted method of measurement will be treated as a variation under the contract, such as, for example, in SBC/Q clause 2.14.3. Unfortunately those drafting the flawed bills of quantities are often those asked to address the valuation of a resulting variation claim and it seems may be reluctant to admit their original error or be keen to unfairly limit the financial effects.

In remeasurable contracts such as the FIDIC Red Book, items that were missed from the bills of quantities should be picked up in the remeasure. However, the authors have seen it argued by those defending bills that they inaccurately prepared that the remeasure is only of those items set out in the bills. This may be related to provisions occasionally inserted into international contracts to the effect that, notwithstanding the requirement for tenderers to price bills of quantities prepared in accordance with a stated standard method of measurement, tenderers still have to allow for all items on the drawings and specification, even though they are missed in the bills of quantities. This can lead to endless argument as to where a missed item should have been priced and the effects of the express terms of the contract and of the applicable law, custom and practice. It overlooks the fundamental intention of bills of quantities and why they (and the quantity surveying profession) came into the construction industry in the first place. This is to provide consistency and certainty for those tendering for works and for employers comparing their bids and managing their budgets. They should not be abused as a sort of lottery as to which bidders will spot the errors and price accordingly and/or raise tender queries.

1.4.8   Unforeseeable Occurrences

That some unforeseeable occurrences will be met on large and lengthy construction and engineering projects may seem so inevitable that they could be described as 'known unknowns rather than unknown unknowns', to paraphrase a past United States Secretary of Defence. The likelihood of such events being met is high on construction and engineering projects given such factors as:

- the susceptibility of some operations to changes or extremes of climatic conditions;
- the susceptibility of other activities to unforeseen ground conditions and obstructions;
- the susceptibility of all operations to changes in resource availability perhaps because of overheating of the market or strike or other civil action;
- the susceptibility of all operations to changes in resource costs perhaps because of the state of the local economy and/or volatility of the local currency;
- the arguably unique location of all projects;
- the unique nature of the design and/or location of most projects;
- the susceptibility of many types of project to changes in the economic or political environment that might render the project unwanted or uneconomic; and
- the long durations of projects that can run into years, which can also increase the likelihood of a project being affected by unforeseeable occurrences.

Given the propensity for such events, contracts will usually anticipate the type of such occurrences that might occur by allowing time and/or recompense to the contractor for those matters that it is considered should not be at the contractor's risk. This generally means those events of a nature that the contractor can neither predict when pricing its tender so as to price the risk with any certainty other than by a conservative contingency allowance, nor control to avoid their occurrence. Some such occurrences might be cost-neutral, so that the contractor gets an extension of time but not money, for example 'exceptionally adverse climatic conditions' under the FIDIC Red Book. Others might give an entitlement to both time and money, such as unforeseeable 'adverse physical conditions or artificial obstructions' under the Infrastructure Conditions.

A further consequences of some unforeseeable events may be that works have to be suspended or contracts terminated. An example of that is how the last economic crash affected projects, particularly in markets that had been overheating prior to the crash, which saw projects paused while employers took stock of the situation and then terminated. Contracts therefore have to provide for the rights of the parties in those circumstances, including suitable remuneration. Claims in relation to the termination of contracts are the particular subject of Chapter 7 of this book.

For client or contractor, and preferably both, the most appropriate means of monitoring such matters is the establishment of a risk register backed with a risk management and mitigation strategy that will enable events that occur to be managed both physically and contractually with the minimum of disruption and effect on both time and money, resulting in the best prospect of avoiding a dispute over claims that arise.

1.4.9 Breach of Contract

Whilst this discussion focusses on changes that are foreseeable and instigated by the parties or unforeseeable as part of the exigencies of the construction and engineering process, there are other contract claims that require evaluation that simply arise out of a breach by one party or other of its obligations under the contract. Provision of such as late design or information have been mentioned elsewhere, but we are concerned here with breaches that do not go to such as extension of time or delay and disruption claims, but result in a direct and recoverable financial consequence of their own. For example, the employer's failure to pay certified sums within time or its wrongful calling of a bond that the contractor has provided. Such other sources of contractual claim are considered in Chapter 8.

1.5   Summary

Paramount to the proper presentation and consideration of a contractual claim is an understanding of the relevant terms of the contract and the applicable law. Good legal advice should always be taken, particularly where operating in a foreign jurisdiction.

The terms of contracts vary, standard terms can be the subject of heavy amendment and bespoke terms are often applied. The discussions and ideas in this book are illustrated by reference to particular standard form examples. SBC/Q is an example of a lump sum contract in use in the UK. The Infrastructure Conditions and FIDIC Red Book illustrate the approaches of remeasurable contracts in use in the UK and internationally respectively. NEC4-ECC illustrates the more innovative approach of that suite of contracts.

The necessary standard of substantiation of a contractual claim will vary depending on the terms of the contract and local legal requirements. There may also be issues of the remoteness of the damages and an obligation to mitigate the costs. The SCL Protocol gives a guide to good practice in many areas of quantification related to delay and disruption.

On construction and engineering projects change is almost inevitable. How those risks are allocated is a key element of contract drafting. How they are managed will be essential to the resulting quantum. Risk management is an all too often overlooked discipline in some countries. Certain causes of change repeat between construction and engineering projects no matter where in the world they are carried out. Understanding the common risks and causes of change will help parties to plan for them and manage their effects and resulting claims.

In this book, the evaluation of the consequences of change is considered by reference to their direct consequences on the programme (Chapter 4) and money (Chapter 5) and then their time consequences in terms of, for example, prolongation, disruption and acceleration costs (Chapter 6). Chapter 7 looks at claims arising out of the termination of contracts. Chapter 8 looks at some other possible heads of claim. However, first it is necessary to consider how to establish the base from which such claims for additional payments are assessed.

2

Establishing the Base

Chapter 1 considered the underlying reasons that mean change may be anticipated and may need to be accommodated in a construction contract and result in claims for additional payment. In undertaking contracts with a large work scope over protracted periods change may not be inevitable but is likely to some extent. The precise extent or type of change encountered will vary with the circumstances of particular projects but can usefully be considered under two broad headings of planned change and unplanned change.

It is, however, essential that, when considering any type of change, consideration is given to the base from which change is quantified. If the base in terms of time, resources, cost and value is not understood then there will be a high risk that the analysis of any change and additional payment will be flawed. The tender submitted by a contractor will be deemed to incorporate everything that would be anticipated from the contract documents by a reasonable and competent contractor, which may be different from the tender actually submitted. However, the deemed allowances will not be the only source of information for establishing the base from which claims for additional payment will be judged. The base, whether it is a planned change or an unplanned change that is being considered, has to be the construction contract between the parties, and that does not vary with differing post-contract circumstances. The analysis of change, and evaluation of payments, must therefore always refer back to the base of the construction contract and its terms and provisions, including such documents as specifications, drawings, bills of quantities, tender correspondence and even a programme, depending on the extent to which, and how, they were incorporated. In the example of the 8506 pages of contract documents for a tram system, cited in Chapter 1, establishing detailed elements of the base can be a daunting task.

Planned change is that which can be identified in advance and managed in an orderly and timely manner, the obvious example being variations to the scope of the works, which should be detailed and instructed in advance of the need to execute them. By contrast, unplanned change encompasses those events that cannot usually be predicted but occur as the works progress, a common example being the encountering of ground conditions different to those anticipated before the works commenced.

It would, of course, be very convenient if all changes neatly fell into one or other of these categories and could be dealt with accordingly. Regrettably, the way of construction projects is not always so neat and tidy. Variations cannot always be predicted and detailed conveniently ahead of the need to instruct them and on occasions such matters, which would seem to be a planned change, may take on the characteristics of an

Evaluating Contract Claims, Third Edition. John Mullen and R. Peter Davison.
© 2020 John Wiley & Sons Ltd. Published 2020 by John Wiley & Sons Ltd.

unplanned change. However, that does not mean that there is no merit in understanding how changes may occur and need to be identified and assessed for payment, if relevant. Indeed, without such understanding it is difficult to recognise when the boundaries of change have passed from planned to unplanned, with consequences for the evaluation of any payments arising.

2.1 Planned Change

Most contracts for construction projects of any size will incorporate provisions for certain changes to occur. For example, from an employer's perspective the most significant example may be those that empower it to instruct variations and safeguard its right to damages if the completion date is not achieved as a result. Without such provisions the contractor might cease to be bound by the contract completion date if variations were required to the works, thereby preventing the achievement of the date. This may leave time 'at large' and the employer without the safeguard of the contract's delay damages provisions for late completion. It may be that the employer would still have a remedy in unliquidated damages, but subject to proof that the contractor had failed to complete within a reasonable time (with the difficulties of establishing what that time is) and of the employer's actual losses, etc., resulting from that failure. This is a less than satisfactory fall back for the employer and illustrates the need to plan for change and provide for its effects.

Further, contracts may, but do not always, contain provisions whereby the quantities of work provided to the contractor at the time of tender will be adjusted wholly or in part to reflect the actual quantities of work executed, such as the FIDIC Red Book and the Infrastructure Conditions as remeasurement contracts. Even lump sum contracts such as SBC/Q may include provisional quantities for this against which the quantities cannot be accurately determined when the bills of quantities are prepared. Such adjustments of course raise the issue as to whether, or to what extent, the tendered unit rates and prices continue to apply to the adjusted quantities.

However, planned change rarely exceeds the scope of ordered variations or quantity changes, as most other changes contemplated by contracts will to a greater or lesser extent be unplanned. Indeed, ordered variations themselves may occur in an unplanned manner, leaving them to be treated as unplanned change, the treatment depending upon the particular facts and circumstances of any instance.

The compensation events regime of NEC4-ECC contemplate that variations will be pre-priced based on a forecast of actual costs. The Infrastructure Conditions (clause 52.1), the FIDIC Red Book (clause 13.3, where it is part of its 'proposal') and SBC/Q (clause 5.3.1 and Schedule 2) all provide for the contractor to present a quotation in relation to proposed variations. The circumstances of such a planned change will be very different to those of variations that are unplanned. All too often, variations arise from, for example, correspondence, verbal directions, the engineer's abuse of shop drawing approval procedures, materials or subcontractor approval procedures, or from responses to contractor's requests for clarifications of the design. They may not be formalised by a proper instruction under the terms of the contract until much later and after the varied work is long completed. Such issues give rise to problems both in the circumstances of carrying out variations and getting their values agreed and are common causes of disputes on construction projects. It might be, for example, that the variation is of a nature that would best have been carried out under the dayworks regime of the contract, such as that in FIDIC Red Book clause 13.6, with daily records of resources kept and submitted to the engineer. If a change is not even recognised as a variation, let alone instructed to be carried out as daywork, the opportunity to record the resources on it, as a basis of ready and uncontroversial valuation, will be lost.

It should also be recognised that not all ordered variations, or changes that the contractor considers to be variations, may give rise to an entitlement to additional payment. For instance, a change to the location of a site establishment from a position shown on the pre-contract information within the boundaries of the work site to another position remote from the work site, and requiring the workforce to cross a narrow footbridge over a road to pass to and from the works, was not a variation under the MF/1 Form of Contract because such a change is not within the definition of variations in that contract. That contract contained an 'entire agreement clause', which operated to exclude any matter outside those expressly contained in the contract, thereby denying any argument that this was indeed a variation, as the pre-contract information had not been incorporated into the contract itself (*Strachen & Henshaw v Stein Industrie (UK) Ltd* (1997) CILL 1349. It is therefore important, as ever, to consult the terms and conditions of any particular contract when trying to ascertain if a particular change can be regarded as a variation under the contract. This neatly illustrates the need for three steps required in any circumstance to establish an entitlement to additional payment (setting aside any issue as to notice as a possible condition precedent):

- Confirm that the circumstances are within the variation provisions of the contract, express, or implied.
- Confirm that the variation, if within the variation provisions, gives rise to an entitlement to additional payment.
- Determine the rules, if any, for the evaluation of the additional payment.

In considering the first step the limitations of variations provisions will need to be borne in mind. The ordered work may clearly be an addition or change to the contract works but the scope of the additional work, or the nature of the change, may be too extreme to fit within the contract. Most contract forms have widely drafted clauses setting out the extent of variations that can be ordered, but that does not mean the architect or engineer has the right to order variations that are unreasonable in the circumstances of the contract. In *Blue Circle Industries PLC v Holland Dredging Co Ltd* (1987) 37 BLR 40, it was decided by the Court of Appeal that the construction of an artificial island in Lough Larne using dredged material from the contract, instead of disposing of the dredged material within the Lough in areas to be approved by the Local Authorities, was outside the scope of variations that could be ordered under clause 51 of the ICE Conditions, Fifth Edition.

There may also be a variation requiring additional payment where the circumstances giving rise to the change were known to the employer's representative prior to the contract but were not treated in accordance with the contract provisions. In *C. Bryant and Son Ltd v Birmingham Hospital Saturday Fund* [1938] 1 All ER 503, the architect for a convalescent home knew that rock might be encountered in the excavations. The contract bills referred the contractor to the drawings and a block plan and required it to satisfy itself as to the nature of the site and conditions but contained no separate item for the excavation of rock. The contract was under a forerunner of the SBC/Q and JCT Standard Forms, the RIBA Form, and contained at clause 11 a provision that the quality and quantity of work to be included in the contract sum would be that which was set out in the bills of quantities. The bills of quantities were stated to have been prepared in accordance with the then current standard method of measurement for building work prepared by the Royal Institution of Chartered Surveyors. That method of measurement required that excavation in rock should be given separately.

It was subsequently held that the contractor was entitled to be paid for the excavation in rock, measured in accordance with the method of measurement, as a variation to the contract sum, calculated as the extra cost of excavation with the addition of a fair profit.

The SBC/Q Form still contains a similar provision at clause 2.13, that the bills are measured in accordance with the standard method of measurement and clause 2.14, which provides that any correction, alteration or modification required as a result of error or inadequacy in the bills will be treated as a variation.

2.1.1   Ordered Variations

The three steps required to establish an entitlement to additional payment in relation to variations conclude with the need to ascertain the rules, if any, set out in the contract for the evaluation of variations. In many instances the rules will seem straightforward, but care should be taken in reading and considering them as many rules raise issues that may not be immediately apparent on a first reading.

In Chapter 5 we consider in detail typical provisions for the valuation of instructed variations. However, such valuation must start with a definition of what has changed and the details of how it has changed. This in turn must start with establishing the base of what was allowed in the contract before that variation. In theory that base ought to be capable of ready determination by reference to the drawings and specifications. However, in practice, that is not always an easy task. Contract drawings and specifications may be unclear, incomplete, contradictory or to a lesser level of detail to the construction designs to which comparison is being attempted. Referring again to the example of a contract comprising 8506 pages plus drawings, the difficulties can be extreme.

A further complication is where the definition of 'variation' includes aspects of work other than their physical design. For example, the FIDIC Red Book sets out in its clause 13.1 a definition of that variation that may include numerous types of change, including as item (f) 'changes to the sequence or timing of the execution of the Works'. Thus a variation under the FIDIC Red Book clause 13.1(f) will require establishing the baseline of the sequence and timing of work as they were before that change. Similar provisions are contained in SBC/Q clause 5.1.2 and NEC4-ECC clause 60.1. NEC4-ECC requires an Accepted Programme. The FIDIC Red Book clause 8.3 requires the contractor to issue a programme and revisions to the Engineer. SBC/Q clause 2.9 provides for a master programme. In theory these ought to provide a baseline against which a variation that involves changes to the sequence or timing of the works can be assessed. In practice this often proves to be more difficult than might be hoped.

The Infrastructure Conditions contain a series of rules for valuation of ordered variations. The first rule, clause 52(1), contains the provision for the engineer to request a quotation from the contractor for any proposed variation, including any consequential delay, for agreement before the order is given and work commenced on the variation work. This has to be good practice wherever possible, regardless of any provision in the contract for such a step. In an ideal world many disputes over valuation of variations would be avoided by the adoption of such a procedure or similar ones, such as those in SBC/Q and the FIDIC Red Book.

In practice it is not always possible to fully define the scope or agree the costs arising and the Infrastructure Conditions clause 52(3) recognises this by setting out the rules for valuation if agreement has not been reached as provided by 52(1). Part (a) of 5(3) provides essentially for a continuation of the quotation and negotiated agreement contemplated in 52(1). If all attempts at agreement fail, the rules for valuation by the engineer appointed under the contract are contained in clause 52(4) where the valuation is required to be based on the rates and prices in the bill of quantities, but in two distinctly differing circumstances:

- Where the varied work is of similar character or is carried out under similar conditions to the work in the bill of quantities then the varied work is to be valued at the bill of quantities rates and prices 'as may be applicable'.
- Where the varied work is not of a similar character or is not carried out under similar conditions (or is executed during the defects correction period) then the bill of quantities rates and prices are to be used as the basis of valuation 'failing which a fair valuation shall be made'.

There are further provisions in part (5) of clause 52 of the Infrastructure Conditions for the contractor or engineer to challenge the bill of quantities rates and prices as being unreasonable or inapplicable because of a variation. The onus again falls upon the engineer to fix the rate it deems proper in the varied circumstances. Finally, under part (6) the engineer can order work to be executed on a daywork basis if it deems that to be the proper means of evaluation.

These rules, however, raise two issues on careful reading:

1. How does one determine what the character of the work is or the conditions under which the work is performed are? This is a fundamental matter that needs to be resolved before it is possible to decide whether or not the bill of quantities rates and prices apply, but the contract gives no guidance on where the character or conditions are to be found.
2. What is a 'fair valuation' as contemplated by clause 52(4)(b)? Fair to whom?

The issue of how to determine the character and conditions for the work in the event of variations occurs in other forms of contract, notably in the SBC/Q Form where, in clause 5.6.3, the valuation rules provide that if work is not of a similar character to that set out in the contract bills then it is to be valued at 'fair rates and prices'. Work of a similar character, but executed under differing conditions, is to be valued by using the bill rates and prices as a basis but shall include a 'fair allowance' for the differing conditions. In the valuation rules in clause 5.9 it is required that if a variation substantially changes the conditions under which other work is executed then that work is to be treated as varied and valued accordingly.

All these expressions raise the issue of how to determine the 'character' and 'conditions' on which the bill of quantities rates and prices are based. Without being able to determine that issue there is no basis on which to proceed with an evaluation.

The first part of the answer to this issue is, as with many other matters, to be found in the contract documents.

In *Wates Construction (South) Ltd v Bredero Fleet Ltd* (1993) 63 BLR 128 the Judge was asked to consider an appeal from the award of an arbitrator who had determined how the 'similar conditions' and 'character' of clause 13.5.1 in the 1980 JCT Form (a forerunner of the SBC/Q Form, which uses identical terms in its clause 5.6.1) could be ascertained. The arbitrator had decided that the information to be considered went further than that contained in the contract documents and that the contractor had obtained further information in relation to the conditions of the work during a period of protracted negotiations over the tender for the project. The Judge stated:

> In my opinion, the material extrinsic to the contract documents ... to which the arbitrator expressly had regard in interpreting clause 13.5 of the contract, includes at least the following:
>
> (i) the state of Wates' actual knowledge of conditions when pricing the cost plan;
> (ii) the knowledge of conditions which it is supposed or assumed that Wates gained as a result of negotiations;
> (iii) the 'perceived' changes in the conditions resulting from certain proposals which had been accepted by Wates ...

The Judge concluded that the arbitrator was wrong to consider such 'extrinsic' matters when determining the contract 'conditions':

> I am thus satisfied that the words of clause 13.5.1.2 clearly and unambiguously express the parties' intention that the conditions under which the contract works were to be executed (i.e. 'the contract works conditions') are those conditions which are to be derived from the express provisions of the contract bills, drawings, and other documents. In my judgement, extrinsic evidence was not admissible to affect or modify the clear and unambiguous language of the contract by means of which that intention of the parties was expressed.

This is useful guidance on the interpretation of contract valuation rules and firmly indicates the effect of phrases such as those used in clause 13.5.1.2 of the 1980 JCT contract. That clause is replicated in SBC/Q clause 5.6.1.2. Similar words are used in clause 52(4)(a) of the Infrastructure Conditions. The FIDIC Red Book clause 12.3 is set out somewhat differently, but the phrase 'not of similar character, or is not carried out under similar conditions' is in subclause (b)(iii). Applying the approach in *Wates v Bredero*, in such contracts the 'character' and 'conditions' of work will be determined by reference to the contract documents alone. That would be the position in England and Wales, but the position could be quite different elsewhere and demands local legal advice. If either party wishes matters considered in pre-contract negotiations to be included in the formation of base 'conditions' or 'character' then those matters will need to be incorporated in the contract documents. In practice this is sometimes achieved by the inclusion of pre-contract meeting minutes or correspondence in the contract documentation, reinforcing the need to consider all aspects of the contract documents carefully, but also meaning that those documents can become lengthy.

Although this is helpful in establishing that the base 'conditions' and 'character' of the work can be found from the contract documents, it does not assist in deciding what constitutes the conditions and character of the contract work. Unfortunately, although allegations of changes in conditions and character often occur in construction contracts, there are no authorities to provide helpful guidance.

This can lead to disagreement as to what the terms 'conditions' and 'character' actually mean. However, in a construction context how do these expressions assist in determining the base conditions and character of the contract works?

It might be that the most useful approach is to consider matters of 'conditions' as being matters affecting the circumstances in which the work will be executed. Therefore, an examination of the contract documents would be needed to ascertain relevant factors, which may include, for instance:

- If the work is described as being executed in abnormal circumstances, for example outside the boundaries of the site or in existing buildings, in tidal areas or areas susceptible to flooding, or is subject to other extraneous factors that will affect the cost of executing the work.
- If, from the contract documents, it can be established that work will have to be undertaken at a time that will affect the cost of the work, for example external work in winter conditions.

The character of the work might most usefully be considered to be factors intrinsic to the work itself rather than the conditions under which it is executed, for instance:

- The general arrangement of concrete work that dictates how many repeat uses of formwork will be anticipated or
- The amount of detailing in brickwork that will affect the overall rates paid to subcontractors for bricklaying.

In conclusion, whenever the pricing of variations is being considered, and the applicability of the contract rates considered, it will usually be the case that an examination of the contract documents will be necessary to decide the 'conditions' and 'character' priced into the contract rates and prices. It is from this base that changes to rates in the contract will be considered, if necessary.

It was stated that it is from a consideration of such factors, based on the contract documents, that the 'conditions' and 'character' can be established as the base, which is deemed to be included in the contract rates and prices. This base comprises the economic circumstances of the work and should entail a proper assessment at the contract tender stage of the effects of all factors that will affect the costs, and value, to the contractor of the work being priced.

Where the contract terms allow for reconsideration of prices on changes in 'conditions' or 'character' then an analysis will be required, not only of the base circumstances but also of the change, to determine the effect on the economics of the operations as they were before and after the change. The means of making such adjustments is considered in more detail in Chapter 5.

2.1.1.1 Fair Rates and Prices

SBC/Q clause 5.6.1.3 refers to 'fair rates and prices'. The Infrastructure Conditions clause 52(5) refers to rates and prices that are 'reasonable and proper', which is the phrase used in earlier ICE drafted contracts on which the Infrastructure Conditions were based. Expressions such as 'fair rates and prices' do not imply that a contractor can submit a price for the total scope of a variation subject to such pricing without providing identification of the constituent items within the variation scope and particularising the rates and prices applied to the items. Judge John Lloyd QC supported the need for such analytical pricing when, in giving judgment on an appeal from the award of an arbitrator in *Crittall Windows Ltd v T. J. Evers Ltd* (1996) 54 Con LR 66, he stated in regard to a valuation under clause 13.5.1.3 of the old JCT Form:

> ... where a valuation of this nature falls to be made it is necessary for the individual items which are to be valued to be particularised and, once they have been identified, to be priced.

The level of particularisation will depend upon the nature of the variation scope and the measurement rules adopted in the contract. However, even in instances where there are no stated rules of measurement in the contract a reasonable itemisation of the work, and rates and prices applied, should be required. It is perhaps part of the 'fair' element of such clauses that both parties can see the detailed build-up to the total amount claimed and therefore have the opportunity to assess the reasonableness or otherwise of the prices used.

A subjective phase such as 'fair rates and prices' is inevitably going to lead to doubt and disagreement. Are fair rates and process to be judged by reference to the levels of rates and prices set out in the contract? Thus, if the contract rates include unusually high allowances for overhead and profit, should the same allowances be made? What if the contract would actually have been loss making? Alternatively, must the valuation of fair rates and prices be looked at more objectively, to avoid a valuation that has passed the use of contract rates and prices straying back into that basis?

Some guidance is available from two cases heard by HHJ Humphrey Lloyd in the TCC relating to the Sixth Edition of the ICE Conditions (forerunner to the Infrastructure Conditions and from the same roots as the FIDIC Red Book). In *Henry Boot Construction Ltd v Alstom Combined Cycles Ltd* [1999] BLR 123 he decided that 'A fair valuation when used as an alternative to calculation by reference to contract rates and prices generally means a valuation which will not give the contractor more than its actual costs reasonably and necessarily incurred plus similar allowances for overheads and profit'. This judgment is considered in more detail in Chapter 5.

In *Weldon Plant Ltd v The Commission for New Towns* [2000] BLR 496, he decided that 'In my judgment a fair calculation must, in the absence of special circumstances ... include an element on account of profit'.

A further consideration is whether such pricing should be fair to both parties. This issue is considered in more detail in Chapter 5. If so, then applying these judgments would achieve that. It would provide the contractor with an economically viable price, covering costs and adding margin, while not requiring the employer to pay an unreasonably excessive amount by way of unreasonable or unnecessary costs and margin. Provision of a detailed particularisation makes the establishment of fair rates and prices more attainable as the cost of the elements can generally more readily be compared to other data than the price for the whole. For example, labour rates can be compared with published and other available information on rates for the particular trades, costs of materials checked with suppliers, etc.

Such changes can, however, have an effect beyond that of the direct economic consequences of the change to the particular construction operation. If the operations are critical to the progress of the works towards the contract completion date there may be consideration of delays as a result of the changes. If the changes to conditions or character of particular operations affect other construction operations there may be disruption to those operations, with economic consequences with or without effect upon the completion date.

2.1.2 Changes in Quantities

Chapter 1 of this book commented on bills of quantities, their purpose and the origins of the quantity surveying profession. Not all contracts provide quantities of work for the contractor to price. Contracts based on a 'design and build' philosophy will leave quantities to the contractor, and some contracts where design and construction are separate may still require the contractor to take responsibility for determining the quantity of work to be performed, the employer providing only drawings and specifications for pricing. There has developed over the years resistance among some employers to incurring the time and expense of having their quantity surveyor prepare bills when contracts such as SBC/Q (clause 2.14.1) expressly make the employer liable for any errors. Some employers see this as a basis for nefarious claims by contractors and unexpected increases in costs on what they thought was a fixed price, lump sum contract.

What is critical is the examination of the contract documents to ascertain the purpose of any quantities provided, and what, if any, rules apply to the measurement of the quantities and the rates and prices set against them. If quantities are not incorporated in the contract then there will be no entitlement to any adjustment of price consequent on changes in quantities other than to the extent that they are required by instructed changes. If the quantities define the work in the contract then there will usually be an entitlement to adjustment of the contract sum consequent on quantity changes, subject to the terms of the contract and any rules contained therein in relation to the quantities.

Where quantities are not incorporated in the contract there may be an issue on the calculation of any rate to be applied to subsequent variations in the scope of the works where the variations are required to be valued at the same rates and prices as the contract works. For instance, if an item for pricing requires the contractor to provide a lump sum for painting of concrete soffits and it includes the sum of £5,000 (having assessed the requirement as being $500 \, \text{m}^2$ at £10 per m^2) but subsequently discovers that the actual requirement based on the contract drawings was only for $400 \, \text{m}^2$, what rate applies to any variations for further painting of concrete soffits? Is it the £10 per m^2 assessed by the contractor or the £12.50 per m^2 it is recovering based on the actual quantity of work in the lump sum item? The answer should be the £12.50 per m^2, as the contractor is bound by any errors in its pricing and gets the benefit or disadvantage depending on whether it overpriced or underpriced. It would still be bound if the actual quantity for the lump sum item were discovered to be $1000 \, \text{m}^2$, thereby halving its rate to £5 per m^2.

Essential to establishing the base where contracts provide quantities will be understanding what the items against which the quantities and the contractor rates and prices have been set are to cover. To this end there are usually, but not always, detailed rules for the measurement of the quantities and the definition of the matters that are included within each item description. This is often referred to as the 'item coverage'. The most appropriate means of establishing such rules is to use a standard method of measurement suitable for the type of work in the contract. In 2013 the RICS's New Rules of Measurement ('NRM') replaced the Standard Method of Measurement of Building Works, which had been first published in the UK in 1922. The Civil Engineering Standard Method of Measurement ('CESMM'), the Standard Method of Measurement for Industrial Engineering Construction ('SMMIEC') and the Pipeline Industries Guild Method of Measurement are standard methods devised for the measurement of particular types of construction work. The first two of these sometimes also appear overseas, particularly where British quantity surveyors are involved. The RICS's Principles of Measurement International ('POMI') are particularly popular internationally. Further afield there are such as Qatar's Method of Measurement for Road and Bridgeworks, South Africa's Standard System for Measuring Building Works and the Australian Method of Measurement of Building Works, among others.

It is also possible, and not unknown, for members of the client's team to measure work on their own rules or scheme of measurement. There is nothing intrinsically wrong in this but if undertaken it must involve setting out the rules or scheme adopted in the contract documents, including a detailed and well thought out approach to 'item coverage', so that the measurement and coverage of the work items will not themselves become contentious and the seedbed of unnecessary dispute.

Sadly, a systematic and consistent approach to the preparation and measurement of bills of quantities is not always adopted, even in large-scale contracts, where the costs would be relatively insignificant but the price of error can be significant. The use of a contract term requiring the contractor to take the risk of items missed from the bills but shown on drawings and specifications has been highlighted in Chapter 1. As an additional example, it has been known for the civil works for a new build large power station to be measured on the basis of notes in a preamble to the bills stating, among other matters, that 'only major items have been measured, minor items are deemed to be included with the major items'. There was no definition of 'major' or 'minor' items in the notes; it had to be assumed that everything not measured was a 'minor' item. It does not take an expert in measurement of construction work to anticipate that such an approach will almost inevitably result in dispute as to:

- Where major items of construction had not been measured, which were larger than many 'minor items' which had been measured and were these errors rather than omissions?
- Where new construction items were required by variations and instructions, were these all 'major' or 'minor' items or were they a mixture of the two? If a mixture of the two, which were 'major' and which were 'minor'?

Such situations are unnecessary and disruptive to the working relationships on a project and only require a modicum of forethought and expertise to obviate. It was reported that, in the case of *Priestley v Stone* heard in 1888, a witness for the contractor stated:

> If we do not think much of the quantity surveyor, and entertain any doubts about him, then we take care to have the quantities put into the contract.

With modern quantity surveying systems there should be no need for such distrust!

2.1.2.1 Quantities and Conditions

There will of course be a difference in the treatment of quantities provided for tendering purposes, depending upon the nature of the contract. Lump sum contracts such as the SBC/Q Form use the quantities to establish the lump sum and changes in the quantities do not automatically affect the contract sum. By contrast, remeasurement contracts such as the Infrastructure Conditions anticipate that the tendered quantities will be completely remeasured to reflect the actual quantities of work undertaken and the contract sum adjusted accordingly.

SBC/Q states in clause 4.1 that the quantity of work is deemed to be that set out in the contract bill of quantities, and further provides in the ensuing clause that there shall be no adjustment other than expressly provided in the contract conditions. There is therefore no question of wholesale remeasurement, only adjustments as provided elsewhere in the contract. Errors or departures from the stated method of measurement can be corrected under clause 2.14.1, by remeasurement at either party's instigation, but otherwise the scope for adjustment of quantities is limited.

SBC/Q includes in its definition of the term 'Variation', at clause 5.1.1.2 the alteration of the kind or standard of any materials or goods for use in the works, but this means that adjustments to quantities set out in the bills of quantities will only be made as a result of a valid variation or correction of errors.

By contrast the Infrastructure Conditions contemplate in clause 56(1) that the value of the works will be ascertained by the engineer by measurement, as the quantities set out in the bill of quantities are estimated quantities and are not to be considered as the actual quantities of work to be undertaken by the contractor (clause 55(1)). The expression used in this particular instance is that the value of the contract will be determined by 'admeasurement'. This can only mean that, as the quantities are expressly stated to be estimated, the final value is to be determined by a remeasurement of the works.

The Infrastructure Conditions are not the only contract conditions to contemplate a complete measurement of the work executed and, as usual, only an examination of the particular conditions or amendments to standard conditions will establish the extent of measurement in any specific contract.

From the discussion in the previous section, it will be appreciated that remeasurement of the contract work raises two issues:

- Do the bill quantities contribute to the establishment of the 'character' of the work covered by the rates and prices in the bill?
- Can changes in quantities alter the character of the works, or otherwise require the rates and prices to be adjusted?

The answer to the first question would logically seem to be yes. It is difficult to envisage circumstances on a civil engineering contract when the extent and mix of quantities would not be considered by the contractor in forming a view as to the nature and form of the required construction operations, i.e. the 'character' of the works. The anticipated requirements for plant, equipment and labour will be based on the various quantities of the work to be executed, and those same quantities will influence the working methods to be employed. They will also be relevant in determining the sequence, duration and timing of the works. It therefore seems that the bill quantities are an intrinsic part of the character of the works.

This relationship between the quantities and the character of the works is expressly recognised in the Infrastructure Conditions by clause 56(2) and by FIDIC Red Book clause 12.3(a). The former allows the adjustment of the bill rates and prices if, in the opinion of the engineer, any increase or decrease in actual quantities from the estimated quantities renders such an adjustment applicable. The answer to the question as to whether changes in quantities can alter the character of the works, or otherwise require the rates and prices to be adjusted, is therefore also yes, where terms such as those in the Infrastructure Conditions are concerned. However, in the absence of express provisions in other remeasurement contracts for such an adjustment of rates and prices, the contractor may have to examine the contract conditions to decide if such an adjustment is permissible, can be implied or is expressly ruled out under the contract terms.

For 'lump sum' contracts such as the SBC/Q Conditions the contractor will need to recognise that it will only be able to claim changes to rates where the provisions of clause 5 allow, as a result of variations, or where correction of errors under provisions such as clause 2.14.1 of the SBC/Q Form is allowed. Clause 4.1 has the effect that the quantities in the contract bill of quantities will be those that are considered in establishing the contract rates and prices, and changes in quantities will not otherwise be a mechanism for adjusting the rates and prices tendered.

2.1.3 Preferential Engineering

Ordered variations and remeasured changes in quantities should be reasonably clear changes to the contract, capable of being established by reference to the baseline of the contract and comparison with what has been instructed to be actually carried out. Unfortunately, not all changes are so easy to define or agree.

Two areas of frequent disagreement that cause difficulty in evaluating any entitlement to additional payment are:

- Preferential engineering: changes arising from agreement of design details between the members of the project team that incorporate differences to the content of the contract.
- Variation by stealth: changes made to the design of a project during the approval of contractor design by the client's project team, a process not infrequently encountered on projects of a design and build nature.

There is a similarity between these two processes in that, in many cases, the participants involved in bringing about the change will often genuinely not be aware that they have changed the contract works. The common basis for such changes is that they were 'thought to be within the contractor's existing obligations'. This has already been raised as an issue in Chapter 1, particularly where the contractor has an obligation to produce shop drawings for approval, but complains that the process is being abused by the engineer or architect by using it to require what are actually changes to their design on which the shop drawings were based and that the contractor is only required to detail. The designer's retort may be that it considers that it has properly required a submitted drawing to be subject to revision and resubmission because the contractor has its details wrong against those designs.

We are concerned here with the evaluation of such changes, rather than establishing a contractual liability, but the basis for both processes will be to establish exactly what was within the contractor's obligations under the contract and then define the change so that it can be evaluated in the same manner as any other similar change. A difficulty that often arises with such changes is that, because of the way in which they arise, they are not recognised as such until some time after the event. This can exacerbate the not unusual problems of inadequate records of the changed work, or costs incurred, and that may be detrimental to a contractor who much later establishes an entitlement to payment for such changes but does not have adequate records in relation to its valuation.

Often, the progress of a project can be such that the contractor in particular is too concerned with getting on with progressing the works and does not have time to properly sit back and assess the situation. However, the only sensible policy is for both, or all, parties to a contract to:

- ensure that all personnel involved in the agreement of, or obtaining approval for, construction works and operations have a good appreciation of the basis of the contract and the work scope contained in it, and
- reinforce the never-ending need to keep proper records of the works and costs.

Such precautions will reduce the incidence of disputes that commonly arise as to the evaluation of changes on projects with a contractor design element.

2.1.4 Value Engineering

Value engineering is a process by which the required function of a project is established and the costs managed so as to achieve that function at the lowest cost. It was born in the manufacturing company General Electric Company of the United States during the Second World War as a method of looking at alternative material components that could achieve their required function but at a cost saving. It has since, slowly, spread around the world and into other industries, including the construction and engineering sectors.

There is clearly much potential benefit for an employer in harnessing the knowledge of contractors in relation to materials and methods that can lead to savings in costs without unacceptably affecting the quality of the end project in terms of such things as speed of delivery, functionality, aesthetics and life cycle costs. The contractor can be encouraged to this end by a sharing approach to any saving achieved. Value Engineering is particularly popular in the public sector, where achieving value for money for the taxpayer and under limited budgets are major considerations. The contract provisions of Concession Agreements and Design Build and Operate Projects, for example, should particularly be expected to contain such a term.

Of the standard forms considered in this book, only the FIDIC Red Book contains a Value Engineering provision, at clause 13.2. This was new in the 1999 edition of the FIDIC contracts; earlier editions did not include value engineering provisions. Clause 13.2 is in a typical form to that of bespoke Value Engineering clauses that are included in many ad hoc contracts.

Typically, the contractor may issue a proposal to change the design in ways that will accelerate completion, reduce the employer's costs, improve the value or efficiency of the end project or otherwise benefit the employer. The proposal includes a financial proposal. If it is accepted, then the contractor designs the change. Commonly the resulting saving in the contract price is shared between the contractor and employer. However, this approach only addresses the saving in costs and takes no cognisance of the fact that there might be some reduction in value or functionality of the end project to the employer. This should be a key consideration of the employer and its advisors on receipt of a contractor's variation proposal. Under the FIDIC Red Book this is addressed in the calculation of the contractor's resulting fee under clause 13.2, which is calculated as half of the difference between the saving in the contract price and the reduced value of the varied works to the employer.

Problems that arise with such an approach include the resetting of the baseline for the consideration of changes to the contract scope of works. This includes the requirement that the design of the value engineered works is by the contractor and leads to the question as to how this interfaces with the design by the employer's consultants of other works. It can also be that change arising from the value engineering exercise becomes entangled with changes instructed under the contract on behalf of the employer such that disagreement arises as to the scope of changes.

In terms of evaluation, under the FIDIC approach, assessing the reduction of value of the project to the employer includes 'reductions in quality, anticipated life or operational efficiencies'. Such issues are outside the scope of this book, but they can each give rise to difficulties of their own. They can be particularly subjective issues.

A particular feature of the FIDIC Red Book approach is that the contractor pays for the costs of making its proposal. With no guarantee that the proposal will be accepted and no provision for the costs to be included in its fee, contractors may seem reluctant to expend such costs. In practice, proposals under clause 13.2 to accelerate completion of the works are especially rare.

In the NEC suite, one innovation of NEC4 was to introduce the benefits of value engineering into the priced contracts, Options A and B (it was always part of the target contracts). The new provision is in clause 63.12 and relates to any change in work scope that reduces the Defined Cost and was proposed by the contractor:

> If the effect of a compensation event is to reduce the total Defined Cost and the event is a change in the Scope provided by the *Client*, which the *Contractor* proposed and the *Project Manager* accepted, the Prices are reduced by an amount calculated by multiplying the assessed effect of the compensation event by the *value engineering percentage*.

2.1.5 Unconfirmed Instructions

In addition to the problems of preferential engineering and variations by stealth there may be instances where there is genuine disagreement as to whether or not an instruction issued by the engineer or architect does in fact constitute a change under the contract, giving rise to an entitlement to additional payment.

In most circumstances this will arise where the contractor contends that work required by an instruction constitutes a variation, but that is denied by the engineer or architect who respond that the contracted works remain the same. Many contracts do not make specific provision for such occurrences but some do, for instance clause 14.7 of the CRINE General Conditions of Contract for Construction issued in 1997 for use in the oil and gas industries.

Such clauses generally do no more than spell out what would be the reasonable course in any event for parties in such circumstances. That is, they should ensure that the situation is properly notified and full records kept and regularly submitted so that everyone is aware not only of the nature of the disputed variation but also the resources and any other elements involved in the execution of the variation. It is also sensible for the party disputing the variation, usually the engineer, architect or employer, to set out what measures they require to be taken and what specific records they require, without prejudice as to liability as to the alleged entitlement or the contractor's right to keep and submit further records it considers relevant and necessary.

It could be argued that the inclusion of similar provisions in other standard forms of contract might be a useful spur to encouraging better recording of disputed or contentious instructions or variations, but there is little doubt that where such provisions exist, if properly followed, they will eliminate the seedbed of many disputes that arise over the evaluation of claims on construction projects. Effectively they would limit the disagreement over whether or not there is a variation, and if it is later determined that there was, the records will avoid a second disagreement over its valuation.

2.2   Unplanned Change

Changes to contract works do not always occur in a planned manner, or even a semi-planned manner, as in the case of preferential engineering and unconfirmed instructions. In contrast to those situations there may be occasions when changes occur or are required as a result of extraneous circumstances rather than a decision of the employer's team to change the design or the contractor's submission of a value engineering proposal. A common occurrence of an unplanned event that is likely to change the works is encountering unforeseeable physical conditions, as envisaged by, for example, FIDIC Red Book clause 4.12. It should, however, be remembered that, depending upon the terms of the contract, less tangible factors such as changes in economic circumstances, for instance in the shape of tax changes or exchange rate fluctuations, may require consideration.

When considering physical factors causing unplanned change, the first consideration, after definition of the physical condition, has to be the assessment of the factors in the contract that have been impacted by the change. As stated previously, one has to look to the contract to establish what was contained within it at the outset in order to conduct an analysis of the impact of any unforeseen or unplanned changes. In practice this will require the establishment of such as the intended programme of work and working methods, always bearing in mind that it is what should have been the interpretation and requirements of the contract documents by a reasonable and competent contractor that will be the starting point, and not necessarily the contractor's actual tender or programme.

2.3   Programmes and Method Statements

2.3.1   The Status of Programmes

It is unusual, but not entirely impossible, to find that the contractor's programme has been incorporated into the contract. There are several related reasons why this is both a usual and undesirable approach:

- At the time of entering into the contract, the programme may be only in outline form, suitable for estimating and tendering purposes. It is very unlikely to be to the level of detail required of a construction programme for project management and control, incorporating the full detail required. For example, the authors have seen a 65 km dual carriageway highway programmed in the successful contractor's tender submission on one A4 sheet of paper using spreadsheet, rather than programming, software. Quite how the employer's tender adjudication team decided based on such scant planning that the contractor understood what it was to build, knew how to construct it and was the best tenderer was never explained. Any such tender programme will have been prepared based on assumptions made during the tender period that require further proper consideration.

- The programme at that stage will have been prepared by the contractor's bid team. Its construction team may have significantly different views on methods, sequences, and durations. It rarely makes a good early impression with employers and their advisors, when the first thing a contractor's project team does when it arrives on site is to issue a significantly revised programme, but if the previous programme was made a contract document, then this can be even more of a problem.

- If the tender programme was to be incorporated into the contract it could, depending on the terms under which it is incorporated, become an obligation on both parties. Thus, the contractor would have to perform in accordance with the details of that programme and the client would have to fulfil its own obligations, such as the release of information, so as to ensure that the contractor is not prevented from so doing. Thus failure of either party in relation to that programme could become a breach of contract.

There may be occasions when the parties wish the exact sequence and timing of operations to be part of the contract between them, but the dangers and difficulties of such an approach can easily be imagined. It is also fundamental to most construction contracts that the contractor should retain the responsibility for the management and organisation of the works, without unnecessary dilution of that responsibility. For that reason most construction contracts do not incorporate the contractor's programme but restrict the contract terms to dates by which the whole, or parts, of the works have to be completed. Some contracts, particularly contracts for 'minor' works, do not refer to or require a programme at all. Other contracts, for works of a more substantial scale including the SBC/Q, the FIDIC Red Book and the Infrastructure Conditions, do require a programme but in differing terms.

That does not mean, however, that the contractor's programmes are irrelevant for the analysis of changes. Far from it. As with the establishment of the 'character' and 'conditions' of the works, there must be a means of establishing base sequences, duration and timings of works from which to measure change to those.

Clauses 2.9.1.2 and 2.9.2 of the SBC/Q Form require the contractor to provide the architect with its 'master programme' for the works and any revisions upon decisions under part 2.9.1.2 (delays to completion), but specifically states in 2.9.3 that nothing contained in the programme shall impose any obligations beyond those in the contract documents. Importantly, the contract does not specify what form the programme should take or what should be incorporated in it. There is therefore no explicit requirement for the contractor to identify the critical path, or paths, through the programme of work leading to the required completion date(s). What will constitute an acceptable master programme will depend upon the circumstances of the project and the nature of the work to be undertaken, but there is no obligation to produce the information in any particular form and nor does the architect have to approve the programme. This is a situation that, in practice, can result in the tabling of an inadequate or incomplete programme as the master programme, with no power to require its resubmission in an acceptable, and usable, form. The only recourse for the employer would be to refer to those inadequacies and its concerns in any ensuing dispute as to the causes of delays to progress and completion dates.

In contrast, clause 14 of the Infrastructure Conditions not only requires the contractor to submit to the engineer its programme for execution of the works within 21 days of being awarded the contract, but the engineer has to approve or reject the programme. If rejecting the programme the engineer has to give reasons for so doing, or the contractor can be required to supply further information, clarify or substantiate the programme. In essence the engineer is able to satisfy itself that the programme is reasonable in the circumstances. As with SBC/Q, there is no specification as to what form the programme should take, what should be incorporated in it including identification of the critical path(s), etc. However, the engineer is in a much stronger position than the architect under the SBC/Q Form in obtaining the information required to identify critical activities if the submitted programme does not already identify them.

Problems often arise where the contractor considers that the employer's representative has unreasonably refused to approve a programme. In this regard, provisions such as FIDIC Red Book clause 8.3 give the engineer power to notify the contractor that the programme does not comply with the contract. That clause provides some statement as to what a programme shall include, but is quite superficial and considerably less detailed than, for example, the model programme provision of Appendix B of the First Edition of the Society of Construction Law (SCL) Protocol. Internationally it is not uncommon for construction contracts to add provisions to FIDIC terms that further detail the requirements of the contractor's programmes in terms of software requirements and such like.

Whilst the contractor may argue that the rejection of its programme was unreasonable, the engineer may argue that the programme was genuinely non-compliant with requirements of the contract. This is particularly so where the engineer considers the programme is intentionally slanted to emphasise and bring forward requirements for design and other information from the engineer. Conversely, contractors often complain that the programme has been unfairly rejected because it identifies or clearly relies upon certain prerequisites from the engineer, which, although genuinely critical of the contractor's contractual obligations, the engineer cannot actually achieve and so does not want the requirement so clearly recorded.

Such provisions as FIDIC Red Book clause 8.3 require the contractor to submit its first programme within 28 days of receiving a notice to commence work under clause 8.1, and the engineer has 21 days to respond. Thus the intention, quite properly, is that the baseline programme is agreed early in the works. However, in practice the process of submission, rejection, and resubmission of programmes under such terms can drag on for months and even never reach a point at which a programme is agreed. As this unfolds, an issue may be the extent to which the next attempt should incorporate actual progress as it has passed. Generally, the approval of a contractor's programme can become a fraught procedure during which contractor and engineer accuse each other of incompetence or of artificially creating a basis for an unmerited claim or for denying a valid claim.

A baseline programme should provide a basis for the assessment of change to the sequence and timing of works following a change. However, a programme is merely an expression of intent, i.e. it shows an intended sequence and timing and the assumptions and intentions inherent in the programme may become invalid for a number of reasons, including:

- Errors in the contractor's analysis when calculating the programme.
- The contractor's inability to obtain the resources it considered it needed as and when it thought it would need them.
- The impact of changes made as the contract progresses, including ordered variations, etc.
- The impact of extraneous factors such as weather conditions.

This raises the issue of the validity of the 'contract programme' when considering the impact of any particular change. In the vast majority of contracts for works of a substantial nature, the programme will be impacted and changed itself by many other factors before the impact of a particular change that needs to be addressed for compensation.

Under the Infrastructure Conditions clause 14(4) and the FIDIC Red Book clause 8.3, the engineer can require revisions to the programme if at any time it believes that the progress of the works does not conform to the accepted programme. Under the SBC/Q clause the requirement for an amended master programme is limited to when a decision has been made on an extension of time (clause 2.9.2). This means that there is likely to be a difference between these two systems of programme requirements that will apply to other contracts with similar requirements, in that under such a system as the Infrastructure Conditions system there should be a reasonable chance that the engineer, as well as the contractor, should have an updated and relevant programme from which to assess change. Under the SBC/Q system there is much less chance that the architect will have a relevant updated programme part way through the contract period, unless the need for it happens to follow the occurrence of a delay to completion.

NEC4-ECC, typical of its ethos, takes a more proactive and detailed approach than that of more traditional standard forms. It provides for an 'Accepted Programme'. Clause 31.1 provides for a programme to be identified in the Contract Data, or submitted by the contractor within the period stated in the Contract Data. Clause 31 then continues through a process for the acceptance of the contractor's programme by the Project Manager. This can be especially helpful towards the aims of allowing the employer to consider whether the contractor has fully understood the project, its methods and is capable of achieving the required completion dates, and also understands what the contractor will require from the employer and its team and when. These are important components of a typical tender adjudication process.

Alternatively, NEC4-ECC clause 31.1 provides for the contractor to submit a first programme for acceptance within the period stated in the Contract Data. Clause 31.2 sets out detailed requirements of the form and contents of each programme submitted by the contractor. Clause 31.3 allows the Project Manager just two weeks to accept a programme or not with reasons, and sets out valid reasons for not accepting a programme. Clause 32 sets out further requirements in relation to revised programmes such as showing provision showing for float and the date when the contractor will require information from others.

Unusually among the standard forms considered in this book, and many others, NEC4-ECC contains a very real sanction for the employer in the event that the contractor fails to submit a first programme for acceptance. This is at clause 50.5 and involves retaining one quarter of amounts due under interim payments until the programme is issued. This is a significant weapon. It recognises the importance of a programme to the successful management of construction and engineering projects and, for the employer, the importance of achieving due completion dates and its potential losses if they are missed. Whilst the other standard forms of contract considered here contain no such express sanction, a common alternative approach among some certifiers is to take account of the failure by reducing the amount certified. This can be in the form of reduction of amounts due against preliminaries and general items that are considered to represent the allowance for preparation of a contractually compliant programme. However, the resulting reduction is usually relatively small, especially compared to NEC4-ECC's quarter of the value of work done.

A further notable feature of NEC4-ECC is at clause 31.1. This does not provide for the project manager to approve the programme if it was included as part of the Contract Data, or what happens if it considers that programme unacceptable. According to the definition in clause 11.2(1), the programme in the Contract Data is the 'Accepted Programme', whether it is accepted or not.

A potential lacuna in NEC3-ECC was where the contractor issued a programme for acceptance under clause 31.3, but the project manager did not respond within the prescribed two weeks. It was not said to become the 'Accepted Programme', but neither did that contract say what the position is or how it is resolved. One criticism of the NEC contracts among certain commentators has been that their terms that the parties to them would comply with their obligations did not detail what happened if they did not. However, others saw such assumptions as one that is to be applauded as part of a co-operative and proactive environment. This issue in clause 31.3 of NEC3-ECC has now been addressed in NEC4-ECC.

The proper basis for analysis of the impact of a change in terms of effect upon the sequence and timing of the works must be the programme in position immediately preceding the change to be analysed, incorporating all known relevant information and revisions at that time. It is important therefore that the contract administrator, for both client and contractor, is in a position to establish that programme with some certainty and any shortcomings in the contract programme requirements may need to be addressed by a system of progress monitoring and reporting to enable any deficiencies to be overcome. In the case of NEC4-ECC this basis for analysis of the impact of a change is repeated through clause 60 'Compensation Events', where the employer's failure to provide access (clause 60.1(2)) or to provide something it is to provide (clause 60.1(3)) or to work within times shown on the Accepted Programme (clause 60.1(5)) are all judged against the time or date shown on the Accepted Programme.

From the updated programme the impact of a change or other event can be determined in terms of its *likely* effect on the completion date, using the logic and information in the revised updated programme. Unfortunately, in practice it is all too often the case that such updated programmes are not maintained because of a lack of proper project management discipline. Alternatively, it may be that there are so many changes, being issued on such a constant basis, that the process has become impossible even for a well-managed contractor.

2.3.2 Programmes and Resources

Whilst on the subject of programmes, an often overlooked issue in relation to establishing the baseline for a change is that of the resources behind them. Too often programmes are issued without details of the resources underlying them. This lack of linking often seems to also affect those preparing programmes and it is surprising how on projects subject to later forensic audit it is found that the contractor's programme seems to have been assessed without reference to the required resources.

The requirements of programmes under NEC4-ECC set out in its clause 31.2 expressly require that it shows 'for each operation, a statement of how the Contractor plans to do the work identifying the principal Equipment and other resources which will be used'. Similarly, the FIDIC Red Book clause 8.3(d)(ii) requires each programme to be accompanied by a supporting report that includes 'details showing the Contractor's reasonable estimate of the number of each class of Contractor's Personnel and of each type of Contractor's Equipment, required on the Site for each major stage'. There is, however, no such requirement in either SBC/Q or the Infrastructure Conditions.

The SCL Protocol suggests at paragraph 1.48 that the details within a proposed programme should include 'key resources such as labour, staff (including that which relates to design where relevant), tradesmen, major plant items, dedicated resources, major materials and work rates should be indicated for major activities (or otherwise explained in the programme software)'.

It may be that, in the absence of an express requirement to do so, many contractors are reluctant to show their hand in relation to the resources on which their programmes are based for fear that, should actually resources be lower, the employer and its consultants will cite that as a failure affecting the contractor's progress. In preparing their programmes it is a reasonable assumption that contractors will have based durations of activities on a view of the resources they will apply to them. Furthermore, the resource functions of programming software such as Primavera should make resourcing and scheduling all part of the same process.

It is suggested that if a programme is resourced, it should form an even better basis for employers and their advisors to check that the contractor fully understands the scope and requirements of the works and has a proper plan to achieve completion dates. This should also work for the contractor, in allowing it to ensure that its programme is soundly based and how it is to be realised.

A resourced programme is a very useful, but too often absent, tool in establishing the baseline against which construction and engineering claims can be assessed. This is particularly the case in relation to variations, but also for such claims as disruption and preliminaries thickening considered in Chapter 6 of this book.

2.3.2.1 Method Statements

Like resource schedules, method statements can add further detail to those of a programme to assist in establishing the baseline from which changes can be identified and evaluated.

The FIDIC Red Book includes limited requirements for method statements from the contractor. Clause 4.1 allows the engineer to require details of the arrangements and methods proposed for the execution of the works. Clause 8.3(d)(ii) requires that programmes issued by the contractor shall include a supporting report that includes 'a general description of the methods the contractor intends to adopt'. Neither of these clauses specify particularly detailed requirements.

The Infrastructure Conditions are even more limited in this respect. They allow the engineer, under clause 14(6) and (7), to request information relating to the contractor's proposed working methods for the purposes of assessing if they will have any detrimental effect on the permanent works, and the engineer can require the contractor to make whatever changes are necessary to conform with the contract or avoid detrimental effect on the permanent works.

There are no similar provisions in the SBC/Q Form, apart from a provision in its Schedule 2 that an instruction for a variation quotation might require 'an indicative statement on … the methods of carrying out the Variation'. Apart from that, working methods are left entirely in the hands of the contractor, save for the programme requirements of clause 2.9.

These approaches are representative of the approaches in many standard forms of contract in relation to method statements, which either opt for no contractual requirement or a limited right to information for specific purposes.

Even where standard forms of contract do not require method statements from the contractor, they may be requested as part of the bidding contractors' tender submissions. This is particularly understandable on complex engineering projects where the design requires complex or unusual construction methods. Here it is understandable if the employer and its team want to check that bidders understand the problems and have suitable planned procedures and processes to solve them.

There is, as always, no substitute for reading and considering the contract documentation in order to establish the base contained within the contract from which the impact of change is to be measured and evaluated. It is, however, very important to appreciate the impact of information relating to working methods if such information has clearly been incorporated in the contract.

Even though clause 4.1 of the FIDIC Red Book and clause 8(3) of the Infrastructure Conditions expressly make the contractor responsible for all site operations and construction methods, the consequence of incorporating a method statement into the contract itself can serve to override such a provision. As with the inclusion of a programme, the result may be that if the method contained in the contract becomes impossible, the contractor may be entitled to a variation, the stated construction methods having been incorporated into the contract and hence becoming part of the base from which change should be measured and valued.

This situation was considered in *Yorkshire Water Authority v Sir Alfred McAlpine & Son (Northern) Ltd* (1985) 32 BLR 114, where the contract specification stated:

> Programme of Work: In addition to the requirement of clause 14 of the Conditions of Contract, the Contractor shall supply with his tender a programme in bar chart or critical path analysis form sufficiently detailed to show that he has taken note of the following requirements and that the estimated rates of progress for each section of the work are realistic in comparison with the labour and plant figures entered in the Schedule of Labour, Plant and Sub-Contractors.

Sir Alfred McAlpine submitted a tender including a bar chart programme and method statement, which was subsequently approved at a meeting with the client and incorporated into the construction contract. The method statement supplied at the tender stage and subsequently incorporated into the contract provided for the construction of a water tunnel commencing at the lower level and working upstream. In the event it proved impossible to work upstream and the contractor contended that working downstream constituted a variation.

In deciding that it was indeed a variation the Judge, on appeal from an arbitration award, stated in his judgment:

> In this case the (employer) could have left the programme methods as the sole responsibility of the (contractor) under clauses 14(1) and 14(3). The risks inherent in such a programme or method would then have been the (contractor's) throughout. Instead, they decided they wanted more control over the methods and programme than clause 14 provided. Hence clause 107 of the specification; hence the method statement; hence the incorporation of the method statement into the contract imposing the obligation on the (contractor) to follow it save insofar as it was legally or physically impossible. It therefore became a specified method for construction by agreement between the parties, who must be taken, in my judgement, to have had the provisions of clause 8 in mind as relevant to any programme subsequently submitted under clause 14(1). No such programme was submitted or demanded, presumably because the (employer) was content with the control over the programme afforded by the specified method statement.

There is, however, an important distinction to be drawn between the contractor's entitlement to a variation because the specified method was deemed to be physically impossible and the situation where the contractor would have retained its responsibilities under clause 8 of that contract in the absence of physical impossibility. In the absence of physical impossibility the contractor would remain liable for any damage or additional cost resulting from the specified method.

It is thus vital not only to ascertain and understand what the base of the contract is when measuring and evaluating change, but it is equally important to appreciate that the base may be subject to variation in some circumstances but not in others.

2.3.3 The SCL Delay and Disruption Protocol

The SCL Protocol sets out detailed requirements for the preparation, approval, and revision of the contractor's programme. The model clause appended to the First Edition was eight pages long. In the Second Edition the Programmes section at paragraphs 1.39 to 1.64 spans five pages. These are significantly longer than the programme clauses of the standard forms considered in this book. The Protocol anticipates that the contract administrator should accept a properly prepared programme that should illustrate the major sequencing and phasing requirements of the project. Under the First Edition of the Protocol, if its model specification clauses for the treatment of programmes from its Appendix B were adopted in the contract, then paragraph 1.4 of Appendix B stated that acceptance of the contractor's programme did not make the programme a contract document or relieve the contractor of the responsibility for the construction of the works. In the Second Edition, this general approach is stated in paragraph 1.57. There is therefore no mandate that the works will be constructed strictly in accordance with the programme.

This may seem a similar approach to that of the Infrastructure Conditions regime but in practice great care was needed with the use of the programme provisions of the First Edition and incorporation of the lengthy and detailed model clauses into a contract, as there were potential conflicts with the standard provisions in a number of standard form contracts.

The most sensible approach to the Protocol is to regard it as a guide to good practice in programming and the analysis of delay and disruption where it can be applied in harmony with the contract provisions, but not to incorporate it into a contract. In both editions it expressly states at paragraph B of the Introduction that it is not intended that it should be a contract document. Indeed, it is difficult to imagine how a Protocol with such a wide range of coverage could be compatible with more than one form of contract at a time, without extensive work on drafting to harmonise their terms.

2.3.4 A Partial Programme

A sensible, and in some cases the only realistic, approach to the submission of programmes on major contracts of long duration is for only the early part of the programme to be fully detailed and with the adoption of only milestone dates and activities used to define the programme thereafter. This may avoid disputes generated by the submission of fully detailed programmes showing what might, on a major project, be thousands of activities over a period of several years. Such programmes are almost bound to be flawed as it is practically impossible for a contractor to predict all the activities in precise detail over such a period, especially at the early stage of a contract when its analysis of the work in the later stages may yet be incomplete or lacking in some details. This approach was recognised in the First Edition of the SCL Protocol, where paragraph 2.2.3 recognised that the contractor may wish to change or develop its programme or expand the details of activities it had not previously fully planned. In the current edition this is referred to as 'rolling wave programming'. As described in paragraph 1.46:

> … common sense should be applied and reasonable summary bar activities incorporated in the programme that are then detailed as the time to execute them draws nearer.

2.3.5 Limitations on Liability

When undertaking any assessment of quantum for an event under a construction contract it may be necessary to observe any limits placed on liability by the terms of the contract. This may be desirable to avoid deterring potential contracting parties if the extent of unlimited liability would be out of proportion to the potential rewards arising from the contract.

An example was detailed in Chapter 1, and is considered in detail in Chapter 6, as the capping of employer's delay damages as a percentage of the contract price. For example, in the Infrastructure Conditions this approach is set out in clause 47(4).

The NEC4–ECC contract has a potentially extensive optional clause, Option X18 'Limitation of Liability', which allows the parties to incorporate all or parts of the provisions therein in order to place a limit on potential liability for matters such as the client's indirect or consequential loss, loss or damage to the client's property, and the client's loss arising from defects in any design provided by the contractor. The limitations are expressed in the Contract Data as individual limits for each category, but there is also a cap on total liability, albeit subject to some excluded matters.

The FIDIC Red Book provides in clause 17.6 a limit on the contractor's liability under or in connection with the contract in an amount stated in the 'Particular Conditions' or otherwise at the 'Accepted Contract Amount', effectively the contract price.

Other contracts may have similar provisions, but the nature and extent of limitations will vary with differing circumstances.

The practical implication of such provisions is that they can effectively curtail a quantum assessment, as there is little value in compiling detailed evaluations for sums that exceed individual limits or aggregate caps on liability, although some work may be required to demonstrate that the limits set by the contract have been achieved or exceeded. On the other hand, such caps sometimes lead to arguments as to whether they are an 'exclusive remedy' and as to which types of default they cover. This is considered further in Chapter 6.

2.4 Summary

The foregoing discussions highlight some of the aspects and difficulties that might be encountered in establishing the base from which claims for additional payment will be made. Whether the changes to the works that generate such claims are of the type antic-ipated by the contract, and can be 'planned' and accommodated in the contract regime, or are 'unplanned', such as those that occur from breaches of obligations and there-fore are more difficult to anticipate and quantify, it will always be necessary to examine the contract documentation to determine precisely the base for adjustment. It is easy to make assumptions about the effect of apparently changed conditions or character, etc., of the works or about the status of programme information or method statements. However, such assumptions should always be checked before being used as the basis for calculating additional payment or assessing a delay event and its effects.

One of the more difficult areas of analysis is considering the impact of changes on the intended programme of works, a subject that is considered in more detail in the following chapter.

3

Effect of Change on Programmes of Work

Chapter 1 explained how change can occur during the course of a construction or engineering project, whether that change is planned or unplanned. Most contracts include provisions for the employer's representative to make changes to the works during the course of the project, be they changes in scope, quality or even aspects of the timing of the works. The one-off nature of the vast majority of construction or engineering projects mean that unplanned events are highly likely to occur: the larger the value of a project and the longer its duration, the greater the likelihood to the point of inevitability. Contracts should provide for such planned and unplanned change, apportioning the risks for them and containing machinery for evaluating their effects and amending the contract in terms of both price and time periods.

Chapter 2 considered how the baseline from which the effects of change on price and time might be assessed. As to how changes encountered in the lifetime of a construction project might impact on the programme of work, this chapter considers some salient points that recur when considering claims for payment related to changes to programmes of work and how those changes can be quantified.

Changes to work will often have an impact on the direct cost of the work affected, which may in some circumstances have a consequential effect on the cost of other, unchanged, work. In this book the expression 'direct cost' is used as meaning the unit cost of the work affected by the change, i.e. the labour, materials, plant and equipment, and related site overhead costs of the construction operation affected. These direct financial consequences of change are considered in Chapter 5. However, the direct consequences of such change may not be the only effect upon the contract works. The contractor's working methods may be affected by the change and/or its programmed sequences and durations might be affected, and the contractual completion date(s) may no longer be achievable. It is also possible that the change, in addition to its direct impact and cost, might cause a disturbance to the contractor's site organisation and costs without having an impact on the completion date. In this book, these latter costs are categorised as the 'time consequences of change' and are considered in Chapter 6.

As will be explained in Chapter 6, there is not always an inextricable link between claims for time resulting from change and claims for the time consequences of change. Either can occur without the other. However, many effects of change on a construction programme will have financial consequences.

Before considering how to evaluate such financial impacts of change it is necessary to appreciate the appropriate means of analysing the time and delay consequences of change. There are many alternative approaches applied by those calculating the effects

Evaluating Contract Claims, Third Edition. John Mullen and R. Peter Davison.
© 2020 John Wiley & Sons Ltd. Published 2020 by John Wiley & Sons Ltd.

of delays on the completion dates of projects. They vary widely in the details of the methods (including both prospective and retrospective analyses) and in their results. This chapter sets out the main approaches and considers some of their advantages and disadvantages. Delay analysis is a very large subject on its own and is the subject of a number of lengthy dedicated texts. It cannot be covered here in detail, but this chapter provides an outline, including the guiding principles that need to be established before any reasoned consideration of evaluation can be made.

Programmes, their preparation, updating and use are significant parts of the SCL Protocol's guidance and they must be at the heart of any analysis of the time, and associated financial, effects of change on construction and engineering projects. There is, however, an important point to bear in mind with any delay or disruption analysis based on programming material. The prime purpose of the contractor's programmes is to provide a tool for the management of the project. Before that, they have another purpose for the contractor in being an important part of its pricing and tendering calculations and for the employer as part of its evaluation of competing tenders. The use of programmes to analyse delay and disruption is, or should be, a subsidiary purpose. Before using it as a basis for any analysis of the effects of change, it is essential to ascertain as far as possible that the programme is the valid and accurate management and pricing tool that it should be. This includes the fact that it has not been doctored or structured with delay and disruption analysis as an intended purpose. Such abuse of the programme is sometimes found in the hiding of float, manipulation of logic links, extending activities beyond their true durations, or by artificial timing of activities. These distortions may be intended to hide the effects of culpable failure and/or to create or exaggerate the effects of failures by the other party. This is done in the hope of founding an otherwise unmerited claim. An example of such distortions includes the early delivery of employer provided materials or equipment, before they are reasonably required, in the hope of being able to produce a 'late delivery' based claim.

3.1   Use of Programmes

In the preceding chapter the status of programmes under various standard forms of contract was considered in the context of establishing the basis of the contract, from which analysis can be conducted.

There is usually only one certain factor common to the programmes at the outset of any significant construction project and that is that each programme will contain errors requiring it to be updated and amended periodically to maintain it as a viable means of managing and monitoring the construction works. The errors in the initial programme might mean that it is more optimistic or pessimistic than necessary and it might well contain a number of such errors, both optimistic and pessimistic, in terms of the achievable periods of activities. If the contractor is fortunate, these may cancel each other out, leaving the overall completion date realistic. The reliability of initial programmes tends to vary considerably and will depend to a large degree upon the following factors at the time that it is prepared:

- The extent to which the design of the works has been completed.
- The extent to which major subcontract or supply packages have been defined and the relevant subcontractors and suppliers have progressed their pricing and any related design information.

- The degree of reliability of information used relating to site-specific factors such as weather conditions, ground conditions, etc.
- The reasonableness of assumptions made in the programme as to the likely outputs to be achieved by labour and plant resources.
- The reasonableness of assumptions made in the programme as to the likely availability of resources such as materials, labour and plant.
- The accuracy of lead-in periods required for materials or manufactured items required.
- The amount of time and effort expended in ensuring the programme has considered factors such as these as far as reasonably possible.
- The intended use of the programme, particularly where it was part of a contractor's tender submission and intended for no use beyond that, and one that can be replaced by a detailed construction programme such as that which FIDIC Red Book clause 8.3 requires to be issued within 28 days of a notice to commence work.

The need to take sufficient time and effort to consider and detail a construction programme may seem an obvious requirement, but it is often the case that the initial programme contains errors that could be eliminated by careful consideration of these factors. It is also the case that the initial programme may be unreasonable by being too detailed. This may seem an unusual, or even unreasonable, criticism of a programme. However, for major projects with a site period of 18–24 months or more it may not be possible to predict all these factors with sufficient accuracy for a fully detailed programme for the full period to be compiled at the outset. To prepare such a long-term programme in great detail may be unrealistic and a waste of programming resources whose time would be better spent detailing the shorter term more accurately and in greater detail. In such circumstances it would seem much more sensible to produce detailed programmes for, say, the first six months, with a less detailed programme for, say, the following six months and a series of planned 'milestones' thereafter. The degree of reasonableness in the milestones will need to be assessed in the light of the operations needed to achieve them, but such an approach obviates the need to put forward fully detailed programmes for works of two years or more in the future, which cannot be predicted with reasonable accuracy to the degree suggested by some programmes.

This approach would not, of course, mean that the contractor would be any less bound by the completion date(s) contained in the contract, but it might avoid unnecessary arguments about the accuracy of predictions which in the prevailing circumstances could never have been reasonably accurate.

The accuracy and details of programmes were discussed in Chapter 2. In addition to the unavoidable limitations of the time and information that are available for the preparation of tender programmes, it sometimes appears that they were prepared only to 'tick a box' within the instructed requirements of the overall tender submission. In such cases they are superficial and bear little resemblance to how the works will actually be constructed. For example, the authors have seen a tender programme that entirely ignored the fact that the tender enquiry set out sectional completion dates for different parts of the works. The authors have even seen a tender programme for one project that was clearly just an edited photocopy of the programme from another earlier project. Such an approach from a tenderer should cause loud alarm bells for those involved in the bid evaluation, in terms of the capabilities of that tenderer. However, if that contractor is successful, then the dangers might be avoided if the contractor then complies with the requirements of the contract in terms of agreeing a construction programme.

For example, in FIDIC Red Book clause 8.3, the contractor is required to submit a programme to the engineer for approval within 28 days of a notice to commence work. However, such express contractual requirements are sometimes neither followed nor imposed. Again, the authors have seen a project on which the contractor's construction management relied on no more than an inadequate tender programme as their construction programme, even basing their eventual delay and costs claims on it. Such wholly inadequate procedures are those for which the contractor and the employer's representative (in not demanding better programming from the contractor) equally share responsibility for both the resulting failure of the project due to poor programme management and the unnecessarily problematical resolution of the claims that followed.

As with any document used to determine the baseline for considering a claim or entitlement, no matter how detailed a programme is, it needs to be assessed for its accuracy, completeness and reasonableness. Impacting delay events into a programme that would not have been achieved in any event, or comparing such a programme with the as-built record, is likely to provide misleading results. Similarly, a programme that does not include proper logic linking and dependencies will not produce an accurate delay analysis.

3.1.1 Provisional Sums in Programmes

The problem of accuracy in initial programmes can sometimes be compounded by the use, or misuse, of provisional sums in tender documents. This is particularly under contracts let on a lump sum basis that ought to more honestly and realistically have been on a remeasurable basis. There may often be legitimate reasons why an element of the works cannot be fully detailed at the time of tender. Works may truly be 'provisional' at the contract stage because it is then not yet certain that they will be required, either in part or at all. Alternatively, it may be known that the works will be required, but their scope and detail may not yet be capable of being detailed and measured, perhaps because that work will be below an existing structure or depends on ground conditions that are as yet uncertain. A further uncertainty as to work scope may be that they are specialist works, such as IT systems, to be carried out by a specialist subcontractor that cannot yet be priced. However, such sums need to be used with care and in a manner that allows the contractor to understand the scope of work being tendered. The extent of such sums should be limited as far as practicable, rather than using them to relieve the need to properly finalise the design at the tender stage. Failure in this regard can lead to disagreement as to what was allowed for in the contractor's programme in relation to provisional sums and what adjustment should be made in relation to the actual works eventually instructed against them. The resolution of such disagreements will depend on the terms of the contract and how they address work covered by, and instructed against, provisional sums in terms of risk and time.

Contracts such as the FIDIC Red Book, the Infrastructure Conditions and the ICE contracts that preceded them appear to require the contractor to include provisional sum work in its programme but to provide for an extension of time should the work instructed against a provisional sum merit it. However, their terms are not particularly clear in this regard.

In the case of the FIDIC Red Book, clause 4.11 expressly deems that 'unless otherwise stated', the 'Accepted Contract Amount' covers all of the contractor's obligations including those under provisional sums. Thus it could be said that the price includes the time-related costs associated with carrying out works against provisional sums and by the 'Time for Completion' stated in the contract. Furthermore, according to definition 1.1.4.1, the 'Accepted Contract Amount' is 'the amount for the execution and completion of the Works'. Combining clauses 4.11 and 1.1.4.1, 'the Works' include work under provisional sums. Since clause 8.2 requires the contractor to complete 'the Works' in the 'Time for Completion', it therefore appears that the contractor is required to complete work under provisional sums in the 'Time for Completion'.

Regarding the FIDIC Red Book's approach to extensions of time in relation to provisional sums, it is firstly notable that clause 8.4 does not list provisional sums among its causes of delay, giving an entitlement to an extension of time. This seems to support the view that the contractor has the risk for completing provisional sum works within the original contract period. However, the question then arises as to what scope of work instructed against a provisional sum is covered by that risk. Is it limited to work to a value equal to the amount of the provisional sum?

What FIDIC clause 8.4 does list among its causes of delay giving an entitlement to an extension of time is 'a Variation'. That term is defined at 1.1.6.9 as 'any change to the Works, which is instructed or approved as a variation under Clause 13'. Clause 13.5 says that work against provisional sums shall be instructed by the engineer and valued under clause 13.3 'Variation Procedure'. Accordingly, provisional sums are adjusted as variations instructed by the engineer and the contractor would be entitled to an extension of time under clause 8.4 to the extent merited by that adjustment. How that extension is calculated is a subject of regular debate between contractors and certifying engineers under terms such as FIDIC. A simplistic approach often applied in contractor's claims sees a pro rata adjustment of the programme activity bar(s) for the value of the provisional sum. For example, if the provisional sum is adjusted, resulting in a 50% increase compared to the amount included against that item in the contract or bills of quantities, then the activity bar(s) is/are extended by the same percentage. Of course, this simplistic approach may under- or overcompensate for the effect of an increased value of work. If a provisional sum for granite stone tiles increased in price because stone from Italy is specified in lieu of stone from India, the material cost might rise but it might take the same time to install the same area. Furthermore, if each new tile is larger in area or on shorter supply period, then time might actually be saved. The answer to this is that each such event for adjustment needs to be looked at on its details and its merits of the effects of the adjustment of each provisional sum on the programme.

This latter point can lead to difficult issues in practice. It needs one to establish what a reasonable programme period should have been for the works in the provisional sum and the same for those actually instructed against the provisional sum. The two periods are then compared. Whereas the first of these requirements might rely on the period allowed in the contractor's accepted programme, it has been explained that there may be no such programme. Even where there is a programme, it may be that durations for the provisional sum works cannot be identified among the activities and bars. Furthermore, it is not at all unusual that contractor's programmes miss works covered by some provisional sums. The authors have seen a tender programme that missed provisional sums entirely, notwithstanding that the tender instructions expressly required that the contractor was to include in its programme for those works. That those advising the employer missed this point in their tender adjudication led to significant issues after that contractor was appointed to the works.

This analysis is based on contracts that do not draw a distinction between 'defined' and 'undefined' provisional sums. This approach, followed by SBC/Q in its incorporation of SMM7, was mentioned in Chapter 1. It appears to be unique to methods of measurement drafted by the Royal Institution of Chartered Surveyors (RICS) based on UK practice, such as their NRM and the predecessor SMM7. However, it does not feature in the RICS's own POMI Rules. Internationally, it also does not appear in methods of measurement produced by such as the Association of South African Quantity Surveyors, Australian Institute of Quantity Surveyors, Society of Chartered Surveyors Ireland, Institution of Surveyors Malaysia, or Singapore Institute of Surveyors and Valuers.

For a provisional sum to be 'defined' under SMM7 and NRM rules, information as to the nature of the work, its construction and relationship to the building is required to be given in the contract, together with indicative quantities and any specific limitations. For provisional sums so defined the contractor is deemed to have made due allowance for the required work in its planning and programming of the works and the pricing of its preliminaries.

It is not difficult to anticipate that, whether a contract incorporates a distinction between 'defined' and 'undefined' or not, a great deal of uncertainty can be introduced into the definition of the works that the contractor is required to plan and make a programme as part of the contact. The greater the proportion of the works covered by provisional sums, the greater the uncertainty. A salutatory lesson in this regard can be found in a read of the judgment in *Walter Lilly & Company Limited v (1) Giles Patrick Cyril Mackay (2) DMW Developments Limited* [2012] EWHC 1773 (TCC) (12 July 2012), which is considered in more detail elsewhere in this book. In the authors' experience, such uncertainties are heightened by the NRM and SMM7 approach.

An inevitable issue arising out of the distinction between 'defined' and 'undefined' provisional sums may be that of whether sums that are labelled in the contract as the former ought in reality to have been labelled as the latter. The requirements for a provisional sum to be defined have been summarised. In practice it may occur that 'defined' provisional sums do not actually satisfy all of those criteria. NRM and SMM7 both address this risk. In NRM the criteria for a 'defined' provisional sum are detailed in paragraph 2.9.1.2 as follows:

> A provisional sum for defined work is a sum provided for work that is not completely designed but for which the following information shall be provided:
>
> (1) the nature and construction of the work;
> (2) a statement of how and where the work is fixed to the building and what other work is to be fixed thereto;
> (3) a quantity or quantities which indicate the scope and extent of the work; and
> (4) any specific limitations identified.

Paragraph 2.9.1.5 then says that a provisional sum given for 'defined' work that does not comprise the information required under paragraph 2.9.1.2 shall be construed as an 'undefined' provisional, irrespective that it was stated in the bills of quantities to be for 'defined' work. However, in practice this can lead to issues as to whether the criteria were met. Some of them are quite subjective. For example, to what extent was 'the nature and construction of the work' set out and was this sufficient? Similarly, were any 'specific limitations' sufficiently described?

In some instances a 'half-way' situation may be encountered where the so called 'defined' provisional sum does not provide all the information required under, for example, paragraph 2.9.1.2, but it is patently clear that the work has been included in the contractor's programme anyway. In such a situation, is the contractor deemed to have included for the full effect of the works in the provisional sum? It seems only reasonable that the contractor should be deemed to have made due allowance in its programming, planning and pricing preliminaries for the works covered by that provisional sum. However, as with the FIDIC Red Book analysis, that would only be to the extent required by the provisional sum so far as the information provided reasonably allowed.

In contrast to 'defined' provisional sums the use of similar 'undefined' sums is allowed for work where the information required for a defined sum is not available. In this case the contractor will be deemed not to have allowed in its programming, planning and pricing of preliminaries and general items for the work covered by the sum. There is therefore little to be gained, from the employer's point of view, by the use of such sums other than to have the work content included in the contract. However, when all the required details are issued to the contractor it will be able to revise its programme and preliminaries, pricing accordingly, in accordance with the contract conditions. This means that such sums, if substantial, are a guaranteed source of future requests for extensions of time and related additional payments.

There may be legitimate reasons for the use of provisional sums in either instance, and these two systems are typical of other standard and ad hoc contracts, but the need to update programmes in the light of further and better information as to the work required under such sums should be understood. If the use of such sums is abused to any extent, by their overuse whereby they begin to constitute an unreasonable portion of the total works, the accuracy of any programme can obviously become almost unachievable, with likely undesired consequences in terms of completion date and claims.

As was noted in Chapter 1, NEC4-ECC avoids such issues by not providing for provisional sums at all. This is on the basis that if an employer cannot clearly define part of the works at the time the contract was formed then it should not be included in the contract because it is not reasonable to expect the contractor to include work without a clear indication of its cost and programme implications. Under NEC, the preferred approach is to deal with such work by use of the early warning provisions of clause 15 and with the work being valued as a compensation event in accordance with the contract terms. However, this approach is not always popular with employers who consider there are good grounds for works to be provisional at the contract stage and that it is detailed to an extent that ought to provide them with certainty as to price (subject to adjustment of the provisional sum) and programme (with the contractor taking the risk).

3.1.2 The Base Cost

Time and cost are invariably entwined on construction and engineering projects. More time will usually mean more cost. The time that an activity takes, or is planned to take, will usually be related to the resources and costs allocated to it. In Chapter 2 we commented on the all too familiar absence of a link between the programme and the resources that should relate to it.

Underlying any evaluation of change in terms of its impact on time and progress of the works is the need to appreciate how the base cost was determined. This involves not only what it does contain but also what it is deemed to contain, whether or not the contractor actually made any, or any adequate, allowance. Once the matters discussed have been considered and any necessary amendments made to the initial programme, then it is possible to ascertain if the contract allowances made by the contractor were likely to be reasonable and adequate.

A degree of caution is required in such exercises in ensuring that 'standard' information is used appropriately when assessing the reasonableness of the contractor's programme and cost assumptions. For instance, the contractor's assumptions of outputs to be achieved by resources in various construction operations need to be tested against independently tested information applied in the light of appropriate experience. However, the application of standard 'S' curves and other statistical devices for the distribution of general costs, such as the setting up, operation and dismantling of site establishment facilities, should be modified to reflect the actual manner in which costs for a particular contract will be incurred. It is not unusual to find that expenditure is incurred in steps rather than in smooth curves. An initial mobilisation cost will be followed by a curve of cost as the early operations commence and proceed, but the opening of further work fronts as the early work makes the introduction of other resources possible will cause steps in cost, i.e. there will be a relatively high expenditure in a short period to establish the new operations, followed again by a cost curve commencing from the top of the step. This process will apply to each element of the total costs, usually categorised as labour, plant, materials and site establishment. The concept of the 'S' curve should therefore be understood to be a graphical representation of underlying costs that generally change in steps, with the steps for each element not necessarily occurring at the same time.

3.2 Use of As-Built Programmes

In many instances there will be a need to consider the use of as-built programme information, i.e. a programme that demonstrates the actual sequence and duration of the various operations on the site rather than the sequence and timing anticipated and programmed by the contractor at the outset. Strictly speaking these are not programmes in the sense of the contract programme. They do not predict the durations and sequences but show factual information. As such it is often accepted that as-built programmes are an accurate representation of the progress of the works, but in reality there can be significant difficulties with the compiling and interpretation of such programmes.

As will be detailed elsewhere in this chapter, some methods of delay analysis particularly rely on the as-built record as a basis for analysis of the contractor's entitlement to extensions of time. The extent of that reliance ranges from those that are solely based on retrospectively collapsing the as-built programme to those that just use the as-built record to update the planned programme before prospectively impacting it with delay events.

3.2.1 Sources of Information for As-Built Programmes

By definition, an as-built programme is compiled after the works have been executed and the common source of information is the contract progress reporting system. Such systems vary enormously depending upon the sophistication of the system, the degree to which it is actually applied in practice, the competence of those applying it, the requirements of the contract and the nature of the work being undertaken. Many such systems suffer from a common shortcoming in that they record an operation, or section of work, as complete only when every piece of work required in connection with that operation or section has been finished. This may seem reasonable when the objective is to report to management the progress of the various parts of the project, for example in monthly reports to management. However, it can introduce distortion into the representation of the works' progress if used unamended for as-built programmes to be used in delay analysis. For instance, the construction of a reinforced concrete retaining wall may be a critical activity in the programme of works. It might be that a section of the wall is brick faced, but the brick facing is not critical in that it can be carried out after completion of the reinforced concrete elements but with an activity float that means it does not hold up succeeding activities. In such a case, if the brickwork is included as part of the retaining wall construction, completion of the wall might be recorded when the brickwork is complete and not at the critical point when the reinforced concrete work was complete, so releasing following activities. In such circumstances the sensible approach is to ensure that the non-critical element of the construction is shown as a separate activity.

Among project reporting system records, photographic, video and webcam evidence can be particularly helpful in relation to establishing the as-built programme. All can suffer where the record would benefit from a contemporaneous narrative of what is being shown. The use of permanently fixed cameras is an increasingly common approach on construction and engineering sites. Taking snap shots, say, every 30 seconds, they can also be web based and thus accessible from a computer anywhere. Particularly in relation to the preparation of as-built programmes, they have been known to remove any doubt as to when progress was achieved, particularly external activities such as groundworks, the structure or envelope of a building. The main limitation is in relation to internal trades. Sometimes (where the fixed cameras are there to monitor the site rather than its security perimeter) staff and labour can be suspicious of being 'spied upon' by such permanent equipment, a reaction that one can only smile at. For a detailed understanding of how technology can be used to supplement and improve data collection and analysis, allowing more detailed and closer to real time risk management, see Construction Risk Management – Technology to Manage Risk (ConTech) by Rob Horne, in C. O'Neil (2019), *Global Construction Success*, Wiley, p. 205.

However, problems can still arise when using progress recording and reporting systems as the source of information because many activities, even if critical, release following activities before they are 100% complete, when the outstanding work preventing reporting of completion is minor or insignificant. Such instances are often identified in progress reporting systems by the recording of an activity being, say, 97% or 98% complete for some period before completion is recorded on the execution of the outstanding minor element.

In compiling programmes, including as-built programmes, links are often made between operations. These links are often between the completion of one piece of the works and the start of an ensuing operation. This is a finish–start relationship in planning terminology. However, it is often the case that a preceding operation, or section of the work, does not need to be absolutely complete before the ensuing work commences. For instance, in works of a mechanical nature it is possible to show a finish–start relationship between the completion of the installation of a particular piece of equipment and the commencement of the erection of the pipework running away from that equipment. The completion of the equipment installation operation for progress reporting might include the 'bolting on' of an ancillary piece of the equipment, perhaps a meter or other measurement device, which in practice does not have any influence on the erection of the related pipework, that can commence when the main body of the equipment is installed and secured.

In this example, by using the progress reporting of 'equipment installation 100%' as the trigger point for the earliest possible commencement of the following pipework, the as-built record will be incorrect. The pipework could, in this instance, have commenced before the installation of the meter allowed the equipment installation operation to be reported as 100% complete. If the period between installation of the main equipment and the meter is significant, then the distortion in the as-built record will be similarly significant.

Only by a detailed analysis and appraisal of the progress reporting data against the assumptions contained in any programming system can such distortions be eliminated to allow the as-built record to be used as a reasonable analysis tool. In this instance, it would be necessary to modify the finish–start relationship between equipment erection and commencement of the related pipework to one that reflected the appropriate progress reporting of the equipment installation prior to the connection of the meter, say 97% equipment completion, or whatever deduction of the completion the meter element represents, as a trigger for commencement of ensuing pipework. Alternatively, the meter works should be shown as a separate programme activity to the equipment to which it is to be attached.

3.2.2   Constant Resource/Continuous Working

There is also a danger in using as-built records, or any programme where the works are represented in bar chart form, that the duration of a bar will be taken to indicate that resources were employed throughout the duration of that bar and at a consistent level. In fact that can be a very misleading impression and it is often the case that the resources employed on a particular operation, or section of the works, will vary considerably at various stages as it progresses. The resources represented by a bar will often not be constantly employed and nor will the operation or works necessarily be continuous. As-built records will show the commencement of the work and the completion, and often represent the intervening duration as a continuous bar, but this may be far from the true representation of how the work was executed and the related costs incurred.

Where there is a break in the as-built duration of a programme activity, there could be important reasons for this in terms of either employer failures or contractor culpability. If this picture is lost because the as-built record fails to show the break, then this can distort analysis of any related claims or even hide claims that could validly have been made.

3.2.3 Recording of Completion

The dangers of the use of completion of operations in the progress recording system as the trigger for the commencement of ensuing works has been discussed elsewhere but cannot be overemphasised. There is a further obvious pitfall of such systems, where the completion of operations, or sections, will usually not be shown as complete until all work is finished.

If the final, say, 2% of a section of the work is the painting of an installed piece of equipment, then that operation will not be reported complete until such time as it is painted. If the painting is delayed, intentionally or otherwise, the reporting of the completion of the operation will be similarly delayed. The as-built record, by using a bar to represent the installation period, may suggest that the particular operation continued for, say, 20 weeks, when in fact 98% of the work was complete in eight weeks and nothing further occurred until, say, two days of painting 12 weeks after the main installation.

Such distortions in reporting systems have to be detected and eliminated when considering the distribution and incurring of related costs to be analysed as part of the calculation of the costs incurred by changes or disruption to the sequence or timing of activities.

3.3 Change Without Overall Prolongation

It is, of course, quite possible that change to the scope of the works, either in quantity or specification terms, can cause change to the sequence or duration of programme activities without having an effect on the date for completion of the project, or dates if the contract sets out sectional completion obligations. Such an effect is often termed 'activity delay' or alternatively 'disruption' where the anticipated sequence of working has been disrupted and another effected in its place. The revised sequence, and/or durations, of activities may not impact on the completion date because the affected activities were not on a critical path through the programmed activities to the completion date. The effects might also better be described as 'local prolongation' as local activities and costs are extended but there is no effect on contractual completion dates. Alternatively, the delays might lie on a sequence of activities where there is 'float' in the programme. If the critical path is the longest sequence of activities from commencement to completion, float is the amount of time on non-critical activities that can be absorbed by the activities, over and above their intended duration, without impacting on the critical path. Thus there is delay, whether on critical or non-critical activities, but again no effect on completion dates. Alternatively, change can affect activities, with additional costs but not delay to those activities, because they require more resource but do not take longer to carry out. This latter example may be pure 'disruption' or alternatively may involve acceleration, two subjects considered in detail in Chapter 6.

It should be remembered that there may well be more than one critical path through a programme, and in detailed programmes for large and complex projects this will often be the case. It is equally important to remember that the critical path, or paths, may change as soon as there is a change in the timing or duration of any activity on site. On this basis, when evaluating the programme impact of change under most of the delay analysis methods considered elsewhere, it is therefore essential to consider that impact against the programme current at the time the change took place so that its critical path is considered. This is a fundamental principle of the 'Time Impact Analysis' approach that was recommended by the First Edition of the SCL Protocol. Furthermore, in the Second Edition, the first Core Principle is that 'the programme should be updated to record actual progress.... If this is done, then the programme can be more easily used as a tool for managing change and determining EOTs and periods of time for which compensation may be due'. It is also notable that in relation to float, the ninth Core Principle of the current edition states that 'The identification of float is greatly assisted where there is a properly prepared and regularly updated programme...'.

The concept of float in programmes raises particular issues in the evaluation of claims for 'disruption', i.e. claims for additional payment as a result of changes to activity durations or sequence that are alleged to incur additional costs but have not affected the contract completion date. Float is considered in detail elsewhere in this section.

'Local prolongation' on non-critical activities that do not affect overall completion, as explained elsewhere, is particularly important where a project is of such a size or nature that the contractor allocates time-related resources or time-related costs of such as preliminaries and general items and/or on-site overheads to parts of the project but not the whole project. It is particularly an approach required on long linear transportation projects. For example, on the construction of a new 85 km railway line, the turnkey contractor divided such costs as follows:

- Engineering, procurement and construction management ('EPCM') resources were housed in what was effectively a project head office, serving the whole project.
- Preliminaries and general resources were located in two geographical centres reflecting the length of the project with one at each end – 'North P&Gs' and 'South P&Gs' – plus a separate establishment for tunnelling works, reflecting their specialist nature.
- At a more local level, preliminaries and general resources were allocated to specific parts of the works, such as the construction of bridges or viaducts.

These separations did not reflect separate contractual completion dates.

The consequence of this approach was that when claiming time-related costs related to delays, the contractor had to carry out the analysis at three different levels. Claims for EPCM costs were related to delays and extension of the overall contract completion date. Claims for such as 'North P&Gs' required separate analysis of delays to the north section of the project as a whole. Delays to local preliminaries and general items required analysis of delays locally to, for example, a specific bridge to which they were related. All claims for preliminaries and general items could therefore be said to be claims without overall prolongation, with only the claim for EPCM costs being related to overall completion.

3.3.1 Who Owns the Float?

In paragraph 8.1 of its Guidance Part B, the SCL Protocol defines float as 'the amount of time by which an activity or group of activities may be shifted without causing Delay to

Completion'. In relation to assessing the effect of change on a programme of work, float in a programme therefore determines whether, and to what extent, a delay to an activity will have an effect on a contractual completion date. However, this depends on whether the float in the programme is maintained or used to absorb the delay as far as it can. The question resulting from this, which is especially relevant to claims for extensions of time, is who owns the float in the programme?

Contractors have historically argued that, since it is they who create float in what is their programme, the float in it should be for their benefit, just as any risks in it are theirs. The employers' position has been that if an event for which they are responsible causes activity delay but no real delay to completion because the programme contains float, then that event should not give the contractor more time to complete the works because as a matter of fact the activity delay will not cause a delay to completion.

Traditionally it has been rare that engineering and construction contracts expressly state who owns the float in the programme. Exceptions include some international Concession Agreements, where provisions such as 'any float in the development Programme shall be for the sole benefit of the Concessionaire' reflect the particular nature of such contracts and the allocation of risk over the whole of their development and concession periods. However, in practice such a clause can lead to understandable disagreements over the details of how they are to be operated. A more balanced approach is seen in contracts where float may be allocated to whoever takes it first, through provisions such as 'Float shall not be for the exclusive use of the Employer or the Contractor'. At the other end of the spectrum, some overseas government bodies seem to have been persuaded as to the efficacy of provisions making any float in the programme for the exclusive use of the employer. For example, 'All float in the programme belongs to the Employer'. Contracts that adopt this approach may also seek to address the fact that it is the contractor's planning team that creates float in a programme, such that a clause to that effect is likely to lead to no float being shown by the contractor. They attempt to do this by also setting out detailed programme approval procedures that allow the employer and its engineer authority to require changes in logic and activity durations in a programme before it is approved. One national government's adoption of this approach is widely considered a 'dog's breakfast' and leads to many of its contracts ending up in dispute in the local courts regarding the detailed operation of those provisions.

In the absence of express terms, for whose benefit does the float in the programme exist? Is it for the benefit of the employer such that a certain amount of change can be accommodated without the completion date being changed? Is it for the benefit of the contractor, who created it in the first place, and who will usually be responsible for the programme and progress of the works, so that it can suffer some difficulties of its own making without running the risk of damages for late completion? Is it for the benefit of the contractor's suppliers and subcontractors so they too have a measure of protection against their defaults? Or is it a combination of all these, with benefit distributed on a 'the first in need gets the benefit' basis?

Float is often referred to as the contractor's 'time risk allowance' in that it is seen as being built into programmes to provide the contractor with a cushion for any unforeseen difficulties or problems it may encounter, and for which it is responsible. For that reason many contractors jealously guard the float as being theirs, and theirs alone, on the basis that they created the programme, and hence any float in it, and that they did so to create a contingency against their own culpable activity delays and therefore the float should be for their sole benefit. For a contractor an analogy might be a financial contingency that it included in pricing its tender, which, should it be successful in securing the contract at a price including that contingency. No employer would argue that, if the contractor allowed a $100,000 pricing contingency for ground water which was never met, then that contingency should be for the employer's benefit and set off against the first $100,000 of instructed variations. On the basis that they should have the benefit of the float, but might lose it if it is to be dealt with on a 'the first in need gets the benefit' basis, some contractors actually instruct their programmers to show no float in their programme, effectively by filling any gaps with extensions of otherwise realistic activity durations.

In England and Wales, the issue of ownership of the float came before the Technology and Construction Court in 1999 in the case of *Ascon Contracting Ltd v Alfred McAlpine Construction Isle of Man Ltd* (1999) 66 Con LR 119. The case concerned claims by a subcontractor, Ascon, against the main contractor, McAlpine, where the subcontractor had a claim for additional time for reasons that were the responsibility of McAlpine, but had also required additional time for reasons that were its own responsibility. Ascon claimed the benefit of float in the main contract programme to demonstrate that its own delays had not caused delay to the main contract works, i.e. the delays for which Ascon were responsible had used up the main contract float but caused no delay. McAlpine had rejected this approach stating that the float was theirs to deal with as they chose. In his judgment Judge Hicks QC said this:

> Before addressing those factual issues I must deal with the point raised by McAlpine as to the effect of its main contract 'float', which would in whole or in part pre-empt them. It does not seem to be in dispute that McAlpine's programme contained a 'float' of five weeks in the sense, as I understand it, that had work started on time and had all subprogrammes for subcontract works and for elements to be carried out by McAlpine's own labour been fulfilled without slippage the main contract would have been completed five weeks early. McAlpine's argument seems to be that it is entitled to the 'benefit' or 'value' of this float and can therefore use it at its option to 'cancel' or reduce delays for which it or other subcontractors would be responsible in preference to those chargeable to Ascon.
>
> In my judgement that argument is misconceived. The float is certainly of value to the main contractor in the sense that delays of up to that amount, however caused, can be accommodated without involving him in liability for liquidated damages to the employer or, if it calculates his own prolongation costs from the contractual completion date (as McAlpine has here) rather from the earlier date which might have been achieved, in any such costs. He cannot, however, while accepting that benefit as against the employer, claim against subcontractors as if it did not exist. That is self-evident if total delays as against subprogrammes do not exceed the float the main contractor, not having suffered any loss of the above kinds, cannot recover from subcontractors the hypothetical loss he would have suffered had the float not existed, and that will be so whether the delay is wholly the fault of one subcontractor, or wholly that of the main contractor himself, or spread in varying degrees between several subcontractors and the main contractor. No doubt those different situations can be described, in a sense, as ones in which the 'benefit' of the float has accrued to the defaulting party or parties, but no-one could suppose that the main contractor has, or should have, any power to alter the result so as to shift that 'benefit'....

I do not see why that analysis should not still hold good if the constituent delays more than use up the float, so that completion is late. Six subcontractors, each responsible for a week's delay, will have caused no loss if there is a six weeks' float. They are equally at fault, and equally share in the 'benefit'. If the float is only five weeks, so that completion is a week late, the same principle should operate; they are equally at fault, and should equally share in the reduced 'benefit' and therefore equally in responsibility for the one week's loss. The allocation should not be in the gift of the main contractor.

It was clear from this judgment that the court considered that float in a programme was not the property of the main contractor and was not to be used for its exclusive benefit, or as it directed it should be allocated or used.

In passing, Judge Hicks also said, as quoted, that 'six subcontractors, each responsible for a week's delay, will have caused no loss if there is a six weeks' float'. This remark was made as an aside, but merits further discussion. There may be no loss in terms of delay to completion, but there may be loss in terms of local delay. The essence of a disruption claim in such a case would be that 'six subcontractors have each caused a week's delay to progress, because of float the completion date is not extended, but as a direct result additional costs have been incurred'. There may be a distinction to be made between project prolongation costs, i.e. those costs incurred as a result of an extension to the contract completion date, and disruption or local or activity prolongation costs, i.e. those additional costs incurred by extended periods of activities without any effect on the completion date. If six subcontractors all overrun on their works by one week, then the contractor will have to continue its attendances on those subcontractors for an additional week at additional cost to it.

The SCL Protocol also addresses the subject of float, its identification and ownership, at some length in its Guidance Part B. In its Core Principle 8 it says this:

> Unless there is express provision to the contrary in the contract, where there is remaining total float in the programme at the time of an Employer Risk Event, an EOT should only be granted to the extent that the Employer Delay is predicted to reduce to below zero the total float on the critical path affected by the Employer Delay to Completion (i.e. if the Employer Delay is predicted to extend the critical path to completion).

This approach is further detailed in Guidance Part paragraph 8.5:

> Core Principle 8 (and 9) set out the Protocol's position on float where the parties in their contract have not made clear provision for how float should be dealt with. This is consistent with current judicial thinking, which is that an Employer Delay has to be critical (to meeting the contract completion date) before an EOT will be due. It has the effect that float is not time for the exclusive use or benefit of either the Employer or the Contractor (unless there is an express provision in the contract).

This approach applies that adopted in *Ascon*. The effect is that, in the absence of an express provision, float in a programme is not to be regarded as for the exclusive use or benefit of either the employer or the contractor. The project is the owner of the float and it would be available for all parties as required. Thus if float existed in a programme and a delay event occurred, then the float would be available to reduce or eliminate the effect of that delay regardless of liability for the delay. On that basis, the benefit of the float in a programme goes to whoever first uses it, although an occasional problem in practice is determining whose delay occurred first, particularly where built records are lacking.

This approach was also applied by HH Judge Humphrey Lloyd QC in *The Royal Brompton Hospital National Health Service Trust v Hammond* [2002] EWHC 2037 (TCC). However,

> Under the JCT conditions, as used here, there can be no doubt that if an architect is required to form an opinion then, if there is then unused float for the benefit of the contractor (and not for another reason such as to deal with p.c. or provisional sums or items), then the architect is bound to take it into account since an extension is only to be granted if completion would otherwise be delayed beyond the then current completion date.

The reasoning here seems to have been based on the facts of the effects of the delay and the terms of the contract in that case, rather than consideration of the (debatable) broader merits of who should own the float in a programme. If a delay event does not cause delay to completion because as a matter of fact float in the programme absorbed the effects, then there is no basis on which to grant an extension of time. Preceding the quoted passage the Judge said this:

> What is required is to track the actual execution of the works. On a factual basis this part of the case requires no further discussion. In addition clause 25 refers to 'expected delay in the completion of the Works' and to the need for the Architect to form an opinion as to whether because of a Relevant Event 'the completion of the Works is likely to be delayed thereby beyond the Completion Date'.

However, the Judge tempered the position with the following:

> This may seem hard to a contractor but the objects of an extension of time clause are to avoid the contractor being liable for liquidated damages where there has been delay for which it is not responsible, and still to establish a new completion date to which the contractor should work so that both the employer and the contractor know where they stand. The architect should in such circumstances inform the contractor that, if thereafter events occur for which an extension of time cannot be granted, and if, as a result, the contractor would be liable for liquidated damages then an appropriate extension, not exceeding the float, would be given. In that way the purposes of the clause can be met: the date for completion is always known; the position on liquidated damages is clear; yet the contractor is not deprived permanently of 'its' float.

This appeared to suggest that if the contractor causes culpable delay after the float in its programme has been used up by an employer delay event, then the contractor would be entitled to an extension of time, not exceeding the float, for the effects of its own culpable delay.

As noted previously, the possibility is that, with float and its ownership being so contentious an issue, contractors will attempt to hide or disguise float in their programmes, particularly any 'end float' that may be available between the completion of the works and the contract completion date, although many will argue that in practice 'end float' rarely if ever occurs. The SCL Protocol approach in paragraph 8.7 of its Guidance Part B is that contractors should include in the activity durations for any perceived risks and leave true float available for the project.

The danger of all of this is that, instead of being a tool for the management and organisation of construction works, and the prediction of time scales for logistics, etc., the programme becomes a contractual battleground from the outset, with both parties attempting to construct or interpret it to their advantage. Such a situation is not a sensible basis on which to commence a project.

Whichever approach is taken to float ownership, it may be necessary to consider how it affects extensions of time and entitlement to additional costs separately. It should be borne in mind when considering additional payments that before any evaluation can take place the effect, if any, of available float in the programme should be determined. The extent of available float will be relevant to considering the heads of cost recoverable.

One of the reasons that float is such a contentious issue, in addition to resolving employer's claims for delay damages, is that parties often believe that an extension of time is also necessary in order for the contractor to be able to claim compensation for delay. The concept of local prolongation that does not affect the completion date has been explained elsewhere. Contracts such as SBC/Q set out their provisions for extensions of time (clauses 2.26 to 2.29) quite separately from those for loss and expense (clauses 4.23 to 4.26). In SBC/Q even the events are given separate headings. For extensions of time clauses 2.26 to 2.29 call them 'Relevant Events', whereas for loss and expense clauses 4.23 to 4.26 call them 'Relevant Matters'. This lack of dependency is often overlooked. An example is where under FIDIC Red Book terms the employer's quantity surveyors preparing interim payment recommendations exclude any delay-related costs because the engineer under the contract has granted no extension of time.

If the available float means that extensions to the period of particular activities do not extend the completion date for the project, then the main contractor is unlikely to be able to recover payment for contract management resources and support facilities that would be in place for the duration of the contract period. The same situation would apply where the delays are off the critical path and only cause local delays, such as in the case of the turnkey contractor's EPCM establishment for the whole of a railway project identified elsewhere. That turnkey contractor will be deemed to have included this in its pricing for the cost of such management, etc., for the contract period, and will be restricted to recovery of the direct costs of the prolonged activities. In that example, this would be its 'North P&Gs' and 'South P&Gs' and local activity related preliminaries and general costs so far as they can be established as having been prolonged.

NEC4-ECC clause 31.2 sets out detailed requirements for the contents of each of the contractor's programmes. Those requirements include provision for float. The NEC User Guide says this regarding clause 31.2:

> Float is any spare time within the programme after the time risk allowances have been included. It is normally available to accommodate the time effects for a compensation event to mitigate or avoid any delay to planned Completion. However, in accordance with Clause 63.5, float attached to the whole programme (i.e. any float between planned Completion and the Completion Date) is not available. Any delay to planned Completion due to a compensation event therefore results in the same delay to the Completion Date (see further explanatory notes on Clause 63.5).

That clause 63.5 says this:

> A delay to the Completion Date is assessed as the length of time that, due to the compensation event, planned Completion is later than planned Completion as shown on the Accepted Programme current at the dividing date.
>
> A delay to the Key date is assessed as the length of time that, due to the compensation event, the planned date when the Condition stated for a Key Date will be met is later than the date shown on the Accepted Programme current at the dividing date.
>
> When assessing delay only those operations which the Contractor has not completed and which are affected by the compensation event are changed.

Furthermore, the NEC User Guide says this regarding clause 63.5:

> By taking this approach, if planned Completion is delayed, the Completion Date is delayed by the same period. If planned Completion is not delayed, the Completion Date is unchanged. If the adjusted programme shows that achievement of the Key Date will be delayed, the Key Date is delayed by the same period.

Contractors will say that the NEC approach places the benefit of float in the programme where it belongs – with the contractor who created it as part of its provision against the risk of its own failures. Under that approach the end or terminal float in a programme is maintained until, and if, it is used up by the contractor's own later culpable delays. In practice, where difficulties arise is when the contractor has failed to properly update its programme, particularly for its own earlier culpable delays.

3.4   Prolongation of the Works

There will of course be many occasions when delays to particular activities in a programme do impact on the critical path (or paths where there is more than one) in the programme and thereby cause the contract completion date to be extended from that originally intended.

It should be recognised that the purpose of the provisions in most construction contracts for the completion date to be extended for defined events or circumstances is twofold. Firstly, to protect the contractor against claims from the employer for damages due for failure to complete by the contractual completion date(s). Secondly, to preserve the employer's right to deduct damages for late completion notwithstanding the defined breaches of contract by the employer. Particularly under English law, without machinery to allow the extension of the contractual completion date(s), in the event of any breach of the employer's contract obligations by the employer or its agents, such breach might enable the contractor to claim that it had been prevented from completing in accordance with the contract. This may leave the contractor with the obligation only to complete within a 'reasonable' time, but with the difficulty for the employer of determining what a reasonable time was. It may also render the agreed rate of delay damages in the contract obsolete, with the employer left to establish its actual damages. Extensions of the completion date provisions therefore serve to preserve the contract mechanism for determining the date of completion and the employer's right to delay damages.

It should equally be recognised that entitlement to an extension of the completion date does not automatically entitle the contractor, or subcontractor, to additional payment for the extended period of the works. It has been highlighted elsewhere how contracts such as SBC/Q keep their provisions for extensions of time and 'loss and expense' separate. The grounds for an extension to the completion date that will also entitle the contractor to additional payment will depend on the terms of the contract.

To expand this by reference to a particular type of delay event, the SBC/Q form lists the matter that will enable a contractor to claim an extension to the contract completion date at clause 2.29, where they are called 'Relevant Events'. The entitlement to payment of additional costs for matters affecting the progress of the works is set out in clauses 4.23 and 4.24, where they are called 'Relevant Matters'. The events/matters under these headings do not fully overlap. Thus, for instance, clause 2.29.3 entitles the contractor to an extension of time for the employer's failure to give possession of the site of the works or a section thereof, and the effect of clause 4.24.5 is to give the contractor an entitlement to loss and/or expense in such circumstances. By contrast clause 2.29.9 gives the contractor an entitlement to an extension to the completion date if 'exceptionally adverse weather conditions' cause critical delay, and clause 2.29.11 provides entitlement to an extension of time for civil commotion, but there are no similar provisions in clauses 4.23 and 4.24 in relation to loss and expense arising from those events. This leaves the contractor with an extension of time in such circumstances but with no entitlement to additional payment. It is therefore protected from damages for late completion caused by such as exceptionally adverse weather, civil commotion, strikes, lockouts and the employer preserving the contract mechanism for fixing the date for completion and its entitlement to damages for late completion, but the contractor cannot claim associated loss and/or expense. Such delays are sometimes referred to as 'cost neutral delay events'. A similar approach is followed in contracts such as FIDIC Red Book and the Infrastructure Conditions, although FIDIC covers weather conditions under the broader heading of 'exceptionally adverse climatic conditions'.

Once the principle of entitlement to additional payment for an extension to the contract completion date has been determined, financial quantification can be addressed. The quantification of prolongation claims is considered in detail in Chapter 6 of this book. However, there are two matters of principle that are worth highlighting at this stage in the context of delay analysis:

1. Whether it should be based on rates or prices or the contractor's actual damages, costs, loss or expense. Unless the contract expressly provides otherwise, the contractor's entitlement to additional payment for prolongation will normally be based on the actual additional cost or expense to the contractor of the extended period.
2. The evaluation should be made by reference to the period of the works, when the relevant event impacted on the progress of the contract works, and not by reference to the period between the original completion date and the extended date, i.e. the extended period at the end of the works.

These principles highlight two difficult areas of evaluation that affect delay analysis as part of consideration of a contractor's financial claims.

There is a need for detailed delay analysis, not only to determine which events caused the delay to the contract completion date but also to identify the periods of the works when the delay occurred. Only very rarely will the delay actually have occurred at the end period extending beyond the original completion date; for example:

- Where the delays occurred because of interference with tests on completion, entitling the contractor to an extension of time and 'Costs plus reasonable profit' under FIDIC Red Book clause 10.3, for example.
- Where contractors sometimes contend that the employer's agent unfairly failed to certify completion, or Taking Over as it is termed in FIDIC parlance. The FIDIC Red Book appears to contain no provision for an extension of time or compensation in such a situation. However, given that under clause 3.1(a) the engineer is 'deemed to act for the Employer', such a failure may amount to a breach of contract by the employer entitling the contractor to damages for that breach, including any prolongation costs as it retained time-related resources on the project for longer than should otherwise have been necessary. Given that the contractor would retain liability for such as insurance and security of the works until the issue of a completion certificate it can be envisaged what additional expenditure the contractor would incur in such a situation.

Notwithstanding such examples, in the vast majority of cases the delay(s) will have occurred during the course of the works and quite usually before the original completion date. The accurate identification of these periods is crucial to a proper and accurate evaluation of the costs incurred by prolongation, as costs will obviously vary with the timing of the delay and extent and state of the works at that time. A delay at the very outset of a project, by perhaps a lack of access to the site of the works, will probably incur relatively low costs as activity has not started at that stage. The main costs may be the abortive cost of staff allocated to the project but unable to commence, together with any site plant and accommodation ordered but unable to be delivered. Financial and 'head office' costs may also be incurred in addition to these direct costs, but the scale of expenditure is very different at commencement compared to a delay period at the height of site activity when main and subcontractor resources are at a peak.

The detailed analysis of delay to identify the relevant periods is not the subject of this book; it is only necessary to emphasise the importance of the analysis and appreciate the implications for evaluation purposes. For guidance as to the techniques that might be employed the SCL Protocol is an excellent starting point, but it should not be thought to be prescriptive or applicable in all circumstances. Internationally, the First Edition gained increasing currency over the years following its publication in 2002, as a persuasive source of guidance on how to address delays and related claims. There is no reason to assume that the Second Edition will not be similarly greeted, particularly as it addresses some of the criticisms of the first. On the two matters of principle identified, Core Principles 20 and 22 of the SCL Protocol respectively state:

(1) '.... the compensation for prolongation caused other than by variations is based on the actual additional cost incurred by the Contractor'; and
(2) 'the evaluation of the sum due is made by reference to the period when the effect of the Employer Risk Event was felt, not by reference to the extended period at the end of the contract'.

However, this is always subject to the express terms of the contract and any principles or rules arising from the substantive law of the local jurisdiction. There is a wealth of information, and differing opinion, on delay analysis and it is important that the techniques used achieve the objective with the information available and address the principles that have been set out.

It is important to understand the limitations of the often simplistic approaches adopted in some calculations of activity durations. The most common approach is to derive the amount of additional work in an activity and divide by the available resource to produce an alleged period of extended activity. Taking a deliberately simplistic example, if 20 additional doors need to be hung, each requiring two hours of labour, then the total labour requirement of 40 hours could be divided by the available resource of, say, two joiners working eight hours per day, to produce an activity extension of two and a half days (40 hours of work divided by 16 hours available per day). This may not, however, be the actual effect on the activity duration if the additional doors are not immediately available, or are only available in stages. It should also, of course, go without saying that the insertion of the additional work into an activity duration does not necessarily impact on the contract period. This may lead to prolongation overall, local prolongation only or no prolongation costs at all, such that the only financial effect is the additional cost of the joiners and materials.

Whilst the detailed analysis of delays and prolongation are not subjects for this book, for those carrying out financial quantification it is important when conducting any evaluation that a number of aspects of delay analysis that might arise are understood and their implications taken into account in the evaluation. These are discussed in Section 3.5.

3.5 Analysis of Time and Delay

3.5.1 Introduction

There can be few issues under the heading of evaluating contract claims as extensively written about and analysed as that of the analysis of claims for delay. There seem to be a number of reasons for this including the following:

1. Time means money and prolongation claims are often the largest of a contractor's claims on a problem project.
2. The analysis of programmes and delays can be a highly technical area given the complexity of programmes and issues such as float.
3. The facts of delay are seldom simple on complex projects, and usually have some competing causes with liability spread between employer and contractor and perhaps also subcontractors and suppliers.
4. There are a number of alternative approaches to delay analysis that can give rise to very different results on the same facts, and each of which has their proponents, opponents, merits and disadvantages.
5. Certain technical issues arising in the analysis of delays have either not been settled by the courts or have been the subject of authorities that appeared, to say the least, to sit uncomfortably with each other.

Typical headings of expenditure and losses that can be incurred by the employer or contractor in the event of delay to a project are scheduled in Section 6.1 of this book. In addition, for both parties, the costs associated with assessing and analysing the causes of delays and reaching a conclusion on culpability are also not to be underestimated. This area is complex and subjective in many aspects, and as a result consultant costs, management time and legal fees can be large for both parties. Indeed, on more complicated disputes there may be several parties. Their costs may include preparing the case, analysing each other's case and taking the matter through to a conclusion as to a decision on responsibility in such as arbitration or the courts.

Why is it that claims for delays on construction projects are so common? Firstly, there are so many potential causes of delay to the progress and completion of construction works. As stated elsewhere, a review of any of the provisions of the major construction contract forms shows the number of events that can impact upon progress and completion of a project. For example, the SBC/Q contract form in clause 2.29 lists 14 headings of 'Relevant Events', any one of which can contain within it a number of subheadings and which can recur many times during the course of the project. Similarly, NEC4-ECC lists 'compensation events' in its clause 60.1 that run to one and a half pages. These events cover a variety of issues, some of which are outside the control of either party, others of which are in the control of the parties but still have plenty of potential to occur.

It is a truism to say that all construction projects are different. The 'one-off' nature of construction means that the scope for learning lessons and using those lessons in subsequent similar projects is reduced, certainly when compared with most other industries and manufacturing processes. On the other hand, even where projects of similar type are carried out by the same or similar teams, lessons may not have been learnt and the same mistakes can be made on a serial basis by the same employer, contractor and consultants. This is a particularly striking feature of public sector clients internationally, with even government ministries that have a continuous pipeline of construction projects making the same mistakes from project to project.

The uncertain nature of construction and the circumstances in which projects are carried out are other factors that are unique or unusual in construction processes compared with other industries. Uncertainty and unpredictability of ground conditions, weather conditions, the performance of third parties such as statutory undertakers or state owned utilities companies, and similar matters that are outside the parties' control, add to the frequency of delays and claims for delays. This leads on to the matter of human fallibility. Whilst sectors such as housing are beginning to increase the use of mechanisation, automation and off-site pre-assembly, the fallible nature of human beings means that mistakes are made, which on construction projects can lead to claims for time. Late information from architects, engineers, other consultants or specialists is the most obvious manifestation of human fallibility and its ability to delay construction projects. Equally, even where information is provided on time, wrong information will subsequently have to be amended or become the subject of query or the seeking of clarification. This is another regular cause of delays.

Such incorrect information and its subsequent correction often leads to change and variation. However, the construction industry is perhaps relatively unusual in its acceptance that change will be made during construction. Designers and/or more particularly the employer itself will want to make changes to the original design as covered by the contract and will expect the contractor to carry out the subsequent varied work. Accordingly, construction contracts are usually drafted in such a way as to make provision for the instruction of variations, for valuation of the work that arises and the assessment of the time effect of such events. The scope for change and the need for change to the design, as with so many of the other causes of delay that occur on construction projects, is exaggerated of course by the time that a project may take – the span of the period of construction. Projects spanning several years are not unusual and it is perhaps inevitable that during the course of such a project legislation and/or technology and/or client requirements and/or market requirements will change such that it is clearly desirable that contracts make provision for the instruction of such change and therefore variation to the scope of work as originally contracted for.

The lengthy nature of many construction projects also increases the extent to which those projects may experience a variety of climatic, weather and seasonal conditions. Construction contracts vary in the terms in which they deal with such conditions and allocate the risks between the parties. Many bespoke contracts lay the risk of weather conditions at the door of the contractor, as do a few standard forms such as the General Conditions of Contract issued by the CRINE Network (Cost Reduction in the New Era) in June 1997 and the LOGIC Marine Construction contract Edition 2 in 2004. However, as noted elsewhere, the standard forms considered in this book make weather conditions a cost-neutral event; that is, they are an event for consideration for extension of time, but not for financial reimbursement of the contractor's related costs. Furthermore, the terms in which standard contracts define the weather conditions that are to be considered are subjective. Traditionally the JCT contracts used the term 'exceptionally inclement weather', although SBC/Q now refers to 'exceptionally adverse weather conditions' as the 'Relevant Event' in its clause 2.29.9. That reflects the approach traditionally adopted by ICE contracts and the Infrastructure Conditions, the latter referring to 'exceptional adverse weather conditions' in their extension of time clause 44(1)(d). FIDIC uses the broader concept of 'exceptionally adverse climatic conditions'. On the other hand, the form of the lump sum contract published by the Institution of Chemical Engineers 2001 uses the term 'exceptionally severe weather conditions'. Such subjective terms give some regular difficulty in defining, for example, what is 'exceptional', 'inclement', 'severe' or 'adverse'. This is even before considering what the effect of any such weather that is experienced has been on the construction process and on the completion date(s).

NEC3-ECC took a more scientific approach to the issue of what weather conditions should count for consideration. As its Guidance Notes 2005 state against clause 60.1(13):

> Rather than rely on the subjective generalisations about 'exceptionally inclement weather' or the like sometimes included in standard forms of contract, the ECC includes a more objective and measurable approach. The idea is to make available *weather data*, which is referred to in the Contract Data, and which is normally compiled by an independent authority, establishing the levels of selected relevant weather conditions for the Site for each calendar month which have had a period of return of more than ten years. If weather conditions more adverse than these levels occur it is a compensation event. Weather which the *weather data* show is likely to occur within a ten-year period will not result in a compensation event arising and the *Contractor* should have made due allowances for such at tender stage.

A further delaying event whose potential tends to become exaggerated by the long duration of construction projects is that of resource shortages. Most standard forms make inability to obtain labour and materials the contractor's risk. Entitlement to extension of time (and particularly financial compensation) rarely appears in bespoke forms. The basis of this approach must be that obtaining such resources is in the sole power of the contractor, who should therefore take the risk for it. However, construction contracts might make exception where the cause of resource shortage is one that is outside the contractor's foresight and control. For example, the FIDIC Red Book has this as grounds for entitlement to extension of time at clause 8.4(d): 'Unforeseeable shortage in the availability of personnel or Goods caused by epidemic or governmental actions'. Therefore the contractor gets an extension of time if a labour shortage was unforeseeable and caused by an epidemic or the government. There is, however, no provision for payment of the contractor's associated costs. This is therefore a 'cost-neutral' event under the FIDIC Red Book. However, where resource shortage does count for extension of time its likelihood of occurring is clearly greater on longer construction projects. Particularly in periods of over-demand for construction work, labour shortages can be a real problem. Further, if the market changes from periods of relatively low activity to periods of very high activity, even during the course of a single construction project, availability of labour can change very quickly. One issue that is often problematic is where periods of labour and material shortages occur due to earlier delays but for which the shortages would not have been met. Contractors in overheated markets such as those in south-east England or the Gulf States 10 years ago, often argued that earlier delays on construction projects have lost them their 'window' for resource availability from their subcontractors. By being pushed out of that 'window' into periods where their subcontractors have other work commitments, they incur further delays. This knock-on effect of earlier delays is typical of the kind of complexity that can occur when considering delays to construction projects. It is a good example of one reason why the analysis for claims for delay is so complex and contentious. The topic of such consequential delay events is considered in detail in Chapter 6 in the context of financial claims.

Construction and engineering projects of any size and length therefore invariably involve an acceptance that change will happen and there is a likelihood of unplanned change, such as that suggested already, meaning claims are often inevitable. However, their complexity is in practice often added to by the following factors:

- The consideration of claims for delay often starts (although all too regularly it starts much later) with the arrival of a notice of delay issued under the contract. Additional issues may arise as to whether or not it complies with notice requirements in the contract. However, the complexity in dealing with the claim subsequently can often be exaggerated by the absence, lateness or quality of such notices.
- An inability or failure to maintain sufficient or appropriate records. This is often the case even after notice has been given and both parties are aware that they will have a claim to address.
- A lack of awareness at the appropriate time that delay has actually occurred. Lack of notice sometimes results from ignorance or oversight as to contractual requirements. However, that an event will have a notifiable delaying effect is not always immediately apparent or spotted.
- Inadequacy or even a lack of an original programme. This may mean that, with no baseline to compare the actual timing of something like the issue of a drawing against,

the lack of a programme may mean that it may not be apparent that there is a delay. Lack of a programme is also likely to mean that analysis of the effects of a delay will be unclear and open to disagreement.

- Inadequacy or shortage of subsequent updated or revised programmes that would enable a delay event to be put into its true context.
- Difficulties in separating causes and effects. This particularly involves 'chicken and egg' analysis of whether an event that has happened late was the cause of the delay or happened late because of delay resulting from other and earlier events. This particularly arises with design information, apparently issued late against an original programme but in fact issued later because the works, and hence its requirement, were already in delay due to other causes. The theoretical approaches to delay of analysis, such as that adopted in *Great Eastern Hotel v John Laing Construction Ltd and Another* [2005] Adj. LR 02/24, which is considered further elsewhere in this book, do not take account of the fact that the architect or engineer, aware that the project is already in delay, may time the issue of information to fit actual progress. This concept of 'pacing' activities is particularly relevant to contractors, who slow non-critical activities in light of delays by the employer to critical works. However, the contractor may later find a defence to its claim being that it had concurrent delays on those deliberately slowed activities.
- Concurrent causes of delays, particularly where these involve a mixture of employer risk events and contractor culpability issues.
- Concurrent causes of delays, particularly where some are compensable and others are cost-neutral.
- Differentiating between critical and non-critical delays.
- As noted previously, the existence of a number of alternative delay analysis techniques, each of which is capable of giving a quite different result from the same set of facts. For example, design release by the engineer may be late against an original programme and therefore give an entitlement to extension of time if an 'impacted as planned' method is adopted. However, delays elsewhere that have affected the programme may mean that it has no effect at all on the actual programme if a 'collapsed as built' method is adopted. This will be illustrated further elsewhere in this when considering some of these techniques.
- How the delays should be evaluated in terms of their financial consequence. This is the detailed subject of Chapter 6 of this book. However, for now it is noted that most heads of delay-related financial claims have more than one possible approach to quantification and some are particularly controversial (for example disruption, global claims and formula approaches to head office overheads).
- Contract provisions, which can vary, and contribute to the complexity of delay analysis. Bespoke amendments to standard contracts are often poorly drafted. There is a reason why the industry has standard forms of contracts, which is to use standardised terminology and provisions, based on an industry consensus as to what is fair between the different interests of contracting parties and have stood the test of use in practice. Amendments particularly to extension of time clauses often cause issues: for example, attempts to define 'exceptionally inclement weather' under the terms of the 1963 edition of the JCT contracts. More recently, some attempts to draft a 'time impact analysis' approach into contracts have appeared to be written by advisors with no experience of actually applying such an approach. Float ownership clauses

may cause disagreement as to their workings and given their unusual nature will lack precedent and other guidance as to their meaning.

The matters discussed in this section are particularly common problems in assessing the delays and additional payments that may be due as a result of delaying events, and they need to be understood so that a proper evaluation of delays and related costs can be made.

3.5.2   Basic Requirements

The basic requirements for any analysis of delay and its effects should start with an original plan. On all but very simple minor projects this should take the form of a programme. If such an original plan can be identified then it must be considered whether it is in a form and level of detail sufficient to enable analysis and furthermore whether the plan was actually achievable such that analysis of it will derive the true effects of the delay event being considered.

Tender programmes are a starting point for such a plan, but must be treated with caution, as explained elsewhere. From the point of view of employers considering contractors' tender programmes, or for that matter contractors considering tender programmes received from subcontractors, it is all too often the case that sufficient consideration is not given at the tender stage to the nature and adequacy of such programmes. The first question that a tender programme should answer is whether the party that has prepared the programme has understood the scope and complexity of the project or work that it is undertaking to carry out and whether it has illustrated that it can indeed build or construct it. Such questions should be asked not only for the purpose of the programme's subsequent use in any analysis but most obviously as essential matters of tender evaluation. The programme should be considered to see that it illustrates sufficient detail and logic linking as well as activity durations. Thus, for example, an A4 single page programme with no logic links, as part of a tender submission for a multimillion pound roads project spanning several years, should immediately sound alarm bells for the recipient. Whether a tender programme has been produced using a recognised planning software package should also illustrate the capabilities of the tenderer. A tenderer that presents a programme prepared using spreadsheet software suggests that it may not have the programming sophistication to manage and control a project of anything other than a simple scope. Other issues that should be apparent in a tender programme include whether the tenderer has understood the sequencing of the work to be carried out and in particular any sectional completion requirements in the enquiry documents. Clearly a tender programme that has not understood or does not properly identify such issues is not only a matter of concern in tender evaluation but is also going to become a false basis for a subsequent delay analysis, should it become necessary to use it for such purposes. Equally important to these questions of understanding and practicality shown by a tender programme is whether the tenderer has shown that it has the resources to carry out the work in accordance with that programme. This means not only sufficient resources but also the resources available at the appropriate time. It was explained in Chapter 2 how often tender programmes do not set out the resources that they were based upon and were prepared without consideration of what resources would be required and would be available.

Whilst these matters of tenderer selection can subsequently be the basis for delay analysis, the contractor's construction programme, which should be seen as an essential project management tool, is even more important to subsequent delay analysis.

Construction programmes should illustrate in greater detail many of the issues first tested by the tender programme. These include identifying how the contractor intends to build the project or work. This programme should give more detail in terms of such as: numbers of activity bars that the work is broken down into, logic links to show interdependencies and the critical path(s). Additionally, the contract programme should become a tool for monitoring progress. It should therefore be in an updatable format using recognised and available software. Of equal importance both to the project management processes and subsequent analysis of delays and their effects is whether the contract programme identifies programme information requirement dates or indeed other requirements that the contractor has, such as dates for access or the completion of work by other parties and provision of design information. Fortunately, as the software available has improved over the last 30 years, so it has become increasingly common for contract terms to set out express requirements of the form and contents of the contract programme. For example, the clause 8.3 requirements of the FIDIC Red Book were not set out in clause 14 of the 1987 edition, which only required a programme 'in such form and detail as the Engineer may reasonably require'. In addition, it is increasingly common for bespoke forms or amendments to standard forms to set out requirements of the programme in terms of its form, extent of detail, the software to be used and, in particular, obligations regarding the updating and revision of that programme during the course of the work. Some of these were either based on, or incorporated, the terms of the model clause of the First Edition of the SCL Protocol.

Alongside programmes should be related method statements, although the requirement for these is often overlooked. These should be required and provided both at tender and contract stage further to set out the methodology for work to be carried out. From a tender assessment and project management viewpoint, method statements should further illustrate how the work is to be done: is the party aware of what is required, is there a clear plan, is the plan practicable? From a delay analysis viewpoint, method statements can often provide an invaluable record of planned intent that can be compared with actual practice. They can often be used to debunk claims made as to supposed planned intent conceived retrospectively for the purpose of creating or exaggerating a claim.

Underlying methods should also be apparent from the programme itself. For example, a programme might show the excavation of a shaft and tunnels for a station and underground railway, but does it show how spoil from the tunnels will be removed to ground level? If the programme does not show that suitable access shafts will also be dug, has the contractor adequately considered the practicalities and how will the effects of delays in relation to those shafts and the land that they occupy be assessed?

Having established a planned intent, the next requirement for delay analysis is notice of events that are said to have caused delay. The need for notices is usually a contractual requirement. Whether it is a condition precedent to entitlement is a legal issue depending on the express terms of the contract and the local substantive law and is not considered in this book. Even where notices have been given, issues often arise as to their sufficiency. Do the notices sufficiently detail the event said to have caused delay and also the effect or likely effect of that delay? Furthermore, were notices given early enough to allow the other party not only to assess and monitor that delay but also potentially to take action to prevent or reduce it?

The absence of notice is often linked to a lack of awareness on projects that delays have actually occurred – not only that there have been delaying events but that those events have had an effect on progress and an effect on completion. It is striking how often delay analyses are carried out on projects spanning long periods of time when for much of the period of delay one or all parties were unaware the delay to completion had actually occurred. Often this can result from a failure or refusal of the party to admit to what is obvious, with management staff in denial that their project is slipping. Equally, contract administrators such as the architect or engineer can refuse to accept that the delay is actually occurring. Usually this denial is linked to culpability, or at least a fear of being proven to be culpable. The practical effect of a lack of notice coupled with a refusal or failure to admit to delays is often that delays are not monitored and records are not kept. As a consequence, subsequent delay analysis can be made more problematic.

Records can contain various pieces of information and vary in quality and form. The records required for subsequent delay analysis need to identify such facts as: what happened, when it occurred, what resources were impacted and for how long, what its local effect was and what its project effect was. As a tool for proper consideration of related financial claims this means both the effect on activities and on completion as a whole. Furthermore, even when records are kept there are often issues as to the quality of those records. Records need to be regularly and consistently kept. Records of limited periods or parts of a project can be of little use unless the delays being considered were of similar limited periods or parts. The need for consistency also applies to their format. Records kept in a variety of formats, by time or part of a project, can be of use but add to the complexity of subsequently attempting to apply them to a delay analysis. On large sites with a number of individuals or teams being responsible for keeping records in their parts of the site or works, it is surprising how often they will maintain records in different formats, level of detail and regularity. A first management task on such a project should be to set these as project-wide standards to ensure best practice and consistency.

Finally, the accuracy of records is of paramount importance. A delay analysis which is based on inaccurate records is likely to be of little value and potentially damaging for the party involved. Inevitably, some records may be inaccurate and analysis may need adjustment. However, where records are shown to be extensively inaccurate, not only can the costs of delay analysis be unnecessarily wasted but credibility can be lost. Little else discredits a claim better that showing that there are inconsistent records as to events it is based upon as, for example, where a monthly management report records one date for the start or finish of an activity, but an updated programme or labour or plant record show a different date. This can happen for various reasons. The records may be for different purposes, with the internal management reports intended to paint a better picture than the reality. It may be simply that they were prepared by different people and were not coordinated. Those people may work to different levels of thoroughness and attention to detail.

It is often surprising which records are quite useful in the analysis of delays but have been overlooked by a party. A party may believe that it has limited suitable records available, when actually it has rather more than it thinks. Experience of delay and disruption analysis shows that the following categories (listed in no particular order) can provide useful records that may be put to good use:

Tender programme	Quality assurance record	Payroll
Draft programme	Health and safety file	Subcontractor payment
Programme update	Shipping record	Supplier payment
Tracked programme	Delivery note	Letter
Drop-line programme	Photograph	Warning notice
Method statement	Video	Delay notice
Revised method statement	Webcam	Instruction
Clerk of Works report	Labour return	Confirmation of Instructions
Taking over certificate	Daywork sheet	Authorisation to proceed
Snagging list	Labour allocation sheet	Memoranda
Punch list	Staff allocation sheet	Request for Information
Weekly report	Plant allocation sheet	Request for Approval
Monthly report	Plant return	Quotation
Management report	Wage record	Order
Meeting minutes	Project accounts	Requisition
Staff diary	Cost report	Subcontractor correspondence
Pour record	Cost/value reconciliation	Supplier correspondence
Inspection sheet	Invoice	Drawing revisions
Testing and commissioning data	Applications for payment	Working drawing
Safety record	Interim valuation	Sketch
Non-conformance report	Surveyors' book	Drawing register
Cube test	Bonus calculation	BIM record (see elsewhere)

This list is not intended to be comprehensive but is to illustrate the range of useful documentation and perhaps some of the more unusual sources of record that can be used in a delay analysis. Creating an as-built record from such documents, or even from witness testimony, can be a very laborious and risky business. Much better, but not included in this list and probably of most value, are updated programmes issued during the course of the project, starting with the tender and construction programmes already discussed. Regular drop-line records against the construction programme, that can be rescheduled to project an effect on completion, are particularly useful not only as a management tool but also in the subsequent delay analysis. However, an occasional difficulty with rescheduled programmes based on a drop-line against an original programme is that they can become little more than theoretical. Whilst such theoretical results may be acceptable under some contract terms, those terms need to be checked in each case. They project the theory of an effect against an original programme when the reality of what is happening on site shows a change in the construction approach. Revised programmes that reflect such change are therefore often more helpful. However, care must be taken to understand the status of such a revised programme; for example, is it a mitigation or a target programme? The terms in which it was submitted and presented at the time is of importance. Such programmes provide two things: firstly a timely warning of delays and their effect and secondly a useful record and basis for later delay analysis.

Having considered the basic requirements for an analysis of time and delay, such as programmes, notices and records, some of the common problems encountered in carrying out the resulting analyses are discussed in the following sections.

3.5.3   Float and Acceleration

The problem of float in contract programmes generally and the issue of ownership of float were discussed in Section 3.3 of this book in the context of float being used up by change and thereby causing disruption to the sequence or duration of non-critical items. However, if all such float is used up and further delay occurs, or delay occurs to an activity on the critical path, then delay to the contract completion date will seem inevitable.

It should be understood that there can be two different types of float in a programme – 'end' or 'total' float and 'free' or 'activity' float. 'End' or 'total' float applies to the total contract works activities and appears as a difference between the completion of the last activity on the programme and the contract completion date. 'Free' or 'activity' float applies to an individual activity only, within the total network of activities. In practice, 'total' or 'end' float is rarely seen in contractors' programmes and even where it might be available is often hidden by artificial extension of activities so that the contractor can hope to have the benefit of such float without the knowledge of other parties, although contractors would say that it is just a sensible provision by it against risk of its encountering culpable delays, by leaving a buffer at the end of its programme. This has been highlighted, where the contract does not expressly allocate the benefit of float to the contractor.

Where the contractor has planned and programmed to complete before the contract date(s), there is, of course, no obligation on the employer generally to fulfil its obligations so as to allow the contractor to achieve that early completion, any more than the contactor is contractually obliged to finish early. It was determined over 100 years ago that the contractor has the right to the contract period in which to complete the works (*Wells v Army & Navy Co-operative Society Ltd* (1902) 86 LT 764. In relation to a contractor's claim against the employer for delays preventing it from finishing early, see *Glenlion Construction Limited v The Guiness Trust* [1987] 39 BLR 89. This was an appeal from the award of an arbitrator on a contract under the 1977 revision of the 1963 JCT Standard Form of Building Contract with Quantities. The court held that, whilst the contractor was entitled to complete before the contractual completion date, there was no obligation on the employer, its servants or agents to perform their obligations so as to enable the contractor to complete the works early. This was the position even though clause 21(1) of that contract required the contractor to complete the works 'on or before the Date for Completion' stated in the contract.

There is, of course, a potential answer to the problem of maintaining a completion date or dates when further delay occurs, but no float exists to accommodate it, and that is to reduce the time required for the remaining activities so as to prevent or reduce the delay to the completion date. This involves acceleration, a topic considered in detail in Section 6.5 of this book, with focus on the evaluation of its financial effects. As noted there, the issues often start with a disagreement as to what acceleration actually is and what acceleration measures are. It is suggested that the latter term covers those measures that mean that the works, or a section or part thereof, are completed earlier than would have been the case had the measures not been taken.

Acceleration might be because of delays for which the contractor is culpable and wishes to avoid delay damages for late completion. In this case it will not generate entitlement to payment from the employer, although it could generate an entitlement against a culpable subcontractor or supplier, depending on the relevant facts and their contract terms. However, if the delay, or potential delay, is the responsibility of the employer, then the contractor is likely to claim compensation for any additional expenditure from the employer.

The concept of 'accelerating' the remaining works raises a number of issues that need to be addressed, starting with what the term actually means. The subject was considered by Judge Hicks QC in the *Ascon Contracting Ltd v Alfred McAlpine Construction Isle of Man Ltd* decision, which is also discussed in Section 6.5 of this book. The Judge's thoughts on the meaning of the term are worth repeating here:

> 'Acceleration' tends to be bandied about as if it were a term of art with a precise technical meaning, but I have found nothing to persuade me that that is the case. The root concept behind the metaphor is no doubt that of increasing speed and therefore, in the context of a construction contract, of finishing earlier. On that basis 'accelerative measures' are steps taken, it is assumed at increased expense, with a view to achieving that end. If the other party is to be charged with that expense, however, that description gives no reason, so far, for such a charge. At least two further questions are relevant to any such issue. The first, implicit in the description itself, is "earlier than what?" The second asks by whose decision the relevant steps were taken.

The topic of acceleration and how it comes about is detailed in Section 6.5 of this book. There all of the potential acceleration claims are explained as arising on construction and engineering in the following alternative ways:

1. On the basis of an express provision of the contract that either:
   (a) directly refers to acceleration, including clauses that allow the employer to instruct or direct acceleration, or
   (b) indirectly either refer to acceleration or can give rise to it.
2. As an extra-contractual agreement between the parties, that effectively agrees an amendment to the original contract.
3. As part of the contractor's claim for the loss, expense, costs or damages arising from delay and disruption.
4. Where the contractor considers that it is entitled to extensions of time, but is denied these by the employer's certifier, and instigates acceleration in order to reduce its potential liability for delay damages. This is commonly described as either 'induced' or 'constructive' acceleration.
5. Where the contractor is in culpable delay and accelerates in order to reduce its risk in relation to its failure to complete on time and potential liability for the employer's delay damages. The associated costs of such measures will not be compensatable by the employer, but where the contractor has instigated acceleration in order to reduce the effects of its own failures, this can have significant effects on the evaluation of its entitlements in relation to other failures by the employer.

3.5.4 Concurrent Delays

The identification of the cause, period and timing of delay is often the first step in moving to evaluation of additional payment to the contractor and will be essential to the employer's entitlement to deduct delay damages by defining the contractually due completion date or dates from which those damages should run. Sadly, in practice it is rarely the case that there is one delaying event causing one extension to the contract period. On large and complex contracts there are likely to be several causes of delay, perhaps dozens, perhaps hundreds. Some of these may be on the critical path activities and others on non-critical activities, with the possibility that their delays cause non-critical events to become critical. All of these may require to be identified both as to their timing and effect.

Further complication will usually appear in that some, and sometimes all, of the various delays will overlap or cause delay to different activities in the programme at the same time. In other words, they will be overlapping in time. The situation can then become further complicated in that some delays may be the responsibility of the employer, some the responsibility of the contractor, some cost-neutral events (such as exceptional weather) and yet others the responsibility of subcontractors or suppliers.

In such situations it will be necessary to separate the effects of different delays on progress and completion. This whole task particularly becomes more complex where there are multiple sectional dates for completion set out in the contract, or the employer exercises a discretion under the contract to take possession of the works in parts (for example under clause 10.2 of the FIDIC Red Book or clause 2.33 of SBC/Q), such that several separate analyses are required.

Where two or more delay events occur at the same time, or their effects occur at the same time, they are sometimes referred to as 'concurrent' delays, although the use of that term needs care given the legal contexts in which it has been considered. This also leads into the further concept of 'true concurrency'. The importance of these concepts relates to what is to be done to the contractual completion date or dates (and to the contractor's financial entitlements) where delays occur at the same time that are both critical, but one is the employer's responsibility and the other the contractor's. The provisions of the contract must be the first source of reference in this conundrum. However, in practice these rarely address concurrency. In the absence of some contractual prescriptions, several potential outcomes might be suggested:

- The contractor is entitled to an extension of time, perhaps on the basis of the prevention principle (under such as that in English law) or because it is fair and reasonable for it to do so. This is on the basis that it would be unfair for the employer to deduct delay damages from a date that matters for which it was liable would have made impossible.
- The contractor is not entitled to an extension of time because it would have completed late anyway because of its own culpable delay. This might be seen as applying the 'but for' test of putting the contractor into the position it would have been but for the event for which the employer is liable.
- Some form of apportionment of time between the effects of the two delays could apply some basis of comparison, such as their relative potency, so that the contractor gets an extension of time for part of the effects of the employer delay.
- Some combination of these outcomes.

- A further consideration might be what to do with the parties' financial claims, particularly the contractor's prolongation cost. Where it is considered to be entitled to an extension of time for a concurrent employer delay, does the 'but for' test mean that it is not entitled to additional costs that it would have incurred anyway?

Along with the ownership of float considered elsewhere, the issue of concurrency of delays is probably the most common cause of dispute in relation to claims for extensions of time and also for compensation of time-related costs. An employer may complain that a contractor's claim for extension of time and related compensation emphasises only events that are the employer's responsibility under the contract and conveniently ignores the contractor's own culpable delays. Alternatively, compensable events are emphasised when concurrent cost-neutral events were the real cause of delay. A contractor may argue that cries of 'concurrency' from an employer and its consultant are a diversion intended to avoid their responsibility for what really caused delay to completion, the contractor's own slow progress being on activities that were not on the critical path and had no effect on completion. Sometimes it is even said by contractors that slow progress to those activities elsewhere was the result of critical delays by the employer, with the contractor having responded by 'pacing' its other activities. In such circumstances the contractor will say that, whilst it was in overlapping delays elsewhere, there never was concurrent delay. This leads to a requirement to define exactly what concurrent delay is.

In his paper 'Concurrent Delay Revisited' (February 2013, SCL Paper no. 179), John Marrin QC defines concurrency as 'a period of project overrun which is caused by two or more effective causes of delay which are of approximately equal causative potency'. That definition was approved by Mr Justice Hamblen in *Adyard Abu Dhabi v SD Marine Services* [2011] EWHC 848 (Comm.) [2011] BLR 384. It was also adopted by Lord Justice Coulson in the recent case *North Midland Building Limited v Cyden Homes Ltd* [2018] EWCA Civ 1744 (see further discussions elsewhere in this chapter). Thus, true concurrency in this sense requires that the events cause delay to completion (that is, they are not just activity delays to non-critical parts of the works) and that they are of roughly similar potency.

Before considering how to address the problem of concurrent delays, it may help to set out the following series of very simple charts to illustrate the issue of what is, and is not, concurrent delay. Figure 3.1 shows a very simple sequence of activities for the construction of a wall, with a critical path running through design, the ordering of bricks, laying of bricks and copings. It also has non-critical activities in excavation, ordering concrete and concreting of footings.

Figure 3.2 identifies two periods of delay to this sequence of activities: firstly, a one-day excusable delay (perhaps for a change in specification) to the ordering of bricks; and, secondly, at the same time, a one-day non-excusable delay to the construction of the concrete footing (perhaps because of breakdown of the mixer).

In the scenario in Figure 3.2, both delays occur at the same time, but only the delay to ordering of bricks is on the critical path. The contractor's culpable delay to construction of the concrete footing may be occurring at the same time, but it is not on the critical path and would not cause a delay to completion. It is therefore not truly a concurrent delay. The one-day excusable delay in the ordering of bricks would cause a one-day delay to completion and would rank for an extension of time.

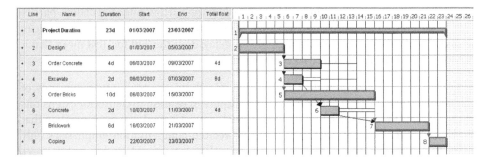

Figure 3.1 Sample programme sequences.

Line	Name	Duration	Start	End	Total float
1	Project Duration	24d	01/03/2007	24/03/2007	
2	Design	5d	01/03/2007	05/03/2007	
3	Order Concrete	4d	06/03/2007	09/03/2007	4d
4	Excavate	2d	06/03/2007	07/03/2007	6d
5	Order Bricks	11d	06/03/2007	16/03/2007	
6	Concrete	3d	10/03/2007	12/03/2007	4d
7	Brickwork	6d	17/03/2007	22/03/2007	
8	Coping	2d	23/03/2007	24/03/2007	

Figure 3.2 Sample programme sequences with non-concurrent delays.

Figure 3.3 identifies the same two causes of delay, but of different periods. Firstly, there is the same one-day excusable delay to the ordering of bricks. However, the non-excusable delay to the construction of the concrete footing is now five days in duration.

In the scenario in Figure 3.3, the delay to completion is again one day, but it is now being caused by two concurrent events. The five-day culpable delay to construction of the concrete footing has made that activity also critical to completion. There are therefore now two concurrent delays with the same one-day critical delay effect on completion.

Line	Name	Duration	Start	End	Total float
1	Project Duration	24d	01/03/2007	24/03/2007	
2	Design	5d	01/03/2007	05/03/2007	
3	Order Concrete	4d	06/03/2007	09/03/2007	
4	Excavate	2d	06/03/2007	07/03/2007	2d
5	Order Bricks	11d	06/03/2007	16/03/2007	
6	Concrete	7d	10/03/2007	16/03/2007	
7	Brickwork	6d	17/03/2007	22/03/2007	
8	Coping	2d	23/03/2007	24/03/2007	

Figure 3.3 Sample programme sequences with concurrent delays.

The problem of how to solve the problem of concurrent delays such as those simplistically set out in Figure 3.3 was considered by many to have been clarified by the decision of Mr Justice Dyson in *Henry Boot Construction (UK) Ltd v Malmaison Hotel (Manchester) Ltd* (1999) 70 ConLR 32, in which he stated, when considering a dispute under the JCT conditions of contract:

> Secondly, it is agreed that if there are two concurrent causes of delay, one of which is a relevant event, and the other not, then the contractor is entitled to an extension of time for the period of delay caused by the relevant event notwithstanding the concurrent effect of the other event. Thus, to take a simple example, if no work is possible on site for a week not only because of exceptionally inclement weather (a relevant event), but also because the contractor has a shortage of labour (not a relevant event), and if the failure to work during that week is likely to delay the works beyond the completion date by one week, then if he considers it fair and reasonable to do so, the architect is required to grant an extension of time of one week. He cannot refuse to do so on the grounds that the delay would have occurred in any event by reason of the shortage of labour.

The approach adopted in *Malmaison* was therefore based upon an agreement between the parties to the action, but it was recorded without criticism by the Judge. The result is that, in the scenario in Figure 3.3, what is since commonly referred to as 'the *Malmaison* approach' would see the contractor granted a one-day extension of time for the delay to the ordering of bricks. In that award, the architect cannot effectively apply the 'but for' test by refusing the extension on the basis that there was a five day delay to concreting activities that would have also caused a one day delay to completion in any event.

Whilst the parties in *Malmaison* agreed that a contractor is entitled to an extension of time where there are concurrent contractor culpable and employer caused delays, they disagreed as to whether, in determining under clause 25 of the then JCT form of contract, if an event is likely to cause delay to the works beyond the completion date, the architect is precluded from considering the effects of other events. The Judge accepted the employer's contention that the architect is not so precluded. Referring to the *Balfour Beatty Building Ltd v Chestermount Properties Ltd* (1993) 62 BLR 12 case that is considered elsewhere in this chapter, he stated 'It seems to me that it is a question of fact in any given case whether a relevant event has caused or is likely to cause delay to completion…'. He concluded that:

> In my judgment, it is incorrect to say that, as a matter of construction of clause 25, when deciding whether a relevant event is likely to cause or has caused delay, the architect may not consider the impact on progress and completion of other events.

The scenario in Figure 3.3 is unrealistically straightforward. In practice events and their delay periods are rarely so conveniently parallel. A more common scenario is that there are multiple delaying events, some the employer's responsibility and others the contractor's culpability, with the events occurring at different times and the periods of delay being different for each event, with some having an immediate effect and others causing delay at times after the event itself. The situation can become further complicated where some of the activities are critical whilst others initially have float.

The Technology and Construction Court considered more complex situations in the case of the *Royal Brompton Hospital NHS Trust v Fredrick A Hammond & Others* [2000] EWHC Technology 39. The court considered the matter as an appeal and the case arose from a complex and long-running saga over major building works for the Royal Brompton Hospital in Chelsea. The contractor had undertaken a major arbitration against the employer as a result of delays causing an overrun of the contract completion date by some 43 weeks and 2 days. After the settlement of the arbitration, resulting in an additional payment in excess of £6 million to the contractor, the employer commenced an action against its professional team, alleging that they had been too generous in awarding extensions to the contract completion date under the JCT Standard Form of Building Contract Local Authorities with Quantities 1980 Edition.

HHJ Seymour had to consider the proper approach to the assessment of extensions of time under the old JCT clause 25, as applied in the *Malmaison* case. He was particularly concerned with concurrency and the JCT requirement that, for an extension to the completion date to be granted, there had to be a delay caused by a 'Relevant Event', i.e. an event that was the responsibility of the employer, and that delay had to cause the completion of the whole works to be delayed beyond the contract completion date. He considered the *Malmaison* decision, and also that in *Balfour Beatty v Chestermount*, and concluded:

> ... as a matter of impression it would seem that there are two conditions which need to be satisfied before an extension of time can be granted, namely:
>
> (i) that a relevant event has occurred; and
> (ii) that relevant event is likely to cause the completion of the works as a whole to be delayed beyond the completion date then fixed under the contract, whether as a result of the original agreement between the contracting parties or as a result of the grant of a previous extension of time.

This analysis does not strike me as particularly exciting or novel, but I felt it was necessary at least to consider the question because it seemed to me that, in his submissions [Counsel] was rather glossing over the second element. He cited, helpfully, ... the recent decision of Dyson J in *Henry Boot Construction (UK) Ltd v Malmaison Hotel (Manchester) Ltd* ... in support of a submission that those decisions 'confirm ... the approach taken by (the project managers) in this case, where relevant and non-relevant events operate concurrently ...'.

However, it is, I think, necessary to be clear what one means by events operating concurrently. It does not mean, in my judgement, a situation in which work already being delayed, let it be supposed, because the contractor has had difficulty in obtaining sufficient labour, an event occurs which is a relevant event and which, had the contractor not been delayed, would have caused him to be delayed, but which in fact, by reason of the existing delay, made no difference. In such a situation although there is a relevant event, 'the completion of the works is [not] likely to be delayed *thereby* beyond the completion date'.

The relevant event simply has no effect upon the completion date. This situation obviously needs to be distinguished from a situation in which, as it were, the works are proceeding in a regular fashion and on programme, when two things happen, either of which, had it happened on its own, would have caused delay, and one is a relevant event, while the other is not. In such circumstances there is a real concurrency of the causes of the delay. It was circumstances such as these that Dyson J was concerned with in the passage from his judgment in *Henry Boot Construction (UK) Ltd v Malmaison Hotel (Manchester) Ltd* ....

On this basis, the approach to granting an extension to the completion date might be different for two delaying events that occur at the same time, as opposed to two delaying events that occur in sequence but whose effects run concurrently. In the former case, the *Malmaison* approach would apply. However, if, in the latter case, the first event is a contractor culpable delay, and the period of delay is the same or greater than the succeeding event, then, even if the succeeding event is the employer's responsibility, there will be no extension of time because that succeeding event is of no effect.

Malmaison and *Brompton v Hammond* are English authorities. However, the latter's approach of narrowly distinguishing 'concurrency' was subsequently considered under Scottish law by Lord Drummond Young in the Outer House of the Scottish Court of Session. The long-running saga of the case of *City Inn Ltd v Shepherd Construction Ltd* [2007] ScotCS CSOH 190 considered the use of computer programs in delay analysis (considered elsewhere in this chapter) and the problem of concurrent delays under the JCT terms. The judgment agreed with the position in *Malmaison*. If a delaying event could be shown to be the 'dominant cause' of the delay, then that event should be considered as the cause of the delay, and there was in fact no concurrency of delays. However, where there was concurrency, Lord Drummond Young disagreed with Judge Seymour's distinction between concurrent delays that follow one another and delays that occur more or less simultaneously. Lord Drummond Young described this as an 'arbitrary criterion' and then stated that:

> It should not matter whether the shortage of labour developed, for example, two days before or two days after the start of a substantial period of inclement weather; in either case the two matters operate concurrently to delay completion of the works.

Further, he went on to say:

> Where there is true concurrency between a relevant event and a contractor default, in the sense that both existed simultaneously, regardless of which started first, it may be appropriate to apportion responsibility for the delay between the two causes; obviously, however, the basis for such apportionment must be fair and reasonable. Precisely what is fair and reasonable is likely to depend on the exact circumstances of the particular case.

Applying these principles to the facts of the particular case Lord Drummond Young found that the contractor's claim for 11 weeks for extension of time for late instructions should be reduced by two weeks for contractor risk events. An appeal to the Inner House of the Scottish Court of Session ([2010] Scot CSIH 68) saw a majority agree with his approach.

As noted previously, but cannot be emphasised enough, whether such as 'but for' or apportionment or any other approach is to be applied to concurrent delays depends on the express terms of the contract and also the applicable law. Apportionment may be of particular relevance in some Middle East jurisdictions, for example. However, subsequent English judgments suggest that the *City Inn v Shepherd* approach of apportioning responsibility for delay is not applicable there.

In *De Beers UK Ltd v Atos Origin IT services UK Ltd* [2010] EWHC 3276 (TCC) 134 ConLR 151, Mr Justice Edwards-Stuart considered claims arising out of the termination of a contract for the development of computer software required for De Beers' movement of part of its operations to Botswana. Those claims included issues as to responsibility for delays and compensation for those delays. At paragraphs 177 and 178 the Judge said:

> The general rule in construction and engineering cases is that where there is concurrent delay to completion caused by matters for which both employer and contractor are responsible, the contractor is entitled to an extension of time but he cannot recover in respect of the loss caused by the delay. In the case of the former, this is because the rule where delay is caused by the employer is that not only must the contractor complete within a reasonable time but also the contractor must have a reasonable time within which to complete. It therefore does not matter if the contractor would have been unable to complete by the contractual completion date if there had been no breaches of contract by the employer (or other events which entitled the contractor to an extension of time), because he is entitled to have the time within which to complete which the contract allows or which the employer's conduct has made reasonably necessary.
>
> By contrast, the contractor cannot recover damages for delay in circumstances where he would have suffered exactly the same loss as a result of causes within his control or for which he is contractually responsible.

The Judge's reference to 'The general rule in construction and engineering' appeared to be a reference to the *Malmaison* approach. His statement that it 'does not matter if the contractor would have been unable to complete by the contractual completion date if there had been no breaches of contract by the employer' similarly makes the 'but for' test irrelevant to a consideration of entitlement to extensions of time for employer delays.

In *Adyard Abu Dhabi v SD Marine Services* [2011] EWHC 848 (Comm), [2011] BLR 384, Mr Justice Hamblen considered a dispute regarding the termination of two shipbuilding contracts. The employer had rescinded the contract on the basis that the shipbuilder was culpable for its failure to have the vessels ready for sea trials by the contractual dates. The shipbuilder argued that it had been prevented from completing the vessels by the due dates by the employer's acts, or alternatively it was entitled to extensions of time for those acts.

The contractor argued that it was sufficient to look at the employer's delay events that it relied on and how they related to the contractual completion date, without the need to prove that they caused actual delay to the progress of the works. As the Judge put this approach, it was one that asserted that 'Notional or theoretical delay suffices'. He disagreed both in principle and based on the legal authorities.

On the principle, the Judge disagreed that there was no need to prove causation in fact. In relation to the contractor's prevention argument he considered that 'necessarily means prevention in fact; not prevention on some notional or hypothetical basis'. In relation to extension of time, he considered the contractor's argument also contrary to the provisions of that contract, which required that the extension was to be 'that occasioned by or resulting from' the event.

On the authorities, the Judge referred to the English cases of *Balfour Beatty v Chestermount*, *Brompton v Hammond* and *Malmaison*, considered elsewhere, as requiring that the shipbuilder establish that the employer's acts were at least a concurrent cause of delay. On the facts, he found that the shipbuilder was unable to prove actual delay to the progress of the works.

Mr Justice Hamblen also referred to Lord Osborne's summary of the position in *City Inn v Shepherd*: 'In the first place, before any claim for an extension of time can succeed, it must plainly be shown that a relevant event is a cause of delay and that the completion of the works is likely to be delayed thereby or has in fact been delayed thereby'.

In *Walter Lilly & Company Limited v (1) Giles Patrick Cyril Mackay (2) DMW Developments Limited* [2012] EWHC 1773 (TCC) (12 July 2012), Mr Justice Akenhead considered various issues arising out of a contract for construction of a new building incorporating the JCT Standard Form of Building Contract 1998 Edition Private Without Quantities. The Judge considered the question of concurrency of delays and the difference of approach in the English and Scottish authorities, including those already summarised. He confirmed that the apportionment approach adopted in *City Inn v Shepherd* was not applicable in England. He referred to the English cases of *Malmaison*, *De Beers v Atos* and *Adyard v SD Marine*, and concluded that:

> I am clearly of the view that, where there is an extension of time clause such as that agreed upon in this case and where delay is caused by two or more effective causes, one of which entitles the Contractor to an extension of time as being a Relevant Event, the Contractor is entitled to a full extension of time.

In conclusion, the first essential when resolving issues over concurrent delays is to understand exactly what it really is. The second consideration, where it actually exists, is how to deal with it. The first port of call in this must be the express provisions of the contract and then the applicable law. Where the contract does not expressly address the issue, in some jurisdictions it may be that an apportionment approach should be adopted. Under English law the contractor will be entitled to an extension of time and hence the employer will lose its right to delay damages.

One approach to both define concurrent delay and to prescribe how it is dealt with is to incorporate the approach adopted by the SCL Protocol. In its Core Principle 10 this says:

> True concurrent delay is the occurrence of two or more delay events at the same time, one an Employer Risk Event, the other a Contractor Risk Event, and the effects of which are felt at the same time. For concurrent delay to exist, each of the Employer Risk Event and the Contractor Risk Event must be an effective cause of Delay to Completion (i.e. the delays must both affect the critical path). Where Contractor Delay to Completion occurs or has an effect concurrently with Employer Delay to Completion, the Contractor's concurrent delay should not reduce any EOT due.

Although this approach is tempered by the SCL Protocol's approach to compensation for prolongation where there are concurrent delays (see elsewhere in this chapter) and might be considered an even-handed approach, it is not popular with all employers.

It has become increasingly common in some jurisdictions for employers to attempt to 'contract away' concurrency by having the contract expressly address the issue. Typical of the clauses attempting this is the following:

> Notwithstanding any other provision of the Contract, if two or more delay events occur, one of which is an event described within [the extension of time clause] and the other(s) constitute events that are not described within [the extension of time clause], the effects of which are felt at the same time, then during such period of concurrency, the Contractor shall not be entitled to an extension to the Milestone Date or the Time for Completion, as applicable, nor any additional Costs.

Thus employers using this clause are attempting to deny the contractor both its protection from delay damages and its related prolongation costs. In practice, the extent which such clauses are successful depends on arguments regarding such as:

- whether, and to what extent, the effects of the competing causes were felt at the same time;
- whether 'effects' in the clause means that the events have to both be on critical path activities; and
- how the provision sits with the applicable law, particularly in relation to such as what under English law is the doctrine of 'prevention'.

The question of the effectiveness of a clause that attempts to deny the contractor an extension of time where it has concurrent delays, and the relationship of that to the principle of prevention, was addressed by the England and Wales Court of Appeal in *North Midland Building Limited v Cyden Homes Ltd* [2018] EWCA Civ 1744. In that case the contract incorporated the JCT Design and Build 2005 Standard Terms and Conditions, but subject to numerous bespoke amendments. These included an amended clause 25, which read as follows:

> If on receiving a notice and particulars under clause 2.24:
>
> 1. any of the events which are stated to be a cause of delay is a Relevant Event; and
> 2. completion of the Works or of any Section has been or is likely to be delayed thereby beyond the relevant Completion Date; and
> 3. provided that
>   (a) the Contractor has made reasonable and proper efforts to mitigate such delay; and
>   (b) any delay caused by a Relevant Event which is concurrent with another delay for which the Contractor is responsible shall not be taken into account; then, save where these Conditions expressly provide otherwise, the Employer shall give an extension of time by fixing such later date as the Completion Date for the Works or Section as he then estimates as to be fair and reasonable.

The contractor challenged the validity of this clause on the basis that it was contrary to the 'prevention principle' in English law. That principle goes to the point as to why extension of time clauses in construction and engineering contracts are there for the benefit of the employer as much as the contractor, and the potential for time under the contract becoming 'at large'. In Chapter 5 we explain how an excuse to make a claim on the basis of a 'quantum meruit' valuation has often been seen by contractors as a 'silver bullet' solving all of its problems. A claim that time is at large is often seen as a solution of similar immediacy and effect.

It has long been the position under English law that it would be wrong to hold a contractor to a contractual completion and a related liquidated rate of delay damages where the employer was at least in part responsible for the contractor's inability to achieve that date (*Holme v Guppy* (1838) 3 M&W 387). As it was put in *Dodd v Churton* [1897] 1 QB 566:

> ... where one party to a contract is prevented from performing it by the act of the other, he is not liable in law for that default; and accordingly a well-recognised rule has been established in cases of this kind, beginning with *Holme v Guppy*, to the effect that, if the building owner has ordered extra work beyond that specified by the original contract which has necessarily increased the time requisite for finishing the work, he is thereby disentitled to claim the penalties for non-completion provided by the contract.

It was to avoid time becoming 'at large' in such a manner that contracts began to include provisions for the employer to extend the time for completion in the event that they caused delay. This was to maintain a contractual completion date and the employer's right to levy the delay damages under the contract.

However, such extension of time provisions had to be drawn up in broad enough terms to address all potential delays by the employer. If an employer delay occurred that was not covered, time could still be come 'at large'. For example, in *Peak v McKinney* (1970) 1 BLR 111, the contractor was responsible for defective piling works and the necessary rectification works took six weeks. However, the overall delay to the project was 58 weeks because the employer delayed in deciding how the problems should be rectified. The employer sought to rely on the contractual term 'any other unavoidable circumstance' as a ground for granting an extension of time under the contract. However, the court held that the delay was avoidable and that, as a result, time was large and the liquidated damages under the contract could not be levied. It is on the basis of cases such as this that contractors have sought to establish that an employer delay event has prevented it from completing but is not covered by the extension of time provision in the contract as a way to avoid liquidated damages.

In *North Midland Building v Cyden*, the contractor argued that the amended clause 25 as quoted was contrary to the prevention principle and therefore ineffective. At first instance, Mr Justice Fraser concluded that:

> ... there is no rule of law of which I am aware that prevents the parties from agreeing that concurrent delay be dealt with in any particular way, and [the claimant's counsel] could not direct me to any. *Multiplex* and the doctrine of prevention are so far off the point, with respect, as to be dealing with something else entirely.

In the Court of Appeal, Lord Justice Coulson summarised his decision on this issue as follows:

> I can see no basis on which clause 2.25.1.3(b) could be struck down or rendered inoperable by the prevention principle. The clause is clear and unambiguous and it does not cut across clause 2.26.2.5 (which prima facie entitled the contractor to

an extension of time for anything that might be considered an act of prevention by the respondent). The only thing the clause does is to stipulate that, where there is a concurrent delay (properly so called), the contractor will not be entitled to an extension of time for a period of delay which was as much his responsibility as that of the employer. That was an allocation of risk which the parties were entitled to agree ....

This leaves the issue of whether the contractor will be entitled to compensation of its financial damages related to the delays, which is considered further elsewhere. However, first, a further topic related to entitlement to extension of time in periods of concurrent culpable contractor delay which should be considered is the principle referred to as 'dot on'.

3.5.5 'Dot on'

The 'dot on' principle addresses what entitlement the contractor has if, during a period of culpable delay after the due completion date, it is delayed by an excusable event that is at the employer's risk. For years contractors had argued that, notwithstanding their own culpability for still progressing the works after the due completion date, if a delay event then arose for which the employer was responsible, this would give rise to either time becoming 'at large' (because the contract contained no facility to extend time in those circumstances) or an entitlement to extension of time on a 'gross' basis. Thus the extension of time would be calculated from the date at which the excusable event occurred and including both the effects of the event itself and the period of culpable delay that preceded it. Contractors sought to justify this approach by reference to the 'prevention' principle explained elsewhere. Employers and their consultants considered that the approach sought to excuse the contractor for its own culpable delay. The position is illustrated by the following charts.

In Figure 3.4 the contractor can be seen to be in a period of culpable delay, overrunning beyond the due completion date on both activities D and E. During that culpable delay period a one-day employer risk event has occurred – perhaps an instruction to carry out a variation. Under the 'gross' method the contractor would claim an extension of time up to the time of the one-day 'Employer Risk Event' and its effect, almost up to its actual completion date, notwithstanding that it had until that event been in a much longer period of culpable delay.

The approach suggested in Figure 3.4 was considered in *Balfour Beatty v Chestermount*. Mr Justice Colman held that 'the contention that the "gross" method is "fair" to both parties cannot stand up'. He concluded that 'it would be wrong in principle to apply the "gross" method, and that the "net" method represents the correct approach'. Applying the 'net' method to the example in Figure 3.4, the extent of the contractor's entitlement to extension of time arising out of the one-day employer risk event is illustrated in the impacted chart in Figure 3.5. The one day of employer delay is 'dotted' on to the due completion date to give an entitlement to one day of extension of time as shown.

The *Balfour Beatty v Chestermount* decision is, of course, in the context of the JCT Standard Form contract. However, it is suggested that the same principles would apply to many other construction contracts where grounds for an extension of the contract

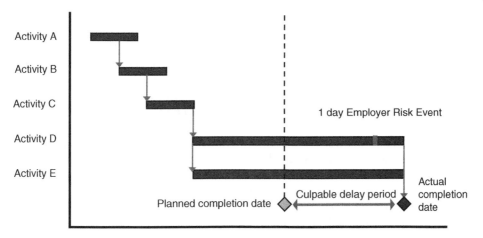

Figure 3.4 Delay in a period of culpable delay.

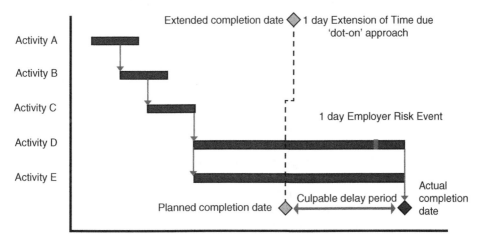

Figure 3.5 Delay in a period of culpable delay 'dotted on'.

completion date are set out and the contractor takes the risk of other delaying events, in anything other than unusual terms.

In *Ascon Contracting Ltd v Alfred McAlpine Construction Isle of Man Limited* (1999) 66 Con LR 119, His Honour Judge Hicks QC considered an extension of time provision in McAlpine's bespoke subcontract conditions, which was in similar terms to clause 11.1 of the JCT's then standard form of domestic subcontract, DOM/2. He found that Ascon was entitled to two extensions of time of six days and eight days and given by adding the total of 14 days to the date for completion under the subcontract, that is, by dotting the delay on to the existing completion date.

More recently, these cases were considered in the context of another subcontract that incorporated an extension of time clause similar to that in DOM/2. There the 'net' method was referred to as a 'contiguous' approach to extensions of time and the 'gross' method as a 'non-contiguous' approach.

In *Carillion Construction Ltd v Wood Bagot Ltd & Others* [2016] EWHC 905 (TCC), Miss Recorder Jefford's judgment was that the extension of time should be on a contiguous basis on the basis of: the natural meaning of the clause; that the alternative may lead to an unsatisfactory result; that the contiguous approach accords with commercial common sense; and that the decision in *Balfour Beatty v Chestermount* supported how a reasonable person would apply the clause. That decision was upheld by the Court of Appeal in *Carillion Construction Ltd v Emcor Engineering Services Ltd & Another* [2017] EWCA Civ 65.

The considerations in these cases were as to the provisions of particular contracts but also such as the fairness of the result. The outcome might have been different in other jurisdictions and under other forms of contract. The Infrastructure Conditions set out in subclause 44(1) the grounds for granting an extension to the completion date, and in subclause 44(2) the engineer is required to assess the delay upon receipt of notification from the contractor that a delaying event has occurred. However, subclause 44(2) also requires the engineer 'to consider all such circumstances known to him at the time'. The engineer might consider that this should include the consideration of any concurrent contractor delay events, or earlier contractor delay events whose effect was, or was likely to be, concurrent with the notified employer delay. Subclause 44(2) further requires that the engineer shall 'make an assessment of the delay (if any) that has been suffered by the contractor as a result of the alleged cause'. The engineer might therefore also consider that in the example set out in Figures 3.4 and 3.5, the delay suffered by the contractor as a result of the '1 day Employer Risk Event' is that one day and not the culpable period preceding it.

To those for whom the 'net' approach seems to be a fair and reasonable solution to this issue and the 'gross' approach would offend any requirement of the applicable law that required a fair and reasonable interpretation and application of contract terms, one argument relates to how the timing of instructions often arise out of the progress of the contractor's works. Standard forms of engineering and construction contracts provide for change to be instructed by the employer by way of variations. In practice many such changes arise as the works as originally designed unfold, because the need for change only becomes apparent at that stage. For example, this is because the employer or its designer sees the physical work in progress for the first time. One can envisage, for example, that in the simplistic activity sequence in Figures 3.1 to 3.3, it might be that the employer decides to change the specification of the last activity in that sequence, the copings, when seeing them as installed against the brickwork. If that stage is only achieved one month after the contract completion date entirely because of the contractor's culpable delays, the employer might consider it an unreasonable outcome that the contractor would be entitled to an extension of time not just for the period required to change in copings, but also the month of culpable delay that preceded it.

3.5.6   Concurrency and the Contractor's Financial Claims

Elsewhere in this chapter we considered the complex issue of concurrency and how it might be resolved in relation to a contractor's entitlement to extensions of time. In relation to the financial effects of concurrent delays, we noted how where there are concurrently both cost-neutral or culpable delays and compensable delays, then the question becomes particularly important. The employer's right to its time-related costs, through delay damages, will be automatically addressed by the resolution of the issue of concurrency and the fixing of a new contractual completion date or dates. However, this leaves the question of whether, even having been granted an extension of time, a contractor should also be compensated for its time-related costs for that extended time where it had its own concurrent culpable delays. We have explained elsewhere in this book how the extensions of time and financial compensation clauses of such as SBC/Q, the FIDIC Red Book and the Infrastructure Conditions are not linked and are set out in separate clauses, such that an entitlement to extension of time does not automatically mean compensation of the contractor's time-related costs.

Of course, the correct approach to a contractor's costs and losses where it has been delayed by employer events, but at the same time was suffering its own culpable delay, will depend firstly on the legal position as set out in the express provisions of the contract and on the basis of the applicable law. However, realistically, there seem to be just two alternative approaches:

1. By adopting an apportionment approach. Under this, the relative potency or significance of the two competing causes of delay is assessed and the contractor's time-related costs are allowed in a proportion that reflects that assessment.
2. By applying a 'but for' test. Applying the principle of damages, that is to put the contractor into the position it would have been had the employer's delay not occurred, this recognises that the contractor would have been in the position of incurring the time-related costs anyway. Under this, any costs and losses that the contractor would have incurred anyway, as a result of its own culpable delays, even if the employer delay had not occurred, would not be compensatable. The question of concurrency in relation to the contractor's time-related costs may also be relevant to non-critical delay. However, under this approach, where it involves critical delay, if the contractor gets an extension of time, then the employer will lose its right to delay damages and the 'but for' test would deny the contractor its financial claim for the same period. Thus, the parties' respective costs and losses could be said to lie where they fall.

The apportionment of financial damages in this context is particularly relevant under some Middle East jurisdictions. It is also the approach adopted in the *City Inn* case considered elsewhere. There Lord Drummond Young followed his approach of apportionment when considering extensions of time, by also allowing apportionment of the contractor's loss and expense. Thus, even where a cause of part of the delay was a matter for which the contractor was responsible, the associated loss and expense claim would not entirely fail. Having awarded the contractor a nine week extension of time out of an 11 week claim, on an apportionment basis, the court also awarded it nine weeks of loss and expense.

As noted elsewhere, subsequent English authorities suggest that the apportionment approach taken by the Scottish courts in *City Inn* will not apply there, and reference is made again to the decision of Mr Justice Edwards-Stuart in *De Beers v Atos*:

> The general rule in construction and engineering cases is that where there is concurrent delay to completion caused by matters for which both employer and contractor are responsible, the contractor is entitled to an extension of time but he cannot recover in respect of the loss caused by the delay.
>
> ...

By contrast, the contractor cannot recover damages for delay in circumstances where he would have suffered exactly the same loss as a result of causes within his control or for which he is contractually responsible.

This 'but for' approach under English law is also that adopted by the SCL Protocol. In its Core Principle 14 it says, in terms similar to those of its First Edition in 2002:

Concurrent delay – effect on entitlement to compensation for prolongation: Where Employer Delay to Completion and Contractor Delay to Completion are concurrent and, as a result of that delay the Contractor incurs additional costs, then the Contractor should only recover compensation if it is able to separate the additional costs caused by the Employer Delay from those caused by the Contractor Delay. If it would have incurred the additional costs in any event as a result of Contractor Delay, the Contractor will not be entitled to recover those additional costs.

In practice, the statement that the contractor's time-related costs fail the 'but for' test where it is in concurrent culpable delays is far too simplistic. It is a feature of many commentaries on this issue that they fail to consider this point in detail. Just because the contractor is in culpable delay does not mean that all of the costs caused by a concurrent employer delay would have been incurred anyway. This can be illustrated by a simple example related to the contractor's site staff, as follows.

Assume that the contractor is in delay in completing its works because its operatives have all gone on strike. Assume that at the same time the contractor cannot complete the works because there are outstanding design details still awaited from the architect. Assume that analysis shows that these are both critical delays to completion and the contractor is granted an extension of time. Does it follow that none of the contractor's time-related costs over the period should be reimbursed because they would have been incurred anyway because of the labour strike? Among its site staff the contractor should have supervisory personnel waiting to manage the operatives on completing the remaining works when their strike ends. Clearly the costs of such supervision would have been incurred in any event because of the strike. However, the contractor may have such employees as document controllers and engineers awaiting the architect's design so as to process it and prepare such things as shop drawings. These would have to be retained whilst the design information was awaited, but would not have been retained for the labour strike. Accordingly, in this example, the time-related costs of the design-related staff ought to be recoverable notwithstanding the contractor's culpable delay, because they were only caused by the employer's delay. Many other examples could be given in relation to site staff, but also in other areas of time-related costs such as the contractor's plant, equipment and temporary facilities. The point is that the analysis of time-related preliminaries and general costs where a 'but for' approach is being adopted to concurrent delays requires more detailed analysis that a simple assertion that they are all not recoverable. Each requires detailed consideration of the question 'would they have been incurred anyway?'

Of course, proponents of the NEC suite of contracts will say that their approach to the valuation of compensation events on a prospective basis based on a forecast of the change in the contractor's actual costs, including time-related costs, prevents such issues occurring. An effect of the NEC approach will be that, where the contractor subsequently falls into concurrent culpable delays, perhaps because of the labour strike in the example set out in this section, the fact that it would have incurred some of its time-related costs anyway becomes irrelevant. As this was very directly put in clause 65.2 of NEC3-ECC:

> The assessment of a compensation event is not revised if a forecast upon which it is based is shown by later recorded information to have been wrong.

In NEC4-ECC, that provision now reads as follows as clause 66.3:

> The assessment of an implemented compensation event is not revised except as stated in these *conditions of contract*.

The effect of this provision is stated as follows in the NEC User Guide:

> This clause emphasises the finality of the assessment of compensation events. The circumstances in which an implemented compensation event is revised are only those in the chosen dispute resolution process.

3.5.7   Delay Analysis Techniques

There are a large number of alternative methods for the analysis of delays. These can be variously carried out prospectively, contemporaneously or retrospectively and are sometimes categorised by these terms. They can also be variously categorised as either 'cause and effect' or 'effect and cause' approaches. They rely on different planned and/or as-built programming and progress information. They include, but are not limited to the following:

- impacted as planned;
- time impact analysis;
- rolling programme analyses;
- windows analysis;
- time slice analysis;
- snapshot analysis;
- watershed analysis;
- milestone analysis;
- as planned versus as built;
- collapsed as built;
- longest path.

None of these are terms of art and they may mean different things to different practitioners, particularly in diverse parts of the world. Equally, there is some degree of overlap between them, in that some are particular applications of others or alternative terms for others. The full details of these methods, and the merits, etc., of each, are set out in a number of extensive works on the topic of delay analysis of engineering and construction projects. This book does not set out to provide further such extensive detailed analysis. However, given the obvious link between delays and the subject of this book, being the topic of *Evaluating Contract Claims*, it would be wrong not to provide some introduction to the different methods. This includes their respective details and merits, but particularly to show how it is that different methods can reach different conclusions using exactly the same facts.

As to which method should be applied in a particular case, the first consideration must be to the terms of the particular contract, to the extent that this gives guidance on the topic. However, many construction and engineering contracts (with the exception of the NEC suite) are silent on the details of how extensions of time are to be calculated and therefore lack guidance on methodology. Furthermore, the fact that the different methods can derive different results on the same facts means that parties often deliberately use different approaches that produce the outcomes that suit their particular interests.

Whilst there are many more delay analysis techniques available and in use, they are all broadly based on just four basic techniques. These can be summarised as follows:

1. Impacting the as-planned programme.
2. Impacting the as-planned programme, but after updating it.
3. Analysing the as-built programme.
4. Comparing the as-planned programme with the as-built programme.

In order to illustrate these techniques, the four most commonly adopted methods of delay analysis are considered. The first two can be carried out prospectively, before the delayed work is carried out, because they only rely on the planned programme (updated in approach 2). The second two can only be carried out retrospectively, after the delayed work has been carried out, since they rely on as-built data on the delayed work:

1. impacted as planned;
2. time impact analysis;
3. collapsed as built;
4. as planned versus as built.

These approaches and their merits are outlined as follows, together with an illustration of the differing possible results of analysis that applies to any one of them on the same facts. This is based on a very simplistic example of the planned sequence of activities A to E set out in Figure 3.6. This also shows the logic links between each activity in the sequence.

3.5.7.1 Impacted as Planned

This technique takes the as-planned programme and impacts into it the delay events. This is illustrated by Figures 3.7 and 3.8, which input an example delay in the planned sequence.

If the sequence in Figure 3.6 is the subject of delay to the duration of activity C, then this is shown in Figure 3.7. By applying an impacted as-planned approach to the extension of the activity duration shown in Figure 3.7, the resulting impacted programme is shown in Figure 3.8. As set out in Figure 3.8, the conclusion of the impacted as-planned approach illustrated in this simple example is that the delay to completion is by a period identical to the delay to the duration of activity C, that activity having been on the critical path.

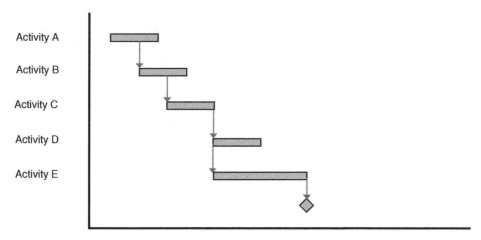

Figure 3.6 Sample activity sequence for delay impacting.

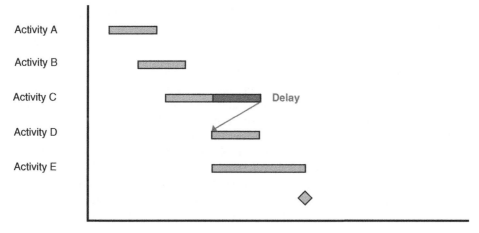

Figure 3.7 Sample activity sequence showing delay to activity C.

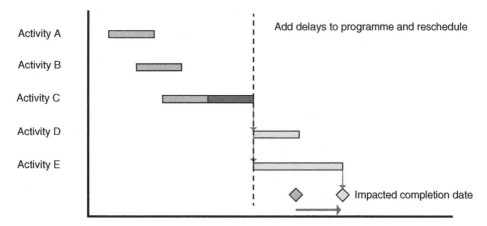

Figure 3.8 Result of impacted as-planned approach.

As with all of the available delay analysis techniques, this approach has its claimed advantages and disadvantages. The main advantage is that it is very quick and simple both to carry out and to understand and the result is therefore relatively transparent. In addition, the approach does not require any as-built data and can therefore be followed in the absence of as-built records, prospectively at any stage, including before the delayed work is carried out.

The main disadvantage and criticism of the impacted as-planned approach is that it takes no account of what actually occurred in terms of progress or the as-built programme or changes to the planned intent on the impacted programme. It therefore results in theoretical conclusions. In practice, the resulting completion date arising from such an approach is very often a date later than that upon which the work is in fact actually completed and those defending a claim based on such an approach will say that this proves that it is wrong. Advocates of the approach will then often respond that the analysis correctly proved the delay but that it also shows acceleration and mitigation of the delays. In other words, the entitlement arising from the delay is somewhat longer than the actual completion delay that resulted from the activity delay and that therefore the approach can be used as a basis for both a prolongation and an acceleration claim! Critics of the impacted as-planned approach may respond that the result is entirely theoretical and proves neither prolongation nor acceleration. Where there is no prescription that an impacted as-planned approach must be adopted, such arguments need to be tested on the detailed facts in each case.

Where this debate particularly comes to a head is on the standard forms that can be read to require such a prospective, and therefore a theoretical, approach. Throughout the many years of their versions, editions and revisions, the JCT contracts, when requiring consideration of claims for notified delays during the course of the work, have talked about whether delay is 'likely' to occur. Thus, the test in SBC/Q clause 2.28.1.2 is whether 'completion of the Works or any Section thereof *is likely to be* delayed thereby beyond the relevant Completion Date' (with emphasis added). In *Leighton Contractors (Asia) Ltd v Stelux Holdings Ltd* HCCT 29/2004, the court considered an appeal from the decision of an arbitrator in relation to a claim for extension of time by the contractor under the terms of JCT 1980 edition, clause 23, which provided the following:

(1) Upon it becoming reasonably apparent that the progress of the Works is delayed, or is likely to be delayed, the Main Contractor shall forthwith give written notice to the Architect of the material circumstances including the cause or causes of the delay. The Main Contractor shall give particulars of the expected effects of the delay, or potential delay, and shall estimate the extent, if any, of the expected delay to the completion of the Works beyond the Date for Completion . . . .

(2) If, in the opinion of the Architect, upon receipt of any notice, particulars and estimate given by the Main Contractor under sub-clause (1) of this Condition, the completion of the Works is likely to be or has been delayed beyond the Date for Completion . . .

  . . . then the Architect shall so soon as he is able to estimate the length of the delay beyond the date or time aforesaid make in writing a fair and reasonable extension of time for completion of the Works.

The contractor argued that this did not limit its entitlement to extensions of time to only where relevant events had delayed completion. It argued that, looking prospectively by standing in the Architect's shoes at 'time slices' over the project's history, if information was issued late by the employer and as a result completion was 'likely to be delayed', then an extension of time should have been granted on that prospective basis. This would be even where, with the benefit of hindsight, it became apparent that the event did not in fact actually delay completion. The Judge disagreed. He found that clause 23(2) requires some causal link between the late information and potential delay to the completion date of the works. It was not enough for the contractor to show that the information was late, it must also show that, on the date when the information ought originally to have been provided, delay to completion was 'likely' to result from its late supply. In this case, substructure works were so delayed that the late information was of no effect.

More recently, the NEC contracts, through their 'compensation events' provisions, specifically require the analysis to be prospective, with no provision for a subsequent back-check against what actually turns out to happen in practice. The effect of clause 66.3 of NEC4-ECC has been explained elsewhere in terms of its 'finality' as this is put in the NEC User Guide.

NEC4-ECC clause 63.5 requires that assessment be based on the 'Accepted Programme', as is required by clause 31 and is to be updated as required by clause 32. However, in practice the NEC provisions can be particularly problematical where their expectation that the programme will have been updated in accordance with clause 32 has not been satisfied. The debate that then ensues is as to whether, following the apparently strict letter of the provisions, the impacted as-planned approach is to be applied to what may be the most recent programme, but is in fact out of date. Impacting a long out of date programme with delay occurring much later can result in an effect that is far removed from the reality of actual progress on site. In the Second Edition of this book it was suggested that, since this is a common problem on NEC-based projects, some authority from the courts on this issue would be desirable. Ten years on that guidance is still awaited. The extent to which revised wording in NEC4-ECC helps to reduce the occurrence of this problem is awaited with interest.

The philosophy behind such a theoretical approach may be that the contractor's original programme itself was just theoretical. It was what the contractor planned would be achieved and when, not what was actually achieved and when. As such, an argument for such an express approach in the contract is that impacting the as-planned approach maintains the basis of the original programme. This is the ethos that underpins the NEC's approach to compensation events, also requiring their financial effects to be assessed on the basis of a forecast, just as the tender was.

An impacted as-planned approach was particularly criticised in *Great Eastern Hotel Company Ltd v John Laing Construction Ltd* [2005] EWHC 181 (TCC). Judge Wilcox set out a number of criticisms of the defendant's expert's analysis in that case, but in particular noted that the analysis took no account of the actual events that occurred on the project and gave rise to a theoretical result.

The further disadvantages of the impacted as-planned approach include that, whilst as-built records may not be required, there is a need for a reasonable and robust as-planned programme. That programme needs to be adequate and properly logic-linked. In the absence of such a programme problems arise. If the programme used is not reasonable, robust, adequate or properly logic-linked then the analysis can give rise to highly misleading results. The parties may in that case seek to create such a programme retrospectively. However, this can be very subjective and can lead to further disagreement as to whether this is being carried out properly and which of any alternative approaches is correct. Reference is made to the *City Inn* case, considered elsewhere in relation to the use of computer software, where the absence of an electronic version of the original programme and the dangers of attempting to reproduce it retrospectively were fully recognised.

By ignoring actual progress the impacted as-planned method can particularly hide concurrent delays. Therefore, it is often used by contractors or subcontractors, where they have their own concurrent culpable delays, to derive an entitlement to an extension of time that hides these failures. The complex matter of resolving concurrency in delays has been considered elsewhere. It may also be used by either party where there are concurrent employer risk events, some of which are cost-neutral, to either hide or emphasise the effects of those cost-neutral events.

As stated elsewhere, proponents of an impacted as-planned approach will particularly commend its simplicity and the relative speed and cost efficiency with which it can be applied. It may be that where the programme and issues are very simple, and there have been no previous changes to the planned intent, this approach has its place. This is particularly so where delays occur at the start of a project, perhaps in mobilisation or possession of the site. Alternatively, it may be that there are no records of actual progress or updates of the programme that will permit an alternative to impacting the planned programme. On the face of it, this might be expected to only apply to building projects that are very simple and have little management and reporting. However, in practice, internationally, it is surprising how much larger engineering and construction projects may lack records and information for either their whole duration or particular periods or remote sections of the works where delay events occurred but with no records of actual progress.

As noted previously, contractors sometimes defend the extreme result of an impacted as-planned approach by claiming to have accelerated to mitigate the effects of an activity delay on completion. This approach can actually be useful where a contractor asserts that the costs of acceleration measures were justified by the amount of delay to completion that was saved by those measures.

3.5.7.2 Time Impact Analysis

This technique can be regarded as a more sophisticated development of the impacted as-planned approach. Again, it takes the planned programme and impacts into it the delay events, but only after having updated the programme for progress just before the delay events occurred. It is therefore an iterative approach, updating the programme and impacting each event in turn. The approach is illustrated by the following simple example.

Taking the same as-planned sequence of activities A to E with links as set out in Figure 3.6, and assuming that activities A to C were already in delay, the programme to be used for impacting would be updated to add progress as it actually stood at the time of the delay event, as shown in Figure 3.9.

Figure 3.9 shows that even before activity D is delayed, activities A to C have already been the subject of delay. Before impacting this programme with the effects of the delay

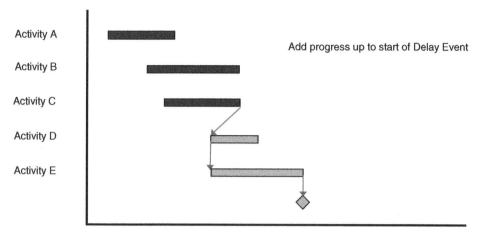

Figure 3.9 Sample activity sequence showing existing delays to actual progress.

being considered, this sequence of activities would be rescheduled to update it for these earlier progress delays to activities A to C to derive a new base line programme for impacting. This updated programme is set out in Figure 3.10.

From this rescheduled programme updated for actual progress, it can be seen that there was already a delay to completion as a result of delays to activities A to C, even before the delay that is being considered is impacted. If that delay is now impacted into the rescheduled programme the new completion date resulting from the delay event is as set out in Figure 3.11. With this time impact analysis approach the net effect of the delay event being considered is just the difference between the new completion date caused by the impacted delay in Figure 3.11 and that which had already resulted from the existing delays to activities A to C, as established in Figure 3.10.

One of the advantages of this approach, when compared with the impacted as-planned approach, is the fact that the time impact analysis takes account of actual progress. It

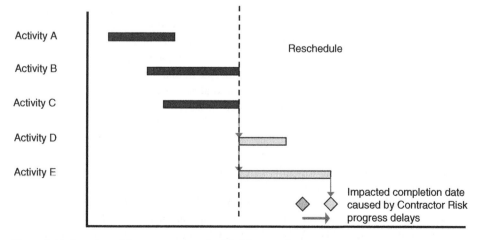

Figure 3.10 Sample activity sequence rescheduled for actual progress.

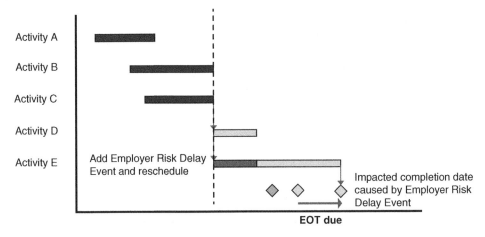

Figure 3.11 Sample activity sequence rescheduled for delay events.

therefore removes much of the theoretical aspects of the impacted as-planned approach. There is still some element of a theoretical result in that a prospective approach time impact analysis still looks forwards to what is likely to happen rather than at what actually happened to the delayed activities. However, the theoretical aspect is clearly significantly reduced. A time impact analysis can also be used to demonstrate acceleration and mitigation, but on a far less speculative basis and without the same extreme results as an impacted as-planned analysis. If a further exercise is carried out comparing its results with what actually happened, then more realistic and robust analysis of acceleration and mitigation can result.

One of the disadvantages of a time impact analysis approach is the need for a reasonable and robust as-planned programme, which includes the issues that were discussed when considering impact analysis as a planned approach. Furthermore, it also requires updating the programme by as-built records of actual progress prior to the delay event. In addition, with the need to keep updating the programme for actual progress before impacting each event, time impact analysis can be extremely time-consuming. In large cases many hundreds of iterations are required for updating the programme and impacting for events. This can lead to criticism of what is known as 'black box syndrome'. The method can result in a huge number of updates and impacted programmes to be looked at sequentially and if at any stage an error is made, particularly in the updating for actual progress, then that error can affect subsequent analyses. The method is therefore also highly dependent upon the existence of complete and accurate as-built progress records as well as the accuracy of the input and analysis. All of the methods considered in this book are subject to the 'garbage in, garbage out' danger. This is especially so with a time impact analysis approach, and its complexity and lack of transparency means that spotting 'garbage in' the subsequent results can be difficult. It can therefore sometimes be used to reduce rather than enhance the clarity of a delay submission or report by masking a flawed analysis behind what is generally considered a legitimate methodology.

The approach was particularly criticised in the judgment in *Skanska Construction UK Ltd v Egger (Barony) Ltd* [2004] EWHC 1748 (TCC). Judge Wilcox was critical of the evidence of the defendant's expert, particularly the complexity of his analysis. That expert's report had over 200 charts attached and was considered too complicated and lengthy for the court to easily understand. The Judge also noted how the defendant's expert had converted an original programme in Powerproject software into Primavera software, and stated that he was satisfied that this led to errors in that expert's analysis. With over 200 charts, this is an extreme example of 'garbage in, garbage out'. In this case the Judge preferred the evidence of the claimant's expert, which he referred to as 'accessible'.

The First Edition of the SCL Protocol said of time impact analysis, at its Guidance Section 4 paragraph 4.8:

> It is also the best technique for determining the amount of EOT that a contractor should have been granted at the time that an Employer Risk Event occurred.

In the Second Edition, the recommendation is not put in such direct terms. However, its Core Principle 4 is headed 'Do not wait and see' regarding impact of delay events (contemporaneous analysis) and this starts:

> The parties should attempt so far as possible to deal with the time impact of Employer Risk Events as the work proceeds ....

In the Guidance Part B on that Core Principle, the Protocol sets out what it refers to in paragraph 4.2 as:

> ... a recommended procedure to be followed in order to deal efficiently and accurately with EOT applications during the course of the project.

It concludes at paragraph 4.12 that:

> The methodology described in this section is known as 'time impact analysis'.

3.5.7.3 Collapsed As-Built Analysis

We have considered the two principal methods of prospective delay analysis that can be used before the effects of a delay event are actually felt. We now turn to consideration of the principal approaches that can only be used retrospectively, after the events have had their effect and the as-built record has unfolded. This explanation starts with the collapsed as-built technique.

This approach takes the as-built programme and collapses it back to establish the programme as it would have been in the absence of the delay events being considered. A simple example is set out in Figures 3.12 to 3.14.

Figure 3.12 sets out the as-built durations and critical path through a sequence of five activities, which sets out this sequence of linked activities and their actual periods. The technique then identifies a delay event whose impacts are to be considered, for example as in Figure 3.13. The technique of collapsing this programme in order to remove and hence isolate the effect of the identified delay event is shown in Figure 3.14.

The effect of the delay event identified in Figure 3.13 has now been established in Figure 3.14 as being the difference between the actual completion date from the as-built programme in Figure 3.13 and the new collapsed completion date in Figure 3.14. Figure 3.14 has created a theoretical programme in which the collapsed completion

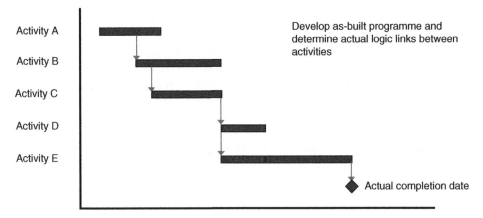

Figure 3.12 Sample as-built sequence of activities.

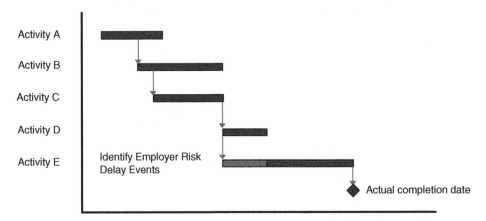

Figure 3.13 Sample as-built sequence of activities identifying delays.

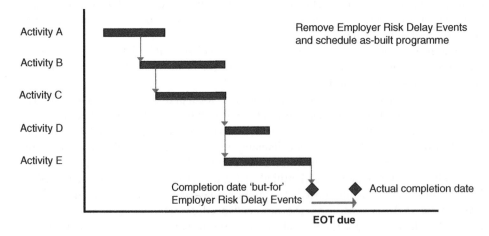

Figure 3.14 Sample as-built sequence of activities – collapsed.

date is the 'but for' date by which it is said the work sequence would have been completed had the delay event not occurred. If the delay event was an employer risk event, then the difference between the two dates is said to be the extension of time due to the contractor.

Such a collapsed as-built technique has the benefit of being simple to understand and relatively simple to carry out, particularly when compared to such as the time impact analysis approach. It could be described as the converse of the impacted as-planned approach also discussed elsewhere, working backwards from what actually occurred rather than forwards from what was planned to occur. It has the particular advantage in comparison that it does take account of actual progress.

However, whilst a fully logic-linked and robust as-planned programme is not required, the technique does require a fully logic-linked as-built programme. This will rarely if ever be immediately available. The technique therefore requires sound as-built records and the creation from those records of an accurate, robust and properly logic-linked as-built programme. The preparatory work for this can be very time-consuming, even if sufficiently detailed records of actual progress are available. Common problems with as-built records include the fact that they are incomplete or are contradictory. Gaps may have to be filled, perhaps through witness statements or flagged assumptions. Inconsistencies can be a particular issue, where different records have the same activities starting or finishing on different dates and having different durations. Often this highlights the fact that the record keepers have different agendas. More likely it may be that they take different interpretations; for example, where one considers work completed but another recognises that it still has minor aspects outstanding. The simplicity of a collapsed as-built approach can be lost where such problems occur with the records and much time is needed to resolve an as-built programme.

The retrospective logic-linking of the as-built programme can also be very subjective and the subject of contention between the parties. The approach can again be criticised on complex projects as giving rise to a 'black box syndrome', although not to the same extent as a time-impact analysis.

A particular criticism of the collapsed as-built approach is that it relies on the critical path as it exists in the final as-built programme, and not at the time that the events occurred. These paths can be very different. As noted already, it results in a theoretical depiction of what would have occurred 'but for' the delays considered. This is true in relation to the critical path, the sequencing of activities and the durations of those activities.

3.5.7.4 As-Planned Versus As-Built Programmes

The last of the four principal techniques for delay analysis briefly outlined in this book, and the second of those that can only be carried out retrospectively at the end of the works, is a comparison of the as-planned and as-built programmes. This approach is illustrated by a simple example shown in Figures 3.15 and 3.16. It starts with a programme setting out for each activity both their as-planned and as-built durations. A simple example is shown in Figure 3.15.

By comparing the as-planned and as-built periods in this programme it can be seen where individual activities have slipped or extended. By the addition of logic links setting out the critical path of the as-built programme, those activities with slippage that were

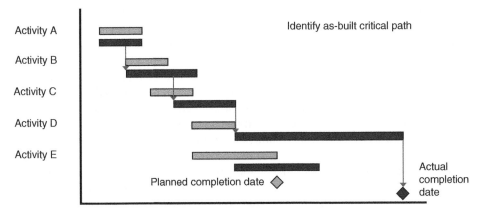

Figure 3.15 Comparative as-planned and as-built programmes.

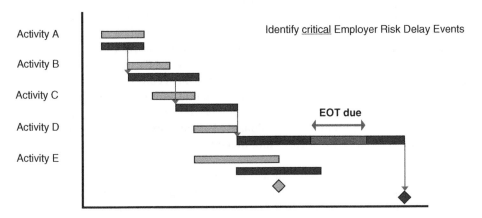

Figure 3.16 Comparative as-planned and as-built programmes – delay identified.

critical to completion are brought into focus and reveal how some activities, although delayed, were not critical and are therefore irrelevant to the analysis. Consideration of the comparative periods shows that the delay between the as-planned and as-built pro-grammes occurred during activities B, C and primarily D. Analysis can then focus on those three activities and finding from the records what may have caused them to be extended, whether it be culpable failure in the contractor's performance or such as vari-ations, additional work or information release by the employer. Hypothetically it may be that the considerable delay in the duration of activity D contains a period of delay caused by an employer risk event. On that basis, the resulting entitlement to extension of time is illustrated in Figure 3.16.

Whilst this example has been made necessarily very simple, it is representative of how an as-planned versus an as-built approach can be easy to understand and is transparent. Although, in practice there will be many more activities than depicted in this illustration. The further advantage over the two described prospective methods is that it does rely on actual progress and therefore avoids theoretical results.

The main disadvantage of this approach is that it requires both a robust and complete as-planned programme and sufficient records of as-built progress to create a logic-linked as-built programme with a critical path through it. The difficulties of assembling these have been set out above. One approach to reducing the extensive research and preparatory work that is required is to start with the as-built programme and its critical path activities and limit the analysis just to those critical activities. However, as noted when considering the collapsed as-built approach, this focusses on the final critical path and not that which stood at the time that the events occurred and the work was being done. As with other approaches involving examining the as-built critical path, the subjective nature of the assessment of that critical path can be problematic. This has also been commented on elsewhere. However, this comparative method was given support by Lord Drummond Young in the *City Inn* case (see elsewhere in this chapter).

3.5.7.5 Windows Analysis

We have not set out simplistic charts illustrating this approach to delay analysis, because windows analysis is not a single method but can be carried out in various forms. In fact, each of the methods considered in this section could be carried out in a 'window', in order to focus the analyses, deal with it in manageable chunks and potentially reduce the extent of work, particularly in creating a robust as-built and/or as-built programme.

Each window will be a manageable period in terms of the extent of analysis required within it. Periods might be chosen because they are bounded by particular programming events, such as monthly updates in the contemporaneous programme or the boundaries of the delayed events being considered. The advantages are obvious in that it can reduce the amount of work required and focus the analysis on convenient periods.

The approach is sometimes limited to analysing only those windows that are of particular interest in terms of encompassing the events or delays being considered. However, this methodology was criticised in *Mirant Asia-Pacific Construction (Hong Kong) Limited v Ove Arup & Partners International Limited and Ove Arup & Partners Hong Kong Limited* [2007] EWHC 918 (TCC). There both parties' experts applied a windows analysis approach. The Judge observed that 'It is, of course, obvious that the analysis is only valid if it is comprehensive and takes account of all activities'. He continued that 'if a retrospective delay analysis is being conducted on a Project, the analysis must include the time to the end of the Project, otherwise activities may occur which will take them on to the (or a) critical path after the date of the final window or watershed'. In this respect he considered that the analysis of one expert was 'seriously flawed' because it did not continue to the end of the works. In particular, windows analyses often conclude that some windows saw delay to completion, some an actual saving against the completion date and others a mixture of both. If all windows are not analysed, then it may be that an early window concludes delays equating to the overall actual delay to completion, when in fact the effects in that window were later mitigated but later windows also saw further delay.

The most common delay analysis approach carried out in 'windows' is to compare the as-planned and as-built programmes, the 'as-planned versus as-built' approach outlined, but dividing that analysis into windows. This therefore focusses the analysis on to each individual window, establishing the actual critical path through each. In this the windows can be selected on any of the bases explained.

A popular alternative windows approach is 'time slice' analysis, which can be particularly based on contemporaneous updates of the programmes and comparing what changed between them. These might be the programmes attached to monthly reports or revisions of a contract programme required by such as FIDIC Red Book clause 8.3. In this approach it is particularly important to check that the programme updates are accurate and complete. In practice programme updates are all too often inaccurate. This can be for a variety of reasons, not least where those issuing the programme have an agenda that might include exaggerating the effects of employer risk events or internally under-reporting culpable delays that have already occurred. How an updated programme portrays future progress and criticality to completion should also be an area for close checking.

3.5.7.6 Software

A further complicating factor in delay analysis is the choice of software to be used. Whilst three software packages (Powerproject, Primavera and Microsoft Project) have accounted for some 90% of the programs in use on construction projects in the UK, there are also a number of other packages available and in use. The difficulty that arises is that different packages have different features and available functions, e.g. in the numbers and type of links available, their calendars and the maintaining of logic. This means that analyses based upon different programs can also give rise to different results. The position is exacerbated where parties seek to transfer the data from an original program in one package into a different package. We have already identified elsewhere how in *Skanska Construction UK Ltd v Egger (Barony) Ltd* [2004] Judge Wilcox was satisfied that an expert's conversion of an original program in Powerproject into Primavera had led to mistakes in that expert's voluminous analyses. A feature of the reported case *Great Eastern Hotel v John Laing Construction Ltd and Another* [2005] EWHC 181 (TCC), discussed elsewhere in this book, is that whereas the original program was in Hornet software, the claimant's expert used Powerproject and the defendant's used Primavera. This is often the result of personal preference of the analysts. The authors have particular experience on petrochemical and ship building projects with original programs in Artemis software, which was traditionally preferred in those industries, being transferred into Primavera. This can immediately result in distortion of the program, particularly loss of logic links, which inevitably distorts any analysis unless corrected. However, making the necessary corrections can become a very subjective, contentious and costly exercise.

Such difficulties also give rise to a note of caution for those using a computer-based method of analysis. This caution arises out of the *City Inn* case previously discussed in relation to concurrent delays. Commenting on computer-based methods Lord Drummond Young said:

> The major difficulty, it seems to me, is that in the type of program used to carry out a critical path analysis any significant error in the information that is fed into the program is liable to invalidate the entire analysis. Moreover for reasons explained by [the defendant's expert], I conclude that it is easy to make such errors. That seems to me to invalidate the use of an as built critical path analysis to discover after the event where the critical path lay, at least in a case where full electronic records are not available from the contractor.

He went on to say (with emphasis added):

> I think it is necessary to revert to the methods that were in use before com-
> puter software came to be used extensively in the programming of complex
> construction contracts. That is essentially what [the defendant's expert] did in
> his evidence. *Those older methods are still plainly valid, and if computer-based
> techniques cannot be used accurately there is no alternative to using these older,
> non-computer-based techniques.*

Whilst this decision, and a reversion to non-computer-based methods, may seem alarm-
ing for those involved in computer-based forensic delay analysis, and run contrary to the
acceptance that projects are now usually managed on the basis of computerised planning
techniques, the key to this decision appears to be in this final sentence emphasised in the
above quote. The defendant's expert did not carry out a critical path analysis as he had no
access to an electronic version of the original programme. He advised the court that he
had tried to replicate the programme but explained that he could not be confident that it
would be correct and hence that his evidence to the court would be accurate. Instead he
therefore relied upon a broader comparison of the planned and actual programmes. The
Judge clearly found this delay expert very convincing. On the other hand, the claimant's
expert did carry out a computer-based critical path analysis, but the defendant's expert
found many errors in that analysis. It is in these circumstances that Lord Drummond
Young preferred the defendant's approach based upon 'factual evidence, sound practical
experience and common sense'.

It seems therefore that there may still be a place for non-computer-based techniques
for delay analysis, but only where computer-based techniques cannot provide an accurate analysis.

At the extreme end of such non-computer-based approaches are 'impressionistic' ones
of the type that was particularly criticised by the Judge in *John Barker Construction Ltd
v London Portman Hotel Ltd* [1996] 83 BLR 31. The contractor claimed a full extension
of time, based on a detailed, logical analysis of the progress on site and reasons for the
delay. That analysis was carried out by a consultant delay analyst appointed by the contractor. In contrast, the employer's architect, appointed under the 1980 edition of the
JCT standard form, had made an impressionistic determination, albeit in good faith, for
a shorter period. Mr Recorder Toulson QC said:

> ... in my judgment [the architect's] assessment of the extension of time due to the
> [claimants] was fundamentally flawed in a number of respects, namely:
> [The architect] did not carry out a logical analysis in a methodical way of the
> impact that the relevant matters had or were likely to have on [the claimant's]
> planned programme. He made an impressionistic, rather than a calculated assess-
> ment of the time which he thought was reasonable for the various items individ-
> ually and overall ([the respondents] themselves were aware of the nature of [the
> architect's] assessment, but decided against having any more detailed analysis of
> [the claimant's] claim carried out unless and until there was litigation). [The archi-
> tect] misapplied the contractual provisions .... Because of his unfamiliarity with
> SMM7 [the standard method of measurement of building works] he did not pay
> sufficient attention to the content of the bills, which was vital in the case of a JCT

contract with quantities. Where [the architect] allowed time for relevant events, the allowance which he made in important instances bore no logical or reasonable relation to the delay caused.

I recognise that the assessment of a fair and reasonable extension involves an exercise of judgment, but that judgment must be fairly and rationally based.

3.5.7.7   Building Information Modelling

The subject of this chapter, the effect of change on the programme of work, is a convenient place for a limited introduction to Building Information Modelling, or 'BIM'. If this system for creating, modelling and managing information on construction and engineering projects lives up to its billing, then it may have profound implications for the extent to which change occurs on engineering and construction projects, particularly unplanned change, and also the methods adopted to evaluate the time consequences of such change where it does occur.

The extent to which BIM has been used on engineering and construction projects has been limited to date, particularly internationally outside the USA. However, the SCL Protocol recognised its growth at paragraph 1.17 of its Guidance Part B. Mega-projects such as the Doha Metro and Abu Dhabi Airport's Midfield Terminal have seen BIM applied. However, the cost of implementation, together with the industries' traditional conservatism may be a factor in its slow take-up. Such conservatism and the slowness of practitioners to adapt to the new disciplines required by BIM are perhaps similar to the drag on the early growth of the NEC contracts, which coincidentally also had the stated aim of 'increasing collaboration'.

In the UK, the Construction Industry Council published its Building Information Model (BIM) Protocol in 2013. Furthermore, with effect from 4 April 2016, the UK Government mandated that BIM be applied for all centrally procured government contracts. The market is also adapting to such change, with, for example, programming software such as Asta Powerproject now having a BIM module that allows the project programme to be linked to multidimensional models in a single application. Others are developing software that can work with BIM to enhance its capabilities, such as extracting claim-related records. Such initiatives will inevitably build a momentum to the application of BIM over coming years that will see it become increasingly important.

In terms of the take-up by standard contract forms, the publishers of those that are particularly focussed on in this book have been slower than proponents of BIM had hoped to amend their contracts to incorporate it. This said, it can be adopted on any construction project whatever its contract terms. In this regard, the CIC Protocol asserts that 'This Protocol is intended to be expressly incorporated into all direct contracts between the Employer and the Project Team Members'.

FIDIC has committees working on BIM and how to fully deal with it. In the meantime the new editions of such as the FIDIC Red Book published in 2017 have high-level guidance on the application of BIM.

The Infrastructure Conditions and SBC/Q are both silent on BIM. However, in January 2016 the JCT released a practice note, 'Building Information Modelling (BIM), Collaborative and Integrated Team Working'. This is a guide to BIM and its integration into the contract process.

Unsurprisingly, the NEC contracts have perhaps been the most proactive. In 2014 the NEC published a guide 'How to use BIM with NEC3 Contracts', which included additional contract clauses in relation to BIM in 2017. The new NEC4 suite includes a new Secondary Option (Clause X10) that addresses 'Information Modelling'.

The claimed benefits of BIM include: greater control of quality; closer co-ordination; greater, quicker and more accurate sharing of common data; tighter design; earlier information input; and increased team co-operation and collaborative working between disciplines. These benefits should in theory reduce the scope for both planned and unplanned change and resulting claims. Further claimed benefits are: provision for early warning and joint risk management; in relation to recording and information; and clearer presentation of complex situations. These benefits should make easier the evaluation of any claims that do arise, from a common data base, and reduce the number that become contentious and the subject of dispute.

The potential benefits are particularly obvious in relation to claims for extensions of time. Some events that might cause delay should be prevented, through such as the early identification of design clashes, for example. Where delays are not prevented, they should at least be highlighted at an earlier date, so as to avoid or reduce their impact. The linking of design to programming activities should enable easy modelling of the effects of delays as they are identified. From an employer's perspective BIM can be particularly used to model the as-planned against as-built position to identify design errors and clashes, resulting variations and their effects on the planned model and revised temporary works. It has already been seen to prove very useful to employers in tracking delay cause, link and effect and as a defence to inflated contractor claims. Properly implemented, its as-built recording should leave no reason to resort to theoretical methods of delay analysis such as the impacted as-planned approach. Similarly, its requirement for a proper baseline programme should mean that the as-built record is not the only tool, which would have left only a collapsed as-built approach available. In practice, proper implementation and maintenance of BIM processes should enable a robust comparison of the as-planned and as-built programmes in terms of analysing events and their delaying effects, and allow that to be done contemporaneously as the project progresses.

As suggested elsewhere, costs can be a factor and much has been written on the relative costs and benefits of BIM. However, to realise those benefits, the project must be planned and set up in a manner that allows the full benefits of BIM to be realised, which may mean significant investment costs. This also requires the education of staff to understand those benefits and how to implement BIM successfully. There is much of the 'old dog new tricks' in this. Also mirroring the requirements of the collaborative aims of the NEC suite of contracts, the implementation of BIM relies on the participants to it both understanding and buying into its ethos and working in an integrated and collaborative manner. As with many such new and innovative processes, early experience has seen some complaints that much has been spent on management and controls infrastructure that provided little benefit in return. Furthermore, as paragraph 1.17 of the SCL Protocol Guidance Part B states 'The effective use of BIM requires specific agreement between the parties regarding its content, use and ownership'.

No doubt the use and success of BIM will increase significantly over the coming years in the UK and internationally, as will the practice, training and familiarity that will enhance its use to its full potential. Adds-on that enable, for example, searches of BIM to find dispute related records will no doubt further increase its benefits and use. It is hoped that by the time a Fourth Edition of this book is warranted, it will particularly require significant amendment to recognise the benefits of BIM in both the avoidance and evaluation of claims for both time and money on construction and engineering projects.

3.5.7.8 Case Law

The UK courts have widely considered the various different methodologies that can be applied to the analysis of delays. This book is not a vehicle for a full analysis of those cases and their effects, but some have been touched on. Some were identified when outlining those different delay analysis methodologies explained. However, the following series of cases are relevant to consideration of the general principles that need to be followed. These cases generally involve terms of the traditional SBC/Q style of wording, and therefore may not be appropriate to the more prescriptive sets of conditions such as bespoke contracts and particularly the NEC suite.

In *Balfour Beatty Construction Ltd v The Mayor of the London Borough of Lambeth* [2002] EWHC 597, His Honour Judge Humphrey Lloyd QC described an approach that compared as-planned and updated programmes, including recognising the critical path as it changed during the progress of the works:

> By now one would have thought that it was well understood that, on a contract of this kind, in order to attack, on the facts, a clause 24 certificate for non-completion (or an extension of time determined under clause 25), the foundation must be the original programme (if capable of justification and substantiation to show its validity and reliability as a contractual starting point) and its success will similarly depend on the soundness of its revisions on the occurrence of every event, so as to be able to provide a satisfactory and convincing demonstration of cause and effect. A valid critical path (or paths) has to be established both initially and at every later material point since it (or they) will almost certainly change. Some means has also to be established for demonstrating the effect of concurrent or parallel delays or other matters for which the employer will not be responsible under the contract.

In *Royal Brompton Hospital NHS Trust v Hammond and Others* [2002] EWHC Technology 39, His Honour Judge Richard Seymour described an approach that only considered events that actually had an effect on completion, as opposed to theoretical approaches that include events with no actual effect:

> ... as a matter of impression it would seem that there are two conditions which need to be satisfied before an extension of time can be granted, namely:
>
> (i) that a Relevant Event has occurred and
> (ii) that the Relevant Event is likely to cause the completion of the works to be delayed.
>
> This analysis does not strike me as particularly exciting or novel, but I felt that it was at least necessary to consider the question because it seemed to me that Mr Taverner QC was rather glossing over the second element. He cited helpfully ... the decision in *Balfour Beatty* v. *Chestermount* and the recent decision in *Henry Boot v Malmaison* in support of a submission that those decisions confirm the approach ... where Relevant and Non-Relevant Events operate concurrently.

However, it is necessary to be clear what one means by events operating concurrently. It does not mean, in my judgement, a situation in which, work already being delayed, let it be supposed, because the contractor had had deficient labour, an event occurs which is a Relevant Event, and which had the contractor not been delayed, would have caused him to be delayed, but by reason of the existing delay, made no difference. In such a situation, although there is a Relevant Event, the completion of the works is not likely to be delayed thereby beyond the completion date. The relevant event simply has no effect upon the completion date.

On appeal the Court of Appeal referred to the judgment as 'exemplary'.

A similarly theoretical approach was rejected in the Australian courts in *Leighton v Stelux*, outlined elsewhere.

In *Great Eastern Hotel Company Ltd v John Laing Construction Ltd* [2005] EWHC 181 (TCC), His Honour Judge Wilcox criticised a delay expert whose analysis did not consider contemporaneous evidence as to what was actually critical to completion of the project:

> 117. There was a body of contemporary documentation available that would have assisted [a witness] had he chosen to conscientiously recollect the precise course of events that he gave evidence about. It would have run counter to the view taken by his Employers as to these events had he taken account of these documents. [The Planning Expert] who was charged with the duty of independently researching and analysing these events singularly failed to take account of this documentation and the photographic evidence in his written report for the court and presented a view of the course of the critical path which was clearly wrong.
> 128. [The Planning Expert] ultimately, in cross-examination, as he had to, revised his opinion as to the criticality of the protection of the Rail-track services to the project. His failure to consider the contemporary documentary evidence photographs and his preference to accept uncritically Laing's untested accounts has led me to the conclusion that little weight can be attached to his evidence.

On the other hand, in the recent case of *City Inn*, to repeat the quote from the Scottish Supreme Court, where the kind of critical path-based analyses suggested by the preceding cases cannot be carried out:

> The major difficulty, it seems to me, is that in the type of programme used to carry out a critical path analysis any significant error in the information that is fed into the programme is liable to invalidate the entire analysis. Moreover, for reasons explained by [one of the experts], I conclude that it is easy to make such errors. That seems to me to invalidate the use of an as-built critical path analysis to discover after the event where the critical path lay, at least in a case where full electronic records are not available from the contractor.

This need to look at the whole project rather than those events and periods identified by the parties has been explained by reference to *Mirant v Ove Arup*, and a similar approach was taken more recently by Deputy Judge Mr Richard Fernyhough QC in *Costain Limited v Charles Haswell & Partners Limited* [2009] EWHC 3140 (TCC). The parties' delay experts had agreed that the appropriate methodology to assess delay was time impact analysis, or 'windows slice analysis' as it was also referred to in this case. However, he noted that the experts had concentrated their attention on the four months during which the effects of the employer risk delays subject of the claim were taking place. He considered that 'Simply because the delaying event itself is on the critical path does not mean that in point of fact it impacted on any other site activity save for those immediately following and dependent upon the activities in question' and found that Costain had called no evidence to establish that the events in question caused wider delay to the project as a whole apart from the particular works on which they occurred.

In *Walter Lilly & Company Limited v (1) Giles Patrick Cyril Mackay (2) DMW Developments Limited* [2012] EWHC 1773 (TCC) (12 July 2012), Mr Justice Akenhead considered, as well as the thorny issue of concurrency, as explained elsewhere, the relative merits of prospective and retrospective delay analysis techniques. He stated:

> The debate about the 'prospective' or 'retrospective' approach to delay analysis was also sterile because both delay experts accepted that, if each approach was done correctly, they should produce the same result.

It is arguable that this is fine in theory, but that in practice it is rarely the outcome. Prospective approaches are generally based on what was planned to occur and will usually produce a longer entitlement to extension of time. That is why they are so often favoured by contractors. For the same reasons, employers tend to favour retrospective approaches, which are based on what actually occurred and will usually produce a shorter entitlement.

The judgment in *Lilly v MacKay* also confirms the view that delay analysis should be based on the facts, and not the theory, of what caused delay. This might give some support to those who consider that delay analysis, as an issue of fact, should not be for expert evidence but determined by the facts. However, again in practice, the reality is that tribunals, in most cases, will find it impossible to determine those facts without the sort of computer-based analysis explained elsewhere and which was provided to the Judge in *Lilly v MacKay*. He said:

> However, both delay experts' approach … involved in reality doing the exercise that the Court must do which is essentially a factual analysis as to what probably delayed the Works overall.

A further feature of the decision in *Mirant v Ove Arup* was the Judge's acknowledgement that, whilst the causes and effects of delay to construction projects are an issue of fact, without critical path analysis carried out by suitably qualified experts the parties may be mistaken as to what was actually on that critical path.

The case of *Fluor v Shanghai Zhenhua Heavy Industry Co Ltd* [2018] EWHC 1 (TCC) is considered in more detail in Chapter 6, particularly in relation to claims for overhead costs. However, it is also relevant to the issue of prospective versus retrospective and the suggestion in *Lilly v Mackay* that they should come to the same result. The Judge, Sir Antony Edwards-Stuart, said this:

The court has been provided with detailed and careful reports by the two delay experts, for which it is grateful – not least because they provide a ready source of reference for the dates of key events. There has been an extensive debate about the correct approach to delay analysis. [The defendant's delay expert] said, and I would accept, that a prospective analysis – in other words considering the critical path at any particular point in time as viewed by those on the ground at that time – does not necessarily produce the same answer as an analysis carried out retrospectively. The former is the correct approach when considering matters such as the award of an extension of time, but that is not the exercise with which the court is concerned in this case. I agree that some form of retrospective analysis is required.

If this approach is adopted, then contemporaneous consideration of an extension of time would apply a prospective approach, with the programme brought up to date (noting the reference to the critical path at any particular point), but thereafter a retrospective approach would apply.

3.5.7.9 Conclusions

There are a large number of available techniques for the analysis of the causes and effects of delays on construction and engineering projects. In this section we have set out a simple review of the four principal methods of delay analysis from which other methods tend to be derived. Hopefully, it is apparent how the different techniques can derive strikingly different results based upon the same facts.

At issue is which of the techniques should be used in any particular circumstance. In this discussion we have set out the respective advantages and disadvantages of each technique. In practice, which technique a party chooses often depends upon no more than whether it is the claimant or the defendant! Contractors, for reasons that are hopefully now obvious, tend to prefer impacted as planned approaches. These are likely to give theoretical results, but be claimed by their authors to also show the benefits of acceleration or mitigation measures instigated by the contractor. Contract administrators, such as architects and engineers, tend to prefer the retrospective approaches that take into consideration what actually happened and in particular the actual rather than a theoretically derived completion date. This inevitably leads to conflict where contracts provide for a prospective approach, but the contract administrator prefers to wait and see what actually happens. Furthermore, the parties will have different views on the extent to which they want an analysis to expose or hide concurrent delays.

It is suggested that the consideration of which of the alternative approaches should be adopted in any particular case should properly involve the following issues:

- What the contract conditions expressly provide for. As identified elsewhere, contracts such as those of the NEC suite are to varying degrees prescriptive as to which approach is to be followed. Furthermore, following the publication of the First Edition of the SCL Protocol, it was not unusual for bespoke contracts to expressly require that its provisions be followed, and in particular that delays were analysed on a time impact analysis approach. The effect of the changes in the current SCL Protocol, in terms of its support for time impact analysis, are awaited with interest.

- In addition to its express provisions, the contract does of course exist within a legal framework depending on the jurisdiction to which it is subject. The UK has seen extensive consideration by the courts of the various approaches.
- What records are available will of course dictate which approach can in practice be carried out. Therefore on a project with limited as-built records the retrospective approaches may be impossible, and the only practicable option may be a prospective approach based on the as-planned programme, such as impacted as planned.
- Similarly the availability of programmes may dictate the approach. If there was no as planned programme then there may be no basis upon which to carry out anything other than a retrospective approach such as collapsed as built.
- The time available for analysis may be another consideration as some of these techniques can be very time-consuming. A time impact analysis approach to a large and complex project can become a huge and extremely expensive exercise and recognition of 'black box syndrome' has seen many such huge and expensive exercises the subject of some criticism.
- In this regard the value of the project and the sums in dispute may be another factor. On the basis of proportionality it may be that a large time impact analysis or an as-planned versus as-built approach may be inappropriate where the dispute is quite small. It may be that a much cruder approach will suit the parties in achieving an impressionistic or even theoretical outcome, if they both consider it reasonable in the circumstances, and at little cost.

Finally, an illustration of the scope for different delay analysis techniques to reach different conclusions on the same facts and the problems that can result is taken from a recent Middle East arbitration. This related to infrastructure works for a new industrial and commercial zone in a capital city. The works were completed in 2011, but the dispute ran until decided by an arbitral award some years later. This lengthy process and its associated substantial costs was in no small part because the parties took polarised views of the contractor's entitlement to extensions of time and delay costs. That position changed little through the evidence of the parties' respective delay experts in the course of the arbitration, such that the award, by three very senior and well-respected international arbitrators, recorded that:

> In many cases, a Tribunal will find itself in the position of being able to prefer the methodology of one expert over the methodology adopted by another. Unfortunately, this is not such a case. It is the Tribunal's very firm conclusion that the methodologies adopted by [the experts] contained different but nevertheless fundamental flaws which meant that each of the analyses could not be relied upon without more to demonstrate what the actual delays and causes of the delay were on the Project.

The contractor's expert had applied a modification of a time impact analysis approach. It impacted a limited number of events into the baseline programme, or a revision thereto, maintaining the logic of that programme. These events were selected by the contractor and ignored the effects of any preceding events, particularly culpable ones. He had then put actual progress into the programme. Where the actual delay was less than the result of his impacted event, the contractor's expert had attributed to that event the lesser of the two periods, asserting that any saving in time was the result of mitigation or acceleration by the contractor.

The employer's expert applied a modified planned versus an as-built approach, but taking activity progress as a percentage based on overall quantities achieved, rather than by location following the progress of the work in its planned sequence. In this he assumed that the contractor was not constrained to work in particular locations or in a particular sequence such that progress was related solely to the quantities of work completed and not where the employer's access delays had allowed it to be carried out.

The award concluded as follows:

> … the Tribunal does not accept the delay analysis of either expert as leading to a conclusion which it can rely upon alone to determine the delays that were caused to the Works, leading to an ability to assess the cause of delays that occurred on the project and/or any extension of time to which [the contractor] may be entitled. But where does that leave the Tribunal?
>
> The question of whether an event caused a critical delay is ultimately one of fact. Although it is common for Courts and Tribunals to be assisted in their analyses of the relevant facts by programming experts, the fact that the Tribunal has found significant problems with parts of both experts' methodologies does not mean that the Tribunal cannot reach conclusions on the issues of what caused the delays that undoubtedly occurred on the present project. Inevitably such conclusions will not be based upon sophisticated computer programmes, and will not be based on an analysis of detailed logic links and precise impacting that has become a common feature of modern day delay analysis. The conclusions will be based upon the facts, as found by the Tribunal, and the Tribunal's assessment, made in a chronological manner, of the events that occurred on site which had an effect on the progress of the works. In reaching its conclusions the Tribunal will have regard to all the factual evidence that it has heard (and read), much of which has also been analysed by the experts.

The tribunal saw the parties' computer-based delay analyses as no more than a helpful 'sense check' on their own analysis based on the facts. This represented scant return for the parties' expenditure of several $million on those experts' fees.

There continues to be some cynicism in certain quarters as to the usefulness of computer-based delay analyses and the need for delay experts in litigation and arbitration. Much of this seems, understandably, to be the result of the lack of efficacy of many very expensive analyses, as illustrated by some of the judgments and the award mentioned elsewhere. A further factor may be the unrealistic outcomes that particularly result from prospective approaches where these are carried out retrospectively and bear little relationship to what actually happened.

3.5.8 Assessment of Productivity

As explained elsewhere, analysis of the effect of change on programmes of work, and in particular establishing a contractor's entitlement to extension of time, will inevitably involve analysis of a programme. These variously involve a planned or as-built programmes or both. The durations of activities on such programmes will depend on labour and plant productivities. Underlying any such programme, and much analysis of delay and disruption, is therefore the assessment of the productivity that was assumed in the planned programme and that which could have been, or was being, actually achieved on site in an as-built programme.

The implications of these productivity assessments are central to much evaluation of the time impacts of change and need to be understood in terms of the productivity anticipated in the tender programme and pricing, that which could have been achieved on site had the change not occurred and that which was actually achieved or should have been achieved after the impact of the change. For example, in an analysis that compares as-planned and as-built programmes, productivity may be required to be assessed in two areas. Firstly, whether a planned duration could have been achieved in the absence of the delays being considered or should have been longer or even shorter than its planned duration. Secondly, whether the as-built duration was reasonable or should have been shorter than its actual duration. It is also relevant to understand that generally productivity is directly related to cost.

3.5.8.1   Tender Productivity

Whichever planned programme is used to analyse delays, it will most commonly be based on the contractor's tender assumptions. This is true whether this is by reference to an original programme (such as that required under FIDIC Red Book clause 8.3 to be issued within 28 days of a notice to commence works) or a later revision of that programme (as also required under FIDIC Red Book clause 8.3). Whilst such data, if it has been preserved in a retrievable and usable form, may be of interest to illustrate the contractor's intentions, it must be subject to the test of reasonableness. The question is not 'what productivity did the contractor include in its calculations?' but rather 'what productivity should a reasonable and competent contractor have included in its tender?' The contractor will hope that there is no difference between the two, and in a perfect world that would be so. The reality is that tender calculations often contain errors and/or overoptimistic assumptions as to how the works would be executed and the productivity that would be achieved. If that is the case and this infects a planned programme, then, for example, impacting that programme for delays under an impacted as-planned approach, or comparing that programme to an as-built approach, will both give misleading results.

On the other hand, it is also possible, though in practice less common, that productivity assumptions underlying a contractor's planned programme are over-pessimistic. If as a result a planned programme overstates the durations that would have occurred in the absence of a delay event, then that can also give rise to a misleading analysis of the effects of delays. In a planned versus an actual analysis the difference between them may understate the effects of delays. In an impacted as-planned approach, doubling the duration of an activity whose quantities had doubled might overstate the effect of that variation if the planned duration was unrealistically long.

Where this is an issue, the contractor's intentions must be subject to a test of reasonableness both in total and in terms of the outputs underpinning critical activities and their programme durations. This of course begs the question of how the contractor's output assumptions can objectively be tested, given that most contractors will have their own in-house estimating data or will use published data with adjustments to meet their own organisational requirements and/or previous experience.

The answer can only be to build up comparable outputs from published data and/or experience and expertise of reasonable outputs for such work, making such adjustments as are necessary for the particular circumstances of that project and in the light of experience of similar works.

3.5.8.2 Achievable Productivity

It is, of course, entirely possible that the contractor's assumptions and calculations underlying its programme were reasonable and sensible at the time it was prepared, but have changed due to other factors that are not related to any delaying event being considered. As a result, apparent underestimation of an activity duration may not be because it was wrong at the time but because something has changed since. It is therefore usually necessary to consider the circumstances of the works as they progress, prior to any alleged delaying events, and determine the effect, if any, of such factors.

This can require consideration of extraneous factors such as the wider economic circumstances, including such factors as labour availability, etc., as well as factors peculiar to the contract, including, for example, any differences between anticipated and actual site conditions, etc.

3.5.8.3 Actual Productivity After a Change

Where an actual programme is to be used, either for comparison with a plan or, for example, a collapsed as-built approach, it is equally important to establish the reasonableness of that actual programme. It is not enough just to assume that, for example, a varied activity has doubled in duration, when that was the result of an instructed variation, if its actual duration was in part caused by the contractor's own inexcusably poor productivity. It may also be that a doubling of work content in an activity sees its duration more than double in length. One of the most difficult challenges in analysing time matters is that of demonstrating that an instructed change to the works has caused a change in productivity. In practice, that change will be, almost without exception, a reduction in productivity. The difficulty arises from the need to isolate, or account for, all other potential causes of change in productivity. Furthermore, the problem can arise both before the commencement of the affected activity so that it affects all of it, and, sometimes, after the activity has commenced on site so that it only affects part of it.

Where there is a question mark over actual productivity after a change, the ideal situation is to have a reasonable period of the activity undertaken on site without the change that is alleged to cause the productivity drop. It is then possible to provide a further analysis of the activity subject to the alleged cause of the productivity change by comparing it with a period that was without that change. This approach is sometimes referred to as the 'measured mile' approach, as it anticipates the analysis of a 'baseline' or 'control' period or activity or 'measured mile' against which future performance, or performance elsewhere, is to be tested. The approach has been explained in detail elsewhere. As explained there, in practice, life is seldom so simple that one is provided with a period of the unaffected activity that can be directly compared to the affected activity or period. Other factors often tend to be present both before and after the change. Of course, no such 'measured mile' assessment of productivity without the change is possible where there is no such comparator at all.

The answer is usually no more than a reasoned analysis of the type of data discussed elsewhere in relation to tendered productivity. This includes comparable outputs from published data and/or experience and expertise of reasonable outputs for such work. This must also take into account the actual circumstances and potential effects of the change to the works. This will not usually provide a precise answer, but at least should provide an answer that is subject to a range of accuracy, thereby enabling an assessment to be made of the time impact, and payment implications, of the claimed change.

3.5.9 Sources of Productivity Data

This is a subject that is not only relevant to the assessment of the time impact of changes but is often also relevant to the pricing of rates for varied or additional work, issues for Chapters 5 and 6 in this book. It is therefore appropriate at this point to give some thought to some of the problems encountered when using 'source' data to establish production outputs, whether they are for time and programme purposes or for the calculation of unit rates. There are, of course, a host of potential different trades and activities that may be encountered in the course of a major construction or engineering project, but the type of considerations that can be required can be illustrated by reference to an activity that is common to many large projects, such as bulk earthworks.

3.5.9.1 The Fundamental Principles

There are four principal pieces of information required when assessing the time required, or price, for any substantial construction activity:

1. Relevant quantities of work required for the activity.
2. Construction equipment and methods best suited to the task in the circumstances.
3. Outputs that can be expected from the resources required on a sustained basis.
4. Level of resources required to complete the activity in the time required, or the time required to complete the activity with available or optimum resources.

Whilst the considerations applicable to different activities can vary, the thought process required is similar for all activities and can be outlined as follows using the bulk earthworks example.

3.5.9.2 Relevant Quantities

There is often a substantial difference between quantities measured for payment purposes and those needed to calculate the time and resources required for the same activity. A principle of many common standard methods of measurement of construction works, including those of the NRM for building works and CESMM4 for civil engineering works, is that the works are measured 'net' from the drawings of the final permanent works, rather than 'gross' in terms of the volume of the materials after it has been removed. This has two implications:

- The 'net' measurement means that quantities shown in bills of quantities or other tender documents are the amounts measured from drawings of the works as they are to be constructed; they are not a measure of the amount of work required to achieve the final constructed works.
- The measured quantities take no account of bulking or shrinkage of the materials required for any activity during the course of the activity, or of any wastage incurred, whether as a result of transportation or conversion to the final works.

As a result, in the example of bulk earthworks, a number of considerations would be required:

- Where, as is usual, the excavation quantities are measured in situ there will be a difference between the quantity excavated from the ground and the quantity that will be carried away from the excavations by trucks. For instance, a truck may be rated to carry $20\,m^3$ of loose material but may only be carrying $15\,m^3$ or less of earth when it was measured in situ prior to excavation, when fully loaded due to the bulking of the material as it is excavated from the ground. The actual extent of bulking will vary with the type of material being excavated. For example, there can be a difference of up to 50% in bulking between sand and sandstone. This in itself will be a factor in the assessment.

- The capacity of the excavation and haulage units would need to be established to ascertain how much 'bulked' material they can excavate and haul per hour or day. Care may need to be taken as published data can be in units different to those used in the UK. In particular some excavators and dump trucks may be rated in American units and conversion may be needed; the American 'short' ton of 2000 lb needs to be identified and converted where necessary.
- The 'fill' factors for the excavator buckets and the haulage units will need to be established and converted back to the 'as dug' or 'net' quantities. The factors will depend upon the nature of the material. Fine material compacts better and achieves greater filling of the bucket or truck than coarse material that will not compact and will bridge voids, resulting in a greater proportion of air voids and creating a greater volume.

From such information, some analysis can be made of the output of excavators and the carrying capacity of the haulage equipment to enable an estimate to be made of the number of trucks required to service the excavators. This will of course depend on further factors such as the length of haul and travel speed of the haulage equipment itself.

It should, of course, be borne in mind that the reverse situation will arise where it is the filling of an earthworks void that is being assessed. The void will be measured net and will be the volume to be filled, but the amount of 'loose' or 'bulked' fill required to fill the void after compaction will need to be calculated in much the same way as the volumes of excavation need to be calculated. It is the greater volume of 'bulked' fill that will be transported, deposited and compacted, so depending upon the type of material and compaction regime, etc., it will be necessary to have available transport for a greater volume of fill than the final void volume. These considerations apply to earthwork excavations but a similar process of analysis has to be contemplated for any activity to ensure that the correct quantities are used for any evaluation procedure.

3.5.9.3 Equipment and Methods

Based on the identified factors, some consideration can be given to the appropriate equipment and outputs that they will provide. With earthwork operations there will need to be a balancing of resources to ensure that excavators are not standing awaiting trucks or stockpiling material that will unnecessarily incur double handling. Alternatively, trucks should not be standing too long awaiting filling because there is not enough equipment to load them or excavated material to be loaded. These circumstances can be referred to as 'under-trucking' or 'over-trucking' respectively and both mean inefficiency.

Once this basic resource analysis has been done, a method statement can be developed for the activity. As the method statements for all the major activities are developed it will be possible to determine an overall method statement for the whole of the works, or section of the works as appropriate. These method statements will underpin the programme and pricing of the works.

There are, however, further considerations before the programme and pricing can be completed.

3.5.9.4 Sustainable Outputs

The outputs used for excavators, etc., can be obtained from a number of sources including published data from the manufacturers, published pricing books, or data compiled over years by a construction company, practice or individual, based on experience of past similar work.

In some cases the output data will refer to continuous working, assuming 100% efficiency, i.e. the resource is working to its capacity all the time it is employed. This is on the basis that there is little real alternative to publishing data on this basis as it would be impossible to predict all factors that might affect productivity, and to what degree, on all projects. This leaves the estimator to apply suitable adjustments for the circumstances of the specific project. Other data will take a view on locational and other factors and make an allowance for them, stating what those assumptions are. The data may also set out factors that can be applied to allow for varying factors such as location, site constraints and other conditions.

In practice it would be most unusual for a resource to be working at 100% efficiency throughout an activity. Time will be lost in excavation operations for matters such as: 'start-up' and 'wind-down' at the commencement and completion of each working day and at meal breaks, etc.; moving from one operation or part of the site to another; obstructions to excavation or hauling of excavated material; cleaning up after bad weather; and breakdowns of the machine or supporting equipment. Other such factors may apply to other activities.

The way this is dealt with, both for time analysis and pricing, is to calculate the output and costs of the resource at a reduced percentage efficiency or by a reduced working time per hour employed. Thus the expected output may be assessed as being 80% and the cost of one hour's resource will then be attributed to 80% of the nominal output of the resource at 100% efficiency. This is sometimes expressed in terms of, say, a 50-minute hour, i.e. the resource is calculated to be producing at 100% capacity for 50 minutes and at nil capacity for the other 10 minutes in each hour.

The application of such efficiency factors should not be confused with inefficient or poor management of the resource. The factors represent the reality of working on site as opposed to the potential output of resources in ideal working conditions with no interruptions, etc. In practice, such conditions do not apply in the field. A particularly problematical area for practitioners may therefore be agreement as to what to allow for local conditions. If, for example, bulk earthworks are being carried out in a harsh environment and by inexperienced local labour, then there may be little specific data on productivity. In such circumstances, the assessment of suitable factors to apply to published norms or those experienced elsewhere may become a very subjective issue.

3.5.9.5 Recalculation Using Efficiency Factors

When an assessment of likely sustainable production has been made, the time and cost calculations can be rerun to produce the time periods and costs that will be behind the programme of works and the unit rate pricing.

To give some idea of how this looks in practice, consider the case of an excavation operation to remove $200\,000\,m^3$, measured in situ, of loam and fine clay soils, with the excavated material being hauled three miles to a spoil heap. Assuming a large-scale cut excavation on an open site with no restrictions on machine movements, the contractor will research output data from, for example, excavator manufacturers' data, its own recorded data or a mixture of such sources. The contractor's basic calculation, assuming an excavator with a nominal bucket capacity of 3.1 m³ discharging direct into waiting dump trucks, will then look something like this:

Theoretical excavation cycle time (using a productivity factor of 1.2 for the type of material being excavated):

Dig 9 seconds × 1.2	10.8
Slew loaded 4 seconds × 1.2	4.8
Dump	2.5
Slew empty	5.0
Total	23.1 seconds

23.1 seconds = 155.8 cycles per hour
Bucket nominal capacity = 3.1 m³, with fill factor of 0.90 = 2.79 m³
Efficiency factor = 85% = 0.85
Output = 1 × 155.8 × 2.79 × 0.85 = 370 m³ per hour
This is loose material per hour, i.e. after bulking; therefore the amount of in situ material excavated, assuming a bulking factor of 0.72 (0.72 m³ in ground = 1 m³ bulked):
370 m³ × 0.72 = 266 m³ per hour

The machines can then be expected to remove 266 m³ of in situ material per hour and will require trucks to carry 370 m³ of bulked material three miles loaded to the tip and return three miles unloaded after tipping. A cycle would then be worked out for the trucks to determine how many trucks will be required to keep the excavators working at the anticipated sustained capacity. When that calculation has been made, the cost of excavation and cart away can be made for 266 m³ of in situ material per excavator per hour.

It can also be calculated that using one machine the excavation will take 752 hours, or 15 × 50 hour weeks. Doubling machine numbers would halve the total period if that were operationally feasible.

It may well be that the norms or other data used by the contractor will consist of the above information in summary form, i.e. the machine in this example will require 0.225 minutes (60 minutes divided by 266) per m³ of excavation in soils of this type. An appreciation of the underlying analysis is often essential, however, when changes to the rate are to be considered or additional payment due to disruption of the working cycle is to be contemplated.

3.5.10 Effect on Contractor's Plant and Equipment

The productivity considerations set out in the example in Section 3.5.9 will be fundamental to the calculation of durations, costs and prices at the tender stages, and similar techniques should be employed in evaluating the effect of changes on the scheme of working, for whatever reason. It is, however, necessary to consider, in the context of time and delay, that the costs incurred by a contractor for plant and equipment do not always vary in proportion to the time expended. Such costs can be considered under two broad headings of 'working plant and equipment' and 'site facilities and equipment'.

3.5.10.1 Working Plant and Equipment

The term 'working plant and equipment' is used to describe contractor's plant, etc., used for the permanent and temporary works and which is usually costed into the unit rates and prices for the work. Such plant will usually include items such as excavators and dump trucks, concrete floats, compressors and welding sets, etc. However, care should be taken, as some categories of plant will sometimes be priced into unit rates and in other instances will be in the general 'site facilities and equipment' or 'preliminaries and general' items and priced in bills of quantities items for those. For instance, cranes used for a specific lift, to place a piece of permanent plant or equipment in its final location, might be priced in the unit rates for that permanent plant or equipment. On the other hand, other cranage (such as a tower crane) that is used for multiple purposes may be priced in the 'preliminaries and general' items section of the bills of quantities. This principle can extend to many categories of plant and equipment and a careful analysis is required to establish how and where the plant and equipment costs have been incorporated in the tender.

One of the general 'principles' of the effect of time on plant costs, when the plant is employed on unit rate work, is that the costs do not vary directly with the time expended.

A simple example is that of an excavator employed on the digging of trenches. If, as a result of changes, the excavator has to work longer hours, the average unit costs do not vary directly. The cost of the machine, assuming it is hired or depreciated by the hour, will be the same as will the fuel and lubricant cost. The cost of the driver may vary, however, if the extended hours result in premium time working. In such circumstances the result of extending the working hours from eight to ten per day may be as follows:

Original cost per hour (plain time working only)	$
Excavator depreciation/hire	30.00
Fuel/lubricants	5.00
Operator	12.00
	Total 47.00

Revised cost per hour (10 hours working per day)	$
Excavator depreciation/hire 10 × $30.00	300.00
Fuel/lubricants 10 × $5.00	50.00
Operator 10 × $12.00	120.00
Operator premium time 2 × $6.00	12.00
	Total 482.00
	Per hour 48.20

It is only the operator costs that are varied by the premium time working; excavators do not get paid overtime! When the effect of the premium time is spread over the full 10-hour working day, then the increase in the average rate per hour is only some 2.5% despite an increase of 25% in the length of the working day. If the extended day is required to do the same or less work than the tender expectation, then the unit rates will increase, but if the extended day is required to undertake increased amounts of work, then there may be no increase in the unit rates or even a decrease if the total amount of work increases by more than the 2.5% increase in the average hourly cost. If the hourly average production remains constant, as would be expected in the absence of other factors, then an increase in working time of 25% has produced an effect on the unit cost of production of only 2.5%. This may seem an obvious point but in the analysis of time effects on cost, the simple fact that the two do not necessarily vary in the same proportions is often overlooked or ignored.

In other circumstances changes in time may result in disproportionately high costs. This is particularly so where high capital cost equipment is being used for works paid or priced on a unit rate basis. For instance, works such as dredging require equipment with extremely high capital and running costs if, for instance, sea-going suction dredgers are employed. If time is extended without corresponding increases in the volume of dredging required, then unit costs will rise rapidly. It is therefore vital that the make-up of the tender costs and pricing is analysed properly before attempting to undertake any evaluation of the effect of prolongation and delay. Not only must the make-up of the unit rates be understood but also the approach to pricing, in terms of where particular plant is priced, must be available to enable any change evaluation to be undertaken on a consistent basis, avoiding any misleading assumptions.

3.5.10.2   Site Facilities and Equipment

The term 'site facilities and equipment' is used to describe plant and equipment used generally on the site, not for particular construction activities, and usually priced into the site overheads or 'preliminaries' section of the tender or the 'preliminaries and general' items section of bills of quantities. Such plant and equipment will often include scaffolding and access equipment, general cranage and site concrete batching plants as well as site offices, messing and welfare and safety facilities.

One of the important aspects of the costing and pricing of such facilities and equipment is that the costs of the relevant plant or equipment rarely change in a smooth curve in proportion to changes in time or resources. The more common experience is that such costs, whilst they may vary a little in response to any time and resource changes, tend to incur substantial changes in costs as 'steps' when critical points are reached.

The easiest example is that of messing facilities, the basic cost of which may vary in direct proportion to time but whose cost will vary in 'steps' with the need to service a greater number in the workforce, it being a characteristic of such facilities that a given size of facility will service a resource up to a certain level, after which an increase in facility size and cost is incurred, which will then remain the same until a further maximum capacity is exceeded. For instance, if messing facilities costing $500 per week are provided, which will accommodate a workforce of up to, say, 100, but numbers above that will require an additional facility at a cost of $250 per week enabling a further 50 to be accommodated, then an increase in resources from 90 persons to 120 persons will incur the 'step' cost of $250 per week as the resource level exceeds the original 100 capacity. Similar types of variation in costs can be experienced with plant costs such as cranes, where a tower crane may be servicing the general lifting and distribution needs. If the capacity of the tower crane is exceeded other mobile cranes may be required to supplement capacity, resulting in steps in the cost of the distribution service.

As with the working plant and equipment the first requirement is to understand how the costs are incurred, what triggers changes in costs and at what point.

3.5.11   Duty to Mitigate

The duty of parties to a contract to mitigate their losses was covered elsewhere in this book, particularly under English law and by reference to the judgment in *British West-inghouse Electric & Manufacturing Co Ltd v Underground Electric Railway Co of London Ltd* [1912] AC 673. From that and other cases, such as '*The Solholt*', *Sotiros Shipping Inc and Another v Shmeiet Solholt* (1983) Com LR 114 and *Banco de Portugal v Waterlow & Sons Ltd* [1932] AC 452 506, a number of principles can be said to apply:

- There is a duty on a party suffering from a breach by another party to take reasonable measures to mitigate its loss resulting from that breach.
- That duty does not limit how the injured party acts, it is free to act as it judges to be in his best interests, but the defendant is only liable for such part of the loss as is properly caused by the breach.
- That duty does not extend to carrying out measures that a prudent person would not normally take in the course of its business.
- The injured party cannot recover that part of its losses that was the consequence of its failure to comply with this duty.
- The costs of reasonable measures that are taken to mitigate loss are recoverable.
- The cost of such measures will not be disallowed merely because the party in breach can suggest that other measures less burdensome to him might have been taken.
- The onus of proving that reasonable measures were not taken is on the party defending the claim.
- Whether the measures were reasonable is to be judged at the time and in the circumstances under which they were decided upon, and not with the benefit of hindsight.
- A party cannot recover any loss that was avoided by its actions, even where those mitigation steps went beyond what is reasonable.

These cases were in relation to financial damages arising from a breach of contract and the duty to mitigate is particularly relevant to financial claims under construction and engineering contracts for such as defective and incomplete work and disruption. However, the principles would equally apply to the actions of a contractor in relation to delays caused by employer risk events.

To translate these principles into the context of claims for delays on a construction or engineering project, it is suggested that the effect is that a contractor can take whatever measures it judges to be in its best interests in overcoming the effects of an employer risk event and it is not open to the employer to criticise on the basis that other less burdensome measures could have been taken. However, the resulting extension of time will only be that which can be demonstrated to have been reasonably incurred as a result of the event. The sanction for the employer will be that, in the event that the contractor reacts unreasonably to a delay, such that the actual delay to completion exceeds its entitlement to extension of time for the event, then the employer will normally be entitled to its delay damages for that overrun.

It seems quite reasonable that recovery of time should not be limited to the minimum that could have resulted from the breach. When faced with an employer risk delay event during the course of a project it is not always possible, or desirable, to take time assessing every possible alternative course of action that may be followed to overcome the effects. Those effects will be both delay, and since time means money, financial consequences. Thus, for example, lacking a final design specification for the wearing course of a road, a contractor will have to consider how best to react to deal with the effects on both the programme and on its costs of such plant and equipment. This may mean balancing the need to retain resources such that when information is made available the effects on the programme are minimised, with mitigating the wasted costs of having resources avoidably standing around awaiting such information.

It is quite proper to expect that any course of action, and the delays incurred, will be reasonable in the circumstances. However, whether the course of action was reasonable must be assessed at the time that the decision was taken, and not with the benefit of hindsight. If measures are taken, which at the time were reasonably expected to reap a benefit in mitigating the effects of a delay event, but they actually turn out to have no such effect, then the contractor cannot be criticised for taking those actions purely based on hindsight. A prudent contractor would, of course, notify the employer of the measures being taken, the reasons for them and any cost implications. This may help to pre-empt any retrospective attempt to deny their reasonableness, but may also satisfy any notice requirements under the contract.

This potential restriction means that the measures taken to overcome an employer risk delay event must always be examined with a test of reasonableness in mind. For example, if a contractor is imminently awaiting a final design specification for the wearing course of a road, but uses the delay as an excuse to remove all of its related plant to another project for which it culpably lacked the necessary plant, where it commits it for the next two months, it cannot then claim from the employer the two months lost due to lack of plant on the original road. On the other hand, if the plant could be readily returned when required and at the time that it was transferred there was a reasonable expectation that the costs of transporting it and the time lost in returning it would be less than the costs of leaving it idle, then the contractor ought to be entitled to an extension of time, not just for the delay awaiting the information but also a reasonable period to get the plant back on site. Reasonableness in such circumstances will be a question of fact depending on the circumstances of the case at the time.

3.6   Summary

In this chapter we have considered the effects of change on a programme of works and how that change can be assessed. There are two particular tools used as a basis for such analysis. These are the as-planned and as-built programmes. As-planned programmes are an essential component of all but very minor construction projects. All standard forms of contract require one to be prepared and updated during the progress of the work. As-built programmes are often maintained during the course of the works by way of updates of a planned programme. Their forensic production on a retrospective basis as a basis for delay analysis can draw on a range of project information.

Change does not always give rise to prolongation or entitlement to extension of time. The ownership of float in a programme, and how it should be taken into account in calculating extensions of time, is a particularly thorny area. Similarly, concurrency where there are delays the responsibility of contractor and employer or delays caused by cost-neutral events. Concurrent delays can have different implications for a contractor's entitlement to extensions of time compared to its entitlement to related financial compensation. Change can give rise to only local delays, with associated local running costs. It can also result in disruption costs without effect on completion.

Where delay to completion does arise, the analysis of its effects generally relies on the as-planned and/or the as-built programme. However, there are a number of techniques in use and these can give very different results even based on identical project facts. Which should be applied in any given case depends on such issues as: the express terms of the contract; the applicable law; the available records; the available programmes; the time available for the analysis; and the parties' wider relationship and commercial interests. There are similarly various different programming software packages and these may give different results where data are transferred between them.

Underlying any programme is the productivity of labour and plant that underpins the duration of each activity. Planned programme durations, in particular, may be too short having been based on overoptimistic outputs. Actual durations may similarly be too long because the contractor culpably failed to achieve reasonable progress. If programme durations used to analyse the effects of change are the result of such flawed productivities, then they may be challenged and need to be investigated. There are various sources of suitable data for such an analysis.

Contractors may have a duty to mitigate the effects of delays. However, it is important to understand that duty and its limits, for example that mitigation measures or the failure to apply them should be judged in the light of circumstances and knowledge as they existed at the time the measures were taken.

Once the programming consequences of delays are established, the next issue may be assessing the financial consequences. This is the subject of the following chapters, starting with consideration of the sources of the types of financial information that can be used in the evaluation of claims for change on engineering and construction contracts.

4

Sources of Financial Information for Evaluation

Having established the basis from which change can be evaluated and considered the effect of change on programmes of work, it is then necessary to examine the sources of information that will facilitate the financial evaluation of a construction or engineering contract claim itself.

As with almost any contract analysis, the starting point must be the terms and conditions of the particular contract. This includes both the express provisions of the agreement and any terms applying from the applicable law, particularly where the express provisions leave gaps in terms of how change is to be valued.

Notwithstanding that they might be the traditional approaches applied in many jurisdictions, it would be wrong to assume that variations are to be valued based on rates and prices in the contract, but the time consequences of change, such as delays, are to be valued on the basis of actual costs. This is not always the case in bespoke contracts, as illustrated by the example detailed in this chapter.

The contract should also be checked for any express definitions of specific expressions. It should never be assumed that a contract expression or term (for example 'cost', 'price' or 'rate') has a particular traditional meaning without checking that it is not expressly defined in the contract. Where a standard form of contract is being used it should be checked that any definition contained in that standard form has not been altered or qualified for the particular contract being considered.

The contract may also place express limits on the extent to which such 'costs', 'loss' or 'expense' are recoverable. For example, it may refer to 'costs directly referable to the variation' or only allow such costs as are traceable to invoice records (thus leaving little room for assessment of those costs). Again, such express terms may limit the sources of financial information that can be applied to the evaluation of change.

Generally, the sources of financial information for the evaluation of a claim may be either the contractor's actual costs, loss and expense or the contractor's original pricing of the contract price. Where the contract terms require the latter approach, considerations may include the examination of the documents produced in the course of obtaining the tenders for the contract. These will usually include the client's invitation to tender, the successful tender itself and the project specification documents, which may extend to bills of quantities or schedules of rates, depending upon the type of contract. The next step may then be to examine the tender itself, including any available information relating to the calculations underlying the contract pricing. Care needs to be taken in this process as it is only that which amplifies and allows a proper understanding of

Evaluating Contract Claims, Third Edition. John Mullen and R. Peter Davison.
© 2020 John Wiley & Sons Ltd. Published 2020 by John Wiley & Sons Ltd.

the financial content of the contract that is relevant. In some jurisdictions, material extrinsic to the contract documents that suggests interpretations contrary to the contract content will generally be inadmissible in a formal dispute, even where material such as bid meeting minutes, not incorporated in the contract documents, show a joint approach by the parties. Local legal advice should be taken on issues such as this.

Where actual costs are the required basis, the next potential source of information for evaluation of additional payments is the, usually voluminous, amount of orders, invoices, and other cost records for the project works, culminating in the final source of such information – the financial accounting records of the contractor.

The potential sources of information can then be considered under discrete headings, as follows:

- The contract provisions.
- Contract and/or tender documents and correspondence, etc.
- Tender and/or contract price calculations and assumptions.
- Invoices and cost records.
- Accounting information.
- External information.

On some occasions some of these categories of documents may not be relevant as they may be excluded by the contract as sources of information for the administration or interpretation of its terms and provisions. Similarly, in other circumstances, there may be a hierarchy of contract documents set out in the contract; i.e. in the event of contra-dictions or anomalies it might be stated which documents take precedence, for instance terms and conditions, specification, drawings, bills of quantities, etc.

In some contracts, such as the SBC/Q Form clause 1.3, there may be a provision to the effect that nothing in the contract documents (such as the bills of quantities, Contrac-tor's Design Portion Documents or any Framework Agreement between the parties) is to override or modify the terms and conditions of the contract itself. Such clauses need to be considered very carefully by the parties at the outset to ensure that, if any of the other contract documents or tender documents, etc. have the effect of modifying the standard terms and conditions then effect is given to that modification by a con-tract amendment. This may be particularly relevant in the context of commercial terms or arrangements contemplated in the tender documents, or other contract documents, but which are potentially in conflict with the standard terms and conditions. Commonly this occurs in relation to payment periods or retention provisions. The effects of such an 'override or modify' clause may come as a surprise to those operating in jurisdictions where the applicable law states that the bespoke terms expressly written for that contract take precedence over contradictory terms of the applicable standard form. The assump-tion may have been that painstakingly drafted amendments to the standard form will be of effect, but an 'override or modify' clause in the contract documents means that the amendments are of no effect at all. The result may be a dispute between the parties as to the correct interpretation. As ever, local legal advice is essential on such matters.

It may, of course, not be necessary or appropriate to examine each or all of the poten-tial sources for every evaluation process, depending upon the type of evaluation being considered. Under some forms of contract some sources of potential valuation infor-mation may be completely irrelevant. For example, where a contract requires that all changes are valued on the basis of the contractor's actual costs, with no reference to the contractor's original pricing, how it priced its tender or any rates and prices set out in the contract documents may be irrelevant.

4.1   The Contract Provisions

The contract terms and conditions provide the framework for all quantum assessments and should be the first source of information to be examined. Not only should the terms and conditions be used as a framework but also the full implication of any expressions used should be considered carefully in the context of the particular contract.

The first priority in examining the contract terms and conditions should be to ascertain if any common words or expressions have been given particular defined meanings for the purposes of the contract. Expressions such as 'cost', 'contract price', 'contract sum', 'tender sum' or 'tender total' will often be defined in ad hoc contracts as well as in standard forms, together with definitions of 'prime cost sum' and 'provisional sum', where such terms are used. While many individuals commencing an evaluation process may be familiar with such terms it is always important to check that the accepted or assumed meaning of the term has not been changed, or in a standard contract has not been qualified in some way. It should never be assumed that a traditional meaning has been left without express amendment or that the applicable law does not have relevance to interpretation.

More fundamentally, many contracts will define terms such as 'loss', 'expense', 'cost', 'profit' or 'margin', and 'overhead' so that they have a particular meaning when they appear in the contract provisions and include some components of valuation but not others. On the other hand, failure to define such terms within the contract may be an invitation for confusion and conflicting interpretation of the contract and what the contractor is entitled to be paid. This is an example of bad practice that should be avoided if at all possible.

4.1.1   Cost

The term 'cost' is often used in contracts and is equally often pertinent to the evaluation of some types of claim for additional payment, both for the direct consequences of a variation and the indirect effects on the programme of works and productivity. *The Concise Oxford Dictionary* gives a definition of cost as being 'the required payment of a specified sum in order for something to be bought or obtained or the amount that something costs'. If there is no express definition available then the word will be interpreted as having its usual meaning under the relevant jurisdiction, with the dictionary definition being an excellent starting point. However, is this what was meant or intended in the contract? If the dictionary definition is adopted then, when evaluating something for which the contract states that 'cost' is relevant, the sum claimed may be everything paid in order for the subject of the evaluation to be bought or obtained.

However, what about the reasonableness of the amount paid out? Dictionary definitions do not concern themselves with such issues, which may be of particular relevance in the context of valuing change under an engineering or construction contract. Contract definitions of 'costs' will therefore often include the qualification that they are 'reasonable' and incurred as a direct consequence of the change that is being evaluated.

Even where this qualification is not expressly made, it may be that the applicable law implies it.

Another issue in relation to the term 'cost' is whether this includes off-site overheads and profit, and if so how to address them. In a contract where the contractor is working to make a profit on his works, is it entitled to include an amount for the profit element in the definition of cost, and an additional element to cover any off-site overhead charges that may indirectly be part of the costs? A definition within the contract such as that found at clause 1(5) of the Infrastructure Conditions is the sensible approach to avoiding such a potential argument. That definition is as follows:

> The word "cost" when used in the Conditions of Contract means expenditure properly incurred or to be incurred whether on or off the Site including overhead finance and other charges properly allocatable thereto but does not include any allowance for profit.

In this definition, 'cost' therefore expressly includes overhead and finance charges, as well as 'other charges properly allocatable thereto', but expressly excludes any allowance for profit. As is explained when considering different types of claim under the Infrastructure Conditions elsewhere in this book, profit is then added to that defined 'cost' in evaluating some claims but not others.

However, even within this Infrastructure Conditions definition, there is scope for contention. There is no guidance as to what the 'other charges' are intended to cover. Since overheads and finance charges are separately stated to be included, presumably such charges would cover matters such as taxes and levies, royalties or license fees that might be payable on certain types of operations and might not be considered as part of the direct costs or the overheads or finance charges.

The definition of 'cost' set out in clause 1(5) of the Infrastructure Conditions will then apply to any use of the word elsewhere in those conditions of contract. For instance, this applies in clause 12(6) in relation to adverse physical conditions, which also provides for the additional of a reasonable percentage addition for profit, and in clause 26(4) for claims arising from the employer's failure in relation to planning permissions, where there is no addition for profit.

When discussing the direct consequences of change in Chapter 5, we explain how the NEC4-ECC contract applies 'costs' to the evaluation of all types of change and their effects. This includes an explanation of how NEC4-ECC defines that term.

The situation is somewhat different under the standard SBC/Q Conditions. As also explained in detail in Chapter 5, there the conditions of contract generally apply the more traditional approach of anticipating that the contract pricing will be used as the basis for evaluating entitlements arising from variations to the works. On the other hand, where there is a need to ascertain the entitlement to payment for matters such as delays caused by the 'Relevant Events' as defined in the contract, the expression used in clause 4.23 is 'direct loss and/or expense' rather than 'cost' or any similar expression. That might bring in losses in relation to such as head office overheads and profit, as discussed in Chapter 6.

The need to not assume that some common terms in relation to evaluation of change have their usual industry definitions may be best illustrated by way of an actual example. An easy assumption under a traditional contract approach would be that the financial effects on the contractor of compensatory delays are compensated on the basis of its actual costs (or 'loss and expense' under JCT terminology), with or without profit, whereas variations are primarily valued on the basis of the contractor's rates and prices set out in the contract. However, in a contract for the construction of a power generation facility the opposite approach was adopted. Variations were to be priced on the basis of 'actual cost' but that 'cost' was defined to exclude any allowance for both off-site overheads and profit. In a final account dispute, the contractor argued that such an approach was contrary to common practice and common sense, whilst the employer said such considerations were irrelevant and relied on the express terms that had been agreed between the parties. In the same contract delays were to be valued on two bases depending on the cause of the delay. The financial effects of delays caused by 'Force Majeure' were subject to agreed hourly rates set out in an annex to the contract. The financial effects of all other delays, such as late information and access, were to be evaluated on the basis of defined 'Cost', but that capitalised term was defined to include both off-site overheads and profit. Such an approach may be unusual, but they illustrate the need for care both in agreeing contract terms and later assuming what they mean when addressing the financial effects of any changes.

Another unusual approach is sometimes found in concession agreements, where the contractor is engaged for both a development period to construct the project and a following concession period during which it operates the resulting facility and takes the profit (or losses) from its running. Agreements for the development period of such projects have been seen to require that all changes be valued subject to a 'no better and no worse' limitation on the financial consequences for the contractor of the valuation of any change. Apart from the unusual nature of this approach, the lack of explanation of what the term 'no better and no worse' actually meant led to inevitable dispute as to whether it meant that the contractor was entitled to additional profit for additional work or was limited to the profit allowed in its original price and therefore carrying out any additional works at no additional profit.

4.1.2 Loss and Expense

The term 'loss and expense' has been used by the JCT throughout its many contracts and subcontracts and their editions over the decades in relation to the evaluation of contractors' and subcontractors' claims for the financial effects of what those contracts refer to as 'relevant matters' affecting the regular progress of their works. It is therefore the equivalent of 'cost' in the Infrastructure Conditions and 'Cost' in FIDIC contracts. It is a particularly favoured expression for practitioners from the UK and it is not uncommon to hear them use it (or the abbreviation 'L and E') inappropriately in relation to contracts in other parts of the world where the specific contract and the applicable law do not recognise it. Just like the term 'cost', it is, rather obviously, a term that could be interpreted in different ways by different people. In the absence of any guidance as to what it means, it could pose a number of questions, particularly:

- Does it include profit?
- Does it include finance charges incurred on other costs and payments?

It is often suggested that profit cannot be covered by the word 'expense' under JCT parlance. If that is correct, then, under a contract that allowed 'loss and expense', profit would have to fall under the word 'loss' in order to be recoverable. A successful claim would therefore have to show that profit was lost due to the breach. In such circumstances a reduced turnover for the contractor due to delay in the works might result in a claim for loss of profit. This is considered further in relation to claims for overheads and profit as part of delay-related claims, discussed in Chapter 6. The expression 'loss and expense' would seem potentially to include such a claim but the contractor will need to show that it would have made a profit, or reduced his loss, if the delay had not reduced its turnover.

Under English law, the question of finance charges was decided by the Court of Appeal in *F.G. Minter v Welsh Health Technical Organisation* (1981) 13 BLR 1, when it considered the expression 'direct loss and/or expense' in the context of the 1963 Standard Form of Building Contract. The court decided that finance charges could be added to the cost of other expenses and payments. Interestingly, in opting not to follow earlier legal precedents against allowing interest as damages, the court stated:

> I do not think that today we should allow medieval abhorrence to usury to make us shrink from implying a promise to pay interest in a contract if by refusing to imply it we thereby deprive a party of what the contract appears on its natural interpretation to give him.

Effectively interest/finance charges incurred by a contractor on other heads of 'loss and expense' were themselves another head of that 'loss and expense'. The decision was based upon 'what the contract appears on its natural interpretation' to give the contractor, but it required the Court of Appeal to decide this was what the natural interpretation meant! How much easier would it have been if the contract terms had spelled out what was intended to be included as 'direct loss and/or expense'? This still leaves the position to be considered where English law does not apply, and in some jurisdictions where there may be still an 'an abhorrence to usury', that leaves the position subject to argument where there is no contractual definition.

4.1.3   To Ascertain

Another expression that causes some difficulty and is often found in the JCT and other ad hoc building contracts (particularly where based on JCT terms), is the requirement that a payment or sum due in relation to claims for such as delay and disruption shall be 'ascertained' or the person responsible is charged 'to ascertain' the sum due.

The Concise Oxford Dictionary defines the word 'ascertain' as to 'find out for certain' and this dictionary definition has led some commentators and practitioners to suggest that for a sum to be ascertained it must be proved with certainty. Such reasoning may seem understandable where one of the rules for the interpretation of contracts is that words shall be given their normal or ordinary meaning. On the other hand, such a definition would make a contractor's damages for delay under SBC/Q somewhat more restricted than the usual civil standard of proof of damages, being 'the balance of probabilities'. It could even be said that on such a basis the standard is nearer to that under criminal law of 'beyond all reasonable doubt'. In practice it may not always be possible to prove all sums beyond the balance of probabilities and it may be disproportionate in terms of administration time and costs to do so. Notwithstanding this, a consequence of such attempted restrictions of entitlement was that some claimants submitted their claims for the financial effects of delay and disruption also on the alternative legal basis of damages for breach of contract. This was on the basis that such a claim would have the lesser burden of proof of the 'balance of probabilities'.

The Technology and Construction Court (TCC) considered this problem, not for the first time, in 1999 in the judgment in the case of *How Engineering Services Ltd v Lindner Ceilings Floors Partitions PLC* (1999) EWHC B7 (TCC). This judgment arose from challenges to two arbitration awards made in respect of disputes concerning claims for 'loss and expense' for work carried out by Lindner as sub-subcontractor to How Engineering on the redevelopment of Cannon Street Station in London. The comments of the Judge are interesting to any person charged with the responsibility to 'ascertain' a sum such as that for loss and expense.

It was agreed that, apart from some discrete issues, the claim was a total cost claim, i.e. it consisted largely of Lindner's incurred costs with a credit for the sums that would have been incurred if the intervening events had not occurred. However, counsel for How Engineering argued that the arbitrator had erred in his award in that in his approach to quantification he accepted a series of assessments put forward on behalf of Lindner which lacked the necessary precision to support a valid claim. The contract between How Engineering and Lindner required that How Engineering should 'ascertain' the sum for loss and/or expense, and the arbitration clause in the contract required that the arbitrator should 'ascertain' and award any sum that ought to be due.

The submissions for How Engineering referred to the earlier judgment in the TCC of *Alfred McAlpine Homes North Ltd v Property and Land Contractors Ltd* (1995) 76 BLR 59, in which the Judge said:

> Furthermore to 'ascertain' means to 'find out for certain' and it does not therefore connote as much use of judgment or the formation of an opinion as had 'assess' or 'evaluate' been used. It thus appears to preclude making general assessments as have at times to be done in quantifying damages recoverable for breach of contract.

These remarks were made in the context of a claim in respect of plant charges, and suggested that it was the actual loss that needed to be ascertained and not a hypothetical cost calculated by assumed or typical hire charges, etc. However, the Judge in *How v. Lindner* did not understand the quoted passage from *Alfred McAlpine v Property and Land* to infer that there was no room for judgment in the process of ascertainment. He said:

> I respectfully suggest that the phrase 'find out for certain' might be misunderstood as implying that what is required is absolute certainty. The arbitrator is required to apply the civil standard of proof.

He then went on:

> In my view it is unhelpful to distinguish between the degree of judgement permissible in an ascertainment of loss from that which may properly be brought to bear in an assessment of damages. A judge or arbitrator who assesses damages

for breach of contract will endeavour to calculate a figure as precisely as it is possible to do on the material before him or her. In some cases, the facts are clear, and there is only one possible answer. In others, the facts are less clear, and different tribunals would reach different conclusions. In some cases, there is more scope for the exercise of judgement. The result is always uncertain until the damages have been assessed. But once the damages have been assessed, the figure becomes certain: it has been ascertained.

The conclusion of this discussion is that there is room for judgement and opinion in 'ascertainment' of 'loss and expense' under a building contract such as in JCT terms, in the same way that there is room for such judgement and opinion in the calculation by a judge or arbitrator of damages for breach of contract.

The 'certainty' implied by the requirement to 'ascertain' in such clauses therefore arises once the contract administrator has calculated the amount of loss and expense. It does not require, as some practitioners asserted when assessing contractors' claims, that the evidence produced in support of that claim had to give certainty as to the actual loss incurred by the contractor.

This is entirely consistent with the approach of the courts to the assessment of damages. In *Murphy v Stone Wallwork (Charlton) Ltd* [1969] 2 All ER 949, Lord Pearce stated:

> ... the assessment of damages for the future is necessarily compounded of prophecy and calculation. The court must do the best it can to reach what seems to be the right figure on a reasonable balance of the probabilities avoiding undue optimism and pessimism ...

In *Mallett v McMonagle* [1970] AC 166 HL, Lord Diplock said:

> ... in making an assessment of damages ... a court decides on the balance of probabilities. Anything that is more probable than not it treats as certain.

How the scope for judgement and opinion should be applied is a matter for succeeding chapters.

The nature of 'ascertainment' was further considered in *Walter Lilly and Company Ltd v Giles Patrick Cyril MacKay* [2012] EWHC 1773 (TCC) in the context of a JCT 1998 Standard Form, Private Edition Without Quantities, where the court had to consider what information needed to be provided to fulfil a contract requirement for the contractor to submit to the architect such details as should reasonably enable it to 'ascertain' the appropriate amount of loss and expense under clause 26.1 of that form of contract. The Court decision included:

- In considering the contractor's obligation to provide information the extent of information already in the architect's possession, for instance through site meetings, should be taken into account.
- The contractual obligation was to provide such information as was 'reasonably necessary' and this obligation may be met by an offer to allow the architect, or other client representative, the opportunity to inspect records at the contractor's offices.

- The standard required of the architect was that he or she needed to be satisfied that the claimed loss and expense was likely to have been incurred. The wording of the clause did not require the architect to be certain that any particular expense had been incurred.

In relation to clause 26.1 of the 1998 JCT form, Mr. Justice Akenhead said this in paragraph 468 of his judgment:

> Clause 26.1 talks of the exercise of ascertainment of loss and expense incurred or to be incurred. The word 'ascertain' means to determine or discover definitely or, more archaically, with certainty. It is argued by DMW's Counsel that the Architect or the Quantity Surveyor cannot ascertain unless a massive amount of detail and supporting documentation is provided. This is almost akin to saying that the Contractor must produce all conceivable material evidence such as is necessary to prove its claim beyond reasonable doubt. In my judgement, it is necessary to construe the words in a sensible and commercial way that would resonate with commercial parties in the real world. The Architect or the Quantity Surveyor must be put in the position in which they can be satisfied that all or some of the loss and expense claimed is likely to be or has been incurred. They do not have to be 'certain'. One has to bear in mind that the ultimate dispute resolution tribunal will decide any litigation or arbitration on a balance of probabilities and at that stage that tribunal will (only) have to be satisfied that the Contractor probably incurred loss or expense as a result of one or more of the events listed in Clause 26.2. Bearing in mind that one of the exercises which the Architect or Quantity Surveyor may do is allow loss and expense, which has not yet been incurred but which is merely 'likely to be incurred'; in the absence of crystal ball gazing, they cannot be certain precisely what will happen in the future but they need only to be satisfied that the loss or expense will probably be incurred.

While it is possible to include wording in a contract to require certainty in ascertainment and to require the contractor to submit physical copies of records in support of a claim, that was not what the wording in the *Lilly v McKay* contract required. As always it is imperative to read the contract carefully and to understand its terms and their meaning and any implications of the local substantive law, and to have accurately included in the contract the standards and procedures the parties require to be satisfied and followed in ascertaining a contractual claim.

4.2   Tender Documents and Information

The status of documents provided for the purposes of obtaining or submitting a tender for a project may have a variety of standings, if a contract is subsequently entered into, depending upon the terms of the contract and the circumstances. For the purposes of assessments of sums due as additional payments, it is of course often necessary to understand the status of the various contract documents and the discussions and exchanges that inevitably surround the tendering process.

4.2.1 Entire Agreements

The first port of call, as always, is the terms and conditions of the contract itself. It is common for tender and enquiry documents to be incorporated into a contract by reference, without any thought as to their relevance or relationship to other documents also incorporated into the contract, particularly those produced much later in a lengthy pre-contract period. The result can be contradiction and inconsistency. In particular there is a not uncommon habit of contracts including every relevant preceding document through the tendering and selection process, seemingly on the assumption that it is better to ensure that nothing is missed than carefully consider what is actually required. Where the contract is the result of lengthy negotiations, offers, counter offers and revisions to scope and details, this can result in contracts that are far longer than they should be. This will almost inevitably result in disagreements between the parties on those details and unnecessary difficulty in seeking to resolve issues such as what the final contracted design or specification was for a particular part of the works. In a striking example of this, the contract documents for a major new railway system comprised over 8000 pages, excluding drawings, and spanned a four year pre-contract period of enquiry and negotiation. The result was to create a huge burden on anyone investigating those contract documents in relation to any contentious issues as to price, design or programme. Inordinate amounts of time and costs were required to search and find all documents relevant to any point at issue, and when such exercises were completed they usually unearthed ambiguity, contradiction and confusion.

It is not unusual to find that a clause has been inserted to the effect that the signed terms and conditions, and any other referenced documents, comprise the whole agreement between the parties and no other documents, exchanges or discussions concerning the tender and contract are to have any relevance. Such clauses are often referred to as 'entire agreement' clauses as they have the effect that the entire agreement between the parties is to be found in the referenced documents. An example can be found at clause 28.8 of the General Conditions of Contract issued by the CRINE Network (Cost Reduction in the New Era) in June 1997, superseded by the LOGIC Marine Construction contract in Edition 2 in 2004:

> The CONTRACT constitutes the entire agreement between the parties hereto with respect to the WORK and supersedes all prior negotiations, representations or agreements related to the CONTRACT, wither written or oral. No amendment to the CONTRACT shall be effective unless evidenced in writing and signed by the parties to the CONTRACT.

NEC4-ECC clause 12.4 puts this much more succinctly as 'The contract is the entire agreement between the Parties'. It is notable that the FIDIC Red Book, the Infrastructure Conditions and SBC/Q do not include such a provision, although parties are often advised to amend those standard terms to include one.

In section 2.1 of this book mention was made of the decision in *Strachen & Henshaw v Stein Industrie (UK) Ltd* in which pre-tender representations relied upon by the contractor were excluded from the contract between the parties by the entire agreement clause included in MF/1. The subsequent cost implications of the pre-contract representations being changed were onerous for the contractor, but the terms of the contract meant that it could not, and strictly should not have, relied on information excluded from the contract documents. For the pre-contract representations to be effective in that particular contract, and to then be available as information used by the contractor in formulating its tender costings and prices and the basis for quantifying a claim for an increase in these, they should have been incorporated in the documents referenced as being within the 'entire agreement'.

In circumstances where there is no 'entire agreement' clause, such information relating to factual issues provided in connection with the tender process may, or may not, be relevant depending upon the particular circumstances and the underlying law. As with 'entire agreement' situations the information, if not subsequently incorporated into the contract, may have no legal effect at all; alternatively, it may constitute a representation relied upon by one of the parties or it might form a collateral warranty with regard to the information provided. In such circumstances the representation or warranty might give rise to a claim if it can be shown that the information was mistaken or wrong such as to constitute a 'negligent misstatement' which induced one of the parties to enter into the contract. Local legal advice is essential where a claim is being considered on this basis.

4.2.2 Misstatements and Misrepresentation

Negligent misstatement was considered by the Court of Appeal in *Howard Marine and Dredging Co Ltd v A. Ogden & Sons (Excavations) Ltd* (1997) 9 BLR 34. Ogden had wished to hire barges from Howard for use in connection with some excavation works, and Howard's manager misstated the deadweight capacity of the barges, stating it to be 1600 tonnes when in fact it was only 1055 tonnes. The 1600 tonnes was based on the manager's recollection of a figure in the Lloyd's Register of Shipping which was incorrect. The correct capacity could have been obtained from Howard's shipping documents. Howard was held liable for the damages arising as a consequence of the misstated capacity under the Misrepresentation Act 1967.

It is possible that similar liability could arise in connection with information concerning, for instance, the ground conditions or general state of a site, but each instance will depend upon the particular circumstances of the contract. A common difficulty arises in connection with ground condition information such as borehole data provided by an employer in connection with the site of the works. The contractor is entitled to use such information in calculating its tender but to what extent can it hold the employer liable if the general, or particular conditions, of the site prove to be different to those suggested by the borehole data?

An example of such a claim arose out of the construction of a very long linear transport infrastructure project. The contractor had, in the employer's view, accepted the risk of the employer's late procurement of the many parcels of land on which the project was to be constructed. In addition, in disputing that risk, the contractor also argued that, if it had accepted that risk, it only did so on the basis of representations by the employer that it had already procured all of the land at contract stage. In fact, the employer had done no such thing and the late and piecemeal handover of parcels to the contractor to start construction became a major source of claims.

As a general rule it may be the position that the data provided by the employer will only be warranted as accurate in respect of what it actually purports to show. That is, borehole information will be warranted to be accurate in respect of the ground conditions at the location of the relevant borehole, but not necessarily elsewhere. Even if the contract terms and conditions do not expressly require the contractor to interpret information provided by the employer, and to make whatever further enquiries and investigations may be prudent and necessary, there may be an implied term that the information will be used by a reasonable, competent and experienced contractor versed in the vagaries of such data on that basis. It would not therefore be possible for a contractor simply to place total reliance on information provided for tendering purposes in the absence of his own proper enquiries and interpretations. It is for this reason that clauses such as clause 12 of the ICC Conditions should always be read in conjunction with the preceding clause 11. Clause 11 expressly sets out the contractor's obligations in connection with the provision and interpretation of information relating to the site and ground conditions. Any subsequent claim for additional payment as a result of adverse physical conditions that concerns ground or site data under clause 12 may first have to address the extent to which the contractor fulfilled his obligations under clause 11.

4.2.3 Mistakes in Tenders

The effect of a mistake by a contractor in calculating an overall tender sum based on unit rates inserted in a bill of quantities or schedule of rates will differ depending upon the nature of the contract.

If the error is one of arithmetic in extending the unit rates into the total sum(s) then the effect of it may disappear in a remeasure and value contract such as the Infrastructure Conditions and the FIDIC Red Book. This can be rather irksome for an employer when a significant error in the arithmetic of the bills of quantities on a remeasurable project contributed to that contractor being selected on the basis of price. Effectively, even with no change to the design and hence quantities, remeasurement applying the bills of quantities rates to the same quantities, but calculated arithmetically correctly, will lead to an increase in the contract price. Employers might argue at that point that, for example, where an item was priced at US$10 per metre, but extended at US$1 per metre, then the real rate in the contract was the US$1 per metre. The rights and wrongs of such an issue may depend on local applicable law and how it deals with such errors. The best advice that an employer can receive is to ensure that its advisors carry out a thorough arithmetical check on tenders as they are received.

If the contract is a 'lump sum' contract such as the SBC/Q Form, where the rates are used to value variations but the original scope of work is not subject to a remeasurement and revaluation at the contract rates, then the contractor is likely to be held to the error without any adjustment in its final account.

Where the error is not one of arithmetic in the tender but is an error in the calculation of the rate itself, the contractor will generally be bound by his error, as discussed in more detail in Section 5.2 of the next chapter. This issue cuts both ways, of course. If the contractor patently overpriced an item, then it will receive a windfall if the quantities of that item increase significantly. This occasionally leads to allegations of tendering claimsmanship on the contractor's part – i.e. that it spotted that the contract quantity for an item was unrealistically low and priced that item unrealistically high in order to gain a bonanza when the quantity was later corrected upwards. On the other hand, if the contractor patently underpriced an item, then it will suffer if the quantities of that item increase significantly. One approach in practice where this occurs is that the parties agree that it is reasonable that the contractor is held to the erroneous rate only for the quantities in the contract and that additional quantities should be the subject of a corrected rate. The correct legal position is likely to be a matter for the local substantive law, as discussed in Section 5.2.

4.2.4   The Conditions for, and Character of, the Works

The difficulties of determining the contract 'conditions' and 'character' for work required under the contract, and consequent assessment of additional payment when those conditions or character change, were considered in Section 2.1.1 of this book, particularly in the context of the *Wates Construction (South) Ltd v Bredero Fleet Ltd* case.

That decision was helpful in setting out the sources of information that should be taken into account in assessing the contract 'conditions' for the work, and less helpful in defining just what constitutes a change in 'conditions'. The crux of the *Wates v Bredero* decision was that 'extrinsic' evidence as to what the 'conditions' might be was irrelevant and the sources of information to be used were the contract documents and tender information. This avoids the situation of requiring some form of 'mind reading' exercise when undertaking such assessments in that it is not relevant what the contractor might have known, or not known, but it is that which can be ascertained from the documents that is relevant. It will, however, always be assumed that the contractor has applied the knowledge that can be ascertained from the contract documents in the manner that could be expected of a reasonable, competentand experienced contractor.

4.3   Tender Calculations and Assumptions

It is common practice for contractors to assess claims for additional payment by using their tender calculations and inclusions as a basis from which their claims for additional payments are assessed. This is particularly so when addressing disruption and prolongation situations.

In relation to disruption, the philosophy often adopted is that resources required in excess of those contained in the tender submitted and accepted as part of the contract are claimed to be the result of the alleged disrupting or delaying events. Such approaches include 'total loss' claims in relation to labour disruption, comparing the actual total labour costs with that tendered, as considered in detail in Chapter 6. The difficulties with such approaches are set out therein.

In relation to prolongation, tender calculations are often used in two ways. Firstly, there is a need to establish the basis from which additional costs are calculated. For example, by noting that a tender allowed for a tower crane for 50 weeks, but that in the event it was on site for 60 weeks, giving a claim for 10 weeks of its costs. Of course, such an approach lacks any programming analysis to establish that the additional 10 weeks was the result of compensatory delays rather than, for example, underestimating in the tendered 50 weeks or the contractor's own inefficiency in actually requiring the tower crane for 60 weeks. This would need to be established.

A second common use by claimants of tender calculation in relation to prolongation claims is as a shortcut to the more onerous task of establishing their actual costs or expenses. This approach is also sometimes adopted by contract administrators to avoid the time and trouble of a proper audit of the contractor's actual costs. This approach often appears to be adopted by consultants appointed by employers for whom the only selection criteria was price and who have been appointed on a very tight fixed fee, with the result that the consultant avoids any additional work for which it will obtain no further remuneration. Whilst such an approach is likely to be wrong (unless the contract expressly provides for it), it is always open to the parties to agree this as an expedient and pragmatic approach. It might also be favoured by an employer where the actual costs are greater than those tendered.

Leaving aside for the moment the connected issue of global claims, which is dealt with separately elsewhere in this book, there are obvious objections to the contractor's tender inclusions being the base from which additional payment can be calculated. The most obvious objections include:

- The contractor's tender may contain errors of calculation for which the contractor is responsible.
- The contractor's tender may omit resources for matters that it should have included within the tender.
- For whatever reason the contractor may have made over- or underoptimistic forecasts of labour or plant productivity which underlie the calculations in the tender.
- The contractor may have made a commercial adjustment to its tender to reflect the perceived market at the time of tender, which results in the sums included in the tender not reflecting the full amount of the resources required.
- The contractor may have made errors in assessing the programme for the works at the time of tender, which would impact on the resource inclusion in the tender.

For all these reasons, and possibly others, it is not reasonable to accept without qualification the actual tender calculations or inclusions as the starting point for the assessment of additional payments. Simply to adopt the tender calculations runs the risk of overstating, or even understating, the additional amounts being calculated. This is particularly the case in relation to claims that equate to damages for breach of contract and are intended to place the contractor in the position it would have been in 'but for' the breach or event that forms the basis of the claim. In such cases the 'base' should be those amounts that would have been incurred by the contractor 'but for' the breach or event, and these are most unlikely to be the same as those allowed by the contractor in its tender.

The correct approach in any situation where it is considered necessary to use 'planned' or 'programmed' resources from a tender as the basis for an assessment is to conduct an independent and objective appraisal of the tender calculations to consider what would have been included by a reasonable, competent and experienced contractor in a tender for the works as specified and set out in the contract documents. If the result is that the contractor's actual tender calculations and inclusions are shown to be a reasonable and acceptable then all well and good. Where divergences are shown, the contractor's tender calculations should be amended and updated to reflect the differences identified by the objective review, with the modified tender calculations, etc., used as the base. This is a topic considered in detail in relation to disruption claims in Chapter 6.

The process that has been described can be a straightforward and simple task or can require considerable recalculation of the tender resources and assumptions underlying them, depending on the extent of any variances revealed by the review of the actual tender.

The traditional approach that has been adopted by many construction and engineering contracts, of using a contractor's rates and prices from its tender to value changes, particularly variations, means that even where those rates and prices have not been required to be set out in bills of quantities or a schedule of rates, either or both parties may seek to use those same rates and prices, or at least the principles applied in the tender. This is sometimes referred to as 'analogous pricing'. To serve this end, some contracts expressly require that a successful contractor will, even where it has bid on a lump sum basis, in a set time period after award of the contract, disclose the details of its original pricing of the contract price. In practice this can lead to a number of difficulties. It may transpire that the contractor never does submit those details, or does so in a form that is considered insufficient by the employer. If the contract contains no sanction for this, it may be insoluble. In other instances, build-ups are provided, which raise eyebrows among the employer and its financial advisers at the small levels of costs and high percentages for preliminaries, overheads and profit apparently allowed, the contractor then seeking to add those high percentages in valuing any variations based on its costs.

4.4 Cost Records

The principal source of cost information on a construction contract is the cost documents and accounting records of the contractor for that project. The starting point of the analysis of such records may be the project's overall cost ledger or cost reports, recording the costs and accruals that have been allocated to it. Each of those allocated costs should be the subject of primary cost records.

Those primary cost records include such items as requisitions, orders, invoices and payments for materials, plant, equipment and subcontractors and payroll data for labour and staff. Such records are particularly relevant to claims for delay and disruption, where these are to be valued on the basis of 'cost', 'loss' or 'expense'. However, they are also relevant to variations under contracts such as those of the NEC suite, which dictate that all aspects of compensatory events should be valued based on actual costs or a forecast of costs. Other forms of contract may also dictate their use for variation valuation under certain circumstances, rather than the primary approach of applying contract rates and prices. Contracts such as the FIDIC Red Book (clause 13.6), the Infrastructure Conditions (clause 56(4)) and SBC/Q (clause 5.7) all include a Dayworks provision, and these require the production of invoices in support of, for example, the prime costs of materials. In addition, clauses such as FIDIC Red Book clause 12.3 require that if the contract contains no relevant rates or prices, then variations are to be valued on the basis of 'Reasonable Cost', with that capitalised term given its own definition in clause 1.1.4.3.

The contractor's cost records in respect of purchases and other expenditure are sometimes regarded as primary source documents that can be used on their face value, with little or no further consideration. There are, however, a number of points that should be noted first.

4.4.1 Identification of Invoices

It should go without saying that any invoice used as a source of costing or pricing information should clearly identify the project to which it was delivered, its cost allocation to that project and, in an accounting system of any sophistication, its allocation to a cost code within that project's accounts. It should also identify the details of the people, material, goods, plant, equipment or other resource supplied, with a clear description where relevant, and identify the point of delivery either on the invoice itself or by reference to separate delivery or acceptance documentation. It is surprising how often such documents are deficient in one respect or another, but on a project of any substantial size there should be sufficient systems in place to ensure that such proper invoicing and allocation records are available, not just to the project but to a cost code within that project.

Further considerations may be to ensure that the invoices have actually been paid, and properly paid under the terms of the contract between the paying contractor and invoicing subcontractor or supplier. This may mean following the full paper trail through its stages, including initiating site requisition, quotations, negotiations, order, delivery notes, invoice, approval and proof of payment.

However, there is also a point, either in terms of cost or amount, below which it is not always reasonable to expect a complete paper trail of substantiation of every item. Any approach to substantiating claims for additional payment has to keep in mind that the amount of time and money expended in assessing the payment should be proportional to the sums involved. While it will usually be the case that items costing thousands of pounds or more should be capable of full documentary proof of cost it will also usually be the case that lesser sums, or those related to incidental materials and equipment, will not be subjected to the same level of documentary evidence. In such cases it will often be sufficient that, for example in the case of a claim for prolongation costs, items are sample checked through their full documented history and/or only larger items are looked at in such detail. A consideration here may be whether the level of costs looks reasonable, such that there is no need to check that the amount claimed for a cubic metre of concrete matches the amount paid and that this is in accordance with the contractor's contractual obligations to its supplier.

It is also sometimes argued by recipients of claims that invoices for bulk materials should be identified to the precise location of each delivery in the finished project works, e.g. each delivery of ready-mixed concrete should be identified to the pour or section of the structure in which it was used. Unless the terms of the contract expressly require that such specific allocation of cost, say to a variation, is made, this is generally unreasonable and should not be considered unless a 'gross error check' reveals that the total amount claimed plus a reasonable allowance for wastage, etc., significantly exceeds the amount that could be required for the works.

4.4.2 Discounts and Credit Notes

The definition of cost should be, in the absence of any specific definition in the relevant contract, the amount actually paid for the goods and services. Therefore, the amount required in respect of supplies will usually be the invoice cost 'net' after deduction of trade, cash or any other form of discounts if applicable. Most invoices will usually state the amount of any types of discount relevant to that invoice. Thus a trade discount should be deducted as should the cash discount, which is available for commercial purposes. There may, however, be occasions when despite cash discounts being shown on an invoice, they are lost due to non-conformance with the required payment terms. An issue may then arise as to whether that loss of discount ought to be reflected in the contractor's claim for those costs as it is passed up to the employer as part of a contractor's claim. Some enquiry may therefore be necessary on occasion to establish the extent to which discounts shown on an invoice were in fact obtained. Comparison of invoice amounts with payment records will usually establish this. However, if a discount is deducted by the contractor and later disputed by the subcontractor or supplier, and credited to the subcontractor or supplier, establishing this may be more difficult. Since it will be the contractor that wants the loss of that discount to be accounted for in its claim, it should be expected to show such later adjustments to its incurred costs.

It is also not unknown for invoices to reflect a level of charging for materials or supplies with a discount obtained by the issue of parallel credit notes for the discount element. For instance, supply of a material may be shown on invoices at, say, £50 per m³ but a credit note for, say, £2 per m³ is issued separately for each invoiced amount. If instituted from the agreement of the supply contract, such a practice can only be viewed with the utmost suspicion and, if detected, may raise questions as to the veracity of other accounting records. Care should be taken, however, to distinguish between such occurrences and the issuing of genuine credit notes that reflect the agreement of a reduction in supply price for commercial or technical reasons, often occurring after the supply has commenced.

4.4.3 Bulk Discounts

A more difficult area to check is that of bulk discounting against invoiced amounts by a supplier, based on a volume of business over a particular period, often a financial year.

The principle of such arrangements is that the supplier offers a discount, usually refunded periodically or at the year-end as a credit against the contractor's account, providing a specified level of business is achieved. More sophisticated arrangements may have a sliding scale of discount so that the greater the levels of business the greater the amount of discount that is earned.

Such arrangements may reflect reasonable commercial incentives both to the vendor and purchaser but there are two problems with such arrangements when deciding if they should be taken into account when assessing additional payments or pricing of claims by the contractor against the employer:

1. Such an arrangement will usually only be known to the contractor and no mention of it may occur in the initial invoicing or cost record. It is usually a system monitored and operated by the accounting function of the contractor. Given that it will usually be confidential to the contractor there will be many occasions when its effect is not declared or taken into account when assessing payments and prices. On the other hand, the contract between the vendor and purchaser should record that the purchase is subject to such a bulk discount arrangement. This illustrates the need to not take invoiced amounts on their face value, but to look into the full paper trail surrounding them. It might then be that if an invoice refers to an order or contract number, the terms of that order or contract can be reviewed and show the discounting arrangement.
2. Even in situations where the existence of a bulk-trading discount is known it may not be possible to show that it applies to a particular project in isolation. For instance, if a bulk discount of 2% is available to a contractor who purchased more than 5000 m³ of concrete from a particular supplier over a 12-month period across all of its projects,

then, even if the 5000 m^3 threshold was achieved part way through the period, is the discount relevant to a project for which only 1000 m^3 was supplied in the period? At what point was the threshold superseded?

There is no easy answer to either of these problems. They are simply matters that personnel involved in assessing claims for additional payments on behalf on the employer should be alert to if any relevant information becomes available.

It is suggested that the answer to the second problem is that, if the discount applies to all quantities, including the 1000 m^3 supplied to the particular project, then the discount should be taken into account in relation to that amount as well. If the discount was only available for amounts supplied in excess of the threshold, rather than all amounts, then the amount of discount should only be taken into account in relation to such part of the 1000 m^3 to which it actually applied.

4.4.4   Coding Systems

Most contractor's cost systems will adopt a coding system for the allocation of cost amounts to the project accounts. There are various standard cost codings in use, and many more non-standard systems. In the event that cost records have to be analysed in detail then it will often be necessary to have access to the coding system used so that the allocation of costs from the 'raw' data to the accounting system can be understood and necessary adjustments made. Such codings can be particularly helpful in the allocation of a contractor's costs to its variation claims.

It is not uncommon for costs to be wrongly costed to a code, as a result of human error and this may be forgivable. However, it may also be thought tempting, where a cost code is specifically created to cover a major variation, for those posting costs to codes to use that variation cost code, even for items that are not correctly allocatable to it. Alternatively, even where the cost code was not specifically set up to cover a variation, it might through circumstances become the basis for valuation of a major variation. For example, on a major new railway project, one of the stations was subject of such extensive change that an arbitral tribunal agreed with the contractor's assertion that its many variations, disruptions and delays meant that the station could only be valued globally as a whole and based on the overall actual costs of its construction. In that case it was suggested that significant parts of the project's costs had been wrongly costed to that station's cost code. In either of these situations, suitable checking should be carried out. In the latter case, the defendant argued that the station's cost code was rendered unreliable to be used as the basis for a valuation as ordered by the tribunal.

4.4.5   Timing of Costs

One of the obvious, but sometimes overlooked, vagaries of cost accounting systems is that they record the cost usually either by invoice date or by the date that the cost was entered into the system. This can cause problems if expenditure in particular periods is being examined, as the period required will usually be a period relevant to the works, not the invoicing system. This is particularly relevant to claims for prolongation costs, where it is necessary to establish running costs of preliminaries and general items incurred at specific times.

For instance, if the costs incurred in the calendar quarter 1 October to 31 December of a particular year are to be analysed, it is most unlikely that simply referring to the cost records produced for that period by an accounting department will provide the required information for a prolongation claim. The most common examples of these are utility bills for water and electricity. The cost-reporting system will show the amounts booked into the system during that calendar quarter, but not necessarily incurred in the period. Some of the amounts in the cost records will probably have been incurred in the period but other amounts will refer to preceding periods, reflecting the variable time lags between an item being supplied or consumed on site and the cost being invoiced and appearing through the accounting system as booked to the project. Similarly, some of the costs for the resources used in the calendar quarter will not appear in the cost system until subsequent periods, and there may be advance payments made for future supplies and services. Similar considerations apply to some staff and labour costs. For example, project and company bonuses paid on an annual basis may need to be spread across the whole year over which they were earned, and not just considered as a cost of the December payroll in which they were paid.

A further issue that arises out of costs only appearing in a project's costs when they are booked to it arises out of disputed costs claims, particularly from subcontractors. This can be particularly relevant to preliminaries and general items supplied and maintained by subcontractors, with scaffolding being a common example. If the scaffolding company is making claims for extended hire costs (and also adaptations that can also become time related) and the contractor is considering those claims without making provision for them in the months to which they factually apply, then when such claims are resolved and paid, this can lead to a large apparent spike in the contractor's preliminaries and general item costs. This would not reflect the contractor's actual costs and its timing, but only when the costs were booked, and would require detailed attention to properly allocate such costs to their relevant periods.

When using such data for analysis it is therefore usually necessary to fully investigate and correct the records so that the costs refer to the period of the works and not the period in which they were applied into the costing system.

4.4.6   Cost Transfers and Accruals

One of the other distorting features that commonly occur in costing systems is the incidence of internal cost transfers and accruals for amounts anticipated but not invoiced or charged in the period. Usually these will involve sums paid centrally and then allocated to the various projects periodically, or will be for items supplied by one project to another within the same organisation. The former will often involve project staff employed through a central office, but whose costs are distributed to the projects, while the latter will often refer to plant, equipment or material transfers between projects.

The potential problem with such transfers is that, in a similar vein to the example given in relation to hired staff, they may result in substantial sums appearing in a particular period in the costing system although the resources were supplied or used in a different period. In some instances it is not uncommon to find such substantial transfers at the commencement of projects, or conversely in the latter stages, as costs are moved about either in anticipation of usage or to reflect past usage.

As with the timing of cost booking, this is another factor that will require painstaking investigation if such related records are to be used for analysis for a prolongation claim, although this can be similarly relevant in relation to a claim for a major variation.

Another feature requiring attention may be the internal hiring of plant and equipment from another business within the same company or group of companies. The plant hire business may be just a separate division, or a separate registered limited company within the group of companies, that hires plant and equipment either on the basis of internal cost transfers or invoices to other parts of the construction and engineering divisions or companies of the group at internally set hire rates. This may also occur in relation to project staff and labour. This may cause some controversy as to the commercial reasonableness of those internal rates compared to those available in the external market. An example of this is a major UK contractor which contained within its group a consultancy business hiring project staff by internal cost transfers to each project at very high hourly rates, and which was renowned as significantly the most profitable part of that major UK contractor's businesses. Where the costs are the subject of an internal cost transfer, then the response to such an approach will include consideration of whether the hire and construction divisions are not the same company, such that its real cost or expense is the cost to the hire division and not the amounts internally charged to the project. Where the hire comes from a separate limited company, there may be two considerations: firstly, the extent to which they are operating on an 'arm's length' basis and, secondly, whether the rates charged are reasonable. It is likely that the inclusion of such costs in claims under the contract is subject to express or implied terms that they are reasonable. If they are being artificially exaggerated within a group of companies, then they should be adjusted accordingly to reflect reasonable market rates for the same resources, be they plant, equipment, labour or staff, for example.

4.4.7 Final Accounts and Economic Duress

Another prime source of information contained in the records of any project will be the final accounts agreed by the contractor with its various subcontractors and suppliers. Many such accounts are documented to a reasonable degree and may then provide a final figure that is a compromise between the total sum requested by the subcontractor or supplier and the amount considered appropriate by the contractor. The final figure paid may therefore not be the sum of the calculations in the particular account but may be a different figure. This can cause some difficulty if parts of the subcontractor or supplier account have to be used to calculate the contractor's claims under the contract with an employer. Unless the documents provide some clue as to how the final figure was reached, and hence how it should properly be allocated, it may be that there is generally only one reasonable way to proceed:

1. Firstly, identify all the sums included in the subcontract or supplier account on an agreed basis.
2. Identify the total sum paid in excess of the total of the agreed items. This is the sum paid for 'contested' items.
3. Average the amount paid for the 'contested' items over those items in proportion to the sums requested for each item.

In the absence of better information this is the only way to produce a 'costed account' where the final figure has been the result of a 'horse deal', unless there are meeting minutes or correspondence relating to the contested amounts that allow some weighting of the distribution to be made on the basis of sums that were less contentious than others. Such records might even show, or sufficiently suggest, that some items were so contentious that they should be considered at nil cost, meaning that the effects of the 'horse deal' only need to be spread across a smaller number of unagreed items.

Even where an account had not been the subject of an overall 'horse deal', it is common that different parts of the final accounts of subcontractors and suppliers are subject to varying degrees of substantiation and detail. For example, a supplier's final account might schedule at length the quantities and rates for materials supplied, but end with lump sums for such as air freighting, extended warehousing, or standing time of delivery lorries and drivers. Similarly, a subcontractor's final account might schedule at length the quantities and rates for work done, but end with lump sums for such as variations and claims for prolongation and disruption. Problems will arise where these less particularised items are included in a claim by the contractor to the employer. The employer's advisors may consider them so lacking in detail that they are excluded in their entirety from their valuation. The contractor will say that it is unfair that such items should be discounted by someone with no detailed knowledge of the discussions and negotiations that led up to its agreement to pay such items. Also, they suggested that it would not have agreed those amounts if it did not consider them valid. A further reality of the management of accounts from suppliers and subcontractors (unless they are relatively large and sophisticated companies) is that their claims for such as prolongation and disruption are rarely as detailed as those of main contractors. The position taken by the employer's advisor is likely to be that it cannot agree sums that lack details as to what they were for and that they were reasonable and also that the contractor has only itself to blame if its negotiations with its subcontractors and suppliers lack transparency and details.

Contractors are well advised to ensure that when they reach a final account agreement with a subcontractor or supplier, it is the subject of a detailed breakdown into each component item or claim, particularly where any of these might form the basis of a claim against the employer.

A common point of issue is where a contractor passes on to an employer agreed costs from a final account with a subcontractor with little transparency as to the basis of those costs. It may be that the employer regards those costs as excessive and wants details as to how the amounts were arrived at. The contractor's response may be that those are its costs, that it incurred them in good faith applying due commercial controls, and that the whole of those costs are the result of a matter for which the employer is liable. This is an issue considered in detail in Chapter 6.

In extreme circumstances there may be reasons why an apparently agreed and paid account is subject to challenge by a party to the agreement and adjustment of the amount paid, although such instances will usually be notified in the documentation of the account. In *Carillion Construction v Felix (UK) Ltd* (2001) 74 Con LR 144, the contractor had entered into a subcontract for the design, manufacture and supply of cladding from Felix. The contract stated that the delivery of cladding was to be complete by January 2000, but by February deliveries were still outstanding and Carillion were complaining about the delays. At this point Felix informed Carillion that future deliveries were dependent upon Carillion agreeing Felix's final account, although the work was not complete and the account had been submitted by Felix in a sum considerably in excess of that considered reasonable by Carillion.

In the absence of any viable alternative, and faced with substantial employer's damages for delay under their contract, Carillion reached an agreement with Felix to settle the final account. Following this settlement Carillion complained in writing to Felix that it considered the agreement had been reached under duress. It was subsequently held that the threat by Felix to withhold future deliveries pending agreement of its final account was an act of duress and the agreement was set aside.

In some, hopefully very limited, cases it may therefore be necessary to determine if the amount of a settlement figure is accepted and agreed by both and not the subject of ongoing challenge, before being used in any quantification of the contractor's claims up the contractual line to the employer, or even down the line to other subcontractors in the form of contracharges.

4.5   Accounting Information

There will be occasions when it is necessary to examine the wider company accounting records of a contracting party, for example to ascertain the level of relevant overhead expenditure incurred as 'off-site overheads'. In other instances it might be that examination is required to establish levels of overheads for 'reimbursable' contracts, or elements of contracts, or daywork agreements, where the applicable level of overhead has not been agreed in advance or it has been agreed that the overhead will reflect the actual level incurred.

In examining the accounting records there will usually be two types of information available: the financial accounts of the business and its management accounts.

4.5.1   Financial Accounts

These are the accounts that a business has by statute to produce and submit to the regulatory authorities of the country of its registration. Their core purpose is to publish to the wider world (and particularly the company's shareholders, creditors and clients) the health of the business in terms of turnover, profit and assets. They will therefore usually comprise two elements: the profit and loss account and the balance sheet. In some jurisdictions the degree to which the financial reporting of companies is publicly available is limited. Even where reporting requirements are more open, it may be that under local company rules those companies of a smaller size, in terms of turnover and/or assets and/or employees, will only be required to publish limited information, for example a balance sheet but not a profit and loss account.

A profit and loss account shows the income and expenditure for the business in the accounting period together with any adjustments or other charges (such as overheads/administration costs and interest charges) and/or other income (such as from property rental) required by accounting practice. The difference between the two being the profit, or loss, for the period is often referred to as the company's 'net margin' or 'bottom line'.

A balance sheet shows the available funds in the business and their sources and the usage to which they have been applied (for example as cash reserves or in the value of property, plant and equipment held), but it is relevant only to the date on which the balance sheet is produced, not to a period of business activity.

The difficulty with these types of financial accounts for analysis purposes in relation to that contract is that they will not be project specific but for the business as a whole and are, by necessity, somewhat brief and lacking in detail. This is particularly so in a large business where separate trading divisions or parts of the business may be aggregated together. The usefulness of these statutory accounts for analysis purposes is therefore often very restricted, but in instances where a detailed examination of accounting information is required, they may act as the 'proof' for subsidiary accounts that should be capable of being followed through to the company's financial accounts to demonstrate that they are genuine and reflect the statutory accounting data of the business as a whole.

To illustrate the lack of detail available in published financial accounts an illustrative set of figures, such as might be found in such company accounts, is set out in Appendix A.

The principal type of subsidiary accounting information that will be of interest in relation to the evaluation of contract claims is the management accounts explained as follows.

4.5.2 Management Accounts

The management accounts of a business are much more extensive and detailed than the financial accounts and will record the income and expenditure of the business in much greater detail. There will usually also be separate management accounts for different trading divisions or parts of the business. As the name suggests, these accounts are intended to provide a basic management tool for the running of the business and will therefore usually identify with reasonable precision the manner and extent in which the business, or the division, incurs its costs. This can be contrasted with the financial accounts in Appendix A, whose level of details is only intended to serve the regulatory authority's requirement that information is publicly available as to the overall financial performance of a business and its health and changes therein.

However, there is likely to be no statutory requirement to publish management accounts for public access. As a result, the recipient of a claim including such items as off-site overheads and profit may be able to access a company's wider statutory accounts in order to check that claim, but sight of management accounts in relation to other heads of claim will depend on willing disclosure by the contractor. Following the mantra that 'he who asserts must prove', it must be expected that such information will be disclosed to the extent that it is necessary to check a claim, otherwise it will lead to doubt as to the veracity of that claim.

The fact that these management accounts are used by the business managers to run the business means that they are often prepared for a particular financial period by the use of budget sums for some, if not all, categories of expenditure. For instance, categories of expenditure such as office rents, salary and wages costs for technical and commercial staff employed off-site and management personnel such as directors and the administration staff will appear as lines in the management accounts with their own figures. Prior to commencement of the accounting period the business managers will have undertaken a forecasting exercise to predict the anticipated levels of expenditure in each category for the period. These forecasts will be reviewed periodically and adjustments made to reflect the actual levels of expenditure based upon the actual level of activity and resources required. The timing of such adjustments may vary from company to company depending upon the management policy and it might be that adjustments are only made to the management accounting information on an annual basis. Whatever the periods of adjustments, the person using the information for analysis needs to understand that in such instances the figures for individual lines in the management accounts when completed for the period will show the predicted figure for the period adjusted to reflect the actual level of accuracy in the forecast for the preceding period.

For instance, using the figures from the example of management accounts included in Appendix B, the following items can be identified:

Rent and rates	£265,000.00
Administration staff	£751,000.00
Audit and accountancy	£114,000.00

Underlying these figures could be adjustments for the preceding year, so that the figures might actually have been calculated as:

	Forecast £	Adjustment £	Total £
Rent and rates	290,000.00	(25,000.00)	265,000.00
Administration staff	720,000.00	31,000.00	751,000.00
Audit and accountancy	152,000.00	(38,000.00)	114,000.00

Taking rent and rates as an example, this shows that the figure in the management accounts of £265,000.00 is made up of a predicted expenditure of £290,000.00, which has been adjusted by a credit of £25,000.00 for the preceding period. The adjustment indicates that the forecast for the preceding period was underspent by the amount of the adjustment, £25,000.00. The reason for the underspend may not be relevant if the preceding period is not of interest, but the relevant point is that expenditure on rent and rates is forecast to be £290,000.00 and not the £265,000.00 that may be shown in the management accounts.

However, this is not, of course, the whole story, as the forecast of £290,000.00 may itself be subject to adjustment in the succeeding period to reflect the actual expenditure in the period. If, for instance, expenditure in the relevant period is £305,000.00 then the succeeding period will show an adjustment figure of an addition of £15,000.00. The figure of £305,000.00 may not appear as a figure in the management accounts for any period but it is the actual expenditure for the period made up of the management account figure of £265,000.00 with the preceding period underspend of £25,000.00 added back, plus the overspend in the period as an adjustment in the succeeding period of £15,000.00.

Thus, what appeared to be an expenditure figure in the management accounts of £265,000.00 disguises actual expenditure in the period of £305,000.00. Adjustments may not always have such an effect, and the effect can be a reduction in the figure as well as an increase, especially in circumstances where trading and market conditions are reasonably stable and the level of activity is not showing significant variations. However, in periods where there are substantial increases or decreases in activity, both of which seem to occur in all too regular cycles in the construction industry, the possible effect of preceding and succeeding year adjustments to management account forecast figures is something that anyone conducting an analysis for the purposes of a claim needs to be aware of.

Companies may organise their accounts in different ways and not all management accounts will adopt such a system of forecasts and adjustments. Others may 'sweep up' adjustments periodically rather than carry them forward to succeeding periods, but the crucial lesson is to question the figures provided and the manner in which they are compiled in order to ensure they truly represent expenditure in the period under review.

4.5.3 Exceptional Items

In examining any set of financial accounts it should be recognised that the figures set out for any particular accounting period may not represent the true costs of the company's trading activities due to distortions caused by the inclusion of exceptional costs in the accounts. Such items can be in revenue or expenditure but may be the result of matters unconnected with the trading activity in the period. For instance, exceptional revenue may be recorded from the sale of an asset or exceptional expenditure may be incurred by the closing down of a division of the company or simply the writing down of asset values in the period as the result of overoptimistic allowances in preceding periods.

Before using such accounts as a source of information for analysis it may be necessary to determine by enquiry if there are any such exceptional items that require adjustment of the figures before they are used for such analysis.

4.6   External Information

In some circumstances it may be that information unrelated to the particular project is of relevance to the evaluation of changes to it. Such external sources might include, for example, pricing books, records from the contractor's other projects and data from the wider market.

Circumstances where such sources might be of relevance include those where the contract expressly provides for their use. For example:

1. In a simple bespoke contract for the construction of a small building, the variations clause gave no guidance on valuation beyond 'fair rates and prices' and the parties agreed to satisfy that provision by reference to rates and prices from a well-known pricing book for building works.
2. Where a joint venture contractor's claim for prolongation included extensive staff costs, its members considered that the salaries of their respective senior project staff were matters for confidentiality and therefore could not be disclosed and/or evidenced. Those staff were therefore priced and agreed in the claim for prolongation costs on the basis of known typical market rates for staff of their grades at that time.
3. Even with the express terms of contracts that make the primary approach to valuation contract rates and prices or actual costs, there may be some scope for the use of external information, in appropriate circumstances. For example, in the Infrastructure Conditions, clause 52(4)(b), there is a typical hierarchy of methods for valuing

variations in the form described in detail in Chapter 5. This ends by providing that, where the contract contains no rates and prices that can reasonably be used as the basis for valuing variations, 'a fair valuation shall be made'. It might be argued that this includes reference to pricing books, other projects or wider market rates. The test of relevance of these will depend on whether they give rise to a 'fair valuation' on the merits of each case.

A key point in using such external information is to ensure that it is reasonably applicable to the works in question and, if not, to make suitable adjustments to it. Related considerations will include whether the specification, design and circumstances of working are the same or sufficiently similar. This latter issue includes quantities, such that prices from a major works pricing book would not be relevant to the pricing of minor works to a house extension, for example. Timing may also be a factor, in relation to price escalation. A contractor may seek to rely on information from another project, but if that project was carried out at a different time, then a suitable adjustment needs to be made for inflation. If the projects are too far apart it may not be realistic to expect that applying indices to the costs of one to get to the reasonable costs of the other will give a reasonable result. The approach also requires that there are suitable cost indices that can be applied and that is not the case in many jurisdictions, where there are no published indices specifically related to construction works, material, labour and other resources.

Whilst such information from sources external to the instant project may be of relevance, this is usually the exception to the use of actual costs, expenditure, rates or prices from the project itself. Where it is used, great care needs to be taken to ensure suitability and make suitable adjustments to suit the claims and circumstances to which it is proposed to apply them.

4.7   Summary

There are several potential sources of financial information for the evaluation of change under construction and engineering contracts. The contract terms should dictate which sources are relevant to which claims, under which circumstances and whether adjustments are to be made to them.

Such contract terms will usually use terms such as 'costs', 'expense', 'rates' and 'prices'. These are not terms of art and it should not be assumed that typical industry definitions of them apply in each case. The contract may include its own definitions and these should be checked before assumptions are made.

Further definitions may be expressed to limit recoverability through terms such as 'ascertain', 'directly referable' and 'reasonable' and again any definitions of such terms should be understood.

Sources of financial information may include the contract, tender, cost records, accounting records and external information. The contract will dictate which of these are relevant and each requires care to ensure they are applicable and suitably adjusted to the claims and circumstances to which they are being applied.

Having considered in the last two chapters the critical appraisal of programme and financial information, it is appropriate to consider next how that information can be used in the analysis of adjustments to the contract sum in relation to claimed change.

Those adjustments are considered in two parts: firstly, the direct consequences of change, that is the unit rates and prices for work done, and, secondly, the time consequences of change, particularly to the programme of works in the form of prolongation, disruption and acceleration. These subjects are considered in the next two chapters respectively.

5

Evaluation of the Direct Consequences of Change

The discussion in this chapter concentrates on the evaluation of the direct consequences of changes made to the works. These consequences are usually changes in the type, quantity or specification of parts of the works. Such changes may be variously referred to by terms such as 'changes', 'variations', 'extras', 'additions', 'omissions' or 'compensation events', as they are called in NEC parlance. In this chapter we generally refer to them as either 'variations' or 'changes', using these terms interchangeably. Traditionally, the construction and engineering industries have usually addressed their valuation in the form of claims for new unit rates for the varied work, or changes to those contract unit rates to reflect the effect of the changes. Most construction contracts, whether of 'lump sum' or 'remeasurement' type, contain provisions by which changes to the contract unit rates can be made, or new rates introduced, so that changes can be evaluated within the contract mechanism. In addition, many contracts contain provisions for revisions to the unit rates relating to work that is not itself changed, but is affected by changes to other work that has been changed and paid for under other rates. There may also be some mechanism for making related adjustments to the preliminaries and general items or 'indirects' charges, as they are often referred to in contracts for engineering construction works, where such adjustments are justified in addressing a change.

The traditional approach has also sometimes resorted to the use of the contractor's actual costs, rather than rates and prices that have been set out in the contract. This is usually only taken in appropriate circumstances which are explained in this chapter. However, recent years have seen increasing use of actual costs to evaluate the direct costs of changes, with some contracts making these the only basis for valuation.

This chapter starts by looking at the distinction between the rates and prices as opposed to an actual costs approach to valuation and then look in detail at each in turn.

Recent years have also seen increasing provision for the pre-agreement of the value of changes before they are instructed and/or carried out. This may involve a proposal or quotation from the contractor and might be based on either rates and prices or (a forecast of) actual costs or any other basis that the contractor choses, with the employer or its agent able to agree it or reject it, just as it did the contractor's original tender.

As always, the person undertaking the evaluation of the direct consequences of changes to the work must start by thoroughly studying and understanding the contract terms and conditions and the implications of all the contract documents, as well as the local substantive law. There are often alternative methods of evaluation depending upon the circumstances of a particular change. Only by properly understanding the

Evaluating Contract Claims, Third Edition. John Mullen and R. Peter Davison.
© 2020 John Wiley & Sons Ltd. Published 2020 by John Wiley & Sons Ltd.

law, the contract documentation and the valuation mechanisms contained therein will the appropriate method be identified and employed.

There is often a problem of scale in applying logical and reasonable analysis to the totality of activities on site, particularly on large projects with many trades and possibly hundreds of individual activities. The nature of many large-scale projects is that they are of long duration and have substantial parts whose design at the contract stage is significantly different to that at the final account stage. Thus, for example, on a project for construction of a new terminal and concourse to a major international airport, the mechanical, electrical and plumbing packages alone were subject to claims for over 3000 variations. Such problems of scale can occur not only in evaluating changes to the unit rates and prices for the work, but also in the evaluation of time impacts, as discussed in the next chapter. It is this problem that sometimes encourages contractors to adopt the global approach to claims. The global approach is rarely, however, an antidote to the problem of large-scale change, disruption or delay but may be applicable in limited circumstances. This is considered fully in Chapter 6 of this book.

The problems of scale may need to be addressed by identifying the most appropriate level at which to apply the analysis and adopt testing techniques such as the 'but for' test discussed in Chapter 6. If analysis at the individual unit rate level can be shown to be unrealistic or impossible then a higher level of analysis such as a trade activity or section of the works may be appropriate if the law and contract allows it or the parties agree it. It should, however, always be the case that the selection of the appropriate level of analysis should be for pragmatic and common sense reasons and not merely because it is hoped it will disguise any contractor liability issues that might be revealed by a more detailed level of analysis.

5.1 Unit Rates and Prices or Actual Costs?

Having properly established the scope of a variation, construction contracts will usually expressly define how it is to be valued. As this is put in *Keating*[1]:

> Extra work of the kind contemplated by the contract will be paid for in the manner provided by the terms of the contract.

That 'manner' will generally adopt one of two alternative approaches to the valuation of variations. Examples of such approaches adopted by international standard forms of construction contract in common use around the world are set out in turn elsewhere in this chapter. The first and more traditional of these approaches applies a hierarchy of different methods that starts with rates and prices agreed and stated in the contract documents, adjusts those where appropriate to differences in circumstances or conditions and reverts to such as market rates or the contractor's costs, where there are no agreed rates and prices or they are not capable of appropriate adjustment. An example of this approach that is explained elsewhere as an illustration is in the FIDIC Red Book.

---

1 *Keating on Construction Contracts*, 10th Edition, published by Sweet and Maxwell Ltd 2016, paragraph 4-084, with reference to the decision in *Thorn v London Corp* (1876) 1 App. Cas. 120.

The second and newer approach is based on the contractor's actual or forecast costs; an example of this approach explained in this chapter is the NEC4-ECC contract.

A common theme of such contracts is that they require that the parties pre-agree and set out in the contract documents not only the method to be followed in valuing variations but also data that can be used in that process. This is explained in more detail elsewhere, but, broadly, such contracts as in the FIDIC Red Book require agreed rates and prices, whilst contracts such as NEC4-ECC require certain percentages that will be applied to the contractor's actual costs. Thus, whilst they may require different data, such contracts all require some form of agreed data. This gives the obvious benefits in terms of ready agreement as opposed to potential disagreement if such valuation components have not been pre-set.

Whilst this chapter will consider price-based regimes as a traditional approach generally adopted by such as that of the FIDIC (with the example of its FIDIC Red Book), the ICE (with the example of the Infrastructure Conditions that succeeded them) and the JCT (with the example of SBC/Q) and distinct from cost-based regimes as the new approach of the NEC contracts (with the example of NEC4-ECC), the demarcation is not that simple. The FIDIC Red Book, the Infrastructure Conditions and SBC/Q all contain some provisions for the use of actual costs in evaluating the direct costs of change in limited circumstances.

Furthermore, parties are free to agree any combination of rates and prices or cost-based approaches as they see fit. In a contract for a marine project in South America, the contract provided for two alternative approaches to valuing variations depending on their size. Minor variations were defined as those whose value was less than $20,000, which were agreed to be valued by applying rates and prices set out in an annex to the contract. Variations whose value was above the $20,000 threshold were agreed to be valued on the basis of the contractor's actual costs. Potential difficulties with this approach arise as the parties to a construction and engineering contract can, and all too often do, disagree over the value of variations. Thus, they might disagree as to whether a given variation was above or below the threshold and therefore which valuation approach should apply. Even more intriguing is the potential that the different approaches to valuation might result in values on opposite sides of the threshold. In Section 5.6 we set out an example of how the application of different rules to the valuation of the same variation can result in strikingly different values.

5.2   Unit Rates and Prices

Keating also explains how contracts usually contain expressly listed contract rates and prices, and the approach if the contract does not. This can be regarded as the traditional approach adopted by construction and engineering contracts:

> Payment will usually be at or with reference to the contract rates. If there are no relevant rates, it will be a reasonable sum.

As to what constitutes a 'reasonable sum', Keating also puts this as follows, at paragraph 4-037:

The courts have laid down no rules limiting the way in which a reasonable sum is to be assessed.

This continues in paragraph 4-040:

Useful evidence in any particular case may include abortive negotiations as to price, prices in a related contract, a calculation based on the net cost of labour and materials used plus a sum for overheads and profit, measurements of work done and materials supplied, and the opinion of quantity surveyors, experienced builders or other experts as to a reasonable sum.

These possible methods of arriving at a 'reasonable sum' are discussed in this chapter, but it is notable how Keating's approach recognises that this may be based on prices in a related contract (which are sometimes referred to as 'market rates') or the contractor's actual costs.

In Hudson the usual approach, of applying contract rates and prices, is explained as follows[2]:

There are many types of variation valuation clauses, differing somewhat in different jurisdictions, which provide for a method or procedure for valuation in the absence of an agreed quotation or price. In more sophisticated forms of contract, the great majority tend to be initially contract price based, using some sort of schedule of rates and prices, whether in unit price or fixed price (lump sum) contracts, which in lump sum contracts may or may not contain estimated quantities. In a classical English bill of quantities measured contract, the bills themselves fulfil this initial variation valuation function as well as the quite separate remeasurement function.

Like Keating, this quote from Hudson identifies how construction contract documents can expressly set out pre-agreed rates and prices to be used in the valuation of variations, but it also identifies the formats in which such rates and prices can be set out. This can be in the form of a simple unquantified schedule of rates or a bill of quantities, in which quantities are stated against each item and rate. A further feature of bills of quantities is that contracts that include them should state which published 'Method of Measurement' has been used in their preparation. Such methods of measurement direct how items are to be measured and described (for example whether by volume, area, thickness, length, weight or number) and what they are to include (for example if formwork is included within items for the volume of concrete or measured separately based on its area). There are many such standard methods of measurement published around the world, as further discussed elsewhere in this chapter.

The only use to which a contract schedule of rates that does not set out quantities can usually be used is in the valuation of variations. However, a contract bill of quantities, because it provides both rates and quantities, can serve several further purposes including: proper comparison of tenders at the pre-contract stage; calculation of interim payments; measurement of variations; and remeasurement of the works.

---

2 *Hudson's Building and Engineering Contracts*, 13th Edition, December 2010, paragraph 5-050.

A number of difficulties can arise from the use of schedules of rates, rather than bills of quantities, as contract documents setting out rates for the valuation of variations. These are particularly made worse where the schedule of rates is unquantified, and is purely that – a schedule of rates and prices. Examples include:

- How to apply such rates to changes in quantities, as discussed elsewhere in this chapter. If a rate in the contract does not say what quantity it is based on, how can it properly be adjusted if the quantity of the work increases or reduces significantly? Indeed, does a change in quantity mean a rate in a schedule of rates is to be considered for adjustment at all? Is the contractor to bear the risk that the actual quantities will be higher (or lower) than it envisaged when it set out its prices in an unquantified schedule? Such issues should be clearly set out from tender enquiry stage. If the contractor is properly advised of such issues, and is able to allow for the risk or opportunities it might create, then it may have no complaint that it considers a rate or price is rendered inapplicable by its actual quantities as constructed.
- The lack of a stated method of measurement for the items in a schedule of rates is likely to mean that its item coverage is unclear and therefore capable of disagreement between the parties. These range from simple questions, such as to whether an item includes all plant necessary for that work, to more complex issues, such as regarding item coverage, for example, of formwork for concrete works mentioned elsewhere. There are a myriad of potential examples of this, but consider, for example, whether a linear measured item for wrapping and lagging of pipes includes wrapping and lagging of fittings, valve and joints, an issue made clear by coverage rule C4 of The Civil Engineering Standard Method of Measurement, CESMM4, but potentially unclear if such a set of measurement rules are not incorporated.
- There may be no way of knowing whether a rate or price in a schedule of rates reflects the contractor's original pricing of its tender sum. This will be important on a contract where the intention is that variation prices are at similar levels to pricing of the original contract scope. It is an issue that employers and their advisors should be wary of at the tender stage, and a very good reason for them to use a bills of quantities approach, with quantities, rates and prices extended to come to the tender total. It is particularly common in relation to their subcontractors that contractors agree an unquantified schedule of rates as part of subcontract documents without considering this point and whether the rates in the schedule are competitive.

It is important that the contract should also clearly state to which of the various potential purposes any agreed rates and prices therein are to be applied. This is because contractors will usually price such contract documents differently, in terms of both pricing detail and pricing levels, depending on the use to which they are going to be put. It may be particularly unfair and inappropriate to use such a document for a purpose different to that which the contract documents expressly say it is intended. The structure and level of detail of a pricing document should match the use to which it is to be put. Thus at one end of the spectrum, a bill of quantities intended for the remeasurement and revaluation of the works on completion should set out full descriptions, quantities and rates for all items. At the other end of the spectrum, a price document intended only for bid comparison may only need to set out a series of lump sums against parts of the work.

The importance of stating in a tender enquiry document the uses to which a pricing documents is to be applied can be illustrated by considering a schedule of milestone payments in a contract. These are usually intended only for the use implied by their title – simplifying the process of calculating interim payments by reference to defined milestones and amounts against the achievement of each in turn. As a result of that use, the amounts may be front-end weighted by a pricing contractor to improve cash flow early in a project. Whilst such a document is unlikely to be of sufficient detail to be used to price variations, it should never be so used, unless it is stated to be for that additional purpose in the tender enquiry. In a concession agreement for construction of a new railway, the employer was faced with the prospect that an arbitral tribunal would find that the lack of any other basis on which to value variations left it only able to value a variation based on actual costs. It therefore argued that the schedule of milestone payments provided sufficient details. It also argued that a 'Base Case Financial Model', whose true purpose was completely unrelated to variations, included further detail that could be used to price variations. The tribunal disagreed, but not until after lengthy and expensive argument and analysis by the parties' legal teams and quantum experts in an arbitration of the issue. This could have been avoided if the contract had clearly defined and limited the uses to which those pricing documents were to be put.

As stated in the quote from Hudson set out elsewhere, contracts that apply to the valuation of variations pre-agreed rates and prices expressly listed therein usually apply these 'initially', with other methods applied where no such rates and prices are included in the contract documents or they are included but are inapplicable for some reason. This gives a traditional 'hierarchy' of alternative variation valuation approaches as follows:

1. Where there are rates and prices expressly agreed between the parties to be used to value variations and these are listed in the contract documents, they are applied where they are applicable in terms of similarity of specification, conditions and circumstances of the work in the variation to that envisaged by those contract rates and prices.
2. Where there are rates and prices expressly agreed between the parties to be used to value variations and listed in the contract documents, but they are not applicable in terms of similarity of specification, conditions and circumstances of the varied work to that envisaged in the contract, then new rates and prices are derived (with suitable adjustments) from those expressly set out in the contract. The resulting rates and prices are sometimes described as 'analogous rates and prices'. Some forms of contract also place a limit on the amount of change, or size of variation (expressed as a percentage or amount), that can be priced at the contract rates and prices before they should be adjusted to reflect that differing scope.
3. Where there are no rates and prices expressly agreed between the parties to be used to value variations, or those that are listed in the contract documents are not capable of suitable adjustment, a 'fair' or 'reasonable' valuation is applied. What is a 'fair' or 'reasonable' calculation will depend on many issues in each case. This could be based upon such sources as: market rates and prices from other projects; rates and prices from published industry pricing books; the experience of experts; or the contractor's actual costs of carrying out the variation, with suitable additions for overheads and profit. The contract should specify which of these sources applies. The applicable law may particularly also affect what is considered 'fair' or 'reasonable'.
4. In some circumstances, such as for variations of a minor or incidental nature, a 'Dayworks' approach is applied. Here the contractor's actual resources and hours used

on additional works are recorded and agreed contemporaneously. Expended labour, plant and equipment resources are then valued at hourly rates set out in the contract or referenced industry source for their 'prime cost' and materials are valued at their actual costs with the addition of a percentage mark-up for items such as overheads and profit, also set out in the contract.

Whilst in this book such a tiered approach to the methods of valuing variations is referred to as a 'hierarchy', it is sometimes referred to as a 'shopping list' or the different methods are referred to as 'limbs' or 'tiers'. The rules that prescribe which of these applies to any given variation are often referred to as valuation 'hurdles' or 'fences', the latter term being used in this book.

The explained hierarchy is only a typical outline of such rules for the valuation of variations. Different standard and ad hoc forms of construction and engineering contracts that initially value variations based on contract rates and prices will contain variants on these themes, which may be significant or quite subtle. In each case a careful read of the clauses of a particular contract is essential. Some such subtle differences between the FIDIC Red Book, the Infrastructure Conditions and SBC/Q are identified in this chapter when considering how they might each address a variation changing the quality of a bill item and adjusting the bill rate appropriately.

In a fully designed and traditionally procured contract there will often be a bill of quantities defining the works for pricing and setting out the measured units, with reference to a published standard method of measurement or other set of measurement rules contained or referred to in the contract documentation. For other contracts, including design and build projects where the contractor has responsibility for designing and defining the full extent of the works in accordance with a brief set out as the employer's requirements, there will often not be a full bill of quantities, but an analysis of the contract sum is generally provided to form a basis for valuation of the works and any adjustments that may be necessary as a result of changes in the employer's requirements. This analysis may include a bill of quantities or a schedule of rates for use in the evaluation of any instructed changes to the works.

The total of the items in the measured works sections of a bill of quantities generally represents the value of the physical construction works, permanent and temporary, to be undertaken. There is, however, usually some overlap in the pricing of the contractor's on- and off-site overheads, management, supervision and ancillary service charges. These latter sums are often substantially contained in the preliminaries and general or 'indirect costs' section of the contract pricing document or bills of quantities, possibly subdivided into lump sums and time-related items.

In practice there can be some difficulty in accurately defining the extent to which the unit rates for measured physical works contain management and other costs over and above the cost of the relevant labour, plant and materials therein. Practice differs between different contracting organisations and formats of bills of quantities, but it is common to find that working supervision, that is the cost of working foremen, gangers and the like, are included in the unit rates, and that there is a uniform percentage addition to the unit rates for off-site overheads. The remainder of the management, on-site overheads and site support costs are then contained in the preliminaries and general or 'indirects' section of the pricing document. A simpler distinction may be that work-related preliminaries and general items are included in the rates for those items of measured work, whereas time-related or project-wide items are included in their own preliminaries and general pricing document. The distinction is, however, seldom that simple and will vary depending on the circumstances of each case.

An alternative approach to the incorporation into a contract of bills of quantities, quantified and prepared by reference to a published industry method of measurement, is the use of a much simpler, schedule of rates. In particular, in relation to specialised trades such as mechanical and electrical installations these may be required from bidding subcontractors, without pre-preparation by contractors. This often appears to be because of a lack of skills among those preparing bills of quantities in relation to the measurement of such specialist works, such that they consider it better to leave it to the specialist subcontractors to measure the works as they price them and prepare a document that sets this out. A result can be huge variations in the detail and quality of such documents between bidding subcontractors. Some of the problems with unquantified schedules of rates in relation to the pricing of variations are set out elsewhere in this chapter.

5.2.1   The FIDIC Forms of Contract

In international construction contracts the most obvious example of a standard form that includes a pricing hierarchy that starts with contract rates and prices expressly agreed between the parties and listed in the contract documents is published by the Federation Internationale des Ingenieurs-Conseils ('FIDIC'). Published for over 50 years, the FIDIC suite of contracts is in wide use throughout the World. The suite's Conditions of Contract for Construction for Building and Engineering Works Designed by the Employer, 1999 Edition ('FIDIC Red Book') puts this as follows at clause 12.3 'Evaluation', to which emphasis has been added:

> For each item of work, the appropriate rate or price for the item shall be the *rate or price specified for such item in the Contract* or, if there is no such item, specified for similar work. However, a new rate or price shall be appropriate for an item of work if:
>
> (a) (i) the measured quantity of the item is changed by more than 10% from the quantity of this item in the Bill of Quantities or other Schedule,
>   (ii) this change in quantity multiplied by such specified rate for this item exceeds 0.01% of the Accepted Contract Amount,
>   (iii) this change in quantity directly changes the Cost per unit quantity of this item by more than 1%, and
>   (iv) this item is not specified in the Contract as a "fixed rate item";
>
> or
>
> (b) (i) the work is instructed under Clause 13 (Variations and Adjustments),
>   (ii) no rate or price is specified in the Contract for this item, and
>   (iii) no specified rate or price is appropriate because the item of work is not of similar character, or is not executed under similar conditions, as any item in the Contract.

Each new rate or price shall be *derived from any relevant rates or prices in the Contract, with reasonable adjustments* to take account of the matters described in sub-paragraph (a) and/or (b), as applicable. *If no rates or prices are relevant* for the derivation of a new rate or price, it shall be derived from the *reasonable Cost* of executing the work, together with reasonable profit, taking account of any other relevant matters.

In addition, FIDIC Red Book clause 13.6 includes provision for valuation on a dayworks basis for what it terms 'work of a minor or incidental nature'. This is considered further in Section 5.6.

The highlighted passages in the quotation of FIDIC Red Book clause 12.3 relate to what has been described as the traditional valuation hierarchy:

1. The 'rate or price specified for such item in the Contract'.
2. Rates or prices 'derived from any relevant rates or prices in the Contract, with reasonable adjustments'.
3. 'Reasonable Cost'.

FIDIC Red Book clause 12.3 will be noted as particularly detailed as to the extent of a change that can be valued using the agreed contract rates and prices before the parties should move on to level (2) of the valuation hierarchy, which has been explained by suitably adjusting those rates. That aspect may be particularly relevant where the variation is extensive in terms of change in quantities and the nature of the work between omission and addition. This means that where a contract contains agreed rates and prices under an FIDIC Red Book approach, they would require reasonable adjustments to take account of the extent to which they had changed. The FIDIC Red Book is also notable for making what has been referred to as level (3) of the traditional hierarchy and as a 'fair' or 'reasonable' valuation one based on the contractor's 'reasonable Cost'. That capitalised word 'Cost' has its own definition in a clause in the FIDIC Red Book as: '… all expenditure reasonably incurred (or to be incurred) by the Contractor, whether on or off the site, including overhead and similar charges, but not profit'. The addition of 'reasonable profit' to those 'Costs' is then required by clause 12.3.

5.2.2 Measurement of Work

One of the greatest strengths of the British-based quantity surveying profession has been its ability to undertake an analytical measurement of even the largest and most complex of construction or engineering projects, break it down into manageable and understandable measured units of work and facilitate the application of detailed pricing based on production and cost data for each unit of work. Where projects are being undertaken under a contractual arrangement, where bills of quantities are prepared on behalf of the employer and provided to tendering contractors, this ability has enabled the use of measurement, not only for pricing and valuation purposes but also for programming and operational planning purposes. Although originally founded in the building industry, as opposed to the civil engineering and industrial engineering construction industries, the quantity surveying profession has expanded into the measurement of all nature of construction works, especially over the past 60 years, as the value of analytical measurement and pricing has been recognised for all modes of construction. In the UK the profession has long-established, detailed methods of measurement for building works, the current version of which is the New Rules of Measurement ('NRM'), published by the Royal Institution of Chartered Surveyors. Previous versions of this date back to 1922. These methods sit happily alongside similar methods such as the Civil Engineering Standard Method of Measurement for civil engineering works (currently 'CESMM4') and the Standard Method of Measurement for Industrial Engineering Construction ('SMMIEC') for heavy mechanical, electrical and process projects. Other methods have been devised for specific sectors of an industry with particular needs or considerations, such as the Pipeline Industries Guild Method of Measurement, or by trade bodies, where it was felt that the other methods were not perhaps totally adequate for their particular purposes, for instance the Thermal Insulation Contractors' Association ('TICA') Method of Measurement.

For international work, the Royal Institution of Chartered Surveyors has also, since 1979, published its Principles of Measurement (International) for Works of Construction ('the POMI Rules'). British Commonwealth jurisdictions have also seen the publication of methods of measurement by their local professional and trade bodies: for example the Association of South African Quantity Surveyors' Standard System of Measuring Building Works 1999, although the first such standard system was published there as early as 1906. Beyond the British Commonwealth, the Qatari's Ministry of Public Works publishes a Method of Measurement for Roads and Bridges. There are a myriad similar examples. Perhaps as a reaction to the proliferation of different approaches to measurement around the world, in 2015 the International Construction Measurement Standards Coalition was formed. This comprises over 40 interested bodies, from the UK's Royal Institution of Chartered Surveyors, Chartered Institute of Building, Institution of Civil Engineers and Royal Institute of British Architects to the Korean Institute of Quantity Surveyors and the Sociedad Mexicana de Ingeniería Económica, Financiera y de Costos.

There are also consultants and employers who have developed their own methods of measurement for their projects, often as a result of a belief that the standard published methods were not wholly relevant to their projects. There is nothing intrinsically wrong with such an approach provided that it results in a systematic and uniform means of measurement and is not merely an attempt to abbreviate measurement at the expense of certainty in the pricing process. All too often, such ad hoc approaches are intended to create unfair advantages for the employer proffering them, but leading to only doubt and dispute regarding item coverage. Where contractors are unfamiliar with them, they can lead to additional costs and avoidable disagreement.

It is the ability to measure and analyse the work content and economics of construction and engineering work in even the largest of projects and to adapt that skill to novel and unique circumstances, that has made the quantity surveying profession valued far beyond its original home in the UK, particularly though British Commonwealth countries. Those engaged in construction work in many European countries and the USA will say that they have managed quite well without a quantity surveying profession, but closer contact and experience of projects in such countries reveals that the work of the quantity surveyor as practised in the UK is still undertaken, with variations, in these countries by a variety of engineers and contract administrators. They have simply not developed a separate profession to the same extent. The penetration of the UK profession into such countries speaks volumes for the regard in which the profession's skills come to be regarded when they become familiar with them.

If the development of the measurement systems in the UK has had an Achilles' heel it was in the existence of the 80:20 rule, by which was inferred that 80% of the project measured value would be accounted for by just 20% of the measured items. By inference the converse applied, whereby 20% of the value was to be found in 80% of the items. The implication of this criticism was, of course, that there were too many relatively minor items measured that were of little or no consequence when pricing the project works. There is no doubt that there have been times when such criticism was justified and there is always a need to ensure that measurement for any project is relevant and sensible and does not result in measurement for the sake of measurement. The more recent versions of the various standard methods of measurement have addressed this problem to a greater or lesser degree and it is now less of an issue than formerly. It is always open to the employer, of course, to allow adaptation or qualification of any particular method of measurement in appropriate instances, although as with any amendment to a published industry standard, great care must be taken.

As an illustration of the potential benefits of adoption of measured bills of quantities in a contract, rather than an unquantified schedule of rates, as described elsewhere in this chapter, consider a real example. A contract in Africa for a series of pumping stations to be constructed in remote desert locations was let on the basis of a US$ 180 million lump sum supported by a schedule of rates comprising four pages and less than one hundred individual items. The schedule had no quantities so that the relationship of the rates to the lump sum was not clear. To further compound the brevity the items were not measured by reference to any method of measurement and nor was any description of item coverage included. Considerable disputes arose over the valuation of variations and the applicability of the rates among other matters, resulting in an arbitration continuing for years after the completion of the project. The dispute over the valuation of variations and applicability of the rates could have been avoided, or at least substantially reduced, by having a quantity surveyor draw up a bill of measured items and quantities using a recognised method of measurement, stating the coverage of each item, for pricing by the contractor and inclusion in the contract documents. The cost of such an exercise would have been a fraction of the cost incurred by the parties in arbitrating such a matter.

5.2.3 Design and Build/Schedule of Rates

In other instances, the project may not have bills of quantities provided by the employer for tendering contractors but may have a 'lump sum' arrangement accompanied by an analysis of the tendered sum or a schedule of unit rates for use in valuing variations to the works. Again, it is important that this intended use is clearly stated in the tender and contract documents, so that pricing contractors can act accordingly.

It is imperative that if such schedules are to be of maximum effect in the administration of the contract then they are drawn up to fully cover the scope of works, with definition as to how the items are compiled or accompanied by notes on item coverage so that it is clear what each item in the schedule covers, what each excludes, and how the items relate to each other.

While there is no detailed measurement available to both parties in such instances there is usually a detailed measurement undertaken by the tendering contractors in order to produce their tender for the works. In design and build or design and construct type contracts it will be necessary to ensure that the contract sum analysis is in such detail as to cover all significant units of work likely to occur and that the pricing applied is reflected in the tender submitted. This is obviously not as straightforward as it is in the contract with a full priced bill of quantities available to both parties, and will require some element of judgment.

A number of the difficulties that can arise from the use of schedules of rates, rather than bills of quantities, as contract documents setting out rates for the valuation of variations were detailed elsewhere in this chapter and also in Chapter 4.

5.2.4 Status of Contract Rates and Prices

Whatever the type of contract, the rates and prices submitted by the contractor should reflect, when taken in total, the whole of its obligations under the contract. However, it is important to understand that, while the rates and prices contained in the contractor's tender are deemed to include the whole of its obligations under the contract, they will only apply to the contract works and authorised variations under the contract.

Mention was made in Chapter 2 of the decision in *Blue Circle Industries PLC v Holland Dredging Co (UK) Ltd* (1987) 37 BLR 40, which illustrated the limits on the scope of variations that can be instructed under a contract, but another decision involving a dredging operation demonstrates the limits that can apply to the applicability of the contract pricing.

In *Costain Civil Engineering Ltd and Tarmac Construction Ltd v Zanen Dredging & Contracting Company Ltd* (1998) 85 BLR 77, the dispute referred to the execution of works in connection with the construction of the A55 Conwy Bypass and an alleged variation to the works. Part of the project consisted of the construction of a tunnel under the River Conwy. This tunnel was to be constructed using large precast concrete tunnel segments, constructed in an adjoining basin excavated for the purpose, which was then flooded for the segments to be towed out for placement in a dredged channel in the river. After completion of the works the contract allowed for various options for the flooded basin but in the event none of these were ordered so under the terms of the main contract between the Costain and Tarmac Joint Venture and the Welsh Office the basin was required to be backfilled and reinstated to its original condition. This would have been less expensive than carting away the excavated material for disposal so in this event the Joint Venture was to allow the employer a credit of around £1 million.

In the event the Crown Estate, who were not a party to the contract, decided to use the flooded basin to construct a marina. As a result a supplemental agreement was entered into between the Welsh Office and the Joint Venture whereby the contract was changed to allow the basin to remain flooded with the Joint Venture undertaking some modifications to it to ensure its stability at an agreed price of some £2½ million.

Zanen had been engaged by the Joint Venture as a subcontractor under the ICE Conditions of Contract (the 'Blue Form') and the Joint Venture instructed Zanen to carry out certain works in connection with the marina. Zanen protested that the work was outside the scope of its contract with the Joint Venture but that was rejected by the Joint Venture who issued detailed instructions for the work, stating:

For your information and clarification the works in question form part of our obligations with the employer and therefore your concern regarding this work being outside of your subcontract is unfounded. In fact your subcontract document clearly identified work to be done in the casting basin.

Zanen carried out the instructed works but did not accept that they were a proper variation under their subcontract, which had an amended clause 8 defining 'authorised variations' as:

(1) (b) A variation which is agreed to be made by the Employer and the Contractor and confirmed in writing to the Sub-contractor by the Contractor, or

(c) A variation ordered in writing by the Contractor, or

(d) Any other circumstance relating to the Sub-Contract Works which constitutes a variation under the Main Contract.

The case came before what is now the Technology and Construction Court (TCC) in London, as an appeal from the decision of an arbitrator who had found that the marina works were not part of the main contract scope of work and so the works undertaken by Zanen in connection with the marina were outside the scope of the subcontract. In agreeing with the arbitrator's decision, the judge stated:

... both [Counsel for the parties] were able to present to the arbitrator a detailed analysis of the contractual position and he concluded, in my judgment correctly, that the works undertaken by Zanen, the respondent to this appeal, were outside the contract. The finding that it was more likely than not that the employers for the marina works and its development were the Crown Estate Commissioners and Pearce Developments is consistent with the evidence cited in the award .... The instructions from the Joint Venture described in ... the amended points of claim did not constitute authorised variations of the subcontract works within the meaning of clause 8(1) of the subcontract made between the parties on 1 June 1987, because such instructions required work to be done outside the scope of the respondent's obligations under the subcontract.

As a result of this decision Zanen were able to recover sums in excess of those they would have been entitled to if the works had been a variation within the contract as the contract pricing mechanism for variations was held not to be applicable. This case illustrates how it is important to consider carefully the implications of any variation instructions, particularly where the instructions involve substantial additional works, and ensure that they are legitimately part of the contract before deciding to apply the contract regime for pricing changes to the works in question.

5.2.5 Errors in Rates and Prices

Where contractors erroneously calculate or insert their rates and prices in bills of quantities or other contract documents issues may arise as to what is to be done about such errors. This can lead to various claims including any of the following:

1. That where bills of quantities items have not been priced at all or not extended through to page totals and the contract price, this error should be adjusted in favour of the contractor by an increase in the contract price.
2. That where a rate or price has been erroneously priced high by the contractor it should be reduced before being applied to variations or additional quantities, to the employer's benefit.
3. That where a rate or price has been erroneously priced low by the contractor it should not be held to that low rate in respect of variations or additional quantities, and the rate or price should be suitably increased.

The merits of these arguments will depend, firstly, on the express provisions of the contract, including any measurement rules. The applicable law may also have a significance. The facts of each case also require consideration, for example where the apparently erroneous rate actually reflects some distinction between the work item in the bills of quantities and those in a variation or additional quantities (for example where it includes some fixed costs).

It may seem obvious that the parties to a contract are bound to rates and prices that are agreed in that contract and that errors in the contractor's calculations of the contract sum, which could be to its benefit or disadvantage, will not be corrected unless the parties so agree (and it is difficult to envisage a situation where they would) or the contract expressly provides for correction (which is most unlikely). In SBC/Q, this is made particularly clear in clause 4.2:

> The Contract Sum shall not be adjusted or altered in any way other than in accordance with the express provisions of these Conditions and, subject to clause 2.14, any error in the computation of the Contract Sum is accepted by the Parties.

Clause 2.14 of SBC/Q covers errors in the preparation of bills of quantities or 'Employer's Requirements', which are to be corrected, altered or modified as necessary and are to be the subject of a variation under that contract. Thus, errors in the preparation of the pricing document by the employer's team are adjusted, but errors in the contractor's pricing of that document are not.

Similarly, the Infrastructure Conditions say this at clause 55(2), to which emphasis has been added in relation to errors in rates and price:

> No error in description in the Bill of Quantities or omission therefrom shall vitiate the Contract nor release the Contractor from the carrying out of the whole or any part of the Works according to the Drawings and Specification or from any of his obligations or liabilities under the Contract. Any such error or omission shall be corrected by the Engineer and the value of the work actually carried out shall be ascertained in accordance with Clause 52(3) or (4). *Provided that there shall be no rectification of any errors, omissions or wrong estimates in the descriptions, rates and prices inserted by the Contractor in the Bill of Quantities.*

This italicised passage emphasises how contractors take the risk for anything they insert into bills of quantities when tendering, including their rates and prices, but not errors in the preparation of those pricing documents by the employer's team at a pre-tender stage.

As outlined in Chapter 4 of this book, there is a significant difference between the potential effects of an error in pricing in a 'lump sum' contract, where the rates and prices are used only to value variations to the works, and remeasurement contracts, where the actual scope of the completed works are remeasured in their entirety and revalued at the bills of quantities rates and prices as part of the final account.

In a remeasurement contract such as the Infrastructure Conditions or the FIDIC Red Book, any error by the contractor in its extension of its rates and prices in a bill of quantities into the tender sum will disappear on remeasurement of the works as the actual quantities are remeasured and multiplied by the rates and prices, to replace the tender quantities at the same rates and prices. The same correction will be made of any errors in the totalling of pages or collection of page totals into section summaries and grand summaries. This emphasises the importance of employers and their advisors carrying out a complete arithmetical check of all tenders received to ensure that a lowest bid has not been distorted by an error in the extension of quantities at rates and prices in a bill of quantities that will be corrected when the work is remeasured. However, this is not always a straightforward exercise. For example, on a remeasurable infrastructure project for a Middle East government agency under FIDIC 1987 terms, the bills of quantities ran to nearly 2500 pages. Furthermore, although they were prepared as recently as 2011, these bills of quantities were only provided to tenderers in hard copy and were therefore priced in manuscript form. The task of ensuring that all bidders had priced and properly extended all items apparently proved too great for the employer's quantity surveyors (who were known to be on a very tight fixed fee for pre-contract services). The result was that the contract bills of quantities as priced by the appointed contractor contained a number of errors, including the rates and prices for some items being wrongly extended or not extended at all, some page totals being incorrect and totals being carried forward wrongly. The modern practice of providing bills of quantities in their native electronic form, such as in the form of an Excel workbook, could have made that checking task (and also the subsequent preparation of monthly payment certificates) much easier.

Given such arcane approaches, it is not that rare for contractors to actually miss pricing some items, leaving the column for their rate or price blank (without such clarifying entry as 'included elsewhere'). The inevitable question that arises on such items is how they are to be dealt with on a contractual remeasure of the works, and in relation to additional quantities for the same item. Under the FIDIC Red Book, the contractor is likely to rely on clause 12.3.(b)(ii) as providing it with relief from this error against all the remeasured quantities for such items. That subclause provides that a new rate will be appropriate for an item of work for which 'no rate or price is specified in the Contract for this item'. Employers sometimes argue that clause 12.3.(b)(ii) relates only to work items not measured in the bills of quantities and that if the item was measured but not priced, the clause does not apply. Another occasional employer assertion is that remeasurement under the FIDIC Red Book is only of items that the contractor has priced and not any that it has missed in its pricing. The contractor's counter will be that clause 12.1 requires 'the Works' to be measured and under the definition of that term in clause 1.1.5.8 that includes all of the permanent works executed by the contractor, whether they were measured and/or priced in the contract documents or not.

In a lump sum contract such as SBC/Q this remeasurement does not take place and therefore the contractor is likely to be left with the effect of any extension errors, be they beneficial or detrimental to its interests. This was the position in a case under SBC/Q's predecessor of the time, the RIBA form. In *MJ Gleeson (Contractors) Ltd v Sleaford UDC* (1953), the bills of quantities included items that the contractor had missed to price. There were rates for similar works elsewhere in the bills of quantities and the contractor argued that these should be used to value the work it had missed. The court disagreed on the basis that the error was not in the preparation of the bills of quantities (which were to be corrected under the RIBA form just as they now are under SBC/Q), but the contractor's pricing of them. There was therefore no facility under the contract to correct this omission.

This approach of not correcting a contractor's calculation errors in its tender for a lump sum project seems the right one for a number of reasons. It may seem to an employer that a contractor could win a tender competition as the lowest bidder but then have an error corrected that meant it is no longer lowest. It may also be that in preparing its tender the contractor chose not to price an item because it anticipated that the item would be omitted anyway. Such an approach might also be part of a contractor's process of risk and opportunity calculation inherent in its tender. It may even be that the contractor has priced for the apparently unpriced work elsewhere in the bills of quantities.

That may cover the matters of omission to price items or errors in the extension of an item's value into the tender, and subsequently contract sum, but are any apparent errors in the rates and prices themselves to be corrected or otherwise amended when those rates are used to price variations or quantities additional to those in the contract?

Again, the answer to such questions must start with a consideration of the express provisions of the contract. With a lump sum type of contract, such as SBC/Q, the rates and prices are purely for the valuation of variations and changes as the contract scope is not to be remeasured and revalued. The provisions of SBC/Q clause 5.6 say that variations will be priced on the basis of rates and prices in the contract bills of quantities. It does not say that those rates and prices will be corrected if subsequently found to be wrong.

Also under the RIBA form of contract, *Dudley Corporation v Parsons and Morrin Limited* (1959), the contractor had wrongly priced its excavation rate at two shilling per cubic yard, a fraction of the rate it should have allowed. The court disagreed with the claimed approach of valuing the original quantity at that erroneous rate and extra quantities at a reasonable rate. It considered that the contractor had taken the risk of its pricing and that if the quantities increased against a low rate, that low rate should apply to the original and additional quantities.

The same philosophy applies to remeasurement contracts such as the Infrastructure Conditions, where the quoted rates and prices are said in clause 52(4) to apply to the valuation of variations without mention of amendment for any apparent errors. Clause 52(5) provides for an increase or decrease in a contract rate or price, but only where it is '… by reason of any variation rendered unreasonable or inapplicable …'. Under this the variation has to render the rate unreasonable or inapplicable, not the fact that it was wrong in the first place. This caters for the adjustment of a rate or price to reflect differences in the circumstances of the variation compared to the contract works.

Judicial authority for this approach of not otherwise 'correcting' a contract rate was provided in the Technology and Construction Court in the case of *Henry Boot Construction Ltd v Alstom Combined Cycles Ltd* [1999] BLR 123, subsequently confirmed by the Court of Appeal (reported in *The Times*, 11 April 2000). In this case the contractor was arguing that it was entitled to price a variation at the rate in the contract although it was apparent that the rate was exceptionally high. The employer argued that the erroneous rate should not apply but a fair and reasonable rate should be substituted. The judgment supported the contractor's position that a variation could not be used as the pretext for amending and correcting errors in the contractor's pricing. In the judge's opinion such an amendment or correction would 'be completely inconsistent with the wording of such a contract and the philosophy to be derived from it'. The same philosophy would of course apply if the rate in question had been exceptionally low and the contractor, rather than arguing it was entitled to the contract rate, had been arguing that it should not be held to it. In short, the contract rates and prices will apply to the pricing of variations regardless of apparent errors in those rates and prices and good luck to whichever party this falls to the benefit of.

Whilst that is likely to be the legal position, a not uncommon approach in practice where this occurs is to agree that it is reasonable that the contractor is held to the erroneous rate only for the quantities in the contract and that any additional quantities should be the subject of a corrected rate. This is very much a function of the parties, their relationship and attitudes and the degree to which they see an equitable approach to dealings on such issues as more important than a strict contractual one. It is easy to see how either party might alternatively argue with equal conviction either that: agreed contract rates and prices are agreed rates and prices, and they are both bound to them, win or lose; or that a party should not benefit or suffer from a rate or price that was clearly wrong.

In the absence of such an agreement between the parties, it may still be possible to rely on express terms of the contract for rectification of particular types of error. For instance, clause 2.14 of SBC/Q allows correction of errors in the contract bills of quantities in respect of departures from the stated method of preparation, errors in item descriptions or in quantity or omissions of items. Such corrections would not automatically extend to correction of the rates inserted in the contract bills of quantities unless it could be shown that, for instance, correction of a description would result in a different rate being justified. Applying the valuation hierarchy in SBC/Q clause 5.6, the inserted rate would still be used, as far as possible, as the basis for the amended rate to reflect the corrected description. This means that the adjustment would only extend to that necessary to reflect the change in description. It would not extend to putting right errors in the contractor's pricing of the original bill item. If those rates were too high then they stand in the valuation of the variation, to the contractor's benefit. If those rates were too low, then they stand in the calculation of the adjusted rate, to the employer's benefit. In either case, this is to the extent that they are applied to the valuation of the variation before necessary adjustments are made to reflect the different nature of the varied works or the work as it should have been properly described but for the error in the preparation of the bills of quantities. Contractors sometimes use situations such as these to re-write their original inaccurate pricing, with a low rate being entirely ignored in the valuation of a variation, rather than applying it with the necessary adjustments just to suit the differences in the nature of the work. This may sometimes even extend to claiming that the new rate should be entirely based on actual costs rather than due adjustment of an original rate.

The need for a detailed consideration of the facts to see why a contract rate is apparently unreasonably high is illustrated by the Hong Kong Court of Appeal decision in *Maeda Corporation, Hitachi Zozen Corporation, Hsin Chong Construction Co Ltd v The Government of the Hong Kong Special Administrative Region* (2012) CACV 230/2011. The joint venture contractor had priced a bill of quantities item very high because it had identified that the item's quantity was underestimated and would increase. The rate therefore included considerable amounts of fixed costs for items that should otherwise have been priced against items in the preliminaries and general section of the bills of quantities. The contract was in terms materially similar to those of the FIDIC Red Book. The contractor argued that: the build-up to its tender rates was not admissible evidence; it was only the final rate inserted into the bills of quantities that is relevant and not the way the rate was built up; if one could look at how a tender developed and whether costs were transferred from one rate to another prior to the formation of the contract, there would be a never-ending inquiry as to the history of how the tender was calculated. The court disagreed with all of these propositions. The fixed element in the rate did not increase proportionately with the quantity of work, so the rate required adjustment in relation to those fixed elements. With respect, this seems no more than sound quantity surveying practice.

One further regular moot point when preparing 'analogous rates and prices' by reference to those in the contract is the extent to which the resulting rates and prices are to reflect the pricing levels allowed in the contract rates and prices. The issue and the alternative approaches can best be explained by considering a hypothetical example.

Assume that a contractor has priced to supply and install ornate fixings for the glass panels to glazed screens within a building. Assume that the contract specification is for these to be cast iron and the contractor prices the item at a rate of $23 per fixing, which comprises $10 per fixing for supply and delivery of the materials, $10 each for labour in installation (a confirmed subcontractor rate) and 15% mark-up for overheads and profit. Thus the contract rate was built up as follows:

		$
Labour		10.00
Materials		10.00
Subtotal		20.00
Overhead and profit	@ 15%	3.00
New rate		23.00

Assume, that, in the event, this is an underestimate and the contractor could only ever have sourced the specified fixing at a best price of $20 each, leaving it with a built-in loss in its bill rate on the materials component and a negative contribution to its overheads and profit.

However, the employer decides that it wants a change in the materials to stainless steel fixings. These revised fixings are of the same size, dimensions and fixing costs, but the materials cost $60 each, three times the actual cost of the originally specified cast iron fixings and six times the contractor's tender allowance for those original materials. How is an analogous rate to be calculated for the new fixings? It is suggested that there are three alternatives:

1. The contractor's claimed approach may well be to replace the materials costs allowed in its bill rate ($10 each) with the actual costs ($60 each) into the build-up of the bill rate as follows:

		$
Labour (bill rate allowance)		10.00
Materials (actual cost)		60.00
Subtotal		70.00
Overhead and profit	@ 15%	10.50
New rate		80.50

2. An employer's approach to an 'analogous rate' might consider that the pricing of the new fixings must be proportional to the original allowances in the bill rate, including the error in it. Thus, this is an assertion that the contractor should not be able to make up its tender error through the valuation of a variation. In this case, since the actual costs of the stainless steel fixing is three times what it would have been for cast iron ($60 each compared to $20 each), the approach followed might be to replace the materials costs allowed in the bill rate ($10 each) with a proportionally increased materials allowance as follows:

		$
Labour (bill rate allowance)		10.00
Materials	3 × $10	30.00
Subtotal		40.00
Overhead and profit	@ 15%	6.00
New rate		46.00

3. A third alternative might be to adjust the bill rate by the increase in the actual cost of the materials (which had gone from $20 to $60) as follows:

		$
Labour (bill rate allowance)		10.00
Materials	$60.00 less $20.00	40.00
Subtotal		50.00
Overhead and profit	@ 15%	7.50
New rate		57.50

As stated elsewhere, how such questions are to be resolved will depend on the detailed provisions of the contract. Even in standard forms there can be subtle differences in provisions, which can be of significance to such a debate.

Contrary to another occasional contractors' claim in such circumstances, where the contractor has inserted an erroneously low rate in the contract bills of quantities and seeks to avoid its consequences of its being used to value additional work, it is not possible for the contractor to recover the potential loss under provisions such as those of SBC/Q as 'direct loss and/or expense'. Pricing errors are outside the scope on the grounds for which such claims can be made, that is the Relevant Matters listed in clause 4.24. The same situation will apply in most other contracts, but the particular terms and conditions should always be examined before reaching a final decision.

Forms of contract in which a contractor will not be held to an erroneously low contract rate, or earn a bonanza from an erroneously high contract rate, through the valuation of additional work, are those of the NEC4-ECC suite. Proponents of the NEC approach will proclaim this as a major advantage of those contracts, giving rise to an equitable outcome and avoiding the need for debate regarding potential alternative approaches to the adjustment of contract rates. In NEC4-ECC, as explained elsewhere in this chapter, the valuation of variations is based on the contractor's actual or forecast 'Defined Cost' and its 'Fee', rather than its original contract pricing.

Given this discussion of the status of the contract rates and prices it is next necessary to consider in more detail how those rates and prices might be used in the evaluation of claims for additional payment arising from ordered variations.

5.3 The Valuation 'Fences'

Many contractors view the occurrence of variations, be they of variation to the specification or scope of works, their circumstances or conditions, as an opportunity to obtain enhanced rates and prices for the work and thereby improve their commercial position. Such an approach is understandable given the risks that contractors run and in many instances have to suffer. However, a philosophy of 'making hay' at every opportunity, while understandable, can lead to unnecessary conflicts over the pricing of changes. It has also to be added that the employer's team are often not blameless in such situations, seeking to have varied work undertaken at the minimum, or no, additional cost rather than the correct value reflecting changes in the contract scope through appropriate adjustment to the contract rates and prices.

There is often little need for such conflict over the pricing of varied work as most, if not all, of the standard forms of contract in regular use have well-tested and logical regimes for the pricing of such work, if they are properly applied by both parties and their advisors. Sadly the same cannot be said of many 'ad hoc' or 'one-off' contracts where the provisions are often not properly developed and considered in the light of all possible future circumstances. It may be tempting for those advising employers that bespoke terms should be drafted that pin the contractor to its original rates and prices no matter what extent of change in quantities, specification or design occurs. In all cases, the contract regime for pricing variations needs to consider carefully how the contract rates and prices are to be applied, and more importantly how those rates and prices can be changed or varied as the effects of the variation differ.

5.3.1   The 'Fences'

As has been explained, standard forms of contract that adopt the traditional approach of applying contract rates and prices to the valuation of variations usually state that the contract rates and prices apply to the pricing of varied work providing it is of the same specification and executed in the same circumstances as the contract work. As the type, specification or circumstances of the work moves away from that of the contract then provision is made for adapting or replacing the contract pricing to reflect the differences from the contract basis. We have described the approach as one of a hierarchy of valuation methods. The effect of this philosophy is to provide a series of questions to be addressed when considering the applicability of the contract rates and prices to work that is part of a variation. Thus a series of 'fences' are in place that need to be passed if the pricing is to pass to the next stage of the pricing regime.

The valuation rules for variations, and provisional sums, are set out for SBC/Q in its clause 5.6 where the 'fences' set out in subclause 5.6.1 can be summarised as:

1. Where the work to be valued is of a similar character to the work in the contract bills of quantities, and is executed under similar conditions and does not significantly change the quantity of work to be executed, the rates and prices set out in the contract bills of quantities apply (subclause 5.6.1.1).
2. Where the work to be valued is of a similar character but is not executed under similar conditions to the work in the contract bills of quantities, and/or the effect of the variation significantly alters the quantity of the work, then the rates and prices in the contract bills of quantities are to be used as the basis of valuation with a 'fair allowance' for any difference in conditions or quantity (subclause 5.6.1.2).
3. Where the work to be valued is not of the same character as the work in the contract bills of quantities, then the work shall be valued at 'fair rates and prices' (subclause 5.6.1.3).

Further provisions are added in SBC/Q subclause 5.6.1.4 for the valuation of approximate quantities in the contract bills of quantities, to the effect that the contract rates and prices apply if the approximate quantity is a reasonably accurate forecast of the actual quantity executed, but where the approximate quantity is not a reasonable forecast, then the contract rates and prices are to be used as a basis with a fair allowance for the effect of the variation in quantity. An obvious issue arising from this clause is what is meant by 'reasonably accurate', a subjective term that parties to such a provision might have very different interpretations.

5.3.2   'Conditions' and 'Character'

The first general rule that can be determined from the SBC/Q subclauses 5.6.1.1 and 5.6.1.2 is that as long as the character of the work remains the same, the contract rates and prices apply either without amendment or with amendment for differences in conditions or significant effects on quantities. Only if the character of the work is different can the rates and prices in the contract bills of quantities be set aside and 'fair rates and prices' used instead, under subclause 5.6.1.3. The implications of the expression 'fair rates and prices' were considered in Chapter 2.

In Chapter 2 the problem of determining the 'conditions' and 'character' of the contract work, which form the basis for any variation valuation, was discussed. The effect of the *Wates Construction v Bredero Fleet* decision was explained as being that the base conditions and character had to be determined from that which could be gathered from the contract documents, and not from the subjective expectations or interpretations of information by the parties. Similarly, the assessment of any difference in the conditions for, or character of, work that is the subject of a variation order has to be conducted on an objective analysis of the work performed.

There is no definition of the terms 'conditions' or 'character' in SBC/Q, or indeed in the Infrastructure Conditions where they are similarly used, and it is therefore left somewhat to the person undertaking the evaluation to interpret the terms. For work to be considered as being of a different character to that in the contract bills of quantities there must be some significant difference from that contained in the bills of quantities. For instance, concrete in reinforced retaining walls might be included in the contract bills of quantities, and an examination of the contract documents and drawings might show that these walls are reasonably uniform in thickness and height. If the varied work is for concrete in retaining walls with regular variations in the thickness and height then it might be argued that the character has changed, especially in respect of the formwork required, where there might be a significant increase in the remaking of the wall forms and consequent wastage. Similarly, rates for steelwork in the contract bills of quantities might be considered, on examination of the contract documents and drawings, to apply to steelwork where the 'character' is that there is an average of, say, 3 tonnes per structural member, i.e. the structural steel is comprised of reasonably substantial members. If, as a result of a variation, further steelwork is introduced in, say, walkways, which on examination are found to comprise on average some eight pieces of steel per tonne, i.e. the average member weighs only some 125 kg, then it might reasonably be considered that the varied steel is of a different character.

The conditions that apply to the work are the circumstances in which it is executed. For instance, work outside the original site boundary, or in a part of the site where that type of work was not originally envisaged, might constitute a change in conditions for the execution of the work depending upon the circumstances. For instance, using the steelwork in walkways example, if the contract rates were for the erection of steelwork external to buildings or other structures and a variation required the installation of similar steelwork but inside a building or structure, this would be a case of similar work but under different conditions, the potential difficulties of access for erection of the steelwork being an obvious consequence of those different conditions.

The valuation where varied work is not of the same character as the contract work is stated under SBC/Q clause 5.6.1.3 to be at 'fair rates and prices' but, as with the other key terms in this clause, there is no definition of how these are to be determined, or what factors are to apply. It is sometimes suggested that the contract rates and prices in the contract bills of quantities should be used as a basis for the valuation, just as they are for changes in conditions, but if that were the case there would be no difference between the valuation for changes in conditions or changes in character despite their different treatment in the contract. In other words, valuation under clause 5.6.1.3 would be the same as under 5.6.1.2. This may not seem to be logical and it would seem that there is an opportunity for new rates and prices to be determined without reference to those contained in the contract bills of quantities. However, the expression used is 'fair' rates and prices not merely 'new' rates and prices and that must have some implication.

It is suggested that the rates and prices applied in the event of changes in the character of the work should be 'fair' to both parties. One approach to this is that the rates and prices for such work should be 'fair' to the contractor in allowing it to price all elements of the changed work, while being 'fair' to the employer by requiring that base prices used for labour, plant and materials should be the same as those used for the contract rates and prices. In this way the changed character of the work is valued but at a pricing level that should reflect the rates and prices that would have applied had the varied work been contained in the contract bills of quantities from the outset, only amended to reflect the changed character.

This is different from the pricing exercise that would be necessary where there is a change in the conditions for the work, where the analysis of the contract rates and prices would apply with the only change being to the element(s) affected by the change in conditions.

To take the example of reinforced concrete retaining walls, the approach might be as follows.

5.3.2.1 Change in Conditions

Assume that the change in conditions arose from the issue of a variation order to construct further concrete retaining walls but at a part of the site where access to mechanical plant was so restricted as to exclude deliveries of ready-mix concrete lorries from the vicinity and eliminate the use of cranage; in addition, the quantity of concrete in the walls subject to the variation was not sufficient to justify the mobilisation and use of a concrete pump. Assume also that no such restrictions applied to the concrete in the retaining walls, which were included in the contract bills of quantities.

The change in conditions will affect the placing of the concrete, it may affect the erection of formwork if the contract scope anticipated the use of cranage as part of the erection operation and might similarly affect the placing of the reinforcement if the contract scope had also anticipated the use of cranage to place prefabricated reinforcement cages.

To use the rate for concrete as an example, the rate build-up for the contract scope, assuming walls of 150 to −450 mm thickness, might have been:

		$
Concrete	Per m^3	58.00
	Wastage 5%	2.90
Labour	4.25 h @ $15.00	63.75
Plant	Air hose and compressor	4.50
	Per m^3	$129.15

The change in conditions entailing the exclusion of ready-mix lorries from the vicinity might have the effect of increasing the labour content to, say, nine hours per cubic metre, as the concrete has to be transported manually from the nearest delivery position to the point of placement. The manual transportation might also increase the wastage factor by another 2½%. The effect of the change in conditions for the concrete price would therefore be, leaving aside for the moment the issue of potential impact on the cranage requirement:

		$
Concrete	Per m^3	58.00
	Wastage 7½%	4.35
Labour	9 h @ $15.00	135.00
Plant	Air hose and compressor	9.53
	Per m^3	$206.88

Note that the price for the air hose and compressor has also increased in proportion to the increase in labour time on the assumption that the plant will be dedicated to the concreting operation for the time that the labour is engaged.

This example assumes that the concrete would have been placed from ready-mix delivery lorries direct to the retaining walls for the contract scope. Where it had been anticipated that plant such as tower cranes would be used for the placement, i.e. using concrete skips to transport and place the concrete, a further potential problem arises in that such items of plant and equipment are often priced in the preliminaries and general items, or site overheads, section of the bills of quantities and not in the unit rates and prices for measured works items (in this example, concrete) in their sections of the bills of quantities. In such circumstances the question arises as to what is the valuation effect of the reduction in usage of the tower crane? The usual answer is that the crane is being paid for through the preliminaries and general items because it was not specific to the items for the concrete. It was a project-related or time-related item and not related to a specific item of work. Thus, the contractor still gets its payment and provides the crane for the anticipated period. The overall effect of a single operation such as the loss of usage for a single variation operation will usually not be significant to such items. However, each such case must be looked at on its own merits as to the effect on measured works items and preliminaries and general items, as well as to whether, to what extent and how they require adjustment.

Where the converse applies and the effect of varied work is to introduce a further requirement for the use of such plant, then some allowance may be required as far as the effect can be demonstrated.

5.3.2.2 Change in Character

Now consider the evaluation to be undertaken for a change in character of the work in this example of reinforced concrete retaining walls. If, as suggested elsewhere, the concrete retaining walls, which are the subject of a variation, are of differing thicknesses and heights in contrast to the contract scope for such walls, which were of relatively uniform thickness and height, then it is suggested that the rate for the varied work will be a new rate built up to reflect all the elements of the varied work but using the same basic rates for labour, materials and plant as were included in the contract bill rates.

The original rate for formwork to the walls, as envisaged in the contract bills of quantities rates, might have been:

			$
Formwork materials	4 uses	$34.00 m²	8.50
Sundries			1.00
Labour	Make	6 h for 4 uses	27.00
		1.5 h @ $18.00	
	Repair and remake	per ½ h @ $10.00	5.00 use
		Per m² $	41.50

On this basis, the new rate for formwork to walls of varying thickness and height might be:

			$
Formwork materials	3 uses	$34.00 m²	11.34
Sundries			1.34
Labour	Make	8 h for 3 uses	48.06
		2.67 h @ $18.00	
	Repair and remake	per ¾ h @ $10.00	7.50 use
		Per m² $	68.24

For this new rate, every element of the build-up has been examined and revalued to reflect the character of the varied work; only the basic prices have remained unaltered and it is suggested that this is the correct approach to such valuations. The effect of the variation on preliminaries and general items might also need to be considered.

The Infrastructure Conditions adopt a similar regime in their clause 52(4), with one difference. Where the work is of similar character and is executed under similar conditions the contract rates and prices apply. Where the varied work is not of a similar character or is not executed under similar conditions or it is ordered during the defects correction period (which would usually suggest that different conditions applied in any event), then the rates and prices in the bill of quantities are to be used as the basis for the valuation, failing which a 'fair valuation' is to be made.

5.3.3 New Rates

The distinction between a change in conditions and a change in character is not made in the Infrastructure Conditions in the manner adopted in SBC/Q. This seems to be a sensible approach as in practice the distinction between the evaluation process under SBC/Q for a change in character can be little different to that for a change in conditions. Where there is to be a valuation under the Infrastructure Conditions for a change in character or conditions then it is suggested that the process for adjusting the affected elements of the contract rate analysis to reflect the changes should be the same as that set out in the concrete and formwork examples set out elsewhere for valuing a change in conditions under SBC/Q.

Where the change is so dramatic as to render the rates and prices unusable even as a basis for evaluation with adjustment to those rates and prices, the principal contracts considered in this book take subtly different approaches:

- FIDIC Red Book clause 12.3 requires that the valuation '… shall be derived from the reasonable Cost of executing the work, together with reasonable profit, taking account of other relevant matters'.
- Infrastructure Conditions subclause 52(4)(b) requires that a 'fair valuation' shall be made.
- SBC/Q clause 5.6.3 requires that valuation '… shall be at fair rates and prices'.

In the FIDIC Red Book wording it has been explained how 'Cost' is a defined term that excludes profit, but that 'reasonable profit' is added to it for the valuation of variations by clause 12.3. However, no guidance is given as to what 'reasonable profit' and 'other relevant matters' are. Reasonable profit could presumably be either that rate allowed by the contractor in its contract pricing or some external rate of profit generally applying in the local market for the relevant work at the relevant time. Of these, the former approach seems better to reflect the traditional approach of contracts such as the FIDIC Red Book, of applying to variations the similar pricing levels to those in the contract. The approach would of course depend on the contractor establishing what its tendered profit percentage was, either on similar work or overall in the contract price. The second approach, by reference to the wider market, might be thought relevant where the contractor priced the original works marginally and is then required to carry out a very substantial variation as well. It might not be 'reasonable' to hold the contractor to its tendered profit level in such circumstances. Another aspect of considering such FIDIC Red Book provisions must be the law of the local jurisdiction, which might have significant relevance in relation to reasonableness in commercial contracts.

The FIDIC Red Book requirement for 'taking account of other relevant matters' can prove very problematical in practice. It is a very wide phrase that is open to a wide interpretation by contractors in including all sorts of issues within their variation claims under clause 12.3.

5.3.4   Valuation of Variations in Quantity

SBC/Q includes in clause 5.6.2 the opportunity to include the influence of changes in quantity in the assessment of valuations for variations executed under different conditions to those on which the contract rates and prices were based. Presumably the same opportunity is not expressly included in SBC/Q provision for valuations for work of a different character as this is a 'fair rates and prices' valuation, as discussed elsewhere, which presumably will include the influence of the quantity of the work in any event.

The Infrastructure Conditions have provision in clause 56(2) for the engineer to increase or decrease any of the contract rates and prices if, in its opinion, the actual quantities of work undertaken are greater or lesser than those in the contract bill of quantities and the difference is such that an adjustment to the rates and prices is justified. This provision is not tied to the valuation of variations due to the contrast between the nature of SBC/Q and the Infrastructure Conditions, i.e. the former being a 'lump sum' contract where the bill rates and prices are for the measurement of variations and provisional sums and the latter being a remeasurement contract in which all the quantities of work undertaken are fully remeasured.

Some contracts, such as various within the FIDIC suites, contain restrictions on the ability of the contractor to claim adjustments to prices for the effect of changes in quantities. A common restriction is that quantities must change by more than 10% before adjustments can be claimed. This can work both ways in that reductions in a profitable rate will not be applicable until the threshold percentage is reached and similarly an increase in an unprofitable rate will not be available below the stated change in quantity. On the other hand, proponents of such restrictions argue that such a threshold avoids frivolous claims for changes in circumstances of relatively minor changes in the volume of work. The effect in any event will depend upon the contract terms and pricing regime but such restrictions on the ability to claim as a result of quantity changes needs to be considered at the outset.

In FIDIC Red Book clause 12.3, the appropriate rate or price for a variation, where only the quantity has changed, is that specified in the contract, a new rate shall be appropriate if four tests are all satisfied. These are:

(a) (i) the measured quantity of the item is changed by more than 10% from the quantity in the Bill of Quantities or other Schedule,

(ii) the change in quantity multiplied by such a specified rate for this item exceeds 0.01% on the Accepted Contract Amount,

(iii) this change in quantity directly changes the Cost per unit quantity of this item by more than 1%, and

(iv) the item is not specified in the Contract as a 'fixed rate item'.

Whether the measured quantity has changed by more than 10%, for the purposes of subclause 12.3(a)(i), should be readily capable of being established, although it must be recognised that parties often disagree as to the final remeasured quantities of work and the outcome of that disagreement may make all the difference in relation to passing that threshold.

Subclause 12.3(a)(ii) requires that the value of the change in quantity exceeds 0.01% of the defined Accepted Contract Amount, which is explained in the FIDIC Guide as being to ensure rates and prices are not adjusted for changes in quantity that give rise to a minor effect on the contract price. It is suggested that this 0.01% is very low, being, for example, only US$ 10,000 on a contract price of US$ 100 million. However, it should not be forgotten that this test is just one of four in clause 12.3 that all need to be satisfied. The Accepted Contract Amount is defined in clause 1.1.4.1 as being the accepted tender amount. This is quite distinct from 'the Contract Price', defined in clause 1.1.4.2 as being after any adjustments are made under the contract. This is an important clarification. Where bespoke contracts have been drafted based on these or past FIDIC terms, it has occasionally been seen that what price (original in the contract or finally adjusted?) the threshold percentage applies to can be a cause of disputes between the parties if it is not clearly defined.

Subclause 12.3(a)(iii) requires that for a change in quantity that changes an item's 'Cost' per unit by more than 1% gives rise to difficulties in establishing what the unit costs would have been but for the change in quantity. The capitalised term 'Cost' was quoted elsewhere as it is defined in clause 1.1.4.3. A contractor may present a claim on the basis that the actual cost per unit is more than 1% higher that allowed in its tender pricing, but that is the wrong test. The tender allowance may simply have been too low. From an employer's perspective, a significant increase in quantity could quite readily be expected to lead to a costs reduction of more than 1%. In practice, however, it seems that contract administrators are rarely attuned to this possibility and that most assertions that rates and prices should change for a change in quantity are from contractors and are that the rates and prices should be increased rather than decreased for economies of scale.

Subclause 12.3(a)(iv) refers to items specified in the contract as 'fixed rate items' and may be particularly relevant to items in bills of quantities for which no quantity is given, but the contractor is required to state its rate, fixed no matter what the actual quantity becomes.

Such provisions for limits on the percentage change in quantities in contracts such as the FIDIC Red Book and the Infrastructure Conditions reflect that they are both remeasurable contracts and anticipate that quantities of the works as actually constructed will be different from those in the contract. This is clearly different to contracts such as SBC/Q which are not remeasurable and anticipate that any changes in quantities will be the subject of variation instructions under the contract.

Whether the quantities of a particular work item have increased or decreased, the influence of the quantity of work can be substantial or negligible depending upon the type of work, the quantities 'before' and 'after', and in some instances the timing of the work and the instruction to vary it. It is impossible to give a definitive list of all the different factors that might need to be taken into account in such circumstances, but the following examples serve to illustrate the range of factors that might be relevant in particular circumstances:

- Reductions in quantity might result in 'small load' charges being levied by suppliers of materials, thereby increasing the unit cost of materials.
- Increases in quantity might result in supplies of material having to be brought in from further afield or at greater expense. For instance, large increases in the quantities of imported fill material required may result in the exhaustion of local or lowest cost supplies. A project that was priced on the basis of a calculated balanced 'cut to fill' might now need substantial amounts of imported fill.
- Increases in quantity might have a similar effect on labour and/or staff. It may be that those directly employed by the contractor or already present on the site cannot service the increased scope. This might mean additional personnel being hired from an agency at a premium cost to those on the payroll. Further complications arise where the local construction market is short of the relevant skills. Additional costs might accrue if such resources have to be hired from agencies or brought from other projects. On one project in Africa, the contractor was faced with extensive increases in the scope of work and was unable to source more staff locally, so it imported its additional site staff from Europe at a significant premium.
- Similarly, in relation to plant and equipment, it may be that increases in quantity mean that those items owned by the contractor or already present on the site cannot service the increased scope such that additional items have to be hired from external plant hire companies or brought from other projects.

- Increases or decreases in quantities might influence the economics of plant employed on the works. This is particularly a problem in contracts where large expensive items of capital equipment are employed in anticipation of a particular volume of work. Such circumstances may, for instance, involve significant fluctuations in quantities of dredged material affecting the economics of mobilised dredging plant, increases in volumes of work resulting in site cranage or other plant being inadequate for the revised works.

The influence of such factors will require the analysis of the relevant contract rates and prices and subsequent adjustment of the affected elements to reflect the proven influences of the changes in quantities, in much the same way as was set out elsewhere for changes in conditions or character of the works.

Whilst the illustrative examples of factors set out elsewhere focus on how changes in the tendered and contracted quantities of work can have a significant negative effect on a contractor's costs and efficiencies, it is possible that they can have some positive consequences. It sometimes appears that contractors consider that whether quantities of a work item have gone up or down, they are entitled to a re-rating of that work upwards and make a claim for additional monies on that basis. In extreme examples, contractors have even been known, when faced with significant increases in quantities, to claim that the affected work should be valued on the basis of their actual costs rather than adjustment of contracted rates and prices, particularly where those actual costs significantly exceed the contracted rate for no better reason than that the tender was wrong or the contractor was culpably inefficient. However, as illustrated elsewhere, it does not necessarily follow that an increase in quantity means an increase in rate at all. In particular, increases in quantities can give rise to additional efficiencies that save a contractor costs per measured unit of work. This can be true in relation to costs of components such as materials (where more are purchased at a bulk purchase rate) or labour, plant and equipment (which can work more efficiently on a larger scope of the same work). This point generally can be illustrated by simple consideration of how published construction and engineering pricing books, such as Wessex, Spons and Laxtons, tend to have separate sections for 'major' and 'minor' works, with the unit rates and prices for the former lower than those for the latter.

A specific further area that is commonly overlooked in relation to the cost benefits of increased quantities may be on the costs of preliminaries and general items. Where bills of quantities do not contain a separate section for the pricing of such items, or where such a section only covers some items (for example those that are time-related rather than work-related), an increase in quantities of measured works items will give rise to an increased recovery of any preliminaries and general items allowed in the unit rates and prices for such measured work. This may warrant consideration of whether such unit rates and prices should be reduced accordingly. On the other hand, where the preliminaries and general items are priced separately, that may also require consideration for changes in quantities of work. This is considered further elsewhere in this chapter. Each instance of such questions will involve their own merits based on the particular facts, pricing and what the contract says regarding valuation.

An illustration of how an increase in the quantity of any item can render it unreasonable where it includes for cost components not affected by the increase in quantities is the Hong Kong Court of Appeal judgment in *Maeda Corporation, Hitachi Zozen*

Corporation, Yokogawa Bridge Corporation, Hsin Chong Construction Co Ltd v The Government of Hong Kong Special Region, CACV 230/2011. This case was outlined above, but is worth further detailing here.

The bills of quantities had included a separate section for the bidders to price their preliminaries and general items. As explained elsewhere in this book, this will usually be where contractors allow for costs that do not directly relate to a bill items for measured works and are not directly affected by changes in the quantities of such measured works. However, in this case, the joint venture contractor had loaded its rate for an item of measured works with such fixed costs that would normally be included elsewhere, because it knew that the quantity of the measured work item was understated and would receive a windfall when that quantity was corrected on remeasurement under what were terms similar to those of remeasurable contracts such as those of the FIDIC and the ICE, being the 1999 edition of the Government of Hong Kong General Conditions of Contract for Civil Engineering Works. This included in GCC 59(4)(b) that:

> Should the actual quantity of work executed in respect of any item be substantially greater or less than that stated in the Bills of Quantities (other than an item included in the daywork schedule if any) and if in the opinion of the Engineer such increase or decrease of itself shall render the rate for such item unreasonable or inapplicable, the Engineer shall determine an appropriate increase or decrease of the rate for the item using the Bills of Quantities rate as the basis for such determination and shall notify the Contractor accordingly.

The Court of Appeal dismissed an appeal against a refusal to grant leave to appeal against the award of an arbitrator in relation to the application of this clause to the circumstances of this case and the contractor's pricing of the item. The contractor had argued that the change in quantity of the item did not render its rate unreasonable, because the high level of that rate was the result of the inclusion of extraneous fixed items within it and were unrelated to its quantity. The employer had argued that since the fixed items were unaffected by the change in quantity, the rate was rendered 'unreasonable or inapplicable' in relation to the increased quantity, and therefore the engineer should determine an 'appropriate increase or decrease of the rate' for the increased quantity. The Court of Appeal agreed with the arbitrator's award that the bill rate was unreasonable for the additional quantity because it included an element that was not increased in proportion to the increase in quantity.

It goes almost without saying that it should be for the contractor to establish the influence of such factors by providing relevant substantiation in the form of, for example, breakdowns of its original pricing and quotations, invoices, delivery notes, etc., for its actual costs. As noted elsewhere, in the Maeda case, the contractor's failed arguments included that the build-up to its tender rates was not admissible evidence. However, difficulties can arise in some circumstances where not every factor is easily established by such documentation. This is particularly so in the case of large items of plant and equipment, already mobilised and established on site, and which are then claimed to be influenced by changes in the volume of work.

As an example, consider the case of a large dragline excavator mobilised to site for a particular excavation operation, which it is anticipated in the contract bill of quantities will comprise some 40 000 cubic metres of excavation. Assume that the contractor priced its unit rate for the relevant excavation thus:

		$
Dragline – mobilisation	2 days @ $1,500	3,000.00
Dragline – working	40 000 m^3 @ 750 m^3 per day @ $1,500 per day	80,000.00
Dragline – demobilisation	2 days @ $1,500	3,000.00
	For 40 000 m^3	86,000.00
	Per m^3	$2.15

As a result of a redesign and alterations to the required excavation levels, assume that the volume of excavation available for the dragline is reduced to, say, 32 000 m^3. In effect, the mobilisation and demobilisation charges will now be spread over the reduced quantity ($6,000 spread over 32 000 m^3, i.e. 18.75 pence per m^3), instead of being spread over the original 40 000 m^3 (15 pence per m^3). The rate would therefore need to be increased by the difference of 3.75 pence per m^3. This of course assumes that any related effect on associated plant such as trucks for carting away, etc., is dealt with under separate items, and prices, for disposal of excavated materials.

Such adjustments should not raise undue difficulty where the detailed pricing of an item of plant can be readily established. Where the adjustment is based on actual costs, this should also be relatively easy to value. Adjustment on a costs basis may not be so straightforward where the contractor owns the plant and external charges do not apply. Rates used for mobilisation and demobilisation might be simply estimates or allowances of the costs that the contractor will incur for use of its own resources. Some items such as transport and associated labour should be possible to establish, but further difficulties posed by such issues are discussed elsewhere in this chapter.

These examples consider plant or equipment already owned or hired-in by the contractor for a particular task. Alternatively, it may be that an item has been purchased specifically for a particular project. Thus, the contractor incurs a fixed capital cost at the outset of a project with the expectation that this capital cost will be recovered in part (where the plant will be sold or reallocated to another project) or possibly in whole (where the length of project and useful life of the plant or equipment is such that it will be written off against that project). Particular examples of the former might include tower cranes with large capital costs and long useful lives. Particular examples of the latter might often include scaffolding and temporary buildings on a long project, where their residual value and condition are poor and the costs of dismantling and making ready for re-use are high. If the unit rates for works that were intended to fund that capital cost are, in the event, applied to reduced quantities of that work, then there may be an under-recovery of costs arising out of the variation that reduced those quantities. This may merit an increase in the unit rate for those reduced quantities to maintain the recovery of the capital cost. This process may be complicated if the reduced quantity means that a planned 'write-off' is no longer necessary and the plant or equipment is given some new or increased residual value that it would otherwise not have had. Further considerations also include whether the reduced quantity is in proportion to the reduced period of retention of that plant and equipment being required at the site.

In each case the assessment of changed rates and prices for changed quantities need consideration on the merits of each case, their particular facts and the provisions of the contract. What should be clear from the above is that the common contractor view that any change in quantity means an increase in rates and prices, and the contrary employer view that an increase in quantity always means reduction in rates and prices, cannot be assumed. In many cases quite complex calculations can be required to provide a resulting valuation properly in accordance with the conditions of the contract.

5.3.5 Effect of Variations on Other Work

Some contracts expressly provide the opportunity to claim additional payments in respect of other work, not itself the subject of a variation, if the conditions under which that other work is executed are affected by compliance with a variation instruction. Effectively, the other work is regarded as having also been the subject of variation such that the variation valuation rules also apply to that other work.

For example, SBC/Q clause 5.9 provides for adjustment to the valuation of other work as a result of: compliance with a variation instruction; instructions for the expenditure of a provisional sum for undefined work; instructions for the expenditure of a provisional sum for defined work that is different from the bill description of such work; or the execution of work that is covered by an approximate quantity in the bills of quantities to the extent that the actual quantity differs from that anticipated in the bills of quantities. However, the effect on that other work has to be 'substantial change' in the conditions under which the other work is executed. This clause is therefore subject to scope for disagreement as to what is, and is not, 'substantial change'.

The Infrastructure Conditions contain in clause 52(5) a more general provision for adjustments of the rates and prices 'for any item of work (not being the subject of any variation)' as a result of any variation. Should either the engineer or the contractor consider that this has occurred, they give the other a notice before the varied work is commenced or as soon thereafter as is reasonable. In practice, contract administrators rarely if ever give such a notice because they consider it not in their interest, or that of their client, to do so. Contractors rarely do so in advance, or in what employers and their agents consider a reasonable time, because they are not sufficiently aware or resourced for such administrative burdens. The common defence by contractors in such situations, where seeking a retrospective revaluation of work not varied but affected by a variation, much later than when the work was done, usually refers to the difficulties of knowing such indirect effects at the time, especially where many variation instructions are being issued and on a regular basis.

The FIDIC Red Book contains no express provision to adjust the rates and prices for other work as a result of instructed changes in quantities of varied work items. It is, however, clear that if changing one item of work can have an indirect effect on other work that should merit adjustment to the valuation of such other work. Indirectly under FIDIC Red Book clause 13.1, clauses that might enable a contractor to claim such adjustment include those that define what constitutes a variation. Subclause 13.1(b) refers to 'changes to the quality and other characteristics of any item of work'. Subclause 13.1(f) refers to 'changes to the sequence or timing of the execution of the Works'. In appropriate circumstances either of these subclauses may give a contractor entitlement to make a claim for adjustment of the rates and prices for other works indirectly affected by a variation.

The principal problem that then follows in practice is deciding and agreeing what the effect was on the other work and how to adjust its value suitably. This can prove subjective and difficult and often lacks records to form a suitable basis. Contractors may seem to be seeking to exploit it in situations that lack merit and employer's agents may seem to be seeking to deny meritorious claims on the basis of a lack of proof.

5.4   Inclusion of Preliminaries and General Items

Clause 5.6.3.3 of SBC/Q requires that when evaluating the effects of variations allowance shall be made where appropriate for any associated addition or reduction in the amount of the preliminaries and general costs items. The requirement is mandatory, by use of the word 'shall', and the type of preliminaries and general items subject to adjustment is limited to those contained in the Standard Method of Measurement. There is no corresponding clause in the Infrastructure Conditions but clause 52(5) allows the rates for any work, not itself the subject of a variation, to be adjusted if the contract rate for that item is rendered unreasonable by any variation. Such items would include preliminaries and general type items and this therefore provides a mechanism for adjustment of such items.

The difference in approach between these two contracts again reflects the difference in the basic philosophy, i.e. lump sum contract versus remeasurement contract. While both approaches are understandable, the SBC/Q approach, of having a mandatory requirement to include preliminaries and general type items in the evaluation of each variation, gives rise to various potential issues in practice. In particular:

- It may not be possible to accurately identify the effect on preliminaries and general items for an individual variation, but a group or series of variations may collectively have a demonstrable effect.
- The problem of identification of the effect on preliminaries and general item costs is often exacerbated by a lack of detail of description in relation to items in a preliminaries and general items section even if bills of quantities are prepared on the basis of a published standard method of measurement. It is notable how this is allowed under such publications that require, in contrast, very detailed descriptions and enumeration of items for measured works.
- Practice can vary between contractors in relation to where they price all of their preliminaries and general items, particularly between time-related and activity-related items.
- Contractors can be tempted to apply for additional preliminaries and general costs on every variation, without regard to the particular effects.
- Employers and their advisors can be tempted to dismiss claims for such adjustments out of hand, without due consideration of the merits in each case.
- Some employers and their advisors can be quick to propose a reduction of preliminaries and general items for any significant reduction of work scope, often on a pro rata basis, without due consideration of the merits in each case and in particular fixed components.

- Some employers and their advisors can be slow to spot the need to adjust preliminaries and general items for a significant reduction of work scope that affects variable preliminaries and general items.

Some of these issues are often addressed in practice by the contractor including in its claim an allowance for the effect of a variation on one or more preliminaries and general items, and the employer's agent promptly deleting such allowances on the basis that the individual variation cannot demonstrate such an effect on the preliminaries and general items. In the strict terms of the contract the allowance is to be made in the individual valuation of the various variations, but a sensible approach may seem to be for the contractor to indicate where it anticipated a variation would impact on the preliminaries and general items and the employer's quantity surveyor to conduct a reasoned assessment for relevant groups or series of variations. Large variations may of course not require such treatment and may be capable of being dealt with individually. Providing such an approach is not abused by the contractor in adopting a policy of simply indicating that all variations have a potential impact on the preliminaries and general items, it should be possible to include a sensible evaluation within the variation account, based on specifics. Where the wording requires such inclusion of preliminaries and general costs to be 'where' or 'as' appropriate it can be argued that the assessment of their adjustments on a group or series basis satisfies that approach.

Typically the type of preliminaries and general items priced in that section of bills of quantities, rather than within rates for measured works, that may be unaffected directly by changes in the scope of the works will be site supervision and services such as cranage. As the contractor will usually have priced such items on a time-related basis, it is therefore often difficult to argue that an individual variation, or even a group of variations, has resulted in additional requirements in the absence of evidence that the time of those resources was extended. Sometimes employer's agents confuse these two issues by responding to a claim for additional preliminaries and general costs related to variations on the basis that it can only be considered as part of a claim for delay and prolongation costs. For supervision it will usually be necessary to show that additional supervisory staff have been employed as a result of the variations, or the anticipated supervisory staff have had to work additional time at additional cost to the contractor. For cranage it will usually be necessary to demonstrate that the cranes have worked longer hours as a result of the variations and additional costs, such as the operator's costs, have been incurred. This can be quite difficult where the anticipated requirement was priced on a time basis and the site working hours have not had to be extended beyond those defined in the contract documentation. It will also be necessary to demonstrate that the additional costs were incurred at times relevant to the varied work and are not simply the result of more general increases in the contractor's supervisory or services costs over and above those allowed in its pricing of the contract. On analysis, the increase might turn out to be the result of the contract pricing being underestimated or additional time and resources being required to address the contractor's own problems.

It is of course necessary to ensure that time-related adjustments to preliminaries and general items within the variation pricing are considered for the elimination of duplication when pricing any prolongation claim on a project, or claims for additional payment arising from disruption of the works, as there is an obvious potential for duplication between such claims. The need to avoid and adjust for such duplication between claims is considered in detail in Chapter 6 of this book.

One approach by contractors, when pricing variations at contract rates that exclude preliminaries and general costs because they are included in a separate bill of quantities section, is to claim additional preliminaries and general costs within the value of variations by adding a percentage for them. Thus, if the total value of a preliminaries and general bill of quantities is 15% of the total value of all measured works bills of quantities, a 15% addition for preliminaries and general costs will be claimed on all variations priced at bill rates. Whilst it is possible that the contract terms provide for this, and it may provide a pragmatic approach that suits certain circumstances, it is all too often adopted as a simplistic approach to recovery of costs and increasing the claimed value of the variations. Conversely, employers and their advisors may be wrong to dismiss the approach out of hand. A proper approach and reaction to such an adjustment is to look at the details of the case on its individual merits. It may be that, if delays on a project are all driven by variations and the contractor has lost significant amounts on preliminaries and general items that are not recovered through its delay claims and the direct valuation of variations, such an approach has some merit in giving rise to a pragmatic approach and a reasonable financial result if it is calculated correctly with any duplications adjusted.

As noted elsewhere, adjustment of items in a preliminaries and general items bill of quantities is often hampered by the lack of detail in the description and enumeration of them. This reflects the approach dictated by standard methods such as NRM, CESMM4 and POMI, which were explained in Chapter 2. For example, in bills of quantities for construction of a new international airport, an item in the preliminaries and general items section of the bills had no more description than 'Allow for scaffolding and working platforms' and was priced as a lump sum item at just under US$ 14,000,000. In contrast a whole page of the measured works section of those same bills of quantities was devoted to a series of ten items for anchoring masonry block walls to concrete columns or walls, each quantified and with their descriptions referring to a further detail in the specifications. Those anchorage items were all priced at 48 cents each and the total value of the ten items was only US$ 6,000. The bills of quantities also required that such rates for measured works were to be shown split between labour, materials and plant, thus adding more detail to the pricing of the anchorages. This is not to criticise those bills of quantities, which had been prepared generally in accordance with the POMI Rules. Neither is it to criticise those rules, which are not alone in such apparently inconsistent requirements for the level of detail of items in different bills of quantities section. This example is only given to illustrate the difficulties that can arise when attempting to include adjustment of preliminaries and general items as part of the valuation of variations, as a consequence of the lack of detail when pricing many financially very significant preliminaries and general items in bills of quantities.

A further complicating factor mentioned elsewhere is doubt and differences in practice in relation to where contractors allow for preliminaries and general items and where they have their own bill section, but practice may be to allow some of them in the prices for measured works to which they directly relate. For example, a tower crane that is to service a whole section of a site and several trades and activities on it is likely to be priced in an item for 'Cranes and lifting plant' in the preliminaries and general requirements section of bills of quantities prepared under POMI Rules, applying rule A5.1.3. On the other hand, a mobile crane brought to the site for the specific activity of offloading and placing air-handling units might be priced in an item for that equipment in the mechanical engineering installations section of bills of quantities prepared under POMI Rules, applying rule Q4.1. Again, a lack of detail of the contents of such items may hinder efforts to price the effects of variations on preliminaries and general items.

It is as a result of such issues that the adjustment of preliminaries and general costs in the valuation of variations is a common source of disputes.

5.5 Percentage Adjustments

Percentage adjustments commonly occur in the context of claims arising from change on construction and engineering contracts in several separate guises: firstly, the application of agreed percentages as components of the calculation of the value itself; secondly, the use of threshold percentages of change that must occur as a prior requirement to any adjustment to the contract rates and prices; thirdly, the use of percentages of the contract sum or price as the value of the adjustment itself; and, fourthly, as a cap on the amount of claims.

5.5.1 Percentages in Variation Valuation

It seems obvious that if the parties to an engineering or construction contract can agree on components of the calculations that go to the evaluation of the consequences of changes being covered by agreed percentages then this should make the process easier and lessen the scope for disagreement. Obvious examples are matters such as off-site overheads and profit. Other possibilities include attendances on subcontractors, particularly those nominated by the employer.

In Section 5.7, we set out the requirements of NEC4-ECC for percentages to be set out in the Contract Data to cover such indirect costs and mark-ups such as design, overheads and profit. Therefore, we do not focus on that form's use of percentages here.

It has been explained elsewhere how FIDIC Red Book clause 12.3 provides for the valuation of some variations on the basis of 'Costs', defined in clause 1.1.4.3 as excluding profit, but with the addition of 'reasonable profit' under clause 12.3. It would be open to the parties, and might be considered sensible practice, to agree at the tender stage what a 'reasonable' level of profit was and include that percentage in the contract. There might even be different percentages depending on whether a variation was carried out by a subcontractor or directly be the contractor's own labour. The Red Book and most other contracts in the FIDIC suite leave the level of profit only as open to the test of reasonableness. The only exception to this is the FIDIC Pink Book, which adds a further definition that 'reasonable profit' is 5% unless otherwise agreed in the contract.

Regarding percentages in relation to subcontractors, standard methods of measurement often provide that, where amounts are included in the bills of quantities in relation to nominated subcontractors, bill items are to follow that allow the contractor to price for both its overheads and profit and its attendances on each subcontractor. The pricing of these items will then be used as a basis for the contractor's mark-ups on any variations carried out by such nominated subcontractors. There is, however, no reason why this should not be extended to the parties agreeing percentages in the contract to cover attendances, overheads and profit on costs incurred with domestic subcontractors where these carry out variations. In such cases, particular care must be taken in relation to the attendances item to ensure that it is clear what is included therein.

Even where variations are carried out by the contractor using its own direct labour and materials, rather than through a specialist subcontractor, there is no reason why a percentage addition for overheads and profit should not also be part of the tenderers' bids and agreed in the contract.

The use of percentages when valuing variations on the basis of dayworks is considered in Section 5.6.

5.5.2 Threshold Percentages

It is common in many ad hoc contracts for engineering and construction, and occasionally other types of contracts, to state that the contract rates and prices shall apply without amendment unless a stated percentage of change in the volume of work undertaken is experienced. This is a topic covered in detail elsewhere in this chapter, particularly in relation to FIDIC Red Book clause 12.3. However, it is worth considering a few more general points as follows.

Where there is express provision in the contract, the percentage of change stated as a precedent to any amendment to rates is often in the order of perhaps 15% or 25%, either of an increase or a reduction in the amount of work. This is less common in standard building and civil engineering contracts, and the JCT and ICE stables of contracts have traditionally not included such precedents. However, they are not unknown and it is possible to find amendments to SBC/Q or the Infrastructure Conditions being incorporated in the contract documents to have the same effect. Similar provisions can also be a feature of ad-hoc or one-off contracts. On the other hand, the FIDIC Red Book does contain such a provision in its clause 12.3 and the preceding edition also contained one, though it was less complicated. The equivalent clause 52.3 in the 1987 edition set the threshold at 15% of 'the Contract Price excluding Provisional Sums and allowance for dayworks' and this level seems to have been adopted by many ad-hoc forms used around the world that were based on those FIDIC provisions.

The desirability of such devices is debatable but is usually intended to prevent argument over the rates and prices when the change in the amount of work is regarded, at least by one party to the contract, to be not significant enough to warrant such changes. That these provisions usually occur in contracts used by large client organisations that are regularly engaged in such contracts supports the contention that they are seen to be a potential safeguard for the employer to avoid claims. They do, however, raise some significant potential problems where the terms do not expressly provide otherwise:

1. Where the criteria is the change in the amount of work, how is this judged? Is it the change in the contract price, i.e. the total price to be paid for the works? Or is it the measured quantities of work, and how is this calculated across many items with different units? Or is it a change in the amount of resources to be expended on the works?
2. Does the change apply to individual items or groups of items or sections of work? Thus, do the works as a whole have to have changed by the threshold percentage, the rates for all items then being open to adjustment? Or, for example, does it involve just all earthworks items as a group? Or can individual bills of quantities items be

considered in isolation and adjusted if they have individually changed by the threshold percentage?

3. Is the amendment to the rates and prices to include preliminaries and general items priced in their own section of the bills of quantities? Or is it restricted to only the unit rates for the affected measured works items? Or does it apply only to the preliminaries and general items (often alternatively referred to as 'indirects' in some industry sectors)?

4. What are the rules to be applied in assessing any amendments to the rates and prices if such are justified? Are the contract rates and prices to be used as a basis for evaluation? Is actual cost relevant?

Sadly these matters are often left unanswered, or only partly answered, by the express provisions, leaving the gaps to be filled by what should have been unnecessary argument between the parties. Other matters that need to be considered include the following.

Where the criteria is the change in the contract price (the FIDIC approach and especially that followed in bespoke contracts based on its 1987 edition with its 15% threshold in clause 52.3), the adjustment depends on whether or not the contract price has changed by more than the threshold percentage. This inevitably leads to doubt where there are large values of disputed variations and other claims, the success or failure of which will or will not put the value over that threshold. In very marginal cases it has even been the case that the final value of the works is so close to the threshold that it is only the contractor's claim to revalue its rates and prices because of an increase in value that has pushed it over the threshold.

A further feature that has been mentioned elsewhere in this chapter is how contractors often see the breaching of such a threshold as opening the flood gates to making claims for increases in unit rates, whereas the correct result might be that no change in rates is warranted or that the increase in value of work has actually disproportionally increased its recoveries such that reduced rates are merited. This is for such as economies of scale or where measured works rates include fixed elements that are not increased in proportion to their increase in quantity.

5.5.2.1 Criteria for Judging Change in the Amount of Work

It seems obvious to suggest that the criterion for judging a change in the amount of work should be the value of that work, and the work used as a basis for establishing that change should be the measured work items, i.e. excluding the preliminaries and general or 'indirects' element, as it is the volume of the contractor's site resources that is usually intended to be used in judging a change in the amount of work executed. In this context 'amount' is usually understood to be measured quantity of work and adopting the alternative of value as the criterion can sometimes cause a problem.

If it is the value of the work that is used as the criterion, the intention of the restriction imposed by the percentage threshold may be circumvented by relatively small increases in the quantities of high value items. In other circumstances large changes of omission and addition, which cause the nature of the work to change in a manner significant to the contractor, may have the effect of largely cancelling each other out in value terms and thereby resulting in a change in value of less than the percentage required to allow a review of the relevant rates and prices.

In poorly worded clauses the position may also be unclear where the threshold is based on value and there are extensive additions and omissions of different items that cancel each other out in terms of value. In such cases, contractors sometimes argue that it is the combined total of the value of additions and omissions that should go to define the amount of change. Appropriate wording of the related contract provisions should clarify this point.

The intention is presumably to allow the contractor to review its rates and prices in the event that the resource profile of the project differs substantially from that envisaged at the time of tender as a result of ordered changes in the works. If that is so then it would seem reasonable to use a measure of the contractor's resources other than value, if such is available, to judge when and if the threshold level has been achieved. In many contracts containing such clauses the works are largely of a mechanical engineering nature, with many of the materials being provided free of charge to the contractor by the client. These contracts also commonly have contract bills of quantities that indicate not only the unit prices for the works but also the unit man-hours. In such circumstances there is a strong case for using the change in man-hours, being the prime resource provided by the contractor, as the criterion by which the amount of change should be judged.

In other circumstances the parties should consider, before adopting such threshold provisions, how the change is to be measured, and set out the conclusion in the contract to avoid later complications as a result of difficulties such as those discussed elsewhere.

5.5.2.2   Which Rates Are to Be Amended?

Whilst judgment as to whether a threshold percentage has been passed will usually be based on the amount of measured work done, the next question is which rates and prices are to be adjusted as a result? The intention of such threshold restrictions is often that they should apply only to the adjustment of the preliminaries and general items, or 'indirects', which are prices in the event that the threshold for review is exceeded. There is some logic to such an intention. Firstly, the changes in the amount of work that go to the threshold calculation will effect a change in the value of the relevant measured works items. Secondly, if the preliminaries and general items or 'indirects' are where the site supervision and management and general services are priced, then it is they that are most likely to be affected by substantial changes in the amount of work undertaken at the unit rates and prices in the contract.

If it is intended that the unit rates and prices themselves are to be reviewed where the contract threshold for the change in scope is exceeded, then the contract should make this clear and also which rates apply.

Whether all rates are subject to adjustment, or only certain rates, the contractor should be aware of the intended regime when pricing the works so it can ensure that any costs that need to be adjusted if the threshold change occurs are priced in the correct part of the contract.

5.5.2.3   What Rules Apply?

It is not uncommon for such clauses simply to provide that the rates and prices, be it for preliminaries and general type work or the unit rates and prices for the measured works themselves, can be reviewed if the threshold is exceeded without stating what, if any, rules of valuation apply to that review. In the absence of any express provision to the contrary, it is suggested that the review of prices should be by reference to the contract prices so that any adjustments reflect the level of pricing in the contractor's tender. This is how FIDIC Red Book clause 12.3 addresses the issue, by reference to a typical hierarchy of valuation methods that starts with the contract rates and prices.

It has been explained elsewhere how the 1987 edition of the FIDIC Red Book included in its clause 52.3, 'Variations Exceeding 15 per cent', a much simpler approach. This was based on the total value of all variations and adjustments of the estimated quantities of works in the contract bills of quantities expressed as a percentage of the 'Effective Contract Price', that is the 'Contract Price' excluding Provisional Sums and allowances for dayworks. If the resulting percentage was in excess of 15% then adjustment was due to be agreed between the parties or, failing that, to be determined by the engineer 'having regard to the Contractor's Site and general overhead costs of the Contract'. As to how this adjustment should be made, the only guidance in clause 52.3 was that 'Such sum shall be based only on the amount by which such additions or deductions shall be in excess of 15 per cent of the Effective Contract Price'.

Whilst the 1987 edition of the FIDIC Red Book was superseded in 1999, many bespoke forms of contract in use in various parts of the world are still based on that old version of the Red Book. Understandably, contracts containing such a provision see regular disagreement as to how this is to be done.

5.5.3 Percentages for Defective or Incomplete Work, etc.

Some contracts, particularly in the process plant or mechanical engineering industries, adopt the device of adjustments to the contract price, in the event that work is incomplete or is defective, by a proportion or percentage of the contract price. It is not uncommon for such contracts also to include a cap on the contractor's liability for defective work, for instance, by use of a percentage restriction.

In many cases the contract may simply provide that the client is not obliged to pay the portion of the contract price that relates to unfinished or defective works.

The issue of how incomplete or defective works should be valued, where such a percentage approach is not adopted in the contract, is a subject of Chapter 8 of this book.

As ever, the express terms of the contract and the underlying legal position will be central to deciding how such provisions are to operate. If the contract simply provides that the relevant proportion, or percentage, of the contract price is not to be paid then there will be ample scope for disagreement as to what represents the relevant proportion where there is no relevant breakdown of the contract sum. It is preferable that the means of measuring such adjustments is set out clearly in the contract, whether by reference to the progress monitoring and reporting regime or by reference to contract requirements for agreement of the state of the works when handed over from the contractor to the employer.

If the contract progress monitoring and reporting system is to be used to establish relevant percentages then it is essential that the system is capable of sustaining such reliance and is properly operated and kept up to date. Recording of the relevant completion percentages at the time of handover will also require careful attention to avoid later disputes. It is preferable for the contract to clearly set out whether the percentage or proportion is to be calculated as a whole, over all the works and services provided by the contractor, or whether it is intended that the assessment should be done for separate elements of the works. If the latter course is adopted, there will of course be further implications for the progress monitoring and reporting as well as handover documentation and procedures.

The contract also needs to define the value to which the percentage or proportion is to be applied, usually the 'contract price', i.e. the value of the contract after all adjustments except those that are the subject of the percentage or proportion adjustments. However, doubt can occur where there is no clarification that the percentage does not just apply to the original contract price agreed in the contract, unamended, for example, as variations and other changes. Clarity in this regard is important to avoid disagreement.

Such devices can obviate the need for much detailed measurement and pricing of work not executed as contemplated by the contract, and laborious marking up of record drawings with the incomplete elements, which on a major mechanical installation can be a time-consuming and costly job, but if implemented without sufficient prior consideration and clarity, such devices can cause more disagreement than they save. One such potential issue may relate to whether the mechanism gives a financial result that reasonably reflects the actual loss or damage suffered by the employer. If the employer's claim for incomplete or defective work is on the basis of breach of contract by the contractor, rather than revaluation of the works, then the contractor may be able to say that the financial result does not address the requirement to put the employer in the position that it would have been in but for the breach. This suggestion assumes that the applicable law of the contract makes this the test of quantification of damages for breach. This topic is considered in more detail in Chapter 8.

5.5.4 Percentage Caps on Adjustments

Just as percentages are sometimes used to define the amount applicable to a particular circumstance, there are occasions when the amount of certain defined adjustments can be limited by a provision in the contract to a percentage of the contract, or sometimes order, value. For instance, liability for delay damages for non-completion by the due contract date or dates might be subject to a limit of 10% of the contract sum. Such devices serve to limit liability in circumstances where the potential liability, if unrestricted, would be such as to discourage the contractor, subcontractor or supplier from undertaking the contract, or where an unrestricted liability would mean the inclusion by tenderers of substantial 'contingency' amounts in their pricing. Thus, whilst such a cap appears on first sight to be only in the contractor's interests, it may actually be of greater benefit to the employer in terms of getting a good competitive tendering process.

However, this is a particular further area where the parties can fall foul of a lack of clarity in the wording of the contact. For example, it may not be clear whether the percentage applies to the contract price as originally set or that adjusted for variations and other changes. A low percentage cap on a party's recoverable damages under an express clause may also encourage it to seek to circumvent that cap by such arguments that it is not an exclusive remedy. The topic of delay damages is addressed further in Chapter 6 of this book where such issues are considered in detail.

5.6 Valuation Using Daywork Provisions

SBC/Q clause 5.7, FIDIC Red Book clause 13.6 and the Infrastructure Conditions clauses 52(6) and 56(4) all contain provision that the valuation of additional or substituted work can be valued by use of the contract daywork provisions in some circumstances. This involves contemporaneously recording the resources and time expended by items such as labour, plant and equipment on the variation and pricing these at rates and/or additions to their costs, as agreed in the contract.

The criteria for deciding if such a method of valuation is appropriate vary between contracts and, as ever with such clauses, careful reading of the agreed terms is essential before applicability, and the details of application, can be assumed. Generally, it could be suggested that dayworks are relevant to variations of a minor or incidental nature whose scope cannot be pre-established with such degree of certainty as to allow pre-agreement (where the contract provides for pre-agreement) and which do not lend themselves to measurement and valuation after the work is completed.

Even in the main standard forms considered in this book there are subtle differences in defining variations subject to a daywork regime. In the case of the SBC/Q, it is that the work 'cannot be valued' by measurement. In the case of the FIDIC Red Book, dayworks apply to 'work of a minor or incidental nature' in which case the engineer 'may instruct' a daywork approach. In the case of the Infrastructure Conditions, the criteria is simply that the engineer considers such a method of valuation 'necessary or desirable'. This Infrastructure Conditions provision's lack of an objective criteria can cause issues.

In contrast, given their regime of valuing compensation events on the basis of forecast or actual cost, NEC4-ECC contains no dayworks provision.

In practice, whether the contract requires the contract administrator to instruct or decide that a dayworks approach should be applied can be significant, because they can often be reluctant to do so even in the most obviously appropriate circumstances. There seem to be two common reasons for this antipathy towards dayworks on the side of the consultants:

- Firstly, those operating contracts that apply the traditional approach of valuing variations on the basis of the contractor's original rates and prices may consider that the contractor should be 'pinned' to those levels at which it priced its successful tender for all additional works. Clearly this is a misguided approach where the circumstances of a change merit valuation using a daywork provision that the parties have agreed to apply to such situations.
- A more common reason for reluctance to apply a dayworks clause is the belief that it will over-compensate the contractor and/or encourages inefficiency because the contractor gets all of its actual resources recorded and compensated.
- Over-compensation may arise because the dayworks sections of the tender documents were not properly managed by the employer's advisors at the pre-contract stage. When those advisors adjudicate on tenders received and advise on selection, it may be the case that they have not considered the contractors' pricing of the dayworks sections. Good practice in the preparation of bills of quantities is that the dayworks section should be structured and priced by the bidders in such a way that its pricing is reflected in the overall tender total. However, if such good practice is not followed and a bidder has priced such a dayworks section at exaggerated rates and percentages within a tender for measured work that is very competitive, then the consequences of the overpriced dayworks section may be overlooked. Again, this is not a valid excuse where the circumstances of a change merit valuation using a daywork provision that the parties have contracted to apply to such situations.

Such causes of client and consultant antipathy to a daywork valuation are often misplaced. However, it is also the case that a dayworks approach can indeed result in a claimed windfall for the contractor against its actual costs, as illustrated by the following example.

On an infrastructure project in the Middle East, the contractor had to retain its dewatering systems installed to allow construction of a new sewerage pumping station for longer than planned due to a delay by the employer. Under a FIDIC Red Book based contract, the engineer had not instructed the extended period to be covered by daywork records, but the contractor maintained and submitted these on a regular basis and the engineer's representative signed them off. On the basis of those records the contractor claimed AED 4.8 million, correctly calculated applying the daywork rates in the contract to the signed daywork sheets. However, the employer valued the extended hire at less than AED 0.4 million, accurately calculated by applying rates in the contract for that dewatering to the extended period. By way of another alternative valuation, disclosure of the contractor's cost records showed that its actual costs of the dewatering were AED 1.25 million. The decision of an arbitral tribunal on this one item of a large final account dispute depended on issues of fact and law that will vary between cases. Because the extended hire was the result of the employer and caused delays in construction of the new sewerage pumping station, the arbitral tribunal in this example awarded the actual costs of AED 1.25 million. However, the outcome might have been different under different legal and factual circumstances, for example if the overrun was the result of a variation. The example is detailed here to illustrate how extreme the range of results can be for the same variation when it is valued, depending on which approach to valuation is adopted, and how a dayworks approach might give a contractor a windfall. Alternatively, it is also noteworthy how a valuation based on the rates in the contract would have involved the contractor in a substantial loss against its actual costs in this case.

Contracts that set out a daywork alternative for pricing variations often refer to standard published documents as being the basis for daywork evaluations, or its 'prime cost' to which bidding contractors price their percentage additions for such items as preliminaries and general costs and overheads and profit. In the case of the Infrastructure Conditions the reference to the Civil Engineering Contractors' Association Schedules of Dayworks ('the CECA Schedule') at clause 56(4) is a default mechanism used only in the absence of the inclusion of a schedule of daywork rates in the contract rather than pricing percentage additions to such an external source of rates. In practice, the CECA Schedule is much more comprehensive and useful than the Definition of Prime Cost of Daywork issued by the Royal Institution of Chartered Surveyors and the Construction Confederation, which is referenced in the SBC/Q clause 5.7. It is not unusual with daywork schedules, and particularly the schedule referenced in SBC/Q, for gaps to be apparent in the lists of items of plant and equipment covered and these may require interpolation of prices for the resources actually used on dayworks. The alternative to reference to a standard published dayworks schedule is for the contractor to set out its hourly rates for each of items such as plant, equipment and labour in a schedule to the contract, but that approach particularly often leaves gaps against the plant and equipment actually recorded on the daywork records.

The FIDIC Red Book requires a Daywork Schedule to be included in the contract documents. Unless the local jurisdiction has a published source for the local 'prime cost' (such as the CECA Schedule) to which contractors only have to quote their percentage additions, such schedules usually set out detailed rates and prices as follows:

- For labour, the contractor's hourly rates for labour and supervision (at chargehand and foreman levels), with a rate against each grade and possibly additional rates for dayworks carried out in overtime. It is important to detail what is to be included in these rates, such as senior supervision, management, hand tools and non-mechanical plant.
- For plant, the contractor's hourly rates for plant and equipment, excluding hand tools and non-mechanical plant.
- For materials, a percentage to be inserted by the contractor, to be added to the actual costs of materials as evidenced by invoices. That percentage is to cover the contractor's on- and off-site overhead costs and profit.

Several limitations of daywork approaches to valuation should be recognised. Unless expressly stated otherwise, they are generally intended for the pricing of labour, plant and materials incidental to the main contract works, i.e. for the use of resources already on site. Daywork schedules are not generally intended to cover the cost of plant and resources brought to site specifically for a variation. This said, contracts are sometimes drafted to expressly include plant and equipment both already on site and that specifically brought to the site for a variation. This seems unfair, in that the same rates cannot fairly value both sources of plant and equipment. However, it may be considered a pragmatic approach that avoids the need to establish where an operative or item of plant or equipment used on the dayworks came from.

Daywork rates are also not intended to cover the cost of work undertaken after the completion of the main works, for instance during defects maintenance periods. This is, of course, true of variation valuation clauses generally. Although, on a very large building project where the employer intended from the outset that it would likely require changes to be made to the works after handover but whilst the contractor still had a presence on site, a bespoke contract contained a requirement for two sets of daywork rates and/or percentages to cover both circumstances. It is of course an extension of this principle that the published daywork rates are not for the valuation of contracts executed wholly on a daywork basis. The schedules contain detailed explanations of what they cover in respect of labour overheads and supervision. In the case of the CECA Schedule this is done by a percentage addition to the defined cost of wages.

One of the difficulties with the plant daywork schedule is the accurate identification of individual items of plant to the correct item in the schedule. A most useful guide in this respect is the 'Surveyors Guide to Civil Engineering Plant' published by the Institution of Civil Engineering Surveyors.

FIDIC Red Book clause 13.6, the Infrastructure Conditions clause 56(4) and SBC/Q clause 5.7 all require the contractor to submit contemporary records of the resources employed on ordered dayworks in order that the employer's team can verify them. Such a procedure is essential if unnecessary disputes concerning the resources deployed are to be avoided. If the contractor fails to submit records within the appropriate time there is a danger that disputes may be generated in respect of the resources deployed. Those periods for submission vary between contracts. It is: 'each day' in the case of FIDIC Red Book clause 13.6; 'at the times the Engineer shall direct' in the case of the Infrastructure Conditions clause 56(4); or 'not later than 7 Business Days after the work was executed' in the case of the SBC/Q clause 5.7. In such circumstances, where the contractor's daywork records are late, the question becomes what is the consequence of that failure? The contractor losing the entitlement to a daywork valuation seems an unlikely outcome where daywork appears the only reasonable approach to valuation. It is suggested that the contractor would still be entitled to recovery of payment for the dayworks but, in practice, any reasonable doubts as to the accuracy of the records should be resolved in favour of the employer, as it is unreasonable that the employer should be prejudiced by the failure to submit records in such time that proper verification is possible.

Once submitted, the next part of the process will be a signature by the contract administrator and it is also preferable to set a time limit for such a signature, or rejection, or correction as merited. In the Infrastructure Conditions clause 56(4) the time limit is only stated as 'within a reasonable time'. Neither SBC/Q clause 5.7 nor FIDIC Red Book clause 13.6 set any such time limit. It is suggested that, whilst it is clearly important that contractors submit their daywork sheets within a specified and short period, when memories are still fresh and/or the related activities can be viewed live, it is odd that SBC/Q and the FIDIC Red Book do not also require the contract administrator to sign them off within a set and reasonable period. Such a punctual approach by both parties should be in the interest of both parties.

One common habit among contract administrators is to sign daywork records on a 'without prejudice' basis, perhaps with the written rider 'For Record Purposes Only' or the abbreviation 'FRPO'. There may be two reasons for this:

- Firstly, to record that the resources and times are accurate, but without accepting that the works and resources covered are part of a claimable variation under the contract.
- Alternatively, to accept that the resources covered are part of a claimable variation under the contract, but not that it should correctly be valued on a daywork basis. The example detailed elsewhere of dewatering to a new sewerage pumping station illustrates why a signatory might not want to be relied on as approving a daywork approach where that results in an exaggerated financial claim.

Such qualifications are understandable where the site representative of the engineer, architect or contract administrator under the contract does not have the authority to agree that an item is a variation and/or how it should be valued. In such instances the signature against the record of resources at least take that element of doubt out of the equation. However, where such qualifications are used merely to avoid commitment to an approach that ought to be agreed and settled can be very unhelpful and only leave disagreement to be resolved later.

Where the FIDIC Red Book, SBC/Q or the Infrastructure Conditions are not being used the contract provisions should be examined to determine what provision is made for the ordering and valuing of work on a dayworks or 'time and resources' basis. In drafting ad hoc or one-off contracts it is sensible to include such provisions, as they may be required as a last resort for works where measurement and pricing evaluation of the direct consequences of an instruction for further or different work that can be described as being of a minor or incidental nature is not possible.

5.7   Use of Actual Costs

The most obvious forms of construction and engineering contracts that value change on the basis of actual costs are cost reimbursable ones. There the contractor's reimbursement for both the original scope of work and any changes is based on its actual costs plus an agreed uplift for items not covered by those costs, such as off-site overheads and profit. In the depths of time, the experience of one of the authors of this book included working as storeman on a cost reimbursable marine project and that period showed how attitudes to efficiency and economy can relax where lost items such as hand tools are replaced at the employer's expense without due controls. However, such disciplines are often imposed by the use of an agreed 'Target Cost', which sees the contractor penalised if actual costs exceed that target or rewarded if it is beaten. This is often referred to as the 'Pain Share/Gain Share' mechanism. The alternative is to only allow costs that are properly incurred and disallow costs that are not properly incurred. The example of contracts in the NEC suite and its concepts of allowing only costs covered by the defined term 'Defined Cost' and excluding (in some Option) costs covered by the defined term 'Disallowed Cost' when addressing 'compensation events' is dealt with in more detail in this section.

The JCT publishes a set of what it terms 'Prime Cost' contracts, comprising a main contract, subcontract and sub-subcontract form. Of the other suites of standard forms focussed on in this book, there is also the Infrastructure Conditions of Contract – Target Cost Version. The NEC suite also includes, through its Option E, a cost reimbursable contract.

The benefits of such entirely costs-based approaches are said to be that they allow a contractor to be appointed early and on projects where the scope of works is particularly undefinable at the contract stage. An example of this might be in relation to the refurbishment of a very dilapidated building. The alternative may be that contractors will be unwilling to price the works at all or allow disproportionate amounts for risks and uncertainties if such works were procured on a lump sum basis. It also removes the need for detailed provisions in relation to the valuation of contract claims, and hence much of the subject of this book! Therefore, little is said in relation to cost reimbursable contracts herein. The removal of more adversarial aspects of contractual relationships on construction and engineering contracts, such as defining and valuing variations and quantifying delays, is said to promote a more co-operative and constructive environment. The downsides for the employer of a cost plus approach are particularly likely to include paying a premium against a lump sum contract.

As set out elsewhere, contracts that make agreed contract rates and prices the primary basis for valuing variations may resort to actual costs where there are no contract rates and prices or they are not suitable even with adjustment. For example, FIDIC Red Book clause 12.3 provides that if no rates or prices in the contract are relevant for valuation of the works, then its valuation shall be derived from the 'Reasonable Cost' of executing it and it is notable that clause 13.3 provides that an Engineer's instruction may require the recording of those 'Costs'. Under SBC/Q, clause 5.6.3 requires that where a variation is not of similar character to that in the bills of quantities, then valuation shall be 'at fair rates and prices' and whilst this does not expressly refer to costs it may be that the contractor's costs provide the only fair approach to valuation of some variations because of their nature. Similarly, under the Infrastructure Conditions, clause 52(4) has at the end of its valuation hierarchy a requirement that 'a fair valuation shall be made', which again could in some circumstances be best served by reference to the contractor's actual costs.

In addition, it may be on any contract that provides for variations to be valued at contract rates and prices, that it is agreed that the particular circumstances of a change are such that a valuation approach based on actual costs is the only reasonable approach. This may particularly be the case where a change is accompanied by delays, acceleration and disruption of the type considered in Chapter 6, to an exceptional extent such that the various causes cannot be separated between the change in design or specification and the effects of those other causes. To give two examples where a costs basis was resorted to:

- A contract for demolition of an old wharf and construction of an office block just behind the tidal wall of a river in a maritime city's business centre. In the contract the foundations were designed to be piled, bearing on the clay many metres down, with a ring beam supporting the foundations of the new building. In the event, removal of the existing structures exposed that the site was substantially subject to an unrecorded underground stream. This rendered the planned piled design impossible to achieve even with substantial dewatering. It was replaced by the construction of a raft foundation, to effectively 'float' on the unstable subsurface below. The contract was a JCT 1963 edition, in which the hierarchy for the valuation of variations included at its third limb was 'fair rates and prices'. In the circumstances, the parties agreed to value the resulting 'add and omit' variation as the omission of the contract prices for the piled foundations and the addition for the raft based on the contractor's actual costs of constructing it and the abortive works carried out in attempting to achieve the original piled design.
- A contract for construction of a new railway scheme. The design included tunnelled sections that contained a new station in the centre of a city. In the contract, the station was designed to be constructed on a 'cut and cover' basis. In the event, the inevitable disruption that would have resulted to the city centre above that station were deemed unacceptable and the design was changed to construct it as a cavern. Furthermore, constraints on the required opening date for the system meant that construction was required to be accelerated with no effect on completion. In the circumstances, an arbitral tribunal gave an interim ruling that the construction of the cavern be valued on the basis of the contractor's actual costs, compared to its estimates of the costs of the cut and cover design at contract date.

These are examples of a cost plus approach taken for a specific unusual change, notwithstanding that the contract terms took a primary approach to valuation of applying contract rates and prices. The most obvious example of construction contracts that make the contractor's actual costs their primary approach are those in the NEC suite, as explained as follows by reference to NEC4-ECC.

5.7.1   The NEC Suite of Contracts – Introduction

The NEC4 suite published in June 2017 is the latest version of a set of contracts first published in 1993. They are now the only forms of contract fully endorsed by the UK's Office of Government Commerce, which recommends them for all public sector construction projects in the UK. By 2015 the NEC suite of contracts was in use in over 20 countries and was, for example, one of the officially recommended forms of contract for use by the South African government. They have also become the contracts of choice of the UK's Institution of Civil Engineers, whose involvement with the long running series of ICE contracts was ended in 2009, whose successor, the Infrastructure Conditions considered as an example in this book, were taken over by the Association for Consultancy and Engineering and the Civil Engineering Contractors Association.

The NEC suite includes a range of contracts to cover such fields as: engineering and construction; term services; professional services; subcontracts; design/build/operate; materials supply; dispute resolution services and frameworks. It also includes short forms of contract for some of these situations. For engineering and construction contracts the suite provides the following procurement options:

Option A. Priced contract with activity schedule (the alternative considered in this book and referred to herein as 'NEC4-ECC').
Option B. Priced contract with bill of quantities.
Option C. Target contract with activity schedule.
Option D. Target contract with bill of quantities.
Option E. Cost reimbursable contract.
Option F. Management contract.

In the following discussions, reference is largely made to clauses in Option A as an example. All of the Options refer to compensatory changes as 'compensation events', although the following discussion also uses the term 'variations' for consistency with the rest of this book. The provisions of the NEC contracts in relation to the valuation of compensation events could justify a textbook in its own right. Therefore the following only sets out the broad principles applied in relation to compensation events and highlights some of the issues that can arise when valuing them under the NEC4-ECC approach.

In relation to variations to the works, NEC4-ECC gives the parties wide discretion as to how to agree their valuation at clause 63.2:

> The Project Manager and the Contractor may agree rates or lump sums to assess a change to the Prices.

The clause contains no guidance as to the basis or source of such 'rates or lump sums', which could, for example, presumably come from any of the sources identified elsewhere in this chapter, such as: the contract; 'market rates' from other projects; published industry pricing books; or estimates of the contractor's actual costs.

The NEC4 User Guide says this regarding clause 63.2:

> This should be a 'go to' clause for minor compensation events, but it is important to understand that the *Contractor* and *Project Manager* must agree to both the principle of using rates, and the rates themselves. If the Parties are not careful, they can spend as much in professional fees and costs evaluating the Defined Cost as the compensation event is worth, which is plainly not sensible.

This 'go to' reference seems a laudable aim, particularly on minor variations. There was no such provision in NEC3-ECC, but the idea that this approach may particularly apply to small items was also stated in the NEC3 Guide, although it did not suggest it should be a 'go to' approach and the extent to what was attained in practice varied. Whether this will change under NEC4-ECC waits to be seen.

The NEC4 User Guide then gives the following guidance regarding the basis for 'rates or lump sums' under clause 63.2:

> Clause 52.1 requires Defined Cost to be at open market or competitively tendered prices, the *Project Manager* and *Contactor* need to have this in mind when agreeing rates or lump sums.

5.7.2   The NEC Suite of Contracts – Forecast or Actual Cost?

Failing agreement on the basis of clause 63.2, NEC4-ECC directs how variations should be valued as follows, at clause 63.1:

> The change to the Prices is assessed as the effect of the compensation event upon:
>
> - the actual Defined Cost of the work done by the dividing date,
> - the forecast Defined Cost of the work not done by the dividing date and
> - the resulting Fee.

In this, 'Defined Cost' includes both the direct costs of construction and the indirect costs of such items as design and preliminaries and general costs, the Fee largely covers the contractor's mark-up for overheads and profit, although more is given on this elsewhere.

The concept of a 'dividing date' in clause 63.1 is new to the NEC4-ECC edition. It is defined therein as follows:

> For a compensation event that arises from the Project Manager or the Supervisor giving an instruction or notification, issuing a certificate or changing an earlier decision, the dividing date is the date of that communication.
> For other compensation events, the dividing date is the date of notification of the compensation event.

This contains a change from the wording of clause 63.1 in the previous edition, NEC3-ECC. That edition did not refer to a 'dividing date'. Clause 63.1 of NEC3-ECC read as follows, where the previous, now changed, wording is emphasised:

> The changes to the Prices are assessed as the effect of the compensation event upon
>
> - the actual Defined Cost of the work *already done,*
> - the forecast Defined Cost of the work not *yet done* and
> - the resulting Fee.
>
> *The date when the Project Manager instructed or should have instructed the Contractor to submit quotations divides the work already done from the work not yet done.*

Under all editions of NEC-ECC, Core Clause 63.1 has, under terms such as these, made it clear that the valuation of a compensation event is to be based on the contractor's 'Defined Costs'. For 'work already done' (NEC3-ECC wording) or 'work done by the dividing date' (NEC4-ECC wording) the valuation should take information as to the actual Defined Costs into account. On the other hand, for 'work not yet done' (NEC3-ECC wording) or 'work not done by the dividing date' (NEC4-ECC wording), the costs are to be a forecast of those Defined Costs.

In practice, this means that where the compensation event is of an 'add and omit' nature and the actual work has been done by what is now termed the 'dividing date', the additional work can take account of actual costs. On the other hand, the omission, since it will not actually yet have been constructed, and therefore cannot be the subject of actual costs, can only be based on an estimate of it. This inevitable aspect of the different approach can be particularly relevant to allowances for contingencies and risks in valuing compensation events. NEC4-ECC clause 63.8 expressly requires inclusion for these in the assessment of a compensation event where they have a 'significant chance of occurring'. Clause 63.6 of NEC3-ECC was in similar terms. The subjective nature of 'a significant chance' will be noted. The NEC4 Guide seeks to help by explaining that risk is allowed in a compensation event 'in the same way that the Contractor would allow for risk in its tender'. This perhaps underlines the philosophy of the NEC approach to valuing compensation events. This is that they are forecast based on actual costs, just in the same way as contractors price their tenders in the first place. It may help in some cases when considering whether a risk has a 'significant chance' to ask whether the contractor is likely to have priced for it in its tender.

The change in the NEC4-ECC wording to add a concept of a 'dividing date' may seek to address a couple of old chestnuts under NEC terms in relation to compensation event valuation. The first is as to which update of the programme is to be used to assess the effect on completion. The second was the perhaps more obvious question, under the old clause 63.1 wording, as to 'already done' when in relation to forecast or actual cost. Already done at what date? Thus, at what date would the pricing of a compensation event go from a forecast of costs to including the actual costs? Clearly, this could have major implications in relation to the risk taken by the contractor in providing a quotation and the amount of such quantification. Furthermore, given how in practice the processing and agreement of compensation events could sometimes take much longer than the periods envisaged in the contract, the valuation could become a 'rolling ball' that changed over time. If compensation events were left in dispute until the end of the project, and even in formal dispute resolution proceedings thereafter, would this mean that all such compensation events would now be assessed based on their actual costs, given that they would then all be 'work done'? From the respective viewpoints of the parties, on a project that ran into such a formal dispute, and was therefore a 'problem project', it may have seemed inevitable that actual costs would be greater than those that might have been forecast at the time, even with allowance for risk. It might, however, be that a forecast of Defined Cost was exaggerated, in such as allowances for risks, such that the actual costs turn out to be less than those forecast. Either way, this question therefore became most important under NEC3-ECC and earlier editions as, for example, contractors claimed on their higher actual costs and employers argued that the correct assessment was of a lesser forecast made at the time.

This 'actual costs or forecast' or 'retrospective or prospective' debate arose regularly on projects under NEC terms and has long since seemed an obvious one for clarification from the courts. Frustratingly for practitioners, none was provided until 2017 and the judgment in *Northern Ireland Housing Executive v Healthy Buildings (Ireland) Limited* [2017] NIQB 43, and even then only to a limited extent. This was the same year that NEC4-ECC was published.

Healthy Buildings were engaged to provide services in relation to asbestos in buildings owned by the Executive, under two separate contracts both under the NEC3-Professional Services Contract ('NEC3-PSC'). In relation to a compensation event assessment, the terms of that NEC form are similar to those for the NEC's various contractor–employer contracts, such as NEC4-ECC, particularly in relation to the need for timely notices of compensation events, requests for quotations and assessment of their values, which are explained in Section 5.14. One difference is that, under NEC3-PSC the term 'Defined Cost' is replaced by 'Time Charge'.

In January 2013 the Executive instructed Healthy Buildings to carry out additional surveys. This change should properly have amounted to a compensation event under the two services contracts and the Executive should have notified it as such and requested a quotation, but it did not. Four months later, Healthy Buildings gave its own notification that the change was a compensation event. It was not until August and December 2013 that the Executive requested that Healthy Buildings provide quotations for the compensation event for the two services contracts respectively. By this time the additional surveys had been carried out and were therefore 'work already done' per the NEC3 wording. Healthy Building provided the requested quotations. In November 2013 the Executive rejected the quotations and assessed the compensation event as having no effect on Healthy Buildings' costs. It will be noted how all of these actions were carried out much later than the time frames dictated by the NEC contracts. Whilst this is not how NEC expects its contracts to be administered, in practice such failures are not especially unusual.

Adjudication followed and found that the additional surveys were a compensation event and that the Executive should pay Healthy Buildings' costs, the amount of which the adjudicator also found. The Executive challenged the decision by referring it to the court. In relation to the common NEC 'actual costs or forecast' or 'retrospective or prospective' debate identified elsewhere, the issues before the court included:

(1) On the true construction of the contract, and in particular Clauses 60 to 65 of the contract, is the assessment of the effect of the compensation event calculated by reference to the forecast or actual Time Charge?

(2) Are actual costs relevant to the assessment process in Clauses 60 to 65 of the contract?

Healthy Buildings submitted that the assessment should be based on a forecast of its costs and not its actual costs. In particular, it referred to these provisions of NEC3 PSC, but which are all in typical NEC3 form:

Clause 63.1:
'The changes to the Prices are assessed as the effect of the compensation event upon:

- the actual Time Charge for the work already done and
- the forecast Time Charge for the work not yet done.

The date when the employer instructed or should have instructed the consultant to submit quotations divides the work already done from the work not yet done.'
Clause 63.6:
'Assessment of the effect of a compensation event includes risk allowances for cost and time for matters which have a significant chance of occurring and are at the consultant's risk under this contract.'
Clause 65.2:
'The assessment of a compensation event is not revised if a forecast upon which it is based is shown by later recorded information to have been wrong.'

Healthy Buildings argued that because the Executive should have instructed the compensation event in January 2013, the assessment of it should be based on a forecast at that date. The court disagreed and found that the assessment should be based on actual costs.

Regarding clause 65.2, the court held that this applied where there was an assessment by the employer, which was based on the consultant's forecast, the employer cannot subsequently revise that assessment because it proves to have been wrong when compared to actual costs. However, as summarised elsewhere, this was not the situation in this case, where the time frames and processes of the contract had been largely ignored. Furthermore, reference was made more broadly to legal principles for the assessment of compensation and the House of Lords decision in *Bwllfa and Merthyr Dare Steam Colliers (1891) Limited v Pontypridd Waterworks Company* [1903] AC 426. Quoting that judgment, the Judge stated:

> Faced with seeking to award compensation to the consultant here for any cost to it as a result of the instruction of 10 January 2013 why should I shut my eyes and grope in the dark when the material is available to show what work they actually did and how much it cost them?

If other courts follow the approach adopted here by the Northern Ireland courts, the effects could be significant for parties operating under NEC terms and failing to follow its processes and time scales properly. To some observers, however, the belief is that the England and Wales courts will not follow this precedent on the basis that it is wrong, on the express wording of the NEC provision. Whilst NEC4-ECC introduces the concept of a 'dividing date', the issues addressed in *Northern Ireland Housing v Healthy Buildings* related to a contract where the parties had failed to properly administer the compensation events machinery properly. The extent to which the change in NEC4 helps to avoid such issues in practice will have to wait to be seen.

Whilst the parties' positions were somewhat reversed in *Northern Ireland Housing v Healthy Buildings*, a common position under NEC contracts is that it is the contractor that seeks to replace what turns out to be an over-optimistic forecast of its costs with its actual costs as they turn out to be higher. On the other hand, it is also often the case that forecasts are over-pessimistic, especially in their allowances for risks, such that it might be the employer that would prefer an actual costs approach to replace a higher forecast.

To the extent that this decision hinged on the failure to operate the provisions of the compensation events procedures in the contract, the temptation for a party to avoid the presence of a forecast, so that actual costs can be applied instead of a forecast, might be great.

The NEC4 User Guide states this in relation to clause 63.1 and regarding the extent to which compensation events should, in practice, be based on a forecast or actual costs:

> Ideally, the assessment will be the forecast of Defined Cost of work which is yet to be done, but it may, on occasions, include an element of incurred Defined Cost for work which has been done (e.g. for a weather compensation event).

The point about weather as a compensation event is that its effects are unlikely to be predicted and therefore subject of a forecast of their effects on the contractor's actual costs. However, even for a predictable event, experience under earlier NEC editions may suggest that this statement in the User Guide may prove optimistic on some projects, particularly those that continue to use the NEC3 contracts or bespoke forms based thereon. The User Guide's optimism as regarding the timing of when compensation events will be assessed continues as follows:

> On the rare occasions when some or all the work arising from a compensation event has already been done, Defined Cost should be readily assessable from records.

The key aspect of NEC4-ECC in this regard may be the change in wording of clause 63.1 to refer to a 'dividing date' for the use of actual as opposed to forecast Defined Cost when assessing compensation events, as explained elsewhere. The NEC4-ECC therefore seeks to pinpoint much more firmly than previous NEC editions the date at which the assessment of a compensation event switches from a forecast to actual Defined Costs.

The NEC4 User Guide suggests that this pinpointing of the date between forecast and actual is for the benefit of contractors:

> This prevents the practice of a *Project Manager* making a retrospective and selective choice between a quotation and the final recorded costs of dealing with a comparison event. This practice was never intended to be allowed because it clearly disadvantages the Contractor and, if adopted, will inevitably lead to adversarialism and game playing.

Of course, it could equally prove to be a protection for an employer, with some members of all sides of the construction and engineering industries being capable of 'game playing' on occasions.

Clause 63.1's definition of the 'dividing date' has been quoted elsewhere. Of this definition, the NEC4 user Guide says this:

> This supports the intention of the NEC that assessments will usually be forecasts of the cost of work yet to be done. Where work has had to start before the quotation has been submitted or even before the instruction to submit was given it is inevitable that the forecast component of quotation will be influenced by the

cost already incurred. Nevertheless, for most cases, the inclusion in the clause of a dividing date set early in the assessment process reinforces the point that compensation events are not cost-reimbursable but are assessed on forecasts with the *Contractor* taking some risk.

This idea that compensation events are priced on a forecast continues the theme suggested elsewhere, that the NEC idea is that changes in scope should be pre-priced, at the contractor's risk, with due allowance for contingencies and risks, in just the same manner as its original tender for the original scope of work.

As explained elsewhere, important to the debate in *Northern Ireland Housing v Healthy Buildings* was clause 65.2 of the NEC3 contracts and its requirement that the assessment of a compensation event is not revised if a forecast upon which it is based is shown by later recorded information to have been wrong. There is no such expressly equivalent clause in NEC4-ECC. The new edition does, however, continue in clause 61.6 with provisions in relation to assumptions in the assessment of compensation events. This tempers the requirement for compensation events to be assessed based on a forecast of the effect on the Defined Cost, by recognising that certain assumptions might have to be made and that these can be dealt with by way of conditions stated in that assessment. Of this provision, the User Guide says this:

> In some cases, the nature of the compensation event may be such that it is impossible to prepare a sufficiently accurate quotation.
>
> ...
>
> In these cases, quotations are submitted based on assumptions stated by the *Project Manager* in its instruction to the *Contractor*. If the assumptions later prove to be wrong, the *Project Manager's* notification of their correction is a separate compensation event (clause 60.1(17)).
>
> Apart from this situation, the assessment of compensation events shall not be revised (clause 66.3). Since each quotation will include due allowance for risk (clause 63.8) and the early warning procedure should minimise the effects of unexpected problems, the need for later review is minimal, and the benefits to both Parties of fixed time and cost effects far outweigh any such need.

The referenced clause 66.3 is as follows:

> The assessment of an implemented compensation event is not revised except as stated in these conditions of contract.

5.7.3   The NEC Suite of Contracts – Defined Cost

Turning to the concept in NEC4-ECC of the defined term 'Defined Cost' referred to in clause 63.1, it is surprising how often those operating the compensation provisions of NEC contracts assume that the contractor is allowed its costs without any reference to or understanding of the limitations placed on these by the term 'Defined Cost' and its definition. It is explained in the various Options in slightly different terms. Taking the example of the Option A: Priced contract with activity schedule, the term is defined in clause 11.2(23) as follows:

> Defined Cost is the cost of the components in the Short Schedule of Cost Components.

The other priced contract version of NEC4, Option B, contains the same definition. Options C, D and E, the cost reimbursable forms, refer to a longer schedule of cost components and the exclusion of what are termed 'Disallowed Costs':

> Defined Cost is the cost of the components in the Schedule of Cost Components less Disallowed Cost.

Both of the NEC's schedules of cost components are contract documents agreed at the outset, and included in the Contract Data section of an NEC contract. They both set out the costs to be paid to the contractor under headings of 'People', 'Equipment', 'Plant and Materials', 'Subcontractors', 'Charges', 'Manufacture and Fabrication', 'Design' and 'Insurance'. Beneath these headings the details of what is allowed are only by way of descriptions of types of cost, rather than measured work items, quantities, rates or prices for work, although they do cross-refer to rates and prices to be found elsewhere, in some cases. As to how they are to be calculated in the two forms of the schedule of cost components can be summarised as follows. It will be noted how in these provisions in which valuation is based on costs, there are still a number of rates and prices (or sources of them) that need to be pre-agreed and set out in the contract.

- *People.* The short schedule of cost components allows 'Amounts calculated by multiplying each of the People Rates by the total time appropriate to that rate within the Working Areas'. 'Working Areas' is a defined term under the contract and is important to several aspects of the Defined Costs allowed in assessing a compensation event. This therefore requires the contractor to provide such People Rates as part of the Contract Data. The longer form of the schedule of cost components has no such pre-agreed rates. For people directly employed by the contractor it details at length those components of the cost of employing people that are allowed as part of the Defined Cost, such as wages, bonuses, travel, vehicle, pension, etc. For People not directly employed, the longer form says 'Amounts paid by the Contractor'. This exposure of the employment details of some directly employed members of site staff can prove to be problematical.
- *Equipment.* The cost of these are paid in various different ways depending on the circumstances. In the shorter schedule of cost components these are:
  - By multiplying the time that the equipment is required by rates from a 'published list' in the Contract Data, multiplied by an adjusting percentage stated in the Contract Data. In practice parties often adopt the Civil Engineering Contractors' Association's 'Schedule of Dayworks Carried out Incidental to Contract Work' ('the CECA Schedule') as this 'published list'. However, such published rates are often higher than actual cost rates. This may cause some disagreement over the application of those higher rates even where the contract requires those rates to be applied.
  - Where an item of Equipment is in the Contract Data, but not in the 'published list', by multiplying the time that the equipment is required by rates from the Contract Data. This therefore requires that the contractor provides rates to be listed in the Contract Data for such items for the valuation of compensation events.

    – Where an item of Equipment is in neither the Contract Data, nor the 'published list', by multiplying the time that the equipment is required by 'competitively tendered or open market rates'. We discussed in Chapter 4 the issues where a contractor internally hires plant and equipment from another company or cost centre within its group, and not on an 'arm's length' relationship. Clearly this NEC provision would mean that inflated rates paid under such a relationship would not fall to be part of the assessment of a compensation event, without suitable adjustment downwards.

    – Unless the item is in the 'published list' and the rate includes it, the purchase price of Equipment that is consumed.

    – Unless the item is in the 'published list' and the rate includes it, payment for transport, erection, dismantling, constructing, fabricating or modifying.

The longer schedule of cost components has variations on these themes and adds to these alternatives such as payments for the hire or rent of equipment not owned by the contractor and at rates in the Contract Data for special equipment listed therein.

- *Plant and materials.* Under both forms of the schedule of cost components, these are paid for at the amounts of payments made by the contractor for purchase, delivery, removal, providing and removing packaging and samples and tests.
- *Subcontractors.* Under both forms of the schedule of cost components, the Defined Cost includes 'Payments to subcontractors for work that is subcontracted…'. In practice, the auditing of a contractor's payments to its subcontractors as part of 'Defined Cost' under the NEC forms has often been dogged by a lack of detail from those subcontractors. The general issue of the common disparity in the level of detail between main contractor pricing and that of subcontractors has been mentioned elsewhere in this chapter. In relation to compensation events, many subcontractors prove unwilling to provide the same level of detail as that requested from the contractor. Many subcontractors do not appear to understand the NEC mechanisms and the level of detailed scrutiny of costs that contactors are often subjected to, which is particularly often the case internationally.
- *Charges.* Both forms of the schedule of cost components set this out in the same terms. It includes, for example, utilities, payments to public authorities, royalties, inspection certificates, consumables, etc., for the Project Manager and Supervisor. They are allowed at the cost paid by the contractor.
- *Manufacture and fabrication.* Both forms of the schedule of cost components set this out in the same terms. This is at rates provided by the contractor as set out in the Contract Data, applied to the 'total time appropriate' of people on their 'manufacture and fabrication of Plant and Materials outside the Working Area'. The subjective nature of the test 'appropriate' in relation to the time spent will be noted here as a potential cause of disagreement.
- *Design.* Both forms of the schedule of cost components set this out in the same terms. This applies hourly rates by the contractor as set out in the Contract Data to the 'total time appropriate' to design done outside the Working Area. This is with the addition of the cost of travel to and from the Working Area. Again, the subjective nature of the test 'appropriate' in relation to the time spent will be noted here as a potential cause of disagreement.
- *Insurance.* Both forms of the schedule of cost components set this out in the same terms. This allows cost with the deduction of the cost of events that the contract requires the contractor to insure and costs paid to the contractor by insurers.

The schedules of cost components are therefore complex, and the fuller form is particularly lengthy. Furthermore, some of the language, as with much of the NEC contracts' terminology, may be unfamiliar to those schooled under JCT, FIDIC or ICE terminology and is therefore hard for them to follow initially. In practice, some NEC contracts are administered without reference to the detailed contents of the schedule of cost components until a dispute arises as to the valuation of a compensation event. In the meantime, parties might happily muddle through agreeing compensation events on the basis of estimated or actual costs without reference to the term Defined Costs or the schedules of cost components. There may be no harm in this if it collaboratively achieves the aim of the NEC approach.

Common issues in relation to the presentation by contractors of their costs under these rules, and the consideration of them by the employer's team, are the proof of allocation to a compensation event and the degree to which costs have to be proven as having been incurred where retrospectively addressing 'the actual Defined Cost of the work done by the dividing date'. The aim of dealing with compensation events on a prospective forecast basis, rather than on costs actually incurred, will avoid such issues. However, all too often under NEC contracts historically, these have been issues in retrospective valuation.

Allocation to a change can be particularly difficult in relation to labour, where that change is not a wholly new piece of work, but a change in its scope or nature, such that records of the additional hours cannot be isolated. Labour usually gets allocated to a specific piece of work, so that this may just be a matter of keeping good records. Staff time may be even more of a problem, as they are usually managing or administering a number of items or areas of the works. In relation to materials similar issues may arise. The problems of obtaining proper details from subcontractors is commented on elsewhere.

Regarding the proof of actual costs, how far does the paper trail have to run? In relation to people, does this extend as far as seeing their salary or wage slips? In relation to staff, this can often be a fraught issue where senior staff in particular regard the details of their income as confidential. In relation to materials, is proof of payment required, or is a cost entry in the contractor's project cost reporting system supported by an invoice sufficient?

It can be imagined how on a project with many compensation events and of significant size, the time and resource burden on the contractor to provide such information for a retrospective compensation event evaluation, and on those checking it for the employer's team, can be disproportionately burdensome. It is suggested that a degree of common sense and proportionality is required here. Reference is also made again to the NEC4 User Guide as quoted elsewhere in relation to clause 63.2 and the danger that 'the Parties … can spend as much in professional fees and costs evaluating the Defined Cost as the compensation event is worth, which is plainly not sensible'.

5.7.4 The NEC Suite of Contracts – Disallowed Cost

As set out elsewhere, Options C, D, and E allow the Defined Cost with reference to the fuller Schedule of Cost Components, but excluding 'Disallowed Cost'. This term is defined in clause 11.2(26) as follows:

> Disallowed Cost is cost which
>
> - is not justified by the *Contractor's* accounts and records,

- should not have been paid to a Subcontractor or supplier in accordance with its contract,
- was incurred only because the *Contractor* did not
  - follow an acceptance or procurement procedure stated in the Scope,
  - give an early warning which the contract required it to give or
  - give notification to the *Project Manager* of the preparation for and conduct of an adjudication or proceedings of a tribunal between the *Contractor* and a Subcontractor or supplier

  and the cost of
- correcting Defects after Completion,
- correcting Defects caused by the *Contractor* not complying with a constraint on how it is to Provide the Works stated in the Scope,
- Plant and Materials not used to Provide the Works (after allowing for reasonable wastage) unless resulting from a change to the Scope,
- resources not used to Provide the Works (after allowing for reasonable availability and utilisation) or not taken away from the Working Areas when the *Project Manager* requested and
- preparation for and conduct of an adjudication, payments to a member of the Dispute Avoidance Board or proceedings of the tribunal between the Parties.

Rather like not understanding that the contractor is only entitled to costs covered by the defined term 'Defined Cost', it is surprising how often those operating under NEC terms are not aware of the contents and effect of such provisions to exclude 'Disallowed Cost'. It is clearly essential that such a definition of costs that are to be excluded is clear, unambiguous, recognised and understood. The NEC definition is detailed but it is a regular cause of disagreement, perhaps because of the difficulty of making it proof against debate and the often contentious way in which it is sometimes applied.

The disallowing of costs such as those 'not used to Provide the Works' is an obvious protection for the employer from inefficiency by the contractor. However, this can lead to difficulty in relation to subjectivity with regards to qualifications such as the use of the word 'reasonable' in relation to wastage.

The question of reasonableness of the contractor's costs also arises in clause 63.9 of all of the Options as follows:

> The assessment of the effect of a compensation event is based upon the assumption that the contractor reacts competently and promptly to the event and that any Defined Cost and time due to the event are reasonably incurred.

The disallowing under clause 11.2(26) of Options C, D and E of costs that resulted from failure to give a due early warning perhaps echoes what we say in Section 5.14 in relation to the consequences of contractors not giving due notice in relation to their claims and the loss of opportunity for the employer and its advisors to readdress a decision, instruction or action in order to avoid or reduce resulting costs.

The disallowing of costs incurred in 'correcting Defects after Completion' is an interesting, and sometimes moot, point. It is sometimes said that the correction of defects after completion is an inherent part of the vast majority of construction and engineering projects. On that basis, it could be argued that the cost of carrying them out is a legitimate part of the contractor's activities. Furthermore, it is sometimes argued by contractors that the nature and/or timing of a change has led to unavoidable post-completion correction, whilst avoiding delay to the achievement of that milestone. On the other hand, allowing costs, but excluding those that are the result of defective works, is an important incentive to ensure that proper management and quality control are maintained. From an employer's perspective it may seem unfair to be paying for the contractor's costs of remedying its own failures. Under NEC terms, it would always be open to a contractor to make some allowance for such post-completion costs in its fee percentage.

5.7.5   The NEC Suite of Contracts – The Fee and Other Agreed Rates

The final component of valuation of compensation events under NEC4-ECC's clause 63.1 is the addition of 'the resulting Fee'. That term is defined at clause 11.2(10) as:

> The Fee is the amount calculated by applying the fee percentage to the amount of the Defined Cost.

That fee percentage is another requirement of pricing information to be provided by the contractor and included in the Contract Data at an agreed rate. It was said elsewhere that this percentage is for elements that include overheads and profit and, as stated in the User Guide, these are likely to be the major elements of the percentage. However, the Fee is also to cover all other costs (including such as overheads and profit) that are not part of the Defined Cost. Clause 52.1 puts this as follows:

> All the *Contractor's* costs which are not included in the Defined Cost are treated as included in the Fee. Defined Cost includes only amounts calculated using rates and percentages stated in the Contract Data and other amounts at open market or competitively tendered prices with deductions for all discounts, rebates and taxes which can be recovered.

Using the example of plant and equipment which a contractor internally hires from another company or cost centre within its group, and not on an 'arm's length' relationship, the fee percentage might therefore in theory include for any extra over costs that this gives rise to for the contractor. It might also include such as allowance for the correction of defects, as noted elsewhere.

A common problem with the fee percentage is to what extent staff are included therein rather than in the Defined Cost. The shorter schedule of cost components, for example, requires People Rates to be set out in the Contract Data to cover the costs of the contractor's operatives and staff. However, what if the job descriptions of actual operatives or staff do not directly align with those of the people listed in the Contract Data? Does this mean these actual people are included in the fee percentage?

What will also be noted is how the parties to an NEC contract need to agree in advance and set out in the Contract Data certain percentages and other rates to be used in the valuation of compensation events. In summary, depending on which Option and schedule of cost components applies, these may include:

- A fee percentage.
- People Rates for people directly or indirectly employed by the contractor working in the defined Working Areas.

- A percentage adjustment to be applied to rates for equipment from a published list stated in the Contract Data.
- Rates for equipment that is in the Contract Data but not in the stated published list.
- A time-related on-cost charge for equipment purchased and listed in the Contract Data.
- Rates for special equipment listed in the Contract Data.
- Rates for people time spent on manufacture and fabrication done outside of the defined Working Areas of the project.
- Rates for people time spent on design done outside the defined Working Areas of the project.

There is no requirement for any of these rates and percentages to reflect open market or competitively tendered prices.

NEC4 has made one aspect of these requirements simpler in that NEC3 required two different fee percentages – one for subcontracted work and one for non-subcontracted work, directly carried out by the contractor's own resources.

In addition to these agreed rates and percentages, another consequence of the 'collaborative' nature of the NEC contracts in practice is that parties often agree at an early stage of a project a pro forma for the pricing of compensation events. This may consist of an Excel workbook, often with several worksheets, and will set out the headings of 'Defined Cost' to be included and the various percentages agreed in the contract. Furthermore, such pro formas often also include further percentages to cover some of the 'Defined Costs', in addition to those required by the schedule of cost components and Contract Data. This might include items such as 'Design' and the management and supervision parts of 'People' being assessed as a percentage on other costs, rather than being based on hours spent. This approach may be adopted where it is considered that allocating such actual costs to a compensation event might be difficult and time consuming and a percentage addition will provide a pragmatic and reasonable answer to quantification. It is suggested that the more such issues can be agreed to be dealt with on such a pragmatic basis, simplifying the provisions of the standard NEC forms, the more that reflects its 'collaborative' ethos. The potential for savings in administrative costs to both parties are obvious.

A consequence of the pre-agreement of such aspects of the valuation of compensation events is that the contractor will arrange its record keeping to suit. For example, if it has been agreed that the cost of design of a compensation event will be priced by applying a percentage addition to the construction costs, then there will be no need for the contractor to maintain detailed records of the time and allocation of design staff that can be specifically allocated to each compensation event. This may be a further benefit in terms of the administrative burden of administering compensation events under NEC terms.

5.7.6 The NEC Suite of Contracts – Objectives

Throughout the NEC terms, it will be noted how the valuation of a compensation event is predominately based on the costs to the contractor in accordance with a pre-agreed valuation structure. The rationale behind this costs-based approach is explained on page 61 of the NEC4 User Guide as follows:

Assessment of compensation events as they affect Prices is based on their effect on Defined Cost plus the Fee. This is different from some standard forms of contract where 'variations' are valued using the rates and prices in the contract as a basis. The reason for this policy is that no compensation event for which a quotation is required is due to the fault of the Contractor or relates to a matter which is at his risk under the contract. It is therefore appropriate to reimburse the Contractor its forecast additional costs (or actual additional costs in certain circumstances) arising from the compensation event. Disputes arising from the applicability of contract rates are avoided.

This is in similar terms to those for the NEC3 Guidance Notes 2005, although the current words in parentheses 'actual additional costs in certain circumstances' were previously 'actual costs if work has already been done'. The philosophy is in accordance with one that the NEC3 Guidance Note referred to as Objectives – Stimulus to Good Management':

> The ECC is therefore intended to provide a modern method for employers, designers, contractors and project managers to work collaboratively.

The preface to the NEC4 User Guide includes, among the stated objectives of drafting NEC4, to:

> provide greater stimulus for good management.

A final point to make about the NEC approach to evaluating the consequences of change is that both the direct consequences (the subject of this chapter) and the time consequences (the subject of Chapter 6) are swept up in the assessment of each compensation event. As this is described in the NEC4 User Guide:

> If the compensation event increases the Contractor's time related costs, for example because planned Completion is delayed, or additional supervisory staff are needed, the increase in Defined Cost is included within the value of the compensation event. There is no such thing as a retrospective and separate 'delay and disruption claim' in the contract.

This is the theory, and an obvious benefit, where changes are administered properly under the NEC compensation events procedures. However, in practice, it is not uncommon, on 'problem projects' for contractors to present separate delay and/or disruption claims, asserting that the nature and extent of compensation events was such as to make separate allocation of some consequences to specific events impossible.

5.8 Unit Costs

The discussions elsewhere in this chapter centre on the adjustment of the contract price for changes that can be evaluated as having a direct effect on the contractor's unit rates and prices. In many instances the ability to make reasonable adjustments to such unit rates and prices will depend on a proper understanding of the unit costs that go into such rates and prices and how they are incurred.

Most construction contracts will include express provisions stating that the rates and prices submitted by the contractor in its tender will be deemed to cover all the contractor's obligations under the contract. The Infrastructure Conditions contain such a statement at clause 11(3)(b). Clause 4.11 of the FIDIC Red Book is in similar terms. SBC/Q states at clause 4.2 that the contract sum shall only be adjusted or altered in accordance with the conditions of contract, and that errors in the computation of the contract sum shall not be corrected unless they come within the ambit of clause 2.14, covering errors in the measurement and description of the works. This should be read in conjunction with articles 1 and 2 of SBC/Q, which states that the contractor will undertake and complete the works in the contract documents for the contract sum or such other sum that becomes payable under the contract.

Most contracts, including civil engineering and building contracts executed under provisions such as those of the Infrastructure Conditions, the FIDIC Red Book or SBC/Q, will have two distinct parts to the pricing documents in the contract, often in the form of bills of quantities. The first part will be the 'preliminaries' or 'preliminaries and general' section covering general on-site overheads, services and lump sum charges, which apply to the whole of the works. In other contracts, such as many for heavy mechanical engineering contracts, such sections of pricing are often termed 'indirects' as they are considered to arise 'indirectly' from the works, as opposed to the 'directs' section of pricing, which relates to the unit prices for detailed individual construction operations or activities. In civil engineering and building contracts this latter section of pricing is often covered by an individual detailed bill of quantities for each of the various trades and elements of construction work. As explained elsewhere in this chapter, both the preliminaries and general and measured works section of bills of quantities are likely to have been measured under a referenced method of measurement. Such a method will dictate how the bills of quantities are separated into separate sections for each trade, for example in the case of CESMM4: 'Class E: Earthworks', 'Class H: Precast concrete' and 'Class V: Painting'.

The rates for the detailed construction work are usually priced against measurement units for pricing, as also dictated by the stated method of measurement. The units for such measured work items can be of volume, area, length, weight or number of an item of work, depending on the work's nature. If a standard method of measurement is not adopted the contract documents should include a detailed explanation of the methodology employed in this respect. Failure to adopt an appropriate and relevant standard method of measurement, or to provide a comprehensive description of the method of measurement adopted, will run the risk of later arguments over the 'item coverage' included by the measurement with potential claims that further items should be measured to fill gaps in the coverage between the individual measured items.

The method of measurement, whether one of the standard publications or an ad hoc set of rules, will therefore define how the work is to be measured and the 'item coverage' for each item of the detailed measure. As the contract requires the contractor to include in its prices for all its obligations under the contract this will include the supply of labour, plant and equipment, and materials for the works, including any subcontracted elements. There may be variations on this, depending upon the type of contract and the nature of the project. For instance, in many mechanical engineering and process plant contracts the client may supply substantial elements of the materials or permanent plant required, and these and other contracts may require a greater or lesser design input from the contractor.

Pricing documents such as bills of quantities usually require a contractor to include a single unit rate or price against each item. This is the approach anticipated by standard methods of measurement such as CESMM4, NRM and POMI described elsewhere. However, it is common in some parts of the world that such pricing documents require the contractors to declare some detail as to the build-up of their rates and prices. In such cases, the pricing document may set out separate columns into which the contractors price their allowances for labour, materials, subcontractors, and plant and equipment costs for each item separately. There may also be a column headed 'Other' for such as the contractor's head office overheads and profit and anything else it wishes to include against an item, such as risk. That 'Other' column might also include any preliminaries and general items that are not priced in their separate pricing section, but relate specifically against the work items to which they relate. The potential benefits of such detail in the valuation of variations is obvious. What may be less obvious is the potential use of such detail in evaluating the time consequences of change, as discussed in Chapter 6 and an aspect that is returned to there.

Generally, leaving aside the less usual aspects of employer-supplied materials and contractor design obligations, the pricing elements may include any of the following components depending on the location of the project and whether the resources are locally sourced or imported. These items will be included in the unit rates for the measured works to which they relate, although some may be included in the preliminaries and general or 'indirects' items section of the pricing document.

(a) Staff
  - Salary, travelling allowances and all payments prescribed by their employment contracts, including overtime payments where applicable.
  - Bonuses, be they personal, project or company related.
  - Contributions for pensions, sickness, unemployment benefits, National Insurance, etc.
  - Contract works, third party and employer's liability insurances.
  - Annual and public holidays with pay.
  - Industrial training levies.
  - Redundancy payment contributions.
  - Any other obligations under local employment legislation.
  - Protective clothing and boots.
  - Subsistence or lodging allowance or provision of accommodation for personnel working away from home.
  - Welfare and messing facilities.
  - Transport to and from the site.
  - Provision of personal car, and possibly driver.
  - Compliance with all health and safety legislation.
  - Food allowance.
  - Medical and health care costs.
  - Flights, for imported staff.
  - Relocation costs.

- End of service 'Gratuity' pay.
- Visa and work permit costs for imported staff.
- Child education costs.

(b) Labour

- Wages, travelling allowances, tool allowances and all payments prescribed by relevant 'working rule agreements', including overtime payments where applicable.
- Bonuses, be they productivity, personal, project or company related.
- Contributions for pensions, sickness, unemployment benefits, National Insurance, etc.
- Contract works, third party and employer's liability insurances.
- Annual and public holidays with pay.
- Industrial training levies.
- Redundancy payment contributions.
- Obligations under local employment legislation.
- Small tools such as picks, shovels, barrows, trowels, ladders, handsaws, buckets, hammers, chisels and all like items including sharpening and replacements.
- Protective clothing and boots.
- Subsistence or lodging allowance for personnel working away from home.
- Welfare and messing facilities.
- Transport to and from the site.
- Compliance with all health and safety legislation.
- Food allowance.
- Medical and health care costs.
- Flights, for imported labour.
- Relocation costs.
- End of service 'Gratuity' pay.
- Visa and work permit costs for imported labour.

(c) Plant and Equipment

- Provision of owned plant and equipment at an initial capital cost depreciated over its useful life.
- Provision of hired plant and equipment at external hire rates, perhaps also including the driver.
- Provision of internally hired plant and equipment at internal hire rates, perhaps also including the driver.
- Maintenance and repairs, including tyres, etc., mechanics and fitters' time, etc.
- Cost of fuel, lubricants, grease, etc., including distribution of fuel to working plant on the site.
- Contract works third party and employer's liability and motor/plant/equipment insurances.
- Road tax, statutory charges, etc., where appropriate.
- Cost of operators' time including all on-costs as listed elsewhere for labour.
- Compliance with health and safety legislation and statutory requirements for particular items of plant.
- Delivery, erection, dismantling and removal charges (for example for tower cranes).

(d) Materials
  - Cost of supply including carriage, freight charges, etc.
  - Adjustment for any discounts.
  - Import or customs duties on imported items.
  - Demurrage charges.
  - Unloading and distribution.
  - Double handling.
  - Wastage in carriage, unloading and distribution.
  - Wastage and losses in conversion, including bulkage and shrinkage where appropriate to the type of material.
  - Cost of insurances during supply, delivery, etc.
  - Temporary storage off site.
(e) Subcontractors
  - Cost of works or supplies from subcontracted companies.
  - Adjustment for any discounts.
  - Cost of attendance on subcontractors, e.g. unloading, distribution of materials, provision of welfare facilities, etc.
  - Cost of supervision.

All of these costs are likely to be subject to additions to cover the contractor's off-site or head office overheads and profit.

As indicated elsewhere, some of the above cost elements may be priced in the contract pricing preliminaries and general items, or 'indirects', section rather than in the unit rates, and care should be taken to ascertain the demarcation between the two. Practice may vary between different types of contract, contractor practice and employer requirements. Particular countries may also require allowance for other costs not in these lists.

These lists give a guide to the range of costs that will usually be deemed to be included in the contract rates, and which are therefore fundamental to the understanding of claims for additional payment. The treatment of the individual elements may also vary from company to company, particularly with regard to the inclusion of head office overheads and profit within the pricing regime.

Some particular aspects of these cost components are considered as follows.

5.8.1 Labour Costs

Most of the individual elements contained within the make-up of the labour costs are self-explanatory, but there are some that may cause a little confusion when considering claims for additional payment.

5.8.1.1 Gang Rates

It is usual practice when pricing the unit rates in a bill of quantities or schedule of rates for the estimator to use a composite rate for pricing the labour in many of the individual measured work items, rather than the rate for a particular craftsman or labourer alone.

This practice reflects practice on site, where craftsmen and labourers (or helpers) rarely work in isolation but are to a greater or lesser extent deployed in teams. An obvious instance is that of bricklayers who, on UK building sites, commonly work in teams of two bricklayers and one labourer or four bricklayers and two labourers, with the labourer(s) mixing and transporting the mortar and distributing the bricks to the point of use so that the bricklayers can focus on laying bricks and mortar. Such a 'two and one' gang would be priced as:

Bricklayers	per hour 2 × $18.00	36.00
Labourer	per hour 1 × $15.00	15.00
		$51.00
Cost per bricklayer hour		$25.50

In effect, half an hour of labourer cost has been added to the hourly rate of the bricklayer so that when brickwork items are priced using bricklayer 'norms' or constants, the appropriate allowance for labourer time is also included through the rate of cost per hour.

This principle may be extended to other trades such as carpenters and joiners, where it might be considered that a labourer will be required to unload and distribute materials for the carpenters and joiners, at the ratio of perhaps one labourer to every eight carpenters and joiners, depending upon the type of work and circumstances. In that case the calculation would be:

Carpenters/Joiners	per hour 8 × $18.00	144.00
Labourer	per hour	15.00
		$159.00
Cost per carpenter/joiner hour		$19.875

In this instance the effect is to add an eighth of an hour of labourer cost to the cost of each craftsman's hour.

In these examples the ratios of skilled to unskilled operatives are typical for a UK building site. In some countries, the ratio of unskilled labour, or 'helpers' can be much higher.

The way in which such labourer time is included in a particular contract may vary. It can be included in gang rates. Alternatively, the unloading and distribution, plus clearing away of rubbish and other general activities in support of the trades teams, may be priced as 'service gangs'. In the latter case the cost of the service gang may not appear in the rates and prices for the individual measured work items but might be included in the preliminaries and general, or 'indirects', section of the pricing document (such as bills of quantities), against items describing the contractor's general obligations.

It is therefore necessary to understand, when considering claims for additional payment and the adjustment of rates and prices, just where the costs of such general contractor's obligations have been included in the rates and prices in the contract.

5.8.1.2 Supervision

The extent to which the rates and prices for measured works include supervision of the tradesmen is also an area where practice may vary, especially in relation to the categories of supervisory staff that might be included in those rates rather than priced in the management items in a preliminaries and general or 'indirects' pricing section of the contract documents.

The common practice is for 'working' supervision to be included in the unit rates and prices for the measured works to which they relate, and supervision by personnel engaged solely in 'management' to be priced in preliminaries and general or 'indirects' sections. However, in practice the definition of the two categories of supervisory personnel can become blurred, with consequent difficulties in determining who is included in the unit rates for measured works and who is not.

The term 'working supervision' usually covers trade foremen and gangers, working with the trade and labourer teams in the field and responsible for their day to day organisation, including obtaining information and material supplies and raising queries with the site management team. Such foremen and gangers may well also be engaged in assisting with the works themselves, depending upon the type of contract and the contractor's supervisory scheme as well as the extent to which the circumstances of working on the project makes other demands on their time.

Another way to understand this distinction is as an illustration of the difference between 'work-related' or 'activity-related' and 'time-related' preliminaries and general costs mentioned elsewhere in this chapter. 'Working supervision' may be included within the rates for the work to which it relates, because the need for it depends on the required quantity of measured work to which it relates. On the other hand, other supervision, and management, may be included in a preliminaries and general section of a pricing document because it is 'time-related', its need depending on how long the contractor is carrying out the work rather than its quantity.

The actual demarcation between the supervisory element included in the unit rates and prices and that included in the preliminaries and general items section may, as in the case of service labour costs, vary from contract to contract and between different contractors depending upon the type of works and the contractor's policy.

5.8.2 Use of Norms in Evaluation

In compiling the unit rates in a tender the contractor will usually utilise a set of 'norms', or standard productivity outputs, to assess the hours for labour and plant against which to apply its cost rates per hour. These norms will most often be sets of data compiled by the contractor's staff from data recorded on similar projects undertaken by the contractor, with appropriate adjustment for differences in nature, circumstances, conditions, etc. It is relatively unusual to find a contractor using published books of norms or pricing information to compile its tender. However, this does occur. Furthermore, such books of norms can be of particular use in the valuation of variations in some circumstances, as also explained elsewhere in this chapter.

These norms will then underpin the labour and plant element of the unit rates and prices in the contract. In most building and civil engineering contracts the norms will not be apparent from the unit rates, as composite rates are provided for each item covering all the contractor's obligations in respect of that item. It is quite usual, however, in contracts for many heavy mechanical engineering works and process type projects, for the contract bills of quantities to have two 'price' columns, one for the monetary value and the other for the number of man-hours contained within the item. Alternatively, as detailed elsewhere, separate columns may be provided for materials, plant, equipment and labour pricing. In such contracts there is therefore an indication of the labour content of each item, which can be used to produce a reasonably accurate assessment of the labour norm for that item, although the items may require more than one labour activity and therefore include more than one norm. For instance, the labour content of an item for welding pipework might include activities such as the preparation, cleaning and bevelling of the pipe ends, clamping of the pipe and root and fill passes of weld, together with any subsequent grinding, etc. There may be norms available for these individual activities separately, but the item will incorporate them all under most methods of measurement. There is still therefore a degree of judgement required in many work items in converting the item man-hours to unit norms, as the stated man-hours may include more than one activity and/or trade and/or gang make-up within the trades.

The ability to assess with reasonable accuracy the tender labour content of the unit rates is a considerable advantage if it becomes necessary to analyse the rates and prices as a result of claims for additional payment, where, for instance, the contract allows the rates to be adjusted for similar work executed under different conditions and the rates and prices require adjustment for the effect on labour outputs of such difficult conditions.

If the labour content of a rate cannot be assessed from the information in the contract bills of quantities, or elsewhere, then recourse can be made to a variety of published data that can assist. The obvious problem is that the published data will almost certainly be different from any internal data of the contractors used to compile the contract rates and prices.

There is a great range of published information relating to the unit rates and prices in construction contracts, including building 'price books' which provide analysed build-ups of unit rates for works, including their labour man-hour content. These publications include the price books published by Spons, Laxtons, Griffiths, Hutchins and Wessex in the UK for building and civil engineering works, which provide detailed build-ups to unit rates, including outputs for labour. Similarly, the Building Costs Information Service (BCIS) Civil Engineering Cost File 2009 provides outputs for both plant and labour. There are also a series of books of labour norms by John Page, published by Gulf Publishing, often referred to collectively as the 'Gulf Norms' or by the names of the authors as 'Page and Nation'. These include: *Estimator's Piping Man Hour Manual, Cost Estimating Manual for Pipelines and Marine Structures, Estimator's Electrical Man-Hour Manual, Estimator's Equipment Installation Man-Hour Manual, Estimator's General Construction Man-Hour Manual* and the *Estimator's Man-Hour Manual on Heating, Air-Conditioning, Ventilating, and Plumbing*. Further books of norms may be found for specific sectors of the industry, such as the *Data Bank of Estimating Norms* published by the Oil and Chemical Plant Constructors' Association (OCPCA Norms) for process type projects. Specific to a trade, *TSI Luckins Electrical Installation Times Guide* provides suggested man-hour installation times for electrical installations and products. Luckins also produces similar guides for mechanical and plumbing installations.

Internationally, more developed construction and engineering markets have a similar extent of such published data as the UK, particularly the USA. In China, the Beijing Commission of Housing and Urban–Rural Development publishes a set of books for different work trades. These are sometimes referred to as 'the Chinese Norms' and are used widely by Chinese contractors when pricing work both in China and abroad, with adjustments to suit the circumstances of overseas countries in which projects take place.

These are just a few examples of available pricing books. There are many more in use around the world.

The overriding principle when consulting such sources of information in connection with any particular situation is that they should not be regarded as being directly applicable without careful consideration of the basis of the published data as compared with, for example, the location, circumstances and nature of the particular project and work.

In particular it is important to understand how the published data addresses a number of issues that might include:

1. What is the technical basis on which the norms have been established? Is this by reference to a relevant standard such as a British Standard, or those of such as the International Organisation for Standardization ('ISO') or the German Institute for Standardization ('DIN'), or some other criterion?
2. How do the norms deal with supervision? Is it included or excluded? If it is included, to what extent?
3. Do the norms make any allowance for lost time in the working week, such as clocking time, time lost between assembly point and workface at the beginning and end of shifts and at meal breaks, etc.?
4. Do the norms exclude other lost time factors that may affect the works, such as inclement weather losses, training and induction time, periodic leave time, travelling time, etc.
5. Have the norms been averaged over time and, if so, what method has been used?
6. Are the referenced projects adequately similar across all the norms to the subject project, for example in relation to labour skills, technology used, method of construction, resources used, etc.?
7. How old is the data on which the norms are based? Have advances in items such as technology, plant, equipment outputs, etc., rendered the norms obsolete?
8. More broadly, the location, nature, size, etc., of the particular project is compared to those that comprise the source of the published norms.

Practice may differ, but it is usual for norms to include some allowance for matters such as (2) above by including for working foremen and gangers, and item (3) above by including an allowance for lost time. In the case of the OCPCA norms it is stated that an allowance of 9.6% has been included for lost time, based on 3¾ hours lost in a standard 39-hour week. If this sort of information is considered in connection with a particular contract it is obviously possible to adjust the norms to reflect the circumstances of the project, if necessary.

It is equally important to understand the type of project and the conditions of work covered by the published norm. For instance, what size of project is covered by the norm? Does it apply to minor or major works, where in the latter the required labour allowance may be less due to economies of scale? Building price books, such as those published by Spons, usually state the size of project used as a basis for the published data and the location of the project. In the case of Spons *Architects and Builders Price Book* the location is stated to be Outer London. This may have implications not only for the costs when considering projects in other locations but also the productivity norms as a contractor may well consider that higher or lower productivity levels will apply to alternative locations.

On a more detailed scale of analysis, a further question is whether the published norm provide data for the same work under different conditions or for one stated condition only? The OCPCA norms state that the pipework erection data provided relates to work at ground level and factors are given in the pipework erection section if the norms are to be applied to work in other locations, e.g. in elevated positions or in process units.

This process of factoring norms is one that commonly applies to published data, and indeed is commonly applied to the contractor's own data to allow for differences in the circumstances of individual projects. Factors may be provided within the published data, such as the pipework erection factors in the OCPCA norms, or may be calculated by the contractor in the light of previous experience. Whatever the source, it is important to understand that any set of published data can only apply to a particular type of contract and set of circumstances, and cannot be all encompassing. When using such norms to adjust rates and prices in the contract, the first process must be to fully understand the basis on which the published data is presented, and then for that basis to be compared with the particular contract. Only when that process is completed, and any necessary adjustments made, is it possible to use the data to assess changes in the unit rates and prices for a specific contract.

A further factor in relation to the use of published data may be whether, notwithstanding the difficulties of applying it to the circumstances of a particular project and making suitable adjustments, there is any better information available for the evaluation of a change. It may be, for example, that a particular country has no published data in relation to its local construction industry and the contractor is unable to provide any such objective details in support of its claimed rates for variations. If the contract say that a particular variation is to be the subject of 'fair valuation', it may be that the use of data from further afield is the only option and that, with suitable adjustments, it provides the only available valuation. Such a valuation should presumably take a particular conservative approach when adjusting such external rates for local conditions and circumstances.

If properly adjusted, published norms can be particularly useful in establishing the proportions of the constituent cost elements within a rate, allowing those proportions to be applied to a contract rate to establish an estimated value for each element.

5.8.3   Plant and Equipment Costs

Many of the comments made in Section 5.8.2 in respect of the application of norms for labour activities apply equally to productivity data for plant and equipment. It is equally important to ensure that productivity data for plant and equipment is adjusted to the particular circumstances of a project as it is for labour productivity norms.

In many instances, particularly on large-scale civil engineering or similar works, the contractor will resort to the use of an analysis of anticipated productivity using the type of information sources discussed elsewhere. Rather than refer to productivity norms, or constants, the expected production levels will be forecast by the establishment of an anticipated method statement and assessment of the production against time that can be expected using the plant and equipment incorporated in the method statement. It is not unusual for such exercises to include 'what if' scenarios for testing the method statement, or alternative method statements, to ascertain the production levels for possible alternative methods. The final data from these exercises will then underpin the calculation of rates and prices rather than simple reference to a set of output norms.

Adjustment of unit outputs for plant and equipment may therefore entail some understanding of the method statement and deployment of such plant and equipment, rather than simple reference to standard output data. If standard outputs are used, then they will be subject to the same considerations as labour constants with reference to their applicability to the particular circumstances under examination.

Potential sources of published plant data include some of the books identified in Section 5.8.2, some of which contain information on plant outputs. However, rather more detail is available elsewhere; for example, the *BCIS Civil Engineering Cost File* 2009 and the *Caterpillar Performance Handbook* 2017.

A matter that might cause greater difficulty in respect of plant and equipment is that of the hourly costs to be used for each item in an analysis, i.e. what is the correct cost per hour, day, week, or month for a particular piece of plant or equipment?

5.8.3.1   Plant and Equipment Cost Rates

In many instances the detailed build-ups to the contract rates and prices that would detail allowances for the plant and equipment component will not be readily available to others outside the contractor's organisation. Although, as explained in Chapter 4, it may be a requirement of the contract for such detailed calculations to be made available by the contractor, and in any event it would be open to the employer to request it in support of the valuation of change. In circumstances where it is not available, there are several alternative sources of costs that could be applied to plant and equipment in the search for a basis for analysis, including:

1. Daywork rates for plant, either from a contract schedule of daywork rates or from published sources such as the CECA Schedule or the RICS Schedule of Basic Plant Charges.
2. Actual costs of externally sourced plant and equipment hire companies.
3. Internal 'hire' rates or charging rates that represent charges made within the contractor's organisation for plant and equipment owned by the contractor.
4. Actual costs of internally sourced owned plant based on its depreciation, running and maintenance costs (although this is highly record dependent).

The first of these alternatives, the daywork rates, will not usually be applicable for the adjustment of unit rates for measured works items as they are intended for the valuation of work instructed to be executed on a daywork basis and undertaken incidental to the contract works. Those rates will include elements that may be included or excluded within the measured unit rates. For instance, the CECA Schedule applies only to the contractor's own plant already on site and does not include fuel distribution in the rates. In contracts, adjustment of unit rates will usually encompass plant from any source and include the distribution of fuel to it as required. Dayworks rates in a schedule attached to a contract may also include overheads and profit, if these are not set out separately as percentage additions to the rates.

The second of the alternatives, plant and equipment sourced from external hire companies, may or may not be relevant depending on whether the item is hired or owned by the contractor. Such costs should be capable of evidencing from hire agreements and invoices, subject to any discounts. The charges may or may not include a driver or operator. These costs are considered in more detail in Chapter 6.

For the third alternative, internal hire rates, consideration may need to be given as to whether or not the rates are reasonable and reflect the contract pricing regime. It will usually be bad practice to simply proffer or accept rates without being reasonably certain that they only include costs for those elements that should be included in such rates. This then raises the issue of what elements should be included. This is a topic considered in more detail in Chapter 4.

This problem was considered in the case of *Alfred McAlpine Homes North Ltd v Property & Land Contractors Ltd* (1995) 76 BLR 59. The court heard the issues as an appeal from the award of an arbitrator, who had awarded amounts to the claimant calculated on the basis of hypothetical hire rates for plant actually owned by the claimant. The arbitrator's reasoning was summarised in his award as:

> I consider it fortuitous … that the whacker, dumper, Stihl saws and mixer were owned by (the claimant). Had they not owned this equipment then it would have been perfectly reasonable for it to have been hired, as they hired the JCB. The arguments advanced in respect of notional depreciation are not relevant.
>
> As with the JCB, the piecemeal nature of the works occasioned by the random sequencing of operations would justify the need for the equipment to be readily available, and I am satisfied that these items of plant were required to progress the works during the period of the overrun.
>
> I consider that the rates claimed for the specific items of plant are very reasonable. Reference to the RICS Schedule of Basic Plant Charges further supports (the claimant's) evidence. It is also significant that in the explanatory notes to these schedules it is confirmed that 'the rates apply to plant and machinery already on site, whether hired or owned by (the claimant)'. I consider that (the employer) has benefited from (the claimant's) ownership of the tools and that the rates charged are more than reasonable.

The arbitrator considered that, as the plant and equipment were justifiably claimed, the cost of them could be recovered by reference to hypothetical hire rates for such items although the items were in fact owned by the claimant. The arbitrator made no distinction in methodology for the calculation of the cost of the JCB, which had been hired in, and the smaller items of owned plant.

When the appeal was considered, the Judge disagreed with the arbitrator's approach, stating:

> From the award it appears that the arbitrator regarded the fact that (the claimant) owned the small plant as 'fortuitous', and, on the basis that, if the plant had not been owned by (the claimant), the plant would have been hired by (the claimant), dismissed arguments about valuing the claim in terms of depreciation. Instead the arbitrator valued the claim by reference to what he regarded as reasonable hire charges ….
>
> Where plant is owned by (the claimant) which would not have been hired or which was not able to be hired out the ascertainment of loss and expense must be on the basis of the true cost to (the claimant) and must not be hypothetical or notional amounts. An ascertainment needs to take account of the substantiated cost of capital and depreciation but will (or may) not include elements which are

included in hire rates and which are calculated, for example, on the basis that the plant will be remunerative for only some of the time and other times be off hire
....

The Judge's decision was made in the context of what he considered to be necessary for an 'ascertainment' of the amount due as an additional payment under the contract of 'loss and expense' for delay and prolongation of the works, the time consequences of change considered in Chapter 6, rather than in the context of the adjustment of unit rates. However, the Judge's comments in this case have been contrasted, in Chapter 4, with those of the Judge in *How Engineering v Lindner*, where that Judge considered that the term 'ascertainment' left more room for opinion than the Judge in *Alfred McAlpine v Property & Land* apparently considered appropriate.

Here the Judge considered that the assessment of the correct amount for plant owned by the contractor should reflect the capital and depreciation costs incurred by the contractor, and not hypothetical or notional hire rates for similar plant available from plant hire companies, or by reference to published daywork rates.

The inappropriateness of daywork rates as a basis for establishing plant and equipment costs in building up rates for a variation not valued on a daywork basis, has already been commented upon but there are also practical difficulties in adopting the philosophy suggested by the Judge in *Alfred McAlpine v Property & Land*. If the value of owned plant and equipment is to be determined by the actual capital and depreciation costs a number of issues arise:

1. Can the particular plant be identified in the contractor's financial records? In the case of items such as whackers, dumpers, pipe bending machines, electric saws and mixers this may not be easy in a large organisation owning a considerable quantity of such items. Larger items might be better recorded, but, what of plant of very long life, such as a marine vessel, purchased 25 years earlier? Will the capital costs still be available? They should be recorded on the contractor's asset register, but may not be, depending on the extent to which such a register records smaller items rather than just major items. For smaller items, though, even where it is theoretically possible to identify the relevant items in the contractor's financial records is the effort and expense of so doing justified in relation to the cost of the item claimed?

2. Should depreciation be based on that recorded in the company's accounts, based on its accounting policies as to write-down periods or actual depreciation based on an item's actual useful life and residual or scrap value thereafter?

3. Should the pattern of depreciation be based on that recorded in the company's accounts (which will often be on a straight line basis) or a more likely actual depreciation curve, starting relatively steeply following purchase and shallowing out over time?

4. It is likely that the contractor will depreciate plant and equipment in large groups, if not as an entire category, and therefore the depreciation charge is likely to be an average charge applying to a group of items and not an individual charge to an individual item of plant.

5. How can the costs of periodic maintenance and servicing be accounted for? Again it is likely that any attempt to establish such costs for individual items of plant and equipment is going to be, at best, time consuming and expensive and, at worst, practically impossible.

If the contractor borrowed money in order to fund the initial purchase of owned plant, then financing costs may also be a factor in its costs.

It should be a reasonable assumption that any contractor will retain an asset register for all of its assets above a certain cost or significance. This may include not only plant and equipment but also any other owned site facilities and temporary works items, including buildings such as portable offices, containers, stores, canteens, workshops, changing rooms, and also even contents such as printers, computers and boardroom tables. The data set out against each such item may include such information as the following in relation to its cost:

- asset number;
- supplier or vendor;
- description and serial number;
- references;
- purchase, arrival, or start date;
- purchase price or capital cost;
- depreciation period;
- depreciation period to date;
- accrued depreciation; and
- residual price.

Some of these issues are considered in more detail in Chapter 6 when considering claims for the time consequences of change, and particularly claims for plant and equipment in the context of a prolongation claim.

For such owned plant and equipment, it is respectfully suggested that the correct approach for a valuation based on cost is to determine reasonable rates for such plant and equipment, taking into account the capital cost, maintenance and running costs and depreciation charges and use them to calculate the appropriate amounts. Reference to hire charges and/or daywork charges should be for reference purposes only, to consider such as the reasonableness of the calculated rate. It would be usually expected that the calculated rates would be less than those suggested by such comparisons. It may also be necessary in a valuation of change to bear in mind that for the adjustment of unit rates the objective may be to try and establish a cost level that is compatible with the pricing level in the contract rates.

In relation to owned plant, this is the approach adopted in the case of *Norwest Holst Construction v Co-op Wholesale Society Ltd* (1998) EWHC Technology 339. Here the court was asked to consider an evaluation made by an arbitrator in relation to the assessment of damages involving the retention on site of plant owned by the subcontractor. The arbitrator had accepted a view put forward by one of the experts that valuation should be based on hire rates less 10% inflation less 20% discount. It was argued that this did not represent the true loss suffered by the subcontractor. The Judge held that the ascertainment by the arbitrator was not clearly wrong and unsustainable, and that it was also a finding of fact that the loss was best represented by the use of such heavily discounted current market hire rates and was therefore not open to challenge. While this decision was in the context of the assessment of damages arising from delay to sub-contract works, the means of valuing owned plant may be of use in the compilation of unit rates for the valuation of the direct consequences of change.

There is, however, a possible exception to this approach and that is where large individual items of capital equipment are being considered. Such items would include dredgers, large bulk excavators, floating or very large mobile cranes, etc., for which it might be difficult to determine rates by the methodology suggested as there are no readily available market hire rates for such major items of equipment. Items of plant and equipment such as these are usually identifiable in the contractor's management accounts, with a greater or lesser degree of effort, suggesting that it should be strictly possible to produce a calculation along the lines of that suggested by the Judge in *Alfred McAlpine v Property & Land*.

The contractor will usually have an in-house charge rate for such owned items, for use in calculating projected costs for the compilation of tenders and for charging the use of the asset to each of its projects for management accounting and control purposes. These charge rates could be based on the factors mentioned in *Alfred McAlpine v Property & Land*, along with other relevant matters such as maintenance and servicing costs, periodic overhaul and equipment replacement costs, licensing and royalty costs where applicable, insurance and operating costs, etc. The type of information considered by the Judge to be relevant should therefore already be available in some form for these major items and could therefore be adopted, subject to reasonable verification that the costs set out in the calculation have been incurred and are reasonable.

It should, however, be noted that this does not give a rate for even these items that is devoid of opinion or judgment. For instance, if considering the appropriate daily charge for a large trailer suction dredger, opinion will be relevant in considering a number of the inputs to calculation, including:

- Over how many years is the dredger's capital cost to be depreciated and what residual or scrap value might it have at the end?
- Should the depreciation be calculated on a straight line or curve basis?
- At what intervals will major components on the vessel, such as the main engines and dredge pumps, require overhaul or replacement?
- What is the rate of wear on the vessel's equipment generally? Will the rate differ depending upon the actual conditions and location of service?
- Is any allowance to be made at the end of the depreciation period for residual value? If so, is it on a sale or scrappage basis?

These, and many other, issues will need to be considered in judging the reasonableness of any rates proffered by the contractor or plant owner. The final decision will be made in the absence of any available comparable hire rates, etc., and will depend upon the perceived reasonableness of the rate and the amount and detail of substantiation supplied, judged against the requirements for the rate being assessed and any requirements and commercial information in the contract.

Therefore, for a dredger the following information might give a basis for an appropriate charge rate, assuming straight line depreciation over 10 years:

Initial capital cost	$8,000,000.00
Less residual value	$500,000.00
Depreciation over useful life	$7,500,000.00
Annual capital depreciation (over 10 years)	$750,000.00
Annual allowance for overhaul and maintenance	$225,000.00
	$975,000.00
Divided by, say, 322 average working days per year	$3,027.95 per day

To this would need to be added crew costs, fuel and consumables and any sundry costs. A further consideration is interest or financing costs on the initial capital outlay of $8,000,000.00. This can prove complicated to calculate, so is ignored for the purposes of this simple example.

If it were considered that the conditions on a particular contract would cause increases, or decreases, to the charges in the company accounts (for example reduced useful life in a hostile marine or desert environment), then suitable adjustments could be made. The reasoning behind the figures taken from the accounts should be available so the capital depreciation might be made up of a purchase cost of $25,000,000 with an anticipated lifetime with the company of 20 years before being sold on for $10,000,000. Considerations will be whether this period and residue are reasonable. The annual maintenance and overhaul costs should be capable of substantiation from accounts for earlier years.

It is also possible that such a level of detailed enquiry and substantiation of cost would only be appropriate for large and expensive items of equipment where no other cost reference data are available.

5.8.3.2 External Hire Charges

The use of external hire charge invoices to establish the cost of an item of plant is usually quite straightforward where the hire company providing the plant is separate and distinct from the contractor using and paying for the plant. Complications can arise, however, when there are elements of common ownership between the hire company and the contractor: a situation that is not uncommon and which was considered in Chapter 4. The potential problems can take on two different forms:

- The rates charged for the plant may not reflect the true cost of the plant but may be contrived to transfer money between companies for taxation or other purposes. In instances where there is a suspicion that such charges are being artificially inflated to transfer money to a related plant hire company the only recourse is to establish the current market rates for comparable items of plant and adjust the charges accordingly.
- The second problem that can arise is that where, although the charges themselves are just and reasonable, credits or transfers are agreed between the related companies for matters that are not directly related to the contract. Complete transparency is needed for these, including the reasons for each.

If credits are issued that have nothing to do with the charges themselves then there may be a question as to whether the charges should be used in calculating sums due under the contract. This problem was considered in the context of credit notes between related companies in the case of *Floods of Queensferry Ltd v Shand Construction Ltd* [2000] BLR 81.

Floods of Queensferry (FOQ) were the earthworks subcontractor engaged by Shand Construction in connection with the A494 Mold bypass commencing in 1991. FOQ was a family business that had interests in other companies including Floods Plant Hire Ltd (FPHL) and Floods Plant Ltd (FPL). FPHL supplied management, plant and labour resources to FOQ in connection with the works on the A494, having itself hired the plant from FPL.

In June 1992 FPHL went into liquidation and FPL took over the supply of management services and plant direct to FOQ. Subsequently FPL was also liquidated in October 1994.

The works had started in April 1991 but were subject to delay and disruption resulting in claims from FOQ, which included the cost of additional management, services and plant provided by FPHL and FPL. In May 1992 FOQ owed some £160,000 to FPHL for plant and other services, of which almost £100,000 related to the A494 contract. FPHL were faced with insolvency and there was the prospect that a liquidator would pursue FOQ for the debts and subsequently threaten the solvency of FOQ. An agreement was therefore made in May 1992 between FOQ and FPHL that a credit note dated 1 June 1992 would be issued by FPHL to FOQ under which FOQ would owe only five pence in the pound on the debt. The FPHL liquidator did not challenge this transaction and subsequently the same type of arrangement was entered into between FOQ and FPL with further credit notes being issued by FPL to FOQ in July 1993 and June 1994.

In the pursuit of the claims by FOQ the question arose as to whether, and if so to what extent, FOQ were entitled to recover the cost of plant or management services covered by the credit notes.

The court held that FOQ were not entitled to recover the costs covered by the credit notes as damages. The credit notes were not gifts but were transactions arranged to prevent the financial difficulties of FPHL and FPL being transferred to FOQ. The credit notes reduced the loss suffered by FOQ and therefore their amount could not be recovered as part of the damages.

However, it was also held that claims by FOQ for the valuation of additional works were not affected by the credit notes, as such valuation is a contractual entitlement and does not require proof of actual loss. The existence of credit notes, or other cost transfers, in circumstances similar to those described in this judgment, will therefore not affect the valuing of unit rates under a contractual valuation procedure that does not require proof of loss.

5.8.4 Materials Costs

The inclusion within unit rates for the materials element will usually be a relatively straightforward exercise and will be based upon the purchase cost of the material, evidenced based upon such as orders and invoices from suppliers. We have commented elsewhere on the need to ensure that quotations represent actual cost. In addition to those basis costs, allowances may be required for a number of other factors, including:

1. *Wastage.* The actual amount of wastage may vary depending upon the type of material and usage but will usually include losses in transportation for bulk materials, losses in unloading and distribution and conversion waste, i.e. losses incurred in converting

the materials to the finished product. For example, cutting sheet piles to a required length or copper pipes to installed routes will result in wastage of the off-cuts unless they can be used elsewhere. Loose materials such as sand and aggregates will have losses during transporting, delivery, unloading and placing, which can be increased by double handling if the circumstances of the project require it. This may be particularly relevant to a variation whose timing prevented proper pre-planning. Fragile items such as light fittings and lamps may also be broken in transporting, delivery and installation. Typical waste factors can vary between 1% and 10%, or even more, depending on the type of materials. However, the nature, conditions and circumstances of the project and works can add further wastage, increasing these typical percentages.

2. *Transportation.* Particularly on imported products, these can be significant, where they may include port handling fees, demurrage and customs duties and other costs. In some cases air transport may be necessary due to time pressures. This particularly occurs where variations are instructed late in relation to works that are critical and only immediate delivery will prevent delay to the works.

3. *Unloading, handling and distribution costs.* These will usually be included in the unit rates for the measured works to which they relate, but in some civil engineering contracts, for instance those for pipeline construction, handling and distribution of the line pipe may be included in the contract bills of quantities as separate items. Where the unit rates require inclusion of such costs they will include labour and equipment for the task, although some major items such as tower cranes may be included in the preliminaries and general items section of the contract pricing documents. It is also not unusual, as mentioned elsewhere, to find that many building contractors include the cost of service gangs for the unloading and distribution of materials on large projects in the preliminaries and general items. Necessary double handling will be included in unit rates for the work, but if the need for it only arises, or increases, because of some change for which the employer is responsible, then it may become the subject of a separate claim.

4. *Storage.* In some instances storage of material, on or off site, may be required and unless such a requirement only arises as a result of some default by the employer then the cost of such storage will generally be deemed to be included in the unit rates for the measured work items to which they relate. However this is another potential materials cost that can more often occur in claims for the consequences of changes that in original contract pricing because change is likely to reduce the ability to plan to limit storage costs.

5. *Bulking and shrinkage.* It is important to note that due allowance will often be required in unit rates for some items, particularly in respect of earthworks and fill items, for bulking or compaction of the material. The obvious example is that of imported stone or hardcore used for fill, which will reduce in volume, as compared with the transported volume, when placed and compacted. This can be particularly relevant to consideration of unit rates for the placing of fill material stockpiled from excavations on the same project. The usual method of measurement is that of net volume for both excavation and fill requirement, i.e. the contract measurement denotes the net volume of excavation and fill. If, for instance, these volumes are the same then it might be considered that no imported fill will be required (in a 'balanced cut to fill'). In practice this is hard to achieve. Losses in transportation

and the reduction in volume resulting from mechanical compaction of the fill may require an import of fill to make up for any shortfall. Unless the contract terms allow payment for this importation, which is not usual, the cost of the importation to make up such losses will be at the contractor's risk for inclusion in the unit rate. Bulking factors on loose materials can be very high. Typically for soil they are around 25%, but for excavated rock might be as high as 60%.

6. *Buy back or salvage value.* Account may be required for any residual value in surplus materials or the scrap value of any surpluses. There may also be credits available for the return of crates and other packaging.

7. *Import or customs duties.* These can vary widely but can be significant. President Trump's March 2018 announcement of tariffs of up to 25% on steel imported into the USA from some countries has thrown this issue into sharp relief.

8. *Overhead and profit.* The contractor will be assumed to have included an element of overhead and profit in relation to materials and the other elements of unit rates unless the contract requires otherwise.

Analysis of the unit rate will be required, when considering adjustments in valuing the direct consequences of change, to identify the material element and to confirm that all such elements, as required by the contract provisions, have been made.

In this context, it might be a good point at which to mention the practice of construction contracts limiting the international sources of manufactured items that will form part of the permanent works to exclude some specific countries or limit supply to some stated countries. This practice seems to be the consequence of the increasing internationalisation of the construction and engineering markets, combined with concerns regarding the quality of cheaper products from some sources. A typical clause to this end might read as follows, from the contracts of a Middle East government agency:

> Manufactured items, including all parts and components thereof, shall be manufactured and assembled in any of the following geographical locations:
>
> • Western Europe;
> • North America;
> • Australia; or
> • Japan.
>
> Manufactured items, including all parts and components thereof, which are manufactured and/or assembled, either in whole or in part, in geographical locations other than those listed above, shall not be acceptable.

Such clauses will be considered in more detail in Chapter 8 when considering employers' claims for defective work. A moot point in relation to such clauses is the extent to which they are really intended to maintain high quality or to address what is sometimes seen as the unfair competitive advantage of contractors from some countries who can source manufactured items in their domestic market with export incentives and without any local duties.

5.8.5 Overheads and Profit

Unless the contract requires that the unit rates for measured works are to be exclusive of overhead and profit, with such matters addressed elsewhere, the unit rates will generally be deemed to be inclusive of overheads, both on-site (preliminaries and general items) and off-site (including 'head office'), and profit. The extent to which such matters are actually included within the unit rates can vary considerably depending upon the policy of the contractor, current market conditions and type of project. The issue of site overheads and the extent to which they are priced in their own preliminaries and general costs pricing document is discussed in this chapter and also in Chapter 4.

It is not unusual for a contractor to price the unit rates as net of profit, and sometimes net of head office overhead. The reason for doing this at the tender stage is to allow the net estimated cost of the works to be determined prior to its senior management determining the amount of head office overhead and profit the contractor wishes to add to the total cost when converting the estimate into a submitted tender. The amount of head office overhead and/or profit ultimately added to the total cost may not be a simple matter of applying calculated rates for overheads from the contractor's management accounts to the projected cost, and then adding a profit element based on the required future return. Other factors may play a significant part in the final decision as to what level of addition is made. These factors can include:

- *The market.* This may be the market for the particular type of contract or the more general market for construction services and the economy at large. Perceptions of the level of alternative work available elsewhere during the period of the tendered project may affect the amount of overhead and profit priced into the contract. In good times, when alternative work is seen as being plentiful, the inclusions may be relatively high, whereas when times are hard overhead additions may be reduced and profit eliminated completely! The effects of the overheating and crashing of some construction markets in the years 2007 to 2009 are discussed in Chapter 7 in relation to its effect on terminations. As discussed there, the effect on tender prices was dramatic.
- *Workload.* This may or may not relate to the state of the wider market. A contractor that is itself short of work is likely to price overheads and profit allowances much more keenly than one that is already overburdened. Equally, if the market is slow, then bidders are likely to have to bid keenly to be competitive.
- *New market?* Workload may relate to just the particular market, state or country, where a contractor is looking to break into new areas or expand its influence. This may mean marginal costing of new opportunities, or even bidding as a 'loss leader'. The increasing internationalisation of construction has made this more relevant over recent times. A commercial director of one of the larger contractors in a Middle East country once described the newly arrived competition from the Far East to the authors as 'a tidal wave I won't even bother bidding against if only price counts'.
- *Availability of resources.* The contractor may have a surplus, or a paucity, of the required resources for the contract being tendered. Either situation will tend to influence how the contractor includes overhead and profit in its tender.
- *Perceptions of the client.* In some instances the contractor may consider some clients more desirable than others, or the contract may be for a client with whom the contractor has established an ongoing relationship, either formally or otherwise, or want to establish one for the future. In such cases, consideration of the client can be a factor in deciding the precise level of overhead and profit inclusion. This can sometimes also extend to consideration of the client's consultants or representatives for the contract.

If the contract requires overhead and profit to be included in the unit rates and prices for measured work items the contractor will often 'spread' the calculated sums for these elements evenly over its rates and prices for the works. This may, however, not be done evenly as the contractor may consider that there are advantages in adding a greater proportion to some elements rather than others. For instance, a greater proportion might be included on the rates and prices for work early in the contract period to assist in maximising the contractor's cash flow during that period. Alternatively, a greater proportion might be added to elements that the contractor believes to be liable to increases in quantities when the work is undertaken, particularly if the contract is a remeasurable one under such as the FIDIC Red Book or the Infrastructure Conditions. There are obvious dangers in such approaches and it is usually assumed that all unit rates and prices include an equal allowance for overheads and profit unless their analysis shows a different addition and/or there is some requirement in the contract for different elements to be priced on a different basis.

5.9 Subcontractor and Supplier Costs

The use of subcontractors is long established in the construction industry, but in the last three decades or so, the extent of work subcontracted by main, or prime, contractors has increased considerably to the point where in some countries contractors act effectively as what were termed 'management contractors' at the start of that period. This is particularly true in the UK, where at one time the common practice was for the main contractor to undertake much, if not all, of the structural work with its own workforce and to sublet only specialist works such as the mechanical, electrical, plumbing, roofing works, etc., to others. The increase in subcontracting means that it is now not uncommon to find the bulk of the works sublet, with the main contractor providing management and ancillary services and only undertaking minor works as part of the project.

There is nothing intrinsically wrong with a policy of subcontracting, and it can be supported on the grounds that increasing specialisation, by having separate contractors for all the major trades and elements, brings greater quality and efficiency. As projects have become more technically specialised, it may be inevitable that they are carried out increasingly by specialists. The contrary view is that 'too many cooks spoil the broth' and that having so many separate contracting parties working on a project leads to difficulties in terms of both interfacing and their competing interests. Such a view also often includes the assertion that the contractor becomes no more than an observer, post box, and cashflow manipulator, passing responsibility and liability for quality, programme and cost around between the various parties to which it has contracted. In the end, much depends on the quality of the subcontractors and ability of the contractor to manage them.

Many contracts place restrictions on the main contractor's ability to sublet the whole or parts of the works without the consent of the employer, but with some safeguards for the contractor. For example, SBC/Q clause 3.7.1 states: 'The Contractor shall not without the Architect/Contract Administrator's consent sub-contract the whole or any part of the Works. Such consent shall not be unreasonably delayed or withheld …'. Similarly, in the Infrastructure Conditions clause 4(2) requires the contractor to notify the engineer, who has seven days to notify 'good reason' for its objection to the employment of that subcontractor 'accompanied by reasons in writing'. FIDIC Red Book clause 4.4(b) requires the engineer's prior consent to proposed subcontractors who were not named in the contract. In practice, notwithstanding such express and agreed limits, it is surprising how often they are ignored, with contractors engaging subcontractors without necessary approvals, but this never becoming an issue with the employer unless such subcontractors fail, in which case the employer might add the lack of approval to its list of complaints. Such issues are outside the remit of this text and the discussion here assumes that subcontracting is authorised. The effect of authorised subcontracting does need some consideration when examining and analysing the rates and prices in the contract.

5.9.1 Subcontractors

In most construction contracts there are two broad categories of subcontractor: firstly, the 'nominated' or 'named' subcontractor where some degree of selection is exercised by the employer and its representatives and, secondly, the 'domestic' subcontractor who is selected and controlled entirely by the main contractor, with the only involvement of the employer and its representatives perhaps being to give the contractor prior consent to sublet to them as required by the contract.

5.9.1.1 Nominated or Named Subcontractors

Tenders for works by nominated subcontractors will usually be obtained by the employer's consultants and selection of the subcontractor to be nominated and employed under a subcontract with the contractor will be made on behalf of the employer by its team. Nomination under most contracts that contain such provisions can be either as part of the contract documents or later under an instruction under the contract.

Where a contract pricing document includes amounts for a nominated subcontractor, this will usually be by way of a prime cost or provisional sum in relation to that nominated subcontract, with provision for the contractor to price a 'mark-up' (for such as overheads and profit) on the subcontractor's price and to price any attendances or services required to be provided by the main contractor for the works of the subcontractor. Here the employer takes complete control over selection but may also take some liability for failure or default by such nominated subcontractors and also ensuring that they are properly paid by the contractor. Safeguards for the contractor are likely to include a right of reasonable objection to a nomination. The extent of such rights, liabilities and responsibilities will vary between the terms of the contract and may have significant consequences for the valuation of claims by the contractor.

Historically, the JCT contracts contained detailed procedures in relation to nomination of subcontractors and left significant aspects of risk of their failures with the employer. Therefore, for example, the 1998 edition of the JCT Standard Form of Building Contract, Private with Quantities made delay by a nominated subcontractor a 'Relevant Event' for extensions of time under clause 25.4, although it was not a matter for 'loss and expense' under clause 26. Such aspects of the allocation of risk for the performance of nominated subcontractors had, however, been a regular source of dispute and perhaps for that reason, SBC/Q contains no provision for nominated subcontractors.

In the Infrastructure Conditions, provision for the nomination of subcontractors is contained in clause 59, which gives the contractor a right of 'reasonable objection'. Clause 59(3) makes the contractor responsible for the works of a nominated subcontractor. However, in the event that the subcontract is terminated due to the subcontractors' default, clause 59(4)(f) provides that any delay to completion of the Works resulting from such delay shall be taken into account in relation to extensions of time under clause 44.

FIDIC Red Book clause 5 contains detailed procedures in relation to nomination of subcontractors, either as one stated in the contract or instructed under Clause 13 'Variations and Adjustments'. Once that subcontractor is appointed the contractor becomes entirely responsible for its performance. However, clause 5.3 also gives the contractor rights to 'reasonable objection' to the nomination.

Where variations are carried out by nominated subcontractors the contractor will usually be entitled to its mark-up and attendances on those variation works as an addition to the subcontractor's price for them. It has been explained elsewhere that the main contract should include a provisional sum in relation to that nominated subcontract, with provision for the contractor to price a 'mark-up' and to price any attendances or services required to be provided by it for the nominee. Where there are bills of quantities measured by reference to a standard method of measurement, those rules will dictate how sums and the contractor's additions are to be included in the bills of quantities. The POMI Rules include this in GP6.1. Regarding the mark-up, GP6.1 says 'an item shall be given for the addition of profit'. GP6.2 requires a further item for 'assistance by the contractor' and lists various inclusions in this, from 'administrative arrangements' to 'unloading and distribution'. This further item is therefore intended to cover all overheads of the contractor, both on-site and off-site. In practice the approach to these additions can vary among those preparing bills of quantities on this basis. Sometimes a single item is included to cover all of the contractor's 'overheads, profit and attendances'. Other bills of quantities might see two items, the first for both 'overheads and profit' and the second just for attendances. Whilst the POMI Rules only require these to be set out as 'items', practice also varies as to whether the items are set out in the bills of quantities to be priced as lump sums or additions in the form of percentages. The practical importance of this to the evaluation of the direct consequences of change under such contracts includes two particularly common points:

- Contractors might variously price these items as lump sums or as percentages, either because the bills of quantities were prepared to require one or the other or because they chose to adopt one or the other when pricing them. If a lump sum addition is in the bills of quantities, the issue then is what is to be added to the nominated subcontractor's cost when pricing a variation? It might appear right, on common commercial logic, that even if the mark-up was as a lump sum, that should be converted into a percentage and applied to the valuation of variations. However, is the same true of an attendances item? Does it follow that additional works by a nominated subcontractor will involve the contractor in additional costs at the same percentage rate? If so, is this always the case? Also, why then did the contractor price the bill of quantities attendances item as a lump sum?

- Issues can also arise as to the coverage of an attendances item. Standard methods of measurement such as POMI Rules CP6.2 set out the details of what is covered, and contracts are often drafted to provide further details, but what of attendances that are not set out in such detailed lists?

CESMM4 paragraph 5.16 requires sums for nominated subcontractors in bills of quantities to be the subject of a 'Prime Cost Item', followed by further items for the contractor's mark-ups and attendances. Subparagraph 5.16(a) requires an item for the contractor's attendances, listing items such as temporary works, unloading, etc. Subparagraph 5.16(b) then requires 'an item expressed as a percentage … in respect of all other charges and profit'. In practice, CESMM4's use of a percentage addition and having that cover the broad 'all other charges' as well as profit, leads to fewer points of issue than the POMI Rules approach.

In the case of named subcontractors, the main contractor will usually have to obtain tenders from a list of potential subcontractors included for the specific section of the works in the tender documentation. Once tenders are received they are treated as the main contractor's domestic subcontractors. The Infrastructure Conditions do not provide for named subcontractors.

Such processes should result in a reasonably transparent regime of pricing and, where appropriate, the subcontractor may provide a priced bill of quantities or schedule of rates. The unit prices so provided can then be analysed and adjusted in much the same way as the rates and prices in the main contract, with the mark-up and attendances adjusted accordingly. Difficulties can arise if subcontract tenders are provided and accepted without sufficient breaking down of the subcontract price. See Section 5.9.2 in this regard, in relation to package equipment suppliers.

In this regard, there may be a distinction between named and nominated subcontractors arising out of how their prices are procured. With named subcontractors the employer will just list names in its enquiry document and the task of obtaining tenders and ensuring that a resulting subcontract contains sufficient pricing details falls to the contractor. In this respect a named subcontractor is thereafter no different to any domestic subcontractor. With nominated subcontractors, the tender is likely to have been obtained by the employer's team, and its amount the subject of a prime cost or provisional sum in the contract with the contractor's additions. In this case, it may be that the detail of that price is very different to that provided by the contractor in relation to its pricing. In this regard it is a helpful practice in relation to variations if employers and their advisors ensure that nominated subcontractors and the main contractor price on the basis of similar documents, such as bills of quantities, in similar forms, methods of measurement and levels of detail. All too often, subcontract prices are obtained on the basis of specifications and drawings, rather than a bills of quantities that gives details of their works on an itemised basis, perhaps for the reasons explained in relation to specialist works explained elsewhere.

In addition to the evaluation of the direct consequences of change, when considering additional payments for matters such as extensions of time or breaches of the contract, the ability to assess the detailed contents of the measured element of the subcontract works may be vital as part of the analysis. All too often it is not, leading to particular problems in relation to nominated subcontractors.

5.9.1.2 Domestic Subcontractors

Under most forms of contract the domestic subcontractor, that is a subcontractor selected and controlled by the main contractor with only approval being required from the employer for the principle of subcontracting the work (if the contract so requires), performs the work effectively as part of the main contractor's organisation. There is therefore no separate consideration of the subcontractor by the employer, after any initial need for approval to sublet.

This does not pose any difficulty in principle as far as control of the works may be concerned, but can create some tensions when considering the adjustment of unit rates and prices under the contract.

When tendering for the works the main contractor will usually obtain prices from potential subcontractors to use in compiling its tender. A common practice is for the subcontractor to provide a total price for the subcontract works, defined by extracts from the tender documentation, rather than submit detailed rates and prices in response to every enquiry. This practice is adopted to save abortive work in supplying detailed quotations for every subcontract and trade enquiry when only a proportion will be successful and need the full detail of the pricing behind the total subcontract sum. Only when the main contractor has been advised that its tender is under active consideration might it usually request a fully detailed price breakdown from the subcontractor. This practice may of course vary depending upon the type of contract and tendering process adopted for a particular project, and some contractors may insist on the return of fully detailed quotations from subcontractors on all occasions at the tender stage to minimise the possibility of errors or misunderstanding in the basis of the works respectively priced by them.

There is of course scope for many variants on this routine and the variant most complained of by subcontractors is that where a subcontractor's price is used by the main contractor in compiling a tender but, upon being advised that its tender is under consideration, the contractor invites further tenders from other potential subcontractors in the hope of finding a lower price than the one used in its tender calculations. Such situations sometimes result in what are known as 'reverse auctions' or 'Dutch auctions' and there have been concerns that the increased use of tendering by electronic means might increase the incidence of such auctions. Alternatively, the subcontractor whose price has been used may be asked for a discount on its price to ensure it is retained for the contract. This may be hard to resist if it has invested a lot of time and expectation in its bid.

This is, of necessity, only a very brief summary of some of the many vagaries of the tendering process but it serves to illustrate that the calculation of unit rates and prices for work which the main contractor anticipates being sublet will often be the result of a subcontractor bidding process, rather than the detailed calculation and analysis of the appropriate rates by the main contractor. The result is that the main contractor may often not have a detailed analysis of the work content of unit rates for sublet works in terms of labour, material, plant and equipment, overheads and profit content. It may only know the relevant subcontract rate and how that compares with the rate in the main contract. The main contract rate itself might even be based on a rate submitted earlier by the same or another subcontractor.

When adjusting unit rates and prices in the main contract, where a subcontractor executes the work, the correct approach is to treat the relevant rates and prices as being those of the main contractor and analyse them as such. Any regard to the subcontractor's rates and prices, if available, should be for information only and to assist with the analysis of the main contract rates as far as possible. However, the issue of 'buying margins', i.e. the difference between the prices in the main contract and the amounts spent with subcontractors, may become relevant where cost-based claims for delay and disruption to the works are to be considered.

In practice the question of duplication of additions or allowances for on-site and off-site overheads and profit may arise, with either two-stage mark-ups being requested to cater for the additions required by the main and subcontractor, or a larger than normal addition being requested to allow payment of the two from the one mark-up. The highly technical nature of some works, particularly in building projects, may mean that subletting of works passes through various layers or tiers of subcontractor, sub-subcontractors, etc. Not only are each of these subcontractors understandably going to want their own overhead and profit additions, but given the likely high-tech nature of their activities and products, those additions may be much higher than that of a typical contractor.

It is implicit in the scheme of most standard construction contracts that duplication of overhead and profit should be avoided. The contract rates and prices apply to the works regardless of whether they are executed by the contractor's own workforce or by subcontractors. Some ad hoc contracts expressly stipulate that only the main contractor's costs will be taken into account in determining rates and prices, leaving the contractor to buy subcontract works within the rates and prices at its own risk. This probably does no more than make explicit what is implicit in other forms of contract, but it is good practice to try to define terms so that grey areas of definition are not exploited by the unscrupulous or overlooked by the naïve. This said, where subcontracts for very specialised works pass down through many high-tech tiers, the respective overheads and profit at each of those tiers should not be duplications.

This can lead to not only several levels of overheads and profit within one variation valuation but the inclusion of some very high percentages at the lower levels. In practice, the more specialised works are, the higher the percentage is likely to be. This can mean that at such a sub-sub-subcontractor level (or the third tier of subcontractor) the percentage can be considerably higher than at the contractor level. This may be a function of the lesser degree of competition at the lower level, giving them higher profit percentages, and also that such highly specialised companies may have very high costs of overheads such as research and development.

5.9.2 Package Equipment Suppliers

It has been noted how some subcontract works can have little or no detail of the build-up to their prices that will allow ready consideration of the consequences of change in relation to those works and it might be useful to illustrate this with an example as set out in this section.

One of the most difficult areas to address can be changes in package equipment provided as part of the contract works, where the rates and prices in the contract can be substantial lump sums with little or no breakdown of the sum against the component parts of the equipment, or the processes required to deliver it to the project. Such equipment as compression packages or chemical or water injection packages in the energy and process industries, and large heating and ventilation plant in the building industries, are usually obtained from specialist suppliers and subcontractors who may, or may not, be involved with the installation of the equipment on site, although it is very unusual for there not to be at least some involvement for them in commissioning the equipment.

Tenders are usually obtained from the supplier on the basis of specification. This is either by reference to standard manufactured units produced by the supplier or by means of a performance specification setting out the input and output requirements for the package together with any other relevant matters such as environmental conditions, storage capacities, etc. The final price for the equipment can include a wide range of engineering, manufacturing, and ancillary matters, including:

- Engineering design, either complete design or detailed design depending upon the circumstances of the particular equipment and order.
- Manufacture of the equipment at the supplier's works.
- Works testing of the equipment to ensure manufacturing defects are eliminated. This may be a supplier-only process or may involve the client's or third party inspection personnel.
- Dismantling of large equipment into component parts for transportation to site, often referred to as 'piece small' packaging for transport.
- *Insurance of the equipment in transit.* This may be on a number of different bases depending upon the terms of the order and may be at the cost of the contractor or client.
- *Shipping and transportation.* This may vary from a relatively simple journey for a lorry from the supplier's factory to the site or may involve shipping by sea with road transport at either end. In extreme cases air transport may be required if timing so demands.
- Reassembly at site, if necessary, and installation into the permanent works.
- Testing and commissioning of the equipment in its final position. This may of course not be completed until other related works have been completed on the project.

It is not difficult to appreciate the difficulties that can arise if changes occur in the requirements for a piece of package equipment which incorporates all of such components, and possibly other items in the supply, without any breakdown or analysis of the original price. This is further compounded if it is considered that the change may occur, not in the specification of a particular component in terms of 'omit type A and substitute type B', but in the performance specification of the equipment itself so that the parameters are changed without definition of the physical changes required to achieve the required result.

If the change is particularised in terms of the former of the two possibilities suggested, i.e. the substitution of one component by another, then the process of evaluating the required adjustment to the price might be reasonably straightforward, coloured only by the difficulty of establishing the reasonable cost of the omitted and substituted components if there is no market data for such items. In such circumstances the process of evaluation is often one of the supplier quoting its proposed prices for both alternatives for comparison and the reasonableness of those prices being assessed against an objective assessment of the relative changes in the component.

Sometimes the change may simply be in the size of a system or part of a system comprised within the equipment. In such circumstances it should be possible to make a reasoned assessment of the change based upon the price for the original equipment and any available comparisons, bearing in mind that the variation in price will usually not be in direct proportion to the change in size or capacity.

5.9.2.1 Example of Change Calculation for Package Equipment

Consider, for example, the case of an injection system specified as part of a process facility and comprising storage capacity and injection capability for a number of chemicals. Assume that there are two such packages, one comprising six equally sized injection and storage systems, each with a capacity of 2.98 m³ storage, while the second system comprises eight injection and storage systems, with storage capacities varying from 2.98 to 31 m³.

If, for instance, one of the systems, with a storage capacity of 18 m³ in the second package, is the result of an instructed change and the two packages are ordered from suppliers with the change already incorporated in the order, the question of separating the cost of the 18 m³ system in the second package might be addressed as follows.

Assume the first package of six 2.98 m³ systems is priced at £220,950 and the second package, now of eight systems of varying size, is priced at £629,490.

The first package provides an average price for a 2.98 m³ storage and injection system of £36 825. Using this as an average base cost for a system of that capacity the cost of the varying sized systems in the second package can be simply apportioned in proportion to the capacity of the storage system in excess of the base size of 2.98 m³, as is detailed in Table 5.1.

Some differences arise due to rounding in the calculations, but this calculation demonstrates a simple method of analysing a package price where some data are available to allow comparisons to be used as a basis. In this example, the cost of the 18 m³ system, system number 5, is £143,797. This assumes that the various systems are subject to the same specifications, etc. If that is not so, further adjustment may be necessary.

Table 5.1 Apportionment of cost to capacity.

System	Storage volume (m³)	Base cost per system (£)	Additional volume over base	Value of additional volume £629,490 − (8 × £36,825) = £334,890 = £ 7,122 per m³ of total additional capacity	Total value (£)
System 1	31.00	36,825	28.02	199,559	236,384
System 2	2.98	36,825	0	0	36,825
System 3	2.98	36,825	0	0	36,825
System 4	4.98	36,825	2.00	14,244	51,069
System 5	18.00	36,825	15.02	106,972	143,797
System 6	4.98	36,825	2.00	14,244	51,069
System 7	2.98	36,825	0	0	36,825
System 8	2.98	36,825	0	0	36,825

However, in the second of the two possibilities considered, i.e. that of a change in performance rather than particular capacities or sizes, matters will usually be more complex.

If, for instance, all these systems were originally specified to the same standards, etc., and the analysis in Table 5.1 reflected the costs of the systems to be supplied, the difference if a requirement was introduced for one system to operate at a higher pressure than the others could not be analysed in this manner as the specification and characteristics of that system would then be different from the others.

In such circumstances, if only the total cost of the package is known, the only way to proceed is to undertake an engineering analysis of the changes required to the component parts of the package to identify piece by piece the various changes needed, i.e. increased pump ratings, different welding procedures, etc. Price comparisons for these items will then be needed to build up a total price change from first principles. Such an analysis may well require input from specialist engineers.

5.10   Valuation of Omissions

The power of an employer, or its agent under the contract, to order omissions from the original scope of work in the contract is discussed in Chapter 2. Some of the issues that commonly arise in relation to omissions are discussed further in Chapter 8.

The valuation of omissions from the contract works can sometimes cause some difficulty, particularly where the omissions have an impact on the economics or commercial aspects of other work, or have an impact on the contractor's general site services, priced in its preliminaries and general items.

SBC/Q provides at clause 5.6.2 that where a valuation is of an omission from the contract works, then the valuation of the omissions shall be determined by reference to the rates and prices in the contract bills of quantities. The omission then falls to be valued under clause 5.6.3 where reference is made to the adjustment of preliminaries and general items, thereby allowing the proper valuation of changes in these items as part of the valuation of the omission.

The Infrastructure Conditions do not specifically refer to valuation of omissions as the remeasurement basis of the contract anticipates that reductions in the quantities of some elements of the work may occur and are covered by that remeasurement. Clause 51(5) of the Infrastructure Conditions states that no instructions are required to cover increases or decreases in quantities where the increase or decrease is not the result of an instruction issued by the engineer. Clause 56(2) then provides the mechanism for adjustment of any rates and prices rendered necessary by changes in quantities, which would include adjustments to preliminaries and general items.

The FIDIC Red Book contains a subclause in relation to omissions of work, at clause 12.4. This particularly addresses costs incurred by the contractor that would have been covered by the value of that omitted work, but for which recovery has been lost as a result of the omission. Typical of such costs may include: preparation costs such as contractor's design and/or shop drawing preparation, preliminaries and general items more generally and materials procured for which the full cost cannot be recovered by return to the supplier.

There is, however, a danger that the contract terms for omission of work may be abused by application beyond the power to omit work, leading to claims from the contractor for such as loss of profit and/or contribution to overheads on omitted work. Claims on this basis are discussed in Chapter 8. For example, this could be where the employer omits the work in order to give it to another contractor at lower rates and hence a saving in its costs. On the other hand, the omission of work can sometimes be a blessing for a contractor, where it has culpable programming difficulties and the employer agrees to reduce its remaining workload, to be carried out by others at a later date. Such a blessing may be balanced by negotiation of the amount to be omitted to reflect the benefit to the contractor and the disadvantages created for the employer, such as future additional costs.

SBC/Q and Infrastructure Conditions have a logical and well thought out regime for the handling of such matters and similar provisions should be considered in ad hoc contracts. In practice many difficulties are caused by attempts to restrict the ability to adjust rates and prices or by devices to provide a threshold level for changes that must be achieved before such an adjustment can take place. These are areas covered in more detail elsewhere in this chapter.

As explained elsewhere in this chapter, the approach of the NEC contracts is very different. Under NEC4-ECC, clause 63.1 does not distinguish between compensation events that involve omissions from those that involve additions. The valuation is based on the effect on the actual forecast Defined Cost. Since omitted works will not have been carried out, their valuation can only be based on an estimate of what the cost would have been. As ever with the NEC4-ECC approach to valuation, what the contractor allowed in the contract price is irrelevant. A practical consequence of this is that the valuation of omissions under the NEC approach can be far more time consuming and contentious than the traditional approach of, for example, the FIDIC Red Book, SBC/Q and the Infrastructure Conditions, which might just require an amount or group of bill of quantities items to be omitted.

5.11   Add and Omit Variations

Other sections of this chapter described how construction contracts can take two alternative approaches to the valuation of the direct consequences of change. The first is the traditional approach, initially adopting agreed contract rates and prices, as the FIDIC Red Book exemplifies. The second is the more recent regime, based on an agreed application of the contractor's actual or forecast actual costs, such as the NEC4-ECC approach.

Common logic may suggest that on an 'add and omit' variation both the added works and the omitted works should be valued on a consistent, or 'like-for-like', basis. The argument would be that to value them on different bases would fail to accurately value the net effect of the variation to omit one item and replace it with another. This might overcompensate or penalise the contractor against the real value or cost of the change, depending on the circumstances.

For example, applying the traditional approach explained elsewhere, if the addition is to be valued based upon agreed contract rates and prices, in the manner that the FIDIC Red Book initially sets out, then the omission would also be valued based upon agreed contract rates and prices.

Whilst NEC applies the alternative, costs-based approach, the position under that contract is even clearer in relation to a 'like for like' approach. Under the NEC approach of valuing a compensation event on the basis of the change on actual cost or forecast cost, then on an add and omit variation, effectively both the addition and omission are valued based on actual or forecast actual costs.

Take a simple hypothetical example of a simple increase in the quantity of a contracted item of work. If such a change were considered as an add and omit item – omitting the contract quantity and adding back the actual – then one would not expect the omission and addition to be valued on different bases (unless there was some other factor than the increase in quantity). However, what if the omission and addition are not of the same directly comparable item and capable of such consistent valuation? In particular, where the traditional approach to valuation applies, if the addition half of an add and omit variation can only be valued on the basis of actual costs, should the other also be valued on some estimate of what costs would have been on the omitted work?

In *MT Hojgaard A/S v E.ON Climate Renewables UK Robin Rigg East Ltd* [2104] EWCA Civ 710, the Court of Appeal upheld the decision of a judge at the first instance in the TCC. The matter related to construction of the Robin Rigg East offshore wind farm. MT Hojgaard's contract was to design, manufacture, deliver, install and commission the foundations for 60 wind turbine generators for a contract price of over €100 million. Of that some €26 million was for installation. The contract included at Schedule L1.1 a high-level summary of the contract price, including the sum of €22.1 million against 'Installation of the foundations'. Schedule L1.3 provided a Schedule of Rates 'which will be used for the evaluation of Variation Orders' and which was expressly referred to in clause 31.3 'Disagreement on Adjustment of the Contract Price', which read as follows:

> If the Contractor and the Employer are unable to agree on the adjustment of the Contract Price, the adjustment shall be determined in accordance with the rates specified in Part L, Schedule L1.3 Schedule of Rates.
> If the rates contained in the Schedule of Rates (Schedule L1.3) are not directly applicable to the specific work in question, suitable rates shall be established by the Engineer reflecting the level of pricing in the Schedule of Rates (Schedule L1.3).
> Where rates are not contained in the said Schedule, the amount shall be such as is in all the circumstances reasonable. Due account shall be taken of any over- or under-recovery of overheads by the Contractor in consequence of the Variation.

This clause is therefore typical of a traditional hierarchy for the valuation of variations as described elsewhere: firstly, using contract rates, secondly, applying suitable rates reflecting the level of pricing in those contract rates and, thirdly, a 'reasonable' amount if there are no such rates.

MT Hojgaard's installation of the monopiles and transition pieces had been served by its barge, 'The Lisa', which proved inadequate for the task. As a result, E.ON issued variation orders that omitted 'The Lisa' and replaced it with a barge which it supplied free of charge, 'The Resolution'. The parties agreed that MT Hojgaard was due some additional monies in relation to its work with 'The Resolution', but disagreed as to how to value the omission for the work it would have had to do with 'The Lisa'.

E.ON submitted that, given the difficulties MT Hojgaard was having with its use of 'The Lisa', the omission should be on a hypothetical calculation based on the length of time, and hence costs, MT Hojgaard would have incurred had it continued to install all of the monopiles using that inadequate barge.

MT Hojgaard submitted that the valuation of the omissions should be based upon the original contribution of the omitted work to the Contract Price, as summarised in Schedule L1.1.

The Judge at first instance agreed with MT Hojgaard, applying Schedule L1.1. He observed that this schedule was inserted into the contract to show what amounts to the constituent parts of the Works listed as contributed to the Contract Price. He noted that the contribution of constituent parts of the Works to the Contract Price was also reflected in Schedule L1.4 with its provision of stage payments by reference to completion of particular stages of the Works. Whilst the correlation between Schedules L1.1 and L1.4 was not exact, he concluded that, overall, the contract recognised the principle that discrete parts of the Works made discrete contributions to the Contract Price.

The Judge also agreed that E.ON's approach, that would have hypothetically considered the costs that MT Hojgaard would have incurred had it continued to work with 'The Lisa', was an approach that attempted to achieve additional contractual remedies for breach of contract under the guise of adjustment of the Contract Price for a variation.

Agreeing with the Judge's decision, the Court of Appeal said:

> The essential fallacy of E.ON's case is that it ignores the fact that by the Contract MTH (a) agreed to carry out the work for a fixed price and (b) assumed the risk that the price would, in the event, not be enough to cover the work which it had promised to do ('the pricing risk'). Whilst the Contract Price was a single price for the whole of the work it is obvious that some part of it related to the work of installation. If the installation work was wholly omitted from the Contract the whole of the price properly attributable to such work would fall to be omitted; and if part was omitted, there should be omitted a proportion of the price that appropriately reflected the work omitted and which, but for the omission, would have been paid.

> It is neither necessary nor appropriate to work out how many days it would, in fact, have taken to complete the installation with the LISA and apply a rate to those days. Such an approach would relieve MTH of the pricing risk as is indicated by the fact that, if E.ON be right, the variation ends up costing € 62 million when the product of the cost of installation and Wait on Weather allowance under Part L1.1 is of the order of € 25 million. The logical conclusion of E.ON's approach would appear to be that if, pursuant to a Variation Order, half the contract works were omitted the deduction from the price could be 100% if the time that MTH would in fact have taken to fulfil the 50% omitted would have been so long that it justified a valuation equivalent to the whole of the price.

As explained elsewhere, SBC/Q provides at clause 5.6.2 that where a valuation is of an omission from the contract works, then the valuation of the omissions shall be determined by reference to the rates and prices in the contract bills of quantities. If there is an add and omit variation, and the only way to properly value the addition is by an approach other than applying the rates and prices in the contract bills of quantities, then there is nothing wrong with this. As ever, the reasonableness of such an apparently inconsistent approach will depend on the circumstances of each case.

As noted elsewhere, under the NEC regime the situation is different to that under the Hojgaard judgment. As this is put in the NEC4 User Guide:

> Where the Work to be done is changed, it is important that the assessment is based upon the change in forecast or recorded Defined Cost. The clause gives no authority for the price for the originally specified work to be deleted or for the forecast Defined Cost of all work now required to be used as the basis for a new price.

Thus, under NEC4-ECC, where there is an add and omit variation, the valuation does not omit the original price for the omitted work. The valuation of such an NEC compensation event under clause 63.1 would assess the increase in the Defined Cost arising out of the difference between the omitted and added work. To that difference is added the contractor's fee percentage. This said, as explained elsewhere, under clause 63.2 the parties are at liberty to agree to apply rates and prices to value any compensation event, so they could under that clause agree to omit the original price for the omitted work, particularly if they considered that this best reflected its likely actual costs and/or gave a pragmatic approach to a reasonable result rather than speculate and argue over what the actual costs might have been.

5.12   Quantum Meruit

In some circumstances the contractor may be entitled to claim additional payment on the basis of a 'quantum meruit' evaluation, rather than a valuation calculated on the basis of the express rules set out in the contract. The term 'quantum meruit' literally means 'the amount he deserves' and implies an obligation to pay whatever the work or service provided is worth. Whilst the term has a particular resonance under English law and in Commonwealth countries, the concept of a contractor being entitled to a 'reasonable sum', rather than one calculated in accordance with the express terms of what the contract does (or does not) say, is of wider relevance. In particular, in this section we consider four situations in which this approach might most likely be met by practitioners under engineering and construction contracts. These are where: there is a contract but no agreed price; the contract only refers to a 'reasonable price'; work is carried out outside of the scope envisaged by the contact; and where there is 'Cardinal Change'. We will then consider some potential bases for establishing what is a 'reasonable sum' in such circumstances.

Claims for valuation on a quantum meruit basis (rather like claims that the time for completion of the works is 'at large', as considered in Chapter 3) are often seen by contractors (and particularly subcontractors) as a panacea for all their contractual and financial ills. The term is therefore often bandied about by their advisors for that reason. However, the term is often used (rather like claims that the time for completion of the works is 'at large') with little understanding of the circumstances in which such an entitlement can actually arise, the need to prove that those circumstances actually exist, what the term means or how valuation on that basis should properly be made. It also sometimes seems that parties resort to a quantum meruit approach because they have failed to keep the necessary records to enable a proper valuation in accordance with the terms of the contract and are therefore looking for a 'quick fix'. In this approach, it can often be seen as akin to the pursuit of a 'global claim', as explained in Chapter 6.

This said, a cry of 'quantum meruit' has occasionally been said to have achieved the benefit of rather focussing the minds of a party that is in receipt of such a claim and fears that it has a risk of being held to have been responsible for creating circumstances where such an approach is actually merited. In this context it may be that the receiving party is persuaded to take a more constructive and reasonable approach to valuation under the terms of the contract if the alternative is a battle over what might prove, for it, to be a more problematical quantum meruit claim.

5.12.1 No Contract or Agreement as to Price

A Quantum Meruit valuation particularly arises on construction and engineering projects where there is a contract but with no agreement as to price. The most common example of this is in relation to letters of intent. A badly drafted letter of intent may create an obligation to carry out works under it (such as preparatory design or procurement), but set out no express provision for the calculation of that payment. This is an area considered in more detail in Chapter 8, particularly by reference to the judgment in *British Steel Corporation v Cleveland Bridge & Engineering Co Ltd* (1981) 24 BLR 94.

For a further case on a letter of intent, see *ERDC Group Limited v Brunel University* [2006] EWHC (TCC), which is also considered in more detail in Chapter 8. This was a case in which the contractor started work on the basis of letters of intent, the last of which had a fixed expiry date. The contractor continued work after that date, but subsequent negotiations between the parties as to terms of a proper contract failed, and the contractor walked away from the part completed work. It was held that for the work done up to expiry of the final letter of intent, the value was to be based on the contract contemplated by the letter of intent. For works done after that date, this was not to be valued on a quantum meruit basis, but on the basis of the same rates as the earlier work. This was because 'The conditions in which the remaining work was carried out did not differ materially from those which (it must be assumed) were originally contemplated'.

On the other hand, if a contract is successfully concluded subsequently, and it does set out machinery for valuation, it can usually be expected to act retrospectively to cover the valuation of work previously done under a letter of intent (unless the letter of intent or the contract say otherwise). This is a common scenario on construction and engineering projects. See in this regard *Trollope and Colls Ltd v Atomic Power Construction Ltd* [1962] 3 All ER 1035.

Where works are commenced on a simple letter of intent and negotiations on the price and the terms of a proper contract follows, the works might unfold in a manner that leads one party to consider its commercial interests are best served by contending that no subsequent contract was entered into, and hence rely on a quantum meruit valuation based on the letter of intent. In this regard two cases are of interest.

In *Mitsui Babcock Energy Ltd v John Brown Engineering* (1996) 51 Con LR 129, John Brown proceeded on the basis of a letter of intent. When disputes arose regarding tolerances, Babcock argued that there was no concluded contract. The court disagreed, noting that the parties had agreed to enter into a contract and only subsequently disagreed over terms without which the contract could still be enforced. Accordingly, payment for the work done was to be based on the terms of the contract and not a quantum meruit.

In *VHE Construction PLC v Alfred McAlpine Construction Ltd* [1997] EWHC 370 (14 April 1997) the Judge observed regarding the parties' dispute as to whether they had actually concluded a contract:

> It is remarkable that this question is probably the most frequent issue raised in the construction industry. On projects involving thousands and sometimes millions of pounds, when a dispute arises about payment, the first issue very often is to decide whether there was a contract and if so what were the terms of the contract, if any. So it is here. The plaintiffs have done work and there is no complaint about the quality of their work. Are they to be paid pursuant to a contract, and if so what contract, or are they to be paid on a quantum meruit? In money terms there is a considerable difference between the parties.

He concluded that there was a contract. His analysis set out at length the conduct of the parties, which was strongly disputed between them, and how that led him to his conclusion:

> A sub-contract was completed by a telephone call between Mr. Brown and Mr. Brian Thomson in late February or early March when they agreed a reduction in the period for payment being the last issue to be resolved between the parties and was further agreed to by the parties by their conduct.

The full account of the parties' negotiations makes an interesting read for anyone involved in such contract processes and contains aspects that will be familiar to many from their own experiences.

5.12.2 Contract Only Says Reasonable Sum

Further situations in which 'quantum meruit' might arise are where the contract expressly provides for a payment of a reasonable sum, without detailing that sum or its valuation procedure, or where the contract requires work to be undertaken but no price is fixed by the contract at all. This is rare on construction and engineering projects.

5.12.3 Work Outside of Contract

It has long been established that if a contractor undertakes specific work under a construction contract but is requested to undertake further work outside the scope of the contract then it will be entitled to be paid a reasonable sum for the further work undertaken, which being outside the contract is not valued by reference to its terms (*Thorn v London Corporation* (1876)).

The judgment in *Sir Lindsay Parkinson & Co Ltd v Commission of Works* [1950] 1 All ER 208 is often cited as the authority for the proposition that a quantum meruit approach may be implied to works undertaken outside the contract. In that case, the contract placed a financial limit on the extent of variations that the Commission could instruct. The court held that where there were works instructed whose value exceeded that limit, valuation should be on a quantum meruit basis.

More recently, where invalid variations were instructed they were treated as outside the contract machinery in *Costain Civil Engineering Ltd and Tarmac Construction Ltd v Zanen Dredging & Contracting Company Ltd* (1998) 85 BLR 77. The arbitrator had awarded Zanen payment on a quantum meruit basis for work undertaken in connection with the construction of a marina in the flooded casting basin used for the contract works. Zanen had been instructed by the Joint Venture to undertake work in connection with the marina construction despite its protests that the work was outside the scope of its subcontract with the Joint Venture. The amount awarded by the arbitrator in respect of Zanen's work on the marina was the sum of £370,756 plus on-costs for the work done and £386,000 in respect of Zanen's share of the profit arising from the marina works. This latter sum had been claimed (in the alternative to half the Joint Venture's profit from the works in connection with the marina) as 'such sum as Zanen might well have assessed as being commensurate with the mobilisation/demobilisation charges its competitors would have had to allow for had competitive tenders been sought for the marina works'.

In the appeal from the arbitrator's award, the court was asked to consider whether the work was outside the scope of Zanen's subcontract, could the valuation on a quantum meruit basis be valued by reference to any profit allegedly made by the Joint Venture on the works and could such a valuation be made by reference to charges that a competitor of Zanen would have incurred in carrying out the work but which Zanen, by virtue of the circumstances, did not incur? Alternatively, was such a valuation confined to the value of the work undertaken and materials supplied plus a reasonable addition for overheads and profit?

The court decided that in respect of quantum meruit there is a distinction to be made between cases where there is an implied term for payment and those where there is no contract. This was a case that fell into the latter category. Had discussions taken place prior to the marina works to discuss the price for the works, then Zanen would have been aware of their commercial advantage over their competitors in not having to incur additional mobilisation and demobilisation costs for the marina works. In such circumstances it would be unrealistic to ignore the commercial realities of the situation and, given that the basis of assessment of the sums for the marina works had been restitution and undue enrichment, the arbitrator could not be held to be at fault for considering such matters. The appeal by the Joint Venture was rejected and the sums confirmed as payable to Zanen.

In respect of the specific points as to whether the alleged profit of the Joint Venture or the disadvantage of competitors in respect of mobilisation and demobilisation costs could be taken into account, the Judge confirmed that they could. The *Zanen* case was decided on the basis of a quantum meruit in respect of restitution and undue enrichment, and the factors taken into account there may not always be relevant.

5.12.4 Cardinal Change

The adoption by a contractor of a quantum meruit claim as a panacea to its problems usually arises on construction and engineering projects where the contract provides for the price to be paid for the works, but the contractor argues that, as a result of the actual conditions or character of the work being so different from those anticipated in the contract, the pricing provisions of the contract are to be regarded as void or incapable of being used as the basis of valuation. On this basis the contractor might believe that it has a claim for just additional works, or even the whole of its works, being valued on a quantum meruit basis.

Contractors seeking to define their projects as being in such a situation might like to consider these very old words from the judgment in *Pepper v Butland* (1792):

> If a man contracts to do work by a certain plan, and that plan is so entirely abandoned that it is impossible to trace the Contract, and to what part of it the work shall be applied, in such a case the workman shall be permitted to charge for the whole work done by measure and value, as if no Contract had ever been made.

Such an extent of change might be described by what is referred to in some jurisdictions as a 'Cardinal Change'.

Much more recently, the relevance of the change being so great that the provisions of the contract can no longer be applied, was tested in the Canadian case of *Morrison-Knudsen v British Columbia Hydro & Power Authority* (1978) 85 DLR 186. The project was for one of the largest hydroelectric power schemes in the world, the Peace River Dam and Power Plant in British Columbia. There the Canadian Court of Appeal rejected the argument that failures by the employer that put it in breach of the contract allowed the contractor to claim payment for the works on a quantum meruit basis. The employer's breaches of contract were extensive and significant. Difficulties arose between the parties almost from the beginning. Soon after the award of the contract, the drawings were marked *'Not for Construction'* and engineering proceeded on a *'design as you go'* basis. The owner and its engineer throughout the performance of the work made significant changes in the design of the structures and were slow to produce the required drawings. The contract was badly behind schedule in March 1966, largely due to delays caused by the owner. The employer ordered acceleration of the work and continued to insist that the work be completed by the times fixed in the contract. The trial Judge found that the employer's actions amounted to a fundamental breach of contract because it had not paid for the costs of acceleration and because, while agreeing with the contractor that it would process claims for acceleration, it had given the Provincial Government a private assurance that it would not pay such costs. However, the contractor had continued with the works in the face of the employer's breaches, and had received periodic payments certified under the contract. The judgment concluded that in such circumstances the contractor cannot claim on a quantum meruit if it affirms the contract by continuing to perform it in spite of the breaches. If the contractor had treated the breaches as discharging the contract it would have had the option of suing on a quantum meruit basis.

In terms of how the contractor's damages should be evaluated, if not by applying a quantum meruit basis, the Court of Appeal suggested a possible basis as follows:

> If Hydro caused the contractor to speed up its work to overcome owner-caused delays, and we think the trial Judge properly found that to be the case, then the additional cost attributable to acceleration can be assessed. Surely the respondents would be able to calculate that cost if they put the people on it who normally prepare their bids. It could be calculated on the basis, say, of a three-and-a-half

year contract instead of a four-and-a-half year contract. They must consider such questions every day.

If it is difficult to fix the amount for a breach it is no more difficult than to fix compensation for a personal injury. Trial Judges do that every day. It may be that the award will be remarkably similar to the result of a quantum meruit and that some rather arbitrary figures will have to be used. But difficulty of assessment does not justify abandoning the attempt and making an award on another basis.

However, the Court of Appeal referred quantification back to the trial Judge and it is not apparent how that evaluation was actually carried out.

The manner in which the contractor had conducted itself was also critical in the South African case *Alfred McAlpine & Son (PTY) Ltd v Transvaal Provincial Administration* [1974] 3 SA 506. The project for construction of a national road was subject of a large number of variations. The contractor argued that the scope was so changed that the contract was tacitly substituted by a new agreement under which it was entitled to reasonable remuneration for the work. However, the contractor had continued to make its claim applying the terms of the contract until a year after completion of the project, and the court held that it could not therefore be said that the parties regarded the original contract as having lapsed. Furthermore, it was held that there was lack of evidence that the road as constructed was substantially different to the one contracted for. It might therefore be suggested that the impacts on the project of the employer's failures were not such as to amount to what might be termed a 'Cardinal Change'. It seems obvious that for a contractor to succeed in such a radical re-writing of the valuation rules agreed in the contract, it must first prove that the project was subject to similarly radical change.

In London, the Court of Appeal decision in *McAlpine Humberoak Ltd v McDermott International* (1992) 58 BLR 1 may also dampen enthusiasm for claims on a quantum meruit basis as a contractor's panacea for recovery of its costs where there is extensive change to its works. There the Judge at first instance held that the extent of increase in drawings issued to McAlpine (approximated eight times as many as were available at the tender stage) was such as to distort the substance and identity of the contract so that it had become frustrated. He found that the effect of that was that a new contract had come into place that entitled McAlpine to a reasonable price for the work (and a reasonable time in which to complete it). The Court of Appeal disagreed and allowed the appeal. Lord Justice Lloyd said this:

> The revised drawings did not 'transform the contract into a different contract' or; 'distort its substance and identity'.
> And:
> … it is impossible to hold that the contractual machinery between the parties had been displaced.

Again, the conduct of the parties was deemed of significance. It was particularly noted how the parties had concluded a full and final settlement of McAlpine's claims for the direct costs of many of the resulting variations.

It therefore seems that circumstances in which a contract with price and payment provisions can be wholly converted to a quantum meruit payment basis will be rare. It will obviously turn on the facts of the case, but also the conduct of the parties. A situation where such an approach is applicable is more likely to be an instance where an element of work is undertaken outside the contract and that this element of the work should be valued on this basis.

5.12.5   How to Calculate

Whilst the position, and the relevance, of the whole concept will vary between jurisdictions, it is suggested that the general rule should be that, if a quantum meruit is undertaken, then the terms of the contract which governed the original scope, if any, do not have any bearing on the evaluation. The whole point is that the machinery in the contract no longer applies, or there was none in the first place. Rules for such an assessment vary between jurisdictions and the legal basis upon which the claim is made. As this was put in under English law in *ERDC Group Limited v Brunel University* [2006] EWHC (TCC): 'It has rightly been said that there are no hard and fast rules for the assessment of a quantum meruit'. However, the following factors may be relevant to valuation, depending on the circumstances of each case:

1. Evidence of a fair commercial rate for the work done. See in this regard, *Malcolm Taylor v Ranjit Bhail* (1995) EWCA Civ 54. Also, *Benedetti v Sawiris and others* [2013] UKSC 50 and references to 'the price which a reasonable person in the defendant's position would have to pay for the services' and discussion as to whether the subjective views of the party receiving the services was relevant.
2. Evidence in respect of prices in a related or similar contract.
3. Evidence of the conditions in which the works were carried out. See in this regard, *Serck Controls Ltd v Drake & Scull Engineering Ltd* (2000) 73 Con LR 100.
4. Evidence of any negotiations between the parties in respect of the applicable price. See in this regard, *Stephen Donald Architects Limited v Christopher King* [2003] EWHC 1867 (TCC) and *ERDC Group Limited v Brunel University* [2006] EWHC (TCC) as explained elsewhere.
5. Measurement, or other quantification, of the amount of works undertaken.
6. Calculations of the reasonably incurred cost of labour, plant, materials and subcontractors required for the works.
7. Evidence of the overheads actually incurred in respect of the works.
8. Evidence as to the level of profit normally anticipated on such work.
9. The value of the works to the other party, although this may not be relevant to a claim for works done against a letter of intent. See *ERDC Group Limited v Brunel University* [2006] EWHC (TCC).
10. The objective market value of the work.

From this it should in theory be possible for experienced contractors, quantity surveyors or engineers to apply a method to arrive at the reasonable value of the works. The work involved in such a valuation may, however, be extensive, particularly if it requires an audit of the contractor's actual costs. The contentious nature of such claims may also mean that agreement is difficult, particularly as to whether the adopted method is the right one.

There will often be disagreement as to whether the correct approach is one of 'fair rates and prices' or the reimbursement of costs plus overheads and profit additions. This question of fair rates versus a cost-plus approach was considered by what is now the

Technology and Construction Court in the case of *Laserbore Ltd v Morrison Biggs Wall Ltd* (1993) CILL 896. In this case the consideration was in connection with the assessment of a fair and reasonable payment under a contract rather than a quantum meruit claim, but the Judge's comments are still instructive and of interest when considering what constitutes assessment of a fair and reasonable payment. The Judge stated:

> I return to the approaches of the respective experts, the 'reasonable rates' basis on the one hand and the 'costs plus' basis on the other .... In a competitive market, one would expect both approaches to result in much the same figure, particularly if one accepts that someone who competes by providing high quality rather than low cost should receive a higher remuneration on both tests. Tenders are usually built up on a costs plus basis and the acceptance or rejection of tenders sets what can be viewed as the market rate. But one problem for the plaintiffs is that they did not expect to have to prove their claim on a costs plus basis and they have not kept records sufficient to prove their claim in that way.
>
> I am in no doubt that the costs plus basis in the form in which it was applied by the defendants' quantum expert (though perhaps not in other forms) is wrong in principle even though in some instances it may produce the right result. One can test it by examples. If a company's directors are sufficiently canny to buy materials for stock at knockdown prices from a liquidator, must they pass on the benefit of their canniness to their customers? If a contractor provides two cranes of equal capacity and equal efficiency to do an equal amount of work, should one be charged at a lower rate than the other because one crane is only a year old but the other is three years old? If an expensive item of equipment has been depreciated to nothing in the company's accounts but by careful maintenance the company continues to use it, must the equipment be provided free of charge apart from the running expenses (fuel and labour)? On the defendant's argument, the answer to those questions is 'Yes'. I cannot accept that that begins to be right.

These comments succinctly set out the sort of considerations that might render a cost plus assessment incorrect for a quantum meruit valuation, although they were made in the context of an evaluation of a fair and reasonable price under a contract rather than a claim for damages to put the contractor back into the position it would have been in but for asserted breaches of contract. A quantum meruit valuation need not be the sum total of the cost of materials and services supplied plus additions to reflect actual overhead costs and the market rate of profit. On the basis of this approach, the calculation of quantum meruit can involve the use of market rates for the work, adjusted for any special circumstances of the actual works, so that a contractor entitled to payment on such a basis is not penalised for its commercial ability to tightly manage its costs.

The correct approach will differ depending upon the particular circumstances, but this discussion gives useful guidance as to the factors that need to be considered in such evaluations.

There is also the matter of the extent to which a contractor can recover costs, where payment is made on a cost rather than on a measured unit rate basis, in circumstances where rates are not applicable or available and there are suggestions that the contractor has been inefficient or has not managed the works as well as it might. A balance needs to be struck here. The general rule is that the contractor is not expected to be perfect or to be the best contractor in the circumstances. It is expected to be a reasonably competent contractor experienced in the particular type of work, and as such it can be expected that a certain level of efficiency can be expected which will include an acceptable level of 'inefficiencies'. Providing the contractor has not made errors, or managed the works in a way that demonstrates it is not a reasonably competent contractor, then no adjustment is relevant to the assessment of costs in such circumstances.

Too often, employers audit a contractor's costs with an over-sharp pencil, removing all costs that they consider were due to the contractor's own faults or inefficiencies, while ignoring the reality that when contractors price a tender they will usually include some allowance for such 'own goals'. These may be included under headings such as 'risks' or 'contingences' or included in outputs, rates or prices that themselves included an allowance for such things. If a contractor bases its outputs on those achieved on past projects then, unless they had been faultlessly efficient on those past projects, those rates will include some inefficiencies. A common example of this relates to corrective work arising from snagging lists. To remove these from a contractor's costs on the basis that 'the employer doesn't pay for the contractor's errors' ignores that all projects will have some extent of required snagging works and on a lump sum contract the employer will be paying for those through any allowance that the contractor made in calculating the contract price. It is suggested that only if the extent of such works is unreasonable should adjustment be made. Even then, the circumstances under which the works were first carried out needs to be considered. Any costs-based valuation of works under a contract that did not principally make valuation of a change based on actual costs, must involve works whose circumstances were unusual in some manner.

5.13 Valuation in Advance

Traditionally, standard forms of construction contract have required the contractor to comply with any variation instruction issued by the employer's agent and the question of financial valuation of such a variation was left to be considered later. The priority was to get on with the variation as part of 'the Works' as whole, so they could be progressed and completed, and to worry about money later as a secondary consideration. However, this philosophy has changed over the last 20 years or so.

For example, clause 11 of the 1963 editions of the JCT contracts contained no provision for financial quotation or proposal prior to the issuing of an architect's instruction. In contrast, SBC/Q Schedule 2 provides for a Variation Quotation. Similarly, in the case of FIDIC contracts, the fourth edition, of 1987, contained no provision for the engineer to request a proposal from the contractor, prior to instructing a variation, such as that which is now in clause 13.3 of the 1999 edition considered in this book. Also, whereas the 1999 Seventh Edition of the ICE Conditions of Contract had introduced at its clause 52(1) provision for the engineer to request a quotation for a proposed variation, there was no such term in the variation provisions in clause 51 of the previous, Sixth Edition, of 1991.

It would, of course, have been open to the parties to such old forms of contract to agree that variations would be the subject of proposals and pricing in advance. There is clearly much attraction in drafting a clause that requires that all aspects of a variation be agreed in advance, including its direct effects, indirect effects and impact on the programme and completion date(s). However, in practice this was rarely done, and this only encouraged disagreement as to a valuation where there was no pressing onus on the parties to agree a value before the variation works would be done.

The valuation of variations can be a very subjective issue (as should be apparent from this book) and often leads to dispute. As part of the tide away from allowing such issues to develop, over the last couple of decades, standard forms have increasingly included provision for proposals to address the financial (and programming) implications of variations in advance of the formal instruction of such variations. This change perhaps reflects the different approach to the same issue adopted by the NEC contracts since their First Edition in 1990. Clause 61.2 provides for the project manager to instruct the contractor to submit a quotation for a proposed instruction or changed decision before it is put into effect.

The provisions of the Infrastructure Conditions clause 52 for the valuation of variations start with the premise in subclause (1) that the contractor, if requested by the engineer, shall submit its quotation for any proposed variation, together with its estimate of any consequential delay. Clause 52(3) then sets out the procedure to be adopted in the event that the engineer does not make a request, or agreement under clause 52(1) is not achieved. The valuation rules of clause 52(3) and (4) then apply in the event that agreement under both clauses 52(1) and 52(2) is not achieved. The procedure is thus set out in a manner suggesting that the pre-agreement of the costs and consequences of variations is the preferred option, and recourse to the hierarchy of valuation rules is the default to be applied in the absence of agreement based on contractor's quotations. It should be noted that pre-agreement could be of a sum calculated on a completely different basis to any of the valuation provisions in clause 52.

SBC/Q similarly has provision, in clause 5.3, for the contractor to submit what is termed a 'Variation Quotation' in respect of work that is the subject of a variation instruction. This refers to Schedule 2 to that form, which sets out a detailed set of 'Variation and Acceleration Quotation Procedures'. In the event that the resulting quotation is not accepted, then clause 5.1.1 of Schedule 2 empowers the Architect to instruct the variation and it is valued applying the rules of clauses 5.6 to 5.10 of the conditions. Similarly, under clause 5.3 of SBC/Q, the contractor can disagree that the quotation procedure should apply, in which case the architect may instruct the variation and it is valued applying the hierarchy of valuation rules of clauses 5.6 to 5.10.

In accordance with its ethos of proactivity and early tackling of such issues, NEC also anticipates that quotations will be used for the pricing of variations in its Core Clauses in relation to what it terms 'compensation events', where the submission of quotations by the contractor is required for events arising from instructions notified by the project manager or the contractor. The project manager may also request quotations for proposed compensation events. Such a regime does, however, anticipate that the parties will staff the project in anticipation of the level of administration such systems require. However, for a variety of reasons, such prior agreement is not always possible, even where the parties have agreed a contract under NEC or similar terms.

The objectives behind these provisions for the valuation of variations in advance would seem to be:

1. Early pricing of the commercial and timing effects of variations, etc., enables the employer's representative to maintain budgetary and programme control for the benefit of the project and the employer.

2. It can of course be argued that this benefits not only the employer but also all parties to the contract and in many instances the contractor may be keen to have variations priced in advance, to avoid administrative debate and costs and to have certainty.
3. Early pricing of variations, etc., reduces the risk of dispute over the pricing or the consequential effects of such matters, as the relevant circumstances and information are still fresh in the minds of those concerned.
4. It may also be relevant that the agreement of the financial and time effects of changes under such as NEC terms comes at a time when the employer and its team is still proactive in relation to a required change and has an interest in getting it agreed so that work can progress. A common complaint from contractors under other regimes is that the contract administrator instructs a variation, it carries out the works in good faith and for the employer's benefit, but the subsequent process of agreement of the associated money and time show little or no urgency or interest by the employer's team.

There is a distinction in approach between the SBC/Q and Infrastructure Conditions as to the relevance of the contract pricing to the calculation of such a quotation. Clause 1.2.1 of Schedule 2 to SBC/Q requires that 'the amount … shall be made by reference, where relevant to the rates and priced in the Contract Bills'. That clause also recognises that the valuation of such a variation may include '… the effect of the instruction on any other work…'. These are issues discussed in more detail elsewhere. The Infrastructure Conditions clause 52(2) make no such reference to the contract pricing. However, in the case of the Infrastructure Conditions this may be thought unnecessary in practice in that, if the engineer considers that its requested quotation does not reflect contract pricing in the way it believes it should, there may be no agreement and the provisions of clause 52(3) will apply, with their reference to contract pricing as explained elsewhere.

While the desirability of early price and programme agreement in relation to variations would seem to be in the interests of all parties, it seems that prior agreement occurs as the exception rather than the rule (where NEC terms do not apply), except on particularly well managed projects. It occurs especially rarely in some jurisdictions. This may be for a variety of reasons, but the following may be relevant:

1. Contractors, when engaged in large projects with possibly many individual instructions and matters to price and constraints on time and availability of key personnel, may consider that prior pricing and agreement is not possible without risking gaps in the evaluation. Contractors may consider, as a matter of policy, that they are likely to get a better valuation at a later stage when they know all the circumstances and effects of a variation, rather than commit themselves to a pre-estimate of a price at the earliest time. Effectively, they see retrospective pricing as removing any risk.
2. Because of the risk element in advance pricing, a contractor will usually wish to include a contingency element in such pricing that may be unacceptable to the employer. The contractor will argue that the contingency is no more than to cover for the risk it takes in advance pricing (just as in its original tender), but if the resulting price is too high, especially compared to a budget or other indication provided by the employer's advisors, agreement may not be possible.
3. The lack of powers for the employer's representative under some forms may also be a factor.

4. In many jurisdictions it seems that such a proactive approach runs contrary to established culture and habits in the construction and engineering industries. This is often also suggested to have also acted as a 'brake' on the uptake of the NEC contracts in some jurisdictions internationally and difficulties in their implementation where they have been applied.

5. A lack of resources, particularly where engineers have been appointed on particularly tight fixed fees in very competitive international markets, often means that the required resources are not available for them to be willing to be so proactive, even if there was the cultural will to be.

6. On the contractors' side, a notable feature of many when they first experienced NEC projects was that their levels of site administrative resources were more suitable to traditional contractual approaches. As a result they did not have the site overhead resources to keep up with the administrative burden of the pre-pricing and agreement of variations.

7. A further feature of the effect of site resourcing is whether the right types of administrative staff are on site. The prospective pricing of a variation might be considered more a discipline for an estimator than a commercial manager. The latter may be reluctant to commit to pre-pricing as a discipline in which they are inexperienced.

8. Where there are sufficient resources, but they are not used to such proactive approaches to variations. This has particularly been a problem for the NEC contracts where they have been adopted in markets where the project teams are from a FIDIC background.

9. Where there are so many variations being issued that it is impossible to keep up with the regime. The nature of how variations are instructed may also exacerbate these problems. These may include their being issued too late to allow pre-consideration or that they are issued in a format that does not follow the formal procedures of the contract, for example through comments on drawings or replies to the contractor's technical queries.

It might also be considered that early pricing on large civil engineering projects is more feasible as instructions tend to be identified in advance of implementation to a greater extent than on building projects where individual instructions may be smaller in effect but far more numerous. This may be easier on a lengthy project for construction of a new road, but not the fast track refurbishment of an existing hotel.

There is no doubt that early agreement of the commercial and timing implications of instructions can be a significant factor in avoiding unnecessary disputes, with the associated costs and drain on the time of the contractor's and employer's teams. Users of ad hoc and one-off contracts should consider the inclusions of provision for such matters and the enhanced potential for the success of such provisions if the employer's representative is able to instigate early pricing and agreement. There is also, of course, a need to retain flexibility in the adopted regime so that the pricing regime can cover all likely, and even some unlikely, eventualities.

A further potential benefit of pre-agreement of variation values is reflected in the Society of Construction Law (SCL) Protocol Core Principle 19:

> Where practicable, the total likely effect of variations should be pre-agreed between the Employer/CA and the Contractor to arrive at, if possible, a fixed

price of a variation, to include not only the direct costs (labour, plant and materials) but also the time-related and disruption costs, an agreed EOT and the necessary revisions to the programme.

The authors heartily agree with this and particularly the potential to include the time consequences of change that are considered in Chapter 6 of this book as part of a variation's value, thus taking out even more of the potential for claims and disagreement in relation to change.

5.14   Requirements for Notices

In the field of evaluating claims under construction and engineering contracts, contractual requirements to notices may be relevant in three particular circumstances as follows:

- In relation to clauses for the valuation of variations that require the contractor to give notice if it considers that rates or prices in the contract have become unreasonable for some reason. For example, see the Infrastructure Conditions clause 52(5), which also require the engineer to notify the contractor if it considers a rate or price unreasonable or inapplicable.
- Where variation valuation procedures include a requirement for the contractor to provide a quotation or proposal if instructed, but subject to its ability to object to that procedure, by notice. See, for example, clause 1.1 of Schedule 2 of SBC/Q.
- Where the giving of such notice is a requirement of a clause giving the contractor a right to make a claim under that clause, perhaps to the extent of making notice a condition precedent to that entitlement and fatal to it if not given. This is a particular feature of some bespoke contracts internationally, where based on FIDIC terms but with the addition of onerous notice requirements. It involves issues of contract and the law of the applicable jurisdiction and is extensively written about in books on those subjects. Legal advice should always be taken on this topic. It can, however, have profound consequences for practitioners involved in the valuation of contract claims, so is considered to some extent in the following.

Most contracts will include provisions for notice to be given by a party wishing to initiate a process of valuation of changes to the contract works. The obvious reason for such requirements is that the party in receipt of the notice, usually the employer or its representative, or the contractor under a subcontract, may wish to take some action in respect of the change for which payment is being claimed, either in respect of the actual works themselves or in respect of records being kept in relation to their direct consequences.

If the employer and/or its contract administrator are made aware that the contractor considers that circumstances have occurred (or are about to occur), which entitles the contractor to a claim for additional monies, they may be given the opportunity to address those circumstances earlier than they might without such notice. This may mean that the effects can be mitigated or even avoided. For example, a revised drawing could be withdrawn or an amended detail re-amended to a lower specification and cost. If the effects cannot be entirely avoided, then processes can be put in place to control and monitor those effects on time and money. This can include recording resources and costs at the time and thus reducing the risk of dispute later on. It may even be that the employer has a better opportunity of pursuing responsibility for the events and their costs with others, which will be lost without notice. From an employer's viewpoint, being denied by the lack of a contractor's notice the opportunity to prevent, mitigate or record such effects and pursue its entitlements elsewhere can be particularly irksome. It is clearly also likely to be in the mutual interest of the contractor to avoid creating such a situation. Understandably, a party faced with a claim for costs caused by failure of a third party is more likely to pay those costs if it has been put in a position to pursue those costs from the third party than if it has been left, through a lack of contractually required notices, unaware and unable to pursue them. It may even be that the employer is able to rely on the breach of contract by the contractor in relation to notice to counterclaim for costs that it could have avoided, reduced or passed on to a third party had the contractual obligation to notify been complied with.

Conversely, those who are criticised for a failure to give contractual notices often protest that the lack of notice is only a matter of procedural formality or detail that was of no practical effect because the recipient would not have changed its conduct if it actually knew of the issue. In particular, a common contractor's attitude is likely to be that the employer would not, or could not, have changed its approach had it given the notice, for example by withdrawing the issuing of a revised set of drawings, such that its lack had no effect, and that the employer is hiding behind a procedural technicality to avoid the consequences of initial failures for which it is contractually responsible.

It is worth adding that the importance of being notified of a claim so as to enable a party to pursue responsibility for the events and the costs with others is often of particular resonance for contractors in their dealings with subcontractors and suppliers. Where a contractor is in contract with many such parties, if one of them has claims that are the result of failures by another (for example an electrical services subcontractor's claim for delays and damage to its works caused by a plastering subcontractor) the contractor may need to know at an early stage of such issues, to ensure that it notifies and withholds enough money from the plastering subcontractor's account to effect any counterclaims it has as a result. In this example, the passing of such a counterclaim or notice to the plastering subcontractor may also reduce the incidence of damage to the electrical services installations.

It has already been identified that the ethos of the NEC contracts is one of information and co-operation. This means that it has a large number of references to the need for notices, particularly in relation to changes. The general thrust of the NEC contracts is the early identification and notification of what they refer to as 'compensation events' (which term covers all forms of change to the works, their scope, character, conditions, timing, etc., not just variations), and the submission and agreement of quotations for them based on the contract regime. As a result, the NEC contracts contain extensive requirements for notices in relation to both the direct and time consequences of change.

5.14.1 Notice in Relation to Unit Rates

As far as unit rates are concerned, the Infrastructure Conditions contain provision in clause 52(5) for either party to give notice to the other if they believe that any rate or price in the contract, which is not the subject of a variation, is rendered unreasonable or inapplicable as a result of a variation. Furthermore, under clause 53(1), if the contractor wishes to claim a higher rate or price than one notified to him by the engineer under the rate fixing powers in clauses 52(3) or (4), the contractor has to notify its intention within 28 days after receiving notification of the relevant rate from the engineer.

Infrastructure Conditions Clause 53 then sets out in paragraphs (3), (4) and (5) the actions that may be taken upon the receipt of a notice under clause 53. These actions can be summarised as:

- The engineer may instruct the contractor to keep contemporary records of the relevant works, without admitting any liability on the part of the employer that the claim is valid (clause 53(3)).
- The contractor has to keep the records required by the engineer and to permit the engineer to inspect such records and supply the engineer with copies of these if and when requested to do so (clause 53(3)).
- After giving notice, the contractor shall submit to the engineer a first interim account giving full particulars and details of the amounts being claimed under that notice to date and the grounds for the claim (clause 53(4)).
- The contractor shall submit further claims to the engineer at such intervals as the engineer requires (clause 53(4)).
- If the contractor fails to comply with the clause 53 procedures, then the contractor is able to recover payment only to the extent that the engineer has not been prevented or prejudiced in its evaluation by the contractor's failures (clause 53(5)).

It is sometimes contended that these provisions in the Infrastructure Conditions are more appropriate and relevant to situations where the contractor intends to claim additional payments under clause 53(2), i.e. in circumstances other than in connection with unit rates. This is not so; the notice provisions apply equally to claims in respect of adjustments to unit rates.

If the contractor wishes to contend that a higher rate shall apply, or wishes to dispute a rate fixed by the engineer, then not only does it have to give the necessary notice but it also has to comply with the requirements for records and accounts of clauses 53(3) and 53(4) as have been summarised. If the contractor omits to follow these procedures it will be at its own risk as it is expressly stated in clause 53(5) that it will only be able to recover payment to the extent that the engineer's consideration of the claim has not been prejudiced by the failure to keep records and submit accounts. It is this last provision that is at the essence of the notice requirement. It is only reasonable that the engineer should be able to verify the records and data relating to the works to ensure that any valuation reflects the true circumstances. Failure to give notice prejudices the engineer, and therefore the employer, and it seems reasonable that the contractor should suffer from any financial effects of this rather than the employer.

It is not difficult to envisage how records might be necessary, and their absence prejudicial, in evaluating claims for a change in a rate. As an example consider a contract in which a bulk excavation activity is required with large excavators discharging to dump trucks for haulage of the excavated material to a spoil heap some 1500 m from the face of the excavations. At the time of tender the contractor will have estimated not only the rate of excavation that is likely to be achieved by the excavators in the anticipated conditions, but will also have considered the haul distance for the dump trucks and their loaded and unloaded speeds on the site in order to produce a balanced resource that enables the excavators to keep working without having to wait for dump trucks into which they can discharge, while at the same time not 'over trucking' the operation so that dump trucks are left idle and queuing to be loaded.

In this example, in the event that the engineer instructs that the anticipated location of the spoil heap is unsuitable or not available and instructs another location to be used some 2250 m from the excavation face the contractor is faced with two principal options:

1. It can continue with the same team of excavators and dump trucks but accept that due to the increased haul distance the excavators are going to have idle time awaiting the arrival of trucks into which they can discharge.
2. Alternatively, it can increase the number of dump trucks in order to keep the excavators working to full capacity.

The actual course taken will depend upon the circumstances of the project. Considerations may include, for example: the amount of excavation to be undertaken in this operation; its criticality or otherwise to the progress of the works as a whole; the comparative likely costs of different solutions; and the availability of additional plant and drivers. There may be other potential factors. However, it is likely that the contractor will wish to notify an intention to claim a higher rate if it is dissatisfied with any rate fixed by the engineer as a result of the change in the siting of the spoil heap. Critical to the evaluation, depending on which option has been adopted, may be records of idle time for the excavators for option (1) or records of the increased time and costs of additional dump trucks if option (2) has been adopted. Failure to give notice promptly or to keep any records required by the engineer of relevant plant operating and idle times may result in the engineer claiming that he or she has been prejudiced in assessing the contractor's claim for a higher rate.

Furthermore, it might even be that, in this example, the employer could legitimately establish that its decision to move the spoil heap was only out of preference rather than necessity, and that had it known it would lead to additional costs for the contractor and a resulting claim, then it would have quickly reversed that decision.

This highlights the need for rate fixing, and any subsequent challenges, to be processed as soon as possible, not necessarily just within the time limits expressed in the contract. On most large contracts the works are proceeding at a considerable pace and if the evaluation process is delayed then the possibility of disagreement and dispute as to the consequences of a change increases.

It is in the contractor's interests to have good record keeping as a norm so that effects of a change, before and after it occurs, can be readily demonstrated. If the record keeping is only instituted after the change, then there may be some difficulty in establishing how the change affected the resources because there is nothing to compare it to (what might be referred to as a 'measured mile', 'baseline' or 'control' period or location). In this example, if option (2) is adopted it will help greatly if there are records of such items as the number of trucks, their travel time and extent of waiting time, both before and after the change in location of the spoil heap. Sadly, all too often such comparative records are not available.

In contrast to the Infrastructure Conditions, SBC/Q anticipates that rates for varied work will be produced by measurement and application of the clause 5.6 valuation rules. If additional or substituted work cannot be valued by measurement, the daywork provisions of clause 5.7 are to apply. In practice this will result in a very similar process to that of the Infrastructure Conditions, but with the different rules of SBC/Q applying to the adjustments, if any, to the rates and prices. This, of course, reflects the difference in nature between the Infrastructure Conditions and SBC/Q, with the latter using the bill of quantities only to define a 'lump sum' price with the rates and prices therein used as the basis for pricing changes. It is the former's remeasurement approach that may necessitate the requirement for a provision in clause 53 for the contractor to notify and claim where it considers that it is entitled to a higher rate for the remeasured works. It is therefore perhaps surprising that the FIDIC Red Book does not contain a similar provision to that in the Infrastructure Conditions clause 53.

However, the architect/contract administrator under SBC/Q can require the contractor to submit such details as are reasonably necessary for the ascertainment of loss and expense under clause 4.23 if sums are to be claimed in addition to the adjustment of unit rates under the contract. This requirement is more of an 'after the event' provision as it seems to assume that the contractor will have kept the records required. In practice, the contractor may have kept adequate records or may have kept inappropriate records in the opinion of the architect or quantity surveyor, in which case the provision to request records, and to specify which records, in advance or concurrently would be preferable.

Under the NEC contracts there are no specific record requirements in the Core Clauses, but it is significant that the basis of advanced agreement and pricing by quotations was reinforced by NEC3-ECC core clause 65.2, which provided that the assessed value of a compensation event shall not be revised if recorded information later shows that the assessment was based on an incorrect forecast. This is an area addressed elsewhere in this chapter when considering the NEC approach to valuing the direct consequences of change.

5.14.2   Notice in Relation to a Quotation or Proposal

As explained elsewhere, recent years have seen the rise of terms in variations clauses in construction and engineering contracts that promote the pre-agreement of their financial consequences, by requiring the contractor to provide a quotation or proposal before the variation is instructed and/or carried out.

SBC/Q clause 5.3.1 enables the architect or contract administrator to require in its instruction for a variation that the contractor shall provide a quotation in accordance with the detailed provisions of Schedule 2 of that form. This is stated as being dependant on the contractor having received sufficient information to enable it to provide a quotation. However, within seven days of receipt of the instruction (or a longer period if stated in the instruction or agreed between them) the contactor can give notice that it disagrees with the application of this procedure to that instruction. If the contractor gives such a notice then under clause 5.3.2 it is not obliged to give the quotation or carry out the variation unless and until instructed to carry it out. It is to be valued in accordance with the variation valuation rules of clause 5.6.

Similarly, the Infrastructure Conditions clause 52 'Valuation of ordered variations' starts with the power for the engineer to request from the contractor a quotation for a proposed variation. This procedure has been detailed. However, clause 52 does not provide for the contractor to give notice that it disagrees with the application of the quotation procedure to a particular variation.

The most obvious reason for a contractor giving such a notice is that hinted at in SBC/Q clause 5.3.1. This is that it has not received sufficient information to enable it to do so. This may be a failure of the architect or contract administrator or it may be that the work is so indeterminate at that stage that it cannot give further details. The nature of many variations is that they were not included in the original scope because the need for them was unknown and hence their detailing and definition will not be to the level of the works as originally priced. Take, for example, the discovery of old underground structures in the excavation for a building's basement and foundations that was thought to be on virgin ground. The contractor is likely to respond that it knows no more than the employer and its advisors as to the scope and nature of those structures and will not until they are all removed. The contractor is likely to consider that it is too risky to price such works in advance or that if it were to do so then it can only price extensive contingencies at a greater cost for the employer. In such circumstances it is unlikely to be in the interests of either party to apply the quotation approach. It may even be that the only reasonable approach to valuation is on the basis of recorded time and materials and potentially on dayworks.

The detailed requirements of NEC4-ECC in relation to compensation events are set out in its clauses 61 to 65. The notice requirements of these provisions are as follows:

- Clauses 61.1 and 61.2 require the project manager to notify the contractor of a compensation event which arises from it or the supervisor, and instruct the contractor to submit a quotation, unless it arose from a contractor default or is of no relevant effect. The types of compensation events covered by these requirements include such items as changes to the works, instructions to stop work, changes in decisions or the employer using part of the works before formal completion.
- Where the contractor identifies what it believes to be a compensation event, and which the project manager has not notified, clause 61.3 requires the contractor to give the notice, within eight weeks of becoming aware that the event has happened. The condition precedent provision in clause 61.3 is considered elsewhere. This should normally capture compensation events not covered by clause 61.1, such as failures in relation to site access, provision of plant and materials by the employer, and other failures and breaches. It should also capture neutral events such as encountered physical conditions or adverse weather.
- Clause 61.4 requires the project manager to reply to a notification from the contractor within one week, unless the contractor has agreed to a longer period. In that reply, the project manager notifies the contractor if it considers that the event is the contractor's fault, has not happened and is not expected to happen, was not notified in the time required in clause 61.3, or has no effect or is not a compensation event. Otherwise the project manager notifies the contactor that it is a compensation event and includes in that notice an instruction to submit a quotation. Clause 61.4's sanction for the project manager's failure to comply with the timetable is discussed elsewhere.
- Under clause 62.3 the contractor submits a quotation within three weeks of being instructed to do so to and the project manager replies within two weeks of the submission, although they can agree to extend these periods. Whereas NEC3-ECC clause 62.5 required that the project manager should 'notify' the contractor of such an extension, it is now termed 'informs'. Clause 62.6's sanction for the project manager's failure to reply to a quotation within the timetable is discussed elsewhere.

- Under clause 62.3 the project manager's reply can: notify its acceptance of the quotation; instruct a revised quotation; or advise that the project manager shall be making the assessment.
- Clause 64 provides for the project manager's assessment. Clause 64.3 requires it to notify its assessment and its details to the contractor within the same period as was allowed for the contractor's submission of its quotation. This seems sensible, in that a large complex compensation event is likely to take them both similarly long periods to consider. Clause 64.4's sanction for the project manager's failure to give an assessment in the time allowed is discussed elsewhere.

5.14.3   Notice as a Condition Precedent and Other Sanctions

This is a particularly problematical issue, which can become a critical one to many a construction or engineering claim. It is not unreasonable to suggest that most significant disputes as to the valuation of contractors' and subcontractors' accounts that end up in international arbitration involve, in addition to their factual and technical issues, arguments regarding contractual notices, whether they were duly given and the effects on the claimant's entitlements, if any. In addition to careful consideration of the express terms of the contract, the local substantive law may be critical to this question. Different jurisdictions can take quite different attitudes to requirements for notice of a claim and the extent to which a lack of them affects entitlement to that claim.

One consequence of the use of this as a defence to claims is the habit of claimants putting their claims in alternatives outside of the terms of the contract that included the notice requirement. Thus, a contractual claim will be made under the contract terms, but also, in the alternative, as a claim for damages for breach of contract, or perhaps terms implied out of the applicable law, that do not require notice as a condition precedent to entitlement. This may have a significant effect on the valuation of the effects of a change because it may then be required to quantify it on two alternative bases: firstly, applying the valuation rules set out in the contract and, secondly, and alternatively, valuing it on some other basis, even a quantum meruit approach.

Notices can be required under the terms of a construction and engineering contract as either a simple administrative requirement or, more significantly, a condition precedent to other rights under the contract. Detailed consideration of the terms of such a notice provision is essential. It is not uncommon to encounter contracts where the giving of a notice of intention to make a claim is stated to be a condition precedent to a valid claim. Usually such clauses state that the contractor shall give the required notice within a stated time limit from the point at which it knew, or should have known, the claim would need to be made.

The obvious intention of such clauses is to eliminate dilatory claims being made long after the event, and as such can be seen to be an effective means of ensuring the employer is not surprised and/or disadvantaged by late claims. The difficulty that can arise is determining the exact point at which the contractor had, or is deemed to have had, knowledge of the claim. This can lead to sometimes critical debate about the start of the stated notice period.

If the employer has been prejudiced by a late notice, in that it has been denied the opportunity to take corrective or other mitigating actions or denied the opportunity to examine relevant aspects of the work at critical times, then such exclusions can be seen to be equitable to the extent that the employer was prejudiced. There is a counter argument that such clauses can be used to exclude otherwise valid claims simply because they were not notified in the specified time period.

In addition to the timing of notice, its form may also be an issue, including the party and address to which it is required to be delivered. Whether a notice complies with such requirements will depend on the facts of each case, the requirements of the contract and any effects of the applicable law. Contracts should be drafted to state what form a notice must take and who it is to be given to. Clauses such as the Infrastructure Conditions clause 1(6) set out the form of any communications required under the contract to be 'in writing'. Clauses such as the Infrastructure Conditions clause 68 set out the respective addresses to which notices to the contractor or employer are to be served. In the FIDIC Red Book the form of a notice and required address are set out in clause 1.3, with reference to further details in the Appendix to Tender. In SBC/Q the form of a notice and required address are set out in clause 1.7, with reference to further details in the Contract Particulars. It is a common failure of parties to construction and engineering contracts that they fail to carefully read such provisions and ensure they comply with them, particularly under bespoke forms of contract.

It is also not uncommon to find that a party hiding behind asserted lack of contractual notice, or that it is not in the format required by the contract, when defending a contentious claim in formal proceedings, but that it did not rely on such formalities on other claims during the course of the project. It may even have addressed the factual merits or quantification of a now disputed claim, without reference to the alleged lack of proper notice. Whether such conduct is relevant will depend on legal considerations as well as the facts of each case.

An issue relating to the form of notices often includes a situation where, whilst the contractor has not served a notice to the employer's agent in the form, etc., required by the contract, it is apparent that it knew of the issues that ought to have been subject to that notice because such a notice would have told it no more than it already knew. Thus, it suffered no prejudice as a result of the lack of notice. Experience suggests that arbitral tribunals in particular are reluctant to allow those receiving a claim to hide behind such technicalities and will then find a way to circumvent their effects. In this regard, the law of the local jurisdiction may again be of critical importance.

The NEC ethos of timely advice of events such as changes that have a time and commercial impact and the timely processing of these events and their effects has been stated elsewhere. Such an ethos may be of little practical effect if there are not sanctions for failure to follow it. As a result, as indicated elsewhere, NEC4-ECC includes a number of sanctions in relation to compensation events where the provisions of clauses 61 to 64 have not been complied with in the time scales set out. In particular:

- Under clause 61.3, if the contractor does not notify the project manager of a compensation event that has not been notified by the project manager within eight weeks of becoming aware of it, the sanction can be severe. That clause continues: 'the Prices, the Completion Date or a Key Date are not changed unless the event arises from the project manager or the Supervisor giving an instruction or notification, issuing a certificate or changing an earlier decision'. These exclusions from the condition precedent sanction are interesting. As explained elsewhere, under clause 61.1 the Project manager is the party that should be giving the notification of such compensation events. It should not have been left to the contractor and clause 61.3. They also seem to recognise that some compensation events directly arise out of the actions of the employer's team and they should be aware of them. It would not, however, give the contractor relief in respect of compensation events such as physical site or weather conditions (clauses 60.1(12) and (13) respectively).

- Clause 61.4's requirement that the project manager replies to a notification from the contractor within one week, or such longer period as has been agreed, allows the contractor to give notice of that failure. If the failure continues for a further two weeks, 'it is treated as an acceptance by the *Project Manager* that the event is a compensation event and an instruction to submit quotations'.
- Clause 62.6's requirement that the project manager replies to a contractor's quotation within two weeks, or such longer period as has been agreed, allows the contractor to give notice of that failure. If the failure continues for a further two weeks, 'it is treated as an acceptance by the Project Manager of the quotation'.
- Clause 64.4's sanction for the project manager's failure to assess a compensation event within the same period as was allowed for the contractor's submission of its quotation, allows the contractor to give notice of that failure. If the failure continues for a further two weeks, 'it is treated as an acceptance by the *Project Manager* of the quotation'.

5.14.4   Further Considerations

The provisions in construction contracts for notices are there to avoid, wherever possible, both unnecessary costs and disputes as to the consequences of events on site. If one party to the contract believes that a change or instructed variation has rendered some, or all, of the relevant rates and prices invalid and adjustments are necessary, it is simple good practice to alert the other party to the contract so that a course can be changed and records and notes can be kept of any relevant matters to avoid unnecessary argument as to the alleged circumstances and effects of the change or instruction.

Apart from the express requirements for notices incorporated in contracts, it makes good commercial sense to have matters recorded jointly as far as possible in the interests of achieving early agreement of any requested adjustments to rates and prices. Cash flow is the lifeblood of the construction and engineering contracting industry and it is in the interests of the contractor to ensure that early and adequate notice is given. Any belief that some advantage can be achieved by delaying notification and restricting the ability of the other party to make relevant records is more than likely misconceived and it is more likely that the contractor will suffer from late agreement and delayed, if not reduced, cash flow as a result of any delay in issuing notices.

In any event, apart from express provisions such as that in clause 53(5) of the Infrastructure Conditions, it will be reasonable to expect that, in the event that delayed notices result in the other party being unable to record relevant circumstances or consequences, then the evaluation of the change or instruction will be limited to the extent that the other party has been prevented or prejudiced by the failure to give reasonable notice.

A perhaps under-used alternative approach by those complaining at a lack of notice is to counterclaim in relation to the effects of that failure and the discrete financial consequences of it. For example, if an employer considers that its contractor's failure to give notice as required by the contract, or late giving of such a notice, has led to its incurring costs that could have been avoided, reduced or charged to a third party if the notice had been given as specified, then it should be able to establish the additional quantum resulting from this as damages. If they are truly the result of breach by the contractor in relation to notice, then those damages might be capable of being claimed on the basis of that breach. It is surprising how often employers adopt the 'magic bullet' approach of trying to defeat a claim in its entirety on the basis that notice was a condition precedent, when such a counterclaim might provide a viable alternative approach. This may also avoid such consequences as: debate on the legal effect of a lack of notice; the contractor relying on an alternative claim outside of the procedures in the contract; and the reluctance of many tribunals to disallow a whole claim on a procedural technicality. However, before considering making a counterclaim on this basis, local legal advice would be essential.

5.15   Summary

This chapter has considered the evaluation of the direct consequences of change and typical standard form provisions for such evaluation. These have been categorised as a traditional hierarchy approach using rates and prices agreed in the contract, or the more recent approach of relying on the contractor's actual costs.

A rates and prices approach has been illustrated by reference to the FIDIC Red Book as an example. Related issues discussed have included: the 'fences' contained in such a hierarchy approach; what to do about errors in rates and prices; the inclusion of amounts for preliminaries and general items; and percentage adjustments and dayworks valuation.

An actual costs-based approach has been illustrated by reference to NEC4-ECC. Related issues discussed have included: the choice of forecast or actual costs; the concept of Defined Cost; the concept of Disallowed Cost; and the use of agreed rates and percentages in this context.

In order to effect a valuation on either basis, it is necessary to understand the typical component parts of contractors' rates, prices and costs, and these have been identified.

Further aspects of the direct consequences of change have included: dealing with subcontractor costs; valuation in advance; quantum meruit claims; and contractual notices. The legal nature of these last two topics has been emphasised.

It is next necessary to consider the evaluation of the time consequences of change. This includes how changes to the programme of work, impacting on cost through their effect on timing and duration of the works rather than on individual unit rates, can be established and evaluated. These issues are addressed in the following chapter.

6

Evaluation of the Time Consequences of Change

6.1 Introduction

It is an often cited truism on construction and engineering projects that if change has an effect on time, then those time implications will have financial consequences for some or all parties involved. This particularly means the employer and contractor, but may also mean the employer's consultants and the contractor's subcontractors and suppliers. To put it baldly, 'time means money'. Those time consequences of change can take the form of prolongation (which broadly means that resources are required for longer), disruption (which broadly means that more resources are required perhaps for less time) or acceleration (which also broadly means that more resources are required). All of these can have a financial effect that requires evaluation. They all involve areas of difficulty that can lead to disagreement and dispute.

In this chapter these areas of difficulty are addressed by particular reference to the terms of the standard forms SBC/Q, the FIDIC Red Book and the Infrastructure Conditions. In relation to the NEC suite of contracts, as explained in the NEC4 User Guide:

> If the compensation event increases the *Contractor's* time related costs, for example because planned Completion is delayed, or additional supervisory staff are needed, the increase in Defined Cost is included within the value of the compensation event. There is no such thing as a retrospective and separate 'delay and disruption claim' in the contract.

6.1.1 Factual and Legal Background

Claims for the time consequences of change will always be set against a background of their factual and legal basis. These can have a profound effect on quantification.

The factual basis will be essential to establishing if there is any monetary entitlement, in terms of what happened, who caused it and what its effect was. Some causes of delays are cost-neutral. The most widely recognised of these is weather or climatic conditions. For example, in SBC/Q, 'exceptionally adverse weather conditions' is a 'Relevant Event' for extensions of time under clause 2.29.9, but it does not appear as a 'Relevant Matter' in relation to the reimbursement of loss and/or expenses under clause 4.24. Similarly, in the Infrastructure Conditions, clause 44(1)(d) has 'exceptional adverse weather conditions' as a ground for extension of time, but it is not a ground for payment of the contractor's additional cost under that form. In the FIDIC Red Book, 'exceptionally adverse climatic conditions' gives an entitlement to extension of time under clause 8.4, but is not a ground for payment of the contractor's additional cost. This approach to the risk of weather or climatic conditions is widely understood in the industry, but confusion often arises in relation to less common events. For example, under the FIDIC Red Book clause 19, delays caused by such as rebellion, radioactivity and earthquake give an entitlement to an extension of time, but additional cost only if they occur in the country of the permanent works.

Cost-neutral delay events can also foster disagreement as to which cause of delay extensions of time should be granted against. It is far from rare that contractors complain that contract administrators grant extensions for such as climatic conditions under the FIDIC Red Book clause 8.4 in order to hide from the employer the effects of their failures to issue drawings or instructions in good time under clause 1.9. This is on the basis that under clause 1.9 the contractor would be due both an extension of time and payment of its associated costs plus reasonable profit. Such an approach has been seen even where the contractor made no claim for delay associated with the climatic conditions. Whilst such craft is clearly an abuse of the contractual procedures, it has also been seen that contractors have reluctantly accepted such a cost-neutral extension of time for the benefit of relief from delay damages, where their case for delays caused by such as late design might have been met with responses referring to concurrent delays due to matters such as a lack of resources on the contractor's part.

The legal basis of a claim for the time consequences of change involves both the applicable law and the express terms of the contract agreed between the parties. In the absence of express terms, a claim is likely to be for damages for breach of contract. In most jurisdictions, the quantum of such a claim will be such an amount as will put the delayed party in the financial position it would have been in absence of the delay. As this is put in the SCL Protocol:

> Unless expressly provided for otherwise in the contract, compensation for prolongation should not be paid for anything other than work actually done, time actually taken up or loss and/or expense actually suffered. In other words, the compensation for prolongation caused other than by variations is based on the actual additional cost incurred by the Contactor. The objective is to put the Contractor in the same financial position it would have been if the Employer Risk Event had not occurred.

The relevance of the SCL Protocol on delay and disruption is discussed more fully elsewhere in this chapter.

This 'but for' approach to quantification means that evaluation of claims for damages related to delays will usually include costs and, potentially, losses of such as interest and contribution to off-site or head office overheads and profit (if not expressly excluded), provided of course that they can be proven as a matter of fact. Such heads of claim are considered in detail elsewhere in this chapter.

A further legal issue will be as to the test of proof that the quantification has to pass. For example, in the UK this will be the civil test of 'the balance of probabilities'. This principle can be overlooked by both parties. Claimants forget that they have to show that, on the balance of probabilities, the failures they plead led to the financial amounts they claim. Those receiving claims forget that they are not subject to a higher test such as 'beyond all reasonable doubt'.

In stating the nature of what the contractor is to be financially reimbursed in the event of delays, the terms of various standard forms contain subtle differences in terminology. The JCT contracts have always applied the term 'direct loss and/or expense', as is currently used in the SBC/Q clause 4.23. The equivalent in FIDIC contracts is its defined term *'cost'*, which is capitalised in the FIDIC Red Book and defined in its clause 1.1.4.3 as 'Cost means all expenditure reasonably incurred (or to be incurred) by the Contractor, whether on or off the Site, including overhead and similar charges but does not include profit'. As a result of this definition, a further consequence of the factual basis of a claim under FIDIC terms is that some clauses in relation to the cost of delay also add *'reasonable profit'* (for example, clause 1.9(b), in relation to delayed drawings or instructions, refers to 'payment of any such Cost plus reasonable profit'), whereas others do not (for example clause 13.7(b), in relation to changes in legislation, just refers to 'payment of any such Cost'). The distinction here is that events that involve failure by the employer or the engineer will add profit to 'Cost' whereas events that do not involve such failure allow only 'Cost'. The Infrastructure Conditions take a similar approach to the FIDIC Red Book. This reflects their common heritage. The Infrastructure Conditions clause 1(5) defines the uncapitalized word 'cost' as 'all expenditure properly incurred or to be incurred whether on or off the Site including overhead finance and other charges properly allocatable thereto but does not include any allowance for profit'. As with the FIDIC Red Book, the clauses that then allow the contractor such 'cost' do add profit in some cases, although the approach appears less consistent with employer or engineer failure than in FIDIC. For example, if the engineer is late in issuing drawings, specifications and instructions, clause 7(4) gives the contractor 'such cost as may be reasonable', but if the contractor encounters unforeseen adverse physical conditions or artificial instructions, then clause 12(6) gives it 'any costs which may reasonably have been incurred by the Contractor (together with a reasonable percentage addition thereto in respect of profit)'.

A further common feature of the FIDIC Red Book and the Infrastructure Conditions is how their definitions of recoverable cost include those 'to be incurred'. This will cover where the contractor has an established liability to a supplier or subcontractor but has not yet paid it. From a cashflow perspective where a contractor has incurred significant compensable costs with its supplier or subcontractor because of issues for which the employer is liable, this can be a significant help.

Like FIDIC and the Infrastructure Conditions, NEC4-ECC refers to cost, under its 'compensation events' regime, but as 'Defined Cost' in its clause 63.1. NEC quantification for 'compensation events' also entitles the Contractor to its 'Fee' for overheads and profit. A further quirk of NEC (for those used to the more traditional FIDIC, Infrastructure Conditions and JCT approaches) is its prospective approach. The 'Defined Cost' of delay will be that actually incurred on work already done, but 'forecast' on work not yet done (see core clause 63.1).

In practice, particularly in formal proceedings, claimants often plead claims for costs related to delays under express contract provisions, but, alternatively, as damages for breach. Often the alternative claim is intended to avoid the consequences of failure to serve contractual notices that might defeat the claim under the contract. It is also sometimes the case that the law allows some additional remuneration to that provided for in the contract, particularly where there is some express limitation in the contract. This can lead to alternative claims being made and requiring alternative quantifications.

6.1.2 Financial Effects for the Employer

As stated elsewhere, time means money in construction and engineering. Both employer and contractor can be affected, but the nature of the financial effects of time will differ between them.

For the employer, the financial effects of culpable delay by the contractor might include such as:

- delayed income from residential sales, rental income or lost plant production;
- continued rental costs of existing accommodation or plant;
- extended running costs of an existing inefficient or expensive facility;
- extended financing or interest costs, whether on borrowings or lost income on capital;
- additional fees on the project under extended consultancy agreements of such as architect, engineer and contract administrator; and
- extended costs of provision on the project of employer provided services for the site, such as security, insurances or dewatering.

Whilst employers could rely on their actual costs and losses for the quantification of a claim for damages for delay in completion by a contractor, the vast majority of construction and engineering contracts provide an agreed rate of delay damages, often referred to as 'liquidated damages' or 'liquidated and ascertained damages' (LADs) in UK parlance, and sometimes referred to as 'delay penalties' in international parlance. These are usually agreed in the contract at a daily rate. Where the works are not subject to a single contractual completion date, but are broken down into contractual sections, there may be several such rates. The advantages of such a liquidated rate for the employer's delay damages include certainty for both parties and the avoidance of the need for the employer to prove actual damages and the contractor to audit and agree them. Particularly for projects in the process industries, it is common for the resulting damages to be capped, at a percentage of the contract price, to avoid the total damages putting off potential contractors and also to allow contractors to know the maximum limit of their potential liability when pricing their tenders. A similar approach is also applied sometimes in subcontracts, where the potential damages resulting from a subcontractor failure that delays a whole project might be out of all proportion to the value of its subcontract. However, there is occasionally a lack of clarity as to whether that cap percentage is applied to the original contract price or its final account total.

The aim of providing certainty through a pre-agreed rate of delay damages is not always realised in practice. Whilst delay damages are often referred to internationally as '*penalties*', in many jurisdictions argument may ensue as to: whether compensation must equal the harm suffered; how an agreed contractual rate sits with this; whether penalties are enforceable at all; and whether the agreed rate is penal (for a detailed analysis of the position under English law, see SCL paper 195, *Liquidated Damages or Penalty: Cavendish v Makdessi* by Joanna Smith QC). Furthermore, internationally, employers occasionally claim amounts in addition to the contract rate, on the basis of legal arguments that it neither covers all of their losses nor limits their entitlements, being not an exclusive remedy.

6.1.3   Financial Effects for the Contractor

Whilst time usually means money for contractors, it does not automatically follow that an extension of time means compensation. Cost-neutral delay events under construction and engineering contracts, such as weather or climatic conditions, were explained elsewhere. Reflecting this, contracts such as the JCT's cover extensions of time and the compensation of *'loss and expense'* in separate clauses (for example in SBC/Q, extensions of time are dealt with in clauses 2.26 to 2.29 and 'loss and/or expense' is in clauses 4.23 to 4.26). On the other hand, this separation of time and financial compensation also reflects how delay costs can be reimbursable to a contractor without any entitlement to extension of time. This can occur where: there are compensable delays to activities that are not on the project's programme critical path; delays cause disruption but no extension of activity durations; or delays are also accompanied by acceleration.

A significant factual issue in relation to the contractor's entitlement to its financial damages where compensable delays have occurred is that of concurrency. As this is put in the SCL Protocol Core Principle 14:

> Where Employer Delay to Completion and Contractor Delay to Completion are concurrent and, as a result of that delay the Contractor incurs additional costs, then the Contractor should only recover compensation if it is able to separate the additional costs caused by the Employer Delay from those caused by the Contractor Delay. If it would have incurred the additional costs in any event as a result of Contractor Delay, the Contractor will not be entitled to recover those additional costs.

Many years ago an engineer working in the UK public sector named Mr Brown proposed that the financial effects of delay for a contractor should be covered by a rate agreed in the contract, similar to the delay damages or liquidated damages rate applied to the employer's claims for culpable late completion by the contractor, and recommended this be covered by a clause in the contract commonly referred to as a 'Brown's clause'. This approach never caught on, even in the UK public sector, although it is now recommended by the SCL Protocol at its paragraph 20.4. That lack of take-up may have been not least because the financial effects of delays can be very different depending on when they occur. For example, the contractor's costs of early delays deferring the contractor's mobilisation and site setup will usually be much lower than the costs of delays when its site activities are at a peak. This issue is discussed further elsewhere in this chapter.

For a contractor, the financial effects of delays caused by matters for which the employer is liable might include such examples as:

- Prolongation costs, comprising its additional expenditure on time-related items because the works took longer. Such items might include site staff, facilities and equipment and some head office resources. This can also include claims for losses such as profit and interest or finance charges.
- Escalation costs, comprising inflationary rises in the costs of such as materials, people, equipment and preliminaries and general items, because the works were carried out later.

- Disruption costs, in terms of lost productivity, particularly in relation to operatives and equipment. As noted elsewhere, however, there can be disruption without delay and there can also be delay without disruption.
- Thickening costs, often a component of a disruption or an acceleration claim, because additional resources, such as staff, have had to be deployed.
- Acceleration costs, because measures have been taken to attempt to complete activities quicker.
- Similar claims to any of the above from subcontractors and suppliers.
- Financing or interest costs, whether paid on borrowings or lost income on capital, on any of the above.

These various heads of claim for the time consequences of change by both employer and contractor are considered in detail elsewhere in this chapter. The evaluation of such claims is among the most frequently encountered and problematical in construction contracts, and the source of many lengthy and costly disputes. Many of these disputes could be avoided, or at least drastically reduced, if the claims for additional payment were better researched and presented in terms of both their factual and legal basis and their evaluation and substantiation. Too many such claims are presented as simple presentations of asserted facts in terms of 'we intended to do this …, but had to do this …, because … ', without properly establishing the basis of the claim, in contract and in fact. Thereafter, many evaluations are presented simply as 'we priced to spend this …, but actually spent this…'. Furthermore, they then fail to link the claimed effects and financial consequences to the asserted causes. In Chapter 1, the following chain of analysis was discussed in relation to establishing a claim for compensation under a construction or engineering contract:

Duty – Breach – Cause – Effect – Damage

This sequence requires that the person making the claim must firstly establish the *duty* that was owed by someone, usually under the contract, for, for instance, supplying information or issuing an instruction in good time. They must then establish that the person who was under that duty was in *breach* of its duty by establishing in what way the duty was not performed, for instance not performed at all, or performed late, etc. The *cause* of the breach must be established in factual terms, although in practice this may not be as easy as it at first might seem. For instance, if drawings are supplied late the recipient may not know for certain, or at all, why they are late. For the purposes of this text the *effect* is crucial as this is the detailed evidence of the consequences of the breach of duty. From this it should be possible to establish the *damages* that have been incurred as a result of the breach. Typically the effects will be described as the changes in the contractor's purchases, resources, working methods, etc., experienced as a result of the breach with the damages being the financial consequences of such changes.

The first parts of the chain of analysis are outside the scope of this text, but it is essential that the process of establishing the necessary causal links between the parts of the sequence are understood, as without each link being established the whole exercise is undermined. As with any claim, the end product will only be as strong as the weakest link in the chain. Thus, for example, no matter how clearly a claim establishes a legal duty on the employer to provide timely access to parts of a site, and the facts of late possession are clearly documented and indisputable, the contractor's related claim for prolongation costs may still substantially fail if its quantification is hopelessly presented and lacks substantiation and evidence. Furthermore, if the quantum of the effect is to be competently and sufficiently established, then it must be evaluated with a good understanding of the preceding parts of the analysis, to ensure that it is made on the correct factual and legal basis. For example, is the contractor entitled to 'loss and/or expense' per JCT terms or 'costs' per FIDIC Red Book terms, or entitled to damages as defined by the applicable law of contract? What is the entitlement to profit, given the distinctions in different FIDIC Red Book and Infrastructure Conditions clauses that have been identified? Establishing the link between the factual and legal cause and the financial effect is essential, but all too often it is dealt with in a superficial manner, particularly in claims for disruption and acceleration.

6.1.4 The Causal Link

It is central to any claim for additional payment that the breach complained of must have caused the loss claimed as damages. This seems a simple statement but in practical terms it is the source of many of the most difficult problems in evaluating construction and engineering contract claims. Whilst the factual circumstances of construction and engineering projects can be complex, this is not an excuse for the shallow attempt that many claimants make to establish that link.

In addressing the establishment of a link between breach, effect and damage the first step is usually to apply the 'but for' test. This test comprises the adoption of the questioning process of 'is it reasonable that but for the breach that effect and that damage would not have been suffered by the claimant?' This test was used in the case of *South Australia Asset Management Corporation v York Montague Ltd* [1996] 3 All ER 365, where Lord Hoffman gave the following example:

> A mountaineer who is about to undertake a difficult climb is concerned about the fitness of his knee. He visits his doctor who negligently makes a superficial examination so that he pronounces the knee fit. The climber goes on the expedition which, importantly, he would not have undertaken if the doctor had correctly informed him of the state of his knee. During the assent he suffers an injury which is unrelated to his knee but is a foreseeable consequence of mountaineering.

In this example it could be thought that the doctor's negligent advice caused the injury as, had the doctor given the correct advice, the mountaineer would not have gone on the expedition and would not have suffered the injury, although the injury had nothing to do with the dodgy knee! However, the court held:

> A duty of care which imposed upon the (defendant) responsibility for losses which would have occurred even if the information given had been correct was not fair and reasonable as between the parties and was therefore inappropriate as an implied term of a contract or tortious duty ....

In other words, if the doctor's advice had been correct and the knee was sound, the injury would still have occurred. In this case, the negligent advice did not cause the damage.

In construction terms an analogy might be the late issue of foundation drawings on a contract where the contractor is subsequently late in constructing the foundations. If the contractor in fact could not obtain the plant for the foundation excavation until the day before it actually commenced the work, have the late drawing issues caused any delay? The question is 'but for the late drawings, would the foundations have commenced on time?' The answer to these facts is no, and the late drawing issue therefore did not cause the delay.

The 'but for' test is therefore the first stage of the process, but it is not the whole process. It is a useful filter, but it can become difficult to obtain clear answers where there are many overlapping and competing issues to test. The 'but for' test may need to be applied in a modified or extended manner to achieve the objective analysis required. The courts, in *Smith New Court Securities Ltd v Scrimgeour Vickers* [1996] 4 All ER 769, stated:

> The development of a single satisfactory theory of causation has taxed great academic minds .... But, as yet, it seems to me that no satisfactory theory of solving the infinite variety of practical problems has been found. Our case law yields few secure footholds. But it is settled that at any rate in the law of obligations causation is to be categorised as an issue of fact. What has further been established is that the 'but for' test, although it often yields the right answer, does not always do so. That has led judges to apply the pragmatic test whether the condition in question was a substantial factor in producing the result. On other occasions judges assert that the guiding criterion is whether in common sense terms there is sufficient causal connection (see *Yorkshire Dale Steamship Co Ltd v Minister of War Transport* [1942] 2 App ER 6 per Lord Wright). There is no material difference between these two approaches. While acknowledging that this hardly amounts to an intellectually satisfying theory of causation, that is how I must approach the question of causation.

It can safely be said that the courts, in seeking to establish causation, are going to address matters 'in common sense terms'. The reference to Lord Wright in the *Yorkshire Dale Steamship* case quoted elsewhere refers to his summary of the common sense approach, where he stated:

> This choice of the real or efficient cause from out of the whole complex of the facts must be made by applying common sense standards. Causation is to be understood as the man in the street, and not as the scientist or the metaphysician would understand it ... without too microscopic analysis but on a broad view.

There are therefore three potential tests:

1. The 'but for' test.
2. The pragmatic test.
3. The common sense test.

The factor most likely to cause complications in the assessment of causation is that of an act or event that intervenes, being the act of the claimant or a third party, and breaks the chain of causation. The courts have again adopted a common sense approach to the occurrence of intervening acts and events and will determine the issue on the facts of the case rather than by applying predetermined tests. In order to break the chain of causation the court will require something to have occurred that is 'unreasonable or extraneous or extrinsic' (*The Sivand* [1998] 2 Lloyds Rep 97 102).

In construction and engineering contracts there may be multiple, or even a multitude, of potential acts or events with impacts on the progress of the works. Some, or all, of these acts and events and their impacts may well occur simultaneously or in a manner causing overlap of events and impacts – for example, what are commonly referred to as 'concurrent delays' in the context of the analysis of delaying events. In such circumstances it can be very difficult to determine which acts or events had crucial impacts and which were secondary or inconsequential. This has led to the production of various competing, or sometimes complementary, systems of analysing events impacting on progress to determine the critical ones.

In October 2002 the SCL Delay and Disruption Protocol was published in order to review competing methods and provide guidance on suitable approaches to the required analysis. The Second Edition was published in February 2017 and is considered as follows.

6.1.5   The SCL Delay and Disruption Protocol

Many of the Protocol's principles and recommendations are outside the boundaries of this book. However, there are a number of issues directly affecting the assessment of quantum where the views expressed in the Protocol are very relevant. In particular, in relation to: establishing the causal link in a claim; the 'core principles' included in the Protocol in respect of 'float' in programmes; and 'concurrent delay'.

The Protocol states, in paragraph 13 of its Core Principles, that a contractor should be able to recover costs incurred as a result of an employer delay providing the employer delay prevents the contractor from completing the works by its planned completion date, even if this planned completion date is earlier than the contract completion date. The Protocol qualifies this position by stating that the contractor's intention to complete before the contract completion date should be known to the employer at the time the contract is entered into, and the intention must be realistic and achievable. In essence the Protocol is stating that the contractor can suffer an employer delay, which, because of float in the programme, does not affect the completion date but still enables the contractor to recover any loss and expense incurred. This reflects the position identified elsewhere, that entitlement to delay costs is not always dependent on an entitlement to extension of time to the contractual completion, or dates where the contract is subject of sectional completion.

There are two points to be made in respect of this position as set out in the Protocol. Firstly, it is obviously correct that a contractor can suffer delay to various activities within a contract which, while that delay causes him to incur costs, the completion date is not affected because the affected activities are not on the critical path of activities for the project, i.e. they do not lie on the longest sequential path of activities through the project. If the employer causes such a delay then, subject to any contract provisions, it is also obviously reasonable for the contractor to recover its loss, expense or costs associated with that non-critical delay. Such 'local prolongation costs' are considered further elsewhere in this chapter.

However, Protocol paragraph 13 suggests that the same reasoning applies to employer delays that are critical, in that they extend the period for the execution of the works, but do not affect the completion date because of 'end float', i.e. float time between the contractor's intended completion of activities and the contract completion date. In such circumstances the quantum of the contractor's potential losses is likely to be greater as in the former case it would incur only costs related to the extension of the delayed activities but not to the management of the whole project, whereas in the latter case, where the effect is to extend the intended period of contract activities, then the contractor is likely to claim that it has had to retain the whole project management and support services for the period of the extended activities.

This raises the second point in relation to the Protocol's paragraph 13, the advisability of an employer entering into a contract in the knowledge that the contractor intends to complete the works prior to the contract completion date. In practice such a situation is unlikely to occur frequently as most employers require their contracts to be completed as soon as possible. If, for whatever reason, the employer is minded to enter into a contract in the knowledge that the contractor intends to complete its works before the contract completion date, then it would be very prudent for the employer to provide in the contract for the eventuality that employer delays could cause the contractor's works to be extended without affecting the completion date. In particular, the extent to which the contractor can recover project and off-site overheads in respect of such delays should be expressly provided for, as the absence of such provision is likely to result in disagreement and dispute if left for settlement after the event. The extent to which the contractor would incur costs in any event between the intended completion of its activities and the contract completion date would need to be considered as, subject to the provisions of the contract, it is likely that the contractor would still remain responsible for the works until the contract completion date, thereby incurring at least some of the project costs in respect of such matters as fire security, insurances, etc. In circumstances where the employer was not aware of the 'end float' the contractor would be deemed to have included its management and support costs for the full period of the project to the stated completion date. In the absence of any express agreement or term in the contract the contractor would be most unlikely to recover its costs in such a situation.

Paragraph 14 of the Protocol's Core Principles then goes on to consider the problem of concurrent delay and its effect on the contractor's entitlement to compensation for prolongation. This principle deals with the situation where the contractor incurs additional costs as a result of concurrent delay by both employer delay and contractor, and states that the contractor should only be able to recover loss and expense to the extent that it is possible to identify costs incurred as a result of the employer delay separately from those caused by the contractor delay. This reflects the 'but for' test explained elsewhere of putting the contractor back into the position it would have been in had the employer delay not occurred. If the contractor was in its own concurrent delay, then it would have incurred prolongation costs even if the employer delay had never also occurred. This issue of compensation where there is concurrent delay is considered further elsewhere in this chapter.

This is reasonably obvious and accords with recent judgments in the English courts confirming the principle that damage that would have been incurred in any event as a result of contractor delay cannot be recovered as a result of concurrent employer delay. Furthermore, this is a principle that applies in the law of damages in many other jurisdictions.

6.1.6 Conclusions

In considering the impact of any event, whether on the unit costs of an individual activity or on the programme for the whole of a project, it is essential that a logical and reasonable testing of cause and effect is adopted. The English courts have demonstrated that such testing does not have to be in the realms of high scientific analysis but should adopt pragmatic and common sense analysis such as the 'but for' test. Publications such as the SCL Protocol can be very useful in providing schemes of analysis that might be appropriate in differing circumstances, but any analysis must be rooted in fact and logic.

The means of analysing differing causes of additional payment arising from changes to programmes is considered further in the following sections, but when considering the means of analysing the different causes the need for the logical and pragmatic approach discussed elsewhere, and illustrated by the 'but for' test, should always be borne in mind.

6.2 Prolongation

When parties and advisors on construction and engineering projects refer to a 'contractual claim' they are most commonly using that broad term to refer specifically to a claim for additional payment for costs incurred by the contractor as a result of delays, that is a prolongation claim, also including or excluding disruption costs. It has been said elsewhere that time means money for both parties on construction and engineering projects. It is also a fact of life in these industries that delays are all too frequent.

As a result, prolongation claims are common on projects of any size and complexity, and most practitioners understand the typical heads of claim that need to be evaluated in them. There are, however, some common areas of misconception on aspects that lie at the root of claims for prolongation, even before the detailed evaluations are considered. These are in relation to the nature of what is to be evaluated and the lack of direct relationship between the contractor's entitlement to extensions of time and financial reimbursement for delays. These are considered next, before turning to the typical components of a prolongation claim.

6.2.1 What Is to be Evaluated?

In terms of what is to be evaluated, this was explained elsewhere by reference to the standard forms of contract considered in this book. This might variously be defined by reference to terms such as 'cost', 'expense', 'loss', or combinations of these terms and also the alternative of 'damages' under the law.

As ever, the starting point for any evaluation must be the express provision of the particular contract and this is often overlooked for its particular nuances and definition of what is allowable and what is not. For example, SBC/Q allows 'loss and/or expense', whereas the FIDIC Red Book and the Infrastructure Conditions refer to 'cost' and include 'cost to be incurred' as well as those already incurred. Furthermore, the latter add 'reasonable profit' for some events (for example delayed drawings or information) but not others (for example unforeseen physical conditions). Whilst bespoke contracts might occasionally refer to rates and prices in the contract (and this is explained further elsewhere in this chapter), generally, reimbursement under most contracts follows the approach of FIDIC, the Infrastructure Conditions and SBC/Q by allowing what the contractor has actually additionally expended or lost as a result of being prolonged in its activities. This will put the contractor into the financial position it would have been in if the delay had not occurred, which is similar to the likely position under the applicable law of damages for breach of contract. The exception to this may be where the prolongation is the consequences of a variation and falls to be included as part of the valuation of that variation by applying rates or prices in the contract, as discussed in Chapter 5 of this book and further explained elsewhere in this chapter.

However, not all of a contractor's cost, expense, loss or damage may be allowable. In many jurisdictions those claiming damages at law have a duty to mitigate those damages and will be limited to those that were reasonably incurred as a result of the breach relied upon. Furthermore, all of the standard forms considered in this book include express limits, as quoted elsewhere with key words emphasised:

- In FIDIC Red Book, clause 1.1.4.3 defines 'Cost' as:
  '… all expenditure *reasonably* incurred (or to be incurred) by the contractor, whether on or off the Site, including overhead and similar charges, but does not include profit'.
- In the Infrastructure Conditions, clause 1(5) defines 'cost' as:
  '… all expenditure *properly* incurred or to be incurred whether on or off the Site including overhead finance and other charges properly allocatable thereto but does not include any allowance for profit'.
- In SBC/Q, clause 4.23 states:
  'If in the execution of this Contract the Contractor incurs or is likely to incur *direct* loss and/or expense …'.

Thus, under the FIDIC Red Book and the Infrastructure Conditions, the contractor is limited to what it has reasonably or properly incurred as a result of delay. It is suggested that the provision of SBC/Q clause 4.23 comes to the same thing. There the contractor's loss and/or expense is limited to that which is a direct result of the delay, such that if either is unreasonably or improperly incurred, then they would not be allowed. These requirements would have to be tested on the basis of the facts in each case. For example, a contractor faced with compensable delay might require urgent delivery of an additional barge to address the effects. If such equipment is in short supply in the market, it may have to either wait some time for its availability or immediately hire one from its own plant hire company at its standard, but high, hire rate. In such circumstances the expenditure might be considered reasonable and proper in those circumstances. There may also be questions of remoteness and foreseeability, but this has implications in relation to hire of plant and equipment in prolongation claims generally, as discussed elsewhere in this chapter.

6.2.2 The Relationship Between Extension of Time and Money Claims

Addressing the second common misconception, the lack of direct relationship between extensions of time and the costs of delay, is apparent from a consideration of how standard forms set these out in separate clauses, for example:

- In SBC/Q, extensions of time are covered by clauses 2.26 to 2.29, with the latter listing those 'Relevant Events', the occurrence of which can give rise to extension of time entitlement, but that clause does not mention associated financial compensation. Quite separately, 'loss and/or expense' is covered by clauses 4.23 to 4.26, with clause 4.24 listing those 'Relevant Matters', the occurrence of which can give rise to the entitlement to 'loss and/or expense'. Whilst some of the events are common to both clauses, those that are cost-neutral, such as exceptionally adverse weather conditions, are listed in clause 2.29 but not in clause 4.24.

- In the Infrastructure Conditions, clause 44 'Extension of time for completion' is silent as to associated financial compensation. The financial claim must therefore be based on those provisions of the contract that cover the particular circumstances that give rise to delay in the first place. For instance, delays caused by 'adverse physical conditions' or 'artificial obstructions' are covered by clause 12, subclause (6) of which separately refers to resulting delay, at 12(6)(a), and reasonable 'costs' and 'a reasonable percentage addition thereto in respect of profit', at 12(6)(b).
- Similarly, in the FIDIC Red Book, clause 8.4 covers extensions of time, listing the related delay events, but with no mention of the financial effects of those delays. Again, entitlement to financial compensation for those events is set out in the specific clauses relating to them, where they are not cost-neutral. For example, in clause 8.4(b) reference is made to delay events giving entitlement to extension of time under subclauses of those conditions, and this includes clause 1.9 where delayed drawings or instructions from the engineer are grounds for extension of time, at subclause 1.9(a), and 'Cost plus reasonable profit', at subclause 1.9(b). On the other hand, in clause 8.4(c), 'exceptionally adverse climatic conditions' is listed as a cause of delay justifying an extension of time, but it does not appear elsewhere in the FIDIC Red Book as a basis for payment of cost.

Notwithstanding this lack of dependency between contact provisions for extensions of time for delay and the payment of associated prolongation costs, both contractor and employer sides of the industry are prone to misunderstandings. Contractors can mistakenly assume either that they must have an extension of time before they can obtain any financial compensation for delay or that if they have an extension of time, then they will automatically be entitled to compensation. This approach ignores the following:

- the manner in which the standard clauses are structured to keep extensions of time and additional money separated as illustrated elsewhere;
- the potential issue of concurrent delays for which the contractor is responsible itself and which may mean it gets an extension of time, but no additional money, as explained elsewhere in this chapter; and
- that prolongation costs can be incurred on non-critical activities, which may mean additional money but no extension of time – local prolongation as explained elsewhere in this chapter.

Equally, among the advisors of employers, it is not uncommon for those certifying interim payments to appear to be hiding behind a lack of extension of time. In particular, internationally, where a quantity surveyor is separately appointed to issue interim valuations as a basis of interim payment certificates, they often assert that they cannot include any additional costs related to delays in the absence of the contractor being granted an extension of time. Where contractors are faced with this fallacy, an analogy that might be worth considering is that if the financial claim were couched as one for damages at law, rather than under the express contract terms, then presumably there would be no suggestion that an extension of time is required under the contract first.

Even where a ground for an extension of time is also a ground for financial compensation, the issue of concurrency may mean that the contractor has an entitlement to the former, but not the latter. This tends to be an issue that contractors conveniently ignore in their claims, even where they know that the course of the project saw culpable delays on both sides. Equally, some employers and their advisors seem to raise the cry of 'concurrent delays' almost as a conditioned reflex as soon as any claim for time and money is received from a contractor, whether it is a valid counter in the factual circumstances of that claim or not.

The lack of direct interdependency between extensions of time and delay costs in standard forms of construction and engineering contracts particularly reflects the fact that delays can occur, and contractors can incur associated costs, without there having been any associated effect on the completion date of the works (or dates for any contractual sections thereof), such that there is no entitlement to extensions of time. The subject of delays to activities that are not on the critical path, or are on the critical path but have float, was mentioned elsewhere by reference to the SCL Protocol. Such 'local delay' or 'activity delay' can cause the contractor costs. Take, for example, a high-rise office tower that also involves external works in landscaping, access roads, hard-standing areas and planting, but with a single contractual completion date for the project as a whole. Assume the critical path through the programme for the project runs through the tower itself. Delays to the external works may never become critical to completion, but can involve the contractor in prolongation costs, in relation to plant and equipment and supervision specifically related to those external works. It is likely that additional delay costs associated with those resources will be quite independent of any delays to the tower itself and the contractual completion date. Thus the contractor might claim for those additional external works delay-related costs without any claim for an extension of time at all. It is likely, however, that the contractor will be required to prepare a delay analysis in relation to those external works to justify its claim for however many additional days or weeks of plant, equipment and staff costs it incurs thereon. That delay analysis will not be of the critical path for the works, but may be of the critical path of those local activities. Extending this example, if overall completion was also delayed and the subject of a claim for prolongation of site-wide costs, then two delay analyses will be required to establish the entitlement to prolongation both locally (on the external works activities) and to the project as a whole.

Furthermore, there may be a number of such local delay analyses that are required to support a prolongation claim. A high-rise building might be expected to have a single critical path, through its foundations, superstructure, envelope and internal works, but consider the example of a long, linear infrastructure project such as a new railway line or highway. This might again have a single completion date (particularly a railway) but be broken down for the purposes of the contractor's management and resources into a number of lengths, each of which can proceed independently of each other (to some varying extent). The contractor might then break down its preliminaries and general resources such as facilities, equipment, management and supervision into several individual mini-projects, as well as an overall site-wide establishment (say for such as a labour camp, project management, commercial management and design department). In such a case, if any one or more of these mini-projects is delayed then they can each be the subject of a local prolongation claim, independent, and perhaps as well as, an overall claim for delay, extension of time and prolongation costs for the project as whole. They will therefore all need their own delay analysis to support their prolongation claims.

This division of the project, and the need for several independent prolongation claims, may or may not be reflected in sectional completion under the contract. For example, the highway might link several other roads or the railway have a number of branches that can run independently, such that there are a number of contractual completion dates. In that case, not only is there likely to be greater emphasis on the need for local management and other resources associated with each section, but the contractor's claim for prolongation will be for several contractual sections and therefore particularly require a separate delay analysis for each section. This will be required to both support the claim for prolongation of sectional resources and also to address the need for extensions of time for each contractual sectional completion.

The flip side of this is that just because overall completion of the project has been delayed, this does not mean that all of the contractor's time-related resources will be affected and are claimable. This is illustrated by the judgment in *Costain Limited v Charles Haswell & Partners Limited* [2009] EWHC 3140 (TCC). Costain had engaged Haswell to advise on the design of suitable foundations for a project Costain was undertaking. That project included 10 buildings. Costain alleged that the foundations design of two of the buildings was defective and that as a result it incurred considerable costs and delay to the works, for which it sought compensation by way of damages from Haswell. Among those damages were costs of £577,018 for 12 weeks and 4 days of prolongation. A financial evaluation issue between the parties was whether Costain was entitled to prolongation costs for the whole project or just those related to the foundations and any other activities delayed by them.

Deputy Judge Richard Fernyhough QC drew a distinction between extensions of time and damages for delay and explained the need to establish the effect of delays on all activities subject of the financial evaluation of the claim as follows:

> In order to understand and resolve this submission it is necessary to draw a distinction between a claim for damages for delay and a claim for an extension of time of the completion date on account of delay. When an extension of time of the project completion date is claimed, the contractor needs to establish that a delay to an activity on the critical path has occurred of a certain number of days or weeks and that that delay has in fact pushed out the completion date at the end of the project by a given number of days or weeks, after taking account of any mitigation or acceleration measures. If the contractor establishes those facts, he is entitled to an extension of time for completion of the whole project including, of course, all those activities which were not in fact delayed by the delaying events at all, i.e. they were not on the critical path.
>
> But a claim for damages on account of delays to construction work is rather different. There, in order to recover substantial damages, the contractor needs to show what losses he has incurred as a result of the prolongation of the activity in question. Those losses will include the increased and additional costs of carrying out the delayed activity itself as well as the additional costs caused to other site activities as a result of the delaying event. But the contractor will not recover the general site overheads of carrying out all the activities on site as a matter of course unless he can establish that the delaying event to one activity in fact impacted on all the other site activities. Simply because the delaying event itself is on the critical path does not mean that in point of fact it impacted on any other site activity save for those immediately following and dependent upon the activities in question.

He continued:

> Costain has not called any evidence to show the relationship on site between the activity involving the [2 affected buildings] and the other activities going on at the same time or thereafter. It is known that there were 10 structures to be built on the Lostock site of which the [affected buildings] were 2. There is no reason to suppose that, as a matter of course, progress on the other 8 structures would be affected by delays to the [2 affected buildings]. On the face of it, it is hard to see why that should be the case, since there would seem to be no reason why the other structures could not be constructed independently of the [2 affected buildings] at least for part of their construction. If therefore, as seems likely, the other activities on the site were continuing regardless of the delays to the [2 affected buildings], then there is no basis upon which it can be argued that Costain can recover the whole of its costs of maintaining the Lostock site simply as a result of delays to one part of that site.

This case is a graphic illustration of the need to establish in a claim for prolongation costs which of the contractor's time-related costs were affected by a compensatory delay. Where the works are divided into independently constructed parts, whether they are contractual sections or not, separate delay analyses may be required to show the effect on each and hence the prolongation costs on each. Furthermore, where time-related costs are partly project-wide (such as for central management) and partly localised (such as for the supervision of one of a number of independent structures), the fact of an overall critical delay to the overall project does not necessarily prove delay (and hence costs) in relation to localised time-related costs. As explained elsewhere, they will each require their own separate analyses to prove delay and financial effects on them.

It has been explained elsewhere how under the standard forms considered in this book, a contractor does not have to have an extension of time as a precondition to its entitlement to a financial claim for prolongation costs. However, it should be expected that where an extension of time has been refused, the contractor is unlikely to be successful in a claim referred to the same certifier in relation to the financial effects of the same delays. In *H Fairweather & Co Ltd v London Borough of Wandsworth* (1987) 39 BLR 106, His Honour Judge Fox Andrews considered the position under the JCT Local Authorities Edition with Quantities 1963 Edition. In that form, clause 23(f) provided for extensions of time in the event of late delivery by the architect of instructions, drawings details or levels and clause 24(1)(a) provided for the payment of associated loss and/or expense. He quoted from the judgment on the same contractual terms in *London Borough of Merton v Stanley Hugh Leach Ltd* (1985) 32 BLR 51 and considered that:

> Neither this part of the judgment nor the terms of the contract itself points to an extension of time under condition 23(f) being a condition precedent to recovery of direct loss and expense under condition 24(1)(a). However, the practical effect ordinarily will be that if the architect has refused an extension under the former the contractor is unlikely to be successful with the architect on an application under condition 24(l)(a).

As he says, this is likely to be the practical effect, but each case will turn on its particular facts.

6.2.3 Time-Related Costs

As noted elsewhere, the core of any claim for prolongation costs will be the contractor's time-related site costs. These are typically in the nature of preliminaries and general costs items that are usually included in that section of bills of quantities. These are also sometimes referred to as on-site overheads, or field overheads, although in this chapter they are referred to as preliminaries and general costs. There are a number of detail issues to consider in evaluating these, but the starting points are what period of costs is relevant and what sort of costs are usually involved? As explained elsewhere in this chapter, it is possible that time-related site costs can also involve idle labour and activity-related plant and equipment. The same starting points will apply to these.

6.2.3.1 When Did the Delay Occur?

The time at which the prolongation costs were incurred needs to be identified as accurately as possible. Generally, this will be by reference to the levels of time-related resources when the delay occurred, and not the period at the end of a programme or activity and over which completion is extended (although exceptions to this general rule are considered elsewhere and the general statement assumes that there will also be a claim for cost escalation in relation to costs that are pushed back and incurred later, including in a period of overrun). This was a common misconception in the past, but claimants and those receiving prolongation claims have become better aware in this regard over recent years. The key to this is in the nature of what is to be included in a delay claim. As noted elsewhere, this is the cost, expense, loss or damage resulting from the delay event and that must be what was incurred when the delay event had its effect. To put this another way, if in a claim for damages the aim of the quantification for delays occurring in March that led to an overrun of completion in November is to put the contractor into the position it would have been had the March delay not occurred, then it must generally be of the additional expenditure that it incurred in March and not November.

To identify when prolongation costs were incurred, the delay analyses explained elsewhere will therefore need to identify not only how much delay an event caused but also when it happened. For instance, the completion of a project might be extended from 28 February to 7 March as a result of a one-week delay whilst awaiting design drawings from the engineer for first-fix services installations carried out in the previous August. The costs incurred by the contractor will generally have been incurred in the week in August when it was awaiting the design drawings. It will not be the costs incurred in the week between 28 February and 7 March by which completion was delayed.

This can be a significant issue in relation to the levels of costs assessed as arising from delay because of the manner in which contractors generally incur their costs in the form of an S-curve between start and completion of a project. On that basis, in this example, the contractor's time-related costs will have been significantly greater when it was carrying out first-fix services installations than those in the last week on site, when it was removing the last of its resources. The point is graphically illustrated in the charts in Section 6.10, addressing overlaps between claims. It is perhaps as a result of this drop-off of costs at the end of projects that it tends to be engineers under FIDIC contracts that still sometimes wrongly assert that prolongation costs should be evaluated in a period of overrun at the end of a programme or activity period, rather than earlier when the delay events occurred.

The approach of evaluating prolongation costs by reference to those incurred at the time that the delay event occurred is supported by the SCL Protocol at its Core Principle 22 as follows:

> Once it is established that compensation for prolongation is due, the evaluation of the sum due is made by reference to the period when the effect of the Employer Risk Event was felt, not by reference to the extended period at the end of the contract.

It is assumed that in this statement the SCL's drafters expected that contractors will also make a claim for increases in costs because delay pushes costs forwards in time to when they are likely to be increased by inflation. It is sometimes said that advice such as this from the SCL and in this book is wrong and that it is the costs incurred in the period of overrun that should be claimed because they capture any inflationary increases caused by the delay. Thus, for example, it might be said that if a project with one site hut runs over by a month, then the additional cost to the contractor of that overrun is the cost of that hut at the hire rate incurred in that overrun period. Whilst this may be a suitable approach in a very simple prolongation claim on a small project, it misses the following points:

1. That the levels (rather than monthly hire rates) of preliminaries and general costs in the overrun period will almost inevitably be lower than when the delays occurred, as explained elsewhere in this section with reference to the typical S-curve shape of a contractor's costs. Thus, in this example, when the causative delays occurred there may have been three huts on the site. In this example, only valuing the one in the overrun period will not compensate the contractor for its full additional resources.
2. That the calculation of the contractor's costs of delay should include a separate claim for escalation caused by that delay, although duplication between these two claims (prolongation and escalation) must be avoided.
3. That the separate claim for escalation should be for all resources and periods over which their costs are escalated, and not just their increased costs in the overrun period. Thus, in this example, if the hire rate of the site huts was at an escalated rate in the overrun period, it is likely to also have been at an escalated rate in later periods towards the original contract period. The fact of delay in the period before completion will have pushed the dates at which the costs of such preliminaries and general items were incurred forwards in time. In this regard, see the charts in Section 6.10 for an illustration.

In conclusion, trying to capture the escalated costs arising from prolongation by valuing preliminaries and general items on the basis of their costs in the overrun period will under-claim against actual additional resource levels and miss cost escalation before the overrun period. Such as approach would be no substitute for a proper analysis of both prolongation and escalation costs, adjusted to avoid duplication.

Some qualification needs to be made to the aim of identifying as accurately as possible when delays and related prolongation costs were incurred for the practicalities involved in delay and programme analysis. Many analyses are based on periods or 'windows', whether between two dates or two particular events. This may also be a function of the available project records, which show the stage of progress at, for example, each programme update or monthly report. This may in practice mean that all delays cannot be allocated to particular days, but can only be assessed as occurring within a particular period. If all works to the foundations of a building under the FIDIC Red Book are suspended under clause 8 for a clear period of five days, then the related claims for extension of time and the contractor's 'Cost' under clause 8.9 should be based on the exact five days over which that suspension lasted, because the days can be readily identified from the records. However, if those foundation works were delayed for five days by 'unforeseeable physical conditions' under clause 4.12, which hampers progress, then it may be that the related claims for extension of time and the contractor's 'Cost' under clause 4.12(a) and (b) can be evaluated only on the basis of the average costs of time-related items over the period of excavation. This might mean, for example, that if the foundations excavation activities took 20 days, one-quarter (5 days/20 days) of the time-related costs in that period would be taken as being the prolongation costs.

The generalisation that it is the costs incurred at the time the delay occurred that should be included in a prolongation claim also needs some qualification. Especially where the claim is for loss of recoveries of costs, rather than expenditure incurred, it may be that the overrun period is the correct evaluation period. This is a particularly pertinent consideration in relation to claims for loss of contribution to head office overhead costs and profit, as is explained later in Section 6.8.

6.2.4 What Costs Were Incurred in the Delay Period?

Ideally, most of the heads of a contractor's prolongation costs claim can be extracted by an analysis of its cost reports or printouts and relevant cost codes therein for preliminaries and general items and then supported by primary records such as invoices, etc. However, there may be several such codes and there may be further subcodes therein. All told, there might be over a hundred cost codes for preliminaries and general costs items in a contractor's accounting records. It is likely that such cost ledgers will identify when costs were booked to a project, but there are several typical problems in assuming from such cost entries that the amount recorded were truly incurred in that period and should therefore be included in a claim for it:

1. Some costs are booked monthly and are also charged monthly so that the costs booked can be said to be actual costs for that period. For example, this is usually the case for hired plant and equipment for which the supplier submits monthly accounts. However, other charges may be for longer periods. For example, temporary utilities such as water and electricity supply might be billed quarterly such that a cost appears in a month, but is related to the two preceding months as well. This may be apparent on the face of the item's description in a cost report, but often is only found when the underlying invoice is considered. In this case, for the purposes of a prolongation claim, such costs should be spread across the whole period to which they relate.

2. On the other hand, the issue in 1 may also mean that some costs may not appear in a period to which they are actually related, because they were not yet invoiced and therefore only appear in a following period's costs. This means it will be necessary to look at the following months' records to find such costs and spread back those across the relevant preceding period to which they actually related.

3. Some costs only get booked to a project's accounts when they are accepted by the contractor as actual project costs. A common example of this is scaffolding, where accounts can be particularly complicated in terms of attempting to extract the time-related costs and identify the periods to which they relate. It may be that a scaffolding subcontractor charges for extended hire, and on a large project charges for items such as large birdcages can be significant, but the contractor does not agree those extended charges and does not book them when invoiced but until some months later, when it agrees to them in whole or part after a period of debate. This might see a large one-off cost being booked in a month that actually refers to a much longer preceding period to which it will have to be reallocated for the purposes of a prolongation claim.

4. On the other hand, the issue in 3 may also mean that no scaffolding costs (to follow the same example) appear in a month's cost report at all and need to be found in the records of a later period and spread back for the purposes of a prolongation claim.

In such cases, the key is to look at the supplier and subcontractor invoices and accounts that evidence each cost entry to identify what the costs were and in what periods they truly were incurred (rather than when they were booked to the project's costs). Those primary records will also be required to check the accuracy of the project costing system.

On a large project, this can become a very complicated and large exercise, to check the booked costs and to allocate them to the correct period. The onus here must be on the contractor as claimant, but it may be that those checking claims for such costs should use a system of sampling and spot checking. This can benefit from focussing on the types of costs that commonly feature such problems (such as utilities) and also any large or unusual cost entries.

Such checking of the records will also assist in establishing whether costs booked to a contractor's preliminaries and general costs are truly time-related and thus should feature in a prolongation claim because they would be increased as time on the works, a section or an activity increased. Some common issues in this regard include some cost codes gathering costs that are not time-related but one-off purchases. Consider, for example, a cost code for 'temporary services – traffic management'. This might include the costs of some people and other costs of maintaining and moving signs, lights, diversions and such-like as the works progress, but may also include the costs of purchasing traffic signs and lights in the first place. These are unlikely to be time-related, unless for some reason long delays have led to a greater need or their replacement. Otherwise they should be excluded from a prolongation claim.

6.2.5 Typical Heads of Prolongation Cost

The cost codes into which contractors divide their preliminaries and general costs will vary between companies and, as noted elsewhere, their number can run into three figures. A typical list might include the following, although there may be other items:

Salaries	Drawings and prints
Offices	Videos/cameras
Establishment	Photographs
Fence/gates/hoarding	PR/entertainment
Laboratory	Safety
Ablutions	Security
Canteen	First aid
Stores	Rubbish removal and cleaning
Scaffolding – external hire	Insurances
Temporary services – electricity	Bond commission and charges
Temporary services – water	Professional fees
Temporary services – telephones	Office sundries
Temporary services – radios	Internal plant hire
Temporary service – road signage	External plant hire
Survey equipment	Plant purchases
Furniture and equipment	Fuel and oil
Computers and electronics equipment	Transport
Stationery	Power tools and repairs
Tea/coffee/cleaning materials	

The extent to which such preliminaries and general costs items are truly time-related will vary with their facts and this will require consideration of the details of the costs included in each such code. The typically larger headings of these costs are considered in detail elsewhere in this chapter. However, in addition, some particular issues regarding smaller items are as follows:

- Laboratory costs are a good example of a code that will include both time-related and work-related costs. The ongoing maintenance, staffing and servicing of a laboratory will be time-related. However, costs may also include one-off costs such as specialist external testing of particular materials.
- Labour. The sample set of cost codes set out above does not include a separate cost code for preliminaries and general labour, because those costs will usually be included within other cost codes. For example, cleaners will usually be included in a cost code such as 'Rubbish Removal and Cleaning'. However, such staff as a 'multi-service gang' or attendant labour might have its own cost centre. Issues as to whether such labour are time-related are discussed elsewhere in this chapter.
- Rubbish removal is the subject of perennial debate as to the extent to which it is wholly work-related or also partly time-related. As with other preliminaries and general costs, this will depend on the facts in each case and needs a detailed analysis of what had been included. It may be that this cost code only includes the hire and

removal of skips, but this is unlikely and there may also be labour in this cost code, kept on site over time to clean up generally. Furthermore, it is arguable that skips get filled with the passage of time, as well as the amount of work done, and it may be that they are emptied at regular time intervals, rather than when full. This cost code can also be used for the booking of one-off costs such as a deep-clean at the end of a project before handover, which may or may not be affected by delay.

- Bond commission and charges are the subject of particular consideration in Chapter 8.
- Professional fees can include items such as the costs of audit, design costs, legal advice and other consultants and these need detailed investigation of what is included. These two last items are considered in more detail elsewhere in this book in terms of the basis upon which they might be claimable. If a project is audited on a regular periodical basis, then it may be that project prolongation will increase their costs. One common issue in relation to consultants is where they include staff seconded from external consultancies, which is considered further elsewhere in this chapter.

6.2.6   Typical Heads of Prolongation Cost – Staff

Project management, administrative and supervisory staff is usually the largest head of a typical prolongation claim, other than on plant-intensive heavy engineering projects.

Whilst it is generally assumed that staff on a project are time-related, there may be some who are related just to an isolated event or works and who are not prolonged in their involvement by delays. Examples might be specialist engineers visiting to do tests or inspections on particular materials or work. One approach to identifying such people is to note those who are on the site for short periods. A typical staff matrix scheduling each member of staff against time will expose this. However, account also needs to be made for temporary staff who visit the site to fill in for permanent staff who are on leave, sick or otherwise absent. For example, a setting-out engineer might be engaged from an agency, or transferred from another site, whilst the resident setting-out engineer is on holiday. A feature of projects that are subject to widespread delay and disruption is often that they suffer high staff turnover and any resulting short attendances of some staff should not be taken to mean that they were never time-related. This issue needs proper analysis of the records and facts in each case.

With reference to a typical international construction project, staff costs may include such as:

- Basic salary paid in accordance with each employee's contract of employment and before the addition of any allowances such as those set out as follows.
- Overtime payments. This is more often related to labour, but on accelerated projects where significant extended hours and/or weekends are needed, works staff may be incentivised in this manner.
- Pension contributions. It may be that part of the employee's employment package is that the employer contributes to his or her pension provision. Whilst this is uncommon in some countries, it is a legal requirement in the UK since the Pensions Act 2008.
- Payroll taxes and on-costs that are collected by the local government (such as employer's National Insurance contributions in the UK).

- Any other contributions that employers are required to make in accordance with the applicable law.
- Accrual of holiday pay.
- Accrual of gratuity or end of service payments. These are legal requirements in some countries. They are often in amounts that accrue during the course of employment based on the period of employment and in proportion to basic salary, such that they are particularly clearly time-related.
- Bonuses. These can be paid to all site staff or personal to an employee, and can be related to the performance of the individual, or of that project, or of the company as a whole. It is sometimes suggested that a company-wide bonus, that is one not related to the project subject to the claim, but the performance of the company as a whole is not allowable as a part of the prolongation costs of that project. It is suggested that this is wrong, on the basis that such a company-wide bonus is part of the cost of employing each individual who receives it. Furthermore, the project's performance will presumably be a factor in the company's performance. A problem with all bonuses is how they will appear in one lump in the payroll for perhaps the final month of the calendar or financial year. This means that they need to be spread back across the whole period to which they relate. For example, if annual bonuses are paid in the December payroll, then they will need to be spread across that whole year, presumably as one-twelfth of the total spread into each month.
- Visa costs for imported staff. This may also include work permits and local identification papers, as required by local legislation. They will be incurred as one-off amounts at the start of an employee's employment in a country and may require renewal costs periodically, perhaps annually or biannually. There may also be charges for cancellation and an exit permit when that employee leaves the country. Such costs may therefore need to be spread across an employee's employment period. The question will be whether these costs are increased by prolongation of an employee's employment on a project, which will depend on the facts in each case.
- Medical expenses. Visa renewals can also require a medical test.
- Relocation allowances. This is likely to be a one-off payment to imported staff to reimburse their relocation costs. It will then be spread across that employee's payroll cost calculation.
- School fees. These will particularly apply to senior project staff, whether as an allowance or the reimbursement of actual costs.
- Flights for imported staff to and from their home country. These may include both one-off costs at the start and end of an employment, but may also include an agreed number of return flights annually. The number of return visits will vary depending on each employee's seniority and their home location, but might be perhaps three times per year. This cost will therefore involve similar considerations to visa costs.
- Accommodation cost. Where the provision of accommodation, such as an apartment or shared house is not booked to another cost centre, as it may be part of the staff employment cost centre. Alternatively, this may be an allowance paid to an employee, sometimes as a percentage of salary, or it may be the actual costs reimbursed.
- Food allowance. This may be similar to accommodation costs, where the employee is paid to supply his or her own food rather than use a site canteen.
- Transport allowance. Where employees do not use the company's own transport, this may be similar to the accommodation cost, in relation to a supplied vehicle or

reimbursement for the employee's use of his or her own vehicle. It can be in the form of an allowance for provision of the employee's own vehicle or for use of public transport. It can also include items such as fuel, parking charges and toll payments.

- Telephone cost, where this is either provided or reimbursed by the employer, such as a mobile phone, which is not included in a separate cost code for 'Temporary Services – Telephones'.
- Life insurance and general medical costs. Particularly for staff imported into a country, it is common for the contractor to provide insurance coverage for medical check-ups, treatment and hospitalisation.
- Expenses. Staff payroll costs may also include reimbursement of out-of-pocket expenses incurred by a member of staff. Such one-offs are unlikely to be time-related and should usually be deducted from calculations for a prolongation claim.
- Deductions from salary. These might appear occasionally as on-off items. An example would be where an employee is given a loan to rent accommodation, which is paid in part as an upfront payment and which the employer then recovers gradually through monthly salary payments, usually interest free.

Staff costs may also include seconded staff from external agencies or consultancies or the contractor's other projects. Where these are carrying out roles that would otherwise require additional payroll staff, there seems no reason why they should not be equally allowable. Issues are particularly often raised in relation to the employment of external consultancies and often this relates as much to their higher costs as what activities they are carrying out. In one unusual example of this, in the 1990s one national contractor established its own internal claims and commercial management consultancy and cross-charged each project for their employment. Considerations here are similar to those relating to an internal plant division and are covered in detail elsewhere in this chapter. In this case, that division was not a separate company and its rates were such that it was joked that this division was the company's most profitable business, much to the annoyance of project managers who were required by internal procedures to use them. One aim was apparently to create an in-house business that could sell commercial and claims expertise to other contractors. Understandably, that aim was not realised.

6.2.7   Typical Heads of Prolongation Cost – Labour

The extent to which labour will feature as part of a contractor's costs resulting from prolongation can vary greatly. Labour can be generally divided into two types. Firstly, tradespeople, labourers and helpers who are generally work-related, that is their involvement on the project will usually vary only with the amount of permanent work to be done, such that their costs are not increased usually by prolongation. Secondly, personnel who are generally time-related, such that their costs are increased by prolongation. The usual reaction to work-related labour within a prolongation claim is to exclude it, but this is not always correct depending on the factual circumstances of the delays and their effects. Tradespeople, labourers and helpers carry out permanent works, but the time that they take to carry out those works can be increased by prolongation in some circumstances.

Some categories of labour are purely time-related and a key to this may be that they will be both priced in the contractor's tender and also booked to its cost codes, against preliminaries and general costs items. This includes workers such as cleaners, security and maintenance labour. In the case of these examples, if a project is prolonged, then: temporary site facilities such as toilets and offices will require cleaning for longer; the site will require guarding for longer; and offices, temporary services and plant will require maintaining for longer.

However, it is also possible that other labour can be prolonged by delays, including those operatives who would usually be expected to be work-related. The obvious example of this is where the works or a part therefore are suspended. This is likely to mean that a claim under such as the FIDIC Red Book clause 8.8 or the Infrastructure Conditions clause 40 includes such labour. For example, on a project for construction of a new highway, the engineer under a FIDIC-based government contract instructed that works to the wearing course were suspended mid-progress because of an issue over the appropriateness of the specified materials. The works were on the project critical path and therefore the contractor's related prolongation claim included the usual heads of costs such as time-related labour. However, the suspension also meant that a number of the contractor's machinery drivers and road laying labour stood idle while the issue and suspension were resolved. Thus, a number of what might normally be work-related people were also included in the evaluation of that prolongation claim.

One of the problems with claims for prolongation of such work-related people is that contactors often fail to keep records to establish that they were the subject of additional time on the project, and hence additional costs. Delays to completion of the works can usually be assumed to have added to the costs of employing time-related people such as the examples of cleaners, security and maintenance labour, without any particular evidence of the effects on them. However, where work-related operatives are also affected, by such as a suspension, contractors should maintain records of those particular effects. A daily record of time spent idle, submitted to the engineer, is ideal. However, a complicating factor can be the extent to which such operatives can be re-allocated temporarily elsewhere to carry out productive work. Such issues will hinge on their facts in each case, but the onus must be on the contractor to keep records. Of course, delays are seldom as 'clean' as a suspension of works that leaves particular operatives completely idle.

The costs components of employing labour in a prolongation claim may include those set out as the costs of employing staff, although without some of the costs that go with seniority, such as children's school fees and the provision of mobile phones. On the other hand, there are some costs that are more likely to be relevant in relation to labour, such as overtime payments.

6.2.8 Typical Heads of Prolongation Cost – Temporary Buildings

Temporary buildings can include such items as offices, stores, workshops, canteen, mess rooms, medical centre, security huts and laboratories. They could be hired, owned by the contractor or specifically constructed for the project. Where such facilities are hired, their time-related costs will be easily established among hire company invoices for delivery, hire and removal. Where they are owned by the contractor, their costs will be similar to those of owned plant, which is considered further elsewhere in this chapter and also in Chapter 5. They will include the building's depreciation, running costs and maintenance costs.

Where they are constructed for a project, the principal costs will be in the activities of supply, erect, dismantle and remove. Running costs such as utilities, cleaning and maintenance will usually be in other cost codes. There may or may not be a residual value at the end of the project, and given the cost of careful dismantling for re-use, they are often just demolished and removed. This may mean that delay leads to no additional depreciation. However, it has been known for delays to be of such a duration that temporary buildings that were intended to be dismantled and re-used elsewhere at completion were in such an aged state that they were only fit for demolition.

A particularly large item on international projects in remote locations, or in countries to which substantial numbers of operatives have to be imported, is a labour camp constructed and maintained by the contractor for the project. These can be large and can include such items as sleeping quarters, toilets, shower blocks, mess rooms, canteens, internet cafes, cinemas, health centres, places for religious gathering, games and sports facilities, etc. The costs of supply, erection and dismantling can be considerable, but usually not time-related. In practice, projects that merit such a camp are usually so lengthy and the occupant's care and maintenance of them such that they are simply demolished and removed on completion so that there is no additional capital cost (or depreciation) arising from delays. However, running and maintenance costs can also be considerable, and these will usually be time-related. They will particularly include such items as management, catering, maintenance staff and utility supplies.

Such running and maintenance of all temporary buildings will usually be time-related. They may be included in cost centres for the buildings or for staff, temporary services and site cleaning.

6.2.9   Typical Heads of Prolongation Cost – Temporary Services

An obvious point in relation to utilities such as electricity and water supplies is that they will involve both fixed elements, such as installation, connection, and removal, as well as on-going costs of supply and consumption, only the latter being time-related. In a prolongation claim the invoices will require analysis to exclude any costs of such as connection and removal.

However, installation, connection, and removal costs may also repeat as the project unfolds. There may also be also be diversion costs, for example because a route clashes with the next area of permanent works and a new route is now available. Rarely, it does occur that such activities are increased by prolongation, because of the particular facts of the project.

6.2.10   Typical Heads of Prolongation Cost – Temporary Works

Typical temporary works items might include: scaffolding; falsework, shoring; formwork, debris netting; dewatering; temporary piling; cofferdams; trench support; hoardings; fencing; crane supports; etc. Aspects of their costs may be work-related (such as connection and removal) or time-related (such as running, maintenance and hire or depreciation), depending on the particular facts.

As a particular example, on some building projects scaffolding costs can be considerable. They will include elements of both time-related and work-related costs. Erect and dismantle will generally be fixed costs, as will any adaptations or movement to suit the permanent work being serviced by the scaffolding. On the other hand, hire is obviously something that can form part of a prolongation claim. This can also include elements of maintenance and safety monitoring. However, these approaches assume that the scaffolding is hired on a per unit basis (for example, per cubic metre of bird-cages), with discrete charges for erect, dismantle, adapt or move. On some large building projects, particularly where durations and/or scope of change are uncertain, it may be that scaffolding is all hired on the basis of rates per day for quantities of materials and people to erect, adapt, maintain, dismantle and remove as and when required. Thus, the contractor hires-in materials and scaffolders on a time basis and will incur additional costs if they are prolonged.

In the event of delays, some of the typical temporary works items are also prone to increased requirements of repair and adaptation. Hoardings and security fencing are particular examples of this.

Temporary works items can also be owned by the contractor, such that the costs are those of depreciation, running costs, and maintenance, similar to temporary buildings and as detailed in relation to plant and equipment elsewhere in this chapter.

6.2.11 Typical Heads of Prolongation Cost – Plant and Equipment

On major civils engineering projects, the costs of the contractors' plant and equipment can be the largest head of a prolongation claim. Those costs may be from one of two sources: firstly, plant hired from external plant hire companies and suppliers and, secondly, plant owned by the contractor.

Where plant and equipment are hired by the contractor, the costs should be capable of being readily established by reference to the supplier's or hire company's invoices. Those rates might be per hour, day, week or month, depending on the type of item and the particular factual circumstances of its use and hire. However, complications can arise where plant is internally hired from another company within the contractor's group of companies. A common complaint where prolongation claims include plant and equipment that is internally hired is that those inter-company rates are not reasonable and that those parties are not at 'arm's length' in commercial terms. The complaint can be even more valid where the 'hire' is only from an internal division of the contractor's company and particularly where it is in the form of an internal cross-charge from another of the same contractor's sites. It was a fact some years ago that the plant division of one international contractor was its most profitable division. This was its prerogative in terms of internal accounting, but if it gives rise to unreasonable plant and equipment 'hire' rates in a prolongation claim, then that should be addressed.

Where plant is owned by the contractor, and not subject of internal hire rates, the contractor's costs will comprise depreciation, running costs and maintenance. Generally, this should be in line with the discussion in Chapter 5, with, under English law, particular reference to the judgment in *Alfred McAlpine Homes North Ltd v Property & Land Contractors Ltd* (1995) 76 BLR 59.

The largest component of the cost of owned plant and equipment is usually its depreciation. The calculation of this cost requires establishing for each item: its capital cost; its useful life period for the contractor; and its residual or scrap value at the end of that life. Financing the capital outlay may also be a component. These issues were considered in detail in Chapter 5, together with an example of a calculation of a plant depreciation cost.

One occasional quirky claim by contractors in relation to the costs of plant and equipment that it owns is that, had the plant's use not been prolonged by delays, it would have externally hired the plant to other companies at its outward hire rates.

Whilst this is conceivable, it is rarely successfully established as a claim based on loss of external earnings on such items.

In relation to owned, or 'internally hired' items, one or two English authorities are also worthy of consideration, as follows.

The principle that in a claim for damages (and therefore 'loss and/or expense' or 'expense' or 'costs' under standard construction contract terminology), involving extended requirements for owned plant, the evaluation of the damages comprises depreciation, interest and maintenance was established in *Sunley Ltd v Cunard White Star Ltd* [1940] 1 KB 740. The principle was further applied more recently in *Whittall Builders Co Ltd v Chester-le-Street District Council* (1985) 12 Const LJ 256.

Where the depreciation calculation is related to the company's accounting policies, if the plant is of such an age that it has been fully depreciated and no longer sits as a cost on the contractor's books, is it right that the plant can be retained on site for a longer period due to compensatory delays at no cost to the employer where the delays are its fault? In *Laserbore Ltd v* Morrison *Biggs Wall Ltd* (1993) CILL 896, Judge Bowsher QC had to consider a claim for 'fair and reasonable payment' for plant whose costs had been fully depreciated. The defendants argued for a costs-based approach by reference to the claimant's management accounts that would have minimalised the claim. The Judge held that:

> If an expensive item of equipment has been depreciated to nothing in the company's accounts but by careful maintenance the company continues to use it, must the equipment be provided free of charge apart from running expenses (fuel and labour)? On the defendants' argument, the answer to those questions is, "Yes". I cannot accept that begins to be right.

The Judge applied the claimant's rates, which were based upon actual commercial hire rates, but excluded any percentage uplift for insurances, head office overheads, or profit. It is repeated that this was in a claim for 'fair and reasonable payment', but the authors have seen such an approach applied to a claim for 'additional costs' associated with delays. The question is whether such an approach may have its place as part of a value-based approach to quantifying claims but not to claims for 'loss and/or expense' or 'expense' or 'costs' under standard construction contract terminology.

Whether plant and equipment is hired or owned by the contractor, its cost may also include the following components:

- Maintenance and repairs, including new tyres, etc., and mechanics and fitters' time, etc.
- Cost of fuel, lubricants and grease, etc., including distribution of fuel to working plant on site.
- Contract works third party and employer's liability and motor/plant/equipment insurances.
- Road tax and statutory charges, etc., where appropriate.

Whilst these costs are likely to be incurred somewhere, consideration will have to be made as to whether they are part of a supplier's rates for hired equipment, or included in other heads of the contractor's preliminaries and general costs.

The need to establish that plant and equipment is actually time-related has been mentioned elsewhere, citing the common example of cranes.

6.2.12 Example of Additional Activity Costs

The crux of the quantum for time-related costs is generally the calculation of additional costs incurred as a result of the delay to activities affected by the events that give rise to entitlement. As mentioned elsewhere, it is important that the costs are related to the relevant delay events, not least to ensure that there is entitlement to profit on costs where the FIDIC Red Book or the Infrastructure Conditions apply. However, there are two rather more significant complicating factors that need to be carefully considered:

1. It may be that there are other activities being undertaken on site concurrently with the activities that are the subject of the claim for prolongation costs. These activities may themselves have been delayed by matters for which the employer does not have responsibility under the terms of the contract, i.e. there could be concurrent delay as discussed in Chapter 3.
2. There may, in some instances, be costs incurred as a result of multiple events, the effects of which are not readily distinguishable, i.e. there could be a 'global' element of cost incurred as a result of these events.

The approach to claims that have a global element is often much misunderstood and the principles of this aspect are therefore discussed in greater detail elsewhere in this chapter. The practical approach to the quantification of such compensable employer delay costs and contractor concurrent delay costs may, however, be demonstrated by taking a simple example, as illustrated by the sample programme chart extract in Figure 6.1.

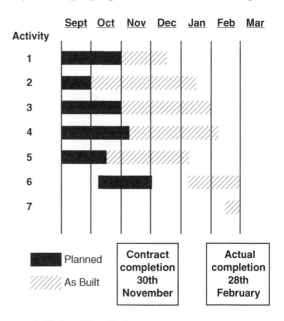

Figure 6.1 Sample programme extract.

Activity 1 Topsoil placement
Activity 2 Final installation – mechanical equipment
Activity 3 Connections to equipment and commissioning
Activity 4 External fencing
Activity 5 Construct parking areas
Activity 6 Test running plant and handover
Activity 7 Install automatic gate barriers

In Figure 6.1, to illustrate the various matters that may arise in evaluating the costs of prolongation, it is assumed the seven activities have been delayed beyond the contract completion date of 30 November for the reasons described elsewhere in this chapter, resulting in an extended completion date of 28 February. The assumed entitlement under the contract for each activity is also explained.

Activity 1: Top soil placement and grading. This has been delayed by earlier exceptionally inclement weather. This is a ground for an extension to the completion date under the contract, but without compensation of any additional costs incurred.

Activity 2: Final installation of mechanical equipment. This has been delayed by the late receipt of performance data required from the employer. This is a ground for an extension to the completion date with compensation.

Activity 3: Electrical connections and mechanical commissioning. This has been delayed as a result of the delay to the preceding Activity 2 (final installation of mechanical equipment).

Activity 4: External fencing. This has been delayed by problems with the contractor's own subcontractor for which no extension of time or compensation is provided under the contract.

Activity 5: Construct parking areas. This has been extended by an instructed variation to the amount of parking space required, for which the contractor is entitled to an extension of time with compensation of additional cost.

Activity 6: Test run plant and handover. This has been delayed because of the delays in the preceding Activities 1 to 5.

Activity 7: Install automatic gate barriers. This is a late instruction of additional works issued and executed in February, for which the contractor is entitled to an extension of time with compensation of additional costs.

From this it will appear that the contractor can expect an extension to the contract completion date to the actual completion date of 28 February, providing it can be shown that the extended periods of activities 2, 3 and 5 are the reasonable result of the matters for which the employer is liable without any compounding factors for which the contractor is responsible. The extension of time in this instance would rely on these activities, although there are concurrent delays to other activities.

The delay to activity 1 is due to delays caused by exceptionally inclement weather, which does not give any entitlement to recovery of associated additional costs. In setting out the quantum of a prolongation claim against the employer the contractor will therefore have to exclude any additional cost element relating to the plant, equipment, supervision, labour and ancillary costs engaged on this activity.

The same will apply to activity 4, as the delay in this case is one for which the contractor's subcontractor is liable. It is assumed that this creates no liability on the employer but whether the contractor is entitled to recover any additional costs it has incurred in respect of this activity from the subcontractor will depend upon the terms of the subcontract and the facts of the matter.

Activity 5 is the subject of a delay for which the contractor can recover associated costs, although it is concurrent with other employer liability delays and is therefore not essential to the establishment of the extension of time to the contract completion date.

Activity 7 is similar to activity 5. The employer is liable for the delay and the contractor can recover associated additional costs, but the activity is not essential to the establishment of the entitlement to extension of time to the contract completion date. Had this item been the only one for which the employer was responsible and all the other delays had been the contractor's responsibility, then it would have only established an entitlement to an extension of time to the contract completion date for the necessary period to execute the instruction for the new automatic barriers. For instance, if the instruction, ordering, delivery and installation had all taken, say, five weeks without any undue procrastination on the part of the contractor, then the completion date of 30 November would have been extended by five weeks, plus any period required to cater for programmed holidays over the Christmas and New Year period. The fact that the instruction had in fact been issued in, say, mid-February, would not entitle the contractor to calculate its additional time from that date under the 'dotting on' principle established in the judgment in *Balfour Beatty Building Ltd v Chestermount Properties Ltd* (1993) 62 BLR 12, as is further explained in Chapter 3.

In this example, the type of costs that may be recovered by the contractor in respect of the activity delays illustrated in Figure 6.1, in addition to the contractor's costs of the overall period of prolongation of the works, will include the following if they pass the testing, i.e. they would not have been incurred 'but for' the employer's cause of delay cited.

Activity 1: Topsoil placement and grading
- This was delayed by the cost-neutral event of weather, so is irrelevant to a prolongation claim.

Activity 2: Final installation of mechanical equipment
- Any increase in the supply cost of the equipment caused by the delay. For instance, if ordering and delivery of some equipment is delayed until after an annual revision of prices by the supplier, where the equipment would have been ordered and supplied before such a price rise without the delay in receipt of performance data from the employer, then the increase in the price will be recoverable.
- As can been seen in Figure 6.1, the installation was undertaken over an extended, as well as delayed, period. This is likely to have involved a reduction in productivity for the labour engaged on the activity. To establish the effect on such work-related resources, it will be necessary to show that no alternative work could have reasonably been used to fill any gaps but in principle the reduction in productivity caused by an extended period will be recoverable.
- If the delay to the activity necessitates undertaking work after wage rates have been increased, where they would have been undertaken prior to such increases without the delay, then the net cost of such increases will be recoverable.
- Any increased delivery or transport costs incurred because equipment is delivered later or in smaller shipments than would otherwise have been the case will also be recoverable.
- If it is necessary to retain specific supervision for this activity on-site for its extended period then the additional cost of that supervision will be recoverable. In this instance it may be that a site mechanical engineer, or engineers, was based on-site for the purpose of supervising the installation of the equipment. If so, their additional cost will be recoverable.

Activity 3: Electrical connections and mechanical commissioning

- The same sort of factors as were considered for activity 2 will need to be reviewed, but in addition there may be a site electrical engineer, or engineers, retained by the delay to this activity and, if so, that cost will also be recoverable.

Activity 4: External fencing

- This is an issue between the contractor and subcontractor and is therefore irrelevant to a prolongation claim against the employer.

Activity 5: Construct parking areas

- This delay is caused by an increase in the scope of the activity as a result of an instruction for additional parking areas. If it is assumed that the additional work is paid for under the contract at the contract rates then it is possible that all additional activity costs are recovered through the contract variation valuation provisions. However, the matters of supervision and engineering, for instance the involvement of setting out engineers, may need to be considered separately if these are not covered by the measured rates that are applied to value the additional works.
- As shown in Figure 6.1, this work was carried out over an extended, as well as delayed, period. As a result, consideration of the productivity of work-related resources used on it may be similar to that for Activity 2.

Activity 6: Test run plant and handover

- This has been delayed due to the late running preceding activities. As the activity has been 'shunted' to a later period, rather than extended to a longer period, it is unlikely that reductions in productivity levels will be relevant. However, the retention of the mechanical and electrical engineer(s) on site for further extended periods will probably generate further recoverable costs.

Activity 7: Install automatic gate barriers

- It is assumed that the cost of this additional work will be recovered under the contract variation valuation provisions of the contract, including allowances for time-related costs.
- However, it is possible that some site supervision for the labour engaged on this activity might have been retained on site after the completion of the preceding activities and may not be covered by the valuation under the variation provisions. If so, the cost of that supervision and similar items will be recoverable in addition.

It may, or may not, be that prolongation, with its resulting increases in the durations of time-related costs, is associated with disruption, with its resulting increases in the levels of all resources, whether time-related or work-related. Thus, in the example based on Figure 6.1, it might be that the late performance data from the employer for Activity 2, or the variations requiring the new Activity 7, were issued on a piecemeal basis. The evaluation of such resulting disruption is one of the perennially difficult issues to face anyone undertaking the quantification of the time consequences of change under construction and engineering contracts and is considered in some detail in the section on the costing of disruption elsewhere in this chapter.

6.2.13 Relevance of Tendered Preliminaries and General Cost Rates

As has been explained, claims for prolongation costs related to compensable delay on construction and engineering projects will almost always be evaluated on the basis of the contractor's actual costs, loss, expenses and/or damages. This is because the provisions of such standard forms of contract such as those of the JCT (and its current SBC/Q), FIDIC (and its current Red Book) and the Infrastructure Conditions (and its predecessor the ICE contracts) use such terms and the law of damages for breach of contract in most jurisdictions prescribe compensation in the form of the actual damages arising from a breach.

In relation to the contractor's tendered allowances generally, the SCL Protocol has this as its Core Principle 21:

> The tender allowances have limited relevance to the evaluation of the cost of prolongation and disruption caused by breach of contract or any other cause that requires the evaluation of additional costs.

It is sometimes asserted by those valuing and certifying prolongation claims for employers that reference to the contractor's tender allowances is relevant because the actual costs are significantly higher. Thus, it is asserted that if the tendered allowances were reasonable at that time, then they limit what is reasonable as the contractor's actual costs at the time of the compensable delay. There often seem to be one or both of two motives for this. Firstly, there is the belief that the contractor has 'got away with' underpricing its tender if it can recover in a prolongation claim actual costs that are not limited by those allowances. However, it is suggested that this is irrelevant to an assessment of the contractor's actual damages, 'cost', 'loss' or 'expense' under the contact. Secondly, there is the belief that because the actual costs are significantly higher than those allowed in the contractor's tender, then they must be unreasonable. Whilst any claimant of damages has a duty to mitigate them, and contracts such as the FIDIC Red Book define compensable 'Cost' as 'expenditure reasonably incurred' (see Definition 1.1.4.3), it is suggested that reasonableness is to be tested on the basis of what was provided, not what was estimated. What was estimated could simply have been wrong, which would be an irrelevant fact, unless it somehow affected the contractor's performance on the project. Such allegations are usually made with no reference to contemporaneous complaints that the contractor had excessive preliminaries and general resources on the project. Reference to tender allowances might be relevant to support and add detail to such complaints, where, for example, the contractor had excessive resources on site because it had its own problems of inefficiency that some resources were addressing or it had no other project to place them on. However, if these facts persisted during the works, one would expect there to be some complaint in this regard from the employer's team.

It has, however, been known for contractors in periods of patent compensable delay on a project, and with other projects completing and their resources awaiting further work, to 'park' equipment and staff on the delayed project as somewhere that they might get some compensation through its prolongation claims, rather than have them idly sitting in a plant yard or head office.

A further point that might be considered by practitioners who disingenuously consider that significantly lower tendered preliminaries and general costs must mean that the actual costs are unreasonable, such that the tendered allowances should be the limit of recovery, is what would their position be if the tendered allowances were significantly higher? Would they then consider this comparison relevant to quantification?

However, in practice there are two principal circumstances in which rates in the contract are sometimes applied to the contractor's claims for delay. These are: firstly, where the contractor and employer and/or the certifier agree to use the contractor's tendered rates notwithstanding what the contract says and, secondly, where the contract contains agreed resource rates to be applied to such delays.

It has been explained elsewhere how in the FIDIC Red Book those clauses that entitle the contractor to the costs of delay refer to its 'Cost' and how that term is defined in its clause 1.1.4.3 as: 'Cost means all expenditure reasonably incurred (or to be incurred) by the Contractor, whether on or off the Site, including overhead and similar charges but does not include profit'. However, and notwithstanding this clear contractual prescription, on occasions it does happen that a contractor's prolongation claim is agreed on the basis of its allowances in the contract price, and particularly by reference to a preliminaries and general bill of quantities. If this approach serves the parties and they are both content to apply it, then there can be nothing wrong with such an extra-contractual approach. It seems to be adopted where there is a reluctance to go to the trouble of establishing the contractor's actual costs and/or a belief that the two approaches would come to similar results in any event. Thus, it saves both parties time and costs and can be less contentious than the contractual approach of substantiation and audit. However, it is not always adopted for such reasons of proportionality and pragmatism. It has also been seen to be adopted where engineers under FIDIC contracts have decided that their fees do not allow for them to spend the time on a proper evaluation of the contractor's claim, and the employer was not aware of the contractual position and/or the short-cut being taken.

The so-called 'Brown's Clause' has been mentioned and is explained elsewhere as a proposal some decades ago that the building industry in the UK adopt a rate (or rates where there are contractual sections) for the contractor's damages for compensable delay, similar to the employer's rate of damages for culpable delay by the contractor. A more sophisticated approach, which addresses failure of the Brown's Clause to consider the varying levels of the contractor's resources at different states of the project, is a schedule of rates and prices to be applied to those actual resources. Those rates can be variously per hour, per day or per month depending on the type of resource. This approach is not common, but is adopted in some countries, particularly in the petrochemical sector and more commonly in subcontracts rather than main contracts. It seems simple enough, on the face of it, and in theory it should allow relatively easy evaluation of prolongation costs, without the need for establishing the contractor's actual costs by reference to project cost reports, invoices, payroll, etc. However, it does have problems in practice. For example:

- Whilst the rate for different resources are agreed, the approach still needs records of those resources, their types, grades, numbers, hours affected, etc. It is suggested that a contractual provision that allows the contractor to avoid having to prove its actual costs by applying agreed resource rates ought to also require it to serve detailed daily records to the employer's engineer, architect or contract administrator for agreement and signature for the record. However, this can create an onerous daily burden. If it is not done, because the records are not kept or are not presented and agreed for that purpose, then there will still be a need to audit the contractor's primary project costs records such as payroll, invoices, etc., to establish how many of each resource were affected on each day.

- Even where records of the actual resources are available (through submitted schedules or through an audit of project cost records), the question arises as to whether those resources were time-related. The classic example of this is certain types of labour. Where this unusual contractual approach is adopted, the contractor will set out in its contract schedule all grades of everything from equipment, to staff, to labour, so that where they are prolonged there is a rate for all eventualities. However, just because the schedule included a rate, for example for electricians, does that mean that electricians on site during a period of compensable delay are allowable as part of the costs of that prolongation, including those usually considered work-related because they are installing permanent works? Alternatively, are only time-related electricians such as those in a maintenance gang allowable?
- A plant item that would have similar issues to that related to electricians as a labour example would be cranes. These are items that can particularly be either time-related (for example, a site-wide tower crane) or work-related (for example, a crane brought to site to unload and place air handling units into their permanent positions).
- Often there are mismatches between the descriptions on the contract schedule and the record of actual resources that need to be resolved to match them together correctly. This can be both laborious and contentious.

In practice, whilst such schedules may work well in some circumstances, particularly by avoiding the need to establish and audit actual resource costs, they therefore often involve their own problems and are far from a panacea.

Finally, on the topic of relevance of the contractor's tendered preliminaries and general cost allowances, it must be remembered that they will be relevant to evaluate the amount of recoveries that the contractor has made against any of its claims for delay and disruption through the value of measured works and particularly variations valued based on those allowances. The topic of duplication of recoveries between such claims is discussed elsewhere in this chapter.

6.2.14   Increases in Costs

If work is carried out later because of delays, then it may seem inevitable that the costs of carrying it out will change with the effects of inflation. This is obvious in relation to such cases as wages of labour, salaries of staff, prices of materials, plant and equipment rates and may also affect subcontractor prices. If the delays were caused by compensable delays, then the resulting inflationary increases in the contractor's costs may form a heading of its claim against the employer, whether the claim is made under the contractual heading of 'cost', 'loss' or 'expense' or as damages.

Particularly in an overheated market, where costs are rising rapidly, if construction activities are delayed, underlying costs of a contractor's resources can rise significantly. Whilst such escalation may be clear, quantification can be difficult. It requires the evaluator to establish what the actual costs were and to compare this with the costs that would have been incurred 'but for' the compensatory delays. The actual costs should be easy to establish through such as the employee payroll and materials, plant and equipment invoices. The '*but for*' costs of such as labour and hired plant and equipment might be established by reference to cost records for the time when those costs would otherwise have been incurred, but this may not always be available.

For materials, items that are required regularly during construction (such as concrete or road layer materials) might be capable of being compared on a similar basis. However, many materials, and particularly permanent plant, are purchased on a one-off basis such that there is no comparator to set against the actual costs. Furthermore, items such as concrete and road layer materials are often pre-ordered or subject to term agreements on a '*call off*' basis, such that it may be that there is no actual increase in costs when delays occur. Other materials might be purchased and paid for at the planned time and then stored for later use, for example pipes and precast concrete units. These may therefore be subject to no change in purchase costs due to delay, although the contractor may incur additional costs such as double-handling, storage and protection.

Whilst utilities charges can also often be readily compared over time, other preliminaries and general costs often contain many individual items, some quite small on their own and with differing inflation rates. It may therefore be that large items (such as temporary works, site establishment, plant, attendant labour, staff, fuel and utilities) can be addressed on their own details, as described elsewhere. However, smaller items may require a laborious exercise of analysis that is out of proportion to the increases in cost that can be found. It may therefore be considered proportionate that such smaller items are addressed together, perhaps by applying construction industry statistics in relation to their costs provided the fact of increases is established.

Whilst delay in activities will generally see inflation increase costs, that is not always the case. The graph in Figure 6.2 plots the cost of gabbro for a road project in Qatar constructed around the last period of economic overheating and ensuing crash. Gabbro is a locally available igneous rock that was specified to be crushed and screened for use in combination with bitumen for the wearing course of the highway.

The original completion date of this project was in June 2007, such that delays pushed gabbro (and other) costs through the overheating of 2008. As a result, in the period from the end of 2007 to the end of 2008 gabbro costs increased by around 70%. However,

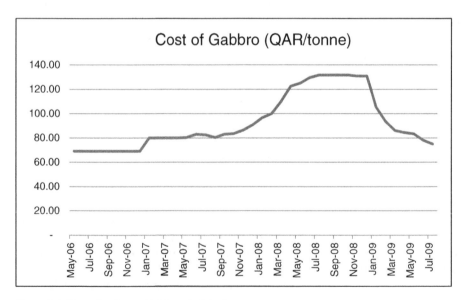

Figure 6.2 Example of a road layer material price.

in 2009, towards the end of the project, prices actually returned to their 2007 levels. The result of this was that the contractor's claim for increased costs in relation to this material required a detailed analysis of when it actually made purchases and when it would have made purchases had the delays not occurred. It will also be noted how, if the original contract period had been through 2008, rather than ending in 2007, the effects of the crash would have meant that delays into 2009 would have saved the contractor a significant part of the costs of this material.

An occasional defence to claims for cost escalation cites contract provisions that the contractor is to bear that risk, for example where the FIDIC Red Book clause 13.8 has been agreed not to apply. However, this should not usually affect the contractor's entitlement to inflation incurred due to the employer's breach or compensatory delays. It may also be that the contract contains a provision for adjustment of the contract price for fluctuations in prices and this, the potential for a claim for increases in costs for delay, the relationship between such claims and approaches to evaluation are all issues considered in detail in Chapter 8.

6.2.15 Off-Site Overheads and Profit

In addition to the costs of prolongation of on-site overheads, or preliminaries and general items, as discussed elsewhere, it is likely that delay on a construction and engineering project will also have an effect on the contractor's off-site overheads, usually those based at its head office, and also its profit. Off-site overheads are also sometimes referred to by terms such as 'head office costs' or 'home office costs'. These are sometimes combined with profit in the term 'gross margin'. The effects on these can be of two forms:

- Additional costs, where the head office provides support during a period of prolongation.
- Loss of overhead and/or profit contribution, where the prolongation of the works prevents the contractor from reallocating its on-site resources to other works on which they can make a contribution through other turnovers.

Off-site overheads and profit therefore feature as a head of claim in most prolongation claims. However, they can similarly be a feature of claims for such items as disruption and acceleration, which are considered in Sections 6.4 and 6.5. Therefore this subject of off-site overheads and profit is considered in full detail in its own Section 6.8.

One particular feature of off-site overheads and profit is that they may be incurred in the period after the original completion date, rather than when the causative events occurred. This is explained in Section 6.8.

6.3 Liquidated Rates for Delay Damages

As explained elsewhere, most standard, and many non-standard, forms of contract for construction and engineering works include a provision for the deduction by the employer of damages at a pre-agreed rate set out in the contract if the due completion date is not met. These are sometimes referred to by the term 'liquidated damages', simply meaning that the rate is fixed and agreed in advance. The intention of such pre-agreement is to provide certainty for both parties and also to avoid the need, cost and potential for disagreement involved in establishing the employer's actual damages. They are often capped at an amount or a percentage of the contract price, for reasons that may benefit both contractor and employer and explained elsewhere in this chapter.

The certainty of an agreed rate of delay damages is important to contractors for several reasons including:

- It is a known level when it tenders for the works, against which it can price any risk allowance it wishes to include in its pricing.
- Where the contractor is considering potential acceleration measures. This subject is covered in detail elsewhere in this chapter, but a part of the equation for a contractor considering whether to accelerate to overcome its own culpable delays is to weigh the costs of accelerative measures against its liability for delay damages if it overruns.
- It can provide certainty where the contractor can pass on the employer's delay damages to a supplier or subcontractor that caused the delay.

Clause 8.7 of the FIDIC Red Book provides for these 'delay damages' to be set at a daily rate calculated at a percentage stated in the Appendix to Tender of the 'Accepted Contract Amount'. Clause 3.32 of SBC/Q refers to 'liquidated damages'. Earlier versions of the JCT's contracts used the term 'liquidated and ascertained damages' and it is still common for British construction professionals practicing around the world to refer to them as 'LADs' (just as they are often prone to refer to claims for delay and disruption as 'L&E claims', by applying the JCT's term 'loss and/or expense'). Under SBC/Q these damages are set at rates per day set out in the Contract Particulars. The Infrastructure Conditions take a similar approach to SBC/Q, referring to 'liquidated damages' in clause 47 and setting these out at rates per day set out in the Appendix to the Form of Tender. Where the applicable law requires delay damages to be a 'genuine pre-estimate' of the employer's loss, the use of a percentage of the finally adjusted contract price might be questionable on the basis that, as that amount would not have been known when the damages rate was set, it cannot have been such an estimate. The requirement for a genuine pre-estimate under English law is a topic considered further elsewhere in this chapter, but legal advice should always be taken on the setting of a liquidated rate for delay damages.

Mention has already been made of the past suggestion of a 'Brown's Clause' but that idea never caught on, not least because of the difficulty of providing for different levels of prolongation costs at different stages of a project, given how delay costs may be incurred by a contractor at very different levels anytime from when it starts until it completes the works. In comparison, for the employer's delay damages, the costs will only run from the due date for completion until when the contractor actually completes the works. Thus, for a 'Brown's Clause' the levels of likely loss would have to be considered over a longer period, with more scope for their fluctuation. It is also tempting to consider that the proposed terms of construction contracts (other than those drafted by cross-industry groups such as the JCT) tend to be proffered by employers rather than contractors, and employers might find the perceived benefits of a liquidated damages provision, such as certainty and no need to prove actual loss, ones that they would want for themselves rather than their contractors.

This approach of including contract rates for the contractor's delay damages is, however, suggested in the *SCL Protocol*, paragraph 20.4, where it states:

> Arguments about proof of loss could be reduced or avoided altogether if the contract contained an agreed amount per day that can be applied to each day of prolongation. This is the reverse of the normal Employer's liquidated damages provision. It may be necessary to have a number of different agreed amounts to be applied depending on the stage in the project where the delay occurs. One method of fixing the figure(s) would be for the Contractor to price a schedule of rates with indicative quantities at tender stage.

Whilst this stated aim of a contractor's delay liquidated damages provision and schedule could have its benefits, but the practicalities of such an approach are questionable. Some of the issues experienced where contracts, unusually, have a schedule of rates and prices per resource for calculating the contractor's delay costs have been set out. If that is to be replaced by a schedule of overall rates to be applied depending on the stage in the project when the delay occurred, other complications may arise. The first is the burden on the tendering contractors to produce such schedules as part of their bids, with sufficient accuracy. The contractors would firstly have to decide how many such stages to price. The type of resource rate schedule explained would have to be married with projections of the numbers of each type of time-related resources at each of those stages. Whilst it could be said this would be part of its tender calculations anyway, it may not be in that form so would still require conversion. From the point of view of the employer and its advisors, to what extent should they consider and investigate these schedules as part of their tender adjudications? How can they be compared if the contractors chose to price different stages? In terms of the application of the schedule when delays do occur, there are likely to be times when delays occur between the stages that have been priced and debate is likely as to how to apply the schedule to such intervening periods. Also, should the contractor's rate of delay damages be required to be reflected in its tender in the same manner that its dayworks rates usually are? If so, against what provisional period(s) of delay?

6.3.1   Challenges to the Rate of Delay Damages

The deduction of liquidated damages by an employer will usually occur when the contractor has failed to complete by the contract completion date (or a section date), or any authorised extension to that date, and the period of the overrun beyond the contract date is not covered by any granted extension of time. It might be expected that the liquidated damages will be an exhaustive remedy, i.e. they will be the total of the amount that the employer is entitled to deduct as compensation for the late completion of the contract. However, this is not always the case, and is often challenged, as explained elsewhere in this chapter.

The rate is intended to be an estimate of the loss that would be suffered by the employer in the event that completion of the works is delayed. The practical problem with this situation is that in many instances the damages, rather than being an excessive amount, are in fact often only a portion of the loss likely to be suffered as a result of late completion. The reason is often not that there was an underestimate but simply that a proper estimate of the likely loss would result in such a high rate of damages that contractors would not accept it as part of the contract. If that occurs then fewer contractors may be willing to bid for the works, reducing competition. Those that do bid may build into their pricing an excessive contingency to cover the level of damages risk they would be accepting, thus increasing the employer's project costs. For the same reason, some contracts, particularly in the heavy mechanical and process industries, include a cap on the total amount of damages that may be deducted for unauthorised delay. The FIDIC Red Book clause 8.7 provides for a maximum amount of 'delay damages', being agreed in the Appendix to Tender as a percentage of the 'final Contract Price'. The Infrastructure Conditions apply a similar limit in clause 47(4)(a), the limit being set at a financial sum stated in the Appendix to the Form of Tender. SBC/Q does not provide a limit on its 'liquidated damages', although it would always be open for the parties to agree to add such a cap to the standard clauses.

As explained elsewhere, capping or limiting the total of damages claimable under a construction contract can be to the benefit of both parties in that one limits its liability and the other does not suffer from inflated bids or the effect of limiting the number of parties willing to tender for the works against a risk of unlimited damages. This latter benefit can be of particular importance in relation to subcontractors, given that even if their works are a small proportion of the overall project, they can cause critical delay to it and the resulting damages in terms of employer liquidated damages, the contractor's losses and expenses, and those of other subcontractors could be out of all proportion to the value of the subcontract. It is therefore often the case that subcontracts place a limit on the subcontractor's liability for any financial claims brought by the contractor. However, as with employers limited by a cap in a main contract, contractors sometimes seek to circumvent the limit in a subcontract.

In *McGee Group Limited v Galliford Try Building Limited* [2017] EWHC 87 (TCC), the subcontract contained two provisions for Galliford to recover the financial effects of default in Galliford's performance:

- Clause 4.21.1:

  .... if the regular progress of the Main Contract Works or any part of them is materially affected by any act, omission, or default of the Sub-Contract ....
  Any sum reasonably estimated by the Contractor as due in respect of any loss, damage, expense or cost thereby caused to the Contractor may pending final determination of the matter in litigation, arbitration, adjudication award or agreement between the Contactor and the Sub-Contractor, be deducted from any monies due or to become due to the Sub-Contractor or shall be recoverable by the Contractor from the Sub-Contractor as a debt.

- Clause 2.21:

  If the Sub-Contractor fails to complete the Sub-Contract Works or such works in any Section within the relevant period or periods for completion, and if the Contractor gives notice to that effect to the Sub-Contractor within a reasonable time of the expiry of the period or periods, the Sub-Contractor shall pay or allow to the Contractor the amount of any direct loss and/or expense suffered or incurred by the Contractor and caused by that failure.

Clause 2.21B then sought to place a cap on McGee's liability as follows:

> Provided always that the Subcontractor's liability for direct loss and/or expense and/or damages shall not exceed 10% (ten percent) of the value of this Subcontract order.

Galliford Try made claims against McGee for delay and disruption and deducted an amount from monies otherwise due to McGee for what was described as 'reimbursement costs loss and expense and costs associated with McGee's failing to regularly and diligently progress their works.' Each time, the figure deducted for these claims was the same, £1,489,733, being 10% of the sub-contract sum. However, Galliford Try's description of these claims changed slightly, and thereafter the amount of the deduction it sought in respect of delay and disruption increased beyond the 10% cap. By the trial, the total amount claimed in relation to delay was £3,318,124.29, of which Galliford Try said that claims worth £2,291,495.53 were not affected by the cap. Galliford Try argued that the cap only applied in relation to claims under clause 2.21 and did not limit McGee's liability in relation to claims under clause 4.21.

Mr Justice Coulson set out the law in relation to clauses that seek to limit liability and concluded that:

> In summary, a clause which seeks to limit the liability of one party to a commercial contract, for some or all of the claims which may be made by the other party, should generally be treated as an element of the parties' wider allocation of benefit, risk and responsibility. No special rules apply to the construction or interpretation of such a clause although, in order to have the effect contended for by the party relying upon it, a clause limiting liability must be clear and unambiguous.

He noted that in this case the cap was not said to be referrable to claims that may be made under particular clauses of the subcontractor or for breach of any express or any implied terms. He considered it to be a cap on McGee's liability for a particular type of claim, namely one for 'direct loss and/or expense and/or damages'. As to Galliford Try's attempts to distinguish between claims for delay and disruption, he said:

> In my view, that argument is also unsustainable. Anyone who has ever put together, argued or been obliged to decide a claim for loss and expense under a building contract, knows that no sensible distinction can be drawn between delay and disruption. One man's delay is another man's disruption. A sub-contractor's failure to complete a particular part of his work may have an adverse effect on the main contractor, but whether the consequential claim is one for delay or disruption, or a mixture of the two, will depend on a raft of factors: whether or not the delay was on the critical path of the main contract programme, what other sub-contractors were affected and how, if others were also in default etc. It is impossible to divide up such claims between delay, on the one hand, and disruption, on the other.

He concluded that:

> I consider that the cap in the sub-contract at clause 2.21B catches all GT's claims for loss and/or expense and/or damages for delay and disruption. I also consider that, on the face of it, it catches all of the claims advanced by GT in their Loss and Expense Claim.

In practice, the aim of providing certainty through a pre-agreed rate of employer damages for delay is not always achieved without some degree of pain through challenge to the rate or amount. Where employers find that changes in market conditions or an underestimate at the pre-contract stage means that the contract rate does not fully compensate their damages, it is not uncommon for them to challenge their own rate. As noted elsewhere, this may involve arguments based on applicable laws asserting that the contract rate neither covers all of the employer's losses nor limits its entitlements arising out of the contractor's failure in relation to timely completion. To avoid such an assertion, clause 35.2 of the CRINE Conditions says that 'Such Liquidated Damages shall be the sole and exclusive financial remedy of the COMPANY in respect of such failure'.

6.3.2   Delay Damages as a 'Penalty'

Inevitably, most challenges to the rates or amounts of delay damages under the contract will come from the contractor. Such a challenge is likely to follow arguments in relation to the contractor's entitlement to further extensions of time. However, in addition, in some jurisdictions the law requires that the damages are a genuine pre-estimate of those likely to be incurred by the employer in the event of delay to completion of the works. If they are excessive, the damages may be regarded as a 'penalty' rather than genuine damages, and may be unenforceable. On the other hand, internationally, some bespoke contracts actually refer to the employer's rate of delay damages as a 'penalty', without the connotations that this term has for those used to working under English law. Particularly in Middle Eastern jurisdictions, arguments can often also follow as to whether the rate of penalties agreed in the contract reflect the actual harm suffered by the employer (rather than a genuine pre-estimate of it) and who has the burden of proving whether they do. These considerations only emphasise the need to consider the position under the local substantive law before either setting or challenging an agreed rate of delay damages.

Concern in relation to the potential contractor's cry of 'it's a penalty' is reflected in how some standard forms seek to circumvent such an assertion. For example, clause 47(3) of the Infrastructure Conditions, under the heading 'Liquidated Damages for Delay' says this under a subheading 'Damages not a penalty':

> All sums payable by the Contractor to the Employer pursuant to this Clause shall be paid as liquidated damages and not as a penalty.

In the CRINE Conditions this is put even more pointedly at clause 35.2:

> All amounts of such Liquidated Damages for which the CONTRACTOR may become liable are agreed as a genuine pre-estimate of the losses which may be sustained by the COMPANY in the event that the CONTRACTOR fails in his respective obligations under the CONTRACT and not a penalty.

Combined with clause 35.2's provision that the liquidated damages rate in the contract is the employer's 'sole and exclusive remedy', the CRINE Conditions can therefore be seen to be particularly keen to aim for certainty in relation to the amounts of the employer's delay damages. They aim to prevent both contractor challenges that the agreed rate(s) are a penalty and employer attempts to claim for further unliquidated damages.

Under English law, until recently, the House of Lords had set out the criteria for judging if a sum was to be regarded as a penalty in *Dunlop Pneumatic Tyre Co Ltd v New Garage & Motor Co* [1915] AC 79. The four principles can be summarised as:

1. If the sum is extravagant and unconscionable in amount in comparison with the greatest loss that could possibly flow from the breach then it will be regarded as a penalty.
2. If the obligation is to pay a sum of money, and failure to do so results in a larger sum being payable, then the larger sum will be regarded as a penalty.
3. Subject to these two rules, if there is only one event upon which the sum is to be paid, then the sum is liquidated damages.
4. If a single lump sum is to be payable on the occurrence of one or more events, some of which are serious and others minor, then there is a presumption that the sum is a penalty.

The judgment as to whether any sum represents a penalty is to be made as at the time the contract was made and not at the time of the breach and with the benefit of hindsight. These principles were confirmed in 2002 in the case of *Jeancharm v Barnet Football Club* (reported in *Building Magazine*, 7 February 2003).

On the basis of these principles, some care therefore needed to be exercised in the calculation of rates of damages and insertion of the rates into a construction or engineering contract. Where such rates were inserted in the contract they needed to be related to an exercise to determine the potential losses resulting from late completion. The rate was unlikely to be challenged by the contractor where it could be shown to be less than the maximum potential damage, but might successfully have been challenged where it could be shown to be 'extravagant and unconscionable' in relation to the employer's maximum potential loss.

However, in a landmark ruling, the Supreme Court has recently set out a new test of what constitutes a 'penalty' and replaced Lord Dunedin's test in *Dunlop Tyres* of a 'genuine pre-estimate of loss'. In *Cavendish Square Holdings BV v El Makdessi* [2015] UKSC 67, [2015] 3 WLR 1317, their Lordships actually heard two appeals together. *Cavendish Square Holdings BV v Talal El Makdessi* concerned a claim for damages for breach of an agreement to sell shares in a company. *Parking Eye Ltd v Beavis* concerned a fine under a consumer contract, for overstaying at a car park. However, both cases involved challenges to the amounts of damages being claimed.

Their Lordships considered the approach taken in *Dunlop Tyres* 100 years earlier, and in particular Lord Dunedin's four principles and the emphasis on financial loss only, as a component of damages for breach. They also set out how subsequent cases had considered how liquidated damages could be set at rates intended to deter breach and taking into account the injured party's interest in timely performance in a broader sense. As Lords Neuberger and Sumption put it:

> In our opinion the law relating to penalties has become the prisoner of artificial categorisation, itself the result of unsatisfactory distinctions: between a penalty and genuine pre-estimate of loss, and between a genuine pre-estimate of loss and a deterrent.

Lord Hodge, put it that the law involved:

> … an over-rigorous emphasis on a dichotomy between a genuine pre-estimate of damages on the one hand and a penalty on the other.

Their Lordships set out a new test of what constitutes a penalty, which recognises that the 'legitimate interest' of the injured party (being the employer in the context of a liquidated delay damages claim) might include issues that go beyond simple financial losses. Lords Neuberger and Sumption held that:

> A damages clause may properly be justified by some other consideration than the desire to recover compensation for a breach. This must depend on whether the innocent party has a legitimate interest in performance extending beyond the prospect of pecuniary compensation flowing directly from the breach in question.

They further set out a test of what constitutes a 'penalty' as follows:

> The true test is whether the impugned provision is a secondary obligation which imposes a detriment on the contract-breaker out of all proportion to any legitimate interest of the innocent party in the enforcement of the primary obligation. The innocent party can have no proper interest in simply punishing the defaulter.

Lord Hodge concluded that:

> …. the correct test for a penalty is whether the sum or remedy stipulated as a consequence of a breach of contract is exorbitant or unconscionable when regard is had to the innocent party's interest in the performance of the contract.

In conclusion, for a liquidated damages provision to avoid being held to be unenforceable as a penalty under the English legal approach to such provisions, the rate no longer needs to be a 'genuine pre-estimate of loss' likely to be incurred by the employer. Whilst it cannot be intended to punish the contractor, it can now take a broader view of the employer's interest in timely completion. In a construction and engineering context, it can be envisaged that these wider 'legitimate interests' might include public interest of a local government authority that has commissioned a new and much needed piece of transport infrastructure. Alternatively, for a home owner, the late refurbishment of a home might involve considerable emotional concern, but little financial consequence. Under the *Dunlop Tyres* test these interests would not have been part of a 'genuine pre-estimate of loss' and a damages provision that sought to allow for them by penalising a late contractor might have been successfully challenged.

The decision in *Cavendish v Makdessi* is likely to have significant consequences where the parties to a construction contract pre-agree the rate of damages for late completion by the contractor. Employers can include in their damages rate allowance for their wider interests in completion, and if the rate is challenged as a 'penalty' by the contractor then those interests will be included as a justifying factor. Contractors can expect it will now be much harder for them to succeed in such a challenge.

As previously stated, the position will vary internationally and in many jurisdictions the pejorative nature of the term 'penalties' and the need for a 'genuine pre-estimate' will not be relevant. As also previously noted, in the Middle East, for example, it is common for contractors to challenge rates of delay damages agreed in a contract on the basis that they do not reflect the actual harm suffered by the employer and that the courts have jurisdiction to vary such an agreement. This may run the risk of the employer claiming its actual damages if they can be established to be greater that those agreed in the contract, particularly where it contains an agreed cap on the total. Furthermore, it will usually be for the contractor to apply for such intervention, with the burden of proving no, or a lesser, harm has been incurred also resting with the contractor. In practice, it may be difficult for the contractor to prove the extent of losses actually incurred by the employer. International arbitration tribunals, in particular, appear reluctant to accept such a challenge to rates of damages that the parties have agreed in their contract, whatever local legal arguments are run by the challenger.

6.3.3 Actual Damages for Delay

If an employer has decided that it does not wish to have liquidated damages as part of the contract, it is under no compulsion to have them. It may decide either that, for whatever reason, it does not wish to deduct damages in any event, or it may prefer to prove its actual loss in the event of delay to completion. However, if the form of contract being used includes a provision for liquidated damages it should be deleted in such instances and not merely have 'nil' inserted as the rate of damages applicable. Insertion of 'nil' as the rate of damages can preclude the employer from pursuing any damages at all in the event of delay to completion. In the English law case of *Temloc v Errill Properties Ltd* (1988) 39 BLR 30, the employer inserted 'nil' as the rate of liquidated damages in the contract, but later argued that this was because it wished to deduct actual damages. The contractor successfully argued that the 'nil' rate reflected a course of dealing between the two firms which had resulted in a desire for there to be no deduction of delay damages, liquidated or otherwise, in any event. Whether the outcome would have been the same if the contractor had not been able to demonstrate the course of dealing on previous contracts is in doubt, but to be sure, the lesson is to delete such provisions completely if they are not required and actual damages are to replace them.

6.3.4 The Date(s) from Which Delay Damages Run

It is important that the contract clearly defines the date(s) from which the delay damages are to be calculated, and any provisions for adjustments to that date or dates. The potential consequences of a lack of such a provision where the employer commits an act of 'prevention' are considered elsewhere in this book. If the contract has provisions for the contract to be completed in sections, then the amount of damages relating to each, and the dates from which damages commence for each, also need to be clearly defined. The Contract Particulars of SBC/Q provide a schedule for the setting out of different daily rates for each 'Section' of the works, those being defined in the Sixth Recital. In the FIDIC Red Book the Appendix to Tender will specify any agreed 'Section', describe it, give its due completion date and state its rate of damages for the purposes of clause 8.7. In the Infrastructure Conditions, the Appendix Part 1 to the Form of Tender follows a similar approach.

Where completion is not planned to be achieved in sections set out in the contract, it may provide for the employer to opt at the time that they are nearing completion to take over the works in parts as they are completed. Contracts vary as to whether the contractor's consent is required for such partial possession. Again, provision must be required to adjust the contract rate of delay damages accordingly. Under SBC/Q partial possession is provided for in clauses 2.33 to 2.37, but it requires the contractor's consent ('which shall not be unreasonably delayed or withheld'). Under clause 2.37 the rate of liquidated damages is then reduced in proportion to the value of the part taken over. FIDIC Red Book clause 10.2 takes a similar approach, including the pro rata reduction to damages rates, although the employer's right to take over parts of the works is 'at its sole discretion'. In the Infrastructure Conditions, partial possession by the employer is covered by clauses 48(3) and (4), but there is no express provision for adjustment to the rates of liquidated damages therein.

6.3.5 Procedure and Prerequisites

The procedure and prerequisites for an employer to deduct delay damages differ between the main contracts considered in this book:

- SBC/Q
  Clause 2.32 contains two prerequisites to the employer's right to require payment, or withhold or deduct, liquidated damages. Subclause 2.32.1.1 requires that the architect or contract administrator has issued a Non-Completion Certificate under clause 23.1. Subclause 2.32.1.2 requires that the employer has notified the contractor before the date of the Final Certificate.
- FIDIC Red Book
  Clause 8.7 makes the contractor's obligation to pay delay damages subject to clause 2.5. Clause 2.5 requires the employer to give notice and particulars to the contractor. Those particulars are required to specify the clause on which the claim is based and include substantiation of the amount to which the employer considers itself entitled. The engineer is then required to proceed in accordance with clause 3.5 to agree or determine the amount of delay damages.
- Infrastructure Conditions
  Clause 47 is comparatively quiet as to procedural preconditions to the employer's right to recover liquidated damages under that clause.

Notable under all of these provisions is that they do not mention the duty of the engineer, architect or contract administrator to operate the extension of time provisions of the contract. In particular, the operation of those provisions is not expressly made a precondition to the employer's entitlement to delay or liquidated damages.

In *Henai Investments Inc. v Beck Interiors Limited* [2015] EWHC 2433 (TCC), Mr Justice Akenhead considered the deduction of liquidated damages under the JCT Standard Building Contract without Quantities 2011, the without quantities version of SBC/Q, but in the same terms in respect of liquidated damages. The issues before the Judge included:

> Would a failure on the part of the CA to make a decision in respect of a contractually compliant application for extension of time render the CA's Non-Completion Certificate invalid or otherwise prevent the Employer from deducting and/or claiming liquidated damages?

The Judge noted how SBC/Q clause 2.32 contained the two conditions in subclauses 2.32.1 and 2.32.2, but considered that:

> It seems odd that, if there was to be a condition precedent that no liquidated damages should be payable or allowable unless the extension of time clauses have been operated properly, it was not spelt out as such.

He concluded that:

> I have formed the view therefore that a failure on the part of the CA to operate the extension of time provisions does not debar the Employer from deducting liquidated damages where the expressed conditions precedent in Clause 2.32.1.1 and 2.32.1.2 have been complied with.

For a discussion regarding the position where delay or liquidated damages are deducted by the employer, but the certifier later grants an extension of time, see Section 6.9 in relation to a contractor's claim for interest.

6.4 Disruption

Disruption may be part of a claim for prolongation costs, or may occur where there is no prolongation either locally or to a contractual section or to the works as a whole. The term 'disruption' generally infers that the contractor's intended sequence and/or duration of construction activities have been rendered impossible, wholly or in part, by extraneous factors such that it has incurred loss of productivity of its resources. Those resources are usually its labour, and/or plant, and/or equipment and and/or supervision. Whether the causative factors are also relevant to the consideration of prolongation will depend on individual circumstances and the relation of the affected activities to the critical path through the project programme. This may also require consideration of local prolongation or non-critical prolongation, as explained in Section 6.2.

The quantification of losses resulting from disruption to the contractor's works is possibly the most difficult area for anyone engaged in quantifying additional payment that might be due as a result of the time consequences of change under a construction or engineering contract. Before considering the often thorny issue of how to deal with disruption and to evaluate its financial consequences, it may be worth pausing to further consider what disruption actually is.

Typical definitions of the word include the following:

'to throw into turmoil or disorder';
'to interrupt the progress of'; and
'to break or split apart'.

Such dictionary definitions sometimes give a key to the issues that have to be considered in evaluating the financial consequences of disruption, for example by quantifying the interruptions or breaks in the continuity of work.

In its Guidance Part A paragraph 5, the SCL Protocol defines disruption as follows:

> In referring to 'disruption', the Protocol is concerned with disturbance, hindrance or interruption to a Contractor's normal working methods, resulting in lower productivity or efficiency in the execution of particular work activities.

A further way to understand what is meant by the term disruption is to consider it as a change in the circumstances or conditions under which the works or parts of them are carried out. This may include where the physical works are not changed at all. In such cases, the only change is in how they are constructed or the circumstances under which they are constructed. This approach to definition is relevant to the particular provisions of some standard forms, as explained elsewhere in this chapter.

The difficulties faced when dealing with disruption claims, whether for the party making the claim, or the party receiving and reviewing the claim and certifying amounts against it, include the following:

- Establishing cause and effect.
- Allocating losses to causes on a particularised basis, where this is necessary.
- Extracting costs that are unrelated to the causes that form the basis of the claim.
- Keeping appropriate contemporaneous records.
- Addressing multiple causes that require their effects to be disentangled, particularly where they also involve mixed liability.
- Addressing the commercial consequences of subcontracting of parts of the works and particularly its effects upon:
  - the extent of records available;
  - the level of detail and sophistication of the submitted claim, particularly in terms of proving cause and effect;
  - the question of whether sums have been paid reasonably (see in relation to this in Section 6.7); and
  - what some contractors consider to be the relative ease with which subcontractors can sometimes obtain money for disruption through the use of so-called 'quick and dirty' adjudication.

6.4.1   Legal Basis of a Disruption Claim

It is often the case that claims for disruption are presented, and even analysed by the recipient, without due consideration of the contract and the legal basis upon which the claim should be made and considered.

Most disruption claims are made under the express provisions in the specific contract. However, even here there are alternative options that could form their basis, some of which are regularly overlooked. There are also alternative terminologies used in different standard forms of contract, and these also need to be considered in order to set a proper basis for evaluation.

Elsewhere in this chapter, the different terminologies used in the claims clauses of the main standard forms considered in this book were highlighted. The relevant clause of SBC/Q is clause 4.23, which refers to 'loss and/or expense' caused by 'matters materially affecting regular progress'. The FIDIC Red Book and the Infrastructure Conditions refer to the defined terms 'Cost' or 'cost' respectively. How these terms can influence what is claimable was explained elsewhere in this chapter.

A regularly overlooked basis for claims for disruption is as a part of the valuation of variations where those variations were the cause of that disruption. This often arises where the contract conditions include clauses that provide that, where variations are instructed, the contractor is entitled to both extensions of time and the payment of its costs or loss or expense resulting from variations. For example, in SBC/Q variations are the first basis for both extensions of time (as the first Relevant Event in subclause 2.29.1) and for 'loss and/or expense' (as the first Relevant Matter in subclause 4.24). As a result, where disruption is associated with delay and both are caused by variations, it is often valued along with delay under such as SBC/Q clause 4.23, as 'loss and/or expense'. However, SBC/Q also says in clause 5.6.1.2 that where additional or substituted work is of similar character to work set out in the bills but is not executed under similar conditions and/or significantly changes in quantity, the rates and prices in the bills shall be the basis of valuation, but with a fair allowance for such differences in conditions and/or quantity.

In addition, clause 5.9 of SBC/Q says that where as a result of a variation there is a substantial change in the conditions under which any other work is executed (including CDP works), then such other work shall be treated as if it had been the subject of a variation and valued as such. Thus, under these provisions of SBC/Q, disruption related to a variation can be valued as if it were a variation, whether it is related to the varied works itself or the effects of a variation on unvaried works.

It seems obvious that such a valuation under Section 5 of SBC/Q should not duplicate amounts included as 'loss and/or expense' under its clause 4.23, but this is expressly confirmed in its clause 5.10.2, which makes it clear that there is to be no duplication between recoveries under Section 5 ('Variations') and clause 4.23 ('Loss and Expense') of SBC/Q. However, the use of the term '… reimbursed by payment under any other provision …', without qualifying that other payment by the addition of the word 'additional' can cause confusion. This could have been avoided by clause 4.23 saying: '… reimbursed by *additional* payment under any other provision …'. It has been explained elsewhere in this book how claims for such as disruption are, generally and put very simply, for the difference between costs incurred as a result of a claimable event and the costs that would have been incurred had that event not occurred. However, what if part of that difference is being paid for as part of the original contract price? Take an example of a simple claim for additional labour time. Say the contractor over-tendered for 1000 man-hours, would have incurred 900 man-hours, but actually incurred 980 man-hours because of disruption for which the employer was liable. Obviously the underlying facts and cause and effect would have to be established, but the contractor's loss due to disruption is 80 hours. On the other hand, the contractor has actually been paid for those additional 80 hours through the contract price (and more – it is actually 100 hours, giving a gain of 20 hours against the actual hours). Where the test is the 'but for' test, it is suggested that it cannot be correct for an evaluator in such a situation to rely on such a recovery, by reference to SBC/Q clause 5.10.2. However, such an assertion has been heard in the past in relation to the similar provisions of clause 13 of previous editions of the JCT contracts. In this hypothetical example, the contractor would have reaped the benefit of being paid for 100 hours of unused labour time had the disruption not occurred. The fact that the contractor overestimated its tender should be no more relevant to the calculation of its claims as it would have been relevant had it underestimated.

A similar clause to SBC/Q clause 5.6.1.2 is provided in the Infrastructure Conditions clause 52(4)(b) as follows:

> Where work is … not carried out under similar conditions … the rates and prices in the bills of quantities shall be used as the basis for valuation so far as may be reasonable failing which a fair valuation shall be made.

Such express contractual provisions as these in SBC/Q and the Infrastructure Conditions therefore require that, in certain circumstances, what amounts to disruption is to be included within the valuation of an instructed variation. Furthermore, the evaluation under such as Infrastructure Conditions clause 52(4)(b) is to be based on the rates and prices in the bills, rather than actual costs, so far as is reasonable. Otherwise, actual costs are likely to be thought to be a proper basis for 'a fair valuation' under that clause.

However, such approaches are limited to variations that have been instructed and require both the valuation of those variations and/or their effects on other work. To this extent the provisions do not cover such other causes of disruption as late information, late access, etc. Furthermore, whereas such provisions might be expected to provide a relatively uncontentious route to value and allow for disruption, this often does not happen. Many administrators of contracts (architects, engineers or quantity surveyors) fail to accept such an approach. This is often because they see 'disruption' as a highly contentious issue that requires consideration quite separate to the relatively simple and non-contentious valuing of variations. Sometimes it is due to inexperience in the use of these provisions for the valuation of disruption (particularly to unvaried work that has been affected by varied work as required by such as SBC/Q clause 5.9), and it is often difficult to persuade such contract administrators to use such clauses in this way. Similarly, contractor and subcontractor staff are often unfamiliar with the possibilities and tend to be slow to use these provisions, seeing disruption as a matter for a separate 'claim' and therefore by reference to such as the 'loss and/or expense' clause 4.23 of the SBC/Q contract form, for example.

As a result of this, it is usual that claims for disruption are made under the 'claims' clauses of construction and engineering contracts, with their references to such as 'direct loss and/or expense' and defined terms such as 'cost' or 'Cost'. How these differing terms can give rise to differing entitlements is explained elsewhere in this chapter in relation to prolongation costs, but the same considerations apply in relation to disruption.

The alternative to a disruption claim under the express provisions of the contract (whether the variations clause or a 'claims' clause, as explained elsewhere) is a claim for damages for breach of contract. In fact, provisions such as clause 4.26 of the SBC/Q contract expressly keep alive both the express contractual and breach of contract routes to a claim by stating that:

> The provisions of clauses 4.23 to 4.25 are without prejudice to any other rights and remedies which the contractor may possess.

In contrast, under the NEC approach, there should neither be doubt as to under which clause disruption caused by a compensation event should be valued, nor a risk of duplication between claims, given that disruption is to be included as part of the valuation of each compensation event including both its direct and indirect financial effects.

Whether a claim for disruption is made under the express 'claims' provision of the contract, as a claim for damages for breach of contract, or both, the principles to be applied usually remain the same. As explained elsewhere in this book, the governing purpose of damages is to put the party whose rights have been violated in the same position, so far as money can do, as if his rights have been observed.

Furthermore, such damages (including disruption evaluation) should fall within one of the two limbs of the judgment in *Hadley v Baxendale*:

> The damages … should be such as may fairly and reasonably be considered either:
>
> [1] Arising naturally, i.e. according to the usual course of things from such breach of contract itself, or
> [2] Such as may reasonably be supposed to have been in the contemplation of both parties at the time they made the contract, as the probable result of the breach.

Broadly, for a claim for disruption, or part thereof (or for prolongation as discussed elsewhere), to succeed, the amount or item claimed must either be the natural result of the breach complained of or have been in the contemplation of the parties as the probable result of that breach when they entered the contract. In practice the effects of this can be more problematical for disruption claims than prolongation claims. Most of the effects of typical prolongation are usually accepted as being covered by one of these limbs, whereas some heads of disruption costs can be less readily recognised as either arising out of the usual course of things or in the contemplation of the parties when they contracted.

Subject to these general legal principles, and converting them into the specifics of how disruption is to be evaluated under construction contracts, it is suggested that the following phrase probably sums up the position as well as any:

> The measure of disruption is the difference between the *production that could have been achieved* without disruption, and the *production actually achieved* as a result of the disruption.

Two clarifications are, however, required in relation to the emphasised words in this suggestion:

- 'production that could have been achieved'
  This is usually the hardest component to establish. As explained elsewhere in this chapter, whilst they are often the basis for a contractor's disruption calculations, this may not be the same as the production that the contractor allowed for in its estimate, which is often the only readily available source of purported 'but for' information.
- 'production actually achieved'
  The vital qualification here is that it must be what was achieved 'as a result of the disruption'. This will exclude the effects of any inefficiencies of the contractor, failures by others such as its subcontractors or suppliers, or matters that are not able to be compensated, such as the effects of exceptionally adverse weather conditions.

Both of these sides of the disruption loss equation can be difficult to establish. Both will be addressed further elsewhere in this chapter.

In making a claim for disruption, a claimant needs to satisfy certain basic principles. It has the burden of proof to establish the following:

- That an event entitling it to make a claim for disruption, be it either a 'Relevant Matter' under the express provisions of the contract such as SBC/Q, or a compensatory event under the FIDIC Red Book or the Infrastructure Conditions and/or a breach of contract, has occurred.
- That the party against whom it is making the claim is factually liable for that event.
- That the party against whom it is making the claim is legally liable for that event.
- That the event has caused it to incur loss.
- The quantum of the loss.
- That it has complied with any contractual preconditions to its entitlement, such as in relation to notices.

Furthermore, the recipient of the claim must be made aware of the case against it in sufficient detail and clarity so as to enable it to respond to that claim. The burden involved in satisfying these requirements can be particularly heavy in relation to disruption claims.

6.4.2   The Factual Basis of a Disruption Claim

Disruption to the contractor's works may arise from a number of causes, including, but not limited to, the impact of the following typical events:

- Ordered variations to the quality or specification of the works.
- Ordered additions or omissions to the scope of the works.
- Late design or other information supplied by the design team or specialists.
- Late provision of access to, or possession of, the site or parts of it.
- Unforeseen physical conditions or obstructions on the site.
- Exceptionally inclement weather or climatic conditions.

- Strikes or lockouts.
- Force majeure events such as civil disorder, war, earthquakes, typhoon or volcanic activities, etc.
- Difficulties or delays in obtaining the necessary labour, plant and/or materials.
- Delays by subcontractors or suppliers, whether they were nominated, named or domestic.
- The opening up of works for inspection.
- The carrying out of works on the site by others, such as other package contractors engaged by the employer on the site.
- Suspension of the works or a part(s) of it.
- Early possession of the works or a part(s) of it.

This list is not exhaustive but gives the prime examples of the potential sources of disruption. It is not difficult to see that in most contracts some of these causes might be the contractor's liability while others will be the employer's, depending on how such risks have been allocated. Where a number of causes occur on a project, with a mix of liability between the contractor and employer, it is not difficult to anticipate the difficulties that can result in accurately separating and identifying the disruptive effects of the various causes. Such situations are common on major projects. This difficulty can perhaps be contrasted with the position in relation to a prolongation claim. Under a prolongation claim, if completion overruns by 10 weeks, but delay analysis shows that 6 weeks were the culpable fault of the contractor and only 4 weeks were able to be compensated, then the time-related costs associated with those compensatory 4 weeks should be readily capable of being isolated. However, separating the associated disruption costs in such a situation is likely to be far less simple, particularly in relation to the loss of production of the contractor's labour, for example.

The situation is further complicated when one considers that the effects of the various causes of disruption may manifest themselves in different ways. For instance, the effect of any of the typical causes of disruption might be to:

- Require the whole, or a section, of the works to be temporarily suspended.
- Reduce labour productivity for a part, section or the whole works.
- Result in certain activities having to be completed on an intermittent basis rather than on a continuous working basis.
- Require the contractor to change the intended sequences of operations and activities for the works.
- Require return visits to working areas to carry out and complete activities.
- Lead to stacking of trades involving increased concurrent working and congestion of working areas.
- Result in late or restricted or out of sequence access to work areas.
- Create uncertainty as to the scope and detail of work to be carried out.
- Reduce morale among operatives, supervisors and management.
- Require the contractor to reduce or extend its working hours.
- Cause the contractor to reduce or increase resources, be they labour, plant, equipment or supervision resources.

- Lead to changes in personnel for reasons such as demotivation or loss of potential bonuses.
- Result in work being carried out under different environmental and weather conditions.
- Shift the balance in numbers between operatives and those supervising them.
- Shift the balance in numbers between operatives and the plant and equipment working with them.
- Cause a combination of the above effects.

In terms of analysis and quantification, it can be very helpful if these effects of the causes of disruption can be separately identified and analyses performed, as explained elsewhere in this chapter.

If multiple causes result in multiple effects, the compounding factor at work on the contractor's site organisation and efficiency can readily be imagined. The first difficulty lies in applying some degree of reasonable analysis and calculation to the process so as to allow the financial effects to be identified to the multiple causes with a degree of reliability. There may then be the problem of separating out the effects to individual causes.

There may be instances where the separation of effects between causes is totally impossible, resulting in the need to produce a 'global' disruption claim, where the effects are not identified to individual causes. Alternatively, it may be that there is left a remaining 'composite' claim, where such particularisation is only partially achieved. Such claims have caused no small degree of controversy in the past and continue to do so; the subject is therefore dealt with in some detail in Section 6.6.

6.4.3 Evaluating the Costs of Disruption – Introduction

There are various potential methods for evaluating the financial consequences of loss of productivity associated with disruption. The method, or methods, adopted depend on such issues as: the documentation available; the size of the claim; the question of proportionality; the stage at which the evaluation is attempted; the law; and any express agreement within the contract as to how such an evaluation is to be carried out. Contractual claims always have to be set against their legal background. Thus, for example, it has been noted elsewhere how the variations clauses of certain construction contracts provide provisions under which disruption might, in certain circumstances, be valued as part of the variation. In addition to the potential benefits of valuing disruption through adjustment for variations, due to the relatively less contentious nature of this approach, it also has a potential benefit in the ease with which the valuation can be made. Variations, including where appropriate the effects of disruption, fall generally to be valued on a 'value' basis, that is generally by reference to prices already set and agreed in the contract, for example the bills of quantities. Thus, for example, a variation that is reasonably assessed as making an activity 50% more difficult to carry out could be the subject of a 50% pro rata adjustment to elements such as labour and plant costs of the allowances in the bills of quantities rates for that work. Clearly such an approach, if it gives rise to a reasonable valuation as required by the express clause of the contract, can make valuation much simpler. In such circumstances, the need to establish actual costs, expense or loss, with the all too common difficulties of records and allocation, can be avoided. Of course this is only possible if the circumstances and provisions of the contract make such a value-based approach applicable. This does, however, reflect a further factor in relation to the appropriate method of evaluation, and this is proportionality, given the high costs of some methods.

As noted elsewhere, under the NEC approach to the valuation of change, disruption will be captured as part of the overall valuation of each individual compensation event. To quote again the NEC User's Guide:

> If the compensation event increases the Contractor's time related costs, for example because planned Completion is delayed, or additional supervisory staff are needed, the increase in Defined Cost is included within the value of the compensation event. There is no such thing as a retrospective and separate 'delay and disruption claim' in the contract.

Furthermore, this means that under the NEC approach, disruption becomes part of the valuation of the direct consequences of change, and it was explained in Chapter 5 that in NEC4-ECC this is normally based on an estimate of the contractor's actual costs on a prospective basis. In that case the example of a change that made works 50% more difficult to carry out should be relatively easy to value as an estimate of the change in the contractor's costs.

However, where there is a need to evaluate disruption claims other than by value and/or a prospective estimate, evaluation methods usually require one or a combination of the following approaches:

- Record sheets.
- Witness statements.
- Expert opinion.
- Planned against actual comparisons.
- Measured mile comparisons.
- Earned value comparisons.
- Project comparison studies. Here the disrupted project is compared to another that was not subject to the same events. The problem here is finding a project that sufficiently reflects the work, scope, specification, location, circumstances, etc., of the disrupted project. It is usually only possible on repetitive housing schemes or linear projects such as phases of a long highway.
- Modelling approaches such as 'systems dynamics modelling'.
- The application of historically based disruption factors.
- The application of historically based cumulative impact data.

As will be explained elsewhere in this chapter, the evidentially difficult and politically controversial nature of disruption claims means that the best ones are often prepared on the basis of a combination of more than one of these approaches, where the amount claimed justifies the expense. This might mean that approaches to evaluation that on their own would not satisfy the burden of proof combine to satisfy that burden because they come to similar conclusions. Alternatively, they might lead to completely different results, leaving neither approach with credibility and proving the contractor's claim wrong.

The direct costs incurred as a result of disruption are predominately in relation to the labour, plant, equipment and on-site supervision and support costs. The components of such costs will usually be the same as those set out in Section 6.2.

The starting point for many claims for disruption costs as presented by contractors are their tender calculations, with the intention of demonstrating that the disrupting factors have caused the contractor to incur costs at levels greater than those anticipated in the tender calculations and thereby incorporated into the contract price. However, on its own, this approach does not establish causation or quantum. It only establishes that the contractor has used more resources than it allowed for, and hence that its actual productivity was less than that estimated. This is often referred to as the 'costs less receipts' approach, for the obvious reason that the contractor is attempting to recover all of its costs in excess of the sums received through the contract. Alternatively, it is a 'total loss' approach as defined in Section 6.6. Core Principle 21 of the SCL Protocol has been quoted elsewhere, but is worth repeating here:

> The tender allowances have limited relevance to the evaluation of the cost of prolongation and disruption caused by breach of contract or any other cause that requires the evaluation of additional costs.

Whilst this has to be broadly correct, the use of the subjective word 'limited' could be debated. As explained elsewhere in this chapter, the contractor's tender allowances can be of some relevance to a disruption claim.

The potential problem of using the contractor's tender calculations as the starting point for the assessment of additional costs was discussed in Chapter 4. If such an approach is valid, then the correct starting point ought to be the reasonable costs that would have been included in the tender by a competent experienced contractor and incorporated in the unit rates and prices for the works based on the full import of the contract documentation and information available to the contractor at the time of tender. Even then, such reasonable tender allowances would require adjustment for such as: any 'own goals' by the contractor; failures of other parties such as suppliers and subcontractors; and cost-neutral events such as weather conditions. The assessment of additional costs must be capable of demonstrating that it is not allowing the contractor to recover inadequacies in the tender calculations as part of the process, or to otherwise recover costs incurred as a result of factors outside the claimed cause(s) of disruption.

6.4.4 Records of Time Lost

The preferred method of productivity analysis should be a records-based approach relying on factual analysis of what additional resources and time were actually required by the contractor, and why. Such an approach requires the keeping of contemporaneous records. The records need to be concurrent with the disruptive event and meet any specified requirements of the contract. It is also advisable that they should also defer to any stated reasonable requests of the engineer or architect or contract administrator, where appropriate. It is even better if their counter signature can be obtained on such contemporaneous records. Such agreement of records may not signify acceptance of liability, but reduces the scope for future dispute regarding the resources and time lost that will be attached to that liability if it is established.

Such contemporaneous records can be of either the specific time lost on an activity or the total time spent on the disrupted activity. Examples of suitable record sheets for each of these alternative approaches, which have been successfully used in practice, are

WORK RECORD SHEET				
Project...................................		Main Contractor / Employer		
Operative's Name(s)		Week Ending Friday		
Location	Description of Work	Quantity of Work	Date	Hours

Representative   Main Contractor's / Employer's Representative

Signature	Print Name	Position	Date	Signature	Print Name	Position	Date

Figure 6.3 Work record sheet.

RECORD OF TIME LOST						
Project...			Main Contractor / Employer			
			Date			
Name of Operative	Location and description of work	Time stopped		Time disrupted		
		Time lost	Cause	Duration	Cause and percentage	

Representative   Main Contractor's / Employer's Representative

Signature	Print Name	Position	Date	Signature	Print Name	Position	Date

Figure 6.4 Time lost record sheet.

given in Figures 6.3 and 6.4. Both are set up to be countersigned by the main contractor (where used by a subcontractor) or the employer or its engineer, architect or contracts administrator (where used by a contractor).

The form in Figure 6.3 is used to record any work that is subject to disruption, the totality of the resources and time spent on it. The intention here is that this can subsequently be used for comparison of the outputs (quantity divided by recorded hours) between activities in different areas, periods or circumstances, with each other (a measured mile approach) or with tender allowances to form a basis for the quantification of a claim for loss due to disruption based on the difference. The particular shortcoming of this type of form is that it does not record specific hours lost or that the work recorded was considered at the time to be the subject of disruption or lost production or to what extent. The form in Figure 6.4 may be more useful in this regard.

The intention of the form in Figure 6.4 is not to record just the totality of resources and time spent on disrupted work, but something more specific. This can be either the hours lost due to disruption or the duration of the period over which disruption is said to have been incurred with an assessment of the extent to which it was being disrupted. In both cases the idea is to record not only the time but also the cause. Recordings in the columns headed 'Time stopped' should speak for themselves in terms of quantification. In the columns headed 'Time disrupted', the recording of periods of disruption is accompanied by an assessment of the degree of disruption in terms of a percentage, being the time considered to have been lost divided by the total time spent. This assessment will be the opinion of the supervisor filling out the form, assessed and recorded at the time.

The specific and rather pointed nature of the form in Figure 6.4, particularly the percentages stated in the final column, is likely to mean that obtaining a counter signature may prove even more difficult than in the case of the form in Figure 6.3. However, in both cases, whether the record is signed by the other party or not, such a record still retains some value, as a contemporaneous, if not agreed, record. An issue in this regard may become whether the refusal of the requested counter-signatory was because of an honest concern that the record was inaccurate, or too speculative, or for no better motive than a desire to avoid providing the contractor (or subcontractor) with such a useful basis for quantifying its disruption claim.

These figures are examples of contemporaneous forms that have been used to record or later calculate the amount of time lost due to disruption, in this case to labour, although a similar approach could be applied to plant and equipment. Where such records are not kept, efforts are sometimes made to produce such allocations retrospectively. This can take the form of allocations supported by witness statements produced subsequently but said to accurately identify what happened at the time. Examples of this include the retrospective allocation of labour and plant to work elements or of hours to specific events. Often such an approach will combine the memories of site staff with some reference to supporting contemporaneous records such as diaries, photographs and video recordings. Such retrospective exercises can be hugely expensive and time consuming, particularly compared with the time that it takes to maintain such records contemporaneously. In addition, such retrospective exercises are open to detailed attack on their credibility. An example of such an approach, and its shortcomings, can be seen in *Bridge UK Com Ltd v Abbey Pynford* [2007] EWHC 728 (TCC), [2007] All ER(D) 156(6), in which Mr Justice Ramsey had to consider a claim for management time based on a retrospective assessment by a witness of the time spent on various activities. Mr Justice Ramsey stated:

> It must be borne in mind that such an assessment is an approximation of the hours spent and may overestimate or underestimate the actual time which would have been recorded at the time.

On this basis he then went on to reduce the hours claimed by 20%. This might be considered by some who have presented unsuccessful disruption claims as a relatively good outcome. However, in this case there was no great doubt over the credibility of the witness giving the estimate. It is to be expected that if the credibility of witnesses carrying out such retrospective allocations can be undermined, then the credibility of the whole allocation might be similarly undermined. For example, if the allocation can be shown to be wrong, not just in a few isolated examples but in significant aspects or to an extensive extent, then it might be that the whole claim will fall rather than being reduced by a margin such as that of only the 20% discount in *Bridge*. The need therefore to ensure a sufficient level of accuracy, and therefore credibility, is great, adding further to the costly and time-consuming nature of such a retrospective approach. It is so much better therefore to maintain contemporaneous records of the types illustrated in Figures 6.3 and 6.4.

It may also be relevant that Bridge UK's claim for management time was only for £7,680, as a consideration of proportionality. In a much larger example of a disruption claim on a new retail development, the contractor based its disruption claim on hundreds of individual events, to each of which it allocated claimed resources and time lost on the basis of witness statements from a large number of its former engineers, foremen and chargehands, as witnesses of fact. The matter settled, so it never reached the lengthy trial at which the credibility and factual recollections of those witnesses would be tested. However, one amusing aspect was that as those witnesses came to understand that the contractor's whole multimillion pound claim depended on them, they started to compete to see who could extract out of the contractor the highest hourly rate for their time!

A few words need to be said about the use of photographic and video records as a method of recording disruption, although such records have been considered in some detail elsewhere in this book. Dated photographs of the state of progress and activities in a location can be useful in recording as-built progress and delays at fixed points in time. However, all too often both photographic and video records are taken intermittently and provide an incomplete record that diminishes their usefulness. It is obvious (but not always observed) that photographs need to be dated at the time, but it also often happens that a photograph (particularly where it is one of many) is taken with no record being kept of precisely where it is, let alone what it is supposed to show. When such records are looked at later they may be of little use unless the cameraman is available and can recall such facts. For this reason, and others, video evidence is often much more useful. It records more than a fixed point in time; however, this too assumes that date, location and purpose are recorded. A further advantage of video is that a voiceover can be made at the time recording these issues. However, such a commentary is often either missing or fails to provide the information that the viewer will later need.

The use of permanently fixed cameras has been mentioned in Chapter 3, in relation to evidence for creating or checking an as-built record. They are much more useful in relation to prolongation than disruption claims, but can still be of benefit.

Video is much more useful as a record of disruption than still photographs. Stills are helpful in recording a state of progress and in relation to extension of time and/or prolongation claims. For a contractor this might be that work is completed to a point, but now stopped, and this is then linked to a lack of information. For the architect or engineer the benefit of a photograph might also be that an area has no work being carried out to it notwithstanding that there is no lack of information in that area. Disruption factors such as overcrowding of working areas can be recorded by still photographs, but this and other disruption factors are usually much better served by video with a voiceover explaining, in addition to the date and location, exactly what the aspect of disruption is.

6.4.5   The 'Measured Mile'

The most effective means of demonstrating the outputs that should have been achievable had disruption not occurred, whether or not those levels of productivity are compatible with the forecasts and assumptions in the contractor's tender, is that of the 'measured mile' approach. This relies on a period or section of undisrupted work that is sufficiently similar to that which it is claimed has suffered disruption and that it can be used as a basis for what the contractor would have achieved without the effect of the alleged disrupting events. The resources, hours and outputs achieved on this undisrupted section can then be compared to a section that was subject to the claimed disruption and the difference used as the basis for a calculation of the lost production and therefore disruption incurred.

This division of the work into disrupted and undisrupted parts for comparison can be carried out on a location or time period basis. Examples by location might be chainages or embedded structures on a roads project or by floors of a high-rise building or individual properties on repetitive housing projects. Division by a time period might be monthly or before or after a particular date, or between dates or milestones. The choice between these alternatives will depend on the facts of how disruption has impacted the project (for example has it hit some culverts or apartments but not others?) and how records have been kept (for example is it possible to separate out resources by chainage or property?). If comprehensive and detailed allocation records are available, then such comparison by location should be possible. However, it often happens that they are not. One advantage of comparison by time is that it can more often be detailed by reference to project records that were not initially kept for the purposes of a disruption claim, but later prove to be useful to that end. This includes reports such as interim payment valuations, cost/value reports and management reports. That these are usually maintained on a monthly basis commonly means that measured mile analyses based on periods of time are by calendar month.

While the measured mile approach may be the ideal approach to a retrospective disruption calculation, in practice they can be difficult to successfully achieve. Often this is due to a lack of sufficiently detailed records that allow a complete and accurate comparison between sections. Sometimes contractors attempt to plug any gaps in the detailed records by making assumptions and approximations. For example, it may be that there are gaps in the allocation records for labour or an item of plant and the contractor uses such as the average allocation for periods that are being analysed. Whether such an assumption is reasonable will depend on their extent and the facts of each case and whether there is some evidence that it is reasonable. However, often such assumptions appear to be no more than conveniences for the contractor to increase the resulting quantum of the claim on an assumption that is too large and significant and cannot be properly tested.

A common complaint from contractors faced with a request from an employer's advisors for a measured mile approach to disruption is that the causes of the disruption affected all the relevant activities throughout their duration such that the possibility of an undisrupted 'control' or 'baseline' section of work did not occur. This relies on the assertion that the extent of the employer's failures were so endemic through the project that they rendered the approach impossible. The contractor will then say that the employer (through its certifying engineer, architect or contract administrator who demands such a measured mile approach) is seeking to benefit from the extent of its own failures.

It may be that no section or period of the work was completely undisrupted, but perhaps one can be found that had the least disruption and this can be used as the baseline. Comparison with the disrupted section or period will not then capture all of the productivity effects of disruption, but this is the best the contractor can do. It might then also consider some other way to address the disruption missed in such an approach.

If the 'measured mile' approach is feasible and there are sections or periods of work unaffected by the alleged causes of disruption, then much will depend upon the quality of records kept of the undisrupted work. It is not uncommon to find that, once causes of disruption have been identified, the contractor has instituted a system of record keeping providing comprehensive records of the circumstances and resources utilised on the disrupted work. Certainly, if the contractor does not institute such a record-keeping regime its omission may seriously prejudice its prospects of compiling a viable claim for compensation later and it only has itself to blame. In contrast, work that was not subject to the alleged disruption factors may not have such good records. The prime reason for this difference is usually one of chronology. The baseline work or period often proceeds ahead of the incidence of the disrupting events and therefore at a period when the potential requirement for detailed records of an undisrupted period or work section is not appreciated or anticipated by the contractor. It therefore does not maintain such records that it later realises might have given it a basis for a measured mile disruption calculation.

When comparing outputs on disrupted and undisrupted works or periods it is essential that the effects of any differences between them that are nothing to do with the disruption are identified and adjusted for. Sometimes the disrupted work occurs at an earlier time than the undisrupted work. For instance, when the cause of the disruption such as late and piecemeal access or design information is overcome, the impact on the relevant activities is thereby removed. In this case, the resources, times and outputs of the later work can be recorded and used to demonstrate what should have been possible had the earlier disruption not occurred. However, it may also be (for example) that during this earlier time, the contractor culpably had insufficient supervision or plant mobilised. It has also been seen in practice that the contractor attempts to achieve exceptional outputs on a small part of the later undisrupted work or period and proffers this as its measured mile in order to inflate the difference in output with the earlier disrupted work. This might include some economy with the truth in relation to recording the resources used on the undisrupted work. Only an objective examination of the working methods, resources and records will be able to determine if such distorting factors are present.

A further consideration should be whether the baseline section or period used in a measured mile calculation is representative or distinct from other parts of the works such that the comparison is misleading. On a road project the contractor proffered as its baseline section of a highway one 100 m length that was constructed alongside its site camp and establishment. The fact that this facility housed staff and resources managing works up to 35 kilometres away was considered to be more of a factor in the lower outputs in those remoter locations than any late information from the engineer. There was also plenty of evidence that the contractor failed to properly manage its resources in those remote locations. There were also doubts as to the accuracy of the records in those locations and the extent to which the resources recorded there were actually doing what the records suggested that they were doing.

This is an example of how, where a 'measured mile' approach is possible, it will be essential that like is compared with like, and any irrelevant influences having an impact on one section or period of work and not the other are identified and their impact removed. An example of this would be where the compared sections are established by reference to time periods and the fact that the allegedly disrupted section was not carried out in a period of less clement weather conditions or even over the winter. Another distorting factor might be comparison with an early period affected by 'the learning curve' of operatives. If it is not possible to remove any such imbalance in the circumstances of the compared sections, the 'measured mile' will not be appropriate as an approach to establishing the extent of disruption and the compensation applicable. The periods or sections of work also need to be representative and large enough to ensure that a meaningful comparison can be made. Attempts to project the results of a small measured mile across a number of other larger sections or periods should be considered carefully. What is an acceptable sample will depend on the facts in each case.

A measured mile calculation compares output, earned value or some other measurable factor in different periods or locations and there are a number of factors that can go to distort such a comparison. Some of these have been identified, but the following also need consideration:

- Who prepared the records on which the analysis is based?
- How have any gaps in those records been addressed?
- The contractor's remedying and re-working of defective work.
- Defaults and failures by other parties such as suppliers and subcontractors.
- Poor coordination of subcontractors and suppliers.
- Poor quality labour.
- Insufficient or poor quality supervision.
- Lack of materials.
- Lack of suitable plant and equipment.
- Do the compared works or periods involve different balances between subcontracting and use of direct labour?
- Do the compared works or periods involve different balances between off-site and on-site fabrication?

There are other potential distorting factors. Those in this list are considered in detail in Section 6.4.6.

Measured mile calculations are usually carried out by converting the outputs in the comparative periods or locations into units. These units can include:

- Earned value achieved per unit of currency spent. An illustrative example of such an approach is provided at the end of this section.
- Earned value achieved per labour hour expended.

- Earned value achieved per plant hour expended.
- Quantity achieved per unit of currency spent.
- Quantity achieved per labour hour expended.
- Quantity achieved per plant hour expended.

A simple example of a comparison applying an earned value per unit approach might be:
Concrete gang recorded as placing 310 m^3 per working week prior to disruption

Cost of concrete gang per week	$2,000.00
Cost per m^3 placed ($2,000.00/310 m^3)	$6.45

Output during disrupted period 260 m^3 per working week

Cost per m^3 placed ($2,000.00/260 m^3)	$7.69

The cost of disruption to the concrete gang is therefore an apparent increase in the labour cost of concrete placed during the disrupted phase of working of $1.24 per m^3. In this very simple example, further adjustments may be necessary for the plant and equipment elements of the operation.

There are two further aspects of calculations such as this simple example that can often cause some difficulty:

1. What if it can be established that the contractor had allowed in its tender for different levels of productivity to that established by the baseline work or period? Also, what if the tendered output was either higher or lower? How does this affect the calculation? For example, the contractor's pricing might have been based on 290 or 350 m^3 per week per gang rather than the 310 m^3 actual performance recorded by the baseline work or period. As explained elsewhere, the answer has to be that it is the effect on the contractor's actual operation that is measurable and is to be compensated. If the contractor had priced for a higher output than that demonstrated by the baseline work or period, it has to stand the loss represented by that difference as an estimating or tendering error. If it had priced for a lower output than was being achieved in the 'measured mile', it should not be penalised by having the lower tender output substituted into the calculation. As has been emphasised several times in this chapter, it will usually be the legal position in relation to such a disruption claim that the contractor is entitled to be put back in the position it would have been in had the disruption not occurred, and that means comparing actual outputs with 'but for' outputs, not those tendered.
2. The other, sometimes controversial, aspect of such calculations is where the engineer or architect, or their representatives, allege that the contractor has not overcome the disrupting factors as well as it should have done, or has been inefficient in its working. Thus, they will say that the output was lower than it should have been had the contractor responded to the disruption more effectively. This will depend on an analysis of the facts. The response to such objections may be that, providing the contractor has not demonstrated incompetence or taken measures that no reasonable contractor would have contemplated, it is not reasonable to object to actual performance in the light of difficulties that are not the liability of the contractor. It is also important

that the assessment of the effectiveness of the measures taken by the contractor is not based on the benefits of hindsight, but by reference to what a reasonable contractor would have done at the time that the disruption occurred. A further consideration is what contemporaneous evidence there is of such inefficiency, including whether the employer or its contract administrator recorded any concerns at the time.

Support for a measured mile approach, which was based on comparative outputs over time, can be found in the judgment in *Whittall Builders Co Ltd v Chester-le-Street District Council* (1996) 12 Const LJ 356. Whittall's claim for disruption was calculated on the basis of comparing earned value during disrupted and undisrupted periods. Mr Recorder Percival QC explained the claim as follows:

> Several different approaches were presented and argued. Most of them are highly complicated, but there was one simple one that was to compare the value to the contractor of the work done per man in the period up to November 1974 with that from November 1974 to the completion of the contract. The figures for this comparison, agreed by the experts for both sides, were £108 per man week while the breaches continued, £161 per man week after they ceased.

He concluded that:

> It seemed to me that the most practical way of estimating the loss of productivity, and the one most in accordance with common sense and having the best chance of producing a real answer, was to take the total cost of labour and reduce it in the proportions which those actual production figures bear to one another – i.e. by taking one-third of the total as the value lost by the contractor.
>
> I asked both Mr Blackburn and Mr Simms if they considered that any of the other methods met those same tests as well as that method or whether they could think of any other approach which met them better than that method. In each case, the answer was no. Indeed, I think that both agreed with me that that was the most realistic and accurate approach of all those discussed. But whether that be so or not, I hold that that is the best approach open to me, and find that the loss of productivity of labour, and in respect of spot bonuses, which the plaintiff suffered is to be quantified by adding the two together and taking one-third of the total.

Noticeable in this judgment is how the Judge considered that the approach was the best available to him. The fact that the resulting award was a figure of only £21,479.35 might also have been significant in terms of the need to adopt a proportionate approach.

However, such authorities for the successful use of a measured mile approach are rare among England and Wales court judgments. The problem with the measured mile approach to disruption claims is rather like most sports fans' experience of supporting their favourite team. There is always a hope of success. Sometimes there is expectation. However, all too often the realities leave such hopes or, particularly, expectations dashed. This reality is perhaps reflected in the slight change in position between the first and current editions of the SCL Protocol. The First Edition had said at its Guidance Note Section 1.19.7:

> The most appropriate way to establish disruption is to apply a technique known as 'the Measured Mile'.

On the other hand, the current edition says this at its Introduction paragraph K(f) when listing the key changes introduced by that edition:

> As in the 1st edition, the preference remains for a measured mile analysis, where the requisite records are available and it is properly carried out.

This change seems to recognise whilst a measured mile is the most appropriate retrospective approach in principle, it is often not possible in practice because of such issues as the available records, and that even where it is possible, the calculations are often distorted by gaps or insufficiencies in those records or flaws in the way in which it is carried out.

Internationally, the approach is the preferred method of the Association for the Advancement of Cost Engineering International, in its Recommended Practice No. 25R-03 – 'Estimating Labor Productivity in Construction Claims' 2004.

6.4.5.1 Illustrative Example of an Earned Value Approach

As has been introduced, earned value is a means of assessing the productivity of a resource such as site labour, although it can be applied to plant and equipment as well. If at the tender stage it is anticipated that, for particular operations that are the subject of a disruption claim, 12 000 man-hours of labour resource were anticipated to be required, and the value of the particular operations at the tender prices was, say, £725,000, then a crude assessment of the value planned to be earned by each man-hour expended was:

$$£725,000/12\ 000\ \text{hours} = £60.42\ \text{per hour}$$

There are, however, some difficulties with such a broad example of an indicator of productivity. Firstly, it takes no account of the mix of trades and labour input in the 12 000 hours. Secondly, it takes no account of the impact of the cost of plant, equipment and materials for the permanent works in the value of £725,000. It is not difficult to anticipate that differences in the labour mix, plant and equipment input or the value of materials for the permanent works, could quickly distort such an analysis when it is rerun with the actual site labour hours and work valuation figures to produce a comparable earned value figure for the actual work.

For instance, say that the actual site hours for the disrupted operations being analysed is 14 260 and the value earned as included in the final account is £683,000. This would derive an actual earned value for each hour spent on these operations on site of:

$$£683,000/14\ 260 = £47.90\ \text{per hour}$$

In this example the earned value per hour is apparently some 20.72% less than that anticipated at the time of tender. The calculation could, however, ignore the impact of, for instance, the contractor using a greater proportion of unskilled, or semi-skilled, labour in the site workforce than anticipated at the time of tender. This may have a potential impact both on the average cost per hour of the employed labour and on the output that could be anticipated in any event, although this might not necessarily be so depending upon the circumstances of the particular work being analysed.

6.4.6 Comparing Tendered and Actual Outputs

Where there are no contemporaneous records of the time lost due to disruption and it is not possible to achieve a measured mile comparison, there may be no real alternative than to compare the actual resources used with those that would have been incurred by an experienced competent contractor 'but for' the disruption. The problem will be in establishing the 'but for' in the absence of a baseline period or section of work for a measured mile approach. This may start with a comparison between the actual resources deployed and the resources anticipated by the contractor's tender calculations and programme. However, both of these sides of the equation – the planned and actual resources – will need some adjustment, the comparison must be 'like for like' and representative and the results should be subject of careful back checking against logic and such issues as the events said to have caused the disruption. This final point is of course true of any quantification of a contractor's claim on any method.

6.4.6.1 Tendered Allowances

Many contractor's disruption claims are based on a comparison of their actual outputs with those allowed in their tender. Whilst those tendered allowances might be a starting point, whether they were unrealistically high or low, they are only that and will require some further objective analysis and may need adjustment.

It may be necessary to examine the planned levels of production and to test them against other objective data and information to ascertain if they are what a competent experienced contractor could have anticipated. This can often lead to differing opinions as to what could, or could not, be achieved without the disrupting factors. This is a frequent cause of dispute. The judgement of what would have been possible has to be made on an objective basis if unnecessary dispute is to be avoided. The contractor must also accept that, if the comparison with that data, etc., reveals the estimates behind its tender pricing to be deficient, it cannot make good any tender errors as part of the disruption calculation.

One source of objective data is published productivity norms. These have to be treated with great care before being used for this purpose: firstly, in terms of their accuracy and reliability and, secondly, to establish that they properly reflect the circumstances and nature of the tendered project. Broad industry pricing books are likely to be too general and approximate. A test of the reliability of broader publications such as those published by Spons, Wessex and Laxtons is that contractors are unlikely to use them to price their tenders. These publications are often regarded as of more use in relation to budget estimating and comparative costs and to be on the conservative side. However, more specialised publications such as those published by Luckins in the UK for mechanical, electrical and plumbing installations might be of use in relation to those trades, particularly given that they are widely used by specialist subcontractors to price tenders for such works. In all such cases, due regard must be made for differences between the project and those upon which the published data were based. Factors such as site location, accessibility, weather seasons, etc., should be considered in this regard.

In respect of plant outputs, publications such as *The Caterpillar Handbook* are a useful source of potential data for excavation plant outputs, and other plant manufacturers publish their own data.

A contractor's tendered allowances might not only be subject to overoptimistic or pessimistic outputs but may also include basic errors such as missing items of work or activities that were required by the tender enquiry documents. This might be the result of a simple error on the part of the estimator or because of a misunderstanding of the scope of the work to be done by the contractor. On the other hand, the tender might include an error that meant it overestimated the resources and times that would be required. Where such errors are identified, they will require adjustment of the tender allowances before they are compared with actual performance.

Given that this part of the disruption calculation should represent what the contractor would actually have incurred without the disruption, it is correct to adjust the tender allowances for the benefit of hindsight to the extent that this means that the tender allowances would not have been achieved in practice. Take, for example, the case where the estimator assumed that the ground conditions would be of a nature that were different to those actually experienced and the contractor had accepted that contractual risk. In that example, it is the actual ground conditions that are relevant to establishing what would have occurred without the disruption, and hence the calculation of the effects of that disruption, not those allowed for.

The use of a contractor's tendered allowances as a comparator to its actual outputs is likely to be criticised on the basis that the estimators may have got it wrong and their prospective calculations were made at a time when they cannot have been completely aware of the actual circumstances and conditions under which the works would have been actually carried out in the absence of the disruption complained of. There are two further ways in which these complaints might be addressed: firstly by looking at other resource planning carried out by the contractor on the project and secondly by the use of expert evidence of outputs and resources for that type of work and project.

It is a common complaint of some contractors' site teams that they inherit project costings and budgets prepared by an estimating team and that they did not properly understand the realities the site team would face in actually constructing the works. It is even sometimes complained that the estimating team's only concern was to win the tender competition. Whether this is fair or not, it does reflect the reality that it is the contractor's site team that is likely to be best qualified and motivated to properly plan what outputs and resources would be achieved in the absence of disruption. The result of this may be that a first task of the site team is to prepare its own resources projections and/or a resourced programme.

In this regard, a programme such as that required by the FIDIC Red Book clause 8.3 may be relevant. In addition to the contractor's detailed time programme, and any revision to it, as required by that clause, subclause 8.3(d)(ii) requires a supporting report that includes:

> … details showing the Contractor's reasonable estimate of the number of each class of Contractor's Personnel and of each type of Contractor's Equipment, required on the Site for each major stage.

It is also not uncommon for particular conditions attached to such standard FIDIC terms to add further requirements for details of the contractor's resource planning and its clause 8 programme.

Where such information is available it may serve a disruption analysis that compares tendered and actual resources in one of two ways. Firstly, they may support the estimators' calculations, such that the two provide mutually supportive evidence for the 'but for' costs. Alternatively, it may be that they are considered a better basis for the disruption calculation and replace the estimate entirely.

A further possibility is for the contractor to obtain expert evidence as to the expected outputs and required resources for the project. However, this can be an expensive exercise and is probably only justified if the amounts claimed as disruption are considerable. Obtaining authoritative expert evidence on issues such as this can be difficult. However, it can add real value where there are issues as to the adequacy of the original tender and a lack of contemporaneous project budgeting from the site team.

6.4.6.2 Actual Outputs

The first issue in relation to actual outputs will be the accuracy and reliability of the records on which they were based. Allocation records should be contemporaneous, but who prepared them and on what basis were they doing so? Was this by reference to the compiler's first-hand accurate knowledge of what resources were working on at any given time? Was the compiler aware of the use and importance to which those records would later be put? It is an occasional theme in relation to contemporaneous site records that those compiling them later admit that had they known of such use and importance at the time, then they would have been more precise in their recording. Repeated identical resource allocations over a period of time are often a clue that the allocation was no more than an exercise in administrative 'box ticking' that required the records to be maintained, rather than providing an accurate basis for a subsequent contractual claim.

The problem of how to address any missing periods of allocation has been mentioned elsewhere. At its simplest, and least forgivable, this has been seen where the member of the contractor's site staff charged with keeping such records went on a two-week holiday and no one was assigned to cover for him over that period. The contractor was left with no alternative than to project the average allocation for the days immediately preceding and following that gap across the unallocated days. This created an unnecessary area of doubt and debate, particularly because it covered a period when many operatives could also have been expected to be taking their main annual holiday.

Typically, the contractor will endeavour to demonstrate that the effect of the cause of the disrupting factor on the particular activity is that less output was achievable by the labour, plant or equipment deployed on that operation. This may, for instance, be a reduction in output in terms of bricks laid per hour by a brickwork gang or a reduction in the amount of material moved per hour by an excavator. It will, however, be necessary to demonstrate that no element of the output reduction is due to factors for which the employer is not liable. If there are other factors affecting the output they will need to be identified and removed from the compensation calculation. Failure to remove any factors for which the employer is not liable may prejudice the whole analysis.

Some of the potential factors that will go to distort a contractor's actual resources used, and hence their calculated outputs, and which will require adjustment to the calculation, are listed as follows. They are sometimes referred to as the contractor's 'own goals'.

- The contractor remedying and re-working defective work. Sometimes this leads to a debate regarding whether the extent of such defects was to some extent the consequences of the disruption pleaded and should therefore be included in the calculation of the disruption, rather than being a subject of correction to it. There may also be debate as to whether some level of remedying and reworks should always be expected and would have been allowed for in the comparative tender outputs.

- Defaults and failures by other parties such as suppliers and subcontractors. These may also be the subject of a separate claim by the contractor against the defaulting third party. Whether such a third party claim is being made or not, the consequences of such a default or failure are unlikely to be a valid part of the disruption calculation.
- Poor coordination of subcontractors and suppliers. This can also become confused by assertions by the contractor that the disruption itself caused its problems with such third parties, for example because their works or supplies had to be overlapped or accelerated or there was ongoing doubt as to what they would have to do or supply and when.
- Poor quality labour, giving rise to lower outputs than ought to have been achieved or the increased need for re-working. A particular complicating factor here is where the contractor addresses the disruption by introducing additional labour, for example from labour agencies. These might be expected to be less efficient than its own payroll employees. There may therefore be some debate as to whether this should be included in the calculation of the disruption and considered an essential component of that, rather than subject of a correction to it.
- Lack of supervision. Poorly supervised labour will always produce lower outputs than it otherwise should. Again, a complicating factor may be where the ratio of operatives to supervisors reduces because labour numbers increase but the supervision numbers cannot be increased in proportion. Alternatively, it might be that the contractor has to open up more work fronts to be worked in parallel, thus spreading its labour across the site but without the necessary additional supervision to cover multiple areas. Even where additional supervision is introduced to manage additional labour, that new resource may also be from sources other than the contractor's own staff. For example, staff from an agency may be of a lower quality than the contractors' own team and it may be relevant that it is unfamiliar with the project and the contractor's internal processes and procedures.
- Lack of materials. Poor scheduling of deliveries by the contractor can leave its labour lacking materials required for them to progress their work. As with other potential 'own goals', if this is the contractor's culpability, then it needs adjustment in the comparison of planned and actual outputs, but this may be only part of the story. It may be that an effect of the disruption is on the contractor's timely receipt of materials. For example, if a cause of the claimed disruption is late design information, then that might leave labour short of materials because they could not be ordered until the specification and scope was clear.
- Lack of plant and equipment. This may involve similar considerations to the lack of labour.
- Uninstructed additional works. This may be unusual, but on one road resurfacing project production on purportedly disrupted locations was found to have been affected by the contractor chairman's habit of offering to various local dignitaries that his plant and operatives would re-grade their local access roads as they passed their villages gratis!

6.4.6.3 Other Potential Distorting Factors

There are a number of factors that can go to distort a comparison of planned and actual labour outputs, even after they have been adjusted respectively for tender errors and construction 'own goals'. These include making sure the analysis is of a representative and sufficiently large section or period and that the comparison of resources is on a 'like for like' basis.

It may be misleading if the analysis is not of a complete activity or operation, but are analyses of just a small part of an activity or operation on site, as there will always be small samples of activities that may show different results to an objective analysis of the whole.

Matters that might mean that a comparison of planned and actual resources is not on a 'like for like' basis include changes in how the contractor procures and manages its resources. For example, it may have planned to subcontract more or less of its works than it actually does in the event, changing the proportion of the works actually done by its directly employed labour. If the claim is only to address this contractor's payroll labour, then this can give it a problem in the calculations. It will have to find some way to adjust either side of the equation to provide a proper comparison. On the other hand, analysis might show that whilst the contractor's plan had been to use its own labour, it was so delayed and disrupted that it had no choice but to subcontract parts of its works. That was essential to a joint venture's claims on one rail project, where it had tendered to use its own labour to construct concreted structures for all of the stations, but in the event had to subcontract some of them because the extent of overlap in their construction increased because of late access for others. This meant that a direct comparison of its planned and actual labour resource outputs on the project needed considerable care. However, it also gave rise to another head of contractual claim related to the enforced additional costs of subcontracting more works.

The amount of work that is subject of off-site prefabrication can also change where delay and disruption occurs. Take, for example, steelwork sections. The tendered plan might have been to have these prefabricated by a supplier or in the contractor's own central workshops, with less time spent in site erection of prefabricated structures. If, in the event, they are brought to site in individual components because of such things as design delays, then the amount of site labour time spent to assemble and erect them will increase. A disruption analysis that is restricted to consideration of on-site labour alone will therefore show increased hours that are not entirely additional, being offset by a reduction in off-site hours. However, consideration would also have to be made of a potential separate claim for inefficiencies or other costs of changing the balance between on-site and off-site work if that was the consequence of disruption.

The ratio of labour to plant might be an important factor in the labour's productivity and if that ratio changes this can also change the output of the labour. If the contractor makes such a change of ratio out of its own strategic volition, then any consequences for its labour productivity would have to be factored out of its claim. On the other hand, if it was the result of the disruption itself, then it might be claimable. For example, the contractor might have had to bring in additional labour to deal with the effects of the disruption, but was unable to similarly increase its plant because of a lack of availability. That additional labour might have achieved some additional output, but its output per man hour may have dropped and be a component of the effects of the disruption.

Ensuring that the comparison of planned and actual resources is on a like-for-like basis also requires adjustment for any other claims made by the contractor that include hours that are in its actual resources. The most obvious area for this is claims for variations. To the extent that variations have absorbed resources and their time, the actual resources in a disruption calculation that compares actual and 'but for' resources would have to be adjusted for those before comparison with the planned resources. This will require a detailed analysis of the build-up to all variations in the period or the works that are the subject of the disruption claim. Complications in this regard can include:

- Where it is difficult to establish when variation works were carried out and adjustment to a disruption calculation needs that timing information.
- Identifying the resources and hours involved in the variations. The resource hours in variations priced on the basis of the contractor's actual time and resources, for example on a daywork basis, might be readily established. However, where variations are valued at contract rates and prices, this will require a detailed understanding of allowances for such as labour and plant in those rates and prices.
- Under remeasurable contracts such as the FIDIC Red Book and the Infrastructure Conditions, the fact of any increase in quantities of work will involve additional labour and plant recoveries and again these will need to be determined by reference to the details of the contract rates and prices applied to those additional quantities. Of course, there may also be a reduction in the quantities as remeasured, which may require an adjustment of a disruption claim in the other direction.
- What to do about claimed variations that the contractor is not actually entitled to. Contractors often conveniently ignore these by only including adjustments to a disruption calculation in relation to variations that are agreed and in their agreed values. However, on a large account there are likely to be works items that were claimed by the contractor as variations, but which were actually part of its original scope and not additional. If the claim for such items was because the contractor missed them in its pricing, then they should be adjusted for in any comparison of tendered and actual resources as a basis for a disruption calculation.

In addition to variations, the contractor may have other claims that include additional resources that are part of its disruption claim. The most common example of this is a prolongation costs calculation that covers the same period as the disruption claim. The issues involved in disentangling prolongation costs from disruption costs are considered in Section 6.10. Other considerations in this regard might be overlaps with claims for acceleration (considered in Section 6.5) and suspension of works (considered in Section 8.4). Where a disruption claim is calculated by reference to comparative resource costs (rather than resource hours), there may also be overlap with a claim for cost escalation (considered in Sections 6.2.3 and 8.3).

6.4.6.4 Back-Checking the Results

It is essential with any approach to the evaluation of disruption costs that the results are subject to 'back-checks', to consider if the financial results of the calculations show a reasonable reflection of the events that are claimed to have caused the disruption being calculated. This is particularly so with calculations comparing actual costs with those tendered or those that should reasonably have been tendered. From the contractor's perspective it may show good correlation that goes to support the approach. However, it is often the case that such consideration of the results of the detailed calculations raises important questions regarding the credibility of the calculations and assumptions made. It might particularly point towards parts of the differences between actual and 'but for' costs being explained by other matters that have no part in the disruption claim at all.

Table 6.1 Disruption factors by resource.

Resource	Unit	Total hours (hours)	Less other earthwork claims (hours)	Less tender recasting (hours)	EO hours (ROs)	Disrupted (%)
Rock	hour	78 875.550	39 807.390	29 673.580	9 394.580	32
Breaker						
Shovel	hour	212 941.100	24 705.620	97 533.750	90 701.730	93
Truck	hour	647 208.200	139 719.670	382 910.320	124 578.210	33
Dozer	hour	161 149.520	26 837.540	54 739.340	79 572.640	145
Water	hour	201 518.660	33 497.150	78 825.710	89 195.800	113
Tanker						
V roller	hour	176 347.200	16 328.380	73 395.760	86 623.060	118
Grader	hour	179 146.400	19 364.100	46 373.420	113 408.880	245
Helper	hour	1 417 409.500	312 201.930	100 857.280	1 004 350.290	996
S labour	hour	582 647.500	164 685.990	5 839.920	412 121.590	7 057
Chargehand	hour	61 045.500	11 287.140	0.000	49 758.360	infinity

Table 6.1 is a real example from the analysis of a contractor's claims on a roadworks project. The contractor prepared its claim by comparison of actual plant and labour hours adjusted for its other claims with its tender 'recast' for its remeasured scope of work. In this case the calculation was for construction of filled embankments.

Table 6.1 shows further analysis by the employer's quantum expert of the contractor's claim, broken down further than the contractor's submission, but using the base information provided by the contractor. This showed the resulting claimed disruption (time lost or 'EO hours') on each type of plant used and each grade of operative involved in the work expressed as a percentage against the recast tender hours. This analysis led to a number of detailed questions regarding the contractor's approach, some of which will be obvious from Table 6.1 and some of which did have a rational explanation and could form the basis of an informed disruption calculation. However, it is provided here to illustrate, by way of a real example, how a detailed back-check on the results of a disruption claim can give rise to fundamental questions for further analysis. The question that needs to be asked of such results is whether the results fit with the disruptive causes and effects pleaded. For example, why was the percentage hours lost on the shovels three times that on the trucks and the loss on skilled labour so astronomically high?

In the same case, further analysis by the employer's quantum expert between different elements of the road's construction showed that the contractor's calculations resulted in claimed disruption factors on each, as set out in Table 6.2. By converting the results of the contractor's disruption calculations into such an analysis by element it was possible to consider whether these different factors reflected the alleged causes of disruption said to have led to them. It will be noted in Table 6.2, for example, how the percentage disruption broadly reduced up the road layers, and this would benefit from analysis of where and when most of the claimed disruption causes were said by the contractor to

Table 6.2 Disruption factors by road element.

Suitable excavation to embankment	89%
Unsuitable excavation to waste	90%
Borrow to embankment	66%
Granular sub-base	104%
Aggregate base course	58%
Bituminous base course	55%
Bituminous wearing course	42%

have occurred. Furthermore, the anomalously high productivity loss on the laying of the granular sub-base would merit detailed consideration as to why that layer was so highly disrupted in comparison to the other layers.

6.4.6.5 Conclusions on Costs Less Recovery Approaches

In the absence of the records of actual lost resources and times or of the allocation records that are required to enable a measured mile approach to disruption, it may be that there is no alternative to a claim that starts by comparing the contractor's planned and actual resources. The planned resources could be supported or adjusted by consideration of planning by the contractor's site team and expert opinion. Suitable adjustments should be made to allow for any errors in such plans that mean they do not reflect what the contractor would actually have incurred had the disruption not occurred. On the other side of the equation, actual costs would have to be discounted for any own-goals by the contractor and adjusted to avoid duplication with the contractor's other claims such as for variations. The results of the comparison of planned and actual costs should then be checked against the claimed causes and effects of disruption, to establish whether the resulting losses reasonably reflect likely outcomes of the causes.

It is easy for the recipient to dismiss out of hand a crude costs less recovery approach to evaluating disruption that has not been the subject of checks on the reasonableness of asserted 'but for' costs, efficiency of the actual costs incurred, adjustments to ensure a like-for-like comparison and back-checks on the results, in their entirety. However, as noted elsewhere, the choice of approach to evaluation of any contractual claim has also to consider two other issues. The first is proportionality, where the amount claimed is such that a more expensive approach to analysis may not be justified. The second is that the test of the claim is likely to be the balance of probabilities. Where the fact and effects of disruption are patently clear it may be that an even broader approach is acceptable, as illustrated by the judgment in *Harvey Shopfitters Ltd v ADI Ltd* [2003] EWCA Civ 1757. There John Uff QC had to consider a costs less recoveries claim for labour disruption in the round sum of £50,000. He held:

> I accept the substance of [the claimant's expert's] analysis showing that the claimant incurred labour costs substantially in excess of those allowed for. The different calculations put forward by the claimant and by [the claimant's expert] all assume that the claimant was blameless and that all established additional costs should be recoverable. Having considered the evidence I am not persuaded

that the whole or even the majority of the claimant's additional labour costs can be attributed to the matters alleged. On the contrary, there was evidence of the unproductive and wasteful use of labour. Doing the best I can on the material presented, I assess that one quarter of the minimum figure put forward on behalf of the claimant by [the claimant's expert] (£50,000) is properly attributable to additional labour costs and I allow the claimant the sum of £12,500.

Several issues are notable here and echo comments made elsewhere. The contractor's comparison of its actual labour costs with those allowed for included supporting expert evidence. The Judge recognised that the contractor's actual labour costs were partly due to its 'own goals', and made a broad assessment of the effects of those on the overspend. That broad allowance of 25% of the amount claimed was his applying his judgment on the materials available and applying a balance of probabilities approach. It is also notable that the amounts claimed and awarded were relatively minor.

In the context of a larger claim, those dismissing a costs less recoveries approach might consider the judgment of HH Judge Thornton in *Amec Process and Energy Limited v Stork Engineers & Contractors BV* 1997 ORB 659. On the broad topic of disruption and its quantification, he said:

> Unusually, the quantification exercise is neither precise nor undertaken using clear cut methods of measurement since human activities, unlike physical phenomena, are not susceptible to precise methods of measurement and the units and process of measurement can be somewhat subjective. By way of example, the speed of liquid flow and the temperature and pressure of gas flow, can readily be measured whereas the rate of fabrication and installation of pipework supports or the extent to which the actual costs of fabrication have been recovered from payments based on a large number of rates made up from historically derived norms are not so readily measurable.

Amec's claim was prepared starting with what number of labour hours was included in its tender. It adjusted this for the additional hours built into the valuation of additional works valued on its contract rates and prices. Deducting the resulting total from the recorded actual number of hours, Amec calculated its disruption as 200 000 hours. He accepted that it was not possible for Amec to measure how many of these additional hours related to each piece of pipework fabrication or erection, in the absence of a contemporary time and motion study. He concluded 'Even then, no clear idea of the disruption involved will be ascertained since the overall disruption is the accumulation of thousands of disrupting events reflected in the overall number of additional hours worked'.

Amec was able to establish that the norms used in its tender calculations were based on its wider experience, where it had used such rates successfully and also published norms. As a result the Judge saw that only four findings were necessary:

1. How many hours did the relevant operatives actually work.
2. How many of those worked hours have Amec, applying standards of fairness and reasonableness, already received payment for under the terms of the contract.

3. How many of the residual number of hours were hours that Amec, again applying standards of fairness and reasonableness, should bear the cost of itself.
4. At what rates should the residual number of hours be remunerated. The answer requires a rate or rates to be fixed, based on the direct cost of providing the labour but subject to any reasonable adjustment upwards or downwards to reflect, again applying the standards of fairness and reasonableness, the circumstances under which the hours were worked.

He concluded that Amec had succeeded in its case for a claim on this basis given the weight of factual and expert evidence regarding the facts of the events and their effects on it:

> My conclusions are that Amec's case succeeds as a result of the mass of primary evidence adduced from the witness statements, oral evidence and the contents of relevant primary documents. This evidence has, in material respects but not wholly, been accurately summarised in Schedule 24. The case is supported and corroborated by, but is not dependent upon, the contents of [its weekly reports] and the schedules, histograms and other documents prepared by or under the direction of [Amec's expert] which subsequently formed the basis of the many experts' agreements reached before and during the trial.

6.4.7 Factors Affecting Productivity

In assessing the disruption that has been caused to labour, a common historical approach was to take the total labour cost actually incurred and apply an assessment, for example that '25% of labour time was lost due to disruption'. Such a broad and global assessment is unlikely to stand scrutiny today. Such factors as the potential large sums involved, increased scrutiny, particularly of public expenditure, and the greater knowledge of practitioners, mean that some more refined approach will be required other than in rare cases. These rare cases might be where the sums involved are small and/or the parties have a business relationship in which they are content to apply a broad brush approach to such issues.

It has been explained elsewhere how loss of productivity can arise out of disruption through a number of effects such as overcrowding or overtime working. One way to apply some detail to a disruption calculation, particularly to the application of percentage factors, is to add further details to the effects of disruption and apply factors to each. To those details it may be possible to apply factors assessed as reasonable for such effects, taking each in turn. This at least adds some particularisation and science to what otherwise may be far too broad a calculation to stand scrutiny, for example where it is based on no more than '25% of labour time was lost due to disruption'.

This approach could be considered as adding an often missing intermediate step between cause and effect. Many contractors present disruption claims that detail at great length, for example, instructions, variations, late approvals, etc., that they say caused the loss and then attach the loss to those causes. The link between the causes and the effect are no more detailed than that and it is this link that can fail. However, it may be possible to further detail the effect. For example, a disruption claim might have been presented in the following summarised terms:

> The following events [listed and evidenced] caused disruption costs as follows [the calculated loss in $].

Analysis of the facts of what the listed causes actually led to might, however, show that they lead to the contractor's losses through several intermediate effects and this claim could have been better presented in terms summarised as follows:

> The following events [listed and evidenced] caused the contractor to incur: greater overlapping of activities [evidenced]; overcrowding of work faces [evidenced]; and overtime working [evidenced]. These effects caused disruption costs as follows [the calculated loss $].

Such a detailed approach might go beyond the vague description of the work as being disrupted to a claimed extent, by looking in detail at the events and breaking down their effects such that each can be considered in more detail than a bald application of such as 'say 25%' for all effects. It may be that an analysis that applies this more detailed approach still leaves some remaining part of the contractor's productivity losses, but this will be significantly less than the 25% in this example. What remains might be referred to as the 'rump' of the contractor's disruption costs.

Some relevant factors that affect productivity are considered as follows. Many of them have been subject of research and statistical analyses dating back several decades. These factors might be applied in a 'bottom up' approach to the evaluation of disruption (see elsewhere in this chapter) and particularly where disruption is calculated on a 'value' basis as part of the valuation of variations (see elsewhere). This latter usage in particular is on the basis that many of these studies are considered by some to be more relevant to preparing a contractor's estimate than quantifying claims for damages. Thus, where disruption is to be included as part of the prospective quantification of a compensation event under NEC4-ECC, or a contractor's quote for a variation under SBC/Q's clause 5.3 and Schedule 2, such data might be considered to be particularly useful. However, with appropriate caution it may also have its uses in relation to a claim for disruption calculated as damages, as is explained elsewhere. Such statistical analyses are often proffered by contractors as a basis for calculating their claim, although they can also be useful to those scrutinising a claim in relation to the effects of a contractor's own failings, for example where its tender missed allowance for overtime working that would always have been required for its allocated resources to achieve the programme.

6.4.7.1 Excessive Overtime, etc.

A particularly common symptom of construction and engineering projects that are subject to disruption is an increase in the working of overtime by labour and its supervisory staff. However, a common problem area is how much of the overtime worked was actually additional and the result of the claimed causes. It may be that the contractor tendered to work some extent of overtime anyway. This is particularly common on remote projects or where labour is imported and housed in a labour camp such that maximising income is of more importance than leisure time. Even if a contractor did not always plan to work some extent of overtime to some degree, it may be that it has had to institute it because of its own failures, such as culpable early slow progress on the works. In such cases separating out the part of the overtime that is the consequence of disruption can be problematical.

The most obvious direct cost of overtime working is the premium hourly rates paid for it. This might be at 'time-and-a-half' and/or 'double-time' or some other premium rate. These rates might be based on a national, local or project working rule agreement or the specifics of a particular employee's employment contract. The non-productive element, i.e. the additional cost of working at premium overtime rates, as opposed to the plain time 'ordinary hours' rate, can usually be easily separated from the total cost.

It is usually a reasonable assumption that there is no additional production achieved for this non-productive payment element of overtime payments. Being paid a time-and-a-half rate to work through an evening will not make an employer work at a greater rate than his or her normal rate of production through those additional hours. In fact, it is likely that those additional hours will see productivity at a lower rate than normal, on the basis of increased levels of fatigue. Indeed it can be argued that where substantial overtime working is introduced the employees' output during their 'plain time' working hours of a normal day may also be reduced due to the increased demands of the longer hours. In practice it is often difficult to identify such an effect on production separately to the effects of the disruption alleged to be affecting the work in question. However, the effect may be demonstrable in circumstances where substantial additional shifts, or six- or seven-day working in lieu of five-day working, are introduced.

This area of the effect of overtime working on productivity on construction sites is one that has particularly been the subject of statistical analyses dating back over the last 40 years (see, for example, the report 'Impact of Overtime on Electrical Labor Productivity', The Foundation for Electrical Construction Inc., Awad S. Hana, University of Wisconsin, 2011). Such studies consider such factors as working of longer days and weekends and also the cumulative impact of these over a number of weeks. Thus, they can show that working extended overtime for one week has a lesser impact than working it for several consecutive weeks. The statistics derived from such analyses are often also combined with the direct cost of the premium overtime rates paid to calculate an overall effect of working overtime.

Such overtime costs might, for example, be claimed to be the result of instructed or induced acceleration, or the result of the contractor's need to mitigate its own delays. Thus contractors' calculations can show that the combined effect of premium time payments and inefficiency said to result from the working of a ten-hour seven-day week is as high as a combined loss of $x\%$.

Alternatively, from an employer's viewpoint, if it can be shown that the working of overtime was not a matter for which it was responsible, then part of larger losses claimed by the contractor to have been caused by disruption factors for which the employer was liable might be deducted from the claim based upon statistics for that level of overtime working. In this regard claims made by contractors that they have been instructed or induced to accelerate are often, at least in part, countered by evidence that the contractor actually always planned to work a degree of overtime, or that its original programme and resources would have always required it. Alternatively, it is sometimes shown that the overtime was being worked already, even before acceleration was allegedly instructed or induced. Acceleration is an area considered in detail elsewhere in this chapter.

Such issues need careful consideration of the facts in each case as well as consideration of the appropriateness and use of such statistical studies in terms of not only their general reliability but also their similarities to the particular nature and circumstances of the project that is the subject of the claim.

6.4.7.2 Weather Conditions

Depending upon the activities being carried out, different weather conditions (and therefore changes in weather conditions) can have a significant effect on construction outputs. It has been noted elsewhere in this book that 'exceptional' weather or climatic conditions are a matter that usually gives rise to an entitlement to extension of time, but not to the reimbursement of associated costs. However, it may be that such 'cost neutral' delay events are not to be confused with the effects of weather conditions that have only been experienced due to other causes for which the contractor is entitled to compensation. It has been explained elsewhere how the resequencing of activities can lead to work being undertaken in less (or more) favourable conditions than would otherwise have been the case. Alternatively, work can be moved into periods of different conditions by earlier delays, and where those earlier delays are the responsibility of the other contracting party there may be grounds for a claim for any reduction in output associated with the changed conditions experienced. This area is considered in Section 6.4.13 as an example of a consequential delay event that only affects the contractor because of some earlier delay event.

The effect of weather conditions on construction outputs is a further area that has been the subject of numerous published statistical analyses in recent decades (see, for example, 'Reducing Seasonal Unemployment in the Construction Industry', Organisation for Economic Cooperation and Development, 1967). Such analyses tend to consider two factors: temperature and humidity. They provide statistics showing the productivity losses said to be associated with low and high temperatures and high humidity levels. In addition, they can show the combined effects of both factors.

Particular care is required when considering statistical data on weather conditions, both as to the reliability and origin of the statistical data itself and its application. Thus, for example, an analysis that is based upon research carried out in North America may be considered inappropriate to consideration of work carried out by labour that is used to conditions in very different climates, such as equatorial Africa. In addition, comparison of work activities that are the subject of the studies with the work activities to which it is intended to apply the resulting statistics is essential. It is obvious that internal and external trades may be affected differently by climatic conditions. However, the method of construction may also be a key consideration. For example, low temperatures may have a very different effect on outputs in hand excavation for underpinning compared with mechanical excavation.

6.4.7.3 Introduction of Additional Resources

Claims based on the contractor having added resources to its planned team to address the effects of disruption are often met with the response that additional resources will be productive at the same level as the tendered resources and so will be compensated by payment for the work executed at the contract rates. For example, it might be asserted that if the contractor adds 30% to its number of pipefitters it will achieve a 30% increase in the output of pipework installed per hour.

The counter to this argument is the obvious one of the law of diminishing returns, i.e. if you increase labour resources beyond the optimum level for any particular operation, then those resources will be productive at a rate that decreases as the resource is increased. There is usually a minimum, maximum and optimum range of resources for the works bearing in mind all the constraints placed on any particular contract. On this basis, enforced increases above, and decreases below, the optimum will affect productivity levels and as the maximum is approached the problems of congestion and servicing of an increased workforce at the upper end of the range will usually mean that production levels reduce.

Calculations and published data on this basis often refer to terms such as 'overmanning' and 'worker density'. This almost inevitably will lead to discussion of the tender resource level and whether it would ever have been adequate, the optimum level required and the extent to which the additional resource has been effective and productive. As with overtime, it may also be entangled with the contractor having already increased its resources in order to address its own culpable failures. It is not unusual to find that such discussions are centred around analysis of labour productivity based on 'earned value'.

This is another area that has been the subject of a number of statistical analyses in the last few decades (see, for example, 'Loss of Construction Labor Productivity Due to Inefficiencies and Disruptions: The Weight of Expert Opinion', Pennsylvania Transport Institute Report 9019, 1990). For example, these analyses consider not only such factors as overmanning and undermanning against optimum gang sizes but also the effect of site congestion and overcrowding. Changes in gang size, congestion and overcrowding can be the result not only of the introduction of additional resources but also increases in the overlapping of work and trades due to delays to some early activities.

Again such statistical analyses need careful use in order to establish that they are appropriate, relevant and suitably applied. Similarly, the same considerations of contractual responsibility for the effects of additional resources apply, as those applied to overtime working considered elsewhere. To what extent was the introduction of additional resources or overcrowding the responsibility both actually and contractually of either of the parties? If it is an issue of a combination of both, then the issue and required calculations can become very complex.

6.4.7.4 Lack of Management

Typical of the areas often raised as matters of contractor inefficiency are lack of management and changes in personnel, both managerial and operatives. It is usually assumed (whether rightly or wrongly) that when a contractor makes an allowance in its preliminaries and general items pricing, or a resourced programme, or a tendered staff organogram, for a certain level of management and supervision, that that level is the optimum level it requires to efficiently carry out the work in accordance with its tendered outputs. All too often at least the last two of these sources can be exaggerated. Where the contractor subsequently provides rather lower levels of management and supervision, any claim by the contractor for the loss of efficiency in the use of labour and plant can usually expect to be faced with a response that at least part of the inefficiency is the result of lack of management and supervision compared to, say, its tender organogram.

On the other hand, and as said before, an apparent lack of management and supervision compared to the number of operatives and pieces of equipment on site can sometimes be the result of delay and disruption that has been imposed upon the contractor and for which it is entitled to be reimbursed because they have changed 'the mix'. This is sometimes referred to as 'dilution of supervision'. In such cases, debate will ensue as to whether management levels were sufficient and, if not, whose fault it was.

As to the financial consequences, it is inevitable that if the level of management and supervision on a project is not sufficient to properly manage them, then the productivity of the operatives and plant that they were intended to manage and supervise will be reduced. The question is how to evaluate the consequences for the outputs of both labour and plant. This is yet another area where historical statistical analyses can be of some use (see, for example, 'Factors Affecting Productivity', Mechanical Contractors Association of America, 1976), but again need to be applied with a large degree of circumspection. They may purport to show such as the optimum ratio of operatives to supervisors and the changes in output if that optimum is not achieved. However, they may not consider such variables as the qualifications and experience of both operatives and supervisors.

6.4.7.5 Changes in Ratio of Labour to Plant and Equipment Mix

This has been mentioned elsewhere, where it is the result of an increase in the contractor's labour numbers due to disruption that cannot be matched by an increase in the necessary plant and equipment that works alongside it, perhaps due to location or procurement problems, or the specialist nature of that plant or equipment. The consequence might be that an increase in labour achieves some increase in overall production, but not in proportion to the increase in headcount.

As is apparent from the discussion on the use of 'earned value' calculations to assess changes in productivity, and the impact of introducing additional resources, it is always necessary when comparing outputs and productivity to ensure that the comparison is being made on a like-for-like basis. Any changes in the make-up of resources between one side of the analysis and the other have the potential to distort the end result. For this reason the approach may sometimes be amended to one of 'earned value per pound spent' rather than 'earned value per man-hour' (or machine hour if appropriate). The reasoning behind this approach is to convert the analysis to one of determining how much is earned for each pound spent in the expectation that the amount spent will reflect the level of competence of the resource in terms of the tradesmen/labourer mix. This may cater for some of the distortion that can arise from any changes in the mix of labour (or plant) hours between the starting point and the actual, but considerations of how reasonable the starting point is will still be appropriate and necessary.

Similar to the discussions on other factors, a key issue will be why the ratio of labour to plant and equipment has changed. It may not always be the natural consequence of disruption for which the employer is liable. For example, where a contractor reacts to disruption by flooding areas with additional helpers, because they are of low cost and available from elsewhere, but cannot supply the necessary associated plant, it could be said by the employer that the contractor has failed to mitigate its losses and/or has been the unnecessary author of those losses rather than their being the consequence of the employer's breaches.

6.4.7.6 Condensing of Activities

One of the commonly claimed effects of disruption is the condensing of activities into a shorter time frame. This may be particularly relevant to activities that are not themselves affected by the causes of the disruption but have their start delayed by disruption to preceding activities. In planning terms, the effect is to delay the activity start but with either no corresponding delay to the activity completion or a completion delay of less than that which is applicable to the start.

The effects of such a condensing of activities may commonly include the following, which are considered elsewhere in this section:

- The working of overtime at premium costs to increase the productivity per day, or shift, of the site workforce.
- The introduction of additional resources to enable completion of the work within the shorter time frame.
- Overcrowding of working areas.
- Change in the ratio of labour to supervision.
- Change in the ratio of labour to plant and equipment.

Whatever approach is adopted, the costs will need to be identified as discussed elsewhere, and evaluation of such costs is no different in practice to the evaluation of acceleration generally, as that is in effect what is being achieved. The quantification of acceleration is considered in Section 6.5.

6.4.7.7 Resequencing Activities

As an alternative to condensing activities, or as part of a wider scheme of reorganisation to overcome disruption, there may be a need to reconsider the contractor's programme and re-plan the sequence of operations and activities through to the project completion date or dates. In such situations programmed activities may be moved in time without their duration being affected, i.e. they are 'shunted' to a later or earlier date to avoid the impact of disrupting factors that may otherwise affect them. Alternatively, they may be moved in the sequence of activities and also have their duration changed in the interest of achieving the same, or another agreed, completion date.

In such situations it is necessary to identify separately, as far as possible, the effects of the 'shunting' and the effects of any change in the duration. Generally the 'shunting' of activities may result in work being increased in cost for factors such as:

- Rates of payment to labour increasing due to cost escalation between the original period and the 'shunted' period.
- Work being undertaken in less (or more!) favourable conditions than would have been the case had the original period been maintained. For instance, excavation work originally planned for August/September in North America, but shunted to the November/December period, will suffer both the impact of less favourable weather conditions (on average) and the impact of shorter working days unless measures are introduced to enable work to continue after dark, presumably at additional cost. The effects of weather conditions are considered further elsewhere in this chapter.
- The cost of holiday periods may be encountered in activities that would otherwise not have been so affected. In the example of excavation work shunted to November/December, the cost of the year-end and other seasonal holidays may have been relevant if the shunting had been to December/January.

- Increases in the overlapping of operations where some are shunted in time, but others are not, with consequences such as spreading resources such as labour, plant and equipment more thinly.

Whatever the effects on a particular activity, the costs will have to be identified to the activity and related to the shunting and/or change of period so that the chain of cause and effect is maintained as far as possible.

6.4.7.8 Breaks in Continuity

This is a particularly likely effect of problems with the release of design information and approvals of such as contractor's shop drawing submittals where they are provided piecemeal and/or too slowly and/or design information is regularly found to be missing or wrong. It especially coincides with projects where there are large numbers of documents such as 'Request for Information' and/or 'Request for Clarification' and/or 'Confirmation of Instruction' forms and resubmittals.

It is obvious that if a contractor's continuity of working is broken and it has to pause or even cease activity in an area, and perhaps relocate to another area to return later, then productivity will be lost. The records-based approach described elsewhere is most useful for such breaks in continuity if records are kept contemporaneously. However, where they are not, then some retrospective method of evaluation may be required. In either event, it is not just the period of stoppage or relocation that may be relevant. The effect of the learning curve, as an area is revisited, and operatives have to re-familiarise themselves with the state of progress it was left in before relocation and even carry out additional preparation work, should not be ignored. It may be that the operatives and supervision that returns to an area is new to it and unfamiliar with the work and the stage of progress it was left in before relocation. It is therefore likely that a break in activity for a defined period will lead to a loss of productivity somewhat greater than the duration of that break. There are some statistical studies that are available purporting to set out the effect on a whole working day of breaks of different durations in that day (see, for example, the Construction Industry Institute, 'The Impact of Changes on Construction Cost and Schedule', 1990). As with all such studies these have to be used with care, as explained elsewhere.

A common approach on projects that are particularly the subject of breaks in continuity is to set these out in the form of planned compared to as-built programmes with broken activity bars to show periods of activity and inactivity in a particular location. Periods for relocation in each break can then be identified and added up. The resources affected by these actions will then have to be established. To be deducted, however, is the extent of revisits that may have been required in any event. It is also essential, of course, to establish that the extent of breaks is the result of the disruption said to have been imposed, and not some failure by the contractor itself to manage the works efficiently. This can be a laborious task. One way to address the topic is to sample the areas or periods, with those samples analysed in detail and the results projected across a larger whole. Key to this is that the sample is representative and this would have to be established.

6.4.7.9 Changes in Personnel

Some similar considerations apply to changes in personnel. This includes both operatives and management/supervision. Both are subject to the 'learning curve' and 'un-learning curve'. Where a contractor experiences excessive changes in personnel this is usually a matter within its control and responsibility. However, where projects are subject to excessive delay and/or excessive disruption it is often the case that one of the effects is increased personnel turnover, particularly managerial or supervisory staff. In a construction market as competitive as that which prevailed in many parts of the world in the overheated period before the last economic crash, grades from helper to site manager were unlikely to be short of options for alternative employment. The issues that arise from this are those of contractual responsibility and evaluation. Evaluating the time lost as a result of the repetition of the 'learning curve' is difficult. Further historical statistical analyses are available (see, for example, 'Factors Affecting Productivity', Mechanical Contractors Association of America, 1976), but the use of these is subject to the same qualifications on their use, mentioned elsewhere.

These costs of changes in personnel are in addition to any premium costs of their employment, for example, where operatives and management are sourced from agencies or are imported from long distances from the site rather than locally to the project. On one project in Africa a European contractor had exhausted all local supplies of supervision and management (and had to as part of local empowerment laws) such that when additional staff were required due to delay and disruption it had to source almost all of them from Europe, at several times the costs per month of local personnel.

6.4.7.10 Contractor Inefficiencies

A very common complicating factor in the assessment of disruption and other contractor claims is the allegation of culpable inefficient working by the contractor. Wherever actual resources or costs are used as part of an analysis or claim there is the potential for a defence that the actual level of such resources or costs is unreasonable, at least in part, because of inefficiencies on the part of the contractor. Such allegations may be specific and identify the resources that have been inefficient and the manner in which that inefficiency occurred, for example by reference to the extent of reworking. Alternatively, there may be a more general allegation of poor productivity due to matters such as a lack of management, insufficient supporting resources or an inadequately skilled workforce.

It is of course important when making any analysis of productivity or the impact of extraneous events on the contractor's resources to ensure as far as possible that other potential causes, such as this type of inefficiency has not had an impact on the analysis. It is good practice to undertake some consideration in this respect irrespective of whether or not specific allegations have been made. On the other hand, it can be too easy (and somewhat lazy) for those seeking to defend a disruption claim to retrospectively assert contractor inefficiencies, when there were no contemporaneous complaints.

One of the more obvious ways to assess if the actual productivity is not affected by such factors is to consider the make-up of the workforce, its management and support services, etc., from the point of view of the experienced competent contractor and compare the resulting resource and management profile with that actually deployed. Published data and work norms may provide assistance in establishing common resource mixes and levels of supervision, etc., for the relevant activities. This will not provide a definitive answer, but if the theoretical results are reasonably compatible with the actuals, then there is a strong likelihood that other factors have not caused any changes in the productivity achieved.

Obviously, if the results are not reasonably compatible, then consideration may have to be given to whether this can still be explained by the disruption for which the employer is liable. The principle as cited in Chapter 1 from the *Banco de Portugal v Waterlow* case, that it ill behoves those who cause a problem to complain about the measures taken to overcome it, will generally be applicable unless it can be demonstrated that the contractor has adopted measures that suggest incompetence or are measures that no reasonable or experienced contractor would have contemplated at the time. While the contractor does not have a blank cheque to overcome disrupting factors for which it is not liable, its duty is to employ reasonable measures in the circumstances. There is no requirement that the contractor must have adopted the 'perfect' scheme of measures to overcome the problems. Neither would it be correct to assess such measures with only the benefit of hindsight, rather than by reference to the circumstances and knowledge that the contractor had at the time that it chose to instigate those measures.

6.4.8 Cumulative Impact Claims

Where alleged disruption is based on instructed changes to the scope of the contractor's work, a common response is that all of the financial consequences of that change should have been covered by the valuation of instructed variations under the relevant provisions of the contract. It has been explained elsewhere how standard forms of contract such as SBC/Q do provide provisions that should allow disruption costs to be included within the valuation of variations in some circumstances. However, in practice, such variation valuation may only cover the direct consequences of each variation, as considered in Chapter 5 of this book.

Particularly problematical causes of changes in circumstances giving rise to grounds for a disruption claim include:

- Widespread late provision by the employer of such as information or site access to allow the works to be constructed.
- Variations to scope that due to such as their timing, extent and/or frequency give rise to a wider effect on the circumstances of work.

In addition to uncertainty, the consequences of such causes can also include matters such as piecemeal working, loss of momentum and deterioration in employee morale.

The direct effects of such events might be able to be isolated and quantified against an individual event. This might also include any direct disruption. However, what about the indirect effects? For example:

- Where access to the route of a linear project such as a highway or railway is divided into hundreds of individual parcels of privately owned land that are publicly procured for the project, but are handed over endemically late, creating ongoing uncertainty as to when works will actually be able to be constructed and in what order.
- Where there are thousands of individual instructed changes to the design of a project, issued at times creating ongoing uncertainty as to what works will actually be constructed, in what quantities, to what design and specification and when outstanding information might be received.

Whilst the direct effect of an instance of late handover of land, or the issuing of a variation, might be capable of due notice and particulars as required by the contract, what about the indirect effects on other parts, or the whole of the works, of lots of such changes? These are likely to become apparent much later, when it is harder to quantify and more controversial to agree.

The indirect effects of lots of change on the wider scope of construction works is often referred to by terms such as 'ripple effect' and 'cumulative impact'. Whilst very few people would argue that these effects cannot exist where there is a substantial amount of repeated changes, the concepts and associated terms tend to be controversial. More importantly, they are extremely hard to quantify and the greater the degree of change, the more difficult quantification is likely to become. Faced with the difficulty of quantification caused by the extent of employer failures, a contractor may complain that the employer is seeking to benefit from the extent of its own failures. An employer may respond that it is just asking the contractor to satisfy its burden of proof.

There are a number of potential approaches adopted in the quantification of the cumulative impact of change. Of the approaches considered in this chapter, Measured Mile Analyses, Earned Value Analyses or Comparison of Planned and Actual Resources should usually be able to capture that impact. In addition, there are specific industry studies of the effects of cumulative impact on past projects. These use a computer model based on past projects to produce a theoretical effect of the events (see, for example, *The Effects of Change Orders on Productivity*, Charles A. Leonard, 1988).

A further computer-based technique is System Dynamics Modelling, where a computer model of the project as-built is created based on the key characteristics that drive its actual performance, such that the claimed changes to those characteristics can be taken out to get to a 'but for' model with the outcome of isolating the effects of the claimed changes.

A general criticism of all statistical or computer-based approaches to the valuation of the effects of change involves the term '*garbage in, garbage out*'. This is particularly said in relation to System Dynamics Modelling, where creating the model and the characteristics on which it is based are essential inputs. This relies on both accurately establishing those characteristics and creating the model on them. The respondent to a claim on this basis, or a tribunal being asked to rely upon it, must be able to understand and check the assumptions made.

A key question in relation to any model will be does the model represent what would have occurred had the events complained of not happened, such that imputing or removing those events accurately identifies their effects? In practice this can be very difficult to test and agree. Furthermore, the accuracy of the resulting effect will depend on the accuracy of the events impacted into or removed from the model. Inevitably such models are very complex to create and to impact. Problems particularly arise where changes are required to what is impacted, for example because it is determined that some of the variations that have been included in the analysis are not claimable. This will mean re-running the model, perhaps a number of times. This can be time consuming and expensive both for the contractor to do and for the employer and its team to carry out due re-interrogation. If this is in the context of a dispute resolution process, the later such re-running is required, the less practicable it becomes. In an extreme position, the tribunal might be being asked to award on the validity of such as whether claimed scope changes were actually variations under the contract and whether the contractor gave necessary contractual notices in relation to some claim events whilst deciding on a cumulative impact claim at the same time. If the tribunal decides 'no' in relation to some of these questions, then it may be too late for the cumulative impact model to be re-run by the claimant and interrogated by the defendant.

Such industry studies based on previous projects can suffer similar accusations of 'garbage in, garbage out', due to the difficulties of testing the study for its accuracy and relevance and the practicalities of changing the analysis for different events. Broadly, there are two types:

- Studies of the effects of specific causes of lost productivity, such as overtime working, overmanning, stacking of trades, reduced supervision ratios, increases in work scope, learning curves and climatic changes, etc. These have been considered elsewhere.
- Specific studies of the cumulative impact of changes. These tend to be limited to increase in the scope of work and its effect on productivity on all work, whether changed or not.

Those defending claims made on the basis of such studies raise a number of criticisms, including the following:

1. The relevance and similarity of the previous projects in the study to the current project. For example, if the model is based on past projects in the energy sector in the United States, are their results applicable to construction of housing in Asia?
2. That such models do not allow interrogation of the details of the previous projects to test their more detailed characteristics and similarity to the project subject of the claim. Some owners of such models are particularly sensitive to attempts to interrogate their models and the details of such as the past projects and assumptions that they rely on. This might be on the stated grounds of commercial sensitivity or copyright, but leaves the suspicion that the real concern is what interrogation would reveal.
3. That the models consider events but not their timing. For example, statistics on the effects of change in work scope might apply the overall percentage of change in such as total labour hours. However, in practice, the effect of variations depends, not only their overall extent but also their number and timing. A major change to a specification or drawing instructed very early in a project might only affect the contractor's early design and procurement, but have no consequence on construction outputs at all. On the other hand, a small change to the alignment of a road after it has been partly constructed could have a significant disruptive effect if resources have been relocated to another chainage of the road.
4. The extent of change also involves their numbers. A single large variation that results in a significant increase of the scope of work to a project is likely to have rather less 'ripple effect' than dozens of small variations of much smaller individual size.
5. That the resulting quantification may be no more than a Global Claim, with all the usual criticisms of such claims. This topic is considered in detail in Section 6.6.
6. That models based on the percentage of change in the scope of work depend on the accuracy of both the asserted original scope and the scope of change. These elements of the equation can be subject to significant challenge that may require a re-run of the calculations that rely on them, as explained elsewhere.
7. That such studies tend to be of the overall change and the effects on a project as a whole. Thus, such studies may be inappropriate to assess productivity loss only on part of a project.

A further issue is that those touting models and programmes for the quantification of such items as a cumulative impact and ripple effect assert that they have been applied successfully, often without supplying appropriate authorities for their use. Much of the support that is cited for the approach is based in the United States. Some observers wonder if their courts' use of juries to try commercial matters lends such approaches a better chance of success than would the rigours of such as a trial in London before the specialist judges of its commercial courts or an ICC arbitration in Paris heard by a tribunal of experienced construction arbitrators.

The aversion of the UK courts to formulaic approaches is reflected in a talk to the Technology and Construction Bar Association and the Technology and Construction Solicitor's Association in 2007 when Sir Justice Ramsey (one of the High Court's leading Technology and Construction judges) spoke of the importance of ensuring that experts' reports in which conclusions are stated clearly set out the analysis that led to those conclusions. He said that such analyses should not be hidden in the 'black box', the expert's head, workings or computer analyses and thus inaccessible to the tribunal. That means that when presenting a cumulative impact claim in particular, the transparency of how the quantum is derived would be important.

It continues to be the case that disruption quantification is a difficult area and claims for cumulative impact or ripple effect are probably its most controversial element.

6.4.9 Top Down or Bottom Up?

Several different approaches to the quantification of the effects of disruption have been identified, one distinction between them being the starting point for the analysis. Some commence with the actual amount of resources deployed and costs incurred and determine how they relate to the activities and events on site. These are often categorised as 'top down' approaches. Examples of these are measured mile calculations. Other approaches start with the planned amount of resources to be deployed and costs anticipated (if established to have been reasonable) and consider adjustment of these in the light of any relevant events. These are often categorised as 'bottom up' approaches. Examples of these are cumulative impact calculations.

The attraction of starting with the actual resources and costs and determining their relationship to events would seem to be that one is dealing with known resources and costs, providing that adequate records of matters on site have been maintained. One disincentive to starting with the activities and events on site and identifying resources and costs to such matters is that this may be a more complex and involved approach requiring more analysis to build the picture step by step. In the absence of adequate records the latter approach may require considerable further research to establish what the reasonably attributable effects of events may have been.

6.4.9.1 The 'Top Down' Approach
There are many variations to this approach to the analysis of resources and costs, but the common feature is that the approach commences with actual resources and/or expenditure and relates it back to activities and events on the project. The employed resources and costs are allocated to the events to provide as factual a picture as possible of the way in which the contractor's costs were incurred.

To illustrate a typical example, using the costs of resources, assume that the following figures may be extracted for three activities, which are alleged to have been disrupted by events on site. In this example the costs have been used to illustrate the analysis, but the same process can be undertaken for individual activities using alternative resource figures such as man-hours, plant-hours, etc., and these are more likely to give a more realistic analysis.

	Tender $	Final A/C $	Cost $
Concrete work	210,000.00	242,000.00	310,000.00
Excavation	160,000.00	170,000.00	190,000.00
Steelwork	95,000.00	85,000.00	100,000.00
Total	465,000.00	497,000.00	600,000.00
Loss on final A/C value			103,000.00

It might be alleged that the extent of disruption in these activities is demonstrated by the loss incurred by the contractor, represented by the excess costs compared with the final account value. On this basis the overall degree of disruption to the contractor would be represented as the loss as a proportion of the value achieved as follows:

$$\$103,000/\$497,000 = 20.72\%$$

On this approach, the contractor may then allege that it has suffered a 20.72% loss of production as a result of disruption. It will then seek to explain that loss by reference to the asserted events that are the basis of its disruption claim.

The flaws in such an approach obviously include that the alleged productivity loss is simply a product of the final cost figure, which may, or may not, be dictated by the impact of disrupting factors on the site activities. It also leaves the defences discussed elsewhere, particularly those where contractor 'own goals' and tendering errors are alleged to have contributed to the increased cost of operations. The same problems apply if the calculation is presented in an alternative to the financial format by using alternative components of the final cost such as man-hours or plant-hours, although at least that removes changes in cost per hour that may not be the result of the causes that are the basis of the claim.

At the very least, if such an approach is considered to be adopted, the analysis should be to each of the activities so that separate factors can be established for each part. In this example, for instance, losses of $68,000 can be identified in the concrete work, $20,000 in the excavation activity and $15,000 in the steelwork. Separate claimed disruption factors for each of these activities would then be:

Concrete work	$68,000/$242,000 = 28.01%
Excavation	$20,000/$170,000 = 11.76%
Steelwork	$15,000/$85,000 = 17.64%

This shows that the various activities of the whole have incurred losses at differing rates to each other and to the whole and may therefore require some examination of each activity, concentrated on the areas of greatest loss, with some prospect of establishing the reasons for the increase in cost attributable to events. This should show, for example, why the concrete work was apparently disrupted to a degree more than double that of the excavation. It may be, however, that only a detailed 'bottom up' analysis will identify the reasons for the difference between the incurred costs and the achieved value on each activity.

There are many variations on this type of 'top down' approach, and this example could obviously be refined by converting the 'final account value' to a cost basis so that comparison with the actual activity cost is on a like-for-like basis. However, it will always suffer from the deficiencies generated by starting with the outcome and trying to work backwards to the cause. The difficulties of allocating actual outcomes to items of causation, separating out non-claimable matters such as contractor's inefficiencies and mistakes, and eliminating the tender/final account value as a source of inaccuracy, all of which has been detailed, will always mean that such an approach can only be adopted with considerable care and painstaking effort to eliminate inherent causes of distortion in the calculations. It will also require a back-check of the results to establish the relationship between the causes claimed and the financial consequences calculated to see if they are reconcilable. This may also include a separate analysis using a different approach to quantification to see if the results are similar, as is explained elsewhere in this chapter.

6.4.9.2 The 'Bottom Up' Approach

This approach starts with two things. In particular, it considers the events on-site, identifies the alleged causes of additional expenditure giving rise to a claim for additional payment and then assesses the impact of those events using as a starting point the resources and costs that would have been incurred had the disruption not occurred. As has been explained, those resources might be based on the contractor's tender, corrected for any errors, and perhaps including consideration of the site team's planning and may also rely on objective expert analysis.

The analysis of events might begin with an analysis of the programmes, both anticipated and as-built, and consideration of the differences in terms of such issues as changed durations, increased overlapping, breaks in continuity and relocations, etc. It will then consider planned and actual resources to identify differences in terms of, for example, increases in labour, overtime working and changes in such as labour/supervision and labour/plant ratios. From this analysis it may be possible to identify the components of the additional cost incurred, be that overtime working, additional resources, condensing of activities overcrowding, lack of supervision and plant or equipment or other factors caused by the disruption or measures adopted to counter the causes of disruption.

From these analyses, it should be possible to build up a level of resource or cost that is generated by the activities and events that are said to have caused disruption, rather than by starting with the contractor's actual levels of resources deployed or costs incurred.

6.4.10 The 'Rump' of the Contractor's Losses

The end result of 'bottom up' approaches in particular is usually that the resulting quantum falls short of the total of actual additional resources or costs incurred by the contractor. In part or whole this might be for identifiable reasons such as contractor's tender errors, construction 'own goals' or other matters for which the contractor remains liable. However, it may also still be that the calculation still leaves a balance of the contractor's expended costs or resources that are not explained. This is often referred to as 'the rump' of the claim.

A similar rump can also appear at the end of a 'top down' approach, where the result of excluding the effects of the various causative events does not reach the contractor's likely 'but for' costs or resources.

The question left at the end of an analysis that leads to such a result is what to do with that 'rump' and in particular have any claimed causes of disruption not been factored into the analysis that might explain it? Is its cause unproven, such that the contractor cannot recover it, or is it explicable by remaining causes of disruption that have not been inputted into the detailed calculations? For example, where a bottom up approach has dealt with factors that are capable of calculation (for example, breaks in continuity and overcrowding) but has not addressed more complex issues such as 'the ripple effect'. Can the rump be associated with that residual cause? The contractor may argue that it can. The employer may argue that the contractor has not established that it can and that it might just as credibly be the result of extraneous issues such as missed contractor 'own goals'.

The presentation of contractual claims in relation to construction and engineering projects is often a process of balancing two forces. The first is the legal requirement to particularise the claim. The second is the practicalities of the factual circumstances of many claims, particularly those in relation to disruption. These practicalities can render the need to particularise more difficult as the causes of disruption become more numerous and the effects more complex.

Particularisation involves putting the party in receipt of the claim in a position such that it can understand the claim being made against it in sufficient detail that it can analyse the claim and defend its position. Where the claim is to be determined by a dispute tribunal, it will require the same detail in order to reach a reasoned award. From the claimant's perspective, a particular benefit of particularisation arises where part of its claim fails. For example, this can occur where disruption is said to be the result of numerous design changes, but one is found to have been part of the original work scope. The danger if this claim were unparticularised might be that the failure of this one cause might imperil the claim as a whole because that part allocatable to the successful causes cannot be isolated with any reasonable degree of accuracy.

The practical problem of particularising disruption claims arises out of the nature of disruption and what causes it. To take the example of construction of a retaining wall, if this is delayed because of the late release of one drawing, then the financial effects, which might be limited to prolongation costs, should be readily capable of being identified and allocated to that single cause. However, say that the same retaining wall is subject to numerous revised, inadequate and late drawings. This might add disruption to prolongation costs and make the allocation of the resulting additional costs, particularly those of the disruption, difficult to allocate between the various events of revised, inadequate and late drawings. Furthermore, say that further disruption occurs because the contractor's working space is limited by late progress by another contractor that is engaged by the employer on the site. In this case the allocation of losses in productivity to the various causes will be even more difficult. As stated elsewhere, a common contractor complaint is that by demanding further particularisation of a disruption claim where the extent of the employer's failures have made it impossible, the employer is seeking to benefit from the extent of its own failures.

The balance between particularisation and practicality in such situations might be struck by particularising a disruption claim as far as is practicable but recognising that full particularisation is not possible and that some amount of the contractor's additional costs will be left. This is what is sometimes referred to as 'the rump' of the contractor's claim. Whether that rump can be said to be the consequence of the causes on which the disruption claim is based will depend on the facts of each case. It will also depend on the legal burden of proof, be it set at 'the balance of probabilities' or some other standard.

A particular consideration is likely to be whether the size of the rump and the nature of the losses therein are commensurate with the unparticularised disruptive events that are said to have given rise to it. Continuing the example of a retaining wall, if the revised, inadequate and late drawings and working space issues were all related to the super-structure, a rump that includes losses on the excavation plant might be confidently said to be unproven as being to the employer's account.

Whether the rump is truly the consequences of the practicalities of the matter, the contractor having particularised as far as possible, rather than being the result of the contractor being too lazy to properly detail its claim, will depend on the facts in each case. It might also be that the failure to particularise as far as possible is because the contractor wants to hide the consequences of its own failures that have contributed to its losses. This is a common complaint from employers and those advising them.

This discussion particularly hints at the broader issue of global claims, a topic that is considered in detail in Section 6.6 of this book.

A worked hypothetical example in Section 6.4.11 illustrates how a rump of a contractor's disruption costs calculation might appear and how it might be considered.

6.4.11 Hypothetical Illustrative Disruption Calculation

It has been explained elsewhere how the best basis for a disruption calculation is contemporaneous records of the resources and times lost, failing which a measured mile analysis is the preferred retrospective approach to a detailed calculation. In the absence of these, other approaches have their problems in terms of proving the contractor's losses with a sufficient degree of certainty, and it may be that applying two such approaches to quantification, which reach similar conclusions, might be a suitable solution that provides a 'back-check' on the results of either. To illustrate this suggestion and some of the other issues of practicality that have been identified, such as what to do with the rump of the contractor's losses, the following very simple hypothetical example is provided. Whether such an approach satisfies the burden of proof and provides a quantification that is sufficient will depend on the facts and legal background in each case.

Assume that a contractor says that its tendered allowance was for a total of $10,000,000 in labour direct costs and that its actual total labour direct costs totalled $18,000,000. That is a loss of $8,000,000. It says that the difference was the result of late, piecemeal and incorrect information from the employer's design team. The fact of the extent of such design problems imposed on the contractor have been established, as has legal liability. The completion date for the project overran by six months and the contractor has been granted a full extension of time and prolongation costs for that full period for late information. The contractor also claims reimbursement of its labour loss of $8,000,000 as disruption. From this the contractor deducts $500,000 as the value of the labour in its agreed variation account. The contractor applies a 'top down' approach to its claim, leaving a balance of a claim for $7,500,000 for labour loss due to disruption.

In this example, the basis of the contractor's disruption claim is that the information release failures that led to delays to the project had four particular disruptive effects on its site labour, as follows:

1. Delayed work activities were carried out in a high temperature and humidity environment of the tropics over summer months that should otherwise have been carried out six months earlier during the much more clement winter months.
2. It instituted overtime working to reduce the effects of the delays and reductions in its productivity. Whilst those additional working hours will have achieved additional output, the contractor asserts that this was at lower levels of productivity than is achieved by working normal working hours, because of the additional effects of fatigue.
3. It increased its labour numbers. This again achieved additional output, but also led to some overcrowding of work faces further reducing productivity.
4. The location of the project meant that it could, and did, increase its labour numbers to reduce the effects on its productivity, but it was unable to increase supervision numbers to the same extent because of the lack of appropriately qualified people in the country of the project. This meant that the planned ratio of labour to supervision could not be maintained, leading to lower productivity.
5. There was a broader 'ripple effect' of uncertainty, learning curves and loss of morale among operatives.

A first area for analysis in this example would be the contractor's approach of comparing planned and actual labour costs, rather than labour hours. Analysis of the contractor's tender shows that its allowance was based on 1 000 000 hours at a rate of $10 per hour, but the payroll shows that its actual cost (excluding overtime payments) was an average of $12 per hour. This rate did not change over the course of the work, even with six months of delay, so it cannot therefore be blamed on compensable delays. Accordingly, part of the labour overspend of $8,000,000 was due to an underestimate in the contractor's average labour cost rate by $2 per hour. In terms of man-hours, the contractor's loss is therefore only 500 000 hours (an actual 1 500 000 man-hours less a tendered 1 000 000 man-hours).

In relation to the contractor's tendered labour allowance, as part of its submitted methods and resourcing of its first construction-issue programme, the contractor's senior site management team prepared a labour histogram that showed that they expected at that very early stage to use a higher than tendered total of 1 200 000 man-hours. The records also show that the contractor started procuring labour to these higher numbers before any of the causes of delay and disruption that form the basis of its delay and disruption claims had occurred. Analysis of the tender shows that the contractor's estimators had missed the labour for some works in their calculations and had applied the same outputs for work at a high level as for work at ground level, which were corrected in the site team's resource projections. In addition, a report from an estimating expert concludes that a reasonable allowance by the estimators should have been around 1 200 000 man-hours, similar to the conclusion of the contractor's site team in their first resourced construction programme. On this basis, the contractor's adjusted loss is therefore only 300 000 hours (an actual 1 500 000 man-hours less a 'but for' 1 200 000 man-hours).

The employer's analysis of the contractor's recoveries of labour costs through the value of variations identifies two points. The first is that the contractor has only based its deduction of $500,000 on those of its variations that were accepted and agreed as variations. However, the contractor's claims for variations had actually been for much more, but included a number of major items that were not part of the agreed variation account. These were work items that were missed by both the contractor's tender and its site team's early planning, but were part of its contractual scope of work. Also, the relevant deduction from the disruption claim is of the labour hours spent on variations, rather than its value. Analysis shows that the total labour hours in the contractor's claimed variations was 100 000 man-hours. On this basis, the contractor's unrecovered labour hours only total 200 000 man-hours (300 000 less 100 000 spent on claimed variations).

Furthermore, the contractor has made other claims that included recoveries of labour hours as follows:

- A claim against one of its subcontractors for defective workmanship that the contractor's labour had to correct at a cost of 2000 man-hours.
- Within the agreed contractor's prolongation costs is some time-related labour that is duplicated in its disruption claim. Excluding that labour removes another 18 000 man-hours from the disruption claim.

Adjusting for these other recoveries, the contractor's loss is reduced to 180 000 man-hours (200 000 less 2000 and 18 000 claimed elsewhere).

The reasonableness of the contractor's actual labour man-hours were never in question. There were no contemporaneous complaints from the employer or its contract administrator that they were working culpably inefficiently, other than to the extent of the need to correct defective works by one subcontractor (adjusted for the above) and a lack of supervision in proportion to the increased numbers of labour on-site (which is part of the basis of the contractor's claim).

In conclusion of this 'top down' analysis, the contractor's labour loss, adjusted for its tender error, other claims and subcontractor problems was 180 000 man-hours on a reasonably planned total of 1 200 000 man-hours summarised as follows:

	Man-hours
Actual:	1 500 000
Less reasonable planned:	1 200 000
Hours lost:	300 000
Adjustments:	
Variation claims:	−100 000
Other claims:	−20 000
Total time lost:	180 000

Expressed as a percentage of the contractor's 'but for' man-hours this is a loss of 15% (180 000/1 200 000 man-hours).

One way of backchecking this result would be to look at the claim by reference to another method of quantification, to see if they reach similar conclusions. For example, looked at on a 'bottom up' basis, the key question would be whether the causes pleaded by the contractor can explain a 15% over-spend on labour-hours (180 000/1 200 000). This could be looked at as follows.

Assume that analysis of the four particular disruptive effects on its site, labour asserted by the contractor shows the following:

1. Summer working. Analysis of the records shows that some 200 000 man-hours were spent on works that were sensitive to the effects of working in the very hot and humid summer months, and which would have been carried out in clement winter conditions had the delays not occurred. The consensus of various sources of published historical data is that the differences in these climatic conditions is a loss of output of about 15% expressed as a proportion of the actual time spent. Within a total of 200 000 man-hours spent, that would be 30 000 man-hours lost.

2. Overtime working. The consensus of various sources of published historical data suggests that the extent of overtime worked by the contractor would result in a loss of output of about 8% as a percentage of the planned hours. However, the contractor's site team's early planning had envisaged a degree of overtime working even before any disruption occurred. Coinciding with this, the project payroll records showed that labour was working some overtime from the very start of its presence on-site. The same published historical statistics suggest that this level of overtime working would have involved a loss of output of about 3%, that is 36 000 man-hours. Thus, the additional effect of the additional overtime was a lost output of only another 5% (8% less 3%). On a reasonable planned 1 200 000 man-hours, that would be 60 000 lost man-hours.

3. Overcrowding of work faces. The records show that not all activities were affected by this. The affected activities accounted for a planned 200 000 man-hours. The consensus of published historical data is that the overcrowding at the level experienced on these activities would reduce their production by 3½%, that is a lost 7000 man-hours.

4. Reduction in labour/supervision ratio. Again, analysis of the records shows that only some trades lacked supervision, where the additional management of the necessary trade experience was not available to be procured. The records also show that the affected activities accounted for a planned 400 000 man-hours. The consensus of published historical data is that the reduced average level of supervision on these activities would reduce production by 10% against plan, that is a lost 40 000 man-hours.

In summary of this 'bottom up' approach, the analysis of the contractor's adjusted loss of 180 000 man-hours would be as follows:

	Man-hours lost
Summer working:	30 000
Additional overtime:	60 000
Overcrowding:	7 000
Labour/supervision ratio:	40 000
Total related to employer:	137 000
Contractor's planned overtime:	36 000
Total time lost:	173 000

The conclusion of this analysis would be that the application of factors addressing each of the causes of disruption asserted by the contractor explains 173 000 of the 180 000 man-hours derived from the comparison of its adjusted actual less planned labour claim. This leaves a remaining balance, or 'rump', of 7000 man-hours.

Regarding this rump, it will be noted that the bottom up approach only addressed four of the five asserted causes of disruption. It ignored item (5), being the 'ripple effect'. Whether this rump of 7000 man-hours is unrelated or unproven, or can be said on the balance of probabilities to be the result of this cause, will depend on the facts. In particular, what detailed evidence there is indicated that this effect existed as a consequence of the employer's failures and that it is likely to have had the effect concluded from the calculations.

In conclusion of this hypothetical claim, two approaches that on their own might not be sufficient to quantify the extent of the contractor's disruption losses on labour substantially provide a 'back-check' to each other's conclusions. It might be considered that together they establish the contractor's loss in relation to 173 000 lost man-hours with a sufficient degree of certainty to satisfy its burden of proof. What happens to the 7000 man-hours rump of the contractor's labour losses is another question.

It will also be noted how one of the two approaches identified part of the contractor's labour loss was the result of matters for which it was liable (overtime worked without the effects of disruption and not allowed for in its tender), and establishes an amount against that which can be discounted from the claim.

It is again emphasised that this is a hypothetical and very simplistic example. It makes a number of assumptions and approximations. In practice, such analyses are rarely this simple, or come to such conveniently similar results when back-checked by looking at the claim on two different bases. It also ignores a number of potential additional claims arising out of such facts. For example, there might be a separate claim for the contractor's cost of premium time payments made to labour working overtime. Another possible example is that it may also be that the additional supervision was employed at a premium cost.

6.4.12   Preliminaries Thickening Claims

This type of claim is considered here, but as mentioned in Section 6.5, it could equally be a potential head of an acceleration claim. Whether preliminaries thickening ought properly to be a head of a disruption or an acceleration claim will depend on its underlying facts. Under either heading, preliminaries thickening claims are particularly common on major projects. They are often a relatively easy heading to quantify, compared to other heads of a typical disruption claim, although they still face a number of difficult hurdles. This is notwithstanding that it is often said that it is easier to show (and to understand) that labour has been disrupted by an employer's failures than to demonstrate a similar effect on, for example, site staff.

The resources that are most commonly claimed under a thickening heading are site preliminaries and general costs such as site staff and plant and equipment. It is, however, possible that they can apply to the likes of temporary works, such as scaffolding, and temporary facilities, such as site accommodation.

A claim for additional preliminaries and general items is often made on the basis that the extent of change instructed to the works has required the introduction of such items as supplementary staff and plant or equipment to serve that additional work. However, in such cases it ought to be possible to address those additional resources in the evaluation of the direct consequences of the change, as considered in Chapter 5 of this book. The more likely footing for a separate preliminaries thickening claim is either change in the contractor's work sequence or the extent of information and change that has been issued to the contractor. This latter cause might have limited direct value for the purposes of valuation of individual variations, as explained in Chapter 5 of this book, but still cause the contractor additional costs.

In relation to plant and equipment, thickening often occurs where the contractor's works sequencing has changed due to events for which the employer is liable, such that it is working on more work fronts than planned and therefore needs its plant and equipment to service more work fronts at the same time. Take, for example, a project comprising one main building and a number of ancillary buildings, perhaps an airport that includes a main terminal but also separate buildings for such things as aviation maintenance, facility maintenance, workshop, emergency services, employee centre, radar and air traffic control, etc. Assume that the main building is where the critical path runs through both the planned and as-built programmes. Assume that the ancillary buildings were planned to have their foundations excavated consecutively, with the same equipment, quite independently from resources for the main building. If construction of some of those ancillary buildings is delayed because of the late release of their design drawings, then it may be that they have to be constructed concurrently, rather than in series, requiring duplication of the necessary construction plant. It might be said that the plant has not increased in its overall cost, because the increase in the number of items is offset by a reduction in the period over which they are required. This may or may not be the case, and will need detailed investigation of the facts. In any event, there may be additional costs in terms of plant mobilisation and removal and extra costs for the additional plant depending on its source. For example, if existing plant is owned by the contractor, but additional plant is sourced from external hire companies, it might well be at premium rates against a calculation based on depreciation and running costs.

In relation to site staff, as with plant, a key consideration will be the extent to which these have been recovered through the valuation of variations, where that is the basis for the staff thickening claim. However, a standalone staff thickening claim usually arises in relation to the extent of changed information received by the contractor, and the quality of that information, rather than the extent of variations. A claim on this basis may look at, for example, the numbers of revised items of information and instructions issued by the employer's design team and the requests for clarification and information that had to be issued by the contractor because of problems with the design team's information. Contractors often submit this in the form of graphs of the numbers of its planned and actual staff on site compared to the numbers of instructions, clarifications and requests, etc., being issued. This is often faced by the response that 'correlation is not causation', i.e. that the fact that numbers of staff are shown as increased and tracking the numbers of pieces of paper being exchanged, does not establish that the latter caused the former. The contractor's counter may be that the link has been evidenced on the balance of probabilities.

The calculation of which staff are additional is often based on a comparison of the contractor's tendered and actual staff numbers. As with other headings of a contractor's disruption claims (such as labour), such an approach begs two key questions. The first is as to whether the claimed planned staff numbers were adequate (rather than over-optimistic). The second is as to whether the actual numbers were the necessary consequences of matters for which the employer was liable (rather than due to the contractor's own inefficiencies).

As discussed elsewhere, a better basis for establishing the staff numbers that would have been required had the employer's failures not occurred may include staff planning by the contractor's site management, rather than its estimators' tender allowances. Sources for the tender allowances can include the detailed build-ups to prices in a preliminaries and general items section of its bill of quantities, and/or staff organograms included in tender submissions. These latter documents are particularly problematical in terms of reflecting what was even expected to be required because they may be inaccurate. They often actually work against the contractor's claim because of the habit among tenderers of exaggerating their tender submission resource organograms in order to make the tender look more attractive. Expert evidence as to what resource ought to have been required may help to establish staff numbers that would have been required had the project proceeded without disruption.

As with labour, early planning by the senior site team is likely to be more persuasive than allowances by an estimating team or histogrammes prepared by a commercial team as part of a bid presentation. On large and lengthy projects a good project manager might start by setting down his or her requirements for staffing the project through to completion, not least so that employees can be relocated from elsewhere or recruited on a planned basis, and also as part of budget setting with consideration of allowances in the contract price.

Rather than just comparing overall numbers, a staff thickening claim should compare the planned and actual staff line by line to establish who exactly is additional, when and why, by lining up job titles and/or personal names. This can prove problematical and require some interpretation to correlate names and titles. A consideration may also be what to do about staff that were planned but were not actually required? Should these be abated from the claim for preliminaries thickening? It may be said that such staff that resulted from an overestimate are irrelevant to a claim for staff that were only required due to failures for which the employer is liable, but it may also be that on detailed investigation there was some indirect effect or offsetting in relation to the staff that were not actually required, such that establishing the true effect of the disruption needs consideration of all of the planned and actual staff. Again, this needs detailed analysis of the facts.

For the contractor, the best evidence in support of a staff thickening claim is contemporaneous records of which members of its site team were additional, and when and why they were introduced. This is unlikely to relate to an individual event of change, but the accumulation of events. For example, where an additional document controller is added on-site, it will not be because of one new or revised drawing, but because the existing team cannot handle the amount of information being issued by the employer's designers. Ideally, senior head office or site management should issue a record every time it makes a decision to add to the site team and why. Often these can show a contemporaneous request for additional staff and when it was made, but not why it was made. However, if this has to be done retrospectively then it will require a statement from whoever made the decision at the time, explaining the reasoning behind it and reference back to contemporaneous records as far as possible.

The reality is that additional staff that are the subject of preliminaries thickening claims are rarely the result of an individual causative event, but the accumulation of a large number of such events. However, this can be controversial. It leads to those receiving such a claim responding that it is unparticularised and global in the pejorative sense of the word explained in Section 6.6. From the contractor's perspective such a complaint may belie the practicalities and the facts. The contractor will then say that to allocate each additional individual to each claimed causative event is impractical and would be contrived and misleading where the cause really is the combination of a group of events.

In conclusion, preliminaries thickening claims, such as for additional site staff, require a comparison of what staff would have been required 'but for' the causative events pleaded, with those that were actually required as a result of those events. The former might start with a consideration of what was tendered, but this may simply have been wrong, and contemporaneous estimates by the site management may be of more value. Alternatively, some objective evidence of the 'but for' requirements of the project may be required. Once the differences are identified, they will require an explanation of why each extra individual was required and relating that to the events pleaded. Adjustment will then be required for any recoveries through items such as the valuation of variations and other claims such as one for prolongation. The potential for overlap with a claim for prolongation and the issues involved in deciding how to avoid such an overlap in quantification are considered in detail in Section 6.10. A recurring problem with many preliminaries thickening claims is that it is impossible to particularise the additional costs to specific employer liability events, such that the claim is labelled a 'global' one.

6.4.13 Consequential Further Delay Events

In the detailed hypothetical example of a disruption claim quantified on two parallel bases, it was assumed that a where a delay caused by a compensable event pushes works into a period of less inclement weather, which causes further labour loss, the contractor is entitled to compensation for the effects on its productivity of that less clement weather. This is notwithstanding the position that under most forms of construction and engineering contract, weather conditions are a cost-neutral event under terms that give the contractor the right to an extension of time but not financial compensation. This has been detailed elsewhere.

This form of event, which is only met as a consequence of an earlier compensable delay, is sometimes referred to as a 'consequential' or 'knock-on' delay. Climatic conditions are a common example of this, but the question applies equally where any other cost neutral delay event is met because of the effects of an earlier compensable delay. An example is a strike among the contractor's site labour where that is a contractor risk event. It is sometime argued that because the later event is of a type that is not compensable under the contract, then there should be no compensation for its effects. The correct answer to such situations will depend on the applicable law. However, if it is a question of causation, it might be said that if bad weather is met because the employer provides late access and work is carried out later, then it is the late access that led to the weather being met and hence the additional costs associated with it. The loss is therefore caused by the late information and not the weather and would therefore be compensatory.

One argument often used by employers and their advisors is that the effects of events such as exceptional weather were a contractor's risk that it allowed for in the pricing of its tender. The typical contractors' response to this is that this misses the point that a tendering contractor only accepts and prices risks to the extent that it can see them and price for them at the time. For example, a contractor pricing work in Northern Europe over the winter months will allow for more lost days due to weather than if those works were planned to span the summer. It can price that risk by reference to the contract period, its programme, its knowledge of when it will be carrying out weather-sensitive works and its experience or data on local weather conditions. It can even price in a risk component against the contingency that its own culpable slow progress will push work back into inclement conditions. That is a risk it can both predict and control. However, how is it to price the risk of the weather effects on a summer project being delayed into the winter by the employer's failures? That is a risk it cannot predict or control. If it were to make allowance for that risk in its pricing would it have any chance of winning the tender competition? An analogous situation is in relation to inflationary increases in the contractor's prices. This area is further considered in Sections 6.2.3 and 8.3 of this book, where it is suggested that on a contract on a fixed price basis, the contractor only allows for inflation in the prices of such items as materials and labour for the duration of the contract period. There it is suggested that if the employer causes compensable delays, then increases in costs caused by that delay are generally recoverable notwithstanding that the contract is a fixed price, although this would be subject to any considerations arising out of the applicable law.

A further feature of weather conditions that are met due to compensable delays is how adverse or exceptional that weather has to be. Standard conditions provide for extensions of time for such as 'exceptionally adverse climatic conditions' (FIDIC Red Book clause 8.4(c)) and 'exceptionally adverse weather conditions' (SBC/Q clause 2.29.9). However, if weather conditions that are met due to compensatory delays are not 'exceptional', but are materially different to those that would have prevailed but for the earlier delay, then is that sufficient for the contractor's claim in relation to that weather? For example, a few days lost due to thunderstorms in central southern Africa in January would not be exceptional, but if the work should have been done in the previous August, then that is a material change in the contractor's circumstances. The basis for saying that the contractor should be compensated for such secondary delay would be that the qualification 'exceptionally' in, for example, the FIDIC Red Book clause 8.4(c) and SBC/Q clause 2.29.9 are irrelevant, as the claim is not being made under those clauses.

Of course, it is equally possible that a delay pushes work into more clement weather conditions, thus reducing the contractor's costs. It is suggested that this should be a factor in assessing the contractor's additional costs arising from the delay. For example, a 12 month delay might push parts of the works from summer to winter, making them less productive, but also push other parts of the works from winter into summer, making them more efficient. The effects of both of these material changes would have to go into the quantification of the effects of the delay on the contactor because the contractor cannot 'cherry pick' only those effects that led to reductions in its productivities.

One judgment that is sometimes cited as a basis for denying the effects of such consequential delay events is that in *Costain Limited v Charles Haswell & Partners Limited* [2009] EWHC 3140 (TCC), which has also been considered elsewhere. One of the issues in that matter was a claim that delays meant that the pipework installation on a project in the UK, instead of being carried out and completed during the summer of 2003, as programmed by Costain, was pushed into October and November 2003, a period of winter working. This caused further delays to Costain resulting from low productivity inherent in working outside during the relatively short days and bad weather of early winter. Richard Fernyhough QC rejected this claim. However, his reasons were as follows:

> In cross-examination, [Costain's programming expert] frankly accepted that this claim and his calculation of it was purely theoretical since he had done no research into the actual effect of winter working on the productivity of works such as pipe work installation. He also accepted that 1 October 2003 was an arbitrary date to commence the calculation since, as we all know, the weather in October can be drier and more settled than in any of the summer months.
>
> …
>
> I have no hesitation in rejecting this part of Mr. Crane's analysis. It is wholly theoretical and based on nothing but the meteorological records for the relevant period and Mr. Crane's experience and hunch. It seems to me to be unlikely that, as a matter of course, productivity of outside building works in October and November is always measurably lower than for, say, the months of August and September. In this country the productivity of outside work depends to a great extent upon the weather which can be changeable at any time of year and there can be no presumption that it will be generally worse in October and November than in any other month. In the absence of hard facts and figures to support such a claim related to the facts of this case, which do not exist, in my judgment, this claim has not been established on the balance of probabilities.

This claim therefore failed on Costain's failure to establish its underlying facts, not on its general principle.

On the other hand, support for the proposition that it is the primary cause of delay that goes to whether associated costs are recoverable can be found in the judgment of His Honour Judge Fox Andrews in *H Fairweather & Co Ltd v London Borough of Wandsworth* (1987) 39 BLR 106. The contract was in the form of the JCT Local Authorities Edition with Quantities 1963 Edition. Clause 11(6) provided for reimbursement for the contractor's direct loss and expense associated with architect's instructions requiring a variation. Clause 23 set out the relevant events that gave the contractor entitlement to extensions of time. Subclause 23(d) covered strike and combination of workmen and subclause 23(e) covered architect's instructions. Clause 24 covered reimbursement of the contractor's loss and expense caused by disturbance of the regular progress of the works and did not include strike and combination of workmen under its relevant matters. The Judge gave a hypothetical example in which a strike occurs during a period of extension of time granted as a result of an architect's instruction requiring a variation. He was of the view that, even if the architect had granted an extension of time for the strike under clause 23(d), he could see no reason why the contractor could not still recover all of its direct loss and expense under clause 11(6), including loss and expense arising as a result of the occurrence of the strike.

In the earlier, Canadian, case of *Ellis-Don v Parking Authority of Toronto* (1978) 28 BLR 98, works to a new car park were delayed by the defendant's breach of contract in failing to obtain a necessary permit allowing excavation works. That permit was obtained 8 weeks late, but the contractor also claimed in respect of consequent further delay events. It was held by the Supreme Court of Ontario that the delay in obtaining the permit caused 7 weeks delay to completion. In addition, it held that consequent delays in commencing the excavation works (1 ½ weeks), consequent delay in obtaining a crane (6 weeks) and time lost by consequently extended winter working (3 weeks) resulted in an overall 17 ½ weeks delay to completion. Furthermore, in addition to the contractor's time-related costs and loss of profit for that 17 ½ week period, it was entitled to the additional costs of pouring concrete in winter conditions (at an estimated rate of $5.50 per cubic yard).

6.4.14 Summary of Disruption

Disruption claims are a perennial and controversial problem within the area of evaluating the time consequences of change on construction and engineering projects. They can be hard to prove in terms of the facts of cause and effect, complex to analyse and expensive to quantify. The greater the extent of disruption that is imposed on it, the harder it can be for the contractor to present its quantification.

The starting point for any evaluation must be the legal basis of the claim and what contract provisions are related to it. Where these allow that disruption can be included as part of the valuation of a variation on a value basis, that is an often overlooked approach and has a number of advantages. This is particularly the case where disruption can be wrapped up as part of a prospective estimate of the financial effects of a change, for example under the NEC's compensation events regime or SBC/Q Schedule 2's procedure for Variation Quotations.

The method of valuation will also involve consideration of proportionality. Where the disruption is localised and readily isolated, claimants ought to be able to maintain contemporaneous allocated records of time lost and seek a counter signatory to those records. Where the sums involved are small, but the facts of disruption and effects are clear, it may be that a broad brush approach should be adopted. However, it might be that in such cases the limited effects on resources ought to be capable of contemporaneous recording.

Where the amounts involved are significant and the effects of the disruption are complex, the preferred approach is a measured mile analysis that compares such items as labour and plant outputs in disrupted and undisrupted or 'baseline' periods or locations. However, the ability to produce a meaningful analysis on this basis is the exception rather than the rule. This is because the available project records are often insufficient. It may also be that there is no undisrupted area or period that can be used as a comparator.

An alternative is a 'top down' comparison of planned and actual resources. However, there are many problems with such an approach. Planned resources may not reflect the resources that the contractor would have needed had the disruption not occurred. They may simply have been underestimated. In this regard, resource planning carried out by the site team is often a better reflection of what would have occurred rather than tender allowances. Expert analysis might also add credibility. Actual resources can be affected by various extraneous issues. They need adjustment for time lost due to contractor 'own goals', recoveries through variations, and resource allowances in other claims, whether against the employer or subcontractors or suppliers.

Alternative 'bottom up' approaches take the resources that would have been required without the disruption, and apply factors related to the various causes of disruption. However, these need great caution in relation to the source and relevance of the factors being applied and the facts upon which they were based compared to the facts of the project that is the subject of the claim. Some claim to provide a panacea for a claiming contractor that uses a computer model to input the claimed causes and output the claimed effect. These can be little more than hypothetical and fail to produce the promised cure in most jurisdictions.

Many of the alternative approaches to evaluating disruption may need a back-check of their results. This can be to see if the results are consistent or explicable across different resources or activities or reflect the details of the contractor's actual losses. Given the difficulties they each involve, it may be that the best approach combines two methods to see if they come to similar conclusions, particularly where the financial amounts that result are significant such that the costs of two approaches are merited.

A hypothetical example of a two-pronged approach to calculation has been set out. Whether that satisfies the burden of proof in relation to what is always a difficult head of claim, particularly where there are no specific records or a 'measured mile', will depend on the facts and the law in each case. However, it will have a better chance of doing so than just one or either of the methods in that example on its own.

Disruption calculations, particularly where carried out on a 'bottom up' approach, often leave a balance or 'rump' of the contractor's losses. Again the success of this will depend on the fact as to whether it satisfies a test such as 'the balance of probabilities'.

Consequential delay events that are only met because of delay caused by earlier events are a particular problem in relation to both delay and disruption costs where the contract has those consequential events as either the contractor's risk or 'cost-neutral'.

Preliminaries thickening claims are a particular form of disruption claim. Increased site staff is a particularly common head of such claims. They can also be a type of acceleration claim, which are considered next.

For examples of the difficulties that parties faced with quantifying disruption claims, and particularly in particularising the financial effects to a number of causative events, as highlighted in some English, Scottish and Commonwealth judgments, see Section 6.6.2 on the subject of Global Claims.

6.5 Acceleration

Financial claims arising from acceleration of the works are often fraught with difficulty, not least because the powers to instruct acceleration at the employer's expense are not common in construction contracts and the basis of the claim may therefore be contentious. The primary area of difficulty is therefore often the contractual basis of the claim and whether the contractor has any entitlement to it at all. It seems understandable that an employer, faced with a delay to its desired delivery of the project for which it is liable, might want to pay additional sums to the contractor to preserve those dates. Therefore some construction contracts cater for this by allowing the employer to seek such an agreement with the contractor, or to instruct it to accelerate. However, there are further ways in which acceleration can occur. These are all considered elsewhere in this chapter. The financial evaluation can also involve some particular problems, for example, in establishing what the contractor's additional costs, as opposed to the bare costs, of the measures were.

Often the issues start with a disagreement as to what acceleration actually is and what acceleration measures are. It is suggested that the latter term covers those measures that mean that the works, or a section or part thereof, are completed earlier than would have been the case had the measures not been taken. The problem with this definition is where, in the event, such measures do not actually result in earlier completion, because of subsequent further problems or causes of delay. Thus the contractor's claim for the costs of its acceleration measures is met with the response that no acceleration was actually achieved. Therefore acceleration claims can also involve measures that were instigated with the intention of achieving a result that is not achieved in practice. This can make the assessment of any associated costs and the contractor's entitlement even more contentious.

The carrying out of work in a shorter time than would have been the case without acceleration can take three different forms:

- where the same scope of work is completed in a shorter period;
- where a greater scope of work is completed in the same period; or
- a combination of the above.

When considering claims for additional payment arising from the acceleration of the works, or part of the works, the two questions posed by Judge Hicks QC in the *Ascon Contracting Ltd v Alfred McAlpine Construction Isle of Man Ltd* case need to be addressed before any quantification can commence. As he put the position:

> Acceleration' tends to be bandied about as if it were a term of art with a precise technical meaning, but I have found nothing to persuade me that that is the case. The root concept behind the metaphor is no doubt that of increasing speed and therefore, in the context of a construction contract, of finishing earlier. On that basis 'accelerative measures' are steps taken, it is assumed at increased expense, with a view to achieving that end. If the other party is to be charged with that expense, however, that description gives no reason, so far, for such a charge. At least two further questions are relevant to any such issue. The first, implicit in the description itself, is "earlier than what?" The second asks by whose decision the relevant steps were taken.

Broadly, acceleration claims can arise on construction and engineering in the following ways:

1. On the basis of an express provision of the contract that either:
   (a) directly refer to acceleration, including clauses that allow the employer to instruct or direct acceleration; or
   (b) indirectly either refer to acceleration or can give rise to it.
2. As an extra-contractual agreement between the parties, that effectively agrees an amendment to the original contract.
3. As part of the contractor's claim for the loss, expense, costs or damages arising from delay and disruption.

4. Where the contractor considers that it is entitled to extensions of time, but is denied these by the employer's certifier, and instigates acceleration in order to reduce its potential liability for delay damages. This is commonly described as either 'induced' or 'constructive' acceleration.

5. Where the contractor is in culpable delay and accelerates in order to reduce its risk in relation to its failure to complete on time and potential liability for delay damages. The associated costs of such measures will not be compensatory by the employer, but where the contractor has instigated acceleration in order to reduce the effects of its own failures, this can have significant effects on the evaluation of its entitlements in relation to other failures by the employer.

Alternative terms for the different types of acceleration include 'prospective acceleration' where those in (1) and (2) are addressed in advance of the measures, and 'retrospective acceleration' where those in (3) and (4) are addressed after the measures have been carried out. One complicating factor in relation to acceleration is how there can a combination of more than one of the circumstances in (1) to (5) at the same time.

6.5.1 Acceleration Under the Contract

The standard forms considered in this book take very different approaches in the extent to which they provide for acceleration. Such provision can be seen as being particularly for the benefit of the employer where the contractor is otherwise not going to achieve the contractual completion date or dates, whether the employer is liable, or values achieving timely completion more that its recourse to delay damages under the contract for the contractor's culpable delay.

In the FIDIC Red Book, clause 8.6 addresses such an event of culpable slow progress by the contractor, and specifically where:

(a) actual progress is too slow to complete within the Time for Completion, and/or

(b) progress has fallen (or will fall) behind the current programme under Sub-Clause 8.3 [*Programme*],

Other than as a result of a clause listed in Sub-Clause 8.4 [Extension of Time for Completion].

This clause then provides the employer with the power to instruct the contractor to submit a revised programme and supporting report detailing the revised methods it proposes to instigate to expedite progress and complete by the contractual completion date(s). The clause requires the contractor to instigate those methods unless the engineer notifies it otherwise. The clause notes that such methods might include increasing working hours or the numbers of the contractor's people, equipment, materials, plant and temporary works and that these measures will be at the contractor's risk and cost. If the revised methods cause the employer to incur additional costs, these are stated as to be reimbursed by the contractor. The most common items of such employer costs include overtime payments in relation to the engineer's staff or those of other consultants to inspect and supervise activities carried out in overtime working instigated by the contractor. Such cost can also include the introduction of additional engineer's staff, where the works to different parts of the project are carried out in parallel rather than sequentially as previously planned. The employer's right to reimbursement of such costs would be in addition to any delay damages that the employer becomes entitled to under the contract where contractual completion dates are not met.

Of course, provisions such as FIDIC Red Book clause 8.6 assume that it is clear that the culpability for the slow rate of progress is the contractor's, and in practice this is often not the case. The contractor may have submitted applications for extensions of time and believe that it is being wrongly denied that entitlement and being blamed for delays for which the employer is liable. In such a situation the contractor has two choices:

- The contractor could refuse the instruction and not adopt any acceleration measures, on the basis that it only appears to be progressing too slowly to complete within the time for completion under the contract because that time has not been properly extended under clause 8.4. However, for this the contractor would have to be very sure of its grounds and entitlement to such extension of time, both in terms of the facts and also that it has fully complied with the procedural requirements of the contract, such as clause 20 of the FIDIC Red Book. In such a situation, the contractor should notify the employer of its reasons for refusing the instruction, in detail and with reference to those of its claims and entitlements that it says have been wrongly denied by the engineer. Such a situation may even involve assertion by the contractor that the time for completion has become 'at large' because of the engineer's failure to properly operate the contract's provisions for extensions of time. See Chapter 3 of this book in relation to time being 'at large'. However, the particular danger of a contractor's refusal to comply with an acceleration instruction may be that it compounds its failures such that it puts the employer in a position to terminate the contract. This therefore requires the contractor to be sure of its grounds in terms of the causes of delay and its right to extensions of time. It would be well advised in such circumstances to take both local legal advice and expert opinion on its delay analysis.
- Alternatively, the contractor could comply with the instruction and adopt acceleration measures, whilst still continuing to pursue its claims for extensions of time. In such a situation, the contractor should notify the employer in similar terms and details. One problem with this approach is where it is subsequently found that the contractor was entitled to extensions of time, was therefore never in default under clause 8.6 and thus the employer never had the authority to issue its instruction under that clause. In such a case there may be doubt as to the contractor's entitlement to compensation against an instruction that was invalid. However, it may still be that the contractor will be entitled to the costs and expense of its acceleration measures as damages for breach of contract by the employer. Again, therefore, there are risks with this strategy and it should be the subject of good advice.

The presence of the two alternative tests in FIDIC Red Book clause 8.6 subclauses (a) and (b) can cause a conflict where the contractor considers that it has been denied its entitlements to extension of time. It may be that it is apparently progressing too slowly to complete within the time for completion as then extended, under subclause 8.6(a), but its progress is not behind its current programme, under subclause 8.6(b). This occurs where the contractor has updated its programme to reflect actual progress and its view of its entitlements, rather that the extensions of time it has received to date. On the other hand, it can also be that the contractor is behind its programme (per 8.6(b)), but it is not progressing too slowly to complete within the contractual time for completion (per subclause 8.6(a)). This may be because the programme was deliberately made tight during its early periods and the contractor knows it can make up that time later.

Clause 8.6 is the limit of the employer's powers to instruct acceleration under the FIDIC Red Book. However, it can also be instigated by the contractor. Acceleration is specifically referred to in FIDIC Red Book clause 13.2, 'Value Engineering'. Under that clause the contractor's proposals for value engineering measures can include those measures that it considers will accelerate completion if adopted. Such a proposal is prepared at the cost of the contractor and is required to include the items listed in subclause 13.3, which includes at 13.3(b) the contractor's proposal for any necessary modification to the time for completion under the contract, and at 13.3(c), its proposal for the evaluation of the acceleration as a 'Variation' (about which more elsewhere in this chapter). Understandably, this provision for contractor instigated acceleration seldom becomes relevant in practice.

Finally, in the FIDIC Red Book, clause 13.1 may indirectly have the practical effect of a form of acceleration. Under that clause, the engineer can instruct, or request the contractor to submit a proposal, in relation to a variation, a term whose definition includes in subclause 13.1(f) 'changes to the sequence or timing of the execution of the Works'. This subparagraph (f) is a more recent addition that was not in older editions of the FIDIC contracts. Whilst this does not appear intended to be used to accelerate completion, it could have that practical effect where works have their sequence or timing changed such as to bring a contractual section forward. However, that does not mean that the completion date of that section can be changed under this clause. In terms of the contractor's costs, these would fall to be valued as a variation under clause 13. It could be envisaged that this might mean, for example, parallel working for some trades, plant and equipment with resulting additional costs for the contractor. A quirk here would be whether those costs fell to be valued as a variation under FIDIC Red Book clause 12, based on the contractor's rates and prices and would also include a contribution to its head office overheads and profit.

The Infrastructure Conditions include provisions at clause 46(1) for the engineer to require the contractor to increase the rate of progress when in the engineer's opinion the rate is too slow to meet the contract completion date. This clause therefore provides a similarly welcome power for the employer to FIDIC Red Book clause 8.6 to address culpable slow progress by the contractor. Similarly, this provision states that the increase in the rate of progress is to be at no cost to the employer.

Infrastructure Conditions clause 46(1) therefore addresses the position where the contractor is proceeding culpably slowly and the employer wishes to ensure that a currently fixed contractual completion date is met. This is similar to FIDIC Red Book clause 8.6. On the other hand, clause 46(3) also allows the employer the possibility of having a currently fixed completion date brought forward to an earlier date, a power that is not in the FIDIC Red Book. In the Infrastructure Conditions this requires the contractor's agreement.

Infrastructure Conditions clause 46(3) is headed 'Provision for accelerated completion'. This allows the engineer or employer to request the contractor to complete the works or a section thereof in a shorter time than the currently extended time. However, the contractor has to agree to do so and the clause requires agreement between the contractor and employer of special terms and conditions of payment before any action is taken. This clause should not therefore generate retrospective claims for acceleration.

In SBC/Q the power for an employer to seek to get a contractor to accelerate is in Schedule 2, Variation and Acceleration Quotation Procedure. This allows the employer to investigate the possibility of achieving early practical completion of the works or a contractual section. The contractor can then either submit its proposal including the time that can be saved, the price and any other conditions attached, or it can explain why it is impracticable to achieve early completion.

Elsewhere in SBC/Q, clause 2.4 requires the contractor to proceed regularly and diligently with the works and constantly use its best endeavours to prevent any delay to the progress of the works and to prevent the contract completion date being delayed, but it is doubtful if this could be construed as being a requirement to accelerate the works in the event of delay caused by matters for which the employer is responsible.

SBC/Q clause 2.28.6.2 is sometimes relied on by employers and their consultants as a basis for instructing acceleration. The clause states:

> in the event of any delay the Contractor shall do all that may reasonably be required to the satisfaction of the Architect/Contract Administrator to proceed with the Works or Section ....

However, this does not grant a power to instruct anything, but is a requirement for the contractor to take measures to proceed with the works and take account of any comments of the architect or contract administrator, not to accelerate because they instruct such acceleration.

Rather like FIDIC Red Book clause 13.1(f), SBC/Q clause 5.1.2.4 does allow the architect to issue an instruction with regard to the execution or completion of the works in any specific order. As discussed elsewhere, it is possible that such an instruction might require some element of acceleration in specific circumstances and valuation might be on the same basis, as a variation. However, as with the FIDIC Red Book clause, this clause does not allow a contractual completion date to be changed and any acceleration would need to be the subject of agreement between the contractor and employer.

This range of approaches in the treatment of potential acceleration of the works in various standard forms of contract underlines the potential difficulties that can be encountered in identifying and costing acceleration claims under the express terms of such contracts.

Those difficulties can be particularly acute for a contractor where it has complied with an acceleration instruction from the employer's contract administrator and has incurred additional costs, but that administrator did not have the contractual authority to issue such an instruction. This would include a purported instruction under FIDIC Red Book clause 8.6 and Infrastructure Conditions clauses 46(1) and 46(3). This may leave the contractor with no contractual entitlement to those costs. The position can become even more complicated where the measures taken do not have the required effect on completion, perhaps because of later delays for which the employer is liable.

An interestingly direct approach to the power to order acceleration in construction contracts is clause 14.1(a)(iii) of the General Conditions of Contract for Construction issued by the CRINE Network. This contract is intended for works of offshore oil and gas development and gives the employing company the right to order a variation to the contract to accelerate the works and, within the limits of what is practical, to reschedule resources to overcome a delay that would otherwise have entitled the contractor to an extension to the date for completion. Clause 14.6 of the CRINE contract provides for the contractor to recover fair and reasonable adjustments to the contract price for an ordered variation. This is in contrast to the 'prior agreement' basis of the common building and civil engineering forms and suggests that the complexities of offshore construction, where many individual contracts need to be coordinated to enable the whole development to succeed, has dictated a different approach. The CRINE approach will, however, generate consideration of additional resources and the effect on price of rescheduling in agreeing the adjustment to the contract price. There is scope for disagreement as to the effectiveness of, or necessity for, particular resources or actions. There may also be fertile grounds for disagreement as to the base rate of progress from which acceleration was required. To avoid such matters being the seedbed of subsequent dispute there will need to be a clear definition of the aims of the acceleration, how it is to be judged and, most importantly, what is to happen if the measures are not wholly, or are only partially, successful.

6.5.2 Acceleration by Extra-Contractual Agreement

In the absence of there being an express power to acceleration in the contract, such as in the standard forms considered in this section, the contractor's obligation will only be to complete the works, or contractual sections thereof, by the date or dates agreed in the contract, or as extended under the extension of time provision of the contract. If the employer wants the contractor to complete earlier than that date or dates (whether to achieve dates earlier than those in the contract or to mitigate the effects of delays for which it is liable), or to carry out additional work in the existing contract period, then it will have to seek the contractor's agreement.

Such an acceleration agreement might be in the form of a new contract or collateral agreement or amendment or supplement to the existing contract. In practice it is usually best to make it an amendment to the existing contract, that keeps other provisions of the contract in place and otherwise the same. Potential risks with making it a new separate agreement include that the dispute resolution procedures may not be the same under both contracts.

On the face of it, the terms of a supplementary agreement or amendment to the existing contract to address acceleration agreed between the parties ought to be simple. In essence, they ought to only need to address just two matters. Firstly, what additional amount the contractor will be paid in addition to the contract price and when. Secondly, what the new contractual completion date, or dates, is or are. However, matters are never that simple and a number of further details are essential.

In practice, such acceleration agreements are fraught with danger. They need as much care and legal advice as the original contract and need to cater for the same possibilities and eventualities as that original contract. In practice, they are often insufficiently thought through and fail to consider all of the potential results of the measures and events that might affect them. Such issues include: what is the effect on the contractor's other claims; which of the contractor's delay events or claims to date are covered; which of the contractor's financial claims to date are now covered; what happens if the measures fail in their effect on completion; and whether the contractor is still entitled to be paid for acceleration measures that do not achieve their required result.

Broadly, such acceleration agreements tend to be made in one of two forms. These are:

1. An agreement setting out the new agreed completion date or dates, and the additional sum of money to be paid to the contractor for achieving those dates. Such a form effectively just agrees revisions to the contractual completion dates and the contract price This can be regarded as an agreement to achieve a defined result, with the contractor to decide what resources and measures it requires to achieve that new contractual obligation.
2. An agreement setting out the measures and additional resources that will be provided and a method of recording them and later determining the price for them. This can be regarded as an agreement to provide agreed resources and measures, rather than an agreed result. This alternative tends to be most common in agreements between contractor and subcontractors for the latter to provide a defined additional resource, such as an increased number of electricians or steel fixers or to instigate overtime working to such extent as it can. This is common where the contractor is in culpable delay under the main contract and needs its subcontractor(s) to help it recover that delay.

There can also be variations of this, for example where there is a defined effect, in terms of new contractual completion date(s), but the resources are to be recorded and compensation worked out later.

Most acceleration agreements are of the type described in (1) above. In such cases the first concern is what is the new agreed contractual completion date or dates? Once they are agreed this should just require the original contract completion date(s) to be 'reset'. However, a common mistake in such agreements is that they do not exactly identify which causes of delay and/or claims from the contractor for extensions of time are covered by that resetting. Sometimes acceleration agreements refer to their covering 'all events occurring up to ... [a date]'. That date can be that of the agreement, or some earlier date. However, what does 'occurring' mean in this context? In relation to an incorrect design, does it mean when the drawing was issued, or when it actually affects the work that it was supposed to detail, or when it was notified? Alternatively, some of this type of acceleration agreement refer to addressing all of those of the contractor's applications for extension of time up to a certain date, or as listed in the agreement, but this can leave the employer a hostage to other events that have already occurred but have not yet been notified by the contractor at that date.

Regarding the contractor's costs, these could be the subject of agreement as a lump sum addition to the contract price or agreed to be valued based on records of the resources and measures taken by the contractor. If based on records, the agreement should specify what form they are to take and the process for contemporaneous agreement of them, together with the rates and prices that are to be applied to value them.

Where an acceleration agreement is to incorporate an agreed sum, the contractor will have to assess what is possible by adopting such measures as overtime working, increased shift working, the introduction of additional resources and the possible reprogramming of the remaining works. In effect it will be in the position of submitting a tender for the accelerative measures and will usually wish to include some contingency to cover risk and unforeseen complications and costs in putting the accelerative measures into effect. Furthermore, the contractor may also consider that it is entitled to add a contribution to its head office overheads and profit to its pricing, on the basis that it would have included such a margin had the measures been part of its original tender for the works. In practice acceleration agreements are usually negotiated on an open basis, with the contractor declaring its allowances for such as risk and margin and these being part of the negotiation with the employer. From the contractor's perspective, it may consider that it should not bear any risk or be expected to mobilise further resources and costs without a margin. In such an agreement it is usually the employer that is keenest to get agreement and the contractor can play the role of 'the reluctant seller'. However, this must be weighed against any benefits to the contractor, such as any doubts as to its case for extensions of time for events to date. Resetting an agreed extended completion date can remove doubt and contractual risk for a contractor.

From the employer's perspective, the costs of the addition to the contract price will be weighed against the financial effects of not accelerating and the project and its being completed later. A regular complicating factor is the time pressure under which such agreements are negotiated, with the project otherwise proceeding towards delayed completion and the benefits of the acceleration perhaps being reduced as the time passes and further work is progressed without the benefit of acceleration measures.

Subject to any express terms of the particular contract, it may be considered that a contingency allowance is particularly appropriate if the accelerative measures can be instructed and required but the payment is only to be made if the required effect on completion of the works is achieved. Depending on the circumstances, the risk might be so great that the contractor either declines an agreement or prices a very large contingencies allowance. An option for the employer then is to agree to pay for the measures but without placing the risk of achieving accelerated completion on the contractor, although this may be very unattractive to an employer other than one in a very difficult position, where it is clearly responsible for delays and cannot countenance the resulting delay to handover.

In this regard, from an employer's viewpoint, a key issue will be what happens to its right to delay damages as set in the contract? If the acceleration agreement sets a new completion date(s), then it should also set out how other terms of the original contract remain in force and unaffected, including the entitlement and rates of delay damages set out therein.

The greatest difficulty in 'prospective acceleration' agreement occurs, as is often the case, when the accelerative measures are adopted with an agreed payment, but they are only partially successful, i.e. they reduce the projected time to completion but not by the amount desired or required. Such a situation needs to be carefully considered before any agreement, or instruction, for accelerative measures is executed and the status of the payment is agreed and set out for all the possible outcomes. It is in essence a matter of who is to carry the risk for the accelerative measures. If the contractor is taking the risk, then the inclusion of a reasonable contingency would seem to be a prudent and acceptable matter.

A further common problem is how acceleration agreements address the contractor's financial claims as they stand at the date of the acceleration agreement. This may include claims for delay and disruption as well as the value variations, which may have been contributory factors to the delays that are now being addressed by the proposed acceleration. Ideally, as previously mentioned, all such claims should be covered by the agreement as closed off. Thus, the new agreed completion date addresses all the extension claims or events, and the agreed amendment to the contract price incorporates all financial claims and variations. In practice this often does not happen because the parties are too focussed on getting the measures for the future in place to worry about closing off what has gone before. However, this can prove to be very short-sighted.

Where an acceleration agreement does not refer to a new completion date(s), but is for the provision of additional measures and resources, it is usually best that the compensation is not an agreed amount, but is to be determined on the basis of what resources and measures are actually provided. This will require an agreed basis for recording the measures and resources and either establishing their costs or pricing them at agreed rates and prices. This retrospective approach would require no inclusion of a continency, but head office overheads and profit are likely to be a factor.

Acceleration agreements can address delays for which the employer is liable, culpable delay by the contractor or a combination of both. It may seem surprising that an employer would reach such an agreement including payment of additional sums in relation to contractor's culpable delays, rather than rest on its right to delay damages, but it may be that the rate at which these have been set in the contract is low, or otherwise insufficient for the real pain that late completion will cause the employer. Take, for example, the construction of stadia, accommodation or transport infrastructure for an event such as a football World Cup or an Olympic Games, where delay damages cannot practicably address the damage to the host nation or city of late delivery. Where such an acceleration agreement addresses culpable delays by the contractor, there may be a potential problem regarding the legal enforceability of that agreement, if the contractor is being paid additional monies to do no more than what it has already contracted to do. It may be that to be legally binding, a contract requires legal consideration to pass from both parties to the other (for example in English law, where the new contract is not entered as a deed).

Under English law, it was long the position that an agreement does no more than what a party had already contracted to do provided there was no consideration from that party. However, in some circumstances consideration can take other forms that are present in the new agreement. In *Lester Williams v Roffey Bros and Nicholls (Contractors) Limited* [1991] 1 QB 1, the Court of Appel considered an appeal from the Judge at first instance. Williams was a carpenter engaged by the Roffey Brothers as contractor on the refurbishment of a block of flats. Williams fell into financial difficulty and it was common ground that this was because it underpriced the work at a total of £20,000 and also failed to supervise its workmen adequately. Fearing that Williams would not complete its works on time and the resulting pain of a delay damages under its main contract, Roffey agreed to pay Williams a further total of £10,300 at a rate of £575 for each flat as it was completed. However, Williams continued to perform badly, only partly completing its works and Roffey engaged other carpenters to complete the works. Even then, Roffey incurred delay damages under the main contract for one week. In the appeal, Roffey submitted that its agreement to pay an additional £575 per completed flat was unenforceable since there was no consideration for it. The Judge at first instance had concluded that the agreement was in the interests of both parties, such that it did not fail for lack of consideration.

The Court of Appeal judgment reviewed the law back to *Stilk v Myrick* (1809) 170 ER 1168. That matter had involved the crew of a ship voyaging to the Baltic and back to London. Two of the crew deserted. The captain agreed with the rest of the crew that if they worked the ship back to London without the two deserters being replaced, he would divide the deserters' pay between them. The court found that this agreement failed for want of consideration as the crew were already bound by the terms of their original contracts to bring the vessel safely back to its destination port. However, in *Williams v Roffey* the court of appeal found that consideration could take the form of a party's promise to do something that it would otherwise fail to do, even though it was already contracted to do it. Lord Justice Glidewell put the position as follows:

(i) if A has entered into a contract with B to do work for, or to supply goods or services to, B in return for payment by B; and

(ii) at some stage before A has completely performed his obligations under the contract B has reason to doubt whether A will, or will be able to, complete his side of the bargain; and

(iii) B thereupon promises A an additional payment in return for A's promise to perform his contractual obligations on time; and

(iv) as a result of giving his promise, B obtains in practice a benefit, or obviates a disbenefit; and

(v) B's promise is not given as a result of economic duress or fraud on the part of A; then

(vi) the benefit to B is capable of being consideration for B's promise, so that the promise will be legally binding.

6.5.3 Acceleration as a Component of Delay and/or Disruption Costs

It has been explained in Chapter 1 of this book that a contractor claiming such as loss, expense, costs or damage under a construction contract is likely to have a duty to mitigate those financial effects, as an obligation under the law. Under English law this was explained with particular reference to the judgment of the House of Lords in *British Westinghouse Electric & Manufacturing Co Ltd v Underground Electric Railway Co of London Ltd* [1912] AC 673. The effect of this principle is that the contractor will not be able to recover damages that could have been avoided by taking reasonable steps. It will, however, be able to recover the costs of those reasonable steps if they were implemented. It can be envisaged that this might include a situation where compensable delays have been caused, and the contractor can take reasonable steps by increasing its resources at additional costs that are less than the resulting reduction in the financial effects of the delay.

Furthermore, or alternatively, the contractor may have a relevant express obligation under the terms of the contract. For example, in SBC/Q subclause 2.28.6, the contractor's entitlement to extensions of time is subject to the caveat that the contractor shall constantly use its best endeavours to prevent delay. Take the common example where there are both problems with the architect or engineer's original designs, in terms of accuracy and completeness, and a large volume of design changes. As a result, the contractor has to issue a large number of requests for information and clarifications and technical queries. It has also received a large number of variation instructions and revised drawings. In such circumstances the contractor might chose to deal with these problems with its existing or planned site resources and claim whatever entitlement to extension of time and associated prolongation costs result. However, and this is the most common response in practice, the contractor might consider that its most efficient response is to increase its resources to deal with this volume of information and change. This might include additional administration, design and engineering staff to process the information as well as additional management and supervision for the varied and affected work itself.

In such circumstance the contractor's claims will include a heading for the costs of these additional staff. Claims relating to such actions and their costs are commonly referred to as 'thickening claims' and have been addressed in Section 6.4.12 of this book under the heading of Disruption. However, they could also be categorised under the heading of acceleration.

Where such costs are challenged, it may be necessary to establish the beneficial effects of the measures taken, in terms of reducing the effects on the completion date(s). A complicating factor in this regard may be where the benefits are lost or reduced by further delays for which the employer is liable. An important principle in relation to assessing the benefit against the cost of such measures is to not apply the benefit of hindsight. The reasonableness of the measures taken must be considered by reference to the knowledge of the contractor at the time it took the decision to add resources.

A further matter to note is that a claim for additional resources on this basis might be subject to the complaint that it is 'global' in the pejorative sense, as explained in Section 6.6 of this book. The counter to that will cite the impracticality of individually allocating each additional member of staff and their costs to each design error, query or change that gave rise to the need for the resources.

6.5.4   Induced or Constructive Acceleration

It is a fact of construction and engineering claims in some parts of the world that the heading of acceleration most commonly arises where the contractor has what it considers valid claims for extension of time that have been unfairly rejected, and asserts that it was therefore forced to instigate acceleration in order to avoid a liability for a period of delay damages that it should already have been relieved of. This is sometimes referred to as 'induced acceleration' on the basis it is induced by the failure of the engineer or other certifier of extensions of time to properly operate the extension of time provisions of the contract.

To succeed in a claim for acceleration on such a basis of breach of contract by the employer, a contractor will have to establish the following:

- where the granting of extensions of time under the contract is in the power of a contract administrator or architect (as in SBC/Q) or the engineer (as in FIDIC Red Book and the Infrastructure Conditions), that the employer is legally liable for the failure of that certifier and any damages that result from it;
- that it complied with any contractual preconditions of its right to extensions of time, including matters such as notices or the provision of supporting details and particulars of the causes and effects of the delays;
- that the certifier under the contract failed in its duties in relation to extensions of time and that failure was clear and sustained;
- that as a result of such failure the contractor incurred acceleration costs;

- that such damages pass any legal requirements in relation to being the foreseeable consequences of the failure; and
- the quantum of its resulting acceleration costs.

Whether the employer is legally liability for the failure of its certifier to grant extensions of time will depend on the contract and the law. This will involve detailed consideration as to the legal nature and extent of the duty of the employer, or its architect or engineer acting on its behalf, in that duty and whether that certifier is acting as the employer's agent in relation to extensions of time. This will be the case if a clause such as 3.1(a) of the FIDIC Red Book applies ('whenever carrying out duties or exercising authority, specified in or implied by the Contract, the Engineer shall be deemed to act for the Employer'). Where there is no such express provision in the contract, the applicable law will be particularly important to this issue.

Where the contractor has eventually received an extension to the contract completion date it may argue that the decision to award the extension came too late to prevent it feeling forced to institute measures to overcome delay for which the employer was responsible. This circumstance is common in practice. It means that the extension of time having been eventually granted, the contractor's case only has to be that it should have been granted earlier, without the need to establish that it is entitled to that extension.

It is sometimes suggested that there should never be claims for induced or constructive acceleration because, if the contractor had an entitlement to extensions of time, then it should have carried on to such date or dates as it considered it was reasonably entitled to and challenged any levying of delay damages by the employer in relation to earlier purportedly due dates. However, this ignores the practicalities that often face contractors when they are in excusable delay, have made extension of time claims and had them refused. The contractor may not be one hundred percent sure of its entitlement. There may be allegations of concurrency in a background where the legal position is unclear. Its legal recourse might be to a dispute forum in which it does not have full faith (consider, for example, the possibility of an overseas contractor working for a national government under a contract where disputes are to be referred to the local courts in a jurisdiction where their independence is in doubt). The potential costs of legal proceedings might be expected to be greater than the costs of the acceleration. The rate of delay damages as set in the contract may be very high.

Even if there is a legal liability for the employer for the certifier's inaction, the contractor will need to establish that the failure to grant an extension of time was a breach under the terms of the contract. Perhaps the contractor gave no notice of the delay event, as required by the contract, such that the employer and/or its certifier was not aware that the event had occurred and/or that the legal duty had been triggered. This may further involve consideration as to whether the notice was a condition precedent to entitlement to an extension of time. However, it may be more likely that a notice was given, but then issues arose as to its adequacy and particularly those of any particulars required in support of the claim for extension of time. FIDIC Red Book clause 20.1 sets out the requirements for the contractor to send to the Engineer 'a fully detailed claim which includes full supporting particulars'. Under SBC/Q clause 2.27.2 the contractor is required to follow its notice of a delay event 'as soon as possible' with 'particulars of its expected effect, including an estimate of any expected delay in the completion . . . .'. Such requirements are subjective. They often lead to dispute as to whether the particulars provided were inadequate to allow the administrator to grant an extension of time or it is just using this as an excuse to deny a known entitlement.

Before a contractor undertakes such preventative acceleration on an induced basis it is also well advised to ensure that it has exhausted all available avenues to getting its claims for extensions of time properly considered. If recourse is available to, for example, a dispute review board or fast-track adjudication then these should be instigated before the acceleration, although there may again be practical constraints here. In this regard, Core Principle 16 of the SCL Protocol advises:

> Where the Contractor is considering implementing acceleration measures to avoid the risk of liquidated damages as a result of not receiving an EOT that it considers is due, and then pursuing a constructive acceleration claim, the Contractor should first take steps to have the dispute or difference about entitlement to an EOT resolved in accordance with the contract dispute resolution provisions.

The contractor should also notify the employer and its engineer or architect/contract administrator of the measures it intends to take, that they have been made necessary by their failures, and that it will be claiming the associated costs from the employer.

The costs of induced acceleration will usually be claimed on the basis that they are damages for breach of contract by the employer. This can be either because it is liable for the certifier's failure or in breach for not procuring that the contract certifier properly operated the extension of time provisions of the contract. As such, they will be based on the additional costs and losses incurred by the contractor as a result of the measures and be quantified on a retrospective basis. Therefore, provision for risk and contingencies will be unnecessary. Similarly, there is unlikely to be any entitlement to profit.

Claims for constructive acceleration have long been recognised in principle in jurisdictions such as the US. Furthermore, some Commonwealth judgments have addressed clear examples of blatant refusal or failure to give consideration to the contractor's extension of time entitlements. Two Canadian cases have facts of sufficient extremity as to merit their own subsection 6.5.5 of this book. In Australia, the judgment in *Perini Corporation v Commonwealth of Australia* [1969] 2NSWLR 530 considered a contract in which the defendant's Director of Works was appointed as certifier in relation to extensions of time. He refused many of Perini's applications on the basis of departmental policy. The court held that the Director of Works had a duty to act impartially in this certifying function and that it was an implied term of the construction contract that the defendant would ensure that he complied with that duty. Thus, the failure of the Director of Works to grant Perini extensions of time to which it was entitled was a breach of contract by the defendant.

In the English courts induced acceleration claims have found more difficulty. In relation to claims for induced or constructive acceleration, the second question posed by Judge Hicks QC in *Ascon v McAlpine*, as to the source of the decision to institute the measures, will be particularly relevant. In that case he rejected Ascon's acceleration claim and said this:

> It is difficult to see how there can be any room for the doctrine of mitigation in relation to damage suffered by reason of the employer's culpable delay in the face of express contractual machinery for dealing with the situation by extension of time and reimbursement of loss and expense. However, that may be as a matter of principle, what is plain is that there cannot be both an extension to the

full extent of the employer's culpable delay, with damages on that basis, and also damages in the form of expense incurred by way of mitigation, unless it is alleged and established that the attempt at mitigation, although reasonable, was wholly ineffective. That is certainly not how Ascon puts its case here; it contends that the work was indeed completed sooner than it would have been in the absence of its accelerative measures.

On the other hand, the judgment in *Motherwell Bridge v Micafil* (2002) CILL 1913 added some impetus to induced acceleration claims in the UK. The Judge accepted that Motherwell were entitled to recover acceleration costs in the face of a refusal of extensions to the contract completion date. Micafil had been the main contractor for the construction of an autoclave with Motherwell as a subcontractor. Motherwell requested extensions of time to cover variations to the works ordered by Micafil, and for restrictions on site imposed by Micafil. The requests for the extension to the date for completion were refused and Micafil insisted that the subcontract completion date must be met. In response Motherwell had to resort to the institution of a nightshift in order to overcome the effect of the changes in their scope of work. It was held by the Technology and Construction Court that the changes to the scope of work caused delay which was on the critical path, and that the delays were not Motherwell's responsibility. Motherwell was therefore entitled to an extension of time and to the acceleration costs of working to overcome the refusal of its requests for an extension of time.

Of course, those promoting the NEC approach to construction and engineering contracts will say that its 'compensation events' approach to such matters will prevent 'induced' or 'constructive' acceleration.

It has sometimes been suggested that constructive acceleration claims cannot be entirely eliminated because sometimes the contractor is not aware until after it has implemented the accelerative measures what the true causes of delay were. To employers and their advisors this may seem to be a somewhat disingenuous line of argument as, if the contractor did not know at the time that it was overcoming employer delay, it does not seem reasonable that the employer should pay for such measures. However, from the contractor's viewpoint, it may be argued that the employer having caused what was eventually recognised as its delay, the employer should not avoid the financial consequences of that delay just because the contractor accelerated to prevent delay to completion. The contractor will say that, had it not accelerated, then the employer would have had to pay for prolongation costs.

6.5.5 The Canadian Cases

As noted elsewhere, two Canadian judgments involve particularly severe examples of contractors being induced to accelerate by failures of certifiers to grant their entitlements to extensions of time. This said, the facts of those failures and the effects on the contractor will be familiar to many practitioners in certain parts of the world, particularly on major public sector projects.

Morrison-Knudsen v British Columbia Hydro & Power Authority (1978) 85 DLR 186 is considered in Section 5.12 in the context of quantum meruit claims. As explained there, among the employer's many failures, it had ordered acceleration of the work and continued to insist that the work be completed by the times fixed in the contract. The trial judge found that the employer's actions amounted to a fundamental breach of contract because it had not paid for the costs of acceleration and because, while agreeing with the contractor that it would process claims for acceleration, it had given the Provincial Government a private assurance that it would not pay such costs. That judgment was quoted extensively by the Supreme Court of British Columbia in *W A Stephenson Construction (Western) Ltd v Metro Canada Ltd* [1987] 27 C.L.R. 113.

W. A. Stephenson Construction signed a contract in September 1983 with Metro Canada Ltd for the construction of a part of the Advanced Light Railway Transit System between Vancouver and New Westminster. The contract involved the construction of 103 large T-shaped columns 30 m apart and the agreed contract sum was $5,865,005. The whole system had to be ready for Expo 86 and potential bidders were warned of the critical nature of the timetable in the original instructions to tenderers. Furthermore, a distinctive feature of the contract was the creation of six contractual milestone events.

During construction there were continuous disputes concerning blocking off streets to traffic and when work areas were delivered (late) they were often not free of obstructions, being occupied by buildings, traffic, overhead and underground utilities. Notwithstanding these problems, the owner failed/refused to grant any extensions of time and insisted on performance by the original completion date(s). It gradually became clear that the City of Vancouver (not a party to the contract) was 'calling the shots' in terms of access and time for completion, but nobody had told the contractor of this in advance.

The background to the litigation includes a wide range of alternative approaches that were adopted to deal with the problems of delay and acceleration. These include: details of concluded agreements on acceleration; verbal promises to accelerate part of the works and address the costs at completion; extended milestones and acceptance of reasonable costs; written requests for increased resources; and formal instructions to accelerate without payment for the contractor. However, the judgment is very specific and arises because there were two separate and distinct provisions for extensions of time:

- A mandatory provision to extend time for third party acts and acts of God.
- A discretionary power to adjust dates and make payments for owner-caused delay or failure to do an act.

Here is what the judgment says about the causes, the breaches of contractual provisions, an analogous case and the results. It is a long quotation, but it embodies the relevant basis for this judgment and the 'analogous case' of *Morrison Knudsen* considered elsewhere:

> Acceleration Law:
> Throughout this job the contractor was told he had to meet the milestone dates in the contract. Conditions of work which delayed the execution of the contract were forced on it by the owner (e.g. non-removal of buildings). Weather conditions plainly impaired the ability to perform. If the owner insisted on performance by the contract date, the slowed work had to be made up with extra resources of capital and labour. This compression in time is one definition of acceleration. Like the Red Queen, the contractor had to run faster just to stay where it was in relation to the time frame of the contract.

The situation here is analogous to that found by the trial judge Mr. Justice Macdonald (as he then was) in *Morrison-Knudsen Company v B.C. Hydro & Power Authority (No. 2)* 1978, 4 W.W.R. 193. He said in his judgment, Part 11.3:

> '...by acceleration I mean speeding up the work increasing the rate of performance of the work – in order to overcome delays and complete by the date specified in the contract work which has fallen behind schedule. Acceleration may be undertaken in order to finish by the contract dates work which has fallen behind schedule due to:
>
> 1) delays for which the contractor is solely responsible;
> 2) delays attributable entirely to the owner; and
> 3) delays which are a combination of the two....
>
> Section 10202 requires completion of the work by specified dates and empowers the engineer to allow extensions of time ... he is not required to grant extensions no matter what the cause of delay. If he should decide against an extension of time, in a clear case of owner-caused delay, the result is that the contractor remains legally bound to complete by the contract dates. That may involve acceleration – at additional cost – to overcome the delay. In such circumstances fair treatment would require the owner to pay that extra cost. The defendant does not contend, that in those circumstances, an owner would not be obliged to pay the cost of acceleration ....
>
> Throughout, the plaintiffs were pressed to complete their work by the dates specified in the contract. The chairman's two letters ... called on the contractor to take all necessary steps to completion by the contract dates, and made it plain that failure to do so was unacceptable to the owner. The second letter was written after a Hydro senior official had accepted the fact that there would be no Peace River power until 1969, but decided that the contractor would not be told of this, and instead, the penalties involved in not completing the contract on time would be pointed out to him ....
>
> ... If extensions of time for owner-caused delays are refused, and a contractor accelerates to overcome those delays plus others which are solely his responsibility, he is entitled to be paid the portion of his acceleration costs attributable to the owner's default ....
>
> In the case at bar, there were two concepts in relation to time: a mandatory contractual obligation to extend time for third party acts or acts of God, and a discretionary power to pay in the case of owner caused delay or failure to do an act together with a date adjustment. In my opinion, the owner failed to do its duty as to the first obligation and is in breach of contract, and having failed to make any proper financial adjustment for the second, must bear proven monetary consequences.

The findings of fact supported the analogy and the judgment continued as follows:

> On all the evidence, I find:
>
> 1. The contractor was not given the work space to which he was entitled under the law and the documents, resulting from the owner's failure to deliver certain properties for occupancy, and failure to prevent undue traffic obstruction.

2. Much obstruction of the work space occurred from utilities, both overhead and underground.
3. For the above and because of weather and labour troubles, time extensions which ought to have been granted under the provisions of the contract were not given by the owner.
4. The owner did not supply timely, accurate information, and because of this the planning of the contractor was affected.

I think the above factors were all serious, and they all affected the planning, execution and cost of the plaintiff's work in every section and in every aspect of the project. The efficiency was interfered with and an assembly-line operation was reduced, on occasion, to custom building with consequent cost effects.

It is my opinion that the above was almost completely the fault of the owner. I think there was one major reason for all this. This was a very large, novel, and complicated project, to be built through a major, urban centre. In a deliberate, anticipatory breach of contract by means of a policy decision, the owner decided that the time for completion would not be extended, no matter what the cause. This inflexible straitjacket seriously distorted the administration of the contract, and I think, executive attitudes. Deprived of the remedy of time extension, but still trying to complete on budget, management appeared to me to be unwilling to face the fact that extra money would be required to overcome the obstacles, even when a principal and obvious one – labour trouble – could not be attributed to either party. The pressure to complete on time for Expo 86, I think, skewed many managerial decisions, and on occasion, they were simply made too late to avoid major strains on the contractor's organization and subsequent monetary consequences.

In its quantification of the claimant's damages, the judgment does not appear to abate any of the claimed values, despite acknowledging that there may be faults with the contractor's performance and that the calculations may not be precise. The Judge excused himself by the following:

It will not have escaped attention that I have awarded the plaintiff damages to the last cent, which might give an illusion of certainty to the reader, but let no such illusion continue. Damage estimates are just that, and in an action this size and complexity, accuracy is impossible. However, where the court can find no good reason for varying what it considers to be a proper estimate, either upwards or downwards, there seems to be no particular reason for altering the numbers. I have therefore left them alone.

And:

In the result I allow the area claims in the amounts originally presented, being of the view that the proven facts support them and the inherent judgmental factors in Gianellia's estimates are reasonable. In one or two cases I could criticize in detail: for example, I think there was insufficient allowance for safety personnel over all, and the break with De Fazio caused temporary inefficiencies. But for each minus, I can think of a plus. I do not think the contractor ever had a fair chance of capitalizing on the obvious economic virtues of his construction train, for which he has the legal right to claim. If he had had an unimpeded run at it, I think he would have completed the whole project a month earlier, according to his plan, and perhaps better than his estimate and a consequent larger percentage profit.

Mr Gianellia was Stephenson's chief executive and project construction manager, who provided evaluations for certain claims. These positions were not given lightly by the Judge and were not merely based on the assessment of a single global claim, with there being three types or categories of claims laid before the Court:

- 4 Area claims, consisting of a claim for acceleration and site constraints for each of 4 main areas of the project.
- 4 Claims based on contract interpretation.
- 3 Claims for loss of profit and productivity (labour, materials/equipment and profit).

The area claims were founded on a combination of invoices that had been submitted on a contemporary basis, a measured mile that was used to establish what had been achieved on an uninterrupted section, and an element of practical judgment that was provided by the contractor's most senior person who had also been on the project for most of the construction period and had been involved in the planning and the original estimate.

The loss of productivity claim was also founded on a wealth of supporting information and three expert opinions, each taking a different approach and broken down as follows:

- Labour: the contractor first took labour overtime premiums, subtracted overtime shown in the area claims, and arrived at a figure. It then took the total number of hours on the job and estimated that 15% of those had been squandered in overcoming the obstacles. It thus arrived at another figure for total loss of productivity in labour. This is in effect a 'rump' type of claim of the type explained in Section 6.4.10 of this book (in addition to the more particularised area claims and contract interpretation claims).
- Materials and Equipment: the contractor took the amount spent on equipment, deducted the equipment cost factors of the area claims and arrived at a figure. It estimated that 3.5 hours per operational day, or just less than half, were attributable to the constraints. Thus it arrived at a total equipment cost, to which he added some materials costs. This is also in effect a 'rump' type claim.
- Loss of Profit: the contractor then calculated the total loss of profit on the entire project by taking 10% of the final contract cost and deducting the profit already incorporated in the other claims.

Two of the parties' expert witnesses used 'measured mile' calculations: one to establish a benchmark for productivity and thereby avoid reference to the tender; the other to establish a total for original project hours in order to decide if the original allowance was reasonable. Both applied their benchmark to calculate a loss of productivity. Experts for both sides were of the opinion that the original work could have been completed with fewer hours than were allowed in the original estimate.

The Judge liked the contractor's own estimate even though it contained practical engineering 'judgement' on the extent of lost production (pages 43 and 50/55). He also confirmed that he did not consider the claim to be what we might now describe as a 'total cost' claim. The contractor had taken the opportunity of a break in the proceedings to assemble what was referred to as the 'Christmas computer analysis'. It assembled the total costs and compared them with the original price and recovery for additional work and other claims. The resulting figure was far greater than the estimates claimed in the action and gave the Judge confidence in his approach to the estimates claimed. When taken together with the wealth of contemporary submissions of invoices and the figures derived by three separate experts using three very distinct approaches, here is what he said about his reasons for admitting the entire contractor's claim for loss of productivity and an adjusted allowance for profit:

> The reasons are much as in the area claims. First, I accept Gianellia's quali-fications. Second, I like the fact that it exposes certain points of engineering judgment, made as a result of experience, and on the territory which can be cross-examined on (and were). The court can then decide as to the correctness and rationality or not, and not just wallow in numbers. Third, it is supported as to magnitude by Smail and Russell. Fourth, it also has some built in moderation in three respects. First, the 46% judgment call concedes a percentage of respon-sibility to the contractor. Second, the Christmas computer analysis, founded on the audited costs, shows that the total of all the claims advanced by invoice still underestimates the actual loss. And last, the practical estimate of Gianellia is less than the experts Smail and Russell,
>
> Second, as to profits. The original estimates allowed for 10% on costs, a standard amount, and this would have allowed a profit of (say) $600,000.00. To allow 10% on the final costs would result in $925,000.00. Is this justified? The owner presses on me that the plaintiff-caused delays and subtrade costly frictions should be con-sidered in reduction of costs and so of profits: I have already stated and repeat that I do not think these made a measurable contribution to the costs. A sec-ond argument, however, points out that a total of $534,558 already paid by the owner in respect of Field Orders and Change Orders already has attracted profit and an allowance should be made for this. I agree that as the contract says that such orders shall include a profit figure consideration should be given to this and assuming that 10% was allowed, will therefore subtract $53,500 from any award. In the result, as I think the costs are justified, so do I the profits. I therefore allow 10% on them and fix the loss of profits at $925,201.00 less $53,500.00 for a total award of $871,701.00.
>
> In making the award for loss of profits it must be remembered that the plain-tiff, at the beginning of this project, had an opportunity of making $600,000.00. I think, unimpeded, it had a good chance not only of doing this, but considering the evidence regarding the measured mile and the experts' evidence, of bettering its performance. In awarding a larger sum it is in part a reflection of the value of the loss of the opportunity of making the original profit.

A notable aspect of these extracts from the judgment is the importance of the impact of the extreme nature of the site's problems. Thus phrases such as:

- 'an assembly-line operation was reduced, on occasion, to custom building with consequent cost effects';

- 'this inflexible straitjacket seriously distorted the administration of the contract, and I think, executive attitudes';
- 'skewed many managerial decisions'; and
- 'major strains on the contractor's organization and subsequent monetary consequences'.

A reading of this may suggest that the claim's success was premised on the court being convinced through factual witness evidence of the great extent of difficulties on the project and of the contractor's suffering. It is also notable how much reliance was placed on the factual evidence of a senior member of the contractor's project team, Mr Gianellia.

In terms of evaluation, the claims were not considered by the court to be global or total cost claims and were not expressly accepted as such by the court. The admission of what appears to be unabated claimed figures for some of the issues is, on detailed inspection, more realistically a confidence call by the Judge based on the 'balance of probabilities', because the magnitude of claims has been justified from a variety of approaches and from separate estimates for damages by three experts. The contractor's estimated values appear to have been allowed because they were smaller than (or in the terms of the judgment seen to be reasonable in the light of) the values derived from more traditional approaches.

It also seems that the court took comfort from the disclosure of the contractor's overall loss and that its award based on the estimates claimed would not over-compensate the contractor.

The judgment in *Stephenson* has been considered in detail here because it is such an interesting case, the facts of which will not be completely unfamiliar to many international practitioners.

6.5.6 Acceleration to Reduce the Contractor's Culpability

It may be that the contractor is in accepted culpable delay which will result in an overrun of completion, knows that it has no prospect of obtaining the protection of an extension of time and accelerates in order to reduce its potential liability for delay damages. This will involve consideration by the contractor as to whether its likely liability for delay damages will outweigh the costs it will incur in accelerating. However, usually they will do so by a significant margin, unless the employer has its delay damages set at a very low rate.

In such circumstances, the contractor's associated costs of acceleration will clearly not be compensatable by the employer. However, the situation can have significant effects on the evaluation of the contractor's entitlements in relation to other failures by the employer. Take, for example, the following situations:

1. The contractor has increased its resources on site to accelerate to recover its own culpable delays, but then events occur for which the employer is financially liable which cause delay. The clearest example of this problem would be where the employer instructs a suspension of the works, under such as FIDIC Red Book clause 8.9. If the contract has already doubled its resources such as people, plant and equipment on-site to address its own culpable problems, is the employer liable for the costs of

all of those resources for the period of the suspension? Or is the employer only liable for the costs of those of the contractor's resources that would have been on-site had it not increased them to address its own culpable problems?

2. The contractor has increased its resources on site to accelerate to recover its own culpable delays, but then is instructed to accelerate to address new and further delays for which the employer is liable, under such as the Infrastructure Conditions clause 46(3). In such a situation what is the starting point for calculating the additional costs to the contractor that will form part of the parties' agreement under that clause? How easily can the disruption costs arising from the two causes be disentangled?

6.5.7 Financial Quantification of Acceleration

The quantification of the financial effects of acceleration can be carried out from two opposing perspectives. The first is a prospective estimate of the likely costs, where it is subject of an agreement required by the contract or otherwise entered into by the parties. The second is a retrospective quantification based on actual costs, loss, expense or damages depending on its legal basis, such as induced acceleration or as part of a delay and disruption claim where that is allowed.

Considering the various contractual and legal bases for an acceleration claim that have been identified, the quantification would variously be made on the following bases:

- FIDIC clause 13.2, as part of a value engineering proposal. Under this clause the contractor's proposal is required to include its proposal for the evaluation of the resulting 'Variation', under subclause 13.3(c). As a 'Variation' it therefore falls to be evaluated in accordance with clause 12. The valuation procedures of that clause are considered in detail in Chapter 5. Forming part of the proposal, valuation would be on a prospective basis. It could be on the basis of rates and prices in the contract, but in practice is more likely to be based on the contractor's estimate of its actual additional costs. As a variation, it could also include allowance for the contractor's head office overheads and profit.
- FIDIC clause 13.1(f). As suggested elsewhere in this section, if an instructed variation under this clause leads to acceleration costs, then they would fall to be valued under clause 12. This gives the possibility of it being valued on the basis of rates and prices in the contract, but in practice is more likely to be based on the contractor's actual additional costs. It could be priced either prospectively or retrospectively, the latter approach being more likely than a prospective estimate of actual costs in these circumstances. Again, as a variation, it could also include allowances for the contractor's head office overheads and profit.
- Infrastructure Conditions clause 46(3). This clause, headed 'Provision for accelerated completion', only says in relation to such a quantification that 'any special terms and conditions of payment shall be agreed between the Contractor and the Employer before any such action is taken'. Costing will therefore be on a prospective basis of the contractor's estimate of its additional costs. It is likely to include risk/contingency allowances. There seems to be no reason why the contractor should not expect to include a margin for both its head office overheads and profit in such an estimate, just as it did in its pricing of the contract price.
- SBC/Q Schedule 2 Acceleration Quotation. The contractor's quotation will be based on its estimate of its costs. This will be done prospectively based on an estimate of

the contractor's planned measures and their associated costs. It is likely to include risk/contingency allowances and head office overheads and profit.

- SBC/Q clause 5.1.2.4. In the same way as FIDIC Red Book clause 13.1(f), an instructed variation under this clause might conceivably lead to acceleration costs. The same comments apply here as made against that FIDIC clause above.
- By prospective priced agreement. Such pricing is almost inevitably going to be based on the contractor's estimate of its costs. This will be done prospectively based on an estimate of the contractor's planned measures and their associated costs. It is likely to include risk/contingency allowances and head office overheads and profit.
- By prospective unpriced agreement. This will require recording of actual additional resources with the payment calculated on a retrospective basis. This might be based on actual costs or agreed rates and prices. There will be no need to include risk/contingency allowances, but addition for head office overheads and profit is likely.
- As a part of delay and/or disruption. Inevitably, the financial consequences of this will be quantified retrospectively based on the contractor's actual expenditure. Since it will be based on the express provisions of the contract for such as 'loss', 'expense' or 'cost', these will dictate whether such items as head office overheads and profit apply, as discussed elsewhere in this chapter. There will be no need to include a risk/contingency allowance.
- Induced acceleration. Inevitably, the financial consequences of this will be quantified retrospectively based on the contractor's actual expenditure. Since it will likely be a claim for damages based on breach of contract, the relevant rules for the quantification of damages under the applicable law will apply. There will be no need to include a risk/contingency allowance. The addition for head office overheads and profit will depend on the considerations explained in Section 6.8.

A controversial point to make about the costs of acceleration measures is that they do not always result in additional costs. It may be that a contractor increases its resources on site but completes the related works in the same number of hours as would have been the case without those additions, just in a shorter period. In such a case there might actually be a saving in costs because time-related costs are reduced. However, this is a rare position and, in reality, acceleration usually does involve additional costs and they can be considerable. Even where it is a simple case of adding work-related resources such as tradespeople, and they are therefore required for a proportionately shorter period, they might be more expensive per hour worked (for example from agencies). There might also be overtime payments and a loss of efficiency from such things as learning curves, longer working hours and crowding of work areas. This said, many contractors' acceleration claims address their additional costs but ignore savings in, for example, time-related preliminaries and general costs. Employers and their advisors should be wary of this.

Judge Hick's first question in his judgement in *Ascon v McAlpine* – 'earlier than what?' – has been discussed, but in relation to the financial evaluation of acceleration, a similar question would be 'greater than what?'. By this is meant, what is the starting point for deciding what the contractor's additional costs are? This is equally applicable whether the acceleration is being priced prospectively or retrospectively. Thus, if the contractor is entitled to payment for additional resources, then evaluation will involve a comparison of actual resources with some lower level of resource. The possibilities for this are:

1. The resources that the contractor allowed in its tender, for the same period of construction activities. This is a common basis for contractors prospectively pricing an acceleration agreement. The contractors' argument for this is often that it is effectively being asked to pre-price its tender on the basis of a shorter construction period and should therefore price the acceleration as it would have been had it been part of its tender. However, if the contractor under-resourced that part of its tender, then part of the increase will not be the result of the acceleration and should have been required anyway.

2. The resources that the contractor had on-site prior to the acceleration measures. However, this misses the point that the coming activities that are to be accelerated might have always needed an increase in the number of resources anyway. Furthermore, the contractor's resources prior to the acceleration might have been inadequate.

3. The resources that the contractor would have had on-site for the accelerated activities, had the acceleration measures not been instigated. Ideally this should be the same as its tender allowance, but it might not be. However, whilst the resources in 1. and 2. can be factually established from the tender and site records, this third alternative will require some judgment as to what would have been required. In some cases an updated resourced programme can help with this, where it is the result of the site team's knowledge of what is actually unfolding on site, rather than the estimating department's application of its norms.

Regarding the types of resources and costs that might be considered in the financial quantification of acceleration, these might include the following:

• The introduction of additional resources, particularly labour. These will be identified to particular activities and operations on-site with the objective of shortening durations. They might be added to work in the same areas as existing resources or to open up new areas for parallel working. One factor for the contractor that is estimating its acceleration costs for a prospective agreement will be the extent to which increases in numbers lead to a proportional increase in production. Such measures as adding resources to any area can reduce efficiency through such things as overcrowding and dilution of supervision. Part of the estimator's skill will be to work out the equation here. As a simple example, it might be concluded that a doubling of the rate of progress needs more than double the number of operatives.

• The working of additional overtime, or adoption of additional shifts, for example the addition of a night shift for pouring concrete structures. In either case the additional cost is the premium paid for work outside normal hours, as the assumption is that additional labour will be productive at the same rate as the original. If it were thought that productivity might be reduced as a consequence of additional overtime or shift working, some basis of assessing the reduction in productivity would have to be built into the calculation. This is an area considered in relation to disruption evaluation elsewhere in this chapter.

• It may be that the introduction of additional resources, or the instituting of additional shift working, requires further labour (and/or plant) resources that are not readily available in the locality. In such circumstances further costs will be incurred in the importation of labour from remote locations, including transport and possibly accommodation costs. It may also be necessary to pay the imported labour a premium to persuade them to travel to the site location.

- Further plant costs may be incurred by duplicating plant already on site in order to service any additional resources and these may be at a premium cost if hired rather than owned.
- Temporary works costs may increase, for instance in circumstances where it is decided that one of the activities to be accelerated is that of formwork to reinforced concrete, and additional resources are introduced to increase the area of formwork constructed and in place for the pouring of concrete at any one time. The amount of formwork material may be increased with a consequent reduction in the number of re-uses achieved on the works. Similarly, increases in the site workforce may require an increase in access resources such as scaffolding and equipment, hoists, etc.
- Payments may have to be made to suppliers to expedite the manufacture and/or delivery of materials and plant, including such items as air freighting and part loads.
- In addition, there may be other such direct works consequences such as an increase in the site supervision and possibly site services to cater for any increase in resources, i.e. additional preliminaries and general costs.
- There may also be additional head office costs and charges if such can be identified, particularly for management time.
- There may also be claims from subcontractors and/or suppliers in relation to any of the above types of cost.

All factors in the build-up to the acceleration cost, such as these, should be identified to the particular activities with an analysis of what is to be achieved. The quantification then becomes a simple case of pricing the various measures.

6.5.8   Sample Acceleration Costing

To take a relatively simple example of acceleration, consider a situation where an analysis of the work to be undertaken to completion shows that the works will overrun due to changes in the fixing of steel reinforcement caused by delays in delivery of the relevant bending schedules, for which the employer is liable.

Assume for the purposes of this example that analysis of the remaining works shows that if the amount of formwork is doubled the reinforcement fixing can be increased now that the required bending schedules are available, but the reinforcement will have to be delivered in small quantities to expedite delivery in order to keep the steel fixers working. The increased formwork area can be erected and dismantled by employing a further six carpenters and six additional labourers in addition to the original formwork labour of eight carpenters and eight labourers, although the whole of the enlarged formwork workforce of 14 carpenters and 14 labourers will need to work an additional 10 hours of overtime per week, for the anticipated remaining period of the formwork of 12 weeks. A further four reinforcing steel fixers are required, but will not need to work any exceptional overtime, although none are available in the locality and they will have to be imported and accommodated for the required eight weeks of the projected period required. In order to increase the rate of concrete pouring to take advantage of the increased areas of formwork and rate of steel fixing, a further concrete gang of four operatives will be required for a period of eight weeks but neither they nor the existing concrete resource will need to work any exceptional overtime, and in order to alleviate cranage bottlenecks caused by the increased rate of concrete placing, a mobile crane will be required for four weeks to supplement the site cranage resource.

In reality all of the identified resources would be identified to specific parts of the formwork, reinforcement and concreting activities, such as columns to be erected and poured 'early' and areas of slab formwork and beams, etc., to be included in the accelerative measures, but this analysis is sufficient to provide a sample build-up of the acceleration costs.

The costing for the above measures would be as follows:

Formwork costs	
Additional resource:	$
6 carpenters × 12 weeks × 45 hours × $18.00	58,320.00
6 labourers × 12 weeks × 45 hours × $16.00	51,840.00
Overtime:	
14 carpenters × 10 hours × 12 weeks × $3.60	6,048.00
14 labourers × 10 hours × 12 weeks × $3.15	5,292.00
Steel fixing costs	
4 steel fixers × 8 weeks × 45 hours × $18.00	25,920.00
Lodging allowance 4 × 8 weeks × 4 nights @ $26.00	3,328.00
Travel time and fares 4 × 8 weeks @ $90.00	2,880.00
Reinforcing steel, small load charge 32 tonnes @ $6.75	216.00
Concrete costs	
Additional resource:	
4 operatives × 8 weeks × 45 hours × $16.85	24,264.00
Mobile crane 4 weeks @ $1,265.00	5,060.00
Total *gross* cost of accelerative resources	$183,168.00

Any increased material costs resulting from, for instance, any reduction in the number of uses of formwork achieved will need to be added to the above resource cost. It will then be necessary to deduct from the gross cost of the acceleration resources and related material costs as calculated, either the costs that would have been incurred had the calculation not includes acceleration measures or the value paid to the contractor, depending on the contractual basis of the claim. If the claim is for an increase in costs, then the additional costs of acceleration can be extracted. If the cost is for loss against a recovered value, then the practical difficulty that may arise in making the appropriate deduction is the identification of the labour element of the concrete work where the contract rates and prices are composite rates for labour, plant and materials, etc. Some analysis of any composite contract rates will be necessary in such circumstances when taking into account the factors relating to unit rates, discussed in Chapter 5.

This is merely a simple example of the type of pricing that is required and would in the real world require related identification of the areas and amounts of permanent and temporary works involved in addition to detailed build-ups of the sums claimed. In many instances the extent of the works to be accelerated and the measures required to achieve the acceleration will be much more extensive than this example, but the 'bottom up' principle would still apply and require each section of the work to be considered, the objective in terms of time set out together with the measures necessary, and the pricing built up for each section, in the manner demonstrated.

6.6   Global Claims and Similar Terms

6.6.1   Potential Definitions

Before considering the vexed subject of global claims in detail, it may help to consider the definition of that term and a number of others that are sometimes used in similar contexts. These other terms include: 'total cost claim', 'total loss claim', 'composite claim', 'rolled-up claim' and also variants on these adding the prefix 'modified'. They tend to be bandied about, often for no better reason than to give contractors' claims a label that has pejorative overtones. They can be applied inconsistently, without consideration of their real meaning and whether the label is fair in the circumstances. None of these terms are terms of art and they lack clear and consistent legal definition, although some are regularly considered in the UK courts. Their definition is a legal matter and may vary between jurisdictions, requiring local legal advice. However, the following are possible definitions based on typical industry usage and by reference to some judgments.

6.6.1.1   'Global Claim'

Sitting in the Supreme Court of Victoria, Australia, Buildings Cases List, Byrne J gave this definition in the Australian case *John Holland Construction & Engineering Pty Ltd v Kvaerner RJ Brown Pty Ltd* (1996) 82 BLR 83:

> (Holland's) claim was a 'global claim', that is, the claimant does not seek to attribute any specific loss to a specific breach of contract, but is content to allege a composite loss as a result of all the breaches alleged.

This seems to suggest that the claimant had not even made an effort to particularise its claim. Quoting this passage, the judgment was described as 'most careful and helpful' by HHJ Humphrey Lloyd in the London High Court, Queen's Bench Division, case *Bernhard's Rugby Landscapes Ltd v Stockley Park Consortium Ltd* [1997] EWHC Technology 374 (7 February 1997). HHJ Humphrey Lloyd also identified one of the reasons that claims are sometimes categorised as 'global' by those defending them:

> Where it describes a pleaded claim it has pejorative overtones as it is usually intended to describe a claim where the causal connection between the matters complained of and their consequences, whether in terms of time or money, are not fully spelt out, but, implicitly, could and should be spelled out.

Again, this suggested that the contractor had not even attempted to particularise its claim.

In the Scottish Court of Sessions case *John Doyle Construction Ltd v Laing Management (Scotland) LTD* [2004] ScotCS 141 (11 June 2004) Lord Drummond Young gave this definition:

> ... a claim in which the individual causal connections between the events giving rise to the claim and the items of loss and expense making up the claim are not specified, but the totality of the loss and expense is said to be a consequence of the totality of the events giving rise to the claim.

In these cases, the courts in England and Wales, Scotland and Australia seemed to apply the same definition under which the globality of a claim was a matter limited to causation, i.e. where an overall sum was claimed against more than one breach. As the Tenth Edition of *Keating on Construction Contracts* puts it, with reference to the judgments in *John Holland v Kvaerner* and also *London Underground Ltd v City Link Ltd* [2007] BLR 391:

> A global claim, however, is one that provides an inadequate explanation of the causal nexus between the breaches or contract or relevant events/matters relied upon and the alleged loss and damage or delay that relief is claimed for.

That the global nature of a claim relates to causation and does not define how it is quantified, seemed to be apparent from this definition from the 1994 Eleventh Edition of *Hudson's Building and Engineering Contracts*, to which emphasis has been added:

> Global claims may be defined as those where a global or composite sum, *however computed*, is put forward as the measure of damages or of contractual compensation where there are two or more separate matters of claim or complaint, and where it is said to be impractical or impossible to provide a breakdown or subdivision of the sum claimed between those matters.

Under this definition a claim may be global, no matter how the sum claimed is computed (it could range from total loss to a measured mile calculation), so long as the sum claimed relies on a number of causes without separately assigning a financial sum to each individual cause.

However, what seemed a simple issue of defining a '*global claim*' as being for a single sum arising from more than one cause, became complicated by this passage from paragraph 484 of Mr Justice Akenhead's judgment in the High Court's Technology and Construction Court in *Walter Lilly & Company Limited v (1) Giles Patrick Cyril Mackay (2) DMW Developments Limited* [2012] EWHC 1773 (TCC) (12 July 2012):

> What is commonly referred to as a global claim is a contractor's claim which identifies numerous potential or actual causes of delay and/or disruption, a total cost on the job, a net payment from the employer and a claim for the balance between costs and payment which is attributed without more and by inference to the causes of delay and disruption relied on.

To some commentators this appeared to confuse the globality of a claim in terms of its assigning one amount to a number of combined causes, with the adopted method of quantification of the sum claimed. On this view, it would be said to combine '*global claims*' with '*total loss claims*' (which are considered elsewhere), although Mr Justice Akenhead did not use that latter term. On this basis it did come in for some criticism. For example, in the paper 'Lilly and Doyle, A Common Sense Approach to Global Claims' by Anneliese Day (Counsel in the case) and Jonathon Cope, it was said that the Judge had 'somewhat controversially' dealt with global claims and total loss claims together. Similarly, in the New Square Chambers paper 'Global Claims: The Law, Where Are We After Walter Lilly?' the authors stated:

> Some commentators have suggested that 'global' claims on the one hand and 'total cost', 'composite' or 'rolled up' claims on the other are separate and distinct concepts but Mr Justice Akenhead deals with them together in his judgment.

Mr Justice Akenhead did, however, give this qualification regarding the use of such terms:

> One needs to be careful in using the expressions 'global' or 'total' cost claims. These are not terms of art or statutorily defined terms.

The current, twelfth, edition of *Hudson* starts its consideration of global claims by quoting HHJ Humphrey Lloyd's definition in *Bernhard's v Stockley* as set out elsewhere. However, *Hudson* then continues by also combining globality as an issue of cause and effect with one dependent on how quantification is carried out:

> A global or total cost claim is simply, as its name implies, one in which the cost of the work incurred by the Contractor in its execution is compared with the tender or contract allowance for that work to arrive at the claimed amount.

This appears to be a significant change since the previous edition of *Hudson*, because it seems to equate a 'global claim' to one for 'total loss'. If so, this may be the result of Mr Justice Akenhead's judgment in *Lilly v Mackay* combining the two different concepts.

This approach is also echoed in a footnote to the 2014 edition of *Emden's Construction Law*. That footnote refers to paragraph 484 of Mr Justice Akenhead's judgment as quoted elsewhere, as follows:

> It has been stated that 'what is commonly referred to as a global claim is a contractor's claim which identified numerous potential or actual causes of delay and/or disruption, a total cost on the job, a net payment from the employer and a claim for the balance between costs and payment which is attributed without more and by inference to the causes of delay and disruption relied on'.

However, as that footnote in *Emden* also warns, echoing the Judge's qualification regarding the use of such terms:

> Care needs to be used when using the expressions 'global' or 'total' costs claims. These are not terms of art of statutorily defined terms.

In the absence of a term of art or statutory definition, the common meaning of the words themselves must surely be of some significance. The word 'global' is defined in the *Shorter Oxford English Dictionary* as:

> Pertaining to or embracing the totality of a group of items, categories or the like.

This dictionary definition of the word 'global' seems to support those definitions of the term '*global claim*' in the England and Wales, Scottish and Australian authorities before *Lilly v Mackay*. This would make globality a question of causation rather than quantification. On this basis, quantification would be a separate issue involving such different terms as '*total loss*' (considered elsewhere in this chapter). Adding to this the recognition that the use of the term '*global claim*' is generally applied for the pejorative meaning identified by HHJ Humphrey Lloyd (as quoted elsewhere), his definition of a '*global claim*' in *Bernhard's v Stockley* might still be preferred:

> … a claim where the causal connection between the matters complained of and their consequences, whether in terms of time or money, are not fully spelt out, but, implicitly, could and should be spelled out.

On this basis, a 'global claim' is not just one that fails to link more than one cause to each relevant part of an overall effect, but one where the claimant could and should have done so. On the other hand, the SCL Protocol gives this definition:

> A global claim is one in which the Contractor seeks compensation for a group of Employer Risk Events but does not or cannot demonstrate a direct link between the loss incurred and the individual Employer Risk Events.

This SCL definition would remove the pejorative element of HHJ Humphrey Lloyd's definition, because a global claim would include one where it was impractical or impossible for the contractor to particularise effects to causes. It does, however, support the view that 'globality' is a question of causation.

6.6.1.2 'Total Costs Claim'

Returning to Byrne J's judgment in *John Holland v Kvaerner*, his description of the claim that he had already defined as '*global*' (see elsewhere) continued as follows, with emphasis added:

> Further, this global claim is in fact a total cost claim. In its simplest manifestation a contractor, as the maker of such a claim, alleges against a proprietor a number of breaches of contract and quantifies its global loss as the *actual cost of the work less the expected cost*.

From this definition, it appears that the Judge considered that a '*global claim*' is not the same thing as a '*total cost claim*'. Furthermore, he appeared to consider that a claim's '*total cost*' nature is a matter of quantification, and (as the term suggests) is based on the difference between the contractor's actual costs (as opposed, for example, to the application of industry norms or factors) and its expected costs.

A similar definition and distinction was given by HHJ Humphrey Lloyd in *Bernhard's v Stockley* (again with emphasis added):

> This global claim was a total costs claim since the contractor had alleged a number of breaches of contract and quantified its global loss as the *actual costs of the work less the expected cost*.

Also, in the Technology and Construction Court, in *How Engineering Services Limited v Lindner Ceilings Floor Partitions plc* [1999] EWHC B7 (TCC), Mr Justice Dyson stated:

> The claim was a total cost claim, that is, one based on actual costs compared with the costs that would have been incurred had delay and disruption not occurred.

However, in *John Doyle v Laing*, Lord Drummond Young followed his explanation of the global nature of the claim in that case with this:

> The expression 'global claim' has normally been used in Scotland, England and other Commonwealth countries to denote a claim calculated in the foregoing manner. In the United States the corresponding expression is 'total cost claim'.

On this basis, it seems that a US definition of *'total costs claim'* would mean the same thing as a *'global claim'* elsewhere.

Returning to Commonwealth jurisdictions, the current edition of *Keating* puts the position as follows:

> A total cost claim is a form of global claim where the contractor has quantified its alleged entitlement to further payment by subtracting the expected cost of the works from the final actual costs.

Again, this suggests that a *'total cost claim'* is different to a *'global claim'*, but might be a type of one, and is a matter of quantification (and not causation), which is based on the contractor's actual and expected costs.

However, a contrary approach appears in Daniel Atkinson's *Causation in Construction Law – Principles and Methods of Analysis* (again with emphasis added):

> The term 'Total Cost Claim' is a claim where a single sum is claimed which is the difference between the *total actual cost and the contract price or valuation of the work*.

In its reference to 'contract price or valuation of the work', this definition appears to be distinct from both the definitions in the Commonwealth cases quoted elsewhere and the US approach described by Lord Drummond Young in *John Doyle v Laing* as also quoted elsewhere. On this view, what Daniel Atkinson is actually describing is a '*total loss claim*', i.e. one whose quantum is calculated as the difference between the contractor's actual costs and the price it received for the work. This is a term that is considered elsewhere in this chapter. This distinction would also seem to align with popular usages of the terms 'cost' and 'loss'. In the construction industry, *'cost'* and *'loss'* are generally well understood terms meaning different, but related, things. '*Cost*' is the expenditure incurred by the contractor on its plant, labour, materials, staff, subcontractors, facilities, etc., in carrying out work. '*Loss*' is the difference between those costs and the amount of payments received by the contractor for that work.

In most jurisdictions, in a claim for a breach or breaches of contract, the measure of damages is the additional cost incurred by the claimant, nett of those costs it would have incurred '*but for*' the breach(es). This is the position where the test of damages is that sum of money that will put the claimant into the financial position it would have been in had the other party not breached. Hence, the references in these cases to such as '*less the expected cost*'. It is not clear how in those cases the '*expected cost*' was to be determined, but the better definition (ignoring the apparent US approach) may be Mr Justice Dyson's in *How v Lindner*, that a '*total cost claim*' is one that is:

> … based on actual costs compared with the costs that would have been incurred had delay and disruption not occurred.

An occasional variation to the term '*total costs claim*' adds the prefix '*modified*'. Practitioners seem to use this for two alternative meanings. Firstly, where the claimant has particularised parts of its claim, as far as it can, and then only claims on a global basis for the remaining unparticularised parts. However, more commonly it is applied to a fully global claim, but with the addition that the claimant would 'modify' its claimed costs to exclude any own goals such as arising from its own errors, inefficiency or mismanagement, etc. Whilst some practitioners use the extended term '*modified total costs claim*' for this second meaning, it is suggested that the addition of the prefix actually adds nothing. This is because the contractor's own goals would have to be excluded anyway to get to Mr Justice Dyson's definition of '*total cost claim*'. They would be excluded by the 'but for' test.

6.6.1.3 'Total Loss Claim'

This term does not appear in any of the judgments considered in relation to the definition of 'global claim', although, as has been explained, Mr Justice Akenhead's judgment in *Lilly v Mackay* was considered by some commentators to have confused the two concepts.

The 2012 Twelfth Edition of *Hudson* heads its section on these issues '*Global or Total Loss Claims*', but the term '*total loss claim*' does not appear again in the text of that section. This may be because *Hudson* followed Mr Justice Akenhead's judgment in *Lilly v Mackay* categorising a '*total loss claim*' and '*global claim*' as the same thing, and repeated here:

> What is commonly referred to as a global claim is a contractor's claim which identifies numerous potential or actual causes of delay and/or disruption, a total cost on the job, a net payment from the employer and a claim for the balance between costs and payment which is attributed without more and by inference to the causes of delay and disruption relied on.

It was explained elsewhere how the construction industry is usually clear on the definition of the words '*cost*' and '*loss*', and the distinction between them, with the latter being the difference between the contractor's costs of carrying out works and the amount received by it for those works. The only other consideration in this respect may be any amounts that the contractor has received through backcharges to its suppliers or sub-contractors for their failures, although these might have gone into reducing its actual costs in the calculation of the additional costs. On this basis, a definition of the term '*total loss claim*' might be:

> A claim for the balance between the contractor's actual costs and the total of payments it has received for those works.

It is suggested that this aligns with what the industry generally understands by the word '*loss*'.

As with the occasional reference to '*modified total costs claim*' explained elsewhere, the term '*total loss claim*' is also sometimes given the prefix '*modified*', usually to recognise that the contractor has adjusted the calculation for recoveries from, for example, its subcontractor and other claims.

6.6.1.4 'Composite Claim' and 'Rolled-Up Claim'

It seems that these two terms are consistently used in the industry to refer to the same thing. However, to some commentators they are also the same thing as '*global claim*'. For example, in *John Holland v Kvaerner*, Byrne J said (with emphasis added):

> The claim as pleaded … is a global claim, that is, the claimant does not seek to attribute any specific loss to a specific breach of contract, but is content to allege a *composite* loss as a result of all of the breaches alleged.

This view that '*composite claims*' and '*global claims*' are one and the same thing is also suggested in the Tenth Edition of *Keating*:

> Global claims are also sometimes referred to as composite claims.

The oldest judgment that is usually cited in relation to the whole subject area of '*global claims*' is that of Mr Justice Donaldson sitting in the Queen's Bench Division of the High Court in *J. Crosby & Sons Ltd v Portland Urban District Council* (1967) 5BLR121. He considered a number of issues arising from the award of an arbitrator, including whether he had been correct to allow a lump sum amount to compensate a contractor for part of its costs arising from numerous delays. Emphasis has been added to this passage from his judgment:

> I can see no reason why he should not recognise the realities of the situation and make individual awards in respect of those parts of individual items of the claim which can be dealt with in isolation and a supplementary award in respect of the *remainder of these claims as a composite whole.*

In short, the duty of a tribunal was to isolate costs to causes as far as possible, but it could thereafter make a '*composite*' award for costs that could not be identified between a number of causes.

A similar approach and terminology was adopted by The Judicial Committee of the Privy Council when hearing the appeal from a decision of the Court of Appeal of Hong Kong in a case considered to be of general significance to construction cases, not only in Hong Kong but also in the United Kingdom. In *Wharf Properties v Eric Cumine Associates* (1991) 52 BLR 1, the judgment, delivered by Lord Oliver of Aytmerton, referred to cases including *Crosby v Portland* and concluded (again with emphasis added):

> Those cases establish no more than that where the full extent of extra costs incurred through delay depends upon a complex interaction between the consequences of various events, so that it may be difficult to make an accurate apportionment of the total extra costs, it may be proper for an arbitrator to make individual financial awards in respect of claims which can conveniently be dealt with in isolation and a supplementary award in respect of the financial consequences of *the remainder as a composite whole.*

This theme of claims being particularised as far as practicable but with a remaining unparticularised '*composite claim*' was put succinctly by Mr Recorder Tackaberry, sitting on Official Referee's Business in the High Court, in *Mid Glamorgan Council v J Devonald Williams* (1992) 8 CLJ 61, as follows (again with emphasis added):

> The claim so put is extremely generalised. There is one clear claim, with specific damages attached to it (the building trades workshop claim, about which no complaint is made); but the rest amounts to relying on all the instructions (and in the case of the complaint about the project architect, on every other event mentioned in the particulars as well) and then applying to all these matters a single *composite* amount of damages.

On the basis of these judgments, one might see a '*composite claim*' as being something different to a '*global claim*', and applying to the unparticularised parts of claims that were otherwise particularised. In other words, the unparticularised remainder or '*rump*' of the contractor's claims after they had been particularised as far as possible.

The reconciliation of the alternative positions that '*composite claims*' are, or are not, the same as '*global claims*' appears to be explained in Daniel Atkinson's *Causation in Construction Law – Principles and Methods of Analysis*, as follows:

> The terms 'Composite Claim' and 'Rolled-Up Claim' are claims where there are a number of events and only some are presented as a group in a Global Claim. In this type of claim, separate sums are claimed for particular events and a single sum is claimed for the remaining group of events that are not so particularised.

This seems a sensible explanation. It would mean that a '*composite claim*' (which is the same as a '*Rolled-Up Claim*') is in itself a '*global claim*', but without the pejorative overtones described by HHJ Humphrey Lloyd in *Bernhard's v Stockley Park*, as quoted elsewhere. Thus, a claimant will have particularised (as far as it can and should) the causal connection between the matters complained of and their individual financial consequences, but is left with remaining items that can only be dealt with on a global basis (although without tagging the claim as '*global*' in its derogatory sense). This is therefore the 'rump' of otherwise particularised claims that the contractor is left with after such particularisation.

On this basis a '*composite claim*' could be defined as follows, paraphrasing Mr Justice Donaldson in *Crosby v Portland*:

> Recognising the realities of the situation, where particularised claims are made in respect of those parts of individual items of the claim which can be dealt with in isolation, a claim in respect of the remainder of the claims as a composite whole.

This definition would equally apply to the term '*rolled-up claims*'.

6.6.2 Global Claims

6.6.2.1 Introduction

As we have detailed elsewhere in this book, a claimant generally has the burden of proving its case for damage through the following stages:

1. That the breach of contract or duty or event giving rise to the claim actually occurred.
2. That the respondent is responsible for loss or damage resulting from that breach or event.
3. That the loss or damage were caused by the breach or event.

On this basis, each breach or event should have its particular effects separately identified and proven. Thus, in a construction and engineering context, the theory would be that a contractor's claim should detail each individual cause of its claim and allocate discrete amounts of loss, expense, costs or damage against each.

However, it is a reasonable proposition that in some instances it will not be possible to specifically allocate some of the costs to a single cause, activity or operation on site, for example because they were generated by a combination of factors. This proposition is particularly relevant in the context of large and complicated engineering and construction projects that can be the subject of thousands of changes. This conflict between what the law requires in theory in order to allow the defendant to properly understand the case against it, and what is practicable for the claimant in the circumstances, has caused much debate and difficulty in the past and continues to do so. It is therefore useful to consider such situations, and the approaches adopted for them, in some detail, by reference to how English law and its courts have addressed it.

Where a party considers that it has suffered losses resulting from a number of different delays and causes, all allegedly caused by the other party, an issue can arise as to the extent to which the claimant can present and pursue its costs, expense, losses or damage in the form of a claim which may variously described as 'global', 'rolled up' or 'composite', as explained elsewhere.

Contractors' claims are presented on a global basis for a variety of reasons and pretexts, some legitimate, others not. From a claimant's viewpoint the following motives might apply:

- Laziness on the part of the claimant, its staff or consultants.
- Their lack of knowledge of the contractual and legal requirements of advancing and proving such claims.

- Where the facts of causes and effects and their nexus are so clear and known to the parties that particularisation is unnecessary.
- Where there are a limited number of causes, all of which are transparently the respondent's fault, such that it is considered unnecessary and/or disproportionate to separate them out.
- A desire to obtain a 'quick fix' negotiated settlement with minimal claims preparation effort and hence costs. This may be fine where the other party is amenable to it or where the claimant will accept a suitably low level of settlement. However, it often leads to later problems where the negotiations fail, and the claim continues to be pursued, unparticularised, and finds itself as part of a pleaded case before a court or arbitral tribunal without proper improvement.
- Where, as suggested elsewhere, it is genuinely impracticable, due to their complexity and/or number, to separate out the effects from the causes, such that the global approach is the only practicable option, any attempted apportionment between causes being contrived and artificial.
- This may be the case where the extent of the respondent's failures is such that there are so may breaches and their interrelationship so complex that particularisation would be impossible. Here the claimant may say that were the global claim to fail it would be the result of the extent of the respondent's failures, such that the respondent would benefit from the great extent of its own breaches.
- That the claim is under an unusual contract provision or legal basis that allows a global, composite or rolled-up approach.

From a respondent's viewpoint the following objections are often raised:

- That the global approach has been adopted to hide the effects of the claimant's own failures.
- That the global approach is seeking to hide the true causes of the loss or damage which are otherwise not claimable, for example because they are cost-neutral events.
- That the approach does not allow it to properly understand the case against it and prejudices its ability to defend itself.
- That the global approach effectively seeks to reverse the burden of proof from the claimant to the defendant, with the latter being left to undertake the detailed analysis of the events and related quantum.
- That the global approach leaves the detail of the claim sufficiently vague, that the contractor can 'change course' part-way through its evidence, for example where one cause of delay fails (perhaps because it is found to be a contractor's risk event), by emphasising others that remain as the real clause of its delays or losses.
- That if the claimed causes fail in part then the approach does not allow those remaining parts to be properly addressed. The question is then, does the whole of the claim fall with the failed parts, and if not how can loss or damage be assigned to those that remain?

The 2002 First Edition of the SCL Protocol stated the following at its Core Principles paragraph 19:

> The not uncommon practice of contractors making composite or global claims without substantiating cause and effect is discouraged by the Protocol and rarely accepted by the courts.

However, in the current, 2017, edition, the SCL Protocol now states at its Core Principles paragraph 17:

> The not uncommon practice of contractors making composite or global claims without attempting to substantiate cause and effect is discouraged by the Protocol, despite an apparent trend for the courts to take a more lenient approach when considering global claims.

This reflects a change in the position taken by the English courts. As a result, an analysis of the position in relation to global claims needs to start with a historical perspective.

6.6.2.2 A Historical Perspective

As reflected by the change in the SCL Protocol, in the UK and Commonwealth jurisdictions, judicial thinking on global claims has fluctuated over recent times. In fact, the approach in both the first and second editions of the SCL Protocol reflect changes to the positions that had existed some time before.

The starting point for an analysis of this historical background is the decision in *J Crosby Limited v Portland UDC* (1967) 5BLR 121. In this case Mr Justice Donaldson considered a number of issues arising from the award of an arbitrator. One of the issues was whether the arbitrator was correct to allow a lump sum amount to compensate a contractor for part of its costs arising from delay. The arbitrator had found that:

> The result, in terms of delay and disorganisation, of each of the matters referred to above was a continuing one. As each matter occurred its consequences were added to the cumulative consequences of the matters which had preceded it. The delay and disorganisation which ultimately resulted was cumulative and attributable to the combined effect of all these matters. It is therefore impracticable, if not impossible, to assess the additional expense caused by delay and disorganisation due to any one of these matters in isolation from the other matters.

The respondent contended that any award of a lump sum amount should only be the result of the arbitrator adding together the individual amounts he finds in respect of each head of claim. The claimant's case was that whilst an artificial apportionment could be made, why should it when it had no basis in reality given that:

> ... the extent of the extra cost incurred depends upon an extremely complex interaction between the consequences of the various denials, suspensions and variations, it may well be difficult or even impossible to make an accurate apportionment of the total extra cost between the several causative events.

The Judge agreed with the claimant that:

> Extra costs are a factor common to all these clauses, and so long as the arbitrator does not make any award which contains a profit element ... and provided he ensures there is no duplication, I can see no reason why he should not recognise the realities of the situation and make individual awards in respect of those parts

of the claim which can be dealt with in isolation and a supplementary award in respect of the remainder of these claims as a composite whole.

In short, the duty of a tribunal was to identify costs to causes as far as possible, but it could then make a composite award for costs that could not be identified between a number of causes.

The *Crosby v Portland* approach was followed in the case of *London Borough of Merton v Leach* (1985) 32 BLR 51, where Mr Justice Vinelott heard an appeal from the award of an arbitrator. The applicant contended in relation to loss and/or expense under the July 1971 revision of the 1963 edition of the JCT Standard Form of Building Contract that:

> … the Architect cannot make an award unless he is in a position to ascertain the direct loss stemming from a specific cause identified in the application and cannot therefore make an award if the loss stemming from the two different causes cannot be separated and each separate part identified as the direct loss stemming from each cause.

Mr Justice Vinelott noted that this contention was at odds with the decision in *Crosby v Portland* and hence that the question for him was whether he should follow Donaldson J's decision in that earlier case. He found:

> I find his reasoning compelling. The position in the instant case is, I think, as follows. If application is made (under Clause 11(6) or 24(1) or under both sub-clauses) for re-imbursement of direct loss or expense attributable to more than one head of claim and at the time when the loss or expense comes to be ascertained it is impracticable to disentangle or disintegrate the part directly attributable to each head of claim, then provided of course that the contractor has not unreasonably delayed in making the claim and so has himself created the difficulty the Architect must ascertain the global loss directly attributable to the two causes.

He continued:

> I should nonetheless say that it is implicit in the reasoning of Donaldson J. that first that a rolled up award can only be made in a case where the loss or expense attributable to each head of claim cannot really strictly be separated and secondly that a rolled up award can only be made where apart from that practical impossibility the conditions which have to be satisfied before an award can be made have been satisfied in relation to each head of claim.

The judgments in *Crosby v Portland* and *London Borough of Merton v Leach* both therefore required that a claimant properly particularise its claim, by reference to the amounts claimed against each breach pleaded, but recognised the practicalities that this might not always be possible.

However, this approval of global claims, albeit with careful qualification as to the conditions that needed to be fulfilled to make such claims viable, appeared to run into some difficulties when global claims were further considered in the cases of *Wharf Properties v Eric Cumine Associates* (1991) 52 BLR 1 and *ICI v Bovis Construction* (1992) 8 CLJ 293.

At the time these judgments were interpreted by many commentators as sounding the death knell of global claims.

Wharf Properties involved a professional negligence claim against the partners of a firm of architects, Eric Cumine Associates. The Judicial Committee of the Privy Council heard the appeal from a decision of the Court of Appeal of Hong Kong, on the basis that the judgment was of general significance to construction cases, not only in Hong Kong but also in the United Kingdom. The decision had been to strike out a statement of case in relation to delays, on the basis that it disclosed no cause of action and that it was embarrassing to the fair trial of the matter and an abuse of process. The judgment's significance was in its impact on the judgments in *Crosby v Portland* and *Merton v Leach*, which were both considered in detail.

The Privy Council dismissed the appeal. It held that the statement of claim did disclose a 'reasonable cause of action' but that it was embarrassing to a fair trial and to permit it to stand would allow an abuse of the court process. The pleading was described in the judgment as 'hopelessly embarrassing as it stands'. This was notwithstanding that it was 'of immense length and complexity' and ran to over 400 pages and supporting schedules. The claimant sought to rely on *Crosby v Portland* and *Merton v Leach* as permitting such an approach to pleading a claim, but the court held that:

> Those cases establish no more than that where the full extent of extra costs incurred through delay depends upon a complex interaction between the consequences of various events, so that it may be difficult to make an accurate apportionment of the total extra costs, it may be proper for an arbitrator to make individual financial awards in respect of claims which can conveniently be dealt with in isolation and a supplementary award in respect of the financial consequences of the remainder as a composite whole. This has, however, no bearing upon the obligation of a Plaintiff to plead his case with such particularity as is sufficient to alert the opposite number to the case which is going to be made against him at trial.

It continued:

> ... the failure even to attempt to specify any discernible nexus between the wrong alleged and the consequent delay provides, to use [counsel's phrase], 'no agenda' for the trial.

This endorsement of the *Crosby v Portland* principle that a tribunal could make a financial award for those matters that could be separately identified in terms of cause and effect, and then make a composite award for the remaining matters where the financial effects could not reasonably be separated, extended the reasoning to expressly state that such a power of the tribunal in making an award did not excuse a claimant from properly setting out the duty, breach, cause and effect for the claimant's complaints. In this regard, Wharf had not served itself well. Firstly, it had consented to provide further particulars as ordered by the Judge in the original trial, but then provided ones that did not comply with the order. It might have been better served by contending that such particulars could not be provided, on the basis accepted in both *Crosby v Portland* and *Merton v Leach*. Furthermore, in the appeal Wharf said that it could and would provide such details and the Privy Council judgment noted that this was some seven years after the proceedings first began. It also noted with regret how long the professional negligence claim had been hanging over the heads of the partners of Eric Cumine Associates.

Finally, it is interesting to note that the Privy Council expressly stated at the start of the judgment that it had become apparent that the case was not actually of general importance and that they did not consider that they were giving a general proposition of law concerning global claims, Lord Oliver stating:

> … their Lordship's Board has been exceptionally concerned with a pure point of pleading peculiar to the particular dispute in which the parties are engaged.

A similar approach to that in *Wharf Properties* was adopted by HHJ Fox-Andrews in the case of *ICI plc v Bovis Construction & Others* 1992 8 CLJ 293. The claimant's case was struck out, leading some to the general proposition that the global claim was no longer viable. However, this case was also concerned very much with the particular manner in which the claimant's case had been pleaded.

ICI claimed against Bovis and the other defendants that their breaches of their obligations to ICI had led to damages of approximately £19 million, out of a total construction overspend of £24,885,097, plus £3,600,000 of related additional professional fees. No attempt was made to link breaches to parts of the claimed losses and these were claimed globally and equally against all three defendants.

In his judgment HHJ Fox-Andrews went to some lengths setting out the inadequacies that remained in the pleaded claim even after ICI had twice accepted its shortcomings and he had twice ordered further particulars and allowed further discovery to enable ICI (on its submission) to plead properly. His second order required production of a Scott Schedule and set out its required format. The judgment identified problems with what was served and repeatedly used terms such as 'inadequate' and 'embarrassing' to describe it.

The global nature of ICI's claim and the effect of that was illustrated by an answer from ICI's counsel to a hypothetical claims outcome put by the defendants:

> The plaintiffs' case is that the various events set out in section 6 all contributed to the sums claimed, with no actual apportionment being possible. However as the total cost claims have in fact been paid, if any of the events is not proven at trial, the only consequence is that the actual sum paid will fall to be distributed between a lesser number of events, not that the total sum recoverable will be less.

The Judge pointed out that this meant that whereas total costs of £840,211 was pleaded against items set out on seven pages of part of the schedules, if only one item on those pages survived then the amount claimed against that one item would still be the whole total of £840,211. Noting that the smallest item on the schedule was for repositioning a fire bell, the Judge described this approach as 'palpable nonsense'. He concluded:

> The history of this matter is quite different from that in Wharf. There is no question here of debarring ICI from pursuing their claim. But the present substitute schedule is inadequate in so many different respects, as I have found, that a fresh schedule must now be served.

Thus, the judgments in both *Wharf v Cumine* and *ICI v Bovis* had seen pleaded global claims struck out or ordered to be amended. However, commentators who saw in these judgments an end to global claims were wrong. Firstly, those judgments were based on the specifics of the particularly inadequate approaches taken by the claimants in those cases. Secondly, at the same time and over subsequent years, other judgments were to apply a softer approach to the pleading of claims on a global basis, again on their specifics and with their particular circumstances and potential consequences set out.

In the months between the judgments in *Wharf v Cumine* and *ICI v Bovis* Mr Recorder Tackaberry had also considered the perceived difficulties with the pleading of global claims in *Mid Glamorgan County Council v J Devonald Williams and Partners* (1992) 8 CLJ 61. In this case the council claimed damages from Devonald Williams as a result of having to pay loss and expense to the contractor for delays in receipt of information on a project where that partnership was the architect. The council's claim was set out in very general terms and did not attribute losses to specific breaches. After a number of attempts by the architect to obtain better particulars of the claim, an application was made to the court for the claim to be struck out on the grounds that it might prejudice or delay a fair hearing of the action, or was an abuse of the court process.

Although the court agreed that the council was creating evidential difficulties for itself at trial, in terms of the prospects for its success, it held that it would be too draconian a step to strike out the claim. Mr Recorder Tackaberry QC summarised the rules on how to plead a case as follows:

> A proper cause of action has to be pleaded.
> Where specific events are relied on as giving rise to a claim for monies under the contract then any pre-conditions which are made applicable to such claims by the terms of the relevant contract will have to be satisfied, and satisfied in respect of each of the causative events relied upon.
> When it comes to quantum, whether time based or not, and whether claimed under the contract or by way of damages, then a proper nexus should be pleaded which relates each event relied upon to the money claimed.
> Where, however, a claim is made for extra costs incurred through delay as a result of various events whose consequences have a complex interaction which renders specific relation between the event and time/money consequence impossible and impractical, it is possible to maintain a composite claim.

In *GMTC Tools and Equipment Ltd v Yuasa Warwick Machinery Ltd* [1994] 73 BLR 102, the Court of Appeal heard an appeal against a decision of HHJ Potter regarding an action for damages arising from the computer-controlled precision lathe purchased from the defendants that the court had found at the liability trial to be '*seriously defective*'.

At that liability trial, GMTC had claimed for management time in dealing with the matter and the hours for which the lathe had been inoperable at a standard rate. It had also pleaded its claim on a global basis in that it did not seek to match each defect for which the defendant was responsible with the specific downtime and loss said to have resulted. HHJ Potter ordered that the claimant set out its case in the form of a Scott Schedule in a specified format. This proved impossible for the claimant such that, having obtained judgment in its favour on liability, the claimant sought to re-amend its quantum. That amended quantum would comprise losses in buying in blanks and finished cutters and lost new business.

The Judge declined that application to amend. In doing so he referred to: the lateness of the application; that the plaintiff's actions had already made the action 'too long, too expensive and unnecessarily difficult'; that the amendment would require further pleadings and evidence and extension of the time for the quantum trial; and that (not knowing their relevance to an amended case on quantum) the defendant had destroyed certain documents. The claimant appealed that decision.

In resisting that appeal, the defendant relied upon four grounds. Firstly, the plaintiff's non-compliance with an 'unless' order. Secondly, the lack of linkage between defect, downtime and loss. Thirdly, that the material facts had not been properly particularised. Fourthly, the prejudice suffered by its destruction of documents.

As to the documents destroyed by the defendant, the Court of Appeal found these to be 'peripheral'.

As to the first three grounds relied upon by the defendant, the Court of Appeal held that the court could not dictate to a party how it presented its case. As Leggatt LJ put it:

> I have come to the clear conclusion that the plaintiffs should be permitted to formulate their claims for damages as they wish, and not be forced into a straitjacket of the judge's or their opponents' choosing.

Relating GMTC's claim to typical global claims under construction and engineering contracts, the following paragraphs from Leggatt LJ's judgment are interesting and may strike a chord with those involved in construction and engineering claims:

> He contends that the amount of downtime caused by a defect found by the Judge to be the defendants' responsibility is at the centre of the case, and that unless the plaintiffs are able to relate to any particular period of downtime damage which they say they sustained, it should be irrecoverable.
>
> ...
>
> He goes so far as to submit that in the event that the machine had not functioned at all, the plaintiffs would have been entitled to present their claim as they wished to do, but that the very fact that it malfunctioned intermittently renders that method of claim inappropriate, if not impossible.
>
> ...
>
> According to their contention, if all causes of downtime were inevitably the defendants' fault it would not be necessary to determine the cause of each period of downtime; but since they were not, it is.
>
> ...
>
> They were criticised on the grounds that some of the heads of claim appeared to fall wholly or mainly outwith the periods of down time insofar as they are specified by the plaintiffs. These submissions, however, appear to me to miss the point so far.

Disagreeing with these submissions, the Judge concluded that they supposed that:

> ... an official referee or a commercial judge, when dealing with the interlocutory stages of an action, is entitled to prescribe the way in which quantum of damage is to be pleaded and proved. I disagree. No judge is entitled to require a party to establish causation and loss by a particular method, especially when the method proceeds, as happened here, on what can only be regarded as an imperfect understanding of the commercial realities of the plaintiffs' manufacturing processes.

However, crucially, making clear that this was only a pleadings point and that it did not mean that a claim on such a global basis would ultimately succeed when tested at trial, he went on:

> That is not to say that either of the claims will necessarily succeed at trial, whether wholly, mainly or in part.

Simon Brown LJ added to this that HHJ Potter had:

> ... sought to force the plaintiffs' damages claim into a straitjacket of his own devising, to impose upon it as an absolute requirement that a direct link be asserted and established between each defect in the machine so found, each period of consequent downtime and each resultant item of loss. But that does not represent the true basis of the plaintiffs' claim. They make no such assertion, indeed they recognise that they cannot. They seek instead to claim damages on an altogether broader basis. Whether that claim will prosper remains to be seen; undoubtedly certain difficulties attend it but those are for the future.

In *British Airways Pensions Trustees v Sir Robert McAlpine & Sons Ltd and Others* (1994) 72 BLR 26, the Court of Appeal allowed an appeal against a decision of HHJ Fox-Andrews QC to strike out a claim for alleged defective work including the costs of remediation and diminution in value of the head lease of the affected property.

The defendants were the contractor, the architect and the supplier of double glazing units. They sought to have the statement of claim struck out on two bases, including that the claimant had pleaded its claim without setting against each alleged defect the amounts claimed either for remediation or diminution. HJ Fox-Andrews QC had struck out the claim and dismissed the action. He described the pleading as embarrassing to the extent that it claimed against all three defendants the same total costs of remedying defects and diminution in lease value and did not ascribe these to each defect individually. However, in the appeal, Saville LJ noted how the defects themselves had been pleaded in detail and that the defendants had been on site for some time, with the knowledge that experience would give them. He stated:

> Thus it seems to me that it can hardly be said that these defendants were in any real fashion placed in a position where they were unable to know what case they had to meet or were facing an unfair hearing. They could, in my view, each prepare to deal with the allegation that they were responsible for the defects and could each assess the cost of remedying any particular defect. They could also

investigate with their own experts to what extent (if at all) any particular alleged defect or class of defects would diminish the sale value of the building. It is true that the pleading does not seek to apportion liability between the active defendants, but this is because it was the plaintiffs' case (good or bad does not matter in this context) that the defects attributable to each defendant caused the whole of the diminution in value.

and:

The basic purpose of pleadings is to enable the opposing party to know what case is being made in sufficient detail to enable that party properly to prepare to answer it. To my mind it seems that in recent years there has been a tendency to forget this basic purpose and to seek particularisation even when it is not really required. This is not only costly in itself, but is calculated to lead to delay and to interlocutory battles in which the parties and the court pore over endless pages of pleadings to see whether or not some particular point has or has not been raised or answered, when in truth each party knows perfectly well what case is made by the other and is able properly to prepare to deal with it. Pleadings are not a game to be played at the expense of the litigants, nor an end in themselves, but a means to the end, and that end is to give each party a fair hearing. Each case must of course be looked at in the light of its own subject matter and circumstances.

At this point he then referred to the then view of the editors of *Hudson's Building and Engineering Contracts*, the Eleventh Edition of which had said at paragraph 8-204, with reference to the judgment in *Wharf v Cumine*:

It is submitted that, in the English and related Commonwealth jurisdictions, claims on a total costs basis, a fortiori if in respect of a number of disparate claims, will prima facie be embarrassing and an abuse of process of the court, justifying their being struck out and the action dismissed at the interlocutory stage. It is further submitted that, even if such a claim is allowed to proceed, it should only be on the basis that, on proof of any not merely trivial damage or additional cost being established indeed any other cause of the additional cost, such as under-pricing for which the owner is not contractually responsible, the entire claim will be dismissed. Any other course places the practical onus of proving the extent of the plaintiff's damage on the defendant or on the court itself.

Saville LJ held that such assertions:

Are not automatically applicable to every case. With regard to the particular pleadings in question, I remain unpersuaded that either McAlpine or PDP were put to any sort of material unfair disadvantage by the way the matter had been set out by the plaintiffs.

The Court of Appeal allowed the appeal.

John Holland Construction & Engineering Pty Ltd v Kvaerner RJ Brown Pty Ltd (1996) 82 BLR 83 was an appeal to the Supreme Court of Victoria in relation to claims for extra costs in carrying out works in relation to a floating production, storage and off-loading facility off the Australian coast. The parties had entered into two contracts. The first was a Pre-Bid Agreement for provision of services at tender stage. The second was a Design Agreement for services at construction stage.

The claimant set out its claim in the form of a schedule of each specific claim. Therein, the loss and damage was set out as the difference between the tender estimated cost of each component and its actual cost. Byrne J stated:

> On behalf of Kvaerner Brown it was put that such an allegation is embarrassing; it does not seek to establish a causal link between the breach and the damage alleged or even to provide any basis for concluding that Holland should be relieved of that obligation; it passes to Kvaerner Brown all of Holland's cost overrun with respect to this component, effectively shifting to the defendant the burden of proof with respect to Holland's loss and damage; it imposes on Kvaerner Brown as defendant the obligation to prepare for trial every aspect of the Project, an obligation which is unfair and unduly expensive; and it imposes on the tribunal of fact an impossible task of determining issues of relevance and an unreasonable burden of sifting through a mass of detail which is not related to defined issues.

The arguments before the court focussed on the link between cause and effect. The causes were alleged breaches of the two contracts by the defendant and the effects were the extra costs incurred by the claimant, and the Judge assumed, for the purposes of the judgment, that these would be established:

> I am concerned with a pleading question; to what extent it is necessary to set out in the statement of claim the causal link between these two asserted facts. I am not concerned with the question, what loss, if any, flowed from the breach.

The Judge described the claim as a global one, in that it alleged a composite loss resulting from all of the breaches alleged. Referring to the judgments in *Merton v Leach* and *Wharf v Cumine*, he said:

> Such a claim has been held to be permissible in the case where it is impractical to disentangle that part of the loss which is attributable to each head of claim, and this situation has not been brought about by delay or other conduct of the claimant.

He further noted that the global claim in this case was calculated on a total loss basis, being the difference between the claimant's actual and expected costs in the form of its estimates. It was also global in that it alleged breaches of both contracts. The claimant was therefore contending that its estimated costs were made too low by breaches of the '*Pre-Bid Agreement*' and its actual costs were made high by breaches of the '*Design Agreement*'. The claim was therefore at the extreme end of practice within the construction claims field being a 'global total costs claim'. Furthermore, both sides of the total loss equation were said to have been affected by the defendant's breaches of two separate contracts.

The judgment provides a helpful analysis of the logic of claims made on such a global and total loss basis, identifying the following stages:

- That the contractor might reasonably have expected to perform the work in accordance with its estimate.
- That the defendant committed breaches of contract.
- That the actual costs were greater than the expected costs.
- That the actual costs were reasonable.
- That the defendant's breaches were the only significant cause of the cost overrun.
- No adjustment is made for variations or extras.

In summary, the Judge put this as two broad assumptions:

- that the defendant's breaches caused extra costs and
- that those extra costs were the amount of the contractor's costs overrun.

As to the first of these assumptions, the Judge surmised that it might be apparent from the nature of the breaches and a general understanding of the project. However, he also identified an intervening stage in relating the extra costs to the breach – how does the cause actually give rise to the effect? This was put as follows:

> For example, it may be that a breach means that work has to be redone, or that work takes longer to perform, or that its labour or material cost increases, or perhaps that there was extra cost due to disruption or loss of productivity.

He continued that, whilst this might be readily apparent in a given case:

> … difficulties will arise for the parties and the tribunal of fact where the global nature of the claim involves the interaction of two or more of these intervening steps, particularly where they and their role are not, in terms, identified and explained.

We explained the importance of claimants detailing this intervening stage when considering disruption claims elsewhere in this chapter. Global claims that assert a series of events as causing a global loss need definition of how those causes led to the loss. Take, for example, a claim for the effect on labour costs of the late and piecemeal release of access to a site. The assertion of late delivery is one of fact that should be readily proven, together with contractual liability for it. That losses were incurred is also an issue of fact that should be easily established from the contractor's factual records. That one led to another can be surmised and in occasional cases may be readily apparent from the facts, but it will usually need detail. One way to do that is to set out examples of the effects of late land that will involve financial loss. For example:

- Standing time whilst the labour lacks available work.
- Down time whilst the labour moves from available area to available area as they are released.
- Increase in periods of learning as areas are returned to.

- Lost efficiency where labour is reduced below optimum levels.
- Lost efficiency where labour is increased above optimum levels to address peaks in access availability.
- Lost efficiency due to changes in plant/labour ratios.
- Lost efficiency due to changes in plant/supervision ratios.
- Premium costs of overtime to mitigate lost time.
- Lost efficiency due to such overtime working.
- Increases in labour costs due to inflation.
- Increased reliance on expensive agency labour over payroll labour.

This list is not comprehensive, and these examples will not all apply in every case, but they are examples of how a claimant might add detail of the link between cause and effect where the factual circumstances fit. The benefits of doing so include:

- The defendant will know the case better that it has to answer, so that it cannot argue that it is being hampered in its attempts to defend itself.
- The issues for trial can be better identified, so that the process can run more efficiently.
- Methods can be found to add particularity of the losses claimed by looking at the intervening individual causes.
- That loss and how the breaches giving rise to it can be recognised.
- Adjustment can be made to the quantification of the claim if a component of its basis fails.

As to the second broad assumption put by the Judge in *John Holland v Kvaerner*, he noted how the claimed extra costs were the amount of the claimant's cost overrun and surmised that this was:

> ... likely to cause the more obvious problem because it involves an allegation that the breaches of contract were the material cause of all of the contractor's cost overrun. This involves an assertion that, given that the breaches of contract caused some extra cost, they must have caused the whole of the extra cost because no other relevant cause was responsible for any part of it.

The Judge also noted how in this case the claimant was asserting that both its estimated and actual costs had been affected by the defendant's failures. He noted how a claim put in this way was even more vulnerable than most global total costs claims if the defendant could show that the underlying assumptions were wrong. He said:

> It may be that a consequence of the formulation of a claim in this way is that Kvaerner Brown, if it can expose the improbability of the assumption which Holland would have the court accept, might achieve the result that Holland, upon whom lies the ultimate burden of proof, will fail to discharge that burden and perhaps fail altogether.

This is a clear statement of the risk that any claimant takes of advancing a claim on a global basis, of its failure at trial and hence of how defendants will usually look to undermine such a claim. However, having explained that vulnerability the judgment then underlined:

First, it is for the parties and not the court, even in a judge-managed list, to determine how their case should be framed. It is not for the court to impose upon them a manner or form of pleading which it thinks better than their own.

...

Secondly, the power of the court to strike out a claim is very limited. So far as is here relevant, it may be exercised where the claim is so evidently untenable that it would be a waste of the resources of the court or the parties for the court to permit this to be demonstrated only after a trial ... or where the pleading is likely to prejudice, embarrass or delay the fair trial of the action.

He cited the Privy Council judgment refusing to uphold an order striking out a global claim on that basis in *Wharf Properties Ltd v Eric Cumine* as authority for the former imposing a very heavy burden on the applicant.

Finally, this case is being managed in a specialist list and is still in its interlocutory stages. One of the advantages of such a list is that the judge, being familiar with the case, can encourage the parties to identify and formulate the issues so that the trial might be conducted in as economical and expeditious a manner as may be consistent with the just disposition of the dispute.

The Judge then considered if the pleading in this case was embarrassing or prejudicial:

... a pleading is not an end in itself: its adequacy must be assessed by reference to its function in the scheme of litigation, having regard to the type of proceeding in which it is delivered and the nature of the dispute with which it is concerned. Each case must be looked at in the light of its own subject matter and circumstances.

In this regard, he referred to, and adopted, the observations of LJ Saville in *British Airways v McAlpine*. He noted how no nexus linking the breaches to the losses had been asserted, but considered regarding such '*conventional pleading practice*' that:

... it appears to be accepted that a departure from this practice may be accepted where it is demonstrated that to do so would be impossible or impractical.

In support of that position he again referred to *Wharf v Cumine* and also to the Australian case *Naura Phosphate Royalties Trust v Matthew Hall Mechanical and Electrical Engineering Pty* [1994] 2VR386.

In the *Naura Phosphate* judgment, Smith J provided several interesting and helpful observations for those seeking either to advance or defend global claims, for example:

It is subject to the obligation to define its case in a manner that discloses a cause of action and to give sufficient particulars to enable the defendant to understand the cases it has to meet and thus to satisfy the principles of natural justice. Subject to those qualifications, however, it is not obliged to give particulars of 'nexus' when it is not part of its case to establish a 'nexus' between each alleged disrupting event, particular disruptions and loss.

It is suggested that circumstances where a claimant might not seek to prove such a nexus are:

1. Where it is impossible or impractical to do so, perhaps because of the number of events and/or the complexity of their interaction.
2. Where the whole of the loss is due to matters that the defendant is responsible for. In such a case, the claimant might say that there is no point going to the difficultly of separately establishing the nexus of each event when the nexus of the whole is established.
3. Where the nexus is obvious from a simple reading of the facts.

Smith J continued:

> On the contrary, a global approach can hide a bogus claim. Nauru argues that this claim is one such claim and points to a failure by Matthew Hall to complain about disruptions during the contract works.

This is a common complaint by those in receipt of pleadings or late claims for loss of productivity of labour set out on a global basis. The complaint in such a case will be that the claim's lack of particularisation is due to fact that there was actually no labour disruption, as reflected by the lack of complaint at the time. The defendant will say that the claim's approach seeks to hide that.

Smith J continued:

> I can see no reason why, for example, a judge controlling a Building Cases List or arbitrator could not require the plaintiff to particularise the 'nexus' or justify its assertion that it is not possible to do so.

This provides an interesting alternative for the Judge and the claimant. This is to either require the claimant to explain why it should not have to particularise, which could be one of the three possibilities listed or some other reason, or if it cannot excuse it, then to particularise.

Finally, Smith J further continued:

> Global claims are difficult for the parties and the court to handle. To compel a plaintiff to give particulars 'of nexus' or justify its inability to do so may reveal the bogus claim.
>
> The issue raised here for decision is whether there is an abuse of process arising from the globally pleaded claim. I consider that, in all the circumstances, there is not.

The *Naura Phosphate* case was one where the arbitrator had allowed the pleading of a global total costs claim, but rejected the claim on the basis that it would substitute for a lump sum contract remuneration on a *quantum meruit* basis. It therefore serves to illustrate the vulnerability of a global claim in that, although it had been allowed through on the pleadings issue at the interlocutory stage, it was lost on its merits. This objection, that the approach would fundamentally change the basis of the contract to a quantum meruit one, is also a regular objection from those receiving global claims that are also quantified on the basis of the contractor's total costs.

Returning to Byrne J's judgment in *John Holland v Kvaerner*, he concluded, referring to the judgment in *British Airways v McAlpine*:

> In my opinion, the court should approach a total cost claim with a great deal of caution, even distrust. I would not, however, elevate this suspicion to the level of concluding that such a claim should be treated as prima facie bad.

However, he continued:

> Nevertheless, the point of logical weakness inherent in such claims, the causal nexus between the wrongful acts or omissions of the defendant and the loss of the plaintiff, must be addressed.

As the claimant found in the *Naura Phosphate* case, that occasion may be at trial, rendering any victory at the interlocutory stage a pyrrhic one. However, in this case the Judge did not strike out the claim but allowed its amendment to correct the bad parts of it, particularly the lack of causal nexus. Byrne J concluded at paragraph 134 of his judgment:

> I put to one side the straightforward case where each aspect of the nexus is apparent from the nature of the breach and loss as alleged. In such a case the objectives of the pleading may be achieved by a short statement of the facts giving rise to the causal nexus. If it is necessary for the given case for this to be supported by particulars, this should be done. But, in other cases, each aspect of the nexus must be fully set out in the pleading unless its probable existence is demonstrated by evidence or argument and further, it is demonstrated that it is impossible or impractical for it to be spelt out further in the pleading. Moreover, the court should be assiduous in pressing the plaintiff to set out this nexus with sufficient particularity to enable the defendant to know exactly what is the case it is required to meet and to enable the defendant to direct its discovery and its allegation generally to that case. And it should not be overlooked that an important means of achieving the result that, once it starts, the trial should be conducted without undue prejudice, embarrassment and delay, is by ensuring that, when it begins, the issues between the parties, including this nexus, are defined with sufficient particularity to enable the trial judge to address the issues, to rule on relevance and generally to contain the parties to those issues.

In *Bernhard's Rugby Landscapes Ltd v Stockley Park Consortium Ltd* [1997] 82 BLR 39, HHJ Humphrey Lloyd heard claims from a contractor including prolongation and other damages arising from delays and which were pleaded on a global basis. The claimant justified its approach by asserting that:

> ... such additional costs were incurred by reason of delay and disruption caused by various events whose consequences have a complex interaction rendering specific relation between the event relied upon and the financial consequences thereof impossible.

The Judge noted, however, that the claim did not differentiate costs attributable to variations from those attributable to other claimed events.

The defendant attacked this approach on the basis that it did not provide a proper causal nexus and the claimant had not explained what the 'complex interaction' was that justified the approach. Furthermore, criticisms were made of the claimed costs in that some were anticipated rather than actual and subcontractor costs were not clarified as being actual costs or liabilities. Referring to the judgment in *Holland v Kvaerner*, counsel submitted that this part of the claim should be 'approached with caution', if not suspicion, and that the financial effect of each alleged delay should be pleaded.

The claimant responded that the claims were not total cost claims in the sense of *Holland v Kvaerner*. It contended that:

- It had pleaded specific periods of delay against specific events.
- The claimed costs could be reduced pro rata if parts of the delay claim failed.

Regarding the practicalities of the position that the contractor had found itself in, the background of the case in *Bernhard's v Stockley* was put by HHJ Humphrey Lloyd as follows:

> … that the project had been prolonged by many years and had been severely disrupted by numerous major and minor changes. In these circumstances it was inevitable that the claim would be presented in the form in which it was.

The claimant referred to *GMTC v Yuasa* as authority for the court not placing a party into a 'straightjacket' or directing how it should plead its case.

HHJ Humphrey Lloyd started setting out his decision on the global claim by distinguishing two forms:

> A global claim can take a variety of forms. Where it describes a pleaded claim it has pejorative overtones as it is usually intended to describe a claim where the causal connection between the matters complained of and their consequences, whether in terms of time or money, are not fully spelt out, but, implicitly, could and should be spelled out. It is to be contrasted with the use of the term where an arbitrator has made an award of a sum which the arbitrator cannot apportion between the various events.

Whilst it may be permissible for an arbitrator to take such an approach, this did not excuse the need to alert the opposing party to the case which he or she will have to address at trial and in that regard he referred to *Wharf v Cumine*. However, the nexus might be inferred.

HHJ Humphrey Lloyd referred to the Australian judgments in *Naura Phosphate* and to paragraph 134 of Byrne J in *Holland v Kvaerner*, which is quoted elsewhere, and considered the position to be identical in the United Kingdom. He put the position as it then stood as follows:

> Whilst a party is entitled to present its case as it thinks fit and it is not to be directed as to the method by which it is to plead or prove its claim whether on

liability or quantum, a defendant on the other hand is entitled to know the case that it has to meet.

With this in mind a court may – indeed must – in order to ensure fairness and observance of the principles of natural justice – require a party to spell out with sufficient particularity its case, and where its case depends upon the causal effect of an interaction of events, to spell out the nexus in an intelligible form. A party will not be entitled to prove at trial a case which it is unable to plead having been given a reasonable opportunity to do so, since the other party would be faced at the trial with a case which it also did not have a reasonable and sufficient opportunity to meet.

What is sufficient particularity is a matter of fact and degree in each case. A balance has to be struck between excessive particularity and basic information. The approach must also be cost effective. The information may already be in the possession of a party or readily available to it, so it may not be necessary to go into great detail.

In relation to the total costs claim, he noted that the claimant had identified in its extension of time claim the periods of delay for each causative event relied upon and therefore he could not see why the nexus between those events and the costs claimed could not also be spelt out. He also considered the work done by Schal, who had been retained as construction manager for the project, as relating to the claims for extension of time, not associated costs. In conclusion the Judge found that the claim did not provide the defendant with sufficient information about the case which it had to meet, but granted leave for its amended statement of claim to be served, subject to provision of further particulars as set out in the judgment.

The *Bernhard's Rugby* case was decided by HHJ Humphrey Lloyd QC in 1997, and the approach he adopted there followed his approach the previous year in the unreported case of *Inserco Limited v Honeywell Control Systems Limited* (19 April 1996). In that earlier judgment he had stated:

> Inserco's pleaded case provided sufficient agenda for the trial and the issues are about quantification. Both Crosby [*Crosby v Portland Urban District Council* (1997)] and Merton [*London Borough of Merton v Stanley Hugh Leach* (1985)] concerned the application of contractual clauses. However I see no reason in principle why I should not follow the same approach in the assessment of the amounts to which Inserco may be entitled. There is here, as in Crosby an extremely complex interaction between the consequences of the various breaches, variations and additional works and, in my judgment it is 'impossible to make an accurate apportionment of the total extra cost between the several causative events' I do not think that even an artificial apportionment could be made – it would certainly be extremely contrived – even in the few occasions where figures could be put on time etc.... It is not possible to disentangle the various elements of Inserco's claims from each other. In my view, the cases show that it is legitimate to make a global award of a sum of money in the circumstances of this somewhat unusual case which will encompass the total costs recoverable ....

6.6.2.3 The Current Position

This tide towards a more supportive approach to global claims was adopted by the Scottish courts in the case of *John Doyle Construction Ltd v Laing Management (Scotland) Ltd* [2004] ScotCS 141, in which Lord Drummond Young delivered the opinion of the Scottish Inner Court of Session upholding a decision from the commercial courts allowing a claim that was both global, and for adjusted total loss, to proceed to trial.

John Doyle had been appointed as works contractor by Laing, as management contractor, to execute various packages of works to a new corporate headquarters building. John Doyle subsequently sued Laing for an extension of time of 22 weeks and recovery of loss and expense arising from delay and disruption to their works. The financial claim was calculated globally, the claimant averring that 'it was not possible to identify the causative links between each such cause of delay and disruption and the cost consequence'. The amount of their loss was calculated by comparing John Doyle's pre-contract estimates of their labour costs with the actual labour costs incurred, with certain adjustments. It also compared labour productivity when work was largely free from disruption with labour productivity achieved when the work was in fact disrupted. That aspect was therefore a 'measured mile' approach as discussed elsewhere in this chapter. The Court of Session's judgment provides a very interesting explanation of such claims and the authorities, including some US cases. It also takes a supportive approach to the pleading of global claims and the effects of their failure as global.

Before referring to the specifics of the case and Doyle's pleadings, the judgment set out the requirements of a pleading in the Scottish commercial courts, in passages including the following:

> All that is required is that a party's averments should satisfy the fundamental requirements of any pleadings, namely that they should give fair notice to the other party of the facts that are relied on, together with the general structure of the legal consequences that are said to follow from those facts.'
> … it is obviously necessary that the events relied on should be set out comprehensively.
> It is also essential that the heads of loss should be set out comprehensively, although that can often best be achieved by a schedule that is separate from the pleadings themselves.
> So far as the causal links are concerned, however, there will usually be no need to do more than set out the general proposition that such links exist.
> Causation is largely a matter of inference, and each side in practice will put forward its own contentions as to what the appropriate inferences are. In commercial cases, at least, it is normal for those contentions to be based on expert reports.
> What is not necessary is that averments of causation should be over-elaborate, covering every possible combination of contractual events that might exist and the loss or losses that might be said to follow from such events.

It was common ground between the parties that a global claim, in which the individual causal connections between the events giving rise to the claim and the items of loss and expense making up the claim are not specified, could in principle be made. Lord Drummond Young referred to the judgments in *Merton v Leach*, *Wharf v Cumine* and *John Holland v Kvaerner*, all of which have been considered elsewhere, and stated that:

The pursuers had averred that, despite their best efforts, it was not possible to identify causal links between each cause of delay and disruption and the cost consequences thereof. On that basis, the defenders accepted that the pursuers were in principle entitled to advance a global claim.

Lord Drummond Young further recognised that:

> Frequently, however, the loss and expense results from delay and disruption caused by a number of different events, in such a way that it is impossible to separate out the consequences of each of those events .... If, however, the contractor is able to demonstrate that all of the events on which he relies are in law the responsibility of the employer, it is not necessary for him to demonstrate causal links between individual events and particular heads of loss.

Later in the judgment he recognised a distinction between claims for delay and disruption as regards the ease with which loss might be separated out against each cause, where delay, occurring in discrete periods of time against which the costs might be allocated, was 'relatively straightforward' to separate out. This distinction has been suggested elsewhere in this chapter.

On the reliance of a global claim on all of the causes of the loss being the legal responsibility of the defendant, Lord Drummond Young gave the examples of bad weather, a neutral event, and also inefficient working by the claimant itself, and continued that:

> ... it appears that a significant cause of the delay and disruption has been a matter for which the employer is not responsible, a claim presented in this manner must necessarily fail.

However, he qualified this with the following, such that a failed global claim may still survive in terms of identifiable parts that fall out of it:

> In some cases it may be possible to separate out the effects of matters for which the employer is not responsible.

Turning to the 'total loss' nature of the claim, the judgment referred extensively to Byrne J's observations in *Holland v Kvaerner* and particularly how such a claim required the effects of any inadequacy in the contractor's tender or causes of cost overrun for which the employer was not responsible to be eliminated. Lord Drummond Young also referred to the US Court of Claims comment on the total cost method of calculation in *Boyajian v United States* 423 F 2d 1231 (1970):

> This theory has never been favoured by the court and has been tolerated only when no other mode was available and when the reliability of the supporting evidence was fully substantiated .... The acceptability of the method hinges on proof that (1) the nature of the particular losses make it impossible or highly impracticable to determine them with a reasonable degree of accuracy; (2) the plaintiff's bid or estimate was realistic; (3) its actual costs were reasonable; and (4) it was not responsible for the added expenses.

At this point it is worth setting out how Doyle had calculated its claimed total loss. Achieved labour productivity was calculated on work that was largely free from disruption, and this was referred to as 'normal work'. Achieved labour productivity was then calculated on work that was disrupted, and this was referred to as 'disrupted work'. Both were calculated over several weeks of the project, to smooth out the results. This approach is the basis of a 'measured mile' approach to a disruption claim, explained elsewhere, and was therefore said to remove from the resulting claim any lost productivity due to the claimant's own failures, thus partly addressing points (3) and (4) in the extract from *Boyajian v United States*. Furthermore, the productivity achieved on the 'normal work' was then applied to the whole of the work to calculate a total labour cost had all the work proceeded as the 'normal work' had. This approach was said to remove from the calculations of loss that followed any errors in the contractor's original tender, thus addressing point (2) in the extract from *Boyajian v United States*.

The calculation then compared the calculated total labour cost had all the work proceeded as the 'normal work' had done with Doyle's actual labour costs. The difference was then discounted for labour in varied and additional works and labour in other claims, to leave a balance said to be the result of matters for which Laing was responsible.

Of particular interest in the judgment is the consideration of how the need to separate out matters for which the employer was responsible, as required in both Byrne J in *Holland v Kvaerner* and the US Court of Claims in *Boyajian v United States*, might (under Scottish law) be addressed by reference to both 'dominance' and 'apportionment'. One possibility would be to establish that a group of events for which the defendant was responsible are causally linked with a group of heads of loss which has no other 'significant cause'. Lord Drummond Young continued that, in determining such significance, the 'dominant cause' approach is of relevance. In this regard he said:

> ... it is frequently possible to say that an item of loss has been caused by a particular event notwithstanding that other events played a part in its occurrence. In such cases, if an event or events for which the employer is responsible can be described as the dominant cause of an item of loss, that will be sufficient to establish liability, notwithstanding the existence of other causes that are to some degree at least concurrent.

and:

> If an item of loss results from concurrent causes, and one of those causes can be identified as the proximate or dominant cause of the loss, it will be treated as the operative cause, and the person responsible for it will be responsible for the loss.

Where it was not possible to identify one of the causes as being the dominant cause of the loss, then he suggested that:

> ... it may be possible to apportion the loss between the causes for which the employer is responsible and other causes. In such a case it is obviously necessary that the event or events for which the employer is responsible should be a material cause of the loss.

These approaches would apply to both claims for the time effect of delay and also loss associated with disruption. In practice either identifying the dominant cause of delay or apportioning the effects of concurrent delay might be relatively straightforward in some cases. However, as Lord Drummond Young recognised, where the claim is one for labour productivity loss due to disruption the matter will usually be 'more complex'. However:

> … we are of opinion that apportionment will frequently be possible in such cases, according to the relative importance of the various causative events in producing the loss. Whether it is possible will clearly depend on the assessment made by the Judge or arbiter, who must of course approach it on a wholly objective basis. It may be said that such an approach produces a somewhat rough and ready result. This procedure does not, however, seem to us to be fundamentally different in nature from that used in relation to contributory negligence or contribution among joint wrongdoers. Moreover, the alternative to such an approach is the strict view that, if a contractor sustains a loss caused partly by events for which the employer is responsible and partly by other events, he cannot recover anything because he cannot demonstrate that the whole of the loss is the responsibility of the employer. That would deny him a remedy even if the conduct of the employer or the architect is plainly culpable, as where an architect fails to produce instructions despite repeated requests and indications that work is being delayed. It seems to us that in such cases the contractor should be able to recover for part of his loss and expense, and we are not persuaded that the practical difficulties of carrying out the exercise should prevent him from doing so.

The Court of Session found that Doyle's approach to calculation was 'relevantly plead', was not a total costs claim comparing estimated and actual costs and 'goes some way towards dealing with the matters set out' in *Boyajian v United States*. On that basis it concluded that the decision to allow the pleaded claim to proceed was correct. However, it noted that in being heard the claimant would still have to establish that:

- their actual labour costs were reasonable;
- neither they nor factors outside the control of either party were responsible for any of the causes of the increased labour costs; and
- it is impossible or highly impracticable to determine the actual additional labour costs arising out of each variation or late instruction.

The judgment in *Doyle v Laing* was a significant point in the history of global claims. In Scotland at least, it confirmed that there was indeed life in an 'old dog' that some had written off as dead after the judgments in *Wharf v Cumine* and *ICI v Bovis*, but which had actually been showing some signs of life all along. In summary the following propositions seemed to result in relation to global claims:

- The pleading of a claim should comprehensively set out:
  - the events relied on and
  - the heads of loss.
- However, the causal links only need be set out to the extent of a proposition that the links exist. That is largely a matter of inference and the evidence of experts.

- Normally it is necessary to establish a causal link between particular events for which the employer is responsible and individual items of loss.
- However, if all the events are events for which the defender is legally responsible, it is unnecessary to insist on proof of which loss has been caused by each event.
- On occasion that may be achieved where it can be established that a group of events for which the employer is responsible are causally linked with a group of heads of loss, provided that the loss has no other significant cause.
- That significance might be established by reference to 'dominant cause'.
- Where a 'dominant cause' cannot be identified it may be possible to 'apportion' the loss between the causes for which the employer is responsible and other causes.

However, *Doyle v Laing* was a case under Scottish law, and some doubt persisted as to the extent to which the England and Wales courts would apply the same principles, particularly in relation to apportionment.

Back in England, in *London Underground Ltd v Citylink Telecommunications Ltd* [2007] BLR 391, Mr Justice Ramsey heard various applications from the parties in relation to an arbitrator's award, including in relation to his granting of an extension of time against a delay claim pleaded on a global basis. Ramsey J noted that the approach adopted in *Doyle v Laing* was not challenged in the application and he accepted the approach taken by the Scottish Court of Session in that case. He defined a global claim as follows:

> The essence of the global claim is that, whilst the breaches and the relief claimed are specified, the question of causation linking the breaches and the relief claimed is based substantially on inference, usually derived from the factual and expert evidence.

He quoted paragraph 20 of Lord Drummond Young's judgment in *Doyle v Laing* as follows:

> In a case involving the causal links that may exist between events having contractual significance and losses suffered by the pursuer, it is obviously necessary that the events relied on should be set out comprehensively. It is also essential that the heads of loss should be set out comprehensively, although that can often best be achieved by a schedule that is separate from the pleadings themselves. So far as the causal links are concerned, however, there will usually be no need to do more than set out the general proposition that such links exist. Causation is largely a matter of inference, and each side in practice will put forward its own contentions as to what the appropriate inferences are. In commercial cases, at least, it is normal for those contentions to be based on expert reports, which should be lodged in process at a relatively early stage in the action.

However, whilst a global claim could be pleaded on this basis, Ramsey J then turned to its being undermined and the effect if it falls as a global claim. The methods by which global claim might be undermined include those set out by Lord Macfadyen at paragraphs 36 and 37 of his judgment in *Doyle v Laing*. These are the standard routes for those defending a global claim:

- That an event played a material part in causing the global loss, but the claimant fails to establish that the event was one for which the defendant was responsible.
- A material contribution to the causation of the global loss has been made by another factor or other factors for which the defendant had no liability.

The question that *Doyle v Laing* had addressed, and Ramsey J then had to consider, was what is the effect of the failure of a global claim on such bases as these two examples? Does the claim die entirely? The defendant might hope so, and will argue that, since no claim has been pleaded that particularises part of the claimed damage against surviving parts of the pleaded causes, the tribunal then has no basis on which to assess part. That this approach would be wrong recognises the practicality of pleading complex construction claims where a large number of interrelated events contributed to a global damage. This might be that hundreds of late design approvals or changes to drawings lead to a global loss of labour productivity or that a number of delay events contributed to an overall delay to completion, as was the case in both *Doyle v Laing* and *LUL v Citylink*. However, the complexity of having to plead such claims on every possible combination of events with their respective combined effects had been identified by Lord Macfadyen and Ramsey J now referred to his paragraph 38:

> The global claim may fail, but there may be in the evidence a sufficient basis to find causal connections between individual losses and individual events, or to make a rational apportionment of part of the global loss to the causative events for which the defender has been held responsible.

As Ramsey J put it:

> The surviving or remaining claim will therefore emerge from the evidence which has been adduced or established from the global claim.

The difficulties that this approach presents to the parties, in particular the defendant, is that such a remaining claim may not become apparent until a late stage in proceedings, once the tribunal has heard the evidence and determined what events have succeeded. What opportunity the parties should be given to address the resulting partially successful case was the subject of one of the applications being heard in *LUL v City Link*. In that case the arbitrator had made his award rejecting the global claim but then granted an extension of time for a surviving delay event. The Judge noted that tribunals frequently have to make decisions in relation to the partial success of a claim and that provided the resulting award is based on the facts in issue in the proceedings there was no need for the arbitrator to seek further submissions from the parties.

The authors of this book have seen this done several times in international arbitration, where a global claim has failed but an award allows part of the claim. An alternative that some international arbitrators follow, and that can raise eyebrows among procedural purists, is to ask the parties for post-hearing submissions on propositions that it gives them as to its likely findings, or asks for the parties' experts to consider and jointly report post-hearing. This has been particularly the case in relation to global claims for labour disruption. In particular, this may occur where the tribunal is satisfied that disruption occurred, that the defendant was responsible for it and has some sympathy for the claimant notwithstanding the fall of its global approach. That might be, for example, because of a lack of claims sophistication on the claimant's part or because the causes of disruption were made inextricably complex by the extent of the defendant's defaults. The important point for practitioners is that for all the risks of presenting a claim on a global basis and the common ways in which defendants seek to undermine them, the legal position may still be that the fall of a global claim does not necessarily mean that the claim is lost completely.

Perhaps tempering this, in *Petromec Inc v Petroleo Brasilerio SA Petrobras* [2007] EWCA Civ 1371, 115 Con LR 11, the Court of Appeal's decision upholding a judgment of Cooke J of a preliminary issue as to the appropriate method of valuing variations was seen by some as fatal to global claims. The case actually hinged on the interpretation of the particular conditions of the contract in the case. However, the court did set out a helpful summary of reasons why the courts '*tended to disapprove of global claims*' and the relevant authorities. Those reasons were:

(i) Dealing with a claim in this way enables a party to avoid the requirements of pleading and proof of entitlement and causation.
(ii) When the case is put in a global way, the other party is not given proper notice of the case it has to meet.
(iii) This approach causes real problems in trial preparation and case management and conduct of the trial itself.

The authorities considered in the judgment were *Bernhard's Rugby v Stockley*, *Wharf v Cumine* and *John Holland v Kvaerner*. The judgment makes no mention of *John Doyle v Laing*. It seems to have particularly focussed on the terms of the contract. Petromec contended that those terms allowed it to claim for extensive changes in specification the difference between the costs it might reasonably have incurred to the original specification and those that it actually incurred, on a global basis. Cooke J and the Court of Appeal disagreed, requiring Petromec to particularise the effects of each variation. Effectively the contract entitled it to the costs of the particularised differences, not its overall difference in costs. In Lord Justice May's judgment:

> ... it would not be fair to Petrobras, nor a practical way of the court proceeding, if Petromec were not required to give adequate particulars of their claim.

6.6.2.4 Lilly and Mackay

The English case of *Walter Lilly & Company Limited v Mackay & Anor* [2012] EWHC 1773 (TCC) (11 July 2012) is widely regarded as an object lesson in how not to procure construction works, how parties to a construction contract should not conduct their relationships and how not to resolve a dispute. The behaviour of the employer was such as to merit a full page exposé in one UK national newspaper. As a result, the underlying facts of the dispute were far from typical, and the judgment of Mr Justice Akenhead reflects this. Mr Justice Akenhead was head of the Technology and Construction Court from 2010 to 2013.

It actually involves three cases before the Technology and Construction Court in London: *Walter Lilly & Company Ltd v Mackay & Anor* [2012] EWHC 3139 (TCC) (11

December 2008); *Walter Lilly & Company Ltd v Mackay & Anor* [2012] EWHC 649 (TCC) (15 March 2012); and *Walter Lilly & Company Ltd v Mackay & Anor* [2012] EWHC 1773 (TCC) (11 July 2012). In addition to global claims, the issues considered in the judgments included legal privilege, record keeping, concurrent delays, dealing with defects, expert evidence and the recovery of head office overheads, claims preparation costs and claim amounts paid to subcontractors. In relation to global claims, it applied similar principles to those adopted by the Scottish court in *John Doyle v Laing*, thus largely aligning the positions in the two jurisdictions.

The project itself appeared simple enough in technical terms. It was for the construction of three houses in central London, with a contract value of just over £15 million and a 78 week programme completing in January 2006. However, the approach to procurement was far from simple. Procured under the JCT Standard Form of Building Contract 1998 Edition Private Without Quantities, the design was not fully developed at the contract stage, with all of the building works being the subject of Provisional Sums. Furthermore, and worryingly for any house builder for a private client, the client expressed a desire for 'a dream home'. The Judge described the project as 'a disaster waiting to happen'. In the event costs overran several-fold and the court awarded the contractor £2.3 million against its claims. Legal costs were estimated at £10 million, with the legal actions running for seven years after the programmed completion date, the employer being denied permission to appeal in January 2013.

Mr Justice Akenhead's review of the law in relation to global claims started with *Crosby v Portland* and continued through most of the preceding judgments, concluding with the Scottish case *John Doyle v Laing*. The judgment considered both 'global claims' and 'total cost claims' and it has been explained how at paragraph 484 Mr Justice Akenhead provided a definition that seemed to some commentators to take an approach that combined both. At paragraph 486 he set out his conclusions, including those paraphrased as follows in relation to global claims:

- Claims for delay or disruption must be proved as a matter of fact.
- The contractor has to demonstrate on a balance of probabilities that:
  - events occurred which entitle it to loss and expense;
  - that those events caused delay and/or disruption; and
  - that such delay and/or disruption caused it to incur loss and damage.
- It is open to contractors to prove these three elements with whatever evidence will satisfy the tribunal and the requisite standard of proof. There is no set way for contractors to prove these three elements.
- There is nothing in principle wrong with a global or total cost claim.
- It does not have to be shown by a claimant of a global claim that it is impossible to plead and prove cause and effect in the normal way or that such impossibility is not its own fault.
- The fact that one or a series of events or factors (unpleaded or which are the risk or fault of the claimant contractor) caused or contributed (or cannot be proved not to have caused or contributed) to the total or global loss does not necessarily mean that the claimant contractor can recover nothing.
- Where a global claim includes events which are the fault or risk of the claimant which caused or cannot be demonstrated not to cause some loss, the overall claim will not be rejected save to the extent that those events caused some loss, in which case their amount would be deducted from the global loss.

- It may be that the tribunal will be more sceptical about the global claim if the direct linkage approach is readily available but is not deployed. But that does not mean that the global claim should be rejected out of hand.
- Arguments that a global award should not be allowed where the contractor has himself created the impossibility of disentanglement are wrong.
- In principle, unless the contract dictates that a global claim is not permissible if certain hurdles are not overcome, such a claim may be permissible on the facts and subject to proof.

6.6.2.5 The Need for a Global Approach

Having set out the evolution of the English courts' approach to global claims, a few practicalities are considered as follows. These assume that any contractual pre-conditions have been satisfied. This includes any requirements for notice and whether these are a condition precedent to entitlement. As ever, the precise wording of the clause under which the claim is being made is a first-read for anyone drafting a claim on any basis, together with an understanding of how the applicable law, particularly to a claim made on a global basis. Bespoke express provisions may exclude any right to make a global claim.

In practice, the necessity of citing a number of events as the cause of a common cost, loss, expense, or damage for the contractor occurs frequently on large construction and engineering projects. The most obvious scenario is where many hundreds of instructions are issued to a contractor. For example, on a new terminal for an international airport, the MEP (Mechanical, Electrical and Plumbing) subcontractor alone had over 3000 variation instructions issued to its works. It would be understandable that such an extent of change had an impact on the subcontractor that went beyond the discrete additional or omitted works content of each instruction; at the least it caused delay to the subcontractor's activities. To require the subcontractor in that situation to allocate that wider impact to each of 3000 instructions would be a monumental task that would also be most likely no more than contrived. Even where there are fewer such events and some can be allocated their individual financial effects, there still may be a 'rump' that cannot be separated out from more than one event.

However, to take a simpler example, consider the situation where a contractor has proven compensable delays caused by a small number of events that are a mixture of late information and changes, say to the construction of a first floor slab. As a result, that slab took 4 weeks rather than 2 weeks to construct. Assume that an extension of time for a 2 week period has been granted, so the employer agrees the case for delay, and that no other issues such as concurrency and lack of notice are being taken. The associated prolongation may not be the result of a huge number of events, but may result from, say, 10 instructions for additional works and 10 late drawings. The contractor is entitled to its prolongation costs as incurred during the 2 additional weeks spent on the slab, but does it have to allocate those costs to each of the 20 events that were the cause of it, individually? Can this be done with any certainty? Is the time and cost required to do so worth it?

Looking in more detail, consider the situation where the contractor has had to incorporate a multiplicity of variations for changed or additional work and has had design information supplied later than required by the terms of the contract. The effects of such matters may manifest themselves in a number of ways:

- The period of some activities affected by the variations or delayed information may be extended beyond that required in the absence of the variations.
- As a result of the extended period of the activities affected by the variations or delayed information, the critical path for the project programme may be extended. Activities that might originally have been on the critical path may become non-critical while other activities, originally non-critical, may become critical as a result of the variations.
- The unit costs of executing some, or all, of the varied works may differ from the unit cost anticipated at the time of tender as a result of the variations.
- The late information may have pushed the related works back in time, so that they are subject of inflationary increases in the contractor's costs.
- The varied works may have an effect upon other non-varied works, thereby affecting the cost or period of those works.
- As a result of the extended period of individual activities, additional supervision and ancillary costs may be incurred that are not recovered through the unit rates for additional work.
- The scope of change and uncertainty may be such that the contractor has to add supervision and management to activities to deal with those issues.
- As a result of the extended period of activities and the effect on the project critical path, the contract completion date may be extended, thereby incurring additional supervisory, management and overhead costs.

It is not difficult to anticipate that, as a consequence of a scenario such as that set out, it may not be possible to identify all the additional costs incurred in each of these bullet points to each individual variation or piece of design information. This will be particularly so where the design information that was supplied late also relates to activities that are the subject of the variations. However, some elements of additional cost should be capable of being identified to specific causes, for instance:

- Differences in the unit costs of activities as a result of the variations should usually be capable of discrete identification.
- Similarly, the cost impact, if any, on non-varied work should be capable of identification.
- The impact on supervisory and ancillary costs of changes in particular activities affected by the variations should usually be capable of identification.

However, in addition to the costs identified to individual activities, there may well be further costs incurred that are attributable to a combination of the variation and information problems. Commonly these costs will be those associated with both disruption affecting productivity and prolongation of the contract completion date. Where activities have been extended by a combination of variations and information delays it may not be possible to attribute these costs to individual causes but only to demonstrate that they are caused by the matters complained of in combination.

However, for such an approach to be convincing and viable in a formal dispute resolution situation, it will be essential to either demonstrate that none of the effects complained of are caused by any shortcomings on the contractor's part, or that any such shortcomings have been identified and the costs of those effects removed from the amounts claimed as a consequence of the variation and information problems.

If, at any time, the recipient of such a claim can demonstrate that either contractor shortcomings are relevant and have been ignored or that the claim is capable of analysis to events beyond those put forward, there will be a real prospect of the claim failing in its entirety if the financial effects of the remaining events cannot be isolated. A global claim is made as a whole and therefore runs the risk of failing as a whole.

Apart from the inherent danger that the whole claim may fail if a part of it can be shown to be suspect, there are other criticisms of the global approach in practice that need to be considered:

- In global claims effort is often concentrated into calculating and supporting the compensation or damage claimed rather than establishing entitlement by as full and detailed an analysis of cause and effect as can be undertaken in the circumstances. This 'puts the cart before the horse'. The cart is going nowhere if the horse cannot pull it.
- In some instances the causal link between the events giving rise to the claim and the compensation or damages can be completely absent. In such situations there is merely a stating of events on the one side and a quantification of damage on the other side, with no apparent link between the two. This is too superficial an approach in any case.
- This failure to establish a causal link between the events complained of and the claimed damage can extend to the lack of any linkage between the events and alleged breaches of the contract, thereby leaving the claim with no proper contractual foundation.
- It is sometimes alleged that a global approach allows the claimant to avoid its duty to prove its claim, i.e. it is able to avoid the 'he who asserts must prove' principle that is fundamental in most claim situations. In most jurisdictions that burden will remain with the claimant.
- Claimants often appear to rely on a risky global approach when, with proper advice, consideration and effort they actually do have the records to produce a properly particularised claim that will give a sufficiently better outcome to more than justify the effort and cost.

It has to be questioned, however, whether these matters are proper criticisms of the global approach to claims or simply evidence of failings to properly research and establish a global claim in circumstances where such an approach would otherwise be acceptable.

No principle of approach will allow a claimant to avoid doing all it reasonably can to establish its claim, nor should the adoption of a global approach allow a faulty or unsustainable claim to be pursued. The basis of a global claim must be established through the chain of duty – breach – cause and effect as far as possible and the global approach only adopted to provide a quantification of loss or damage where it arises from a multiplicity of causes whose effects truly cannot be separated or it is disproportionate and unnecessary to do so.

In this approach, a global claim is replaced by ones that particularise the claims as far as practicable, but leave a 'composite' or 'rolled-up' rump of the remainder against more than one event.

6.6.3   Conclusions

Terms such as 'global', 'total loss', 'total costs', 'composite' and 'rolled-up', with or without the prefix 'modified', are regularly used in relation to contractors' claims. This is done inconsistently, often indiscriminately and with their users unsure of their actual meaning. These are not terms of art and local legal advice is required, although some possibilities regarding their interpretation are set out in this section, particularly by reference to UK authorities.

Such terms are often seen as a criticism of a claim and the term 'global' is particularly used in a pejorative sense to imply that the claim lacks a nexus between cause and effect that could and should have been provided by the claimant and which unfairly prejudices the recipient of the claim, and that the claim should fall at the first hurdle for that reason.

However, the practicalities of major projects that are subject to extensive numbers of changes is that a global approach to all or parts of a contractor's claim might be the only viable method of valuation. Where this is a result of the sheer number of causative events for which the employer is liable, it may seem a harsh outcome that the employer benefits from the extent of such failures. Furthermore, it is often the case that a number of events are proven to have occurred, to have been notified, to have been the employer's liability, and to have together caused loss, costs, expense or damage to the contractor, such that the work of particularisation seems an unnecessary burden. Such separation may also only be possible on a contrived basis.

On the other hand, contractors sometimes take a shortcut to the presentation of their claims by applying a global approach. Motives can include laziness or a desire to hide 'own-goals' or causes for which the contractor has no entitlement under the contract. From the defendant's perspective, the presentation of a claim on a global basis can genuinely prejudice its ability to understand the case against it and to defend itself.

Under English law, the approach to global claims has softened in the last decade or so, and is now substantially similar to that under Scottish law. However, contractors should be wary of applying the approach without proper consideration. This starts with a consideration of the express terms of the contract and the applicable law, particularly whether they have satisfied any related requirements therein. Thereafter, making a claim entirely on a global basis may make it a 'hostage to fortune' if the asserted underlying facts or any part of it then fail in any way. Even where a tribunal considers that it can still allow something for what is left, it may take a conservative approach to such allowance. Unless the particular circumstances dictate otherwise, contractors are better advised to particularise their claims as far as is practicable, leaving a 'composite claim' or 'rolled up' claim for the remainder of their costs, losses, expenses or damages – the 'rump' that cannot be particularised.

Returning to the current approach of the SCL Protocol in its Core Principles paragraph 17:

> The not uncommon practice of contractors making composite or global claims without attempting to substantiate cause and effect is discouraged by the Protocol, despite an apparent trend for the courts to take a more lenient approach when considering global claims.

6.7   Subcontractor Costs

The incorporation into a contractor's claim against its employer of claims it has received from its subcontractors (and its suppliers) can cause a number of problems in practice. These revolve around three main areas:

1. Whether the subcontractor or supply contract is on terms that sufficiently align with those of the main contract to enable ready inclusion of an amount paid to the sub- contractor or supplier in the contractor's claim against the employer. Providing the terms of the subcontract are 'back to back' with those of the main contract as far as the relevant terms for quantification of claims is concerned there should be no real issue. However, for example, there might be some distortion between the basis of items such as subcontract rates and prices and the basis of pricing in the main con- tract, due to the time lapse between the tendering of the main works and the letting of subcontracts. If the contractor is aware of difficulties or changes that have arisen between the commencement of the works and the letting of a subcontract it will gen- erally try to procure a subcontract package on the basis of the changed circumstances, or that incorporates the effect of any difficulties that have arisen. If this has happened it may well be that some analysis of the subcontract pricing will be required to enable the subcontractor's costs to be applied on the same basis as the original main con- tract pricing, i.e. the effect of any 'bought in' changes and difficulties will need to be identified and substantiated, and taken into account in any subsequent analysis.
2. Whether an amount paid to a subcontractor or supplier is recoverable from the employer, and/or is sufficiently detailed and substantiated and/or is wholly in respect of matters that relate to the employer's failures.
3. What happens where the contractor has a claim from a subcontractor or supplier for which the employer is clearly liable but has not yet paid, and the contractor has not incurred any cost by paying the subcontractor or supplier?

Areas (2) and (3) are considered in detail as follows.

6.7.1   Third Party Settlements

It has been suggested more than once in this book that claims from subcontractors and suppliers are often of a relatively low standard of substantiation, certainly lower than the standard expected by many employer's advisors of their main contractor's claims. This can lead to difficulties in practice where a contractor seeks to pass amounts to its employer that it has paid to a subcontractor for events for which the employer is allegedly liable, for example for delay and disruption. Similar issues can arise in relation to claims against contractors from their suppliers. In the other direction, contractor's settlements with their employers sometimes lack details of amounts claimed by the employer that are the result of failures by a subcontractor or supplier, that the contractor then seeks to pass to the subcontractor or supplier that was at fault.

  In particular, the details of subcontractor and supplier settlements can sometimes also:

- Be the subject of broad settlements that appear to contain little detail.
- Wrap a number of claims or items into a single rounded lump sum.

- Contain details for much of the account, but with a final addition to get to a mutually agreed deal, often in a rounded amount, with that final addition having little or no detail at all.
- Address issues such as disruption and prolongation in their totality, without separating them into amounts related to defaults by different parties. For example, where a subcontractor's disruption claim is settled in a single amount that does not separate parts that were the consequence of failures by the employer from those parts that were 'domestic issues' for which only the contractor is responsible.
- Involve the sacrifice of claims that a party would otherwise have made against a third party, rather than the payment of amounts. Thus, for example, the settlement amount with a subcontractor includes allowance for a claim arising from employer delay or disruption, but no amount is recorded against it because it is off-set against backcharges from the contractor to the subcontractor.

A further issue arising out of party C's passing on to a contract breaker, party B, an amount that it has paid to party A for party B's breach is whether that settlement amount is the limit of the amount that can be claimed against party B, even where it is a very low settlement that is clearly much lower than the party A's true damages arising from party B's failure.

Where a recorded settlement amount appears high, it may be contended by a party to which that amount is now being claimed that its amount and description were no more than a convenient construct to manufacture an exaggerated counterclaim. For example, where a contractor that has claims from several subcontractors for delay and disruption caused by both the contractor's own bad management and events for which the employer is liable, and records the settlement of those subcontractors' claims at unusually high proportions of their claimed amounts and against descriptions such as 'delay caused by employer'. This might be accompanied by 'domestic' variations (i.e. those that cannot be passed on to the employer) being valued at nil.

In relation particularly to subcontractor claims, the issues are sometimes complicated by the use of adjudication as a form of alternative dispute resolution and the willingness of some adjudicators to allow subcontractors' claims, particularly for such difficult areas as disruption, in a poorly particularised form that would not normally be sufficient for the contractor to pass 'up the line' to an architect, contract administrator or engineer for certification. From a contractor's perspective, it might expect that a sum it has paid to a subcontractor on the basis of a decision of an adjudicator in such as a UK statutory adjudication, even when that sum is no more than his or her broad assessment of the subcontractor's entitlement, ought to be a sufficient basis for the quantum of that claim being passed on to the employer. This is on the basis that the amount has been tested by the adjudicator, with the contractor making any case for such as the quantum being exaggerated, and has become a genuine cost or liability of the contractor. However, what if the amount awarded was clearly exaggerated when compared to the subcontractors real likely losses?

For an understanding of the position on claims for the amounts of third party settlements under English law, a good place to start is the Court of Appeal decision in *Biggin & Co Ltd v Permanite Ltd* [1951] 2 KB 314. Biggin had accepted liability to the Dutch Government in the sum of £43,000 as a result of defects in the bituminous adhesive which they had supplied. That adhesive had originally been sold to Biggin by Permanite. Biggin then sought to recover the £43,000 from Permanite. The Court of Appeal concluded that the Judge had been wrong to regard the settlement between Biggin and the Dutch Government as irrelevant. It was held that if the settlement was reasonable, even if it was at the upper limit, it should be taken as the measure of Biggin's damage. Somervell LJ said that the plaintiff had to lead evidence, which could be cross-examined, as to whether or not the sum paid was reasonable, and the defendant could endeavour to demonstrate that it was not reasonable. The defendant, he said, 'might in some cases show that some vital matter had been overlooked'. Singleton LJ considered that the only issue was concerned with the reasonableness of the damages and that, if the Judge was satisfied that the damages would be somewhere around the settlement figure, he or she would be justified in awarding that figure as damages. He said:

> The question is not whether the plaintiffs acted reasonably in settling the claim, but whether the settlement was a reasonable one; and, in considering it, the court is entitled to bear in mind the fact that costs would grow every day the litigation continued. That is one reason for saying that it is sufficient for the purpose of the plaintiffs if they satisfy the Judge that somewhere around the figure of settlement would have been awarded as damages.

In *Comyn Ching & Co Ltd v Oriental Tube Co Ltd* [1979] 17 BLR 47, the employer was Queen Mary College, who engaged Minter as main contractors to build two halls of residence. Comyn Ching were nominated subcontractors and were instructed to use a particular steel piping manufactured by the Oriental Tube Company. Comyn Ching were concerned that the piping would not work and so they obtained letters of guarantee from Oriental. Comyn Ching's fears were realised when the piping failed and had to be completely replaced. Queen Mary's College pursued both Minter and Comyn Ching for damages for breach of contract. Those claims were settled by Comyn Ching, who then pursued Oriental for the sums paid out in the settlement. Oriental took the point that there could be no claim against them, because the College's original claim against Comyn Ching was hopeless, on the basis that Comyn Ching had provided precisely the pipes that they had been instructed to provide by the College. On this basis, at first instance, the claim was dismissed. However, the Court of Appeal allowed the appeal, on the basis that the letters of guarantee provided an indemnity in respect of 'claims', which were construed as meaning 'all claims having a reasonable prospect of success'. Thus, the Court of Appeal held that a loss would be sustained in consequence of a claim if it arose from a reasonable settlement of a claim that had some prospect or a significant chance of success.

In *P&O Developments Ltd v Guys and Thomas' National Health Service Trust & Ors* [1999] BLR 3, His Honour Judge Bowsher QC concluded that there were two reasons why an agreement that had been made with a person who was not a party to the action was relevant and admissible. The first was by operation of a rule of evidence, as part of the policy of the courts to encourage settlements. Thus, if a third party's claim was settled, proof of that settlement was some evidence of its true value, although it was not conclusive. The settlement set a maximum value to the claim. Secondly, Judge Bowsher concluded that the settlement was relevant pursuant to the second rule in *Hadley v Baxendale* (1854) 9 Ex 341. The reasonable settlement of claims was a matter that the parties may be held to have had in reasonable contemplation under the second limb of *Hadley v Baxendale*.

The area of third party claims that have been settled and then sought to be recovered as part of a claim to another was considered by HHJ Peter Coulson QC in *John F. Hunt Demolition Ltd v ASME Engineering Ltd* [2007] EWHC 1507 (TCC).

By a main contract, Kier (Whitehall Place) Ltd engaged Kier Build Ltd to carry out the design and construction of commercial office premises. By a subcontract Kier Build Ltd retained Hunt to demolish the existing buildings. By a sub-subcontract, Hunt engaged ASME to construct a temporary steel structure to support the existing facades of the property during the demolition works. While ASME was carrying out welding work it set light to the bitumen felt weather-proofing on the retained facades, which caught fire. It took over 45 minutes for the fire to be extinguished.

The two separate Kier companies, indicated a joint claim against Hunt for the consequences of the damage to the existing facades and the repair work that had to be carried out to them. The total amount claimed was £248,145.04. Hunt notified ASME of the claim and they jointly instructed quantity surveyors to assess the value of it. They advised that it was worth about £151,545 exclusive of interest, on a quantum-only basis, assuming a full liability to the Kier companies. Hunt offered to settle the claims for £152,500 and this offer was accepted by the Kier companies. Hunt sought to recover that amount from ASME.

Of the relevance of the amount in which Hunt had settled with the Kier companies to the quantum of its claim against ASME, the Judge said this:

> In my judgment, the settlement between Hunt and the Kier companies was either reasonable or it was not. If it was reasonable then, prima facie, Hunt can recover the £152,500. As I have said, whether or not it was reasonable will turn on the facts. However, if that settlement was not reasonable on the facts, then, prima facie, the settlement has no evidential value (see *P&O Developments Ltd v Guys and Thomas' National Health Service Trust & Ors* [1999] BLR 3). Moreover, the amount paid pursuant to an unreasonable settlement agreement would not be recoverable under the second limb of *Hadley v Baxendale*, because it would be unforeseeable; to put it another way, the payment of an unreasonable sum by Hunt to the Kier companies would break the necessary chain of causation as between ASME and the sum paid. The unreasonable settlement may well therefore become altogether irrelevant. That is what I take Goff LJ to mean when he said in *Comyn Ching & Co Ltd v Oriental Tube Co Ltd* ([1979] 17 BLR 47) that the settlement was 'either good or bad': if it was bad, it seems to me that it cannot be relied on at all.

This passage from the judgment in *Hunt v ASME* was quoted by Mr Justice Akenhead in *AXA Insurance UK PLC v Cunningham Lindsey United Kingdom* [2007] EWHC 3023. There Mr Justice Akenhead continued:

> I draw from that case and the cases quoted with approval in it that:
>
> (a) If there is no effective causal link between the breaches of duty of the defendant and the need for the claimant to enter into the settlement with a third party or the payment of the sums pursuant to the settlement agreement, there will be no liability to pay the settlement sums irrespective of whether the settlement was reasonable.

(b) The onus of proof in establishing the reasonableness of the settlement is upon the claimant. Thus, there must be some reliable evidence for the court to conclude that it was a reasonable settlement.

(c) The mere fact that the claimant is not liable to the third party either at all or for all the sums payable pursuant to the settlement is not necessarily a bar to recovery or to the establishment of the reasonableness of the settlement. However, the fact that the claimant was not liable to the third party either at all or for anything approaching the sums payable may be a factor in determining that the settlement was unreasonable.

(d) Where a settlement is not established as reasonable, it is still open to the claimant to recover from the culpable defendant elements of the sums paid pursuant to the settlement to the third party to the extent that it can be proved that there is an effective causal link between the payment of those sums and the established breaches of duty. In those circumstances, it is legitimate for the court to consider and establish what was likely to have been payable as a matter of fact and law to the third party as the foreseeable result of the defendant's breaches.

In *Siemens Building Technologies FE Limited v Supershield Limited* [2010] BLR 145, Mr Justice Ramsey reviewed many of the relevant authorities and concluded that:

> In my judgment the following principles can, in summary, be derived from the authorities:
>
> (1) For C to be liable to A in respect of A's liability to B which was the subject of a settlement it is not necessary for A to prove on the balance of probabilities that A was or would have been liable to B or that A was or would have been liable for the amount of the settlement.
>
> (2) For C to be liable to A in respect of the settlement, A must show that the specified eventuality (in the case of an indemnity given by C to A) or the breach of contract (in the case of a breach of contract between C and A) has caused the loss incurred in satisfying the settlement in the manner set out in the indemnity or as required for causation of damages and that the loss was within the loss covered by the indemnity or the damages were not too remote.
>
> (3) Unless the claim is of sufficient strength reasonably to justify a settlement and the amount paid in settlement is reasonable having regard to the strength of the claim, it cannot be shown that the loss has been caused by the relevant eventuality or breach of contract. In assessing the strength of the claim, unless the claim is so weak that no reasonable party would take it sufficiently seriously to negotiate any settlement involving payment, it cannot be said that the loss attributable to a reasonable settlement was not caused by the eventuality or the breach.
>
> (4) In general if, when a party is in breach of contract, a claim by a third party is in the reasonable contemplation of the parties as a probable result of the breach, then it will generally also be in the reasonable contemplation of the parties that there might be a reasonable settlement of any such claim by the other party.

(5) The test of whether the amount paid in settlement was reasonable is whether the settlement was, in all the circumstances, within the range of settlements which reasonable people in the position of the settling party might have made. Such circumstances will generally include:
   (a) The strength of the claim;
   (b) Whether the settlement was the result of legal advice;
   (c) The uncertainties and expenses of litigation;
   (d) The benefits of settling the case rather than disputing it.
(6) The question of whether a settlement was reasonable is to be assessed at the date of the settlement when necessarily the issues between A and B remained unresolved.

This judgment was in the context of a claim for damages for breach of contract. However, in *Walter Lilly & Company Limited v Mackay & Anor* [2012] EWHC 1773 (TCC) (11 July 2012), Mr Justice Akenhead considered that there is nothing to distinguish it in practice from a claim for cost incurred as the result of a settlement under what was then clause 26 JCT Standard Form of Building Contract 1998 Edition Private Without Quantities, and would now be SBC/Q clause 4.23 'Loss and expense'.

The facts in *Lilly v Mackay* have been set out in detail elsewhere. In relation to claims from Walter Lilly's subcontractors, these included claims from its mechanical and electrical engineering subcontractor Norstead. It made claims against Walter Lilly including headings of additional preliminaries and plant, additional labour costs for reduced productivity, under-recovery of overheads and costs escalation. Walter Lilly settled those claims with Norstead and the issue was whether and to what extent the amounts paid to Norstead under that settlement were recoverable from Mackay. The Judge quoted the passages from *AXA v Cunningham* and *Siemens v Supashield*. With reference to the former case, he said:

> It is open to the Court in appropriate circumstances to make an apportionment of the settlement sum if and to the extent that it can be confident that the sum allowed represents a realistic and reasonable allowance which can safely be attributed to the matters for which the defending party is liable.

Mackay's primary position, based on the evidence of its quantum expert, was that nothing was due in relation to Norstead's claim because it was not supported, but that 'if one is to proceed by way of a best estimate some £164,000 is due which includes nothing for the disruption or loss of productivity'. Of this primary position, Mr Justice Akenhead said:

> In my judgement the primary position is simply unrealistic. It is beyond doubt that Norstead was and must have been very substantially delayed and disrupted not only by the simple fact that it was on the site working for 99 weeks longer than it had anticipated (a 236% increase on the agreed sub-contract period) but also by the unavoidably disrupted nature of its own and all the other work.

Of Mackay's quantum expert's evidence, he said:

> His analysis of the loss of productivity or labour disruption claim ignores the overwhelming inference (if nothing else) that there must have been very substantial disruption and loss of productivity for which Norstead is unlikely to have been reimbursed under the measured or variation part of the evaluation.

He concluded:

> In my judgement, the settlement achieved with Norstead was a reasonable one in all the circumstances. WLC was faced with a frustrated and increasingly aggressive sub-contractor which, by and large, had right on its side. It had been seriously delayed for reasons which entitled it to a full extension of time and it was and must have been obvious to WLC that there was a probability that Norstead would recover not only the extended preliminary costs for such delay but also some compensation for disruption and escalation in costs. It is likely that this was the best settlement available and in reaching that view I take into account the fact that this settlement was achieved before there had been any recognition by DMW or its advisers that any further extension beyond February 2007 was due; it was in WLC's interests to keep the settlement as low as it could achieve given that pending the likely future litigation it would have to pay the unpaid element of the settlement to Norstead with no certainty that it would be recovered from DMW. WLC was put in a position in which it faced a substantial and broadly meritorious claim which it was reasonable to settle. The need to settle with those parties was caused by delay and disruption caused by DMW and the settlement fell well within the 'reasonable range of settlement'.
>
> Whilst it is reasonable to take as the starting point the settlement effectively achieved in relation to the delay and disruption cost and loss to Norstead, there is or may be a residual uncertainty as to whether there was an exact or 100% correlation in terms of all the factors attributable to DMW which delayed and disrupted WLC and Norstead and other factors which delayed and disrupted Norstead; for instance, there may be factors which have not been pleaded which disrupted Norstead. In my judgement, it would be safer to allow some amount off the full amount of the settlement and, in seeking to allow an amount which the Court can be confident directly relates to no less than Norstead was entitled in relation to the factors attributable to DMW and for which WLC is entitled to compensation, I fix that sum at £300,000.

The case of *Fluor v Shanghai Zhenhua Heavy Industry Co Ltd* [2018] EWHC 1 (TCC) arose out of the construction of 140 wind turbine generators which were to be installed in the North Sea. Fluor had contracted with an offshore windfarm developer ('GGOWL') to engineer, procure and construct the foundations and structures to support the turbine generators. Fluor engaged the defendant ('ZPMC') to fabricate parts of the structures. However, on testing, ZPMC's work was found to include defective welds. The resulting costs to Fluor included amounts arising out of a settlement agreement it reached of its account with GGOWL. This included £32.325 million against a heading of 'Counter-claim Damages'.

By reference to the judgment in *Biggin v Permatite*, Sir Antony Edwards-Stuart summarised the legal position as follows:

> It is settled law that, in principle, C can recover from a contract breaker, B, sums that it has paid to A in settlement of a claim made by A against C in respect of loss caused by B's breach of its contract with C.
>
> However, C's settlement with A must be an objectively reasonable settlement and, if it is, that sum represents the measure of C's damages in respect of B's breach of contract (assuming that there were no other heads of loss). Even if C can show that its settlement with A was at an undervalue, the settlement sum still represents a ceiling on the amount that it can recover from B.

However, the Judge noted how the position was relatively straightforward in *Biggin v Permatite*, in that it was a claim for damages involving a single head of claim being passed down the contractual chain, but that:

> The position is more complicated where several heads of claim have been settled between, say, an employer and a main contractor but where the defendant subcontractor is alleged to have been responsible for only one of them.

He explained that this type of more complicated situation had been considered by His Honour Judge Thornton QC in *Bovis Lend Lease v R&D Fire Protection* (2003) 89 Con LR 169. Bovis had settled a claim brought by its employer that involved numerous claims and cross claims. Among other claims for failure by the defendant, Bovis sought to recover amounts settled with its employer. However, Bovis produced no evidence of the circumstances surrounding the settlement or as to the apportionment or breakdown of the sum that it had paid. The court held that in a multi-party or a multi-issue dispute, the court would have to decide (if it could) what part of the overall settlement was attributable to the relevant breaches of contract of a particular defendant. Any such allocation had to be reasonable and would then represent a ceiling on the damages payable by the relevant defendant. Bovis took the position that it was entitled to recover the costs of remedying the defective work irrespective of the settlement. Further, it took this position in circumstances where there was no evidence that the defective work in question would be remedied. The court held that, so long as Bovis was contending that it suffered a loss by entering into the settlement caused by breaches of contract by the defendants, the burden was on Bovis to establish what that financial detriment was.

In *Fluor v Shanghai*, Sir Antony Edwards-Stuart noted the case before him was similar to that in *Bovis v R&D*, but that, whilst Fluor had explained how the settlement sum should be apportioned, ZPMC did not accept the explanation. The Judge set out the terms of the settlement agreement, including that: 'Fluor shall pay GGOWL the sum of £32,325,000 (thirty two million, three hundred and twenty-five thousand pounds) (the "Counter Claim Damages")'; GGOWL would assume responsibility for the works and any outstanding works; and the parties would deliver to each other an 'Extended Warranty Deed'. He then set out in detail the history of the negotiations that led to the settlement. He concluded that the settlement included £10 million for historic weld dispute costs and £3.825 million as compensation for a reduced warranty, both of which were claimable as arising from the defects in ZPMC's welds. As to whether this was reasonable, he concluded that:

> Overall, the settlement with GGOWL was the result of hard fought and pro-tracted negotiations in which each side had the benefit of informed legal advice. None, or at least very few, of the claims to which a value had been attributed by either party was so weak as not to be taken seriously. The analysis that I have car-ried out shows that the reasonable settlement value of GGOWL's claims was very close to the sum paid or foregone by Fluor, so on that basis alone that aspect of the settlement was self-evidently reasonable.
>
> Taking all the circumstances into account, in particular the inclusion of the war-ranty, I have no hesitation in concluding that the settlement with GGOWL was a reasonable one. Accordingly, Fluor is entitled to recover damages in the sum of £13.825 million on account of the settlement.

In conclusion of this review of the law in England, it is suggested that, when settling an account, all or part of which it is intended to claim from a third party as damages for its breach of its contract, a party might consider the following, particularly if the amounts are significant:

- Taking local legal advice on such an approach, including on the terms of the settle-ment agreement.
- Recording and agreeing the settlement on the basis of details that ensure that the individual amounts intended to be passed on are readily separately identifiable.
- Where it is impossible to separately identify an individual amount, because it is part of a composite settlement or a composite amount added in order to close a settlement, recording what that composite amount was for and why that settlement, or that part of the settlement, was made on that basis.
- Recording and keeping records of the negotiations.
- Ensuring that the settlement can be shown to have been objectively reasonable when it was entered into.
- That the burden of proving it was reasonable is likely to lie with the claimant.
- That factors that might be relevant to assessing that reasonableness are likely to include:
  - the strength of the claim;
  - whether the settlement was the result of legal advice;
  - the uncertainties and expenses of litigation; and
  - the benefits of settling the case rather than disputing it.
- Taking expert quantum advice on the reasonableness of the quantum before it is agreed.
- That the amount of the settlement is likely to set the upper limit of what can be passed on as a claim.

A further thought is whether it is possible to get a party to a settlement agreement to commit, as part of that agreement, to assisting with the paying party's future attempts to pass on amounts paid or waived as part of the agreement to a third party. This might have the potential to be particularly helpful if the claim against the third party proceeds to formal dispute proceedings and then a witness can be called who is not an employee of either party to the proceedings but can give independent evidence as to how an amount in a settlement agreement was established and what events or issues it was for.

6.7.2 Unsettled Third Party Claims

The modern trend, particularly in Europe, towards contractors employing subcontractors to carry out much of the works, rather than employing their own payroll labour, has increasingly placed contractors effectively in the position of management contractor, paid by the employer to manage subcontractors and in turn paying those subcontractors to do the work. The lack of detail in many subcontractors' financial claims, mentioned elsewhere, leads many contractors to complain that they find themselves 'between a rock and a hard place' when failures by the employer lead to claims from those subcontractors against the contractor as their only contractual route. In such circumstances the 'rock' will be a subcontractor that has made a claim for additional payment and the 'hard place' will be an employer's certifier who refuses to include in a payment certificate amounts against such a claim because it says the subcontractor has not adequately substantiated it. This might be in terms of substantiating its contractual basis or its quantification or both.

Where the subcontract does not include an effective 'pay-if-paid', 'pay-when-paid' (which were prohibited by section 113 of the Housing Grants, Construction and Regeneration (HGCR) Act 1996, from United Kingdom construction contracts as defined by the Act), 'pass-through' or similar provision, then the reaction to such a contractor's complaint might be that this is the nature of its role and the law of privity of contract. On this basis, critics will say that the contractor is being paid to manage its subcontractors, including related contractual and financial issues. If the claim is inadequate, then it should address this with the subcontractor. If it is adequate, it should address this with the certifier. The danger is that if it is only the subcontractor that is out of pocket, then the contractor has little incentive to do the latter. This is one of the reasons pay-when-paid clauses were prohibited in the UK and some other countries.

The position of the contractor can become more difficult where the subcontractor is particularly aggressive and is threatening to refer to claim to third party resolution and where the claim is large. Mention has been made elsewhere of concerns among some contractors that some adjudicators are too lenient on poorly presented subcontractor claims. However, it has also explained how this might give the contractor the benefit of a crystallised claim and quantum that it can pursue from the employer. A further factor in this may be the cost to the contractor of third party resolution, both in terms of legal fees and a drain and distraction on its project management and administration resources.

However difficult the situation that contractors might consider such circumstances put them in, it is suggested that the contractor's approach should be based on its commercial analysis and decision as to the merits and reasonable value of the claim. This might mean either:

1. Concluding that the subcontractor is right and that the claim is worth rather more than the certifier's valuation of it. In this case, it is suggested that the contractor should:
   - Pay the subcontractor what the contractor considers to be the reasonable value of the claim.

- Tell the certifier and employer of the resulting financial shortfall.
- Give notice under the contract that the contractor is incurring interest or finance charges on the amount by which it is out of pocket as a result of the under-certification, which it will claim from the employer. However, see in this regard Section 6.9 of this book.
- Consider its own dispute resolution proceedings against the employer under their contract.
2. Concluding that the certifier is right and that the claim is worth rather less than the subcontractor's claimed amount of it. In this case, it is suggested that the contractor should:
   - Pay the subcontractor what the contractor considers to be the reasonable value of the claim.
   - Explain in detail to the subcontractor the reasons for its lower valuation and that the employer's certifier has reached similar conclusions.
   - Establish with the employer's certifier exactly what further substantiation it requires.
   - Advise the subcontractor of what further substantiation they both require.

The wording of the definitions of compensable cost in the FIDIC Red Book and the Infrastructure Conditions as including those that are yet 'to be incurred' by the contractor can be particularly helpful to contractors where they have claims from subcontractors or suppliers that are a clear liability but not yet payable.

6.8 Off-Site Overheads and Profit

Contractors commonly include amounts for both off-site overheads and profit in the sums they present as claims for compensation for the time consequences of changes. The former of these is sometimes referred to as head office costs or home office costs. Sometimes 'administration' is added in terms such as 'head office administration and overheads costs'. It is distinct from site overheads (also referred to as field overheads or preliminaries and general costs) considered in Section 6.2. Combined allowances for both head office costs and profit are commonly referred to as gross margin. It is that gross margin that goes to pay for the contractor's head office costs, with what is left after those costs going to profit.

The approaches to calculation of an overheads and profit claim reflect the manner in which items such as delay and disruption can potentially affect them. These are:

- Additional costs incurred in relation to the head office where these provide additional support during periods of such as delays and disruption; or
- Loss of contribution to the contractor's gross margin where those of the contractor's resources whose role is to provide such a contribution are prevented from being available for other work on which they would have provided such a contribution.

Contractors apply a variety of methods to the calculation of these claims. It is not uncommon for head office overheads to be included in a claim on a percentage basis related to the quantum of other heads of claim or as a daily or weekly sum calculated pro rata to the contract sum allowances for off-site overheads. Profit is often claimed on a similar basis, often with the two heads being presented as a single calculation for 'head office overheads and profit' or 'gross margin'. Sometimes this is on the basis that a contribution to them has been lost and, sometimes, as an item of costs or expense.

Typical employer responses will include 'prove to me that your overheads increased' and 'you are not entitled to profit on a claim'.

However, the basis for claiming these items, and the most appropriate method of calculation, need to be considered in some detail. What the contract says, and what the applicable law allows as a head of a particular type of claim, are first considerations. Essential to these questions will be the basis on which the claim is made. There are two alternatives:

- By reference to an express term of the contract. In that case it may be a contractual claim for such as 'direct loss/and or expense' under JCT terms or for 'cost' or 'expenditure' under the FIDIC Red Book or Infrastructure Conditions terms.
- Alternatively, it may be a claim for damages for breach of contract and therefore subject to local legal principles for the quantification of damages for breach. As explained elsewhere, this usually means such an amount as will put the contractor into the position it would have been in 'but for' the delay or disruption. Usually this means expenses or costs, but might also include losses.

These alternatives have different implications for head office overheads and profit, so the two are considered separately as follows.

6.8.1   Off-Site Overheads

Taking in turn each of the three standard forms of contract considered in this book, the scope and basis for claiming head office overheads in the evaluation of the time consequences change are as follows:

- Under SBC/Q clause 4.23, for 'matters affecting its regular progress', the contractor's entitlement is stated as being to 'loss and/or expense'. On this basis, it would seem that head office overheads could fall to be claimed either on the basis that the contractor has lost a contribution to them, or that it has incurred additional expense in relation to them.
- In the FIDIC Red Book, various clauses set out the contractor's entitlement to 'Cost', a term defined in clause 1.1.4.3 as follows, to which emphasis has been added:

> … all expenditure reasonably incurred (or to be incurred) by the Contractor, whether *on or off the Site, including overhead* and similar charges, but does not include profit.

This would clearly include head office overheads, but on the basis of expenditure, rather than a loss of contribution.

- The Infrastructure Conditions take a similar approach to the FIDIC Red Book, in that various clauses set out the contractor's entitlement to 'cost', a term defined in clause 1(5) as follows, to which emphasis has been added:

> … all expenditure properly incurred or to be incurred whether on or off the Site including overhead finance and other charges properly allocatable thereto but does not include any allowance for profit.

Like the FIDIC Red Book, this would clearly include head office overheads, again on the basis of expenditure, rather than a loss of contribution.

Where the claim is made as one for damages for breach of contract, it has been explained elsewhere how the quantum of such a claim would usually be that amount which would put the contractor back in the position it would have been in had the breach(es) of contract not been committed by the employer. It is possible to envisage that this might mean a claim for head office overheads on the basis of additional costs or as one for loss of opportunity to earn a contribution to them. As with the alternatives for the standard forms, which alternative applies will depend on the contractor's circumstances in the case and the facts that the contractor will have to establish. Essentially this will mean either establishing that it did lose a contribution to its head office costs or that it did incur additional expenditure on head office overheads.

Applying these alternatives, the contractor's claim for either expenditure or contribution in relation to head office overheads might be as follows.

6.8.1.1   Off-Site Overheads on an Expenditure Basis

Where calculated on an expenditure basis, contractors have commonly presented claims for off-site, or head office, overheads as a percentage based on a simple analysis of the company accounts for the period in question. This will start by setting out: the company's turnover; administration and head office overheads costs; other costs; and its profit. For example:

Total turnover for the financial year	$24,000,000.00
Less total of costs of construction activities	$20,500,000.00
Less total of administration and head office costs	$2,250,000.00
Profit	$1,250,000.00

Where the claim spans a number of years, the calculation may set out the equivalent amounts for all those years and be based on their average. Copies of the audited accounts will be exhibited to support the figures. The off-site overhead costs may then be expressed as a percentage of other costs as follows:

$$\$2,250,000.00/\$20,500,000.00 = 10.9756\%$$

That percentage will then be applied to the quantum of all other heads of claim, such as prolonged preliminaries and general costs, disrupted labour costs or thickened staff costs, to calculate the claim for head office overheads. In this element, this approach has an advantage over the formula approaches considered elsewhere in this section, which are based on time and only serve a prolongation claim.

This approach is attractive in its relative simplicity, but has the major disadvantage that the contractor is recovering monies based on an all-encompassing calculation, which assumes that additional overhead is incurred at the same rate as the total overhead for the year. Some consideration of the offsite activities usually encompassed by the administration and head office costs may lead to the conclusion that the assumption is not a reasonable one as the costs may include matters such as:

- The contractor's estimating and tendering department, which is unlikely to have any further input to the contract after it has been procured, so will not have been increased.

- The contractor's head office contract management and quantity surveying function, which may, or may not, have any greater involvement in the project as a result of the events and causes of compensation being considered.
- The contractor's human resources and payroll department which may, or may not, incur any greater level of expenditure as a result of the events on site, even where further resources are employed on site.
- The fixed costs of such as rental of a head office building, which will not be increased because the costs or time on one project increase.

There are other possible examples, but this is sufficient to illustrate the problem with an all-encompassing percentage calculation covering all of the contractor's head office overheads.

The reservation as to the applicability of a percentage addition to recover head office management time has been demonstrated in the courts in the case of *Tate & Lyle Food Distribution Ltd v Greater London Council* [1982] 1 WLR 149. In this case, which involved the need for dredging operations at the plaintiff's jetties in the Thames River as a result of piling activities by the Greater London Council, a claim for management time was advanced on the basis of a percentage addition. The Judge stated:

> I think there is evidence that managerial time was in fact expended on dealing with remedial measures. There was a whole series of meetings, in which the plaintiff's top managerial people took a leading role, the object of which was to find out what could be done to remedy the situation and to persuade the defendants to do something about it. In addition, while there is no evidence about the extent of the disruption caused, it is clear that there must have been a great deal of managerial time involved in dealing with the dredging required and in rearranging berthing schedules and so on .... I have no doubt that the expenditure of managerial time in remedying an actionable wrong done to a trading concern can properly form the subject matter of a head of special damage. In a case such as this it would be wholly unrealistic to assume that no such additional managerial time was in fact expended. I would also accept that it must be extremely difficult to quantify. But modern office arrangements permit of the recording of the time spent by managerial staff on particular projects. I do not believe that it would have been impossible for the plaintiffs in this case to have kept some record to show the extent to which their trading routine was disturbed by the necessity for continual dredging sessions. In the absence of any evidence about the extent to which this occurred the only suggestion [Counsel] can make is that I should follow Admiralty practice and award a percentage on the total damages. But what percentage? .... While I am satisfied that this head of damage can properly be claimed, I am not prepared to advance into an area of pure speculation when it comes to quantum. I feel bound to hold that the plaintiffs have failed to prove that any sum is due under this head.

It is notable how the Judge considered that contractors' managerial staff ought to be able to record their time spent on each project. Contractors may argue as to the practicality of this. Even where there is no such blanket requirement on staff, it is often said that they ought to be required to record time spent on a delayed project as one on which they are incurring unusual levels of involvement, as the basis for the claiming of compensation for their additional time. A problem with that is that it may not be apparent that compensable delays are being incurred. In relation to a prolongation claim, this is particularly true given the explanation elsewhere that what is usually to be compensated is costs incurred at the time that delays occurred, rather than in an overrun period after the original completion date.

The approach in *Tate & Lyle* was adopted by His Honour Judge Humphrey Lloyd in *Babcock Energy Ltd v Lodge Sturtevant Limited* 1994 CILL 981. He allowed Babcock to recover as a head of damage the time and cost incurred by certain members of its staff in dealing with the problems created by the defendant as a sum calculated on the basis of constructive charging rates applied to the total number of hours recorded for those staff at the material time.

The parallels between the situation in *Tate and Lyle* and that of a contractor whose senior management and other head office resources become greatly involved in resolving issues on a contract where there are substantial delays and disruption is not difficult to understand. The Judge's remarks, however, make clear that the application of blanket percentages may not be acceptable where one is left to speculate what the appropriate percentage is. For the reasons set out, that would seem reasonable in the context of the organisation of a construction company. What then is the way forward for such off-site costs if they are to be recovered on the basis of expenditure?

It seems sensible to consider the costs incurred by the contractor in its off-site management and administration in two tranches. The first tranche would include construction-related staff and services such as contract management, quantity surveying, buying and subcontractor procurement, etc. These are the types of head office resources whose involvement with a project is likely to vary with the amount of work and time that project continues. These are sometimes referred to as 'absorbed' or 'dedicated' or 'additional' overheads. In principle, as suggested by the Judge in *Tate & Lyle*, there is no reason why these personnel cannot record the additional time they expend in resolving the issues on-site. This would leave only the agreement of applicable rates and any dispute as to what was additional to their anticipated input to be resolved.

Rarely, the additional involvement of head office staff on a delayed or disrupted project can be isolated by reference to the employment of an additional person specifically engaged because of the problems of that project. In such a case, the contractor must clearly record that the employee is being appointed for that reason and no other. In practice, however, contractors can usually service even the most delayed and disrupted projects with their existing head office staff. This said, in one instance an electrical services contractor had a manager at head office controlling several projects, with the intention that when the largest of them finished he would move to be resident manager of a new large contract. Delays to one of the existing projects meant that he had to be retained at head office with an increasing roll on that problem project, and the new contract required the recruitment of a new manager. Thankfully, the evidence for the factual link between the problem project and additional costs on the new contract was very well maintained and evidenced by the subcontractor in that case.

The second tranche of head office costs would include the general costs of the off-site establishment such as head office rent, heating, lighting, etc., together with the general administration staff such as human resources and payroll and accountancy functions, and the construction-related activities not involved in resolving matters on-site, such as the estimating and tendering department. These costs can be considered as the general overhead as opposed to the construction activity overhead, and are sometimes referred to as the 'unabsorbed' overhead, presumably because they are not absorbed into the construction costs. These are the types of head office resources whose involvement with a project does not vary with the amount of work and time that project continues.

The problem with the bulk, if not all, of the unabsorbed overhead is that it can usually be argued that it would be incurred regardless of any delay or disruption on the contract under consideration. For instance, the contractor may well own their offices or hold them on a long lease, in which case costs in respect of the building would continue regardless of the circumstances on a particular contract, and would not increase just because one project was delayed or disrupted.

The judgment in *Fluor v Shanghai Zhenhua Heavy Industry C Ltd* [2018] EWHC 490 (TCC) was part of long-running proceedings between these parties. The earlier quantum judgment has been discussed in Section 6.7. Fluor had mentioned, but not detailed, a claim for overheads and profit. In his quantum judgment, Sir Antony Edwards-Stuart directed that Fluor was to identify the sums in the quantum judgment to which overheads and profit addition were to apply, the basis for doing so and the amounts claimed. In the event Fluor did not claim profit. He summarised Fluor's submissions in relation to head office overheads as follows:

(a) whenever a company incurs a direct cost, it does not do so in a vacuum, at no cost to itself: it generally incurs head office costs – commonly known as overheads – in doing so;

(b) that overhead cost is just as much a cost of the relevant event as the direct costs; and

(c) it is a cost which should be directly recoverable.

ZMPC accepted that overheads could be claimed as a matter of principle, but submitted that the circumstances in which a contractor could do so were not present in this case. It referred to passages from Emden's Construction Law and asserted that:

… the contractor must show that, had it not been for the breach of contract, its labour force would otherwise have been profitably employed on other work thereby making a contribution to the contractor's fixed overheads (such as head office costs). The loss of the opportunity to recover that contribution to its overheads can be a legitimate head of claim.

The Judge considered that Fluor's submission was incorrect:

A contractor's head office costs, for example, are fixed – in that, in principle, they are incurred at roughly the same rate every month. When a contractor submits a tender to carry out a contract during a specified period, it will include an allowance for its profit and its fixed costs or overheads. It may happen that, whilst the contract duration is not extended, the contractor has to use more plant than he expected as a result of circumstances for which liability rests with the employer. In that situation the contractor is entitled to claim the direct costs of hiring the additional plant, but his fixed costs – such as the costs of maintaining his head office – will have remained unchanged and so there can be no claim in respect of them.

But if the effect of the employer's breach of contract is to delay completion, so that the contractor remains on-site for, say, an additional three months, the contractor may have a claim for a contribution to his fixed costs for the extended period if it can show that it would have been able to redeploy its labour force on other profitable work during the period of delay. This is because the contractor would have been deprived of the opportunity to earn, not only profit, but a contribution to its fixed costs during that period.

However, in addition to such a loss of opportunity type of overheads claim, the Judge identified a further potential basis:

> This is where his head office overheads have been 'thickened' during the period of the contract. I gave an example of how this could happen during the course of argument. Suppose that in its head office a contractor employed six accountants, whose activities concerned the administration of the business generally and were not project related. As a result of the substantial extra administration required to deal with problems caused by a sub-contractor's breach of contract, such as those with Shipment No 1 at Vlissingen, the company's most experienced accountant, Mr Cruncher, is required to spend half his time dealing with those problems. However, fortunately his five colleagues are not overworked and are able to deal with the balance of Mr Cruncher's work in their ordinary working time. But when further problems arise – such as those with Shipment No 2 – the contractor decides that Mr Cruncher will be required to spend all his time dealing with those problems. Now his colleagues are no longer able to cope with the balance of his work and so a new accountant, Mr Bean-Counter, is taken on. Being inexperienced, Mr Bean-Counter is assigned relatively menial tasks, none of which has anything to do with the breaches of contract. In those circumstances, in spite of the fact that Mr Bean-Counter was not involved with the problems caused by the breaches, it seems to me that the cost of employing him would have been a direct consequence of those breaches and would therefore be recoverable as damages. I did not understand Mr White [Counsel] to dissent from this. It represents a 'thickening' of the head office costs.

This approach to a 'thickening' claim has been explained elsewhere in this chapter in relation to on-site supervision and management. Sir Antony Edwards-Stuart noted that Fluor had not advanced its claim on such a basis. Fluor's calculation of its overheads claim was by the addition of 4% to its other heads of claim. This had been calculated by one of Fluor's managers based on its company overheads and turnover. Of this approach, the Judge said:

> … it seems to me that it forms no basis for a claim against ZPMC in respect of increased head office overheads. The figure of 4% is simply a ratio of one set of costs against another: it tells one nothing about the extent to which the former was increased as a result of breaches of contract by ZPMC.

Like Forbes J in the Tate & Lyle case, I suspect that Fluor may well have incurred increased head overhead costs, such as telephone bills or IT costs, as a result of the breaches of contract by ZPMC, but I am not prepared to pluck a figure out of the air. I appreciate that it may well be very difficult for Fluor to put forward a precise figure, but in the absence of any figure there is little that the court can do. For these reasons, I consider that Fluor has failed to establish the facts necessary to support any claim for overheads.

It is because of this difficulty the courts have had to consider the recovery of unabsorbed overheads in some detail, particularly as a head of a claim for prolongation.

6.8.1.2   The Time and Cost of Absorbed Overheads

As an alternative to, or as part of, a claim for overheads it is common to see claims for the cost of identified management time expended in dealing with the problem that is the cause of the claim. Even where such head office staff are established as having additional involvement on a project because it was delayed and/or disrupted, such claims are often resisted by the response that the claimant would have employed the staff concerned in any event and therefore no additional cost has been incurred.

Such a claim was made in *Bridge UK Com Ltd (t/a Bridge Communications) v Abbey Pynford PLC* [2007] EWHC 728 (TCC), when the claimant included the cost of management time expended in connection with resolving difficulties caused by the construction of a concrete base that proved to be inadequate for the printing press sited on it. The court concluded that the correct approach to the recovery of management time was that set out in the judgment in *R & V Versicherung AG v Risk Insurance and Reinsurance Solutions SA* [2006] EWHC 42 (Comm), where it was decided that a company can recover damages in relation to the diversion of internal resources even if it cannot be demonstrated that the company has lost any income as a result of the diversion. The decision suggests that if no loss of income has occurred the claimant has to show that there has been significant disruption to the running of the business and that the time claimed as 'wasted' has been as a result of diverting staff from their normal duties to deal with the subject matter of the claim, i.e. a causal link will need to be established. The staff involved will also need to keep accurate notes on the time spent, activities undertaken and whether any other resources were used.

As far as quantification of the loss is concerned, a charge-out rate for the staff can be used if such exists and can be established; otherwise the claimant is entitled to recover cost of salary plus demonstrable overhead, providing of course there is no duplication with other heads of claim.

For a full discussion of claims for in-house management time and particularly the judgment in *the Liverpool Museums* case, see section 8.8.4.

6.8.1.3   Lost Contribution to Unabsorbed Overhead

A common basis for seeking compensation of the unabsorbed element of head office costs on a project that has been delayed is that the overhead was not allowed for in the contract for the period of any extension to the contract period and cannot be recovered elsewhere as the contractor is unable to take on other work where this overhead would be recovered because its resources are tied up on the subject contract because of compensable delays.

This type of claim was considered in *J.F. Finnegan Ltd v Sheffield City Council* (1988) 43 BLR 124, in which the Judge stated:

> The defendants maintain that the plaintiff's claim is purely speculative and bears no relation to the actual loss. That is true, in one sense, but this claim for overheads during the period of overrun is not related to actual loss but is assessed by allowing the contractor's average yearly percentage for overheads on turnover for the period of overrun, provided of course that expenditure on overheads was incurred in that period. It is a notional figure in the sense that it applies the average values of the company's working figures to the period of the overrun, as if the company had been able to deploy its work-force on another contract during that period.
>
> .... The plaintiffs do not seek to recover their actual overheads expenditure for the overrun period but to substitute the funding of overheads which a notional other contract would have provided.
>
> However, I confess that I consider the plaintiff's method of calculation of the overheads on the basis of a notional contract valued by uplifting the value of the direct cost by [a constant] as being too speculative and I infinitely prefer the 'Hudson' formula ....

The Judge accepted that a contractor delayed on a project due to compensable delays might have a claim for head office overheads on the basis that the workforce, tied up by delay and prolongation on the contract under examination, could have worked on another contract and recovered a contribution to those costs. However, in this case the contractor's approach was considered to be too speculative and a formula approach was preferred. The subject of the adoption of formulae to address the calculation of applicable overhead, and the various alternative formulae, is considered in detail elsewhere in this chapter, together with a number of other judgments of the courts that bring us up to date in terms of the approach of English law to the use of formulae.

6.8.1.4 Management Charges

In the same way that companies within a group can sometimes be funded centrally and financing charges applied at rates decided and charged by the holding company, it is not unusual to find member companies within a group of companies paying management charges to the centre. It is important that such charges, if they occur, are identified separately to, and not included with, any other head of claim, such as for finance charges, to avoid any overlap.

Such management charges may genuinely be a charge for services provided centrally, or may simply be a way of subsidiaries returning profit or contribution to the centre for taxation and other purposes. On occasions these charges may include royalty or licence payments, particularly in companies involved in the process plant construction industry.

Depending on the basis of calculation, such charges to a subsidiary might be increased by such as delay or disruption to one of its projects, perhaps because they are calculated based on the passage of time and/or increase in turnover. The issue as to whether this is an allowable head of claim may start with considerations of remoteness and foreseeability. As to whether they are real cost, questions will include the nature of the relationship between the parties to the arrangement and whether they are 'at arm's length' and separate financial and legal entities. The key is often to establish that the costs are for a real service provided by the centre, which would otherwise have had to be provided by the local business itself and this leads into the approach in some construction and engineering joint ventures.

When major projects are carried out by a joint venture company, set up and shared by several companies, it is generally because they each cannot viably bid for those projects on their own. Each partner brings something to the joint venture vehicle that they individually do not have. For example, an international joint venture for the construction and concession operation of a new railway might comprise: a local civil engineering contractor; a train rolling stock manufacturer; a company with experience of running a railway system; and a foreign contractor with experience in railway design and construction. It may then be that during the construction and concession periods the different parties charge the joint venture company for the skills they are bringing during that phase. For example, during construction, the joint venture company might establish its own management and administration set up for the project, but pay the foreign contractor a fee for its particular expertise in project and risk management of the design and construction of such projects. This might be termed a 'Leadership Fee' and may be calculated as a percentage of the construction contract price. Where claims arise against the joint venture's client, and these increase the calculation of the leadership fee, allowance will be made in the valuation of the claims for that fee as, effectively, a part of the joint venture's overhead costs. The question for the recipient of such a claim will be whether that is a genuine part of the joint venture's costs or just a way of adding further profit for one partner. If the facts are that the fee is for the provision of a genuine service that would otherwise have to be provided and paid for elsewhere, for example, by the joint venture employing additional staff on to its payroll, and it is calculated related to the amount of work being carried out and its duration, then this may be an allowable head of the joint venture's claim in this example.

6.8.2   Profit

Returning again to the three standard forms of contract considered in this book, the scope and basis for claiming profit in the evaluation of the time consequences change are in turn as follows:

- Under SBC/Q clause 4.23, for 'matters affecting its regular progress' the contractor's entitlement is stated as being to 'loss and/or expense'. On this basis, it is suggested that profit could only fall to be claimed on the basis that the contractor has lost a contribution to it, because profit cannot be an 'expense'.
- In the FIDIC Red Book, the definition of 'Cost' in clause 1.1.4.3 is repeated as follows with emphasis added:

> … all expenditure reasonably incurred (or to be incurred) by the Contractor, whether on or off the Site, including overhead and similar charges, *but does not include profit.*

On this basis, a claim for 'Cost' would expressly exclude profit. However, some of those the FIDIC Red Book clauses that entitle to contractor to claim 'Cost' then expressly add profit to it. For example, in clause 1.9, where the works are likely to be delayed or disrupted by a late drawing or information, the contractor is entitled to (emphasis added):

> '(b) payment of any such Cost *plus reasonable profit*, which shall be included in the Contract Price.'

A similar approach is adopted in clauses 2.1 (for late access), 4.7 (errors in the engineer's setting out data), 7.4 (delay or instructions in relation to testing), 10.2 (early taking over of part of the Works) and 10.3 (interference with tests on completion). These are events that involve a degree of default on the part of the employer or its engineer. On the other hand, for delay and disruption events that do not involve such fault, there is no addition for profit, for example in clauses 4.12 (unforeseeable physical conditions), 4.24 (discovery of fossils) and 8.8 (suspension of work). As for what is a 'reasonable profit' under these FIDIC terms, see elsewhere in this chapter.

- The Infrastructure Conditions take a similar approach to that of FIDIC, by using a similar defined term. The definition of 'cost' in clause 1(5) is repeated as follows with emphasis added:

> … all expenditure properly incurred or to be incurred whether on or off the Site including overhead finance and other charges properly allocatable thereto *but does not include any allowance for profit*.

Like the FIDIC Red Book, the Infrastructure conditions also add profit in some circumstances only. However, they do so on two different bases as follows:

- Clauses such as 12(6), where the contractor makes a claim for delay and extra cost in relation to adverse physical conditions and artificial obstructions, adds profit as follows (emphasis added):

> (b)… the amount of any costs which may reasonably have been incurred by the Contractor (*together with a reasonable percentage addition thereto in respect of profit*).

- A slightly different approach is adopted for the addition of profit in some Infrastructure Conditions clauses, which say no more than:

> Profit shall be added to the contractor's cost.

Here there is no explanation of the basis for such profit addition; it is not said to be a percentage or to have to be reasonable. Examples of this are in relation to additional permanent or temporary works in clauses 13(3), 14(8), 31(2) and 42(3). Similarly, in clause 40(1) in relation to suspension of work.

Where the FIDIC Red Book and the Infrastructure Conditions expressly provide for the addition of profit, the issues will be how and at what level they should be added? The alternatives might include:

- The contractor's tendered allowance. This has the advantage of being what the contractor priced to achieve, but the disadvantage of being open to the question as to whether it was actually achievable. It might, however, be considered as appropriate to a value-based approach rather than one based on cost.

- The contractor's general achievable profit levels on its other work, which might come from its audited accounts. This has the advantage of being a real percentage, rather than a tendered hope, but the disadvantage of being open to the question to whether it was actually achievable on the delayed project.
- The contractor's profit level actually achieved on the project. This approach has the attractions of being based on an actual rate and for the project subject of the claim. The problem with it is the obvious one that the fact of the claim being made on it means that the profit being made by the contractor on the project will be depressed by the circumstances on which the claim is based.
- A typical market rate. This has the advantage of objectivity, but the disadvantage of being open to the question as to whether it was actually achievable by the contractor, who might be capable of achieving a significantly different net margin.
- A percentage agreed between the parties. This can be pre-agreed as part of the contract negotiations and inserted into the contract. In practice, the percentage is sometimes adopted that has already been agreed between the parties for the valuation of variations.

As have been set out, several provisions of the FIDIC Red Book provide for the contractor to be paid its 'Cost plus reasonable profit'. Interestingly, the 2000 edition of the FIDIC Contracts Guide suggests that the following sentence may be included in the Particular Conditions:

> In these Conditions, provisions including the expression 'Cost plus reasonable profit' require this profit to be one-twentieth (5%) of this Cost.

Perhaps following that guidance, in practice many FIDIC-based contracts include such a clarification and the use of a 5% rate is not uncommon as an agreed rate for this purpose.

Where the claim is made as one for damages for breach of contract, it has been explained elsewhere how the quantum of such a claim would usually be that amount which would put the contractor back in the position it would have been in had the breach(es) of contract not been committed by the employer. It is possible to envisage that this might mean a claim for profit, on the basis of loss of opportunity to earn it elsewhere. As with loss of contribution to head office overheads, this will very much depend on the contractor proving the lost opportunity, why it was lost and what margin would have been achieved on it.

For any such claim to succeed, the contractor will need to demonstrate that it had other work available for the resources and could reasonably have been expected to earn the profit claimed. It is not sufficient merely to demonstrate that resources have been retained by delay or disruption and thereby claim a profit element in relation to those resources. This, of course, also implies that the profitability, or otherwise, of the contract on which the resources are retained is not relevant to the claim, nor the appropriate rate of profit.

If the contractor can provide the necessary evidence of work turned away for which the retained resources would have been required then a claim for loss of profit may be sustainable, if the contract or the law does not prohibit it. It may not be sufficient merely to show that tender opportunities have been declined, as there would be no way of knowing whether or not the contractor would have been successful, and if successful, at what profit level. Some evidence of the turning away of real work with realistic prospects of profit at a level will be required.

The rate of profit would not necessarily be based on the overall profitability, or otherwise, of the contractor's business as a whole if it can be demonstrated that the declined work had real prospects of a different level of profitability. It may, however, be quite reasonable that the level of profit should not be excessive for the type of work and prevailing market conditions, following any legal principle of foreseeability, and that some discount to the anticipated rate might be expected given that recovery as compensation will remove the risks inherent in any alternative work.

One consequence of how some contracts exclude profit as a head of a claim for the financial effects of delay is that contractors make the alternative claim of damages for breach of contract, where the law does allow the recovery of profit (usually on a lost opportunity basis). This is particularly relevant to bespoke international contracts that expressly exclude profit in claims even for delay events that involve default on the part of the employer and/or its engineer.

6.8.3 Formula Approaches

As has been emphasised throughout this chapter, the evaluation of contract claims arising from delays or disruption on engineering and construction contracts generally calls for systematic record keeping and a considerable amount of detailed presentation of accounting information if it is to be successful. This will also require significant time and effort in analysing the information presented. Even when all the necessary information for analysis is readily to hand there can be differences of opinion as to the relevance of portions of the figures presented. For example, in relation to head office overheads, where costs for an accounting department are included in the figures, there can be argument that only elements (perhaps the payroll section) are affected by any claim for delay.

There is therefore a great attraction in the potential adoption of a formula that could be used to calculate the appropriate amount of the head office overheads without recourse to accounting data and the required detailed analysis of it. Often this is coupled with a similar formula approach to profit. This is usually in relation to claims for prolongation rather than any other claim for the time consequences of change, such as disruption or acceleration.

The benefits of a suitable formula would obviously be the elimination of the accounting analysis and records, with a consequent saving in time and cost, and the avoidance of argument as to the relevance of any part of the costs. Sadly, there are, however, considerable problems in the adoption of formulae to calculate possible levels of head office overhead and profit recovery as a consequence of the prolongation of a contract period. A number of standard criticisms are usually raised, sometimes rightly and sometimes wrongly, including:

- For contractors the attraction of the simplicity of a formula approach is often matched by the attraction of a resulting high claim amount. It is not uncommon, particularly among specialist subcontractors, that a formula overheads and profit calculation is the largest head of their claims for compensation for delay, claimed at levels significantly greater than their costs of labour, plant, equipment and on-site overheads. Therefore, if strictly applied and accepted, a formula calculation may significantly over-compensate the claimant.

- That the calculation will put to the employer's account the costs of peripheral matters such as, for example, the company's box at Wembley stadium, debentures at Twickenham and the annual 'team building' skiing weekend in the French Alps.
- Many formula calculations will need to be based on the contract data for time and money, i.e. the programme and make-up of the contract sum. Where there are errors or deficiencies in either, but particularly the latter, the level of recovery can be distorted.
- It is difficult, if not impossible, to structure some formulae approaches to discount elements of the contract sum that may not be applicable to the extended period of the contract.
- The adoption of a formula will generally require the assumption that the rate of activity, and therefore the rate of head office overhead cost, is uniform throughout the course of the project. Only in exceptional circumstances will such an assumption be valid.
- Many formulae require the assumption that the level of overhead commitment and cost during the period of the prolongation of the contract works is the same as the average commitment and cost during the original contract period as calculated from the contract data. This is again an assumption that will be valid only in exceptional circumstances. For instance, the works may be delayed at the outset by a late possession of the site but in circumstances that incur a low level of commitment and cost. Alternatively, the delay may occur at the height of activity during the construction works when commitment and cost are at their highest.
- The formulae do not allow for recoveries that the contractor has made through the valuation of additional works. This area of double-counting is considered elsewhere in this chapter.
- The formulae do not establish that any additional overhead costs were actually incurred and that the reality is that any under-recovery is no more than delayed into a later period.

Whether such criticisms are fair will depend on the facts in each case, which formula is adopted and how it is applied.

In principle, it seems inevitable that if a contractor is prolonged in its involvement with a project that will prolong the involvement of its head office resources, this will involve it in additional costs or loss in some financial form, together with a potential effect on its profits. Therefore, notwithstanding the problems listed, the England and Wales courts have been attracted by the apparent simplicity of the formula approach and have given some measure of approval. However, that approval has been tempered by some essential factual preconditions that a claiming contractor would have to satisfy first.

There are various potential formulae that may be considered for the calculation of the head office overhead and/or profit element of a prolongation claim. Whilst there are others used in some parts of the world, the most widely applied internationally are the Hudson, Emden and Eichleay formulae.

6.8.3.1   The Hudson Formula
This formula takes its name from the legal textbook *Hudson's Building and Engineering Contracts*, in which publication it first appeared in 1970.

It requires the contract sum to be divided by the contract period, in days, to produce a daily amount of the contract sum per week. This amount is then multiplied by the head office overheads and profit percentage included in the contract sum, to produce a daily rate for head office overheads and profit allowed in the contract sum. This daily rate is then multiplied by the period of delay, in days, to produce a recoverable sum of head office overhead and profit for the delay period. It can be summarised as follows:

$$\frac{\text{Contract Sum (\$)}}{\text{Contract Period}} \times \frac{\text{Tender OH \& P\%}}{100} \times \text{Delay Period} = \text{Claim (\$)}$$

It could be summarised as applying the tendered allowance for head office overheads and profit to the delay period on a pro rata basis, based on time. It therefore could be said to evaluate head office overheads and profit on a price basis rather than a cost basis.

In its text, *Hudson* made no mention of how the head office overheads and profit percentage included in the contract sum was to be obtained, if it was not quoted in the contract or agreed between the parties. However, the Tenth Edition, in which the formula first appeared, suggested that rates of between 3% and 7% of the total cost, including prime cost items and provisional sums, was the range to be expected for competitively tendered projects in the UK at that time.

The Hudson formula suffers from most, if not all, of the potential criticisms of formula approaches generally and the following additions:

- In applying the overhead and profit percentage to the weekly amount of the contract sum the percentage is being applied to a figure *which itself includes an element of overheads and profit*. The recovery being calculated therefore includes some double counting because of this failing. If the Hudson formula is to be applied at all, the formula needs correcting to reduce the contract sum used in the formula to a figure net of overhead and profit.
- It uses the tendered allowances for head office overheads and profit, without establishing if these were realistic or achievable. It is therefore an approach based on value rather than loss, expense or cost of damage. The counter argument to this is that it actually is a 'loss' approach, and reference is made to the factual preconditions that are identified elsewhere in this chapter. Proponents will say that the loss is the margin that was allowed in the tender, because that is what the contractor would have earned. The alternative is to consider what its accounts show that it actually did achieve. This leads the discussion on to a consideration of Emden formula.

6.8.3.2 The Emden Formula

This formula also takes its name from the legal textbook in which it first appeared, in this case *Emden's Construction Law*.

It follows a similar calculation approach as the Hudson formula, but with one important difference in that it uses actual head office overheads and profit rather than tendered allowances. It therefore has a first stage in which the company's total actual overhead cost and profit is expressed as a percentage of the company's actual total revenue in the same period. The formula is then identical to the Hudson formula with the percentage derived from the first stage being used to calculate the weekly overhead and profit amount recoverable. It can be summarised as follows:

$$\frac{\text{Contract Sum (\$)}}{\text{Contract Period}} \times \frac{\text{Actual OH \& P\%}}{100} \times \text{Delay Period} = \text{Claim (\$)}$$

It could be summarised as allocating the contractor's actual company-wide head office overhead costs and profit to the project on the basis of that project's price as a proportion of its total revenue.

This approach has the advantages of:

- Defining how the percentage is to be calculated, on the company's actual overhead and profit as a proportion of total revenue.
- It is based on the contractor's actual margins rather than those allowed by it in pricing the contract sum. As such, when it first appeared, the benefit was considered to be that the use of this formula would give a more realistic result, based on actual overhead and profit figures, rather than Hudson's reliance on the theory of what the contractor hoped to obtain when it tendered, which could have been unrealistically high or low.

To some commentators, Emden was seen as a more realistic replacement for the Hudson formula. However, it introduces two distinct complications that detract from the attraction of simplicity in formulae:

1. The calculation of the percentage as the first stage requires the production of accounting records and data, presumably for at least one relevant financial year, and therefore the avoidance of time and cost involved in such production of data begins to be eroded.
2. Secondly, and more importantly, the overhead and profit have to be identified from the accounting records, thereby introducing an opportunity for disagreement as to which items in the accounts are, or are not, truly head office overhead and administration items. In this regard, accountancy practice varies internationally. In one case, a South American contractor's published audited accounts recorded its costs against an accounting heading of 'overhead and administration' as only averaging 2.09% of turnover, but it made a case that its real costs of such items were over 7% of turnover, because some such costs were allocated elsewhere in the accounts.

6.8.3.3 The Eichleay Formula

This formula takes its name from a decision of the United States Armed Services Board of Contract Appeals in *Eichleay Corporation* 60-2 B.C.A (CCH) 1960. It therefore pre-dates both Hudson and Emden formulae. Its use is usually limited to head office overheads rather than including profit as well.

A calculation of head office overheads in a prolongation claim using the Eichleay formula involves three stages:

1. The total project revenue in the period of the project is divided by the company's total revenue in the same period to produce the proportion of the company's revenue attributable to the project. This is then multiplied by the company's total head office overhead costs in the period to produce an amount of head office overhead costs attributable to the project. This stage can be summarised as follows:

$$\frac{\text{Project Revenue (\$)}}{\text{Company Revenue (\$)}} \times \text{Actual OH (\$)} = \text{Attributable OH (\$)}$$

2. That the amount of head office overhead costs attributable to the project from stage (1) is then divided by the total actual project period in days to produce a daily rate at

which the project needs to earn its attributable head office overhead costs. This stage can be summarised as follows:

$$\frac{\text{Attributable OH (\$)}}{\text{Actual period (days)}} = \text{Attributable OH (\$) per day}$$

That rate of head office overhead costs attributable to the project per day from stage (2) is then multiplied by the period of delay on that project in days to produce a claimed amount for that project. This stage can be summarised as follows:

$$\text{Attributable OH (\$) per day} \times \text{Delay Period} = \text{Claim (\$)}$$

It could be summarised as aiming to estimate an equitable apportionment of the contractor's actual company-wide head office overheads to the project on the basis of that project's revenue as a proportion of its total revenue by applying that apportionment to the delay period on a pro rata basis.

Eichleay does not include profit in the formula and differs in that respect from both Hudson and Emden. Like Emden, it has the advantage over Hudson of being based on actual head office overhead costs and then relating that to time, rather than applying the tender allowance. It still suffers the criticisms discussed elsewhere and, especially, assumes a uniform rate of expenditure, etc., but this often misses the point that it is only intended to estimate a reasonable cost allocation. Whether that approach is allowable in assessing damages or loss is another question. It might be considered as more appropriate to a claim based on a prospective estimate rather than proven actuals.

6.8.3.4 The Adoption of Formulae

The use of formulae for the calculation of head office overheads as part of a contractor's claim is relatively well developed in the United States, particularly the adoption of the Eichleay formula. That said, the approach varies widely between different States. As explained, that formula is intended to achieve no more than to estimate an equitable apportionment of the company's head office overheads to the project. As explained in the judgment in *Eichleay Corporation*:

> The problem out of which this dispute arises is how to allocate home office expenses incurred during a period of suspension of work. These expenses continue during temporary or partial suspensions, and it was not in this case practical for the contractor to undertake the performance of other work which might absorb them. There is no exact method to determine the amount of such expenses to be allocated to any particular contract or part of a contract. It has been held a number of times that it is not necessary to prove that specific amount, but only to determine a fair allocation for the purposes of compensating a contractor for delay by the Government.

In the UK there was a time, a few decades ago now, when formulae were routinely applied and accepted in evaluating prolongation claims. That said, in the England and Wales courts consideration of formula happened later and less enthusiastically than in the United States. Whilst the judgments have affirmed the application of such as Emden and Hudson in principle, they have placed a particular emphasis on the factual preconditions to be satisfied by the contractor before a formula applies. This has meant that whilst formula have earned legal approval in principle, they are now harder to be successfully claimed, because practitioners are more aware of the preconditions and hurdles that many claimants find impossible to jump. As a result of this, it is worth considering the details of some of the court judgments relevant to the position under English law, not just to see how the current position under English law came about but also to understand the factual issues that have to be addressed in a claim on this basis.

From a Commonwealth perspective, the first authorities for the use of a formula to calculate a claim for head office overheads and profit are Canadian. In the judgment of the Supreme Court of Ontario in *Ellis-Don Ltd v Parking Authority of Toronto* (1978) 28 BLR 98, the court accepted that the contractor was entitled to recover overheads and profit in a period of compensable delay on the basis of a tendered allowance of 3.8%. It considered the earlier Canadian case of *Shore and Horowitz Construction Company Limited v Franki of Canada Limited* (1964) SCR9. The Judge found there was ample evidence that the resources of the claimant that were tied up by delays by the defendant could otherwise have readily been deployed elsewhere and could have continued to earn the claimant contribution to its overheads and profit.

In the UK, the adoption of the Emden formula was approved in the case of *Whittall Builders Co Ltd v Chester-le-Street District Council* (1985) 12 Const LJ 256, in preference to the Hudson formula, with Mr Recorder Percival QC stating in his judgment:

> Lastly, I come to overheads and profit. What has to be calculated here is the contribution to off-site overheads and profit which [the contractor] might reasonably have expected to earn with those resources if not deprived of them. The percentage to be taken for overheads and profit for this purpose is not therefore the percentage allowed by [the contractor] in compiling the price for this particular contract, which may have been larger or smaller than his usual percentage and may or may not have been realised. It is not that percentage that one has to take for this purpose, but the average earned by [the contractor] on his turnover as shown by [the contractor's] accounts.

This judgment was therefore quite specific in stating that the percentage that was to be considered appropriate for the calculation of overhead and profit was that which might have been actually earned by the contractor, as demonstrated by its financial accounts, not the percentage that had been included by the contractor in its tender calculations.

Notwithstanding this earlier preference for the Emden formula, the court appeared, at least on the face of the judgment, to approve the Hudson formula in the case of *J.F. Finnegan Ltd v Sheffield City Council* (1988) 43 BLR 130. The contractor had presented a claim for recovery of overhead and profit using a formula basis but not adopting any of the published formulae. Sir William Stabb QC stated in his judgment:

> It is generally accepted that, on principle, a contractor who has been delayed in completing a contract due to the default of his employer, may properly have a claim for head office or off site overheads during the period of delay, on the basis that the work force, but for the delay, might have had the opportunity of being employed on another contract which would have had the effect of funding the overheads during the overrun period.

Having accepted this general principle, he then considered the method of calculation adopted by J.F. Finnegan in that case and concluded:

> I confess that I consider [the contractor's] method of calculation of the overheads on the basis of a notional contract valued by uplifting the value of the direct cost by the constant of 3.51 as being too speculative and I infinitely prefer the Hudson formula which, in my judgement, is the right one to apply in this case, that is to say, overhead and profit percentage based upon a fair annual average, multiplied by the contract sum and the period of delay in weeks, divided by the contract period.

Therefore, although the Judge expressly referred to the Hudson formula, the process he describes was closer to that of the Emden formula, i.e. the use of a fair annual average rate of overhead, rather than the rate included by the contractor in its tender calculations. Whilst this suggests some confusion as to which formula was considered appropriate, the important point was that this was a further support for their use in general principle.

Amec Building Ltd v Cadmus Investments Co Ltd 1995-ORB-1232, was an appeal from the decision of an arbitrator. The parties had agreed that head office overheads were recoverable as a head of damages and that there must be some proof as to the level of damages suffered. They disagreed as to the correct calculation of those damages. Amec had appealed that the arbitrator had been wrong in law to decide that its claim failed because it failed to keep records and to reject its claim for head office overheads based on a formula calculation it had put forward. Interestingly, Mr Recorder Kallipetis QC first explained the arbitrator's approach to Amec's claim for profit:

> .... The Arbitrator concluded that the delays in construction caused on this project was not such as would deter a building contractor of the size and standing of the claimants from tendering for work and he concluded that in his view the onset of the recession in the industry better reflected the tendering opportunities open to the Claimants. Having read and heard all the evidence and considered all the matters he rejected the claim for loss of profit.

Thus, he refused profit because the market was such that there was no lost opportunity to earn it. Turning to the arbitrator's approach to head office overheads, the Judge continued as follows:

> He did not reject a formula calculation overall but applied a formula to those parts of the overheads which he was satisfied on the evidence he had heard fell to be awarded as arising during the extended period. Although no records existed he was prepared to use a formula with a much reduced percentage element ....

In that same year, the use of formulae for the calculation of a claim for head office overheads in a period of delay was further considered by the courts in the case of *Alfred McAlpine Homes North Ltd v Property & Land Contractors Ltd* (1995) ORB 179, where His Honour Judge Humphrey Lloyd QC stated:

> ... the Emden formula ... is one of a number of methods conventionally applied in an attempt to arrive at an approximation of the damages supposedly incurred

by a contractor where there has been delay to the progress of the works whereby completion is similarly delayed. The theory is that because the period of delay is uncertain and thus [the contractor] can take no steps to reduce its head office expenditure and other overhead costs and cannot obtain additional work there are no means whereby [the contractor] can avoid incurring the continuing head office expenditure, notwithstanding the reduction in turnover as a result of the suspension or delay to the progress of the work. This type of loss (sometimes called a claim for 'unabsorbed overheads') is, however, to be contrasted with the loss that may occur if there is a prolongation of the contract period which results in [the contractor's] allocating more overhead expenditure to the project than was to have been contemplated at the date of contract. The latter might perhaps be best described as 'additional overheads' and will, of course, be subject to proof that the additional expenditure was in fact incurred.

Furthermore, the Emden formula, in common with the Hudson formula ... and with its American counterpart the Eichleay formula, is dependent on various assumptions which are not always present and which, if not present, will not justify the use of a formula. For example, the Hudson formula makes it clear that an element of constraint is required ... in relation to profit, that there was profit capable of being earned elsewhere and there was no change in the market thereafter affecting profitability of the work. It must also be established that [the contractor] was unable to deploy resources elsewhere and had no possibility of recovering the cost of the overheads from other sources, e.g. from an increased volume of the work. Thus, such formulae are likely only to be of value if the event causing delay is (or has the characteristics of) a breach of contract.

Again this was support in principle, but the Judge's reservations about the use of formulae particularly emphasised factual preconditions, such as the contractor not being able to reduce its overhead costs or redeploy resources to earn income elsewhere and that market conditions should not have materially changed thereby affecting the contractor's prospects of obtaining alternative profitable work.

The contractor particularly failed to satisfy the burden of proving that it did lose opportunities to earn a contribution to its gross margin from other work, in the judgment of HH Judge Toulmin in *City Axis Ltd v Daniel P Jackson* (1998) 64 ConLR 84:

There was no direct evidence that City Axis turned away other work or of specific projects which City Axis might have been able to bid for successfully but for having resources tied up in this contract. Mr Ridge said that 'inquiries are generated. They don't just happen'. I should have expected at least some evidence that City Axis would have been asked to bid for projects which did take place in this period but were not asked to do so because they were not available, or did not seek to do so because they were involved in completing Mr Jackson's project. The only evidence relating to a specific project is that they were able to obtain one in Bruton Street in September 1995 very soon after the work on Mr Jackson's project had been completed.

The evidence which has been provided is statistical and inconclusive particularly in relation to the period from 16 June 1995 to the end of August 1995. If I had found that Mr Jackson was liable for any consequential delay during this period

I should have found that City Axis had failed to prove any loss and damage as a consequence of Mr Jackson's delay.

In *St Modwen Developments Limited v Bowmer and Kirkland Limited* (1998) Const L.J. 214, Judge James Fox Andrews considered an appeal from the decision of an arbitrator who had awarded amounts to Bowmer and Kirkland in relation to a shortfall in its recovery of overheads and profit because of delays. St Modwen argued that the contractor had failed to prove the claim in fact or in law. The judgment particularly set out how the arbitrator had recognised the caution required before applying a formula approach and referred to the judgments in cases such as *Ellis-Don* and *Babcock Energy v Sturtevant*. As the arbitrator had put it:

> . . . . without evidence that there was a genuine loss and that there was work available on which the contractor would have earned a return contribution towards the overheads and profit but for his prolonged involvement with the project in question.

The arbitrator had also observed:

> ... the construction industry was buoyant/booming at the material time. These are precisely the conditions in which a formula approach may be acceptable.

The Judge concluded that the arbitrator had not erred in law and that his finding of fact was not one that no reasonable arbitrator could have made. The appeal was therefore rejected.

The *McAlpine v Property & Land* decision was also considered in *Norwest Holst Construction v Co-op Wholesale Society Ltd* [1998] EWHC Technology 339 in which, in considering an appeal from the award of an arbitrator, his Honour Judge Thornton QC explained that the arbitrator had found as an issue of fact that as a result of delays to progress caused to works undertaken by CWS:

(1) The delay caused CWS some additional costs. This was represented by time being spent by senior management working on administrative tasks on this contract in the period of delay.

(2) Some loss and/or expense in respect of head office costs occurred because of the delay on this contract. This loss and/or expense was a combination of rates, lighting, heating and the like.

(3) A claim for a 'loss of contribution to overhead recovery' would be justified if CWS could show that it had suffered loss.

(4) The additional time spent on this contract would have been spent productively on other contracts had it not had to be spent on this contract.

(5) CWS suffered a 'loss of contribution to overhead recovery' caused by senior management spending less time on other contracts because, in the period of delay, they were working on this contract.

(6) It was not possible to accept that the 'loss of contribution to overhead recovery' was as much as would be provided for by the Emden formula.

(7) The appropriate way of compensating for both types of loss was to award CWS a composite sum per week for the 19 weeks in question. This loss was calculated by taking one-fifth of the Emden formula weekly recovery.

The Judge then went on to say:

> Thus Emden-style formula is sustainable and may be used as the basis of ascertaining a contractor's entitlement to payment for loss and/or expense in the following circumstances:
>
> (1) The loss in question must be proved to have occurred.
> (2) The delay in question must be shown to have caused the contractor to decline to take on other work which was available, and which would have contributed to its overhead recovery. Alternatively, it must have caused a reduction in the overhead recovery in the relevant financial year or years which would have been earned but for that delay.
> (3) The delay must not have had associated with it a commensurate increase in turnover and recovery towards overheads.
> (4) The overheads must not have been ones which would have been incurred in any event without the contractor achieving turnover to pay for them.
> (5) There must have been no change in the market affecting the possibility of earning profit elsewhere and an alternative market must have been available. Furthermore there must have been no means for the contractor to deploy its resources elsewhere despite the delay. In other words, there must not have been a constraint in recovery of overheads elsewhere.

Some of these factors are the ones that comprise part of the objection to the use of formulae, but the crux of the matter in the appeal was whether the arbitrator was entitled to use the Emden formula and then discount it to one-fifth of the sum produced by the formula. Having been satisfied that the arbitrator's findings of fact overcame the possible objections set out, the Judge held that it was quite proper for the arbitrator to apply the reduction as he had found it impossible to ascertain the elements of the overhead element by element.

It might be thought that the approach adopted by the arbitrator in *Norwest Holst v Co-op*, whereby a formula is adopted regardless of the acknowledged mathematical shortcomings of such formulae, but then is substantially reduced as it is considered over-generous, is merely compounding the objections to the adoption of formulae. That is, that they do not establish actual loss. On the other hand, the arbitrator's approach might be thought to have taken a pragmatic approach to a difficult problem. The contractor having satisfied the factual preconditions listed by the Judge as in (1) to (5), the arbitrator used his judgment to assess its loss, taking a conservative approach to ensure that the contractor was not over-compensated, whilst making sure that it obtained some recovery for the established financial effects of the employer's defaults.

It should also be noted that the types of reservations set out by HH Judge Humphrey Lloyd QC in *McAlpine v Property & Land* did not prevent a successful use of a formula in the Scottish case of *Beechwood Development Company (Scotland) Ltd v Stuart Mitchell* (2001) CILL 1727. The background facts in this case were not typical of a contractor's prolongation claim, in that the claimant, Beechwood, was a house building company generally carrying out one contract at a time. This perhaps made it particularly susceptible to the effect of a delay on its one current project and to the relevance of a formula approach to resulting loss of gross margin. Beechwood had developed a housing site at their own risk and had a site survey carried out on their behalf by the defendant. The survey proved defective and they sued their surveyors claiming, among other matters, loss of overheads and profits using the Hudson formula and referring the court to the judgment in *Finnegan v Sheffield*. In his judgment, Lord Hamilton put the claim as follows:

> The claim under this head is essentially one for reimbursement to Beechwood of contributions to overheads and net profit which would have been made had Beechwood not, as a result of the defender's fault, lost a period of effective working (now assessed by me at 10 weeks).

and:

> The averred approach appears to be the well-recognised 'Hudson Formula' ....

Having considered the evidence, particularly accountancy evidence as to Beechwood's financial performance, he concluded:

> In my view there is a sufficient evidential basis to allow an award of damages under this head to be assessed and made. I am satisfied that Beechwood did sustain a loss by reason of its inability for a period in 1995 to generate income through carrying out construction works. While the dramatic fall in gross profit from that earned in the year to 31 March 1995 to that earned in the year to 31 March 1996 may not wholly be explained by that inability, it is consistent with a situation in which, by reason of enforced idleness, a loss of productive earning is sustained. Not the whole period of unproductive earning can be attributed to the defender's fault but, in my view, a part of it can. The task is then to assess the amount of the relevant loss.

Moving on, he noted:

> Provided the pattern of ordinary trading is established, together with relative information as to the finances of the company, the court may be in a position to make an evaluation.

He concluded by awarding the claimant an amount by express reference to the Hudson formula, but based on the average gross margin of the previous three years. Therefore, it seems that what he actually used was the Emden formula.

For those seeking to make a claim on the basis of the Emden formula, and therefore to satisfy the burden of proving their loss of opportunity, HH Judge Thornton QC's decision in *CFW Architects v Cowlin Construction Ltd* [2006] EWHC 6 (TCC) might give rise to some optimism in that it described the formula as the 'conventional approach' and seemed to readily accept the factual evidence for lost opportunity:

Head office overheads.

The claim is based on the conventional application of Emden's formula for a 10-week period. The sum is agreed, subject to proof that the loss was incurred. The loss is the loss of recovery of profit and head office overheads arising from the inability to earn these recoveries from other work in the relevant period because Cowlin's resources were still employed on non-profitable, non-financial recovering work for DHE.

Mr Spiller gave evidence to the effect that the effects of the repudiation were that he and Mr Brown were much more heavily involved in the project than they should have been. This precluded them chasing other work, being involved in negotiations and tendering and otherwise generating financially rewarding new work.

I readily accept that the heavy additional involvement that these two senior members of Cowlin's management team reasonably became involved in at Tidworth precluded significant additional earnings elsewhere. It follows that the conventional basis for assessing this loss, recourse to the Emden formula for a 10-week period, is appropriate.

Cowlin is entitled to recover the sum of £143,088.90.

Most significantly and more broadly, the use of formulae was considered by Mr Justice Akenhead in *Walter Lilly & Company Limited v Mackay & Anor* [2012] EWHC 1773 (TCC) (11 July 2012). The claimant had used the Emden formula to assess its claim for head office overheads and profit, of which he started his analysis with this explanation:

This represents a well established basis of claim whereby a contractor, which has suffered delay on compensable grounds seeks the losses which it has suffered as a result of not being able to take on other projects as a result of that delay and disruption (here to Unit C), that loss being the loss of its opportunity to defray its head office overheads over those other projects and the loss of profit from those lost jobs.

He then referred to the discussions in *Norwest Holst v CWS*, *Whittall Builders v Chester Le-Street*, *Finnegan v Sheffield* and *Beechwood v Mitchell*, considered elsewhere. He then quoted HHJ Humphrey Lloyd's judgment in *Alfred McAlpine v Property and Land*. He concluded that:

(a) A contractor can recover head office overheads and profit lost as a result of delay on a construction project caused by factors which entitle it to loss and expense.

(b) It is necessary for the contractor to prove on a balance of probabilities that if the delay had not occurred it would have secured work or projects which would have produced a return (over and above costs) representing a profit and/or a contribution to head office overheads.

(c) The use of a formula, such as Emden or Hudson, is a legitimate and indeed helpful way of ascertaining, on a balance of probabilities, what that return can be calculated to be.

(d) The 'ascertainment' process under Clause 26 does not mean that the Architect/Quantity Surveyor or indeed the ultimate dispute resolution tribunal must be certain (that is sure beyond reasonable doubt) that the overheads and profit have been lost. HHJ Lloyd QC was not saying that assessment could not be part of the ascertainment process. What one has to do is to be able to be confident that the loss or expense being allowed had actually been incurred as a result of the Clause 26 delay or disruption causing factors.

Turning to the factual evidence Mr Justice Akenhead was impressed by the factual evidence of one of Walter Lilly's witnesses, regarding how his company ran its business in commercial terms and the effects on it of retention of key staff on a delayed project. Accepting that evidence he stated:

> Mr Corless has given detailed evidence that WLC's 'business model' and mode of operation is and was to use only direct employed (that is not agency) staff in lead roles when carrying out contracts. Its Business Development Department was tasked to identify suitable contracts to tender which would commence on site at a time when the appropriate staff become available, that is following the completion of their current projects or when their expertise is no longer required on a particular project. WLC's directors assisted in this process by carrying out a review of future tendering opportunities and staff availability on a weekly basis every Monday morning. Thus the strategy was constantly under review and allowed the relevant director to accept or reject tender opportunities depending upon resource availability ahead of their receipt in the office. Between January 2006 and September 2008 WLC's tender success rate was in the order of 1 in 4 (explained in evidence to be based on tenders submitted). During that period WLC had to and did decline a number of tendering opportunities: that was not said vaguely, or in a vacuum of support: the opportunities received and declined were precisely detailed on a comprehensive schedule attached to Mr Corless' statement.

He continued:

> It was clear from the schedule (attached to the statement) that the number of opportunities in the relevant period was significant and the market for the type of projects constructed by Walter Lilly was relatively buoyant in the 2006 to 2008 period.

The parties' quantum experts had largely agreed on a without prejudice basis how to calculate the Emden formula, with a minor difference between their resulting weekly rates. Preferring one of those rates, he awarded Walter Lilly 99 weeks of delay at that rate, but with:

> ... credit being given, wholly properly, for the overhead and profit recovered by WLC on the difference between the amount of profit and overheads earned on the works (by way of the 4½% addition to overheads and profit) and the tendered overhead and profit allowance.

The factual explanation and evidencing of Walter Lilly's business model is of relevance to any party considering a claim on the basis of loss of opportunity. It might particularly be relevant to a smaller contactor, or subcontractor, whose model is to use all of its resources to construct one project at a time or one large project dominates its activities and turnover, and where a delay to one project has an inevitable effect on a following project. In this regard the judgment in *Alfred McAlpine v Property and Land* may also be of interest.

6.8.3.5 When to Use a Formula

As ever the answer to this question starts with a consideration of what the contract and the applicable law say. Do they allow recovery of 'cost', 'expense', 'loss' and/or 'damage'. Do they allow a lost contribution to overhead costs and profit that would otherwise be made?

This discussion of the principal formulae and their shortcomings, together with the developing approach of the England and Wales courts, serve to illustrate the substantial factual and evidential difficulties entailed in adopting and using a formula to calculate the amount of overhead and/or profit recoverable for a period of delay where it is recoverable.

For contractors, it is also relevant to consider what type of resource it is that are essential to its 'business model' in financing the costs of its head office overheads and achieving a profit. The nature of contracting in the UK and elsewhere in the modern era is that main contractors have little direct labour and subcontract most work, acting as managers of those subcontractors. This will mean that the essential resources will be its project management and supervision rather than its labour. Alternatively, it may be certain specialist plant and equipment that the contractor owns that are essential in this regard.

At best, the use of a formula will only produce an average overhead and/or profit figure based on a number of assumptions that may, or may not, be realistic in any particular circumstances. That said, if the test is 'the balance of probabilities', then that may be sufficient. Where the basis of the claim is one for expenditure incurred, it may be preferable to establish the actual overhead costs from records and accounting data. This may only be possible in relation to absorbed overheads, leaving unabsorbed overheads to be dealt with in some other way, such as a formula.

If the adoption of a formula is considered appropriate to a particular law, contract, organisation and/or project then the matter should be raised, if at all possible, before the contract is agreed between the parties. If the use of a formula is considered carefully at this stage it is possible for the parties to agree the relevant percentages to be applied at the outset and agree to its implementation in the event of delay requiring compensation. However, this would require considerable thought and should be the subject of local legal advice if it is to be adopted. It may also be that such an approach will be more attractive to a contractor than an employer (in the absence of hindsight in some cases where they spend much time and money arguing over what they could have settled at the outset).

Whilst the United States adoption of the Eichleay formula has been driven by government contracts and what might be considered to be an enlightened approach to public procurement, elsewhere government procurement might be expected to involve a rather more restrictive approach. The same might be feared in some subcontracts. This might mean preparing contract terms that expressly excludes entitlement to overheads and/or profit as a head of claim calculated on a formula basis, or any basis at all on the assertion that delays do not cause additional head office overheads to be incurred.

Any parties to a contract agreeing the incorporation of a formula adjustment for overheads at the outset obviously need to consider carefully the shortcomings of the formula approach and accept its deficiencies in return for the saving in time, costs and possible disputes if the actual costs have to be researched and presented for agreement after the event. In these circumstances the parties will also have the advantage of knowing from the outset what rate of overhead and/or profit is to be paid for any delay for which compensation is recoverable. This is perhaps analogous to agreeing the employer's rate of delay damages.

If, as is mostly the case, there is no such prior agreement, the only way to establish the actual overhead costs of a delay is to produce the management records and accounting data to allow a proper analysis of the way overhead is incurred by the particular organisation, and how any delay affects the recovery of that overhead. If it is decided in particular circumstances that a formula is the only practical way to address the matter then the Emden formula should be preferred as it avoids reliance on the contract tender, has a relationship to the actual levels of cost incurred in the contractor's business and does not require amendment to avoid the double counting problem with the Hudson formula. However, even if a formula such as Eichleay is to be adopted, the accounting data used should be subject to critical examination and not used without considering the trading activities represented by those figures and any adjustments required.

In large contracting organisations the overhead can represent as little as 5% or less of the sales revenue, but large fluctuations in levels are possible both as a result of different trading patterns and varying levels of activity. An organisation that only undertakes major projects, where most staff and resources are site based, may have a relatively small head office establishment, whereas an organisation undertaking a much larger number of smaller contracts may have a higher proportion of head office-based personnel, emphasising the need to identify those that are genuinely overhead and those that, although they might be budgeted and reported in the accounts as 'head office', are actually contract-related personnel priced into the preliminaries and general items costs. Very specialised high-tech subcontractors such as those in the information technology field may have considerably higher head office overhead costs and be capable of achieving much higher net margins.

Whatever the pattern of trading and accounting systems, there should be a process of questioning and adjustment that needs to be undertaken whenever such information is to be used for head office formula calculations. The use of accounting data from the company management system in an unquestioned form should be avoided as far as possible. There will, however, often be points of difference as to what is a genuine overhead and what is not. For instance, the management accounts in Appendix B include 'bad debts' of £143,000. Is this an overhead? Some will argue that bad debts are sadly a fact of life and a charge on the general running of the business, others will say that an employer should not be required to reimburse as overhead the default of another of the contractor's clients. Perhaps the answer lies in the reason for the provision. If it is an average charge based on trading experience then it may be an overhead; if it is a one-off provision for a particular debt then it might not be regarded as a genuine overhead. As emphasised elsewhere, only proper questioning of the accounts will establish the answer and most appropriate route.

Even when such data are refined before use in a formula, it does not overcome the basic objection to the formula approach where it assumes that overheads are incurred at a uniform rate over all activities and over time. The approach to be preferred wherever possible is that of identifying the relevant overhead incurred by the contractor and adding it to the cost of the resources employed.

6.8.3.6 Example of a Formula Claim

In order to illustrate some of the issues that can arise when considering a contractor's annual accounts to establish an effect on its margins to support the use of a formula approach, it may be worth considering a real example. Figure 6.5 sets out the summary accounts of a contractor over the three years of an engineering project. That project was planned for 18 months, from the third quarter of 2007 to the first quarter of 2009. It involved the fabrication and erection of steelwork, pipework and concrete works. In the event, completion was delayed by three months, to the middle of 2009. The contractor's claims included a heading of head office overheads and profit. This asserted that the delay that held it on the project for the three additional months in 2009 meant that its resources could not be allocated to other projects to earn a contribution to its overhead costs and profit over the period. The contractor provided the following details from its audited company accounts.

	2007	2008	2009
Turnover	152,643,598	189,009,348	162,352,505
Costs	115,207,540	137,988,947	108,256,198
Gross Margin	37,436,058	51,020,402	54,096,307
Admin &	1,850,570	2,207,402	2,372,043
Overhead costs	1.61%	1.60%	2.19%
Net profit	35,585,488	48,813,000	51,724,263
	30.89%	35.37%	47.78%

Figure 6.5 Sample accounts data.

Setting aside whether the contractor established its factual assertions regarding its resources and loss of contribution and the other preconditions for consideration of a formula approach, these numbers gave rise to the following points of interest on their face:

- The contract sum was around US$ 110,000,000. Over 18 months this represented around 45% of the contractor's turnover. The project was therefore of great significance to the contractor, such that it could be readily understood that delay on this project could have had a significant effect on its financial performance, particularly recovery of overheads and profit.
- The turnover for 2009 showed a significant 13% fall against 2008. This compared to a 24% increase between 2007 and 2008. This would fit the assertion that delay on the project caused a significant reduction in the company's turnover.
- However, how this fitted with any wider pattern as to the company's fortunes was lost by its failure to exhibit annual accounts for any other years than these three that covered the years of the project. This was notwithstanding that the claim was submitted some years after 2009. It should normally be expected that a contractor making a claim on this basis should provide data for years other than those of the affected

project's duration, to put the effect of that project into the full context of the wider business.

- Furthermore, the years 2008 and 2009 saw a significant slowdown in construction activity in the contractor's markets generally. The contractor's falling turnover in 2009 could have just been a symptom of that wider market malaise.
- Head office administration and overhead costs increased between 2008 and 2009, notwithstanding the fall in turnover, thus the contractor had been unable to reduce head office costs when its turnover fell. That they actually increased, and in a period when the market was in recession, was not explained by the contractor.
- Other costs (excluding head office administration and overheads) fell significantly in 2009, by far more than the fall in turnover. No explanation was offered for this.
- The result of this fall in other costs was that 2009, the year in which the contractor claimed to have suffered an under-recovery of overheads and profit, was actually its most profitable year of the three.
- The 2009 net profit at 47.78% of other costs was extraordinarily good by most standards for work of this project's nature. Taking the average of the three years, the contractor based its loss of profit claim on a rate of 38.01%.
- However, the contractor adduced no evidence that it had lost any opportunities or turned any work away as a result of being delayed on this project.
- Furthermore, the contractor adduced no evidence that its resources on the project could have earned it a net profit to 38.01% of other costs anywhere else. This contractor was also known to be running several infrastructure concessions over the period, which were considered likely to have been the contributors of much higher profit levels than that of the delayed construction project. On that basis, the profit percentage in Figure 6.5 might not be representative of what the contractor might have achieved using that project's resources on other works of the same type.
- The low percentage levels of the contractor's head office administration and overheads costs will be noted in Figure 6.5. The contractor's explanation was that it booked some head office costs under the heading of 'Costs' and that properly adjusted its real head office administration and overheads costs averaged over 7%. This just added to the doubt created by these numbers. There was a lack of evidence for this assertion and detailed explanation of what costs were booked and where.

Considerations such as these may be relevant to most overhead and profit claimed based on a loss of contribution.

6.8.3.7 Concluding Comments on Formulae
Gathering these discussions together, it is suggested that an approach for a contractor making a claim for head office overhead and/or profit on the basis of a formula should be as follows:

- Decide the basis of the claim that is being made; is it under the terms of the contract or for damages for breach?
- If under the contract, what does that allow, in terms of such as 'costs', 'loss,' expense, and allowability of profit in particular?
- Check the contract and the applicable law in relation to formula approaches.
- So far as overheads (particularly absorbed overheads) can be identified in terms of resource, time and costs, address these separately on their detailed evidence.

- Set out how the company earns a contribution to the recovery of its overhead costs (particularly unabsorbed overheads) and profit through the margin achieved by its project resources. If this relies on certain key resources, which are they? This is likely to involve witness statements from very senior management explaining how the company works in this regard. What is its 'business model'?
- Identify the delay period over which it is claimed that recovery was lost. This will be a product of the contractor's programme delay analysis.
- Show that in that period, company turnover fell. This should be apparent from its management or published accounts, depending on what is available when the claim is prepared. Explanatory witness evidence should support this.
- Show that in that period, head office overheads costs (particularly unabsorbed overheads) could not be reduced, at least proportionally to the fall in turnover. This may be for practical reasons, or perhaps because of the uncertain duration of the delay. Again, this should be apparent from the accounts and should be supported by witness evidence to explain the difficulty.
- Show that as a result of the retention of resources on the delayed project, the contractor was unable to obtain additional work that would have obtained a contribution to its gross margin. Ideally, this should be evidenced by documentary records. These might include business plans and their revisions, board minutes, marketing and estimating team reports, and/or returned tenders. Witness evidence that the strategic decision was made and why, and how and when it was kept under review, etc., will also help.
- Show what that gross margin would have been on those other opportunities had they arisen. This involves consideration of whether the contractor is likely to have been successful had it bid, and at what margin. The first point should be apparent if the contractor has a general record of successful bidding. The second point will come from the company's accounts.
- Show that there was no change in the market in the relevant period that would have thwarted it in obtaining other contributory work in any event. This could come from industry statistics and/or records of the numbers of enquiries still being received.
- Remember that the burden of proof on each of these issues will usually be on the claimant and be set as the balance of probabilities. However, that does not mean that each of the above can be satisfied on the basis of a '50/50' assumption. Each stage has to be established with evidence.
- Credit should be given for any recoveries of the claimed lost head office costs or profit for recoveries made through, for example, variations in the period.

One quirk of some formula approaches to a claim for head office overheads and profit arises out its basis being as a claim for loss, rather than for costs or expense. As has been explained, generally, claims for the financial effects of compensable delays should be quantified at the time that the delay occurred. If a contractor incurs costs or expenditure as a result of a delay, then that financial effect will be incurred as and when that delay happens. However, the situation may be different in relation to loss, and this is specifically relevant in relation to a claim for loss of contribution to a contractor's head office overheads and profit. If the basis of such a gross margin claim (for example, the Emden formula, although Eichleay has the same problem) is that the contractor was held on site for longer, beyond its planned completion date, then it is the overheads and profit that would have been achieved in that overrun period that is the contractor's loss. This means that, strictly speaking, a calculation applying Emden or Eichleay formulae cannot be done until after the end of the overrun period at the end of a project, when the necessary figures are all available.

This distinction also highlights a further quirk. Often contractors assume that if their completion date overran by, say, 20 weeks, due to compensatory delay, then their loss of overhead and profit should be calculated on the basis of that 20 weeks. Thus, whichever of the three formulae explained is adopted, the 'delay period' applied in the formula would be 20 weeks. However, if the basis of such a claim is the tying up of resources for longer on the delayed project, it is quite feasible that they were tied up for longer than just that additional 20 weeks. Say, for example, that the contractor's key contribution-earning resource is particularly skilled people or technical plant and equipment for the installation phase of a programme. They might have been planned to move on to another project 10 weeks before the originally programmed completion date, leaving a period for its testing and commissioning by other employees. It is the nature of many delayed projects that such consecutive periods become overlapped in the face of delays, such that in this example the key resources might have been delayed until somewhat less than 10 weeks before the actual completion date. It is theoretically possible that they have to stay at the delayed project until its actual completion. In that case their prolonged involvement would be 30 weeks. However, the costs of the additional weeks in this would have to pass any test of foreseeability and remoteness such as in *Hadley v Baxendale* under English law.

A final point in relation to claims for head office overheads and profit as a head of a claim for prolongation is of the particular danger of overlap between such a claim and recoveries claimed elsewhere. In particular, this is relevant to recoveries of overheads and profit through the valuation of variations. The quoted conclusion of Mr Justice Akenhead in *Lilly v Mackay* is relevant in this regard. This issue of overlap is considered later in more detail in Section 6.10.

6.9   Interest and Finance Charges

It is now some decades since Lord Denning, in his judgment in *Dawnays Ltd v Minter and Trollope and Colls* [1971] 1 WLR 1205, described cashflow as the lifeblood of the construction industry. In the 1980s the authors had as a client a UK conglomerate that purchased several medium size mechanical, electrical and plumbing services subcontractors to create a specialist division whose sole purpose from the PLC's perspective was to create positive cashflow that its brokers could play with on the international money markets. More recently they heard the main-board finance director of another UK public company tell his commercial directors' conference that he was not interested in profit, only in maximising the period between money being paid in by its clients and paid out to its suppliers and subcontractors. Whilst this second approach can seem to echo what used to be called 'subbie bashing' and be destined to prove short-sighted, both illustrate the importance of cash to contractors and subcontractors and why it is that construction and engineering contract claims so often include a heading of interest or finance charges.

The general distinction between interest and finance charges is that the former is limited to the payment of interest on a loan or facility, or interest earned on such as a deposit account, whereas finance costs include such interest and also any other related costs and charges. Such charges can include matters such as facility establishment costs and financial transaction tax in some jurisdictions. For an example in this latter regard, see Section 8.2 of this book of a complex arrangement of facilities and currency swaps to borrow money following (in that example) the calling of a contractor's bonds. Either way, they generally involve either incurring interest or losing the opportunity to earn it. A contractor's entitlement to recover such losses or costs can arise in one of several ways including:

- As an express entitlement under the terms of the contract for interest or finance charges to be paid in a specific eventuality.
- Where the law provides that a financial entitlement under the contract includes interest or finance.
- As damages for breach of contract.
- As a right under statute, for example in the UK under the Late Payment of Commercial Debts (Interest) Act 1998.
- As part of a judgment or award. For example, under a statute such as section 49 of the Arbitration Act 1996, or arbitration rules such as rule 12.8 of the JCT/CIMAR Rules, which incorporates section 49 of the Act.

The last of these is outside of the remit of this book; the others are considered as follows.

6.9.1   Finance and Interest Under Contract Provisions

Typical standard forms of construction and engineering contracts will particularly include express provisions for interest to be paid to a contractor in the event that certified sums are paid by the employer later than the periods prescribed in the contract. Some will extend this to the failure of the certifier under the contract to certify payments in the first place, as required by the contract. There may also be other particular provisions for interest payments, for example in the event of such as the return of delay damages that have proven to be wrongly deducted because extensions of time are granted after the deduction.

Typical of the simpler and more limited provisions for the payment of interest are those of the FIDIC Red Book. Its clause 14.8 enables the contractor to claim interest on any amounts that the engineer has certified, but the employer fails to pay, in accordance with clause 14.7. The interest is stated to be compounded monthly. Unless otherwise stated in the Contract Particulars, the applicable interest rate is stated as 'three percentage points above the discount rate of the central bank in the country of the currency of payment'.

In the Infrastructure Conditions, clause 60(7) is similar to FIDIC Red Book clause 14.8, but extends the right to interest to also include failure by the engineer to certify in accordance with subparagraphs (2), (4) and (6) of clause 60. These subparagraphs respectively deal with monthly payments, the final account and the payment of retention monies. As in the FIDIC Red Book, the interest is stated to be compounded monthly. The applicable interest rate is stated as '2% per annum above base lending rate of the bank specified in the Appendix to the Form of Tender'.

The Infrastructure Conditions also expressly provide for interest at that same clause 60(7) compounded rate in the event of:

- repayment to the contractor of any liquidated damages deducted by the employer to it which becomes no longer entitled because the engineer subsequently grants an extension of time (clause 47(5)) and
- the contractor having paid to a Nominated Subcontractor amounts previously certified by the engineer that are deleted or reduced in the final certificate, where the contractor is unable to recover such amounts and interest thereon from the subcontractor (clause 60(8)).

In SBC/Q, under clause 4.12.6, the contractor is entitled to simple interest if the employer fails to pay any sum, or part of it, that is due to the contractor under the contract by the final date for its payment. The applicable interest rate is stated as 'a rate 5% per annum above the official dealing rate of the Bank of England current at the date that a payment due under this Contract becomes overdue'.

In addition, in relation to the final certificate and final payment, clause 4.15.7 of SBC/Q provides for either party to pay simple interest at that same rate on any amount it should have paid, or repaid, to the other party by the final date for payment. SBC/Q clause 4.15.7 is therefore unusual among the standard forms considered here, in that it provides for the payment of interest by the contractor to the employer, in the event of the particular circumstances set out therein.

It is therefore generally the case that interest is expressly due where the employer does not pay a certified amount, but the Infrastructure Conditions are unusual in addressing the engineer's failure to certify. For the former failure, as payment is due on certificates, any failure of the employer to make a due payment should be easily identifiable. It will be a matter of the contractual and factual positions as to when certification was made, when payment should then have followed and when it was actually paid or that it still has not been paid when the interest is calculated. In contrast, the matter of failure to certify is potentially less easy to determine, depending on whether it is a wholesale failure to certify at all or an under-certification.

On the face of it, wholesale failure by a certifier to issue a due payment certificate at all ought to be another matter that can be relatively simply established based on the contractual and factual positions, in this case as to when a certificate should have been issued and it has not actually been issued. However, problems do arise in this respect, most obviously as to how much should have been certified. It is worth considering relevant parts of the provisions of Infrastructure Conditions clause 60 in detail:

> 60(1) Unless otherwise agreed the Contractor shall submit to the Engineer at monthly intervals commencing one month after the Works Commencement Date a statement (in such form if any as may be prescribed in the Specification) showing
> (a) the estimated contract value of the Permanent Works carried out up to the end of that month
> (b) a list of any goods or materials delivered to the Site for but not yet incorporated in the Permanent Works and their value

> (c) a list of any of those goods or materials identified in the Appendix to the Form of Tender which have not yet been delivered to the Site but of which the property has vested in the Employer pursuant to Clause 54 and their value and
>
> (d) the estimated amounts to which the Contractor considers himself entitled in connection with all other matters for which provision is made under the Contract including any Temporary Works or Contractor's Equipment for which separate amounts are included in the Bill of Quantities....
>
> 60(2) Within 25 days of the date of delivery of the Contractor's monthly statement to the Engineer or the Engineer's Representative in accordance with sub-clause (I) of this Clause the Engineer shall certify....
>
> 60(3) Until the whole of the Works has been certified as substantially complete in accordance with Clause 48 the Engineer shall not be bound to issue an interim certificate for a sum less than that stated in the Appendix to the Form of Tender....

The scope for disagreement in practice here is clear. Typical issues under such provisions are:

- The quality of the contractor's statement required by clause 60(1). The clause is silent as to the level of detail required, which is to be in a form as prescribed in the contract specification. Whether the extent and quality information provided by the contractor is sufficient for the engineer to have certified will be a matter of subjectivity in each case, particularly as to whether the contractor has an entitlement to items included in its application in principle as well as quantification. This issue is considered further elsewhere in this chapter.
- The area of 'all other matters' in subclause 60(1)(d) may be a particular problem in relation to contentious items such as delay and/or disruption claims and disputed variations.
- The threshold value in subclause 60(3) may be a problem in months where the interim value of work done is marginal. Where the contractor considers that its application merits a sum greater than the threshold, but the engineer considers the proper value falls below that threshold, then the contractor may receive no certificate and therefore no payment at all for that month and contest this.

Turning to the issue of a lower payment certificate, rather than no certificate at all, obviously there may be genuine reasons for the engineers making a different valuation in a certificate to that contended for by the contractor. This is a hardly uncommon eventuality on construction and engineering projects. This raises the prospect of difficulties in establishing under such a provision as Infrastructure Conditions clause 60(7) whether an under-certification has occurred and, if so, precisely when. It also raises the question as to whether or not interest is due if an engineer is subsequently said to have under-certified simply because a later certificate increases the amount for the same work, item or claim.

The process of interim certification for payments and the assessment of sums to be included is sometimes not a simple matter. Later adjustments to figures may become appropriate for a number of reasons, including better information becoming available or an earlier estimate of a sum for inclusion, although reasonable at the time, later proving inaccurate. The problems that can arise in the certification process and the consequent implications for the inclusion of interest under a contract provision were considered by the England and Wales courts in the case of *Secretary of State for Transport v Birse-Farr Joint Venture* (1993) 35 Con LR 8 in the context of the clause in the Sixth Edition of the ICE Conditions, one of the forerunners of the Infrastructure Conditions. The court held that there is a distinction between sums included in a certificate by the engineer, as the result of a bona fide assessment of the value of the work at a lower sum than that claimed by the contractor, even if later adjusted upwards, and the undercertification of sums due to a misunderstanding or misinterpretation of the contract provisions:

> A distinction clearly emerges from this case between the issue of a certificate which, bona fide, assesses the value of the work done at a lower figure than that claimed by the contractor and a certificate which, because it adopts some mistaken principle or some error of law, presumably in relation to the correct understanding of the contract between the parties, produces an undercertification. In the latter case there can be said to have been a withholding of the certificate ....
> The opinion which the engineer is required to form and express in his certificate is a contractual opinion. It must be a bona fide opinion arrived at in accordance with a proper discharge of his professional functions under the contract. In sub-clause (3) there is an express reference to 'the amount which in his opinion is finally due under the contract'. It is implicit in sub-clause (2) that the sum certified is that which in his opinion he considers to be due under the contract as an interim payment for that month. If it therefore should be the case that the engineer's opinion is based upon a wrong view of the contract then it can be said that he has failed to issue a certificate in accordance with the provisions of the contract .... Therefore, leaving on one side all questions of bad faith or improper motive ... a contractor who is asserting that there has been a failure to certify must demonstrate some misapplication or misunderstanding of the contract by the engineer. For example, it certainly does not suffice that the contractor should merely point to a later certification by the engineer of a sum, which had earlier been claimed but not then certified ....
> In my judgement the correct construction of the language used in clause 60 having regard to the overall scheme of the contract and the general commercial circumstances which surround any contract of this kind is that the words 'failure by the engineer to certify' refer to, and only to, some failure of the engineer which can be identified as the failure of the engineer to respect and give effect to the provisions of the contract. Those words do not refer to an undercertification which does not involve any contractual error or misconduct of the engineer.

The Judge went on to hold that even the award by an arbitrator of a sum greater than that certified by the engineer was not necessarily evidence that the engineer had failed to certify the right amount, as better evidence might be available to an arbitrator than was available to the engineer. The decision in this case was further considered in *Royal Borough of Kingston upon Thames v Amec Civil Engineering Ltd* (1993) 35 Con LR 39 and the same principles applied. It therefore seems that to apply an interest provision on the basis of the engineer's failure to certify, the contractor will need to establish that the engineer has failed to certify in accordance with the provisions of the contract, not that he has, in good faith, merely valued something at a different figure.

A notable exception among standard forms in use in the UK, and which might be considered to reflect a more modern approach to the need of contractors to maintain good cashflow, is that in NEC4-ECC. Clause 51.3 provides that if an amount due to the contractor is corrected in a later certificate, either by the Project Manager or following the decision of the Adjudicator or 'the tribunal' or a recommendation of the Dispute Review Board, then interest is due on the correcting amount from the date the incorrect amount was certified until the date the correct amount is certified.

In conclusion, the standard forms considered in detail in this book all expressly provide for the contractor to be paid interest to some extent in relation to payments, in particular the late payment of certified sums. However, what about where the contractor incurs interest or finance charges as a part of the time consequences of change? Is it entitled to claim that interest or finance and, if so, on what legal basis?

6.9.2   Finance and Interest as a Head of Claim

The concept of interest or finance charges as a head of claim for the time consequences of change under the contract is quite distinct from a claim for such charges based on an express provision of the contract, such as on undercertification or on a judgment or award. It is based on the premise that the contractor has had to finance (from its own borrowings or capital and other financial resources) the cost, loss, expense or damages it has incurred as a result of matters such as employer-caused disruption and/or prolongation of its resources on site. In such an approach the finance or interest is claimed on the basis that it effectively becomes part of those financial consequences, just the same as the wages paid for the contractor's extended management time or hire charges for its idle plant. In such circumstances the contractual mechanism for the giving of notices of the intention to claim interest of finance may be crucial to such a claim, and careful consideration of the notice requirements and periods during which such charges are incurred will be necessary. This need for notice is explained elsewhere in this chapter.

It has been said how cash flow is recognised to be the lifeblood of the construction industry, and it is certainly remarkable how many substantial construction undertakings maintain impressively high turnover figures on relatively small amounts of employed capital. This implies both a great pressure on the capital in the business and usually a reliance on external sources of finance in order to maintain the business. The other side of this scenario is that many critics would point out that most construction companies rely heavily on deferring payments to subcontractors and suppliers that are used as a source of maintaining the high turnover figures without employing further capital in the business. However, the point at issue here is that since contractors will have to fund their expenditure on a project by either borrowing at an interest or finance cost or using their own capital at a loss of interest, if that expenditure increases as a result of, say, idle working of its people, plant or equipment on a project because of its employer's failures, then that cost or loss will also be increased.

The case for the inclusion of interest as 'direct loss and expense' under what was then the terms of the 1963 editions of the JCT contracts, but would equally apply to the same term in the current SBC/Q form, was summarised in the case of *F.G. Minter v Welsh*

Health Technical Services Organisation (1981) 13 BLR 1, where the Court of Appeal held that such costs are part of the direct loss incurred by a contractor and not, as was argued by the defendant, simply a claim for interest. Lord Justice Stephenson summarised the position as follows:

> It is agreed that … the court should apply to the interpretation of what loss or expense is direct the distinction between direct and indirect or consequential which was discussed by Mr Justice Atkinson in *Saint Line Ltd v Richardson, Westgarth & Co* [1940] 2KB 99, in construing an exclusion of liability clause and should at least recognise that loss of profit and expenses thrown away on wages and stores may be recoverable as direct loss or expense, as he there held. So this court held that the cost of men and materials being kept on a site without work was recoverable as damages and not excluded as 'consequential' in *Croudace Construction Ltd v Cawoods Concrete Products Ltd* [1978] 2 Lloyd's Rep 55. It had held the same in *B. Sunley & Co v Cunard White Star Ltd* [1940] 1 KB 740, making there a small award of interest on money invested and wasted.
>
> It is further agreed that in the building and construction industry the 'cash flow' is vital to the contractor and delay in paying him for the work he does naturally results in the ordinary course of things in his being short of working capital, having to borrow capital to pay wages and hire charges and locking up plant, labour and materials capital which he would have invested elsewhere. The loss of the interest which he has to pay on the capital he is forced to borrow and on the capital which he is not free to invest would be recoverable for the employer's breach of contract within the first rule in *Hadley v Baxendale* (1854) without resorting to the second, and would accordingly be a direct loss, if an authorised variation of the works, or the regular progress of the works having been materially affected by an event specified in clause 24(1), has involved the contractor in that loss. [Note: The reference to clause 24(1) is to the 1963 JCT Standard Form, now clause 26 of the 1980 version.]
>
> On reaching this point the claimants might be thought to be nearly home. Why [ask the claimants], should they not be entitled to be repaid what they have lost as a natural as well as a contemplated result of what has happened to their contract through happenings which are in no way their responsibility, if not the respondents' faults or the fault of their architect acting for them? If the respondents choose to vary the contract, or if their architect fails to give them the necessary instructions and they have to keep men and machinery idle and have to pay for them and their subcontractors, why should they be allowed the cost of the extra work but be deprived of what is part of the same loss? Why should the contract sum be adjusted to include one but not the other? [They submit] that the whole risk of the claimant's loss, or the risk of the whole of their loss, naturally resulting from the happenings necessary to activate clauses 11(6) and 24(1), was contractually assumed by the respondents. Yet if the Judge is right, the claimants are worse off as a result of actions for which the respondents are responsible and that unjust result is reached by limiting the natural meaning of direct loss and/or expense to exclude losses which were just as clearly within that meaning as the sums which the architect has allowed. There is [they submit], nothing in the machinery provided by the two clauses or its operation in this case to break the chain of causation

or to halt the operation of the root causes of these losses, even if there is any concurrent or supervening cause, or to turn an obviously direct business loss into a
loss that is indirect or inconsequential.

This seems a succinct and reasonable summary of the reason for the inclusion of interest
as a head of 'loss and expense' under what was then clause 26 of the 1963 edition of the
JCT contracts, but is now effectively the same in SBC/Q. The judgment went on to allow
such charges subject to qualifications about notice requirements under the contract.

The need for due notice to be given of the intention to claim finance charges, again
under the 1963 version of the JCT Standard Form, was considered further by the Court
of Appeal in *Rees & Kirby Ltd v Swansea City Council* (1985) 30 BLR 1. There it was
held that the contractor's application for payment must include a clear reference that it
includes finance charges and that such notice should generally be given within a reasonable time of the loss or damage being incurred. The requirement for notice will always
be subject to any express provisions of the relevant contract but in the absence of such
provision, care should be taken to ensure that clear notice of the intention to include
finance charges is given within a reasonable period, as without it the recipient of the
claim will not be aware that such charges have been, and presumably are continuing
to be, incurred. In the absence of such notice the recipient may act differently than if
proper notice is given, i.e. it should have the opportunity to restrict the amount of such
charges by dealing with the claim as soon as reasonably possible.

The decision in *Minter v WHTSO* makes the position clear in relation to interest being
a head of a claim under the JCT (now SBC/Q) term 'loss and/or expense', under English
law. The same approach to this term 'loss and/or expense' was adopted by the Scottish
courts in *Ogilvie Builders Ltd v City of Glasgow District Council* 1995 SLT 15.

It seems to be widely accepted that there is nothing in the decision in *Minter v WHTSO*
that would make the position any different under a construction contract under English
law that uses terms such as 'cost' and 'expenditure' in such standard forms as the FIDIC
Red Book and the Infrastructure Conditions, rather than 'loss and/or expense' under
SBC/Q terms. This is because those terms 'cost' and 'expenditure' in clauses for compensation relating to employer delays would be similarly the equivalent to damages for
breach and that finance or interest costs would fall as a head of such damages.

However, that may well not be the case elsewhere. Some legal systems take a different approach to allowing the charging of interest. This is particularly the case where
Sharia law applies. A contractor may therefore have to carefully consider how it frames
a claim for interest or finance charges. This must start with local legal advice. However,
this should involve a simple consideration of the meanings of the relevant terms used
in the contract that define the contractor's entitlement in the event of the causes relied
upon. Taking the three standard forms considered in this book, the following suggestions are made.

As noted elsewhere, under SBC/Q the term 'loss and/or expense' is used. Applying
these two alternative words, it could firstly be envisaged that a claim for loss of interest
on capital that would have been deployed might be argued as a head of 'loss'. Taking the
alternative funding position, finance charges from a lender might be argued as a head of
either 'loss' or 'expense'.

In the FIDIC Red Book, the defined term 'Cost' in clause 1.1.4.3 has been quoted previously and explained as being the basis of the contractor's entitlement in relation to the

various delay and associated costs clauses of that form. However, it is quoted again with emphasis added:

> 'Cost' means all *expenditure* reasonably incurred (or to be incurred) by the Contractor, whether on or off the Site, *including overhead and similar charges*, but does not include profit.

Under this provision, finance costs incurred on borrowings might be considered covered by 'expenditure' and in particular within the term 'overhead and similar charges'. This is on the basis that financing its projects is an integral part of any contractor's overheads, even before it might in any event be captured by 'similar charges'. In relation to financing charges, and also loss of interest that might otherwise have been earned, the FIDIC Contracts Guide 2000 put it as follows:

> Overhead charges may include reasonable financing costs incurred by reason of payment being received after expenditure. In some countries, financing costs might be included within 'Cost', even though funds were not borrowed because the Contractor had sufficient funds at his disposal.

In the Infrastructure Conditions, the defined term 'cost' in clause 1(5) has been quoted previously and explained as being the basis of the contractor's entitlement in relation to the various delay and associated cost clauses of that form. However, it is quoted again with emphasis added:

> The word 'cost' when used in the Conditions of Contract means all *expenditure* properly incurred or to be incurred whether on or off the Site *including overhead finance and other charges* properly allocatable thereto but does not include any allowance for profit.

This is therefore similar to the FIDIC Red Book definition, but might be thought to even more pointedly cover 'finance' charges. Alternatively, as suggested in the FIDIC Contract Guide as quoted, in some jurisdictions financing costs might be included within 'cost' in the Infrastructure Conditions.

These suggestions based on the FIDIC Red Book, the Infrastructure Conditions and SBC/Q are no more than ideas as to how interest or finance costs might be claimed as a head of a claim for such prolongation, disruption or acceleration costs under the express terms of those forms, where the approach of such as English law in equating such claims to ones for damages and therefore including such components is not available. In such other jurisdictions local legal advice is essential.

In addition to interest or finance charges on its other heads of a claim for prolongation, contractors sometimes make specific interest claims in relation to some other consequences of delay. These include items such as interest or finance charges on the delayed release of retention monies or contract bonds. The basis of such claims will be that the release of such items is dependent on a trigger event, such as completion of the works, and if that event is delayed, then the contractor will incur additional costs. Such costs in relation to extended contract bonds are considered in Chapter 8.

Retention percentages on international projects can be as high as 10% or more of the contract price. If the contract price is in the hundreds of millions of dollars, then the retention fund held by the employer at 10% will be in the tens of millions of dollars. Thus, delays in the release of that retention may have a significant financial effect for the contractor's cash flow. Taking the position under FIDIC Red Book clause 14.9 requires the first half of the retention money to be certified by the engineer for payment to the contractor when the Taking-Over Certificate has been issued under clause 10.2. If delays to the works delay that certificate, then the contractor may be stood out of tens of millions of dollars for as long as completion is delayed. This will similarly delay the release of the second half of the retention money following the Defects Notification Period, with similar financial effects.

However, the effects of delay to the works is likely to also mean that the value of the works carried out will accumulate at a slower place, and hence the rate at which retention monies are deducted from interim payment certificates will also be delayed. This means that the calculation of the contractor's additional interest or finance charges in relation to retention monies will require careful consideration. This will compare when retention was deducted and released with when it would have been deducted and released 'but for' the compensable delays. This requires a 'but for' cashflow projection, which might have to be developed retrospectively although it may be that the tender documents include such information. A further complication is where the works have been subject to both compensable delays and delays that were either cost-neutral (such as weather) and/or the contractor's own culpability. Such complications will require adjustment to the 'but for' analysis to exclude the effects of the cost-neutral events and the contractor's own failures. Where the contract is subject to sectional completion and/or the works are subject to partial possession, with the partial release of retention monies, then the calculation can become even more of a mathematical maze.

6.9.3 Damages for Breach

Several situations can be envisaged in which a contractor would not frame its claim for such as delay, disruption or acceleration as one for 'loss', 'expense', 'cost' or 'expenditure' or similar words under the express provisions of the contract, but as a claim for damages for breach of contract. These include:

- Where the contract contains no express provisions dealing with the default of the employer that has given rise to its claim.
- Where the contractor has failed to comply with a condition precedent in the express terms, for example in relation to notice, and wishes to circumvent the negative consequences of that failure.
- Where the express terms place some limit on what the contractor can recover compared to what it can recover under the wider law of damages for breach.

The inclusion of interest or finance charges on borrowing or capital as part of the assessment of such as contractual 'loss and expense' is distinct from the inclusion of interest charges in the assessment of damages. The loss and expense assessment arises from a contractual entitlement, whereas damages are assessed for breaches of contract.

Historically, English law did not allow interest as damages for non-payment of an unpaid debt. This antipathy towards the lending and borrowing of money still affects claims for interest in many other jurisdictions. Under English law, the usually quoted authority for this position was the decision of the House of Lords in *London Chatham and Dover Railway v South Eastern Railway* [1893] A.C. 429. Even then, this was regarded by their lordships as an anachronism, but to have been the position under the law in England for too long for them to depart from it.

Thereafter, the law moved slowly. In the 1980s interest as damages was accepted as special damages under the second limb of *Hadley v Baxendale* in the Court of Appeal judgment in *Wadsworth v Lydall* [1981] 1 WLR 598. There the court took up a suggestion from Lord Denning in *Trans Trust SPRL v Danubian Trading Co Ltd* [1952] 2 QB 297 and held that the *London Chatham and Dover Railway* case was only concerned with a claim for general damages, and not special damages. As Brightman LJ put it:

> If a plaintiff pleads and can prove that he has suffered special damage as a result of the defendant's failure to perform his obligation under a contract, and such damage is not too remote on the principle of *Hadley v Baxendale* (1854) 9 Exch 341, I can see no logical reason why such special damage should be irrecoverable merely because the obligation on which the defendant defaulted was an obligation to pay money and not some other type of obligation.

This decision was approved by the House of Lords in *President of India v La Pintada Comania Navigacion SA* [1985] AC 104.

As a result, in the Second Edition of this book we wrote 'Interest still cannot be claimed as part of general damages but can be claimed as part of special damages'. However, this position has since been brought in line with the realities of modern commercial life by the House of Lords in *Sempra Metals Ltd v Revenue & Anor* [2007] UKHL 34:

> The common law should sanction injustice no longer. The House should recognise the remnant of the restrictive common law exception for what it is: the unprincipled remnant of an unprincipled rule. The House should erase the remains of this blot on English common law jurisprudence.

The House held that loss of interest resulting from late payment of a debt was recoverable whenever it has been pleaded and proven as a matter of fact. Furthermore, that interest could be on a simple or compound basis, depending on the facts. Such interest would, however, still be subject to the usual tests of remoteness and foreseeability:

> To be recoverable the losses suffered by a claimant must satisfy the usual remoteness tests. The losses must have been reasonably foreseeable at the time of the contract as liable to result from the breach.

One form in which contractors sometimes claim interest or finance charges in relation to claims for such as delay, disruption or acceleration is not as a head of such claims, but on the basis that the claims should have been paid earlier than they were. In such a case the basis of claim is not the events of delay, disruption or acceleration themselves, but breach of contract by the employer in not compensating the contractor for their financial consequences in a timely manner as required by the contract.

An important distinction between claiming interest as just another head of the contractor's claim for the time consequences of change, alongside such as its staff salaries and plant hire costs, as opposed to such a claim for interest as damages for breach, may be as to the date or dates at which the interest starts to run. This can make a very significant difference to the interest calculation. This is illustrated by a recent example that was not under English law. In the 2018 arbitration of claims arising out of construction of a liquified natural gas terminal, the claimant subcontractor claimed interest on all of its claims. The principal sums of those claims (such as unpaid variations and delays costs) were mainly for events and costs incurred by it in 2009. It claimed interest, not as a head of those costs, but as damages for breach of contract on the basis that it should have been paid for those claims when reasonable details and substantiation were submitted by it. Its calculation of interest was therefore not from 2009 but from the dates it contended it submitted properly detailed and substantiated claims. That interest start date varied between the various principal heads of claim, but for its claim for delay and suspension of works in 2009 the start date of the interest calculation was in 2016, because it accepted that it had not submitted a properly substantiated claim until then. Thus, the interest on that major head of claim was for two years (2016 to 2018), rather than the nine years (2009 to 2018) it would have been had the interest been claimed as part of the delay and suspension costs when they were incurred. As can be imagined, this had a significant effect on the amount of the contractor's resulting calculations of its interest claims.

In this example, as in any claim for damages based on a breach, the contractor still had to prove the breach. For this claim for interest, this meant establishing when the certifier ought reasonably to have included monies in respect of each principal head of claim in a payment certificate, such that they would have been paid. In any such scenario, having established when amounts should have been paid, the next question will be in what amounts they ought to have been paid. Such analyses are inevitably highly subjective. They involve considering such as what submissions, information and substantiation the certifier had received at the time and comparing that to what the contractor was required to provide by the terms of the contract. The consideration is then what amount ought reasonably to have been certified against what was actually provided.

As explained, this was a claim by a subcontractor. Therefore the role of both certifier and payer sat with one party – the contractor. Where there is a separation of these roles, a further issue will be whether the payer is liable for the failure of the certifier. This is discussed in section 8.8.2.

The difficulties of establishing such issues are another reason that makes an interest claim on this basis usually much less attractive than one based on the interest as a part of the expenditure incurred by the contractor as a result of items such as delay and disruption, at the time and in the amounts that they were incurred.

6.9.4   The Rate and Compounding of Interest

Having established that a contractor can include in a claim for the time consequences of change for costs or losses incurred by way of financing or interest on capital retained or monies denied, the question of the applicable rate arises. In addition, there is the question of whether the interest should be simple or compounded, and if the latter, then at what rest periods?

If interest or finance charges are being included as such as 'loss', 'expense', 'costs', 'expenditure' or damages, it seems logical that the level of the charge should reflect the actual amounts suffered by the contractor, and not some nominal or notional interest rate. The Court of Appeal in *Rees & Kirby* agreed that the rate of such charges should acknowledge the circumstances of the contractor in incurring the charges. Lord Justice Goff set out the means of calculating such charges as follows:

> There remains to be considered the question whether the respondents are entitled to recover their financing charges only on the basis of simple interest, or whether they are entitled to assess their claim on the basis of compound interest, calculated at quarterly rests, as they have done. Now here, it seems to me, we must adopt a realistic approach. We must bear in mind, moreover, that what we are here considering is a debt due under a contract; this is not a claim to interest as such, as for example a claim to interest under the Law Reform Act, but a claim in respect of loss or expense in which a contractor has been involved by reason of certain specific events. The respondents, like (I imagine) most building contractors, operated over the relevant period on the basis of a substantial overdraft at their bank, and their claim in respect of financing charges consists of a claim in respect of interest paid by them to the bank on the relevant amount during that period. It is notorious that banks do themselves, when calculating interest on overdrafts, operate on the basis of periodic rests: on the basis of the principle stated by the Court of Appeal in Minter's case, which we here have to apply, I for my part can see no reason why that fact should not be taken into account when calculating the respondents' claim for loss and expense in the present case. It follows that, in order to calculate the respondents' contractual claim it will be necessary to calculate it with reference to the total sum … in relation to the period I have indicated … taking into account:
>
> (1) the two payments made on account during that period;
> (2) the rates of interest charged by the bank to the respondents at various times over that period; and
> (3) the periodic rests on the basis of which the bank from time to time added outstanding interest to the capital sum outstanding for the purposes of calculating interest thereafter.

On this basis, the rate of interest or finance charges, whether it is compound or simple, and the manner in which it is incurred in respect of rests, etc., in its calculation, would therefore be the same in the claim for payment as those which apply in incurring the charge. There are, however, further situations that may be encountered in practice, but which were not addressed in the *Rees & Kirby* case.

Firstly, there may be circumstances that mean the contractor has had to borrow money or otherwise obtain finance other than by straightforward bank borrowing which have resulted in the costs of the finance being substantially more than could be expected from normal commercial bank borrowing. If such exceptional circumstances have not themselves been directly caused by the events giving rise to the claim it seems there may be an objection to the actual costs being incorporated in the claim on the grounds that such a level of charge was not reasonably foreseeable, and therefore was not in the parties' contemplation as a consequence. Even if exceptional costs have been incurred as a direct result of the events, it would still be necessary to establish that the enhanced level of charging was not caused by difficulties of the contractor that could not have been known, or anticipated, by the other party. If the contractor has a difficult financial situation not caused by the events giving rise to the claim it may not seem reasonable that the exceptional level of charge should be recoverable as the other party could presumably have no way of knowing that the contractor would incur such a rate of charge. In such circumstances it would seem reasonable to calculate the charges at a commercial level of interest and periodic rests, etc., that would reflect a reasonable market level of interest charges at the time.

Similarly, if a high interest or finance rate is the result of the claimant's own bad management, then it should not be allowed to benefit from this (see in this regard *Harlequin Property (SVG) Ltd v Wilkins Kennedy* [2016] EWHC 3233 (TCC) described elsewhere in this chapter).

A further difficulty that may arise is where the contractor incurring the charges is part of a group of companies and financing is provided centrally. In these circumstances it is not uncommon to find notional rates charged by the parent or holding company on funds required by the subsidiaries. These circumstances generally raise two difficulties:

1. The source of financing is remote from the party undertaking the contract, and the obtaining of finance on a group basis might impact on the level of such costs incurred. The impact could be beneficial if the charges are less than would be expected for the contractor's business alone or may be detrimental if the extent of central borrowing, and circumstances of the group, is such that rates are higher than would be expected for the contractor's business alone.
2. The complications outlined in 1. demonstrate that there may be a break in the chain of causation in respect of the finance charges, as, particularly with large groups with other cash rich enterprises, it may be difficult to separate out the impact of funding the contractor's financing of the events giving rise to the claim.

Other considerations here may be similar to those in relation to the hiring of plant and equipment between companies or divisions of the same wider legal entity, which were considered in Section 6.2. Questions will again involve the legal structure of a group and whether the companies are truly operating 'at arm's length. In such circumstances the actual level of charges incurred centrally may be relevant and the effect of any of the matters outlined above will need to be considered in the light of the particular financial arrangements in order to establish the real rate of charge incurred by the group, rather than any notional rate imposed by the parent or holding company.

In any claim for interest or finance charges, there will probably be added complication as the rates change over time, but this is a matter that a simple spreadsheet can usually address, as is the addition of accumulated interest in a compound interest calculation. More complex is the position where the contractor moves between overdraft and being cash-positive, such that it moves between incurring and losing interest. This might also affect the recoverability of the interest if, for some legal or contractual reason, the contractor is only entitled to interest paid and not interest lost.

Some contractors' funding arrangements can particularly add both complexity to the calculations and also issues as to whether the costs are recoverable. In this regard reference is made to Section 8.2 of this book and the example of a contractor's approach to obtaining funding through a mixture of facilities and currency swaps. Whilst that example related to the effects of calls on bonds, the same situation can arise in relation to such things as a delay or disruption claim. In that case arguments ensued as to whether or not the arrangements and associated costs were the natural result of the alleged failures that were the basis of the claim and were in the contemplation of the parties. That claim was not made under English law, but if it were then such a claim might be defended on the basis that the resulting finance costs fell under neither limb of allowable damages under the rules in *Hadley v Baxendale*.

Of course, finance or interest charges can be incurred or lost and are not only incurred when a payee has been underpaid. They can also be incurred where a payer has overpaid. This was the position considered in *Harlequin v Wilkins Kennedy* referenced elsewhere. Harlequin was the developer of a resort in the Grenadines. It engaged Wilkins Kennedy to provide accountancy and audit services. The case was a professional negligence action. Harlequin asserted that Wilkins Kennedy was in breach of contract and other duties and was responsible for cost overruns of the project. There were also allegations of fraud. One of the issues was the applicable rate of interest to which Harlequin was entitled on the overpayments it had made and was entitled to recover from Wilkins Kennedy. The Judge was referred to a number of cases in which such a percentage was used: *Hunt v Optima (Cambridge) Ltd* [2013] EWHC 1121 (TCC) (2% over base); *Ramzan v Brookwide Ltd* [2011] 2 All ER 38 (2% over base); *PGF II SA v Royal Sun Alliance Insurance PLC* [2010] EWHC 1459 (TCC) (3% over LIBOR). He concluded that all of these cases make it plain that:

> ... the object of an award of interest is to compensate a claimant for being kept out of the money that should have been paid to him as damages.

Of Harlequin's claim for interest, he noted that it was not one for special damages and that it did not borrow any money, so it did not incur any interest charges or suffer any actual loss in consequence of the overpayments. He explained how Harlequin had attempted to claim interest as general damages was:

> ... an unsubtle attempt to create a windfall for the claimant by recovering interest at a rate which it never incurred and would never have incurred ....

and that it:

> ... did not borrow money and, on the balance of probabilities ... would never have been lent any money because of its complete lack of financial security and credit-worthiness. Moreover, these difficulties arose because the claimant was operating a questionable, and even potentially criminal, business enterprise. It would be wholly wrong in principle for the claimant now to take advantage of their own sub-sub-prime status to increase the recoverable interest rate.

He considered that for these reasons, any comparison with a hypothetical company who may have been able to borrow money would be artificial and unhelpful.

He referred to Lord Justice Rix in *Jaura v Ahmed* [2002] EWCA Civ. 210 and that the rate of interest should reflect 'the real cost of borrowing incurred'. He concluded that:

Accordingly, taking into account the polarised submissions made by both sides, I have concluded that the real cost of borrowing in this case would be reflected by a rate of 1.5% above base for the relevant period (a figure only slightly above that which was originally proposed by the defendant). I am confident that that reflects the commercial reality of borrowing and investing at the relevant time, and does not reward the claimant for its self-inflicted status as a sub-sub-prime borrower.

6.9.5 Adjustments to the Calculation

The calculation of interest sums in situations where the claim is based upon matters such as delayed progress and subsequently delayed revenues also raises the issue of whether the calculation should include a compensating deduction to account for the contractor's rate of expenditure also being delayed, as discussed elsewhere. This is particularly relevant in circumstances where the contractor is undertaking works involving the letting of substantial portions to subcontractors, and is therefore the conduit for the flow of cash rather than its final recipient. However, it might also be relevant in relation to suppliers. For example, say that a contractor makes a claim for 'direct loss and/or expense' in the form of prolongation costs that includes amounts in relation to plant hire companies and interest. If the contractor is running an overdraft, the point at which it will start to incur interest in relation to those plant costs will be when it pays its hire company, not when the delay occurred. Of course, the hire company might be charging interest under its supply contract up to the point that it is paid late, and this might form part of the contractor's claim up the line. The key here is to establish who was 'out of pocket' at any one time.

There may also be instances where the ultimate payment is calculated, in whole or in part, by reference to the contract unit rates and prices, rather than actual expenditure. These situations have been considered in Section 6.2 of this book and in particular where prolongation is priced as a part of the valuation of a variation or the parties agree to apply rates from a preliminaries and general bill of quantities for reasons such as expediency. In such circumstances the contract prices will usually be deemed to include an element of financing, at least for the anticipated duration of the works, and due allowance may therefore need to be made to avoid duplication of recoveries.

Further potential for duplication with a claim for finance or interest charges is with a claim for head office overheads as considered in Section 6.8 of this book. Where a contractor is using facilities to fund its operations, it seems likely that head office overhead costs will include finance costs on such facilities. The extent of such duplication will depend on how the claim for head office overheads has been calculated. This may require some detailed analysis of the contractor's head office costs.

Calculation of finance charges may therefore require credits to be applied for the related delay in cash flow outward as well as inward, for duplication with the deemed inclusion of a financing element in any contract rates and prices used or the allowance for head office overhead costs in the same claim.

6.9.6 Statutory Interest

In the United Kingdom, in an attempt to reduce the pressures on cash flow, and the consequent pressure on the viability of businesses, especially small and medium sized businesses, the Late Payment of Commercial Debts (Interest) Act 1998 was introduced. The effect of the Act is to imply a term into commercial contracts that any qualifying debt created by the contract will carry simple interest in accordance with the terms of the Act.

The Act applies to contracts for the supply of goods or services where both purchaser and supplier are acting in the course of a business. Some contracts such as consumer credit agreements, mortgage contracts and other security or charge contracts are excluded from the Act.

A debt created by the contract to which the Act applies accrues interest from the day after the relevant day for payment of the debt under the contract. Interest is set to be calculated at a rate of 8% over the Bank of England base rate. The claimant can also recover a sum for the recovery of that late payment.

The Act was introduced in stages so that at first it only applied to contracts involving small businesses, defined as a business with an average of fewer than 50 employees in the financial year prior to the date of the contract, but it has since been extended to larger enterprises.

Under Section 8 of the Act the right to statutory interest, that is interest under the Act, can only be excluded from a contract where the parties have agreed what is referred to as a 'substantial remedy' for late payment of debts under the contract. This requires consideration of what is a 'substantial remedy'.

As noted elsewhere in this chapter, the decision in *Lilly v Mackay* considered the terms of the JCT Standard Form of Building Contract 1998 Edition Private Without Quantities. In relation to interest, its provisions are effectively the same as those of SBC/Q. One of Lilly's assertions was that that form's provision for interest at 5% above the Bank of England base rate was not a 'substantial remedy' for the purposes of the Act. The Judge disagreed:

> This contract rate is to be compared with the statutory rate under the statute which is Base Rate plus 8%; so the statutory rate is 3% better than the contract rate. I have no doubt that the contract rate provides a 'substantial remedy' within the meaning of Section 8 of the statute. Any 'substantial remedy' must be one which at least judged at the date of the contract would provide adequate compensation for late payment. Section 8 is obviously considering at least the possibility that the 'substantial remedy' will be less than the statutory interest remedy. Whilst the statutory 'remedy' is Base Rate plus 8% and that is a 'better' remedy for the Contractor than the contractual remedy for late payment, that does not mean that the contractual remedy is not 'substantial'.

On the other hand, it is not certain whether the provisions of standard forms such as that included at clause 60 of the Infrastructure Conditions would be considered to include such a 'substantial remedy' in order to allow the contracting out from the Act. There is certainly a marked difference between the contractual remedy in the Infrastructure Conditions at 2% above the base lending rate of the bank inserted in the Appendix to Form of Tender and the Act's 8% over the Bank of England base rate. This is especially the case in times when base rates are only 1% or less. However, the temptation to insert excessive interest rates in the contract should be resisted, as it is possible that they will be regarded as a penalty, as demonstrated by the 2003 case in the Court of Appeal in

Jeancharm v Barnet Football Club (*Building Magazine*, 7 February 2003). In this case Jeancharm had inserted an interest rate of 5% per week into a contract for the supply of football shirts to Barnet Football Club in the event that they were not paid in accordance with the contract, i.e. 260% per annum on a simple basis! The court rejected this level of interest on the basis of its being extravagant and unrealistic in respect of the greatest damage that could flow from late payment and the interest provision was therefore a penalty and unenforceable.

6.10   Duplication of Recoveries

It seems obvious that the evaluation of a contractor's various contractual claims, whether for the direct consequences of change considered in Chapter 5 of this book, the time consequences of change considered in this chapter, termination considered in Chapter 7, or the other claim causes considered in Chapter 8, should not overlap such that the same compensation is duplicated. In SBC/Q this is expressly stated in clause 4.23, by limiting the contractor's entitlement to 'direct loss/and or expense' as follows, to which emphasis has been added:

> If in the execution of the Contract the Contractor incurs or is likely to incur direct loss and/or expense *for which he would not be reimbursed by a payment under any other provision in these Conditions* ....

The scope for duplication is apparent from a summary list of the claims considered in this book:

- Chapter 5:
  - Remeasurement
  - Variations
  - Dayworks
  - Omissions
- Chapter 6:
  - Prolongation
  - Disruption
  - Preliminaries thickening
  - Overheads and profit
  - Interest charges
- Chapter 7:
  - Termination
- Chapter 8:
  - Fluctuations in costs
  - Suspension
  - Post-handover costs
  - Claims preparation costs

The most obvious potential for overlap between such claims is in relation to preliminaries and general costs and off-site overheads and profit. This is particularly the case between the claims considered in Chapters 5 and 6 and allowances for such as plant, equipment, supervision and management costs. However, there can be overlap in claims that are considered in the same chapter, for example between prolongation and a disruption claim. Of the claims considered in Chapter 8, fluctuations in costs can overlap with any of the claims in Chapters 5 and 6; post-handover costs can overlap with most of the claims in Chapter 5 where additional works are carried out in that period; and claims preparation costs can overlap with allowances for site and head office overheads in any other claim. There are other possibilities; these are just common examples.

Two particular practical problems that arise in relation to any overlaps between claims are deciding the following:

- in which of the overlapping claims to make the required reduction and
- how to calculate that adjustment.

The first of these questions might seem credulous, on the basis that it must be the issues that caused the additional value, cost, loss, expense or damage to which it should be attached. It cannot have been caused by issues included in both claims. However, in practice this is not always that simple, as is explained elsewhere in this chapter. Approaches to these practical problems can best be suggested by reference to some examples.

6.10.1   Overlaps Between Prolongation and Disruption Claims

In Section 6.2 of this chapter, prolongation costs were explained as usually including allowances for the contractor's management and supervision and it was said that the costs should be those incurred at the time the compensable delay events had their effect. In Section 6.4, disruption claims were considered, including claims for thickening of supervision and management resources. If the thickening of resources occurs over periods during which compensable delays occurred (as is usually the case), then the quantification of these claims will overlap, unless they are properly adjusted. However, the initial question is as stated elsewhere – which of these claims is to be reduced to avoid duplication? This can be considered by reference to a simplified illustrative example as follows.

This is based on a very simple hypothetical four month planned programme, which is delayed by compensable delays of one month to an actual period of five months. In addition to a claim for prolongation associated with this additional one month working, the contractor also claims that its levels of preliminaries and general costs (particularly staff) were increased as a result of disruption also caused by the employer.

The contractor asserts that, but for the delays caused by the employer, it would have completed the works in the planned four months with 10 staff in the first month, rising to 20 in the second, peaking at 30 in the third month and falling to 10 in the final month, a total of 70 man-months. This is illustrated in the staff histogram in Figure 6.6. For reference, the four planned months are labelled A to D.

The contractor asserts one month of compensable delay in month B, giving a new month E. It also claims that disruption increased staff to 40 in months B and E, 50 in C and 20 in D, a new total of 160 man-months. This is illustrated in Figure 6.7.

Therefore, the contractor claims 90 additional man-months as set out in Figure 6.8, with the additional staff highlighted. The conundrum is how these 90 additional man-hours should be split between the contractor's prolongation and thickening claims. The traditional approach addresses prolongation first, such that the claims are

Figure 6.6 Planned staff numbers.

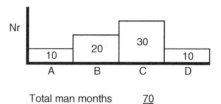

Total man months 70

Figure 6.7 Actual staff numbers.

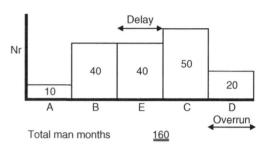

Total man months 160

Figure 6.8 Additional staff numbers highlighted.

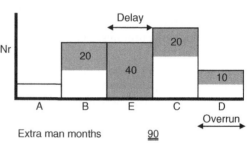

Extra man months 90

Figure 6.9 Claim if prolongation is addressed first.

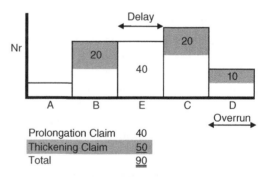

Prolongation Claim	40
Thickening Claim	50
Total	90

as shown in Figure 6.9, with the thickening highlighted. The alternative approach would address the thickening first, such that the claims are as shown in Figure 6.10, again with the thickening highlighted.

Whilst it has been suggested that the traditional approach would evaluate the prolongation claim first, there are other considerations. It will be noted how the thickening occurred first, from month B, and on this basis it might be argued that the thickening should be addressed first. However, if it were not for the prolongation, the thickening in month E would never have been experienced.

Whichever of these alternatives is correct, the practical point relates to what happens if one of the prolongation and thickening claims is successful. It is the nature of

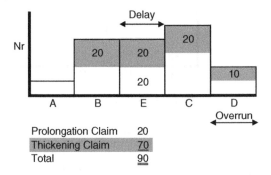

Figure 6.10 Claim if thickening is addressed first.

Prolongation Claim	20
Thickening Claim	70
Total	90

construction claims that those for prolongation tend to be more successful than those for disruption, such as preliminaries thickening. It may be that this has led contractors into the traditional approach suggested. If the thickening claim fails in this hypothetical scenario, the employer might argue that the successful prolongation claim should be limited to 20 members of staff in the delay month E. This is on the basis that the additional 20 in that month were related to disruption. The employer will say that, because that disruption claim has failed, the contractor should not be able to have part of it 'through the back door' of a prolongation claim that includes the full actual 40 staff members in month E.

However, it might be relevant to understand why the thickening claim failed. Two of the possibilities go to why there was a difference between the contractor's planned and actual staff numbers. These alternative reasons and their suggested consequences for the hypothetical claims set out are as follows:

1. Because the planned staff numbers in Figure 6.6 were inadequate. Thus, the contractor could never have completed the project with the asserted planned staff levels and would have always required 40 members in month B, and hence additional month E. In this case, it is suggested that the approach in Figure 6.9 is the correct one, with the prolongation in month E being valued on the basis of the actual 40 staff.
2. Because the actual staff numbers in Figure 6.7 were too high. Thus, those actuals were not the result of the employer's actions, but were the result of problems the contractor was having of its own making, such as inefficiency, own-goals or unnecessary over-staffing. In this case, it is suggested that the approach in Figure 6.10 is the correct one, with the prolongation in month E being valued on the basis of the 20 staff that were planned and would have been there but for the contractor's own problems.

These suggested approaches will not fit all situations and the facts in each case will be essential, Furthermore, such overlap situations are never as clean as those set out in these figures.

6.10.2 Overlaps Between Additional Work and Prolongation

This is the most common area in which duplication arises between claims and for which adjustment is required. As noted elsewhere, it includes remeasurements, variations, dayworks and any items where the value of additional work done by the contractor includes recoveries of items that have gone into the evaluation of prolongation costs. This particularly includes on-site overheads (or preliminaries and general costs), such as people, plant, equipment and site establishment, and also off-site overheads. It also requires consideration of any omissions that have reduced the contractor's recoveries of such items.

Effectively it requires consideration of the details of any valuation of the 'direct consequences of change' as considered in Chapter 5 of this book that might recover allowances in the evaluation of the 'time consequences of change' considered in this chapter.

Regarding where the adjustment should be made, this is almost universally carried by any recoveries through the value of additional works (additional quantities, variations, dayworks, etc.) being abated from the prolongation claim. It is suggested that this must be correct, except in exceptional circumstances where the costs were never really related to the variation.

As explained elsewhere, if the cause of prolongation is variations, then it may be that there is no prolongation claim to abate, if the valuation of those variations includes the full value of their time-related effects as required by, for example, SBC/Q clause 5.6.3.3. Alternatively, it may be that there is a risk of substantial duplication of claims.

The areas of the contractor's account for the valuation of work done that might need to be considered include the following. This is set out by reference to recoveries of preliminaries and general items, but could also include off-site overheads and profit, depending on how these have been calculated in a prolongation claim.

- Increased value of the works through increases in quantities under contracts where the contract quantities are remeasurable, such as the FIDIC Red Book and the Infrastructure Conditions. When the rates and prices in the contract are applied to increased quantities, it may be that they will recover costs included in a prolongation claim. This is particularly the case where those rates and prices are inclusive of the contractor's preliminaries and general costs. In this case calculating the extent of duplication will require details of what allowances there were for preliminaries and general items in the rates and prices for the remeasured works whose quantities have changed.
- Even if the contract includes bills of quantities measured under rules of measurement, such as POMI and CESMM4, which require a separate section for the pricing of preliminaries and general items, the rates for measured work items may still include some allowances for such items. It may be that the contractor has only included in that separate bill section for preliminaries and general items that are time-related, with work-related items included in the rates for the items of permanent work for which they will be required. It might be said that such work-related preliminaries and general costs should not be included in the prolongation claim, but in practice things are usually not that simple and there is some element of duplication. Therefore, detailed analysis is required.
- Variations. As explained in Chapter 5, the approach of contracts such as SBC/Q, the Infrastructure Conditions and the FIDIC Red Book is that these should primarily be valued at rates and prices in the contract, or rates and prices derived therefrom. As a result, similar considerations to those in the above bullet points will apply.
- Variations carried out on a Dayworks basis. Those dayworks rates will be set out in the contract and usually have their own section in the bills of quantities. Obviously, the extent of duplication with its other claims will depend upon what the contractor's daywork rates allow for by way of costs that could form part of such a prolongation

claim. This in turn will require consideration of the conditions of contract and the method of measurement under which the bills of quantities were prepared. Taking an example of each, the possibilities for duplication are illustrated as follows:

- Under SBC/Q clause 5.7 the valuation of Dayworks comprises the prime cost in accordance with the 'Definition of Prime Cost of Daywork carried out under a Building Contract' issued by the RICS, together with the 'Percentage Additions on the prime cost at the rates stated in the Contract Bills'. That 'Percentage Addition' is defined in section 3.1 of the June 2007 Edition of the RICS Schedule as covering 'Incidental costs, overheads and profit' and further detailed in its section 6, where the items to be included in the percentage are listed at length as items (a) to (t). Of these, the following may be particularly relevant to overlap with a prolongation claim:

    (a) Head Office charges.
    (b) Site staff, including supervision.
    (o) Use of erected scaffolding, stages, trestles and the like.
    (p) Use of tarpaulins, plastic sheeting or the like, all necessary protective clothing, artificial lighting, safety and welfare facilities, storage and the like that may be available on the site.
    (t) Profit.

- Under the FIDIC Red Book, it may be that bills of quantities will have been prepared in accordance with international measurement rules such as the POMI Rules of Measurement. This covers Dayworks in its section GP9 and requires at item GP9.7 that 'An item shall be given for the addition of establishment charges, overheads and profit to each of the sums or schedules of labour, materials or plant'. GP9.8 then defines what is included therein as follows:

    Establishment charges, overheads and profit shall include:
    1. Costs related to the employment of labour
    2. Costs related to the storage of materials, including handling and waste in storage
    3. Contractor's administrative arrangements
    4. Constructional plant, except plant employed exclusively on dayworks
    5. Contractor's facilities
    6. Temporary works
    7. Sundry items.

    Clearly, items 2 to 7 in this list may particularly include items in a prolongation claim, but even item 1 might when there is a claim for time-related labour.

- Omissions. It is not suggested that where there are only omissions they should merit consideration in relation to adjusting the evaluation of prolongation claims. It is suggested, however, that where the contractor has an account for, say, US$3 million of additional work items and also US$1 million of instructed omissions, it is the net overlapping recoveries of such as preliminaries and general costs in US$2 million of increased work value that should be abated from any overlapping other claims.

- Provisional sums and prime cost items in bills of quantities against which the contractor has added a percentage. If the values of those sums increase by additional works, for a specialist subcontractor, for example, then the contractor's recoveries through the percentage addition will increase. The question resulting in relation to abating recoveries against a separate claim for prolongation will be what was allowed for in the percentage? For example, measurement rules such as CESSM4 clause 5.16(b) require contractors to price a single addition to such sums for 'all other charges and profit'. This can result in an issue as to how much of the percentage is respectively for site overheads, off-site overheads or profit. Slightly more helpful in this respect is the guidance in rule 2.11.2 of the NRM2 detailed Rules of Measurement for Building Works that 'When required, overheads and profit can be treated as two separate cost items; namely, "overheads" and "profit". Either way, some approach to establishing what is included within the contractor's additions to prime cost and provisional sums will be needed'.

In terms of how recoveries within the valuation of any of these items can be calculated, this can be complex and contentious. It depends on the detail and quality of information as to the contractor's pricing. Possible sources of such information include:

- A split bill of quantities. Where used, this requires all tendering contractors to price not just the bills of quantities in the usual way, by inserting their rates and prices against items and extending them by their quantities, but also to provide a schedule splitting each rate and price into its components. Ideally, this can be done by requiring tenderers to price bills of quantities in the form of Excel workbooks, that have columns for each component of each rate and price. These will obviously include columns for labour, materials and subcontractors, but if they can also have columns for such as plant and equipment, other on-site overheads, off-site overheads and profit separately, then this can provide useful detail of recoveries when these rates and prices are applied to additional works.
- In the absence of a split bill, it may be that the contractor will have to expose its full tender calculations. In some jurisdictions this might be a common requirement in some sectors (such as an 'Urkalkulation' in German engineering industries).
- The valuation of some variations may have included express and detailed allowances for some preliminaries and general items, such as plant and equipment.
- Dayworks should in particular show such express allowances.

6.10.3   Overlaps Between Overheads and Profit and Variations

Claims in relation to a contractor's head office overheads and profit as part of the evaluation of the time consequences of change have been considered in Section 6.8 of this book. Various approaches were set out there, with the main focus on the thorny issue of the use of formulae in a claim for prolongation. Whether the head office overheads and profit claim are part of a prolongation claim, or one for such as acceleration, labour disruption or preliminaries thickening, there is a probability in most cases that there is some degree of overlap with recoveries for that same head office overheads and profit through the valuation of additional works. That might be through variation instructions or the remeasurement of works under, for example, the FIDIC Red Book and the Infrastructure Conditions.

In practice the most risk of overlap is in relation to a formula head office overheads and profit claim. Mention has already been made of Mr Justice Akenhead taking a credit from his calculation of the claimant's head office overheads and profit claim and applying the Emden formula for the additional recoveries the claimant had already made in the value of the work done in his decision in *Walter Lilly v Mackay*. It seems that the parties' quantum experts had agreed the need for such a credit and its amount. The credit was explained in the judgment as follows:

> ... with credit being given, wholly properly, for the overhead and profit recovered by WLC on the difference between the amount of profit and overheads earned on the works (by way of the 4½% addition to overheads and profit) and the tendered overhead and profit allowance.

Thus it seems to have been a credit for recoveries made in relation to all of the works carried out by Walter Lilly, including all variations and additional works carried out by the contractor, no matter when they were carried out. It may be a moot point for consideration as to whether this is the correct approach in all cases, or whether it should only be a credit for those additional works recoveries made during the period over which the formula calculation is being applied.

If the aim of such as an Emden formula overheads calculation is to compensate for the contractor's loss of recovery of gross margin over the extended period that it is delayed on this site, and the aim is to put the contractor in the position it would have been in but for the delay, are variation works that were carried out months previously and that are unrelated to the delay relevant? This may also involve consideration of the relevant events that form the basis of the prolongation claim. If this was additional works, then it would seem that the overlap in recoveries should be credited. However, if the prolongation was all related to late drawings (involving no additional works), should any credit be given? The contractor is likely to argue not.

In terms of how such a credit is calculated, then this will obviously require a detailed understanding of recoveries of head office overheads and profit within the value of variations. Where any variations have been valued on the basis of invoiced costs plus margin, the overlap might be readily apparent, but generally, this process will require transparency as to allowances in the contractor's contract rates and prices, daywork rates, etc.

6.10.4 Overlaps Between Additional Work and Disruption

This is particularly relevant in relation to labour. Clearly it depends on how a labour disruption claim has been quantified, and much has been said about this in Section 6.4. In terms of relevant information as to the contractor's pricing of items that will overlap with such a claim, the points set out against prolongation claims are equally relevant here. However, the following are particularly noted:

- The valuation of additional quantities of work will particularly include additional labour hours, and these also might be in a disruption claim.
- Dayworks will always set out numbers of skilled and unskilled labour hours so that the duplication of hours can easily be established.
- There will often be some variations valued on the basis of assessments of the operative hours spent thereon.
- All split bills will require a column for labour allowances in the contractor's rates and prices which can be invaluable to the work of adjustment.

The topic of duplication of recoveries between a contractor's claims can be a large and complicated one, requiring considerable thought and research to remove it. This discussion just highlights some key examples and ideas on how they might be dealt with.

6.11   Summary

It is a well-worn truism that on construction and engineering projects time means money. Thus, where there is impact on the programme there will be a financial impact for one or both of the parties, and potentially also the contractor's subcontractors and suppliers.

Essential to any claim for the time consequences of change is the establishment of the factual and legal background. It can be a highly complex process to disentangle the effects of matters for which the employer is liable and those that are the responsibility of the contractor, its subcontractors or suppliers. The allocation of risk for time and its effects varies between contracts and the presence of cost-neutral delay events can further complicate the picture.

For the employer, delay can particularly mean loss of income or rent from the completed project, together with interest or financing costs and professional fees. The usual basis for compensation of the employer where the contractor is in culpable delay is by way of an agreed rate or rates of delay damages (or 'liquidated damages') set out in the contract. Generally, the agreement of such a rate should give an easy formula for quantification of the amount due to the employer, but both employers and contractors sometimes seek to challenge the agreed rate as not comprehensive or unreasonably high respectively. A recent judgment of the England and Wales courts means that an agreed rate of damages is less likely to be defeated as a 'penalty' as it previously was in that jurisdiction.

For the contractor, delays can directly lead to a combination or any one of the costs of prolongation, cost escalation, disruption, or acceleration together with associated interest or finance costs. The basis of the contractor's payment in the event of compensable delay effects varies under standard forms between terms such as cost, loss, expense and damages depending on the basis of the claim and the express terms. Generally, the aim of the payment will be to put the contractor as near as possible into the financial position it would have been in had the compensable events not occurred.

To the extent that it can be included in the valuation of a variation that is its cause, the quantification of such items as delay, disruption and acceleration as part of a variation can be a convenient and less controversial route than a standalone claim. This facility in some standard forms is under-used. It is, however, a particular benefit of the prospective approach of the NEC suite of contracts and their requirement that the valuation of compensation events are pre-agreed, provided those provisions are properly followed.

Prolongation is generally the easiest head of a contractor's time-related claims to quantify, although where escalation costs are an element, it can be difficult to establish what the costs of such as materials and labour would have been 'but for' the delay. There may be some role for the use of the contractor's tendered rates and prices in limited contractual circumstances. It is also possible, although not the norm, to agree rates in the contract for the contractor's delay costs. This is more common in subcontracts, particularly in the process industry.

Disruption to labour, plant and/or equipment is a particularly complicated head of claim to establish and quantify. Establishing the fact of disruption, its link to the events that are its basis and its financial effects can all be difficult. Various methods of quantification are possible. Pre-agreement based on estimates such as under the NEC approach would avoid many of the difficulties. The contemporaneous recording of resources and times lost would be ideal but is only possible in limited cases. The preferred retrospective approach is a measured mile analysis comparing a disrupted area of period of work with a baseline area or period that was not disrupted. However, the records for such an approach are often incomplete or missing. Where the records are available it is essential that a measured mile analysis is of comparable and representative periods or areas. Similar considerations apply to such approaches as an 'earned value analysis'.

Many contractors' disruption claims simply rely on a comparison of actual costs with the contractor's tender, but this may not address the aim of putting the contractor into the financial position it would have been in had the disruption not occurred. Both sides of such an equation are liable to criticism. Actual costs may include failures by the contractor, as well as recoveries through variations and other claims. The tendered costs might have been underestimated. Consideration of resource estimates by site management or expert analysis might give a more reliable 'but for' position. Any other distorting factors that render a costs less recoveries approach misleading must be identified and adjusted for, but this can lead to complexity and disagreement.

Approaches to quantifying disruption can be divided between 'top down' and 'bottom up' approaches. One of the latter approaches is to apply factors to the contractor's 'but for' resources for each of the effects of disruption. This requires an intermediate step between detailing the grounds for the disruption claim and its quantification, in looking at how the causes affected the resources subject of the claim. These effects can include such overtime working and overcrowding of working areas. Published data on such effects need to be applied with caution in terms of their similarity to the disrupted project and can only ever give an estimate of the effects. As such, it is perhaps best used in relation to an estimate for an NEC compensation event or as part of a Variation Quotation under SBC/Q Schedule 2, for example.

At the extreme end of theoretical models of disruption quantification are cumulative impact and System Dynamics Modelling. Such computer-based approaches are particularly susceptible to the response 'garbage in, garbage out'. The basis of their original modelling and extent to which they are representative of the disrupted project are key issues. They also suffer from difficulties in analysis of their basis and details and re-running them for changes in those assumptions. In practice, there seems to be little approval of them outside of the United States.

Given the difficulties with most approaches to the quantification of disruption, it is recommended that consideration might be given to the adoption of two alternative approaches to see if they come to similar conclusions. A hypothetical example of this has been set out. However, this can prove very expensive and may only be justified where the amounts claimed are considerable. Proportionality is always a consideration in relation to disruption claims, and it may be that on a very small disruption claim some much broader approach is required when the facts of the disruption are patently clear.

Preliminaries thickening claims are a common head of the results of disruption. A simple comparison of tender allowances and actual numbers of staff, for example, will not on its own establish the actual damage, costs, loss or expense incurred by the contractor as a consequence of failures for which the employer is liable. A better approach to such staff thickening is to identify and to find evidence for who is additional, why they were additional and how this is to the liability of the employer. Whether a preliminaries thickening claim should be made under the heading of disruption or acceleration will depend on its underlying facts.

Disruption claims often feature the 'knock-on' effects of later events that were only met by the contractor as a consequence of an earlier delay. An example could be the effects of weather on work that is pushed back into a less favourable season. Whether the contractor is entitled to be compensated for such consequential further events will be a legal question but may be a matter of simple causation.

Acceleration is a controversial head of claim, particularly in relation to entitlement. Provisions for acceleration vary under standard forms. Where contracts are silent on the issue, it would always be open to the parties to enter into an acceleration agreement, either as an amendment to their existing contract or as a supplement. However, such agreements can be fraught in practice and their terms need great care in their drafting.

Some standard forms give an express power for the employer to instruct acceleration where the contractor is in culpable delay. However, that culpability may not be certain, particularly where the contractor has made applications for extensions of time. Where contractors are denied a proper entitlement to an extension of time and instigate acceleration in order to reduce their potential liability for delay damages, they often make claims for 'induced' or 'constructive' acceleration costs. Such claims are particularly contentious and have to clear a number of high hurdles in terms of the employer's liability for the acceleration costs incurred.

The term 'global claim' is one of a number that are bandied about in relation to contractors' claims, often to give them a pejorative label. Typical definitions of such terms have been set out. The 'global' label is particularly important to construction and engineering claims, so the history of the approach adopted in the UK and Commonwealth jurisdictions to claims that are made on that basis has been set out in detail. Whilst under English law the approach to global claims has softened in the last decade, contractors should be wary of applying the approach without proper consideration, starting with the express terms of the contract and the applicable law, particularly whether they have satisfied any related requirements therein. Thereafter, making a claim entirely on a global basis may make it a 'hostage to fortune' if the asserted underlying facts or any part of it fail in any way. Particularisation as far as practicable may be the key. Where this leaves a 'rump', a tribunal may take a conservative approach to allowance for that.

A problematic feature of many contractual claims is amounts in relation to subcontractors and suppliers. Contractors often find themselves 'between and rock and a hard place' where a subcontractor or supplier has submitted a claim to a certain level of sophistication that does not satisfy the employer's certifier. The advent of 'quick and dirty' adjudication under security of payment legislation in various countries can particularly leave a contractor feeling vulnerable. When settling claims for such as prolongation, disruption and acceleration from subcontractors and suppliers, contractors are advised to take legal advice on the approach and the terms of the agreement as well as quantum advice. Generally, amounts paid to subcontractors and suppliers that reasonably reflect, for example, the strength of the claim, advice taken and the benefits of settlement rather than disputing it, should be recoverable provided other issues such as the link to the employer's failures are established.

Off-site overheads and profit can be a particularly difficult head of claims for prolongation, disruption and acceleration. Entitlement to them will firstly depend on the basis of the claim and the definition of terms such as 'Cost' under the FIDIC Suite. So far as additional overhead costs can be separately identified and quantified, they should be particularised on that basis. Claims on the basis of formulae such as those of Hudson, Emden and Eichleay can particularly boost the size of claim, but require significant preconditions to be met before they form the basis of calculation. Where the claim is for loss in relation to unabsorbed overheads that could not be recovered by securing other works, that loss needs evidence of such issues as that resources held on the project would have been used to cover the contractor's margin and that events on the project for which the employer is liable prevented such opportunities.

Claims for interest or finance charges can arise out of express terms of the contract or as a head of a claim. Late payment of certified sums is likely to be expressly covered. Under-certification is expressly covered in Infrastructure Conditions clause 60(7) and in NEC4-ECC clause 51.3, but under other forms it may depend on whether it was the result of a failure to certify as opposed to a lower valuation that was made in good faith.

Under English law, it is suggested that finance or interest charges would fall as a head of claims for prolongation, disruption and acceleration under most standard forms under the principle in *Minter v WHTSO*. Under express terms, the definition of terms such as 'cost' under clause 1(5) of the Infrastructure Conditions may also give an entitlement. Where a claim for interest or finance charges is for damages for breach, a key question will be when the breach occurred and hence when the interest or finance starts to run. Many claims for interest fail to address this question.

The rate of interest or finance may be set in the contract, particularly where it relates to a clause for interest on late payment of certified sums. Otherwise, if the claim is for contractual terms such as 'loss', 'expense', 'costs', 'expenditure', or is for damages, it seems logical that the level of the charge should reflect the actual amounts suffered by the contractor, and not some nominal or notional interest rate. Security of payment legislation such as the UK's Late Payment of Commercial Debts (Interest) Act 1998 may also be relevant, both as providing a rate of interest and in requiring that a rate agreed under the contract provides a 'substantial remedy'.

A constant consideration when evaluating the financial consequences of change is that there is no overlap between the contractor's different heads of claim. The heads of claim considered in this chapter and in Chapter 5 in particular provide substantial scope for such duplication. This needs detailed analysis and the calculations can be complex and time consuming. An issue may be which of the claims should be reduced and which should retain the amounts that would otherwise have been duplicated.

However, not all contractor's claims are covered by the direct or time consequences of change as considered in Chapter 5 and this chapter, and some other sources of claim are considered in the next two chapters.

7

Termination Claims

7.1 Introduction

It might be expected that when the parties to an engineering or construction contract sign up for the construction of the works thereunder, they will see the project through to its conclusion, the contractor executing the works through to completion and the employer paying for them in their entirety. This would be the case unless one party commits a breach of contract of such a serious nature that the law allows the other party to terminate the contract. However, the grounds for such termination at law are likely to be quite narrow and in practice there are a number of other events that can occur during the course of construction that might alone be thought to justify allowing one or either of the parties to terminate that contract, with the works still incomplete. Accordingly standard forms of construction and engineering contracts will set out express provisions for the parties to terminate the contract on the basis of certain events, some of which are breaches for which the applicable law may or may not allow such recourse in any event, and others that do not involve any failure by a party at all.

Such termination clauses will also set out the rights and obligations of the parties after termination and the quantification of the terminated contractor's final account including allowances for losses and damages resulting from the termination. Thus, it is important to remember that express contract terms in standard forms for termination do not terminate the contract, only the contractor's employment under that contract. The contract will still set out the rights and obligations of the parties post-termination.

Such termination events may involve default or choice by one or other party, so that the related contractual provisions require different approaches to their interests and the financial evaluations that follow. In all cases the first component of valuation will be the value of work completed by the contractor at the date of completion. This will include any accrued entitlements the contractor has under the contract in relation to such items as the value of variations, prolongation costs or fluctuations and any accrued entitlements the employer has in relation to, for example, defective work and delay damages. Thereafter there may be provision for the quantification and recovery damages suffered by the terminating party as a result of the termination.

The events giving rise to a right to terminate in a typical standard form of construction and engineering contract can be categorised into three types:

- Where one of the parties is in default of such a nature as to merit allowing the other party to terminate under express provisions of the contract rather than having to rely

Evaluating Contract Claims, Third Edition. John Mullen and R. Peter Davison.
© 2020 John Wiley & Sons Ltd. Published 2020 by John Wiley & Sons Ltd.

on its rights under the applicable law. Such clauses may also expressly preserve the injured party's other legal rights to terminate at law, by making the express clause 'without prejudice' to those other legal rights. There may be a number of listed failures by a party that give rise to the express right of the other party to terminate, but the most commonly relevant ones in practice are: insolvency or bankruptcy of either party; the employer's failure in relation to payment; or the contractor's failure in relation to progress or suspending work without good reason or refusing to remove and/or make good defective works. In many legal jurisdictions, these 'trigger' events are of a nature that would, in any event, allow the injured party to regard the offending party's conduct as a fundamental breach, forfeiture or repudiation of the contract, which it can accept, relieving itself of further performance and allowing it to claim for damages. In addition, such express clauses may also allow termination for less serious listed breaches that would not count as sufficiently serious as to allow termination under the applicable law. The obvious advantages of an express term allowing termination for default is that it should clearly and expressly define the events that give the right to terminate and also set out the rights and obligations of the parties thereafter. Given that it is the party defaulted against, a clause to this effect will usually place the initial protection on the interests of the party terminating and provide for recovery of its damages, costs and losses arising out of the 'trigger' default and termination.

- Where it is convenient to one of the parties for it to terminate the contract, because its circumstances have changed in some significant way. This can be a wide-ranging right, but will usually exclude the employer electing to complete the works itself by employing another contractor because, for example, that is a cheaper route to completion of the project. Given the absence of default by either party, the initial protection in such a termination for convenience clause will be in the interests of the party subject to the termination and the clause will provide for recovery of its damages, costs and losses arising out of the other party's exercise of its convenience.

- Where an unforeseen event occurs that is neither party's fault, but is of such significance that the parties have provided in the contract that either of them may elect to terminate the contract if such an event occurs. This might include, for instance, the outbreak of war or a terrorist act, and the contract may set out the minimum length of the effect of that event on the works before it becomes an event so serious that it justifies a party being able to terminate. Given the lack of default by either party, a clause to this effect will balance the interests of both parties and provide for payment of such as the value of work completed and both parties' costs arising out of the termination.

In practice, a further feature of the operation of such clauses in general is how often the parties dispute whether the event relied upon for the termination actually occurred to the extent required by the termination clause and/or whether the procedural requirements of the termination clause, such as for notice, were properly followed. This can lead to further complexity in evaluation of the claims arising from termination, with the requirement for alternative valuations depending on who is correct as to the validity of the termination. As will be explained, a period of uncertainty as to the validity of the termination is not uncommon, and may both increase the claimable costs arising from termination and add further heads of claim.

In the UK domestic market, the termination of construction and engineering contracts, by either party, has been, and continues to be, relatively rare. The reasons for this relate to the types of event that are provided for in typical termination clauses and that these have occurred less frequently in the UK than in many other parts of the world. This has been the case even through turbulent economic times for the construction and engineering industries and other industries that rely on them. Reflecting this, in the first two editions of this book, the subject of termination did not merit its own chapter. However, the global financial crisis of 2008 and 2009 led to an increase in the types of 'trigger' event, which in turn gave rise to a large number of terminations in some overseas markets. The resulting financial claims and counterclaims often led to protracted dispute resolution proceedings, particularly arbitration under the auspices of the International Chamber of Commerce and some particular regional arbitration centres. Given the ongoing global uncertainties in some parts, such claims have continued to be more common than they once were and are likely to spike again in any future economic downturn. Hence this new chapter in this Third Edition.

The last global economic crash saw a large number of international arbitrations of claims between employers and contractors for the termination, or wrongful termination, of the contractor's employment under construction contracts. In many parts of the World, that crash was preceded by (and was not unrelated to) a period of particularly high construction activity, and even '*over-heating*'. This led to some quite complex and interlinked claims and counterclaims. Whilst most termination claims will share some common features, they will also depend on their underlying facts, the contract terms relied upon, the contractual basis of the termination and also whether both parties consider it a valid termination. A consequence of this validity question may mean that the terminated party interprets the terminating party's conduct as a repudiation of the contract, or a breach in itself, that gives it the right to terminate. Take, for example, a contractor that is accused of failure to proceed with the works in accordance with the contract, but considers that it has good grounds, and has made valid and adequately substantiated claims, for extensions of time that have been unreasonably rejected. Where such claims are determined by the engineer, architect or administrator under the contract who is by its conduct (such as late design) the cause of that delay, contractors often complain that their claims are being rejected by them to hide their failures and protect their own interests. If the employer terminates the contract on the basis of such unfairly alleged failure to proceed with the works, the contractor may say that there was no valid ground for such termination. It may then be put into a position to terminate itself, for example, where the employer refuses to pay an interim certificate on the basis that its termination allowed it to withhold such payments under the related termination clause of the contract. The result of this may be that both parties make claims and counterclaims against each other for termination based on alternative terms of the contract that give them each such recourse for default.

One feature of termination claims, particularly where they are arbitrated and the full legal and contractual basis of the claims arising are examined, is the lack of precedent or published literature on the details of how the heads of typical termination claims should be evaluated. It might be said that termination claims only comprise heads that are common to those arising from other contractual events, but they do present some particular issues. At the heart of most evaluations post-termination are the value of the contractor's work as completed at the date of termination and the offended party's damages, costs, and losses associated with the termination. In the event of an employer's valid termination, this latter heading will particularly include such additional costs and delays incurred in completing any incomplete or defective works. In the event of a contractor's valid termination, it will particularly include such costs as demobilisation and removal of its resources and loss of the profit it would have made on the works that were outstanding at the date of termination. These claims are covered by other chapters of this book.

It is important that, where a construction or engineering contract gives a party the express right to terminate for specified reasons, the remedies for such a termination are set out in the contract. This is because, where a party terminates under an express provision of the contract, it may only be entitled to the remedies set out in the contract in relation to that termination. The difficulties that this can cause are illustrated by the case of *Thomas Feather & Co (Bradford) Ltd v Keighley Corporation* (1953) 52 LGR 30, where the employment of a contractor was determined by the employer as the contractor had sublet parts of the work in breach of an express provision in the contract that there should be no subletting of work without the employer's approval. The restriction on subletting had been introduced as the employer was anxious to ensure that all persons employed on the works were subject to a 'fair wages' clause that had been inserted in the contract. The court decided, on appeal from the decision of an arbitrator, that the employer would not have been able to determine for such a breach but for the express provision in the contract and therefore was not entitled to any remedy that was not provided for in the contract. The employer therefore lost its claim for some £21,000 incurred as the additional cost of employing other contractors to complete the works because the termination clause did not provide for recovery of such additional costs.

Contracts such as the FIDIC Red Book, SBC/Q Standard Form and the Infrastructure Conditions overcome this problem by ensuring that the employer has wide ranging remedies in such circumstances. However, this may not be the same with all contracts and the extent of remedies available in the event of a termination needs to be carefully considered before any evaluation can be made. Another consideration is the extent to which the contractual remedy is the only remedy that the injured party has in the event of the types of defaults for which these standard forms allow it to terminate. The contract may expressly preserve those rights. For example, in SBC/Q clause 8.3.1 the rights of the employer to terminate for default by the contractor under clauses 8.4 to 8.7 and the rights of the contractor to terminate for default by the employer under clauses 8.9 and 8.10 are stated to be 'without prejudice to any other rights and remedies' that those parties have. Hence, if the breach is sufficiently serious that it would give the right to terminate under the law, then the injured party under SBC/Q would have the choice as to whether to terminate at law or under the express provision of the contract. Whether an express contractual provision replaces the right to terminate at law for the same failures will depend on the provisions of the contract and the applicable law.

A further distinction is that on a termination at law there is unlikely to be any prescription as to how the parties are to act, apart from a likely obligation on the injured party to mitigate its resulting losses if possible. In contrast, the termination clauses of most standard forms of construction and engineering contract will also set out procedures such as possession of the site, use of, for example, the contractor's facilities, equipment and materials, and for subsequent payments and a final accounting between the parties. In this regard, termination under an express term will not be as all-encompassing of the parties' rights and obligations as it may sound. It is perhaps for that reason that JCT contracts, including the current SBC/Q, refer to terminating 'the Contractor's employment' rather than the starker FIDIC term 'terminate the Contract'.

One particular significance of express contractual provisions preserving the terms of the contract is that it may be that a termination under the law of breach of contract may leave the valuation of the works to be based on a quantum meruit or some other approach that is not based on the rates and prices in the contract. This is on the basis that if the contract has been terminated, then so too has its payment provisions. This can have profound effects where the contractor has substantially over-priced or under-priced its works. It is another factor that sometimes comes into play when parties have the option of either terminating by reference to the express contractual provisions or on the basis of the applicable law. For those evaluating claims arising out of termination on that basis, this may require very different approaches to that evaluation.

This book is concerned with the evaluation of claims under express provisions of standard forms of engineering and construction contract. Termination under the law will require legal advice for the relevant jurisdiction. However, as stated elsewhere, an advantage of the termination provisions of standard engineering and construction forms is how they both add definition and introduce additional grounds for termination under the law. Those grounds under the law are likely to be limited to fundamental breaches that go to the heart of the benefit that a party contracted to obtain from the contract, such as an employer's failure to pay for the works or the contractor's failure to construct them. As will be seen in this chapter, express provisions such as SBC/Qs will add to these by making an employer's interference with an interim payment certification or a contractor's failure to proceed 'regularly and diligently' with the works grounds for termination.

As explained elsewhere, before setting out the parties' rights within a financial evaluation following termination, contract clauses will usually start by requiring a valuation of the works completed by the contractor. In most circumstances, where a decision is taken to terminate, relations between the parties will have become strained at the least, and will have broken down almost completely at the worst. This may be even before the act of termination itself makes relationships even more fraught. It can therefore sometimes be difficult to arrange the recording of matters such as the extent of completed works and quantities of unused materials and goods on site. Inevitably, materials and goods stored off site or held by a supplier or subcontractor pending their requirement on site can be subject of even greater difficulty. However, such recording is essential to avoid further dispute. If possible the records should be taken by the parties together and signed off as agreed, but if this is not possible then each party should take a record and forward it immediately to the other for verification. At the least this may allow any areas of difference to be quickly identified.

The manner of recording the progress of the works at termination should be carefully considered and while the traditional method has been to mark up drawings and sign off schedules with completed work and work in progress, it is often, on large projects, difficult to execute this satisfactorily for all elements, particularly in the usually restricted time scale of a termination. Photographs have also long been a tool for recording in such circumstances. However, consideration should also be given to video recording of the works, stocks of materials, plant on site, etc. In either case, some means of date verification of the recording is essential. Voice commentary on video recordings are all too often missed, so that when the video is played back later it is often not apparent what is being shown. Joint recordings are ideal. Scale is a common problem with photographs in particular, with, for example, racks of pipes shown, but it not apparent what diameter they actually were. Properly made, such recordings can provide an enormous amount of information in a relatively short time compared to handwritten schedules and marked-up drawings.

Disputes as to valuation of completed work should be one of the comparatively more straightforward heads to evaluate, except for any allegations of work that is considered to be properly complete by the contractor, but to be incomplete or defective by the employer. Only proper records will resolve the former, while testing may be required to establish the latter and this may require the involvement of engineering expertise.

The greater potential for dispute can arise from matters such as goods and equipment for the works that the contractor has ordered but which are not yet on site. These can be goods for incorporation in the works, or special equipment required to execute future parts of the works and which require adaptation for the particular contract. There may also be instances where the contractor has reserved scarce or specialised equipment required for future parts of the works that cannot be sourced elsewhere or in the required time period. With increasing globalisation, such items might be held anywhere around the world, for example, turbines from Germany or structural steel from China. In such circumstances there will usually be some negotiation required between employer, contractor and supplier to enable the employer to gain the benefit of such arrangements and to minimise cancellation or abortive costs. Provisions in standard-form termination clauses requiring the terminated contractor to assist with this for the employer's benefit are identified in this chapter.

Bearing in mind these issues, this chapter considers some of the particular claims that arise out of contract terminations. However, it starts with a review of termination itself, the typical grounds for termination and the consequences of which party terminated. This is by reference to the FIDIC Red Book, the Infrastructure Conditions and SBC/Q as examples of standard forms. It then looks into evaluation of the different heads of claim that arise out of different scenarios, depending on which party terminated and whether they did so validly. That evaluation section is by primary reference to the FIDIC Red Book as an example, reflecting the reality that terminations are far more common on international projects than domestic UK ones.

7.2   Termination Under Standard Forms

As explained elsewhere, provisions of construction and engineering contracts can commonly provide for termination by one or both parties in the event of four types of event. These four are: default of the contractor; default of the employer; a party's convenience; and the occurrence of major events in neither party's control but which are so significant as to justify either party ending their duty to further perform the works. These types of event are considered in turn as follows.

7.2.1   Termination for Default by the Contractor

The defaults by the contractor that entitle the employer to terminate under the standard forms considered in this book are outlined as follows. The wording differs slightly within the provisions of those contracts, and they need a very careful reading before being applied in any case, including any bespoke amendments, but the following paraphrases them. The employer's termination may require notice both of the intention to terminate if the breach continues beyond a contractually set period and earlier notice in relation to some of these failures themselves.

- The contractor's failure to provide a performance security as required by the contract (FIDIC Red Book clause 15.2(a)). This can be understood as a particularly significant breach from an employer's perspective on an international project where it has concerns regarding the contractor's abilities and wants some insurance.
- The contractor's failure to carry out its obligation under the contract (FIDIC Red Book clause 15.2(a) by reference to clause 15.1, Infrastructure Conditions clause 65(1)(j)). This is a potentially broad provision which is notable by its absence from SBC/Q. However, in the Infrastructure Conditions clause 65(1)(j) it requires 'previous warnings by the Engineer in writing' regarding the contractor's failure. The FIDIC Red Book requires notice at two stages, firstly, from the engineer under clause 15.1 in relation to the breach and, secondly, under clause 15.2 in relation to termination. Furthermore, it seems unlikely that an employer could use it as a pretext for termination on any trivial and isolated contractual failure by the contractor. How serious such a breach must be will require legal advice and might involve reference to what failures would have allowed termination under the applicable law separately to the express provision.
- The contractor having an unstayed or undischarged execution levied on its goods (Infrastructure Conditions clause 65(1)(d)).
- The contractor's abandoning the works (FIDIC Red Book clause 15.2(b), Infrastructure Conditions clause 65(1)(e) (without due cause)).
- The contractor otherwise demonstrating an intention not to continue performance (FIDIC Red Book clause 15.2(b)).
- The contractor's failure to commence the works without reasonable cause (Infrastructure Conditions clause 65(1)(f)).
- The contractor's suspension of the works without reasonable cause (Infrastructure Conditions clause 65(1)(g), SBC/Q clause 8.4.1.1).
- The contractor's failure to proceed with the works (FIDIC Red Book clause 15.2(c)(i), Infrastructure Conditions clause 65(1)(j) (with due diligence), SBC/Q clause 8.4.1.2 (regularly and diligently)). The subjectivity of these provisions and terms such as 'due diligence' is obvious.
- The contractor's failure to comply with a notice to remove faulty materials and goods (FIDIC Red Book clause 15.2(c)(ii), Infrastructure Conditions clause 65(1)(h), SBC/Q clause 8.4.1.3). This is hardly likely to be a basis for termination under the applicable law. Along with that in the next bullet point, in practice this ground for termination is particularly the subject of complaint by terminated contractors that it has been abused by employers as a pretext for termination.
- The contractor's failure to comply with a notice in relation to remedial works (FIDIC Red Book clause 15.2(c)(ii)). The note against the previous bullet point is repeated. In both cases, contractors often complain that their obligation is to complete the works by the contractual due date or dates, and that if in the meantime there are faulty materials or goods or required remedial works, which should be of no concern to the employer.
- The contractor subcontracting the works, or assigning the contract, without agreement from the employer (FIDIC Red Book clause 15.2(d), Infrastructure Conditions clause 65(1)(a), SBC/Q clause 8.4.1.4). This is also hardly likely to be a basis for

termination under the applicable law and is another common source of complaint by terminated contractors that it has been abused. This includes instances where employers have known at the time that the contractor was subletting without permission, and were not concerned to get involved, but then later used it when looking for an excuse to terminate.

- The contractor entering various financial situations such as bankruptcy, insolvency, going into liquidation, becoming the subject of an administration order or making an arrangement with its creditors (FIDIC Red Book clause 15.2(e), Infrastructure Conditions clause 65(1)(b), SBC/Q clause 8.5). Such a 'trigger' event should surely be at the heart of any express termination clause.
- The contractor committing acts of bribery (FIDIC Red Book clause 15.2(j), SBC/Q clause 8.6).
- The contractor's failure to comply with the UK's Construction Design and Management Regulations (SBC/Q clause 8.4.1.5).

In practice, the most common ground for termination by an employer for default alleges failure by the contractor in its obligations to progress the works. There can be a variety of background reasons for this. It may simply be that the contractor over-stretched itself by taking on that project, because it has too much work on elsewhere or that the project is beyond its normal capacity. In practice, this is more likely in a period of over-heating of the market such that contractors have been over-stretched and unable to obtain the resources necessary to enable them to properly service all of their projects. However, the background to such an allegation often includes disagreements as to responsibility for the progress of the works, with the contractor claiming entitlement to extensions of time and costs related to delay and, perhaps, disruption, but those being denied by the engineer, architect or contract administrator determining such issues under the contract. Where contractors are over-stretched by an intensely busy market, so too may be engineers, architects, and contract administrators appointed by the employer, which might affect their ability to provide designs, approvals etc. timeously. As a result, in this context, the claims and counterclaims that follow an employer's termination for purported default by the contractor are particularly likely to also include arguments related to the parties' respective responsibilities for delays to the works and their financial consequences, as well as responsibility for the termination itself. Thus, the employer's claims for its costs associated with a purported termination for failure to progress the works are met by a contractor's claim for delay, related costs, wrongful termination by the employer and its losses and damages resulting from its own purported termination.

Even where there is no dispute as to the contractor's entitlement to extensions of time, the subjective nature of such as Infrastructure Conditions clause 65(1)(j) and its reference to 'with due diligence' or SBC/Q clause 8.4.1.2 and its reference to 'regularly and diligently' has been noted elsewhere. If a contractor is working slowly, is that sufficient? How slowly must it be working? 'Slowly' compared to what? It is suggested that judgement of this issue must particularly consider the contractual completion date (or dates if there are contractual sections) and the likelihood that the contractor will achieve these. This in turn will usually require consideration of the contractor's programme, what rate of progress that required and where the contractor is against it. This may also require consideration of float in the programme, particularly end-float. It may be that the programme has considerable amounts of such float, that the contractor's resources have been detained by another project and it has decided to slow its progress on the current project with no prejudice to its completion because float means it can catch up later. The contractor will say that its primary obligation is to complete by the contractual date(s) and how it gets there is its business. In practice, this can become a central issue in a disputed termination for failure to process with the works, requiring reference to the contract and local legal principles as well as programming expertise.

Following termination by the employer, its interests are then protected by various contract provisions in relation to its continuing the works in the absence of the terminated contractor, such as the following. Again, the terminology varies between the standard forms considered in this book and should be read for its precise wording, in this summary those terms have been paraphrased:

- The contractor leaves the site (FIDIC Red Book clause 15.2, Infrastructure Conditions clause 65(1)).
- The employer takes possession of the site (SBC/Q clause 8.7.1).
- The contractor delivers to the engineer any equipment, materials, plant, temporary works, designs and drawings made by it (FIDIC Red Book clause 15.2, SBC/Q clause 8.7.2.2, in relation to the design documents).
- The contractor complies with any reasonable instructions regarding assignment to the employer of any of its subcontracts (FIDIC Red Book clause 15.2, Infrastructure Conditions clause 65(3), SBC/Q clause 8.7.2.3).
- The contractor complies with any reasonable instructions for the protection of life or property or safety of the works (FIDIC Red Book clause 15.2). It is also likely that the applicable law will place obligations on the contractor in this regard.
- The employer may complete the works and/or engage others to do so, and they may use, for example, equipment, materials, plant, temporary works, designs and drawings, etc., made by the contractor (FIDIC Red Book clause 15.2, Infrastructure Conditions clause 65(2), SBC/Q clause 8.7.1).
- The employer gives notice to the contractor regarding release of any equipment and temporary works that it no longer requires, and either:
  - the contractor removes them at its own expense (FIDIC Red Book clause 15.2, SBC/Q clause 8.7.2.1) or
  - the employer sells them, and the resulting monies are taken into account in the final balance of monies due between the parties (FIDIC Red Book clause 15.2, Infrastructure Conditions clause 65(2)).

In practice, provisions for the delivery of equipment, materials, plant, temporary works, designs, and drawings and assignment of subcontractors are particularly useful express additions to the likely lack of such requirements under the law. This should both help the employer to complete the works as efficiently as possible and is to the contractor's benefit in mitigating the employer's costs and hence termination claim against it.

Regarding payments and the contractor's final account following a termination by the employer for default by the contractor, the position is, very broadly paraphrased as follows, the detailed components of valuation are considered in more detail elsewhere in this chapter.

- The employer can withhold further payment to the contractor (FIDIC Red Book clause 15.4(b), SBC/Q clause 8.7.3). Of course, the employer could make an interim payment if the balance of monies due between the parties was sufficiently clear, but it rarely happens in practice.

- The Infrastructure Conditions clause 65(5)(d) allows that the engineer 'may' issue an interim certificate to the contractor following termination for its default before the works are subsequently completed. In practice, the effect is likely to be the same as under the FIDIC Red Book and SBC/Q clauses, given the use of the word 'may' in clause 65(5)(d).
- The balance between amounts due to the contractor for work done etc. by it up to the date of termination and the employer's additional costs, losses, damages etc. resulting from the termination and completing the works become either:
  - a debt payable by the contractor to the employer or
  - paid by the employer to the contractor.
  (FIDIC Red Book clause 15.4(c), Infrastructure Conditions clause 65(5), SBC/Q clause 8.7.4).
- SBC/Q clause 8.8 also provides for where the employer decides not to complete the works after terminating and the resulting financial accounting between the parties.

7.2.2 Termination for Default by the Employer

The defaults by the employer that entitle the contractor to terminate under the standard forms considered in this book are as follows. The wording differs slightly within the provisions of those contracts, and they need a very careful reading before being applied in any case, including for any bespoke amendments, but the following paraphrases them. The contractor's termination may require notice both of the intention to terminate if the breach continues beyond a contractually set period and in relation to some of the employer failures themselves:

- The employer's failure to provide evidence that it has suitable financial arrangements in place to pay for the works (FIDIC Red Book clause 16.2(a)).
- Failure by the engineer under the contract to issue a payment certificate (FIDIC Red Book clause 16.2(b)).
- The employer's failure to pay against an interim payment certificate (FIDIC Red Book clause 16.2(c), SBC/Q clause 8.9.1.1).
- The employer's interference or obstruction of the architect/contract administrator's issuing of any certificate under the contract (SBC/Q clause 8.9.1.2).
- The employer's substantial failure to perform its obligations under the contract (FIDIC Red Book clause 16.2(d)).
- The employer's failure to enter into a contract within 28 days of the contractor receiving a Letter of Acceptance (FIDIC Red Book clause 16.2(e)).
- The employer assigning the contract or part thereof without prior agreement (FIDIC Red Book clause 16.2(e), Infrastructure Conditions clause 64(1)(a), SBC/Q clause 8.9.1.3).
- Suspension of the whole of the works for more than the period stated in the contract (FIDIC Red Book clause 16.2(f), SBC/Q clause 8.9.2).
- The employer entering financial situations such as bankruptcy, insolvency, going into liquidation, becoming subject of an administration order or making an arrangement with its creditors (FIDIC Red Book clause 16.2(g), Infrastructure Conditions clause 64(1)(b), SBC/Q clause 8.10).

- The employer's failure to comply with the UK's Construction Design and Management Regulations (SBC/Q clause 8.9.1.4).
- The employer having an unstayed or undischarged execution levied on its goods (Infrastructure Conditions clause 64(1)(c)).

In practice, the most common ground for termination by a contractor alleges default by the employer in its obligations to pay certified amounts. This is more likely in a period of severe economic downturn, where reasons extraneous to the particular construction contract have left the employer short of funds. Provisions entitling a contractor to terminate for this cause usually precede that right with a right to suspend work (or to reduce the rate of work). For example, in the FIDIC Red Book, the contractor's termination provisions in clause 16.2 are preceded by its right to exercise either of those preceding actions in clause 16.1. As a result, in this context, the contractor's claims that follow termination are likely to also include costs related to the suspension or reduced rate of work, for example the under-utilisation of equipment, staff and labour.

Following termination by the contractor, its interests are then protected by various provisions in relation to its goods, materials, temporary works, facilities, etc. Again, the terminology varies between the standard forms considered and should be read for its precise wording. In this summary those terms have again been paraphrased:

- The contractor ceases work, except as instructed by the engineer for the protection of life or property or the safety of the works (FIDIC Red Book clause 16.3(a)).
- Whilst the Infrastructure Conditions and SBC/Q contain no express equivalent to FIDIC Red Book clause 16.3(a), this is inherent in their respective clauses 64(1) and 8.9 and the contractor terminating its employment under the contract. Also, under any jurisdiction, it may well be that the applicable law leaves the contractor with some obligations in relation to safety and the protection of life.
- The contractor hands over any documents, plant and equipment for which it has been paid (FIDIC Red Book clause 16.3(b)).
- The contractor hands over any Design Portion documents referred to in SBC/Q clause 2.40 (its clause 8.12.2.2).
- The contractor removes all other equipment, materials, plant and temporary works (FIDIC Red Book clause 16.3(c) (except as necessary for safety), Infrastructure Conditions clause 64(2) (in relation to equipment), SBC/Q clause 8.12.2.1).

Regarding payments and the contractor's final account following a termination by the contractor for default by the employer, the position is, very broadly as follows, while the detailed components of valuation are considered in more detail elsewhere in this chapter.

- Further sums only become due to the contractor as set out in the termination clause (SBC/Q clause 8.12.1).
- The employer releases any performance security to the contractor (FIDIC Red Book clause 16.4(a)).
- The balance between amounts previously paid to the contractor for work done, etc., by it up to the date of termination and the total of: its value of work done; its claims under the contract; its costs of materials properly ordered and for which it is liable to pay; its costs of removing temporary buildings, tools, plant and equipment; and losses and/or damages caused by the termination, become either:

– paid by the employer to the contractor or
– paid by the contractor to the employer.
(FIDIC Red Book clause 19.6 and 16.4(c), Infrastructure Conditions clauses 63(4) and 64(3), SBC/Q clause 8.12.5).

7.2.3 Termination for a Party's Convenience

Some construction contracts also provide the employer with the right to terminate for what FIDIC Red Book clause 15.5 refers to as 'the Employer's convenience'. Not only is an express entitlement to terminate for convenience rare in standard forms, it is usually only an entitlement of the employer and not of the contractor. This is intended to cover significant changes in the employer's circumstances (such as economic or political ones) that mean the employer no longer wants to continue with the project. This was a feature of the last economic crisis, where employers found that sufficient funding to complete a project was no longer available, or the project itself was rendered no longer economically viable, for example, due to collapse in sale or rental values. Such clauses are also sometimes applied when, whilst neither party is in a breach that would allow the other to terminate, relationships have so broken down that the employer wishes to start again with a new contractor. However, such provisions are not usually intended to allow the employer to give the work to a replacement contractor (see elsewhere regarding FIDIC Red Book clause 15.5).

Another consequence of the last economic downturn was that, with substantial falls in tender prices, employers that (for example) procured work in an overheated construction market in 2007 might have looked at tender prices in 2009 and regretted their timing of that procurement, particularly, for example, where the sale or rental value of a building being constructed had also dropped sharply. In these circumstances, a number of termination disputes at that time saw contractors assert that a purported 'termination for default' was actually an abuse of the 'termination for convenience' provision of contracts under FIDIC terms, or to get around lack of a termination for convenience clause, or a qualification such as that in FIDIC Red Book clause 15.5. Contractors in such circumstances then further asserted that the employer's real motive was to re-tender the remaining parts of a project at a significant saving in a depressed construction market.

Termination for convenience clauses are also sometimes resorted to by employers who consider their contractor to be in default but are concerned that the degree of failure might not be provable as justifying termination for default. The scope for such concern can be illustrated by considering the subjective nature of a typical ground for termination for default such as in SBC/Q clause 8.4.1.2 – 'If the contractor … fails to proceed regularly and diligently with the Works'. Terms such as 'regularly' and 'diligently' may be difficult to define objectively. As explained elsewhere, it may also be the case that the contractor has outstanding claims that its progress is delayed because of failures by the engineer, architect or contract administrator under the contract. However, as will be seen, the downside for the employer in a termination for convenience is how the financial consequences will be not as attractive as under a termination for default clause. It may therefore be a difficult issue of judgement for an employer and its advisors as to whether accepting the less favourable financial consequences of terminating for convenience is justified by the greater legal and factual risk of a termination for default.

It is notable that of the standard forms of contract considered in detail in this chapter, the FIDIC Red Book is the only one that provides for the employer to terminate the contract for its convenience. It is suggested that this reflects a different approach internationally to that in the UK domestic market. If that is right, then perhaps the increasing international take-up of the NEC suite of contracts reflects that its clause 90 allows the employer to terminate 'for any reason', which would include its convenience. In contrast, the contracts of the NEC suite limit the contractor's right to terminate to the occurrence of specific events set out in that contract.

Under most jurisdictions, taking work away from an appointed contractor in order to give it to another contractor amounts to a breach of contract. Accordingly, '*convenience*' in this context may not stretch to a desire to give the work to another contractor. However, if the termination for convenience clause is silent on this issue, it might be that the employer can complete the work with a new contractor without putting itself in breach of contract. Under English law, see in this regard *TSG Building Services Plc v South Anglia Housing Ltd* [2013] EWHC 1151.

FIDIC Red Book clause 15.5 expressly states: '*The Employer shall not terminate the Contract under this Sub-Clause in order to execute the Works himself or to arrange for the Works to be executed by another contractor*'. It is qualifications such as this that sometimes lead employers to resort to an unjustified allegation of default instead, allowing the employer to terminate without such a restriction on its ability to subsequently procure a replacement contractor to complete the project at a lower contract price.

Following a termination for convenience under FIDIC Red Book clause 15.5, the contractor is required to proceed in accordance with clause 16.3, which provides it with the protections summarised elsewhere when considering that clause in the context of a contractor's termination for default by the employer. None of the employer's ancillary rights, such as use of the contractor's equipment and materials and assignment of subcontractors, such as that found in a termination for contractor default provision, will apply. Generally, where an employer terminates for convenience, payment to the contractor will include some of the costs, liabilities, loss and damages of the type to which a contractor is entitled in a termination for employer default provision.

Valuation after a termination for convenience under FIDIC Red Book clause 15.5 is by reference to clause 19.6, which requires the engineer to value the work done and issue a payment certificate including:

- amounts payable for any work carried out;
- the cost of permanent materials and plant delivered to the contractor;
- the cost of permanent materials and plant undelivered, but for which the contractor is liable to pay;
- any other liability reasonably incurred by the contractor in the expectation of completing the works;
- the costs of removal of the contractor's temporary works, plant, equipment vehicles, etc.; and
- the costs of repatriating the contractor's staff and labour.

Termination for convenience clauses of the FIDIC Red Book are sometimes used by employers as an alternative ground for termination where their primary basis for terminating is default by the contractor but they are concerned that the default is not sufficiently significant. Thus, they terminate on both bases to 'hedge their bets'. Alternatively, some employers have been known to re-characterise their termination by default as a termination for convenience because it has been challenged by the contractor and they have been advised, or come to fear, that the termination by default might fail.

7.2.4   Termination for Major Events in Neither Party's Control

Given the size and long programme periods of some construction and engineering projects, it may be considered desirable for contracts to provide for the occurrence of unexpected events of such magnitude that it is best to allow either party to walk away from further performance should it so wish, whilst also setting out the rights and obligations of the parties in the event of such a termination.

In FIDIC Red Book, clause 19, 'Force Majeure', includes at subclause 19.6 the option, where an event occurs of sufficient duration, to allow either part to terminate the contract. The definition of 'Force Majeure' is in subclause 19.1 and includes such events as war, hostilities, rebellion, terrorism, riot, strike, lockout, explosives, radiation and natural catastrophes such as earthquakes, etc. Such events have to be beyond a party's control, unforeseeable when entering the contract, unavoidable and not substantially attributable to the other party. The required magnitude of such an event is defined as one that prevents 'the execution of substantially all the Works in progress' for a continuous period of 84 days or multiple periods totalling more than 140 days. Thereafter, subclause 19.6 requires the engineer to value the work done and issue a payment certificate including the amounts as set out when considering that subclause in relation to termination for convenience.

In SBC/Q, clause 8.11 also allows either party to terminate in the event of an occurrence that seriously affects the works for a lengthy period. Here the period is agreed by the parties in the contract particulars, and the effect must be to suspend 'the whole or substantially the whole of the uncompleted Works'. The events listed in its subclause 8.11.1 include force majeure, negligence or default of a Statutory Undertaker (as defined in clause 1.1), Specified Perils (as defined in clause 6.8), civil commotion, terrorism or the exercise of a statutory power by the UK Government. Thereafter, the consequences of termination under SBC/Q clause 8.11 are set out in its clause 8.12.

In the Infrastructure Conditions, subclause 63 refers to 'Frustration', where 'any circumstance outside the control of both parties arises during the currency of the Contract which renders it impossible or illegal for either party to fulfil his contractual obligations'. It also contains a 'War Clause' in sub-clause 63(2). In either event the clauses provide for the works to be deemed abandoned. Thereafter, subclause 63(3) requires the contractor to remove all of its temporary plant, equipment, facilities, etc., required for the works. Thereafter, clause 63(4) entitles the contractor to be paid:

- the value of all work carried out;
- amounts payable for preliminary items;
- the cost of permanent materials and goods delivered to the contractor;
- the cost of permanent materials and goods undelivered, but for which the contractor is liable;
- any expenditure reasonably incurred by the contractor in the expectation of completing the works; and
- the costs of removal of the contractor's temporary plant, equipment, facilities, etc., required for the works.

7.2.5 Competing Claims for Termination

As noted elsewhere, in practice the parties often disagree as to whether the termination was lawful or not. This may relate to the facts asserted as grounds for the termination or whether the termination was properly effected in accordance with the procedures in the contract for such items as notice requirements. They may even disagree as to who terminated. This may occur, for example, where the contractor considers the employer in breach of contract such as to entitle it to terminate and cease work, but the employer disagrees that it was in breach and exercises its right to terminate because the contractor has failed to continue with the works. In such cases the parties will set out their claims and counterclaims on their polarised views of who is correct on the merits of the factual and contractual situation. Should the claims and counterclaims proceed to dispute resolution, the parties will need to also present their claims on alternative bases, depending on the court or tribunal's determination of those merits. This can make evaluation of the claims very complicated as they have to consider a number of alternative approaches to the same heads of claim.

The consequences of all of this is that the quantum practitioner may be faced with a variety of claims and counterclaims, made on a variety of alternative contractual and legal bases as follows:

- The contractor's lawful termination.
- The employer's lawful termination.
- The employer's unlawful termination.
- The contractor's unlawful termination.

These are outlined as follows, addressing each separately, with specific reference to the provisions of the FIDIC Red Book as an example.

7.3 Claims for the Contractor's Lawful Termination

Understandably, given that the provisions relied upon for such a claim will be those that address a fundamental failure of the employer such as insolvency, bankruptcy or failure to pay, here the contractor will broadly be entitled to the value of all work done and its costs or losses resulting from the termination. The FIDIC Red Book sets this out and the components of the contractor's entitlements in clause 16.4, and by reference to its subclauses 19.6(a) to (e), as quoted as follows, following each of which some particular evaluation issues that commonly arise are explained.

7.3.1 'Amounts Payable for any Work Carried Out'

This is in FIDIC Red Book clause 19.6(a). This should particularly involve a complete measure and valuation of all work done, including variations, at the date of termination. Clause 19.6(a) places the duty to do this on the engineer under the contract. Often, for expediency, this will substantially rely on measurement and valuation as it was last set out in the most recent interim payment certificate. In addition, it is likely to require significant joint surveying, measurement and recording between the contractor and the employer's representatives immediately following the termination, although in practice the events of a termination are often such that co-operation of that nature between engineer, employer and contractor is hard to achieve. This can be particularly supplemented by photographic records and even a video recording, as explained elsewhere. The records should be signed off by both as agreed, and a complete copy exchanged, so far as the political climate allows, as noted elsewhere.

Apart from a failure to carry out a necessary survey jointly and to get it signed off by both parties, common problems with this are as follows:

- Issues as to whether works carried out at the date of termination were in accordance with specifications. This may require the input of engineering expertise from both parties. If work is not in accordance with the specification, then there may also be issues as to why it is non-compliant and how to value it. Some items may be of no value and will require complete replacement, for example, stainless steel items that are not of a grade as specified for hostile environmental conditions. Other items may only require some further work, for example a sub-base layer for a road that has not been fully compacted. This illustrates the potential for issues as to why work is not to specification. This could be a simple result of the termination interrupting the contractor's progress. It might also involve issues regarding the interpretation of the specification and the effect of variations that have changed the specification.
- Failure to record the work done in a manner consistent with its pricing in the contract documents. This can mean that, whilst a record has been agreed, it is not in a form that readily allows the resulting value to be agreed. It is obviously essential, but not always the case, that those measuring works completed at termination do so by reference to how the items are measured, and hence priced, in the contract documents, so that rates and prices in the contract can be applied to their quantities to get to a proper value under the contract.
- Photographic records sometimes lack context, scale and location, such that when they are used later it is unclear how they can be converted into a value of the work photographed. The example of racks of pipe lengths with no scale was mentioned elsewhere. In this regard, video recordings may be preferred, or be a useful supplement, because they allow the camera operator to commentate on what is being shown, explain it in detail, and say why it is relevant.
- However, many film makers do not always have a commentator's skills. They can be shy to commentate over the recording. Alternatively, they can fail to appreciate what narrative information will be useful to the viewer sometime in the future. They can also give a biased account that the other party does not agree with. At the end of a joint survey, signing off photographs may in practice prove to be much simpler than agreeing a video recording and its narrative.
- Whether works are being measured or recorded, if they were carried out by a subcontractor it may be important to also involve it in the discussions on a joint basis, so that the contractor is not exposed to costs that it cannot recover from the employer. Many subcontracts contain a provision that they are automatically terminated if the contractor's contract is terminated, but that obvious coordination is often ignored by the contractor and employer when agreeing the value of work carried out under their contract. Alternatively, contractors spend time on detailed surveys of their works and agreeing them with the employer but adopt a blasé approach to their subcontractors' accounts, on the basis that the value of the former will dictate the latter. In practice

this can be dangerous for the contactor, particularly on very specialised trades such as structured cabling systems and computer network equipment, where the subcontractor is likely to have more detailed technical understanding of what it has done than the staff of the contractor and often the engineer.

- The usefulness of interim certificates before termination to a measurement and valuation of works after termination can be affected by their being based on assessed approximate quantities. Such approximations may be acceptable for the purposes of interim payments when it is expected that a proper remeasurement will be carried out when those works are completed, but this is unlikely to be sufficiently accurate for the purposes of a post-termination final account. Such assessed interim certificates are particularly common on major and complex projects, such that where the project is terminated the engineer can find itself having to carry out extensive measurement exercises. This can be complicated where the termination of the contract was due to payment issues that have also become a feature of the engineer's relationship with the employer because of some general financial difficulty the employer is experiencing. Similarly, it may be that the termination related to slow progress and the employer and engineer are having their own squabbles regarding responsibility for such a delay. The reality of many terminated construction and engineering contracts is that relationships between employer, engineer and contractor are all strained to some extent, such that when interim certification has been carried out on a superficial basis, the engineer does not engage with its post-termination duties as required by, for example, FIDIC Red Book clause 19.6(a) with the resources and commitment necessary, but simply relies on the last interim certificate.

'Amounts payable for any work carried out' should also include the valuation of variations. In the context of a termination, these can involve some particular problems. The parties may disagree as to whether their values were agreed before the termination. Relationships that once meant that particular values were considered by one party to have been agreed and were paid through interim certificates are now disputed. The extent of work carried out to a variation should be recorded as part of a survey, the same as for contract work items, preferably jointly.

In addition to the measured valuation of completed works and variations, amounts payable for any works done under such as FIDIC Red Book clause 19.6(a) would also cover other amounts due to the contractor under the contract for carrying out such works up to the date of termination. Under the FIDIC Red Book, this would include such items as claims for delay under clauses such as clauses 1.9, 2.1, 4.7, 4.12 or 8.10, and costs escalation under clauses such as clauses 13.7 and 13.8 for example.

An agreed measurement and valuation can be particularly easily applied to rates and prices in a contract document such as bills of quantities to get to the value of works completed at termination, but this assumes that such a level of detailed pricing is available. Where the contract only sets out a series of lump sums, valuation is more difficult. This occurs, for example, on large housing schemes where the only price breakdown is for numbers of units of different types with a lump sum price against each. The preparation of bills of quantities based on standard methods of measurement as explained in Chapter 2 of this book is a real boon in such circumstances.

In SBC/Q, clause 8.12.3.1 uses a similar term to that of FIDIC Red Book clause 19.6(a) considered elsewhere, in that it refers to 'the total value of work properly executed'.

However, SBC/Q clause 8.12.3.2 also expressly covers 'any sums ascertained in respect of direct loss/and or expense under clause 4.23 (whether ascertained before or after the date of termination)'. That reference to clause 4.23 would cover delay and disruption claims of the type considered in Chapter 6 of this book.

In the Infrastructure Conditions, clause 63(4) refers to 'the Contract value of all work carried out', which perhaps more pointedly requires valuation on the basis of rates and prices in the contract, although the effect appears the same. However, the Infrastructure Conditions clause 63(4)(a) also refers to 'the amounts payable in respect of any preliminary items'. This is a reference to preliminaries and general items, usually measured in their own section for any bills of quantities separately from the measured works items. Clearly, the FIDIC Red Book and SBC/Q would require their inclusion in their 'value of work properly executed' or 'amounts payable for any work carried out', however the separate reference in the Infrastructure Conditions does bring to mind a common problem in relation to valuing preliminaries and general items at termination.

The full provision for inclusion of preliminaries and general items in clause 63(4)(a) of the Infrastructure Conditions reads as follows:

> the amounts payable in respect of any preliminary items so far as the work or service comprised therein has been carried out or performed and a proper proportion of any such items which have been partially carried out or performed.

The previous, 1987, edition of FIDIC contained this same wording in its clause 65.8(a), such that it still appears internationally in many bespoke contracts based on that old FIDIC form, particularly in some government standard construction contracts. The interpretation of what is 'a proper proportion' for preliminaries and general bills items that have been part completed is a perennial issue in termination claims under such terms. For time-related bill items such as provision of fixed cranes or supervision, contractors will often argue that the proportion should be based on the proportion of the contract period that has expired at the date of termination. This can become complicated where there is a sectional completion with a number of completion dates, but the main problem arises where there is dispute as to delay, for example if the contract is, say, 40 weeks into a 50-week programme, but has achieved, by total value of measured works items, only 50% progress. The contractor may say that it should be paid 80% (40 weeks/50 weeks) of its bill allowance for time-related preliminaries and general items. The employer's response may be that the logical extension of the contractor's argument would be that, if it had overrun by 10 weeks, would it ask for 120% (60 weeks/50 weeks) of its bill allowance? The employer will say that this must be wrong and also accuse the contractor of seeking to make a claim for prolongation costs without proving its case for delay and quantifying a claim based on bill of quantities rates, rather than proven actual costs as required by clauses such as FIDIC Red Book clauses 1.9, 2.1, 4.7, 4.12 and 8.10.

Such issues tend to result in an argument over who is responsible for the works only being 50% completed when 80% into their programme (in this example), and whether the contractor's claims for extensions of time have been properly assessed by the engineer under the contract. A further subtlety in relation to the employer's approach of 'a proper proportion' being based on the progress of measured works is which measured works? Is it the whole of the works, or just those to which each relevant preliminaries and general items relates? For example, if the bills of quantities for a high-rise office include a preliminaries and general section, a section for measured works to the high-rise block and another section for low level car parks, and preliminaries and a general item for static cranes is being considered, should its value at termination be based solely on progress of the high-rise works for which it was required? Alternatively, should it be based on progress of the works as a whole?

Dealing with preliminaries and general cost items will also require a consideration of whether each is a fixed or time-related item, and this is made more complex where there are composite items that include not just a number of components (for example staff and supervision without any detail as to how many or for how long) but also include both fixed and time-related components (for example both mobile cranes for offloading and placing permanent plant and static cranes for servicing the whole project for its duration). Take, for example, a simple bill of quantities item for 'provision of welfare facilities', such as toilets, washing, changing, drinking water and eating. This will include: initial activities of delivery, erection, service connection and setup, followed by a period of running and maintenance and a final demolition and removal. A preliminaries and general bill of quantities may have required separate prices for each of these three periods (although it is notable that rules such as the POMI Rules do not require such a split). Alternatively, the parties may have agreed a breakdown for the purposes of interim payments. However, if these are not in place, or it is considered that a split for interim payments was notional and not sufficiently accurate, then there will be a need to establish a proper split between the three stages of 'establish', 'run' and 'remove', in order to value that 'provision of welfare facilities' to the extent that they have been carried out at termination.

Further complexity occurs where a preliminaries and general composite item allows for components some of which have been provided by the contractor before termination and others that have not. To take an example for the contractor's welfare facilities further, say that the bill of quantities item also included a first aid clinic, but that facility had not yet been provided at the date of termination. Valuation at that date of such a composite item will need its price to be broken down between the various components of it, such that its progress at termination can properly take account of what had, and had not, been provided. This may also require consideration of when the different components were actually constructed and a detailed breakdown of the bill item into its component parts and prices against each. The result of such considerations can mean that proper valuation of preliminaries and general items at termination can be a very complex and contentious area.

It is surprising how often a proper valuation of preliminaries and general items at termination shows that the contractor has been overpaid for such items through interim payments prior to termination. This seems to occur where the pricing document in the contract lacks required detail as to the pricing of such items and/or the certifier takes a complacent approach to those interim certifications on the basis that any overpayment will 'come out in the wash' later, as described elsewhere.

The 'amounts payable for any work carried out' under, for example, FIDIC Red Book clause 19.6(b) would also include any entitlements that the employer has accrued at the date of termination such as to counterclaims and set-offs for defective works and delay damages, for example. Since these occur most commonly in a termination by the employer for the contractor's default, they are considered in more detail elsewhere when discussing that situation.

7.3.2 'The Cost of Plant and Materials Ordered for the Work'

This is in FIDIC Red Book clause 19.6(b). The terms 'Plant' and 'Materials' are defined in clause 1.1.5 as being for those items that are intended to form part of the permanent works (for example air handling units, turbines or concrete), rather than being items such as the contractor's excavators and fuel. Unfixed permanent plant and materials should be part of a similar joint survey to that described elsewhere, and are commonly the subject of the same problems as set out there. Apart from a failure to carry out the survey jointly and to get it signed off by both parties, common problems with this are as follows:

- Supplier invoices and subcontractor accounts should confirm the costs and allowance should also be made for delivery charges. However, it might be considered necessary to obtain proof of payment and investigate whether the contractor has any discounting arrangements with a supplier that are not shown on the face of its invoices.
- A further occasional issue is whether materials on a site were properly for that project. It is not unknown for contractors to use large storage areas on such infrastructure projects to store materials that are intended for use on its other projects, which are nearby but on more restricted sites. It is also not unknown for contractors to over-order materials, which may require consideration of waste factors to establish whether the over-ordering simply reflects that. On one infrastructure project the engineer had for months been including unnecessarily high quantities of district cooling pipes under the 'unfixed materials on site' heading of interim certificates. That might have never become apparent had the project been completed and not been terminated. However, the engineer's valuation of the works at termination similarly failed to consider that the total length certified as installed and as unfixed materials on site was significantly more than that required to complete the works, a point that did not become apparent until the parties' termination claims and counterclaims were arbitrated. The contractor had simply over-ordered the pipes, but had been very happy to leave them decaying on site so long as they were included in interim payments, and it feared that a proper adjustment to exclude them would leave it with a negative interim certificate whilst it incurred the cost of removing them but only recovered a partial refund from its supplier, at best.
- An issue that the valuation of unfixed plant and materials particularly shares with the valuation of work done is that of compliance with the specification. It may be that those materials that are left unfixed at termination were unfixed and lying on site for the very reason that they were not to specification. If that was because the contractor had made a mistake in its procurement, then, again, it may be happy to just leave them on site if they have previously been paid for through interim certificates.
- Non-compliance with the specification may also involve consideration of the reasons for that non-compliance. Take, for example, unfixed reinforcing bars or internal grade pipework that has been stored on a long-delayed project and which will require treatment before they can be used post-termination.
- FIDIC Red Book clause 19.6(b) expressly refers to materials and plant to be included in a valuation at termination as including items that are not yet delivered to site. The full clause reads as follows, to which emphasis has been added:

> … the Cost of Plant and Materials ordered for the Works which have been delivered to the Contractor, *or of which the Contractor is liable to accept delivery*: this Plant and Materials shall become the property of (and be at the risk of the Employer) when paid for by the Employer, and the Contractor shall place the same at the Employer's disposal.

Therefore the Materials and Plant can be off-site, but the contractor has to be under a legal obligation to accept their delivery. This may mean that the employer and contractor have to make visits to off-site locations to inspect materials, which can be difficult to coordinate. This is to check that the materials are there, their quantities, that they are to specification, and that they are set aside and secured as the legal property of the contractor. The increased internationalisation of the engineering and construction industries means that this might require overseas trips. Proving ownership, or the liability to accept delivery, can be particularly difficult. It is often complained by contractors that employers and their representatives deliberately make the logistics of such checking more difficult than they need to be, in order to avoid paying for such items in a post-termination valuation. This particularly occurs where relationships have broken down due to the causes and the act of termination. For items stored overseas, the requirement to 'place the same at the Employer's disposal' is often effected by the items being kept set aside and secured, but the labelling being changed to put them in the employer's name rather than that of the contractor. Difficulties can arise in this respect where goods are being manufactured for the works and are part completed at the manufacturer's premises at the time of termination. In such circumstances there can be problems arising from the alternative courses of action that may be available. These are cancellation charges and the cost of abortive work, if the goods are not to be delivered at all, or payment for the full amount on completion of manufacture and delivery or storage at the employer's order. In the latter case, work would continue after the date of termination and special agreements and terms may be required. There will in most cases be a benefit to the employer in taking over such orders, if undue delay occasioned by the need to reorder is to be avoided. It may also be to the employer's benefit to take over the order from the contractor to avoid giving the manufacturer the opportunity to charge a premium to the original order value on the negotiation of a new order. Such matters fall outside the contract mechanism and would need to be negotiated between the parties, depending upon the terms of any relevant orders, but there would be potential benefits to both employer and contractor in working together to achieve this.

- Joint surveys of materials are often less thorough and comprehensive than those for work done. They should not be. Otherwise, the result can be that the contractor follows an agreed survey by sending through supplementary lists of items it says were missed and which it says should be added. That is a risk for the contractor that should be preventable by a proper joint survey in the first place. Such problems seem to arise out of the urgent and pressurised atmosphere under which joint surveys are often carried out after termination.

These are all issues that quantity surveyors valuing plant and materials on site following termination need to be wary of, but all too commonly miss.

In SBC/Q, clause 8.12.3.4 is simpler than the equivalent provisions of FIDIC Red Book clause 19.6(b), considered elsewhere. Reference is to 'the cost of materials or goods (including Site Materials)'. The capitalised term 'Site Materials' is defined in clause 1.1 as materials and goods delivered or adjacent to the site and intended for incorporation therein. It is suggested that the effect would be the same as those of the FIDIC Red Book, the clause's addition of the word 'properly' in its 'materials or goods (including Site Materials) properly ordered for the Works' would particularly preclude payment for materials that had been over-ordered by the contractor (if such express qualification were thought necessary).

In the Infrastructure Conditions, clause 63(4)(b) is in similar terms to those of FIDIC Red Book clause 19.6(b), considered elsewhere. Although reference is the same as that of SBC/Q, to the uncapitalised, undefined, terms 'plant and materials', it is suggested that the effect would be the same as that of the FIDIC Red Book, the clause's addition of the word 'properly' in its 'materials or goods (including Site Materials' properly ordered for the Works' would particularly preclude payment for materials that had been over-ordered by the contractor (if such express qualification were thought necessary). Some bespoke provisions for the valuation of the contractor's progress at termination do not contain a separate and express provision relating to the 'cost' of the contractor's materials. In such cases, argument may result as to whether they should be included under an item for such as 'the value of work carried out' and whether that covers the supply of materials. If so, contractors sometimes argue on this basis that unfixed materials are to be paid for based on the value allowed for them in pricing the contract, including overheads and profit, rather than actual costs incurred by the contractor. Where the valuation at the termination clause contains no provision for including permanent plant and materials, then the contractor will retain ownership and will remove them. If the employer wants to use them for the purposes of its replacement contractor, then it would still have to pay for them.

7.3.3 'Any Other Cost or Liability Which in the Circumstances was Reasonably Incurred … in the Expectation of Completing the Works'

This is in FIDIC Red Book clause 19.6(c). The capitalised term 'Cost' is defined in clause 1.1.4.3 as 'expenditure reasonably incurred (or to be incurred) by the Contractor, whether on or off the Site, including overhead and similar charges, but does not include profit'.

The reference to 'liability' in FIDIC Red Book clause 19.6(c) is clearly a desirable addition in practice to protect the interests of the contractor, to allow it to be paid for items before it has paid a supplier or subcontractor. This recognises the realities of an industry where cashflow is so important to contractors and periods for interim payment from the employer are usually shorter than those of payments to their suppliers and subcontractors. This is particularly important in relation to expensive items of permanent materials and plant. One issue to consider is the extent that such liability might be reduced by the termination, for example where materials have been set aside by a supplier for delivery before the termination, the contractor is due to pay for them, but following termination the supplier agrees a credit for part or all of them because they can be sold by it elsewhere. FIDIC Red Book clause 19.6(c) usually covers costs incurred by the contractor in preparatory activities that will not be covered by the valuation of work carried out under clause 19.6(a) or the cost of permanent plant and materials covered by clause 19.6(b). Often costs claimed under clause 19.6(c) appear to be items that could potentially have been claimed under those other heads, but the 'catch-all' nature of the wording of 19.6(c) means that any issues are avoided as to whether, for example, temporary works not expressly set out in the description of their related measured works items or an item in a preliminaries and general items bill, are covered or have been recorded, in part or whole, by clause 19.6(a). Common examples of the types of costs claimed under clauses such as FIDIC Red Book clause 19.6(c) include such terms as the following:

- The purchase of items of construction equipment and temporary plant to be used on the works. These may involve a significant initial capital outlay, for example items such as cranes for high-rise buildings or barges for marine works. Such capital costs and the extent to which they become unrecovered because of a termination of a contract can be very significant, as illustrated by a simplistic hypothetical example as follows. Say a contractor has to purchase a secondhand 200 tonne crane for the erection of structural steelwork to a new petrochemical plant, and that crane cost US$ 1 million to purchase. Also assume that its estimators expected a project duration of two years and that the crane would remain in use on later projects for another eight years thereafter. Furthermore, that they allowed for straight line depreciation and no residual value at the end of that 10 years of overall use. They would on this basis have assumed that the capital costs that were required to be recovered by that project (excluding other costs such as financing) were US$ 200,000 (US$ 1 million × 2/10 years). Let us also assume that the estimators added a further US$ 50,000 for project ancillary costs such as delivery, running, maintenance, repairs, fuel, spares, consumables and removal. That is a total project cost of US$ 250,000. Their allowance of that US$ 250,000 in the tender could have been through one of two places, depending on the structure of the pricing document and their approach to pricing it. These alternatives are set out as follows along with consideration of how such costs might be recovered in a termination claim. The assumed circumstances are that the project has been terminated when the crane has been on site for 10% of its planned period and has completed 10% by tonnage of the steelwork erection that it was purchased for.
  - The item is priced against a specific item for cranage in a separate preliminaries and general items bill. If the engineer under an FIDIC Red Book valuation considers that clause 19.6(a) is limited to work, that standard form's lack of separate reference to preliminaries and general items might cause the contractor a problem in the absence of clause 19.6(c). Even under, for example, the Infrastructure Condition clause 63.4(a) and 'so far as the work or service comprised therein has been carried out or performed', a catch-all provision such as FIDIC Red Book clause 19.6(c) should prevent issues as to the effect of 'so far as' in the context of this example. If 'so far as' means reference to the 10% period or extent of work done, that might fairly recover 10% of most of the ancillary costs, but would not fairly recover the capital cost incurred by the contractor.
  - The item is priced within the unit rates against the measured works items of structural steelwork for which the crane was required. The problem of recovering only 10% against the progress of the steelwork erection that the crane was purchased for against that of FIDIC Red Book clause 19.6(a) for 'amounts payable for any work carried out' is particularly clear here.

This simplistic example ignores the potential need to make a particular adjustment for ancillary costs under such a provision. It also consistently assumes progress is

10% whether based on time passed or tonnage of work completed, and avoids such debates as discussed elsewhere as to which of these is the correct basis of valuation. However, it is given purely to illustrate the problems of recovery of large plant capital costs at termination through the valuation of progress to that date that clause 19.6(c) is intended to avoid.

- A next consideration in relation to this example of a US$ 1 million crane purchased for a project may be what actually happened to it after termination. If the contractor has no alternative current work for it elsewhere, then it may sit in its plant yard until the following other project starts. The claimable amount of depreciation will only be that over the two years between the dates of termination and when the works would otherwise have finished. However, it might be that immediately following termination the crane is remobilised to another project or is hired-out externally. In such circumstances the parties may differ as to whether the benefits of such other use or hire income are relevant to an assessment of the contractor's claim.

- Similar considerations to the above may apply in relation to hired items of plant and equipment where these are externally hired-in on the basis of fixed or minimum durations, which would have continued beyond the date of termination or include significant delivery and set-up costs that have only been partly recovered by the date of termination.

- The construction of major temporary works, for example a concrete pre-casting yard or a cofferdam. Potential problems against such items in the absence of clauses such as FIDIC Red Book clause 19.6(c) can be the same as for the crane example, in terms of the contractor having a large initial cost that it priced to be recovered through a preliminaries and general item or the value of measured permanent works that such temporary works will service. The only additional issue relates to such items being work that is temporary work, rather than plant and equipment purchases. In FIDIC Red Book clause 19.6(a), 'amounts payable for any work carried out' is notable in not referring to that contract's defined term 'Works', but the full provision reads: 'the amounts payable for any work carried out for which a price is stated in the Contract'. That final, capitalised, word 'Contract' is defined in clause 1.1.1 such that this is likely to mean the bills of quantities or other pricing document. If those do not state a separate price for the temporary works, then catch-all clause 19.6(c) may be very important for the contractor.

- Recruitment costs of staff and labour. This is particularly relevant to major international projects where a contractor brings large numbers of employees into a country for a particular project, incurring such things as agency fees, health checks, transport such as flights, visa costs, etc. Quantification of such items may also be further complicated by the extent to which those costs were recovered on the work carried out before the termination.

In SBC/Q there is no express provision for recovery of 'any other Cost or liability which in the circumstances was reasonably incurred … in the expectation of completing the Works', which is similar to FIDIC Red Book clause 19.6(c). It is not readily apparent where under the termination provisions of SBC/Q the contractor would be entitled to such costs. They might, however, be considered as 'direct loss and/or damage caused to the Contractor by the termination' under clause 8.12.3.5 if the termination prevented the contractor from obtaining payment for them through the valuation of work to which they relate.

In the Infrastructure Conditions, provision very similar to FIDIC Red Book clause 19.6(c) is in clause 63.4(c).

7.3.4 'The Cost of Removal of Temporary Works and Contractor's Equipment'

This is in FIDIC Red Book clause 19.6(d). Contractor's Equipment is defined in clauses 1.1.5 as including 'all apparatus, machinery, vehicles and other things required for the execution and completion of the Works'. It will include such items as barges, cranes, batchers, mixers, earthworks plant, trucks, staff vehicles, buses, etc. Temporary Works are defined in clause 1.1.5 as 'all temporary works of every kind (other than Contractor's Equipment) required on Site for the execution and completion of the Permanent Works' and therefore include such as fencing, scaffolding, falsework, shoring, earthworks support, cofferdams, temporary roads, etc. These headings may also cover a labour camp constructed and maintained for the project.

These removal costs are usually readily capable of being recorded by the contractor for its own resources, or are the subject of discrete and documented charges from such as hire companies and subcontractors.

Contractors should record the time and resources of their people and plant and equipment that they expend on such removal using their own resources, whether on daywork forms or some other form of record that can be presented for counter-signature. The circumstances of the termination by the contractor may be such that it may be unlikely that such agreement will be gained, but it is good practice to present them in this manner in any event. However, it may also be that on a large project the contractor just retains a pool of staff and operatives on site to carry out whatever demobilisation and removal activities are required, as they arise. In this case, the discrete costs of particular activities may not be capable of recording, only the overall cost of the retained resources.

Charges from hire companies (such as for cranes) and subcontractors (such as for scaffolding) may also include charges for the remaining period of a fixed or minimum hire term agreed under their agreement. Whether these fall to be recovered as part of FIDIC Red Book clause 19.6(d) for 'cost of removal' or clause 19.6(c) for 'other Cost or liability … reasonably incurred' may be a moot point, but they should be recoverable under one of these clauses, provided it was reasonable for the contractor to have entered into such a fixed or minimum term agreement in the first place.

In SBC/Q clause 8.12.3.3, the contractor is entitled to its reasonable costs of removal of items under clause 8.12.2, which requires it to remove its temporary buildings, plant, tools, goods and equipment as wells as 'Contractor's Persons', who are widely defined under clause 1.1 as its employees or agents and all other persons employed or engaged on the works or parts thereof, excluding those of the employer and its consultants and statutory undertakers.

In the Infrastructure Conditions, a provision very similar to FIDIC Red Book clause 19.6(d) is in clause 63.4(d).

7.3.5 'The Cost of Repatriation of the Contractor's Staff and Labour'

This is in FIDIC Red Book clause 19.6(e) and is an obvious area of potential termination costs on international projects where the contractor has had to import staff and labour to service a construction or engineering project in the absence of sufficient local resources with the appropriate level of skills and/or experience.

Such repatriation costs usually include such items as transport, flights, exit permits, visa cancellation, end of service gratuity or bonus, and may include compensation payments where (particularly staff) are sent home early against a fixed or minimum term contract. There may also be unused periods of visas and identification papers paid for a period that extends beyond the repatriation date.

A common counter to the quantum of this head of claim is that the contractor would have had to repatriate those employees at the end of the contract anyway, such that it will have allowed for them in the contract price and will hence have recovered those allowances through the valuation of work carried out under FIDIC Red Book clause 19.6(a). This is a matter that varies from project to project. It can give rise to two issues:

- Whilst it is usual, it is not always the case that contractors plan, and price, for the repatriation of their imported employees to their home country at the end of a project. It may be that they expected to reallocate them to further later projects, such that no allowance was made to recover the costs of repatriation. However, if the circumstances of the termination are such that there is no following project that they can be reallocated to, the repatriation costs may be entirely additional. Such circumstances will, most obviously, be early termination at a time so long before the later project starts that it is uneconomic to keep the employees in-country. Less obviously, some terminations can lead to a wider effect on the contractor's local workload. For example, an international contractor claimed as part of its costs of termination that the project was its first in a newly developing country and was planned to be a springboard for securing further projects there and to creating a permanent local presence. It asserted that the wrongful termination of that first highly prestigious project was of such market interest that its reputation was destroyed in that country and it had to withdraw its plans for further work and establishment of a local business vehicle. Thus, it had not allowed in its contract price for the costs of repatriating its people, but incurred those costs when that contract was terminated, also destroying the prospect of future work.
- If the contractor priced the works to include the costs of repatriation at its end, then account may need to be taken of the recovery of those repatriation costs through the value of work completed prior to the termination and included in the valuation at termination under such as FIDIC Red Book clause 19.6(a). This is similar to 'costs and liabilities incurred in the expectation' under clause 19.6(c) and can be a similarly complex exercise. Such recovery may be through the pricing of measured work items (particularly for the labour related to that measured work) or the pricing of preliminaries and general items (particularly for supervision and management staff). The evaluation of such adjustments can be problematical. A superficial approach to recoveries through measured work items might be based on the percentage of those works completed at termination. For example, if $10,000 has been incurred in repatriating steel reinforcement fixers for concrete structures, which are 40% completed (by value) at termination, then the recovery of the repatriation would be said to be $4,000 for 'work carried out' under FIDIC Red Book clause 19.6(a), leaving $6,000 to be allowed under clause 19.6(c). However, this approach is too superficial in most cases. A more accurate approach would require a build-up to the contractor's pricing

of the reinforcement to establish exactly what recovery has been made through its valuation. Similar considerations will be required in relation to repatriation of people allowed for in the pricing of preliminaries and general items, establishing which items covered such supervision and staff, what allowances were made and how much had been recovered through the value of those items at termination.

Neither SBC/Q nor the Infrastructure Conditions contain an express provision for the contractor to recover its costs of repatriation of its staff and labour. This reflects that these forms are intended for domestic use in the UK, unlike the FIDIC Red Book, and therefore that the importation of people from overseas is less likely. However, it is suggested that such costs would be covered by SBC/Q clause 8.12.3.5 for 'any direct loss and/or damage caused to the Contractor by the termination' and the Infrastructure Conditions clause 64(3) for 'any loss or damage to the Contractor arising from or as a consequence of such termination'. Alternatively, SBC/Q clause 8.12.3.3 might be claimed to cover repatriation costs in those related to removing from the site 'Contractor's Persons', as discussed elsewhere.

7.3.6 'A Loss of Profit or Other Loss or Damage Sustained by the Contractor'

This is in FIDIC Red Book clause 16.4(c). The reference to 'other loss or damage' is especially broad, understandable and helpful as a 'catch-all' for the protection of the contractor's interests where it has terminated due to a serious failure by the employer. One item that contractors often put under this heading is loss of contribution to their off-site or head office overheads. This is usually, though not always, addressed along with its loss of profit as a heading of claim for 'loss of overheads and profit'. In the following these are addressed together, but the points made could apply equally to a claim for just loss of profit.

In practice, loss of overheads and profit ('margin') is a particularly problematical head of claim to prove and to quantify and contractor's submissions vary greatly in their level of detail. Such claims that are based simply on the contractor's company overheads and profits stated in its published company accounts beg the question of whether that same level of overheads and profit contribution would have been made on the project, let alone the works not carried out as a result of the termination. A calculation based on the contractor's allowances in its tender build-up may have the advantage of addressing the particular project, but still begs the question as to whether that level of overheads and profit contribution would have been achievable in practice, which usually requires consideration of what the contractor was achieving pre-termination. On that basis, a better approach to loss of overheads and profit claims may be to analyse the actual profitability of work carried out pre-termination and apply that across the remaining work. This approach is particularly credible on repetitive work such as a large housing scheme. However, where a contractor made a certain level of contribution to its overheads and profits on the foundations to a high-rise building, it does not follow that it would have made the same profit on the superstructure, particularly including specialist subcontract trades such as mechanical and electrical services and finishing trades.

It is also often the case that termination of a contract follows a period of delay and disruption to the contractor's progress and that its claims following termination include not only those arising out of the termination but also for losses on such as labour productivity before the termination. Where this is the case, analysis of the contribution to overheads and profit achieved on works before the termination may show an actual loss or level somewhat lower than would have been the case but for the earlier delay and disruption. The employer will then say there was no loss of contribution to overheads and profit on the work stopped by the termination because the contractor had shown its inability to achieve such a contribution on the work it did carry out before termination. In response, the contractor may say that it would have achieved a contribution to its overheads and profit on works in both periods had it not been delayed and disrupted by the employer. On that basis, the contractor might adjust its calculations of margin on the work done prior to the termination for amounts claimed in its delay and disruption claims to derive a 'but for' level of contribution it would have been capable of achieving on the work done and to be applied to the work remaining at termination. However, such calculations can be very complicated and are dependent on the success of the delay and disruption claims. Furthermore, they still rely on the assumption that the work in the two periods, before and after termination, were sufficiently similar.

As noted elsewhere, such similarity can occur on highly repetitive works, such as housing. In one terminated residential project, the works were broken down into numerous phases in adjacent locations and of similar construction and specification. The contract was terminated for the employer's convenience as a result of the crash in the local housing market in 2008 and 2009, halfway through the third phase. The first phase was completed and many of the properties occupied. This allowed the contractor to model its claim for loss of contribution to its overheads and profit for the incomplete properties left at termination on what it had actually achieved on those that it had completed and for which it had detailed cost records. This was particularly helpful in relation to direct labour and equipment outputs, the costs of materials and subcontracted works being areas that could in any event have been relatively easily established by reference to documentation such as supplier and subcontractor invoices. That contractor also set out comparative analyses of the levels of margin earned: by contractors generally in the market at the time, its own company accounts and its tender allowances. All four of these approaches gave similar results, which gave rise to a very persuasive claim for loss of overhead and profit on terminated works.

Where completed parts of a terminated project cannot be used to model lost margins on works taken away by the termination, it occasionally occurs that the contractor has other documentary records of the actual costs it would have incurred on those later works through its procurement records. A contract for the construction of several buildings to an international airport was terminated before any of those buildings were progressed far above ground floor slab. Therefore, only substructure works had detailed cost and value records and these could not be used with any great certainty to represent what would have been achieved on the superstructures. However, by the date of termination, the contractor's commercial team had substantially placed its subcontractor and supplier contracts and orders for those later works. Even where contracts and orders were not in place, it had obtained competitive quotations for other packages. This allowed it to prepare a claim for loss of overhead and profit on the terminated works by comparing its procurement records of the costs of subcontracts and supplies with those included in its tender build-up. This provided a very detailed claim for loss of margin for most works excluded by the termination, except its direct labour and preliminaries and general costs. However, for those preliminaries and general costs the contractor could show that its cost projections and planning at the date of termination envisaged no change to such as its temporary site facilities and which staff it would add to that project to deal with such items as later mechanical and electrical services and finishing trades, when and how long for. This enabled a comparison of likely actual preliminaries and general costs with those in its tender build-up. That only left labour, but most trades were planned to be subcontracted and had already been procured such that this was a small part of its claim for loss of contribution to its overheads and profit.

Loss of margin should usually be considerably easier to establish in relation to works that are covered by prime cost items, where the exact type of product or component cannot be specified, or provisional sums, where building components or items cannot be measured and described in accordance with the applicable detailed measurement rules. These are items against which tendering contractors are required to separately set out their additions for such items as overheads and profit. However, rules such as CESMM clause 5.16(b) only require a single item for the contractor's percentage for 'all other charges and profit' and this can result in an issue as to how much of the percentage is for head office or off-site overheads and profit and what is included in 'other charges'. A point then might become that the 'other charges' would have involved a directly proportional cost that has not been incurred, so should be excluded from the calculation. More helpful in this regard is the guidance in rule 2.11.2 of the NRM2 detailed Rules of Measurement for Building Works that 'When required, overheads and profit can be treated as two separate cost items; namely, "overheads" and "profit"'. Such an approach will allow the profit in particular to be separately established. The overheads might still require consideration of what costs would have been incurred.

If the contractor's percentage for contribution to overheads and profit against a provisional or prime cost sum can be established from the contract documents, then the remaining issue may be the actual cost that would have been incurred against that sum and to which the percentage would have applied had the works not been stopped by termination. Of course, it does not follow that this cost is the amount allowed in the bills of quantities in the contract documents and consideration should be given as to what the real amount is likely to have been. This may be significant where the contract documents appreciably underestimated those provisional and prime cost sums.

As quoted elsewhere, FIDIC Red Book clause 16.4(c) expressly includes loss of profit in its 'any loss of profit or other loss or damage sustained by the Contractor'. However, the contractor's entitlements under SBC/Q clause 8.12, where it has terminated for the employer's default under clauses 8.9 and 8.10, do not mention loss of profit and sub-clause 8.12.3.5 only refers to 'any direct loss and/or damage caused to the Contractor by the termination'. In the Infrastructure Conditions, payment on termination by the contractor for default by the employer under clause 64 similarly includes in subclause 64(3) 'the amount of any loss or damage to the Contractor arising from or as a consequences of the termination'. However, in *Wraight Limited v P. H. & T. (HOLDINGS) Limited* (1968) 13 BLR 27, works were suspended because of unexpectedly difficult ground conditions for a duration longer that the period agreed in the contract, such that Wraight determined its employment. The court considered its claim for 'direct loss and/or damage caused to the Contractor by the determination' under clause 26(2)(b)(vi) of the 1963 edition of the JCT contract, which included loss of profit on the works not carried out as a result of the determination. The Judge held that:

> ... the claimants are entitled to recover, as being direct loss and/or damage, those sums of money which they would have made if the contract had been performed, less the money which has been saved to them because of the disappearance of their contractual obligation.

Whilst this was a case in relation to termination for an event that was in neither party's control, of the type mentioned elsewhere and now covered by SBC/Q's clause 8.11, the wording 'loss or damage' is the same there and under clause 64 of the Infrastructure Conditions.

7.3.7   Demobilisation

A feature of many contract clauses for the contractor's valuation after it has terminated that contract is how they tend to expressly cater for the value of works done up to the date of termination and the costs of the contractor's subsequent departure, but not expressly consider the contractor's ongoing costs in the intervening period between termination and departure. Such periods are a practical inevitability of most terminations, but can be particularly significant on major projects and where there is some uncertainty or dispute as to the validity of the termination.

Under the FIDIC Red Book provisions considered elsewhere, the contractor will be compensated for its temporary works, labour, staff, vehicles, equipment and plant costs up to termination through the 'amount payable for any work carried out' under clause 19.6(a). It will be paid for removing temporary works, such as scaffolding and shoring, etc., under clause 19.6(d). It will be paid for removing its equipment, such as a batching plant and cranes, etc., under clause 19.6(d). It will be paid for repatriating its staff and labour under clause 19.6(c). However, what about the contractor's on-going costs of these things between the termination and their removal?

Claims for such costs are usually made by terminated contractors under a submission subheading of 'Demobilisation Costs', which recognises that such costs will be incurred in a period of demobilisation incurred between termination and removal. Typically they will comprise the following headings:

- Labour costs including such items as ongoing payroll costs, food allowances, accommodation, transportation. Following termination, there may be employment notice periods to be passed or a period before labour can be allocated to another project. As has been explained, some labour will be required to be retained on-site for activities of dismantling and removing the contractor's facilities, temporary works, etc. That dismantling and removing may not be started until issues as to the termination itself have been resolved and operatives will have been retained on-site in the meantime. A further consideration where labour is being repatriated to their home country is the availability of transport, particularly flights, which may take time to organise and to become available where the project has hundreds of imported operatives to be returned.
- Staff costs, which in many respects may be similar to labour costs. Apart from the practicalities of reallocating or repatriating them, some staff will be required to be retained on site to manage both the practical processes after a termination, such as removing temporary works, and the recording and assembly of the contractor's financial account at termination.

- Plant and equipment costs. These particularly include ongoing hire charges. It may be that the employer prevents the removal of some or all plant and equipment at termination, particularly where there is a disputed termination, such that the employer believes it can retain such plant and equipment for its use in completing the works under, for example, FIDIC Red Book clause 15.2, but the contractor considers that termination was wrongful. In such circumstances, the costs may continue until the merits of the termination are resolved or the employer completes the works. Alternatively, costs may be ongoing after termination for reasons of practicality. For hired items, it may be that the terms of their agreements dictate a minimum period or a notice period that needs to expire. For owned plant purchased for the project and whose costs are incurred on the basis of depreciation, it may be that the costs on that project will continue until the plant can be reallocated to another project, sold or externally hired.
- Temporary works costs, which in many respects may be similar to plant and equipment costs. They can continue particularly for reasons of practicality or safety. For example, items of temporary support to works such as shoring or falsework may not be removable immediately for safety reasons. Dewatering systems might be required to be kept in place and running until the replacement contractor's facilities are installed or they are taken over by the replacement contractor.

Under the FIDIC Red Book, for example, such headings of 'demobilisation costs' should variously be recoverable under such subclauses as 19.6(d) or 19.6(e), or the 'catch-all' subclause at 16.4(c).

7.3.8 Subcontractor and Supplier Claims

It is good practice that the terms of a contractor's contracts with its subcontractors and suppliers should include provision for their immediate termination in the event that the contractor's contract with the employer is terminated. For example, in the FIDIC suite of contracts, its standard Conditions of Subcontract for Construction provides at clause 15.1 that 'If the Main Contract is terminated … then the Contractor may by notice to the Subcontractor terminate the Subcontract immediately'. The lack of requirement for a period for the notice, and the immediacy of the termination, are notable here and essential for the protection of the contractor in relation to such issues as the value of work done at the date of termination. Of even more immediate effect are provisions for the automatic termination of the subcontract on the termination of the main contract, without notice. For example, in the JCT's Design and Build Sub-Contract Conditions 2011, clause 7.9 states 'If the Contractor's employment under the Main Contract is terminated, the Sub-Contractor's employment under this Sub-Contract shall thereupon terminate and the Contractor shall immediately notify the Sub-Contractor'.

Claims from subcontractors and suppliers are not specifically mentioned in any of SBC/Q clause 8, Infrastructure Conditions clauses 63 to 65 or FIDIC Red Book clause 19.6. However, they can be covered by the provisions of various of its subclauses, discussed elsewhere, depending on their nature, although they will otherwise usually be caught by 'catch-all' clauses such as FIDIC Red Book subcause 19.6(c) and 'any other Cost or liability'. A difficulty with such claims can be the level of detail and sophistication that many subcontractors and suppliers provide in their claims. The result can be that the contractor is placed in a difficult position where the engineer, in its determination under such as FIDIC Red Book clause 19.6, does not allow such claims because of their poor presentation. For example, under clause 19.6(c), the engineer might consider that 'Cost' arising out of a payment to a subcontractor was not 'reasonably incurred' because it was paid based on a superficial claim from a subcontractor. Alternatively, it might consider that there is no realistic 'liability' under clause 19.6(c), for example because a supplier's invoice includes a superficial claim for escalation in materials prices and there was no escalation clause in the supply contract.

7.3.9 Other Heads of Contractor Losses or Costs

These discussions have focused on the provisions of the 'Termination by Contractor' clause 19.6 of the FIDIC Red Book as an example. Other standard forms can expressly include other headings of items to be considered in such a termination.

For example, the SBC/Q provisions specifically entitle the terminating contractor to recover the amounts of any retentions prior to termination. Under the Infrastructure Conditions the retention would be released under the provision of its clause 63(5), which states that the terms of clause 60(4) apply at the date of termination as if that date were the date of the defects correction certificate.

7.4 Claims for the Employer's Lawful Termination

Where the employer terminates, it will broadly be due to pay the contractor the value of all work properly carried out by it, the value of its materials, equipment, plant, temporary works, etc. (in so far as they are adopted by the employer), and other sums due under the contract (for example see FIDIC Red Book clause 15.3). Deducted from such valuation, the employer will be able to recover its costs or losses resulting from the termination (for example see FIDIC Red Book clause 15.4(c)). Typical clauses, such as FIDIC Red Book clause 15.4(b), also protect the employer by freezing payments to the contractor pending a final accounting of the balance between amounts due to the contractor for the works and the employer's costs and losses in completing the works. In doing so, the employer will usually also be entitled to take over and use such of the contractor's facilities, materials, etc., on site as it sees fit (for example as set out in FIDIC Red Book clause 15.2).

Taking FIDIC Red Book clause 15 as an example, valuation issues that can arise in a final accounting following a valid termination by the employer are as follows.

7.4.1 'The Value of the Works, Goods and Contractor's Documents'

This is in FIDIC Red Book clause 15.3, which is headed 'Valuation at Date of Termination'. The full text requires that:

> As soon as practicable after a notice of termination under Sub-Clause 15.2 [*Termination by Employer*] has taken effect, the Engineer shall proceed in accordance with Sub-Clause 3.5 [*Determinations*] to agree or determine the

value of the Works, Goods and Contractor's Documents, and any other sums due to the Contractor for work executed in accordance with the Contract.

This valuation will be very similar to that which would arise under FIDIC Red Book clause 19.6, where the contractor has terminated. Comments made on FIDIC Red Book subclauses 19.6 (a) to (c) are therefore equally valid here and the components will include all work done, including variations, preliminaries and general items and consideration of entitlements to claims under the contract for such items as delay, disruption and cost escalation accrued up to the date of termination.

Again, ideally, there will be a joint survey of the site to record quantities, etc., of work done. However, in practice it becomes less likely that this will be achieved, where the employer forcibly removes the contractor from the site following termination and the whole event is surrounded in acrimony. It is not uncommon for terminating employers to remove the contractor's staff from the site, place their own security on the site and prevent the contractor's staff from returning even to collect their files, personal belongings and vehicles. This seems to be a particularly common approach on government projects in some parts of the world. It can prevent a joint survey, which is a short-sighted approach given the costs of disputes on this issue that can result. It can also cause the contractor difficulties where it cannot access its project records, files and computers, particularly in relation to the progress, quality and costs of its activities carried out pre-termination. This is especially so where relationships are damaged because the contractor considers the employer's termination to have been unlawful and to have been based on an unreasonable determination of such elements as its entitlements in relation to delays and extensions of time. In addition, it may be that the employer's patience with the contractor is lost to such an extent that it instructs the engineer or quantity surveyor to carry out a unilateral survey. This will, of course, also be necessary for the preparation of tender and contract documents for a replacement contractor and often that is the overriding consideration for the employer. In practice, the priority for the employer and its consultants is often to progress the works with a replacement contractor, rather than set out the account of the terminated contractor, especially where the termination clause has frozen the obligation to pay the terminated contractor as under FIDIC Red Book clause 15.4(b). It is suggested, however, that employers and their advisors should not lose sight of the benefits of getting the survey agreed with the terminated contractor so that it does not become the subject of dispute on this issue later. This may involve 'biting their tongue' in the short term, but save unnecessary dispute and related costs later.

'Contractor's Documents' under FIDIC Red Book clause 15.3, as quoted, are defined in its clause 1.1 as 'the calculations, computer programs and other software, drawings, manuals, models and other documents of a technical nature (if any) supplied by the Contractor under the Contract'.

The 'any other sums due to the Contractor for work executed in accordance with the Contract' on a termination by the employer under FIDIC Red Book clause 15.3 will also include any other entitlements that the contractor has accrued at the date of termination to, for example, the costs of delay under clauses such as 1.9, 2.1, 4.7, 4.12 or 8.10 and costs escalation under clause 13.7 and 13.8.

Where an employer terminates for default by the contractor, valuation under such as FIDIC Red Book clause 15.3 is particularly likely to require allowance for any accrued rights the employer has at the date of termination to counterclaims and set-offs. This is particularly true of defective works and delay damages. It may particularly be that where the employer terminates under subclause 15.2(c)(i) for 'failure to proceed with the Works', the contract date for completion, or more likely a sectional completion date, has already passed. On that basis the employer may have already accrued rights to delay damages under clause 8.7.

7.4.2 'Any Losses and Damages Incurred by the Employer'

7.4.2.1 'Any Extra Costs of Completing the Works'
The employer's right to recover these is set out in FIDIC Red Book clause 15.4(c). In practice some of the types of item covered by these two headings are difficult to separate, so they are considered together here. Typically, they may include the following:

- The costs of securing the site against things such as damage and theft. This is usually taken on by the replacement contractor, but may not be, and there is likely to be a period between the termination of one contractor and arrival of a replacement contractor during which the employer will have to provide such security guards itself.
- The costs of protective measures such as dewatering. Again, this is usually taken on by the replacement contractor, but may not be, and there is likely to be a period between the termination of one contractor and the arrival of a replacement contractor during which the employer will have to maintain such measures itself or ask the terminated contractor to continue with its existing measures and include its costs in the post-termination final accounting. If such measures are not maintained in such an intervening period, issues may arise as to who is to bear any consequences. Is resulting damage the consequence of the termination or of the employer's failure to maintain the temporary works? Dewatering is a particularly common item in this regard and each such instance may turn on its particular facts. A prudent terminated contractor should write to the employer immediately on termination to set out the measures required and the potential results if they are not taken. Consideration here may also include any applicable law on legal duty on claimants to mitigate their losses and safety legislation.
- Alternatively, the costs of protective measures such as dewatering might have already been part of the scope of works of others such as an enabling works contractor, in which case its charges to the employer of dewatering during additional periods resulting from termination should be easily identifiable and claimable on a time-related basis for the period of delay whilst a replacement contractor is appointed and mobilised.
- The costs of surveying, measurement and valuation of the work done, variations and claims at termination. This may involve fees for the employer's engineer, architect, contract administrator and/or quantity surveyor. Terminated contractors sometimes argue that this should require consideration of the extent to which such costs would have been incurred anyway on completion of the Works. However, where a replacement contractor is brought in, the costs will be incurred again when it achieves completion of its works, such that the measurement, valuation, etc., at termination is truly additional.
- The procurement costs of engaging a replacement contractor, including professional fees in preparing new tender and contract documents. This may include: quantity surveyor to measure and schedule the required works, prepare documents and advise

on tenders; engineer or architect/contract administrator to advise on issues such as scope, design, specification and quality; and legal advisors in relation to the replacement contract's terms and their agreement.

- The extra-over costs of the replacement contractor, although, as noted previously, in some economic climates, there might even be a saving. A common complication here is 'scope creep' or 'betterment', where the new construction contract has work included in it that was not part of the terminated contractor's scope. In the termination of a contract for construction of three high-rise buildings, analysis of the replacement contractor's contract, scope of works and specifications actually showed that the costs of the terminated contractor's scope and specification were actually higher than with the replacement contractor. The only reason the replacement contract was at a premium contract price compared to the amount that remained under the terminated contract price was because the scope and quality had been increased in some respects. This is something that a terminated contractor should be very careful to check for when it receives an account for the employer's costs of completing the works. From an employer's perspective, debate on this issue can be most obviously avoided by ensuring that the scope of the replacement contractor is the same as that of the terminated contractor. However, circumstances may be that it is in the interests of the employer to require some change or enhancement now, rather than perhaps by way of post-contract works when the replacement contractor has finished. In this case, the employer is well advised to keep any such changes in scope or specification very clear and with a discrete and readily identifiable cost attached to it in the replacement contractor's contract.

- These extra-over costs of completing with the replacement contractor will also require consideration of any variations instructed to its scope of works, whether they would have also been variations for the terminated contractor, what the comparative costs would have been and whether or not they arise out of its failures.

- Say, for example, a replacement contractor prices the exact same scope of pipework as that which was left by a terminated contractor, but at higher unit rates. If the progress of the replacement contractor's works reveals that additional pipework is required, and it is instructed to carry this out as a variation, then the terminated contractor may argue that since the work is a variation to the replacement contractor's scope, it can never have been part of its scope before termination. This may be correct, but if the necessity for additional pipework is such that it would have always been required from the terminated contractor and would have been instructed as a variation to its scope, then having it done by the replacement at its higher contract rates will involve the employer in additional costs that may be claimable from the terminated contractor.

- Furthermore, additional works to a replacement contractor's initial scope may be because of the failure of the terminated contractor, such that their costs are entirely claimable from the terminated contractor. Common examples include remedying defective work not discovered when the works were surveyed at termination. However, this may require consideration of the extent to which the terminated contractor was actually paid for those works without deduction for defects.

- Other direct costs of completing construction might include such costs as maintaining security or a site camp and/or offices and site security, where these are not part of the replacement contractor's contract. This can occur where the termination takes place when the works are advanced such that what remains is work that is best procured directly by the employer from specialist subcontractors without appointing a replacement contractor. These are likely to be those previously engaged by the terminated contractor, but that the employer now takes on direct. The obligations placed on a terminated contractor by FIDIC Red Book clause 15.2, Infrastructure Conditions clause 65(3) and SBC/Q clause 8.7.2.3 in relation to assigning subcontractors have been noted. In that case it may be most efficient for the employer to provide some items of attendances, etc., that would normally be a main contractor's preliminaries and general items and provide these as attendances to the subcontractors that it takes on to complete the remaining works.

- Further costs of an employer taking on a terminated contractor's subcontractors and suppliers directly may be fees of its advisors in procuring such services. These are similar to those consultants' fees explained in relation to procuring a replacement contractor.
- The resulting subcontractor and supply contracts may also involve price premium for the employer. Often where a contractor is terminated, it is against a background of other problems on the project, such as between the contractor and its subcontractors. These may have a variety of causes and responsibilities, and may even have been a contributory factor in the contractor's failures that gave rise to its default and the employer's termination. Where there were such issues, the employer may find that a subcontractor or supplier requires some premium in its pricing of the direct works to complete the project for the employer to cover its losses or unpaid amounts under its contract with the contractor. This can lead to debate as to whether it was reasonable for the employer to have agreed such terms and the premium payment in the circumstances. This will usually involve consideration of the facts in each case and what practical alternative the employer had in terms of approaching other subcontractor and suppliers. Replacement subcontractors are particularly often wary of taking on the completion of works carried out by others under terminated contracts on anything other than very conservative terms, especially in relation to quality and defects. For employers, such situations are often a balance between mitigating direct and immediate costs with expediting the re-start and completion of the works with indirect costs savings later. Criticisms by terminated contractors of employers' actions in such situations often apply the benefit of hindsight or ignore the practicalities of the situation that the employer found itself in following termination. Where that termination was for the contractor's default, tribunals tend to be particularly unsympathetic to such complaints by contractors.
- Additional professional fees in relation to the replacement contractor and its works. This may be the subject of a new agreement with each consultant, or addenda to their existing agreements. This head of claim usually results in a debate as to why additional fees were incurred at all and if they were reasonable. Terminated contractors occasionally complain that a re-negotiation has been exploited by the consultants, perhaps because they found during the course of the project before the termination that their fees were inadequate. Whether a resulting re-negotiation is reasonable may depend on terms of the consultant's original agreements and their terms in relation to termination of the construction contract. Those agreements may expressly provide a requirement to negotiate a new consultancy agreement between employer and consultant. Again, complaints from the terminated contractor in this regard can be argued as using the benefit of hindsight and ignoring the difficult position that a

termination for default can put an employer in. However, it has also been seen in practice that such situations are exploited by consultants where the employer takes an unusually relaxed approach to additional fees on the basis that they will be charged to the terminated contractor anyway.

- Damages for delay to completion of the works or sections thereof. Obvious delay periods arising out of termination will include time lost whilst procuring the replacement contractor and this will almost inevitably be critical to completion of the works and any contractual sections. This delay will include periods for: preparation of tender documents; submission of tenders; adjudication of those tenders; contract negotiation; and appointment and mobilisation of a replacement contractor. However, they will also require consideration of the reasonableness of the replacement contractor's programme and its relationship to the terminated contractor's programmed periods. This also often involves analysis of: avoidable delays that occur to the replacement contractor perhaps because the employer, engineer or architect use the period to change or enhance their design; responsibility for those changes; and whether they would have also impacted the terminated contractor. It may be that changes in the scope and specification in the new contract led to extra time being required to achieve completion and that these are not the result of the termination and therefore not part of the employer's right to claim the 'extra costs of completing the Works'. The resulting necessary adjustments can be complex.

- It may seem inevitable that a replacement contractor will require longer to complete works that are left by a terminated contractor, not least because of the learning curves that new staff, labour and subcontractors will have to go through. This may depend on the degree to which a replacement contractor can engage, such as the old subcontractors to complete their works. However, it may be for the employer to specify in its enquiry documents the date by which it requires the replacement contractor to finish. This will also involve consideration of the time lost between terminating one contractor and mobilisation and replacement, but often involves acceleration payments to the replacement contractor to recover some, or even all, of the lost time. The financial effects will also be part of the premium paid to it compared to the terminated contractor. Again, questions of mitigation and reasonableness of the costs, in their factual context, may arise.

- A further consideration in relation to delays to completion of works after termination is the relevance of rates for delay damages set out in the terminated contractor's agreement to the calculation of the employer's claim against it for a valid termination. The contract will normally set out agreed rates per day for failure by the contractor to complete the works, or sections thereof, by the date, or dates, set out in the contract and as extended by any extensions of time granted to it. However, say, for example, that a contractor is 10 weeks behind the programme on critical path activities, with no realistic prospect of making up that delay and every possibility of falling further behind. If the contractor had not been terminated, but had continued to completion, then it would have become liable for delay damages at the rate in the contract for those 10 weeks or such other longer period that this actually became. However, if the employer concluded that the 10 weeks by which progress was behind programme amounted to a failure to proceed with the works as required by the contract, it could elect to terminate that contract and appoint a replacement contractor, with a period of selections, appointment and mobilisation, say that the replacement contractor finished the works 15 weeks late against the terminated contractor's contract completion date. The question would then be whether the employer's claim for the 15 weeks of delay to completion should be all based on the 'actual losses and damages incurred by the Employer' per FIDIC Red Book clause 15.4(c), the rate of damages agreed in the terminated contract now being irrelevant. The alternative view would be that only 5 weeks of the delay to completion were due to the termination, with the other 10 weeks due to delay by the terminated contractor. On that alternative basis, the 5 weeks would be at the employer's actual losses and damages with the 10 weeks being at the terminated contract's rate of delay damages. The applicable law may be important here, but the argument will be that the agreement in the contract of a liquidated rate of delay damages was based on the contractor's failure to complete the works on time and that by terminating the contract the employer stopped the contractor from being able to achieve completion such that the agreed damages rate in the terminated contract became redundant.

- A related argument that is sometimes successfully run by contractors is that any maximum cap on delay damages in the contract would still be valid, on the basis that when the contractor priced the works it did so by reference to that capped risk. The employer may counter that the cap is only in the delay damages provision of the contract and is irrelevant to a termination. Clearly, these arguments will depend on legal arguments in each case and the position in the local jurisdiction. However, the practical point is that the replacement of an agreed contractual rate of delay damages by actual loss and damage might influence the employer's decision as to whether to terminate or not. For example, if the delay damages were agreed at a high rate in the contract and a sharp fall in the local property market means that they become significantly higher than the employer's actual loss and damage, it might elect not to terminate, but to allow the contractor to struggle on to late completion and apply its contract rates of delay damages. The level at which any cap on delay damages was set under the contract may also be an important consideration here. If the falling market also means that letting or selling a completed building was going to become more difficult than originally expected, then this might further encourage such an election by the employer. To take the contrary scenario, if it became apparent that the agreed rate of delay damages was too low against the actual loss and damage, this might encourage an election by the employer to terminate, particularly if it considered that this would circumvent any cap on delay damages under the contract. Such decisions will revolve around the factual circumstances and the legal advice that the employer is given at the time.

7.4.3 Bonds

In practice, termination by an employer is likely to be accompanied by its making a call against any bonds provided by the contractor and which have not expired or already been called before the date of termination. The most obvious call is on a performance bond or parent company guarantee, the failure giving rise to the termination being most likely to be of a nature envisaged by the terms of such bonds, such as 'the due performance of all the Contractor's obligations and liabilities under the contract' as is provided for in the Example Form of Parent Company Guarantee at Annex 6 to the FIDIC Red Book. However, it also often happens that the employer makes a call on other bonds that are in place, such as a retention bond and advance payment bond. Whether the employer is entitled to make a call on such bonds in the circumstances of a particular termination will depend on the terms of such bonds, and there may be some dispute regarding this.

The practical point for evaluation purposes is that financial accounting for a termination will need to ensure that amounts recovered by the employer through calling such bonds are taken into account in a payment after termination such as under FIDIC Red Book clause 15.5(c). By way of a further issue, if the termination was wrongful, then there may be a counterclaim from the contractor for its financing of the call on any bonds on the basis that they should never have been called by the employer.

The quantification of claims for the financing of the wrongful call of bonds are the subject of detailed consideration in Chapter 8.

7.5 Claims for the Employer's Unlawful Termination

As noted elsewhere, it is quite common for a contractor to respond to a purported termination by its employer based on an allegation of default in the contractor's performance, by disputing the factual and/or legal basis of the termination and asserting that it was wrongful. This is partly due to the subjective nature of some of the contractor defaults that give rise to an employer's right to termination as they are set out in standard forms. However, another contributory factor is how often both parties and their advisors fall into failures that lead to competing claims and counterclaims and debate as to such as concurrent delays.

Where the contractor considers an employer's purported termination to be wrongful, it may counterclaim against the employer. The basis for this will depend on the applicable law, but may generally include one of two bases, depending on the employer's conduct in its act of termination or in its actions thereafter. These alternatives are:

- By the contractor terminating the contract itself, where the employer's subsequent conduct, such as withholding payment or excluding the contractor from the site, amount to grounds for such termination by the contractor under express terms of the contract. The contractor's claim will then be for whatever costs, losses, etc., the clause that allows it to terminate for employer default entitles it to.
- By the contractor regarding the employer's purported termination as a breach of contract or a repudiation (where the applicable law includes such a concept). In such cases, those contractor damages will not be limited by the express provisions of the contract, but by the applicable law regarding quantification of damages in such circumstances. As a result, the contractor may prefer this route if termination under the express provisions of the contract places some limit on what it can recover. It has also been noted elsewhere how clauses such as 8.3.1 of SBC/Q preserve the rights of the parties by making the termination for default provisions of that contract without prejudice to their other rights and remedies under the law.

Sometimes, contractors will set out their counterclaim on both of these legal bases as alternatives.

In such circumstances a contractor's claims will usually comprise similar heads to those outlined in Section 7.3 of this book in relation to its lawful termination under the provisions of the contract. However, it might also include additional heads of claim in relation to the following:

- Confiscation by the employer of the contractor's plant and equipment. Where the employer considers that it has properly terminated the contract under a provision such as FIDIC Red Book clause 15.2, it will consider itself entitled to retain the contractor's plant and equipment for its own use, or that of a replacement contractor, until it has completed the works. However, if the contractor considers that the termination was wrongful, it may claim for the confiscation of that plant and equipment by the employer. Its damages may be the costs to it of replacing those items, where it has to replace them or the loss in their value over the period during which they were wrongly detained by the employer. Thus, for example, if a contractor places a concrete batching plant on a project and intends to retain it there until the structures are completed and then move it to another project, it may have to purchase another such plant for that subsequent project if the employer wrongly takes possession of it for the use of a replacement contractor. An extreme example of the wrongful detention of items owned by a terminated contractor was an employer who detained 55 staff vehicles owned by the contractor that were parked on site on the day of termination. Some were put to the use of the staff of the employer, engineer or replacement contractor whilst others just sat decaying on-site for the next three years. The employer asserted that it was only liable to the contractor for the residual value of those vehicles when its replacement contractor had completed the project and it no longer needed them, when it would sell them and account to the contractor for their second-hand value at that date. The contractor asserted that they had been wrongly confiscated by the employer and that it was due their value at the date they were confiscated. Such situations will always be resolved on their factual and legal background, but having decided that the termination was wrongful, the arbitral tribunal in that case decided some years later that the contractor was entitled to the residual value as it would have been at the date that they were confiscated by the employer. For a court judgment on this issue, see *Egan v South Australia Railways Commissioner* (1982) 31 S.A.S.R. 481.
- Loss of contribution to off-site or head office overheads and profit ('margin') on the work not done as a result of the termination. This was considered in Section 7.3 in the context of a lawful termination by the contractor. This included reference to the judgment in *Wraight v PH&T Holdings*, which was a case in which the employer was held to have wrongly terminated the contract.
- Loss of, or damage to, the contractor's reputation as a consequence of its being terminated on a project. This means loss of the margin that it would have made on future projects, rather than on the works not carried out to the terminated project. It is a particularly difficult head to establish in terms of causation and prove in terms of quantification. It requires establishing: the wrongful termination; the reputational damage; that work was lost or denied to the contractor as a result; and what profit would have been made on that work. It may also entail establishing that such damages are recoverable under the applicable law as a head of loss arising from such a breach and not too remote, for example. The easiest of these hurdles might be the loss of reputation, by reference to press and news articles, market comment, notification of being dropped from tender lists, etc. However, even this is never straightforward. Proving the resulting financial loss tends to be even more problematical. That loss

should be a function of the value of projects that the contractor would have secured, but for the loss of reputation, and the profit it would have achieved on those projects. However, this involves a degree of speculation as to the position the business would have been in 'but for' the wrongful termination and resulting reputational damage. This said, if the test of proof of quantum of damages is the 'balance of probabilities' then such a test may be passable. Though the case is not related to termination of a contract, readers might examine the judgment in *Aldgate Construction Company Limited v Unbar Plumbing and Heating Limited* [2010] EWHC 1063 (TCC). There the defendant's careless work caused a fire in newly built offices which the claimant was planning to sell and reinvest the proceeds in future developments. The Judge found that the claimant would have carried out further developments and would have earned significant profits thereon.

- As stated at the start of this chapter, disputed terminations often particularly include a period of uncertainty following the purported termination, during which the contractor will retain resources at site (or tries to if it has been locked out) until the contractual situation is confirmed one way or the other. This can particularly mean that a claim for wrongful termination includes not only a section for the costs of removal of resources, but also costs incurred in a period between termination and removal during which costs of labour, staff, and equipment continued. This is often addressed as a claim under the subheading of 'Demobilisation Costs', as exampled in Section 7.3 in the context of a valid termination by the contractor. However, it can be a more significant item in the context of a disputed termination by the employer. The employer may argue that demobilisation should have been immediate. The contractor will argue that this denies the realities of such situations generally and the particular circumstances where alleged termination is not resolved for some time. Such issues and the resulting evaluations will depend on their factual background, which can vary greatly.
- Similar claims are also likely to be received by the contractor from its subcontractors and some suppliers. As noted elsewhere, these can be in varying degrees of detail and substantiation, such that their corresponding claims for such as retention or demobilisation of resources are even more problematical to evaluate.
- A claim in relation to any call on any bonds that were in place at termination. The events of such calls and termination often coincide, so that if the contractor considers the termination wrongful, it will usually claim for amounts received under the bonds and any associated costs that it has incurred, such as financing the amounts called.

7.5.1 Repudiation

Not all legal systems recognise the doctrine of repudiation. However, as noted elsewhere, claims for termination are often involved with claims and counterclaims for repudiation. It may be that a party considers the alleged termination was wrongful and takes the other party's action as a repudiation, accepts that repudiation and makes a counterclaim on that basis. Furthermore, claims for termination are sometimes pleaded on the basis of repudiation as an alternative to a termination under express contractual provisions. It can also be the case that a contract contains no termination provision, such that repudiation may be a party's only recourse if it wants to be relieved of further performance under the contract for a breach of contract such as the employer's non-payment of a certified amount or the contractor's failure to properly proceed with the work.

Where one party repudiates a contract, the valuation of work executed before the repudiation can be an issue in the absence of express provision as to quantification. In the England and Wales courts, in *Tomlinson v Iain Wilson (t/a Wilson and Chamberlain)* [2007] the court accepted that the correct approach was that proposed in *McGregor on Damages*, i.e. payment for the work done should be such a proportion of the contract price as the cost of the work done bears to the total cost of the whole contract works plus, for the remaining work, such profit as would have been made on it.

In another case under English law, *Golden Strait Corporation v Nippon Usen Kubishka Kaisha* [2007] UKHL 12, it was held that the costs of repudiation could be calculated with reference to subsequent events that had not occurred at the time of repudiation but were contemplated by the contract and occurred before the assessment of damage. In this case a contract for the charter of a ship contemplated cancellation in the event of war between, among others, the United States, the United Kingdom and Iraq. The contract was repudiated, leading to a claim for damages, but at the subsequent arbitration the defendant claimed that the contract would have been cancelled by the outbreak of the second Gulf War, although this post-dated the repudiation. The appeals reached the House of Lords, who agreed that the damages should be assessed at the date of repudiation by valuing the chance that the contingency contemplated by the contract might occur and that the value might lie anywhere on a scale between extremely unlikely and virtual certainty. However, by the time of the arbitration the contingency was an established fact and the arbitrator had been correct to take it into account in assessing the loss.

7.6   Claims for the Contractor's Unlawful Termination

As a result of the more objective nature of the most commonly occurring employer defaults that can give a contractor the power to terminate, such as failure to pay a certified sum, terminations by contractors are far less commonly disputed by employers and the subject of counterclaims asserting wrongful termination.

Where an employer does consider a contractor to have wrongfully terminated, its counterclaims are likely to be put on a legal basis similar to those set out in Section 7.4. The heads of claim that follow, and related problems, are likely to be similar to those set out through this chapter.

7.7   Summary

Standard forms of construction and engineering contract typically provide for termination of the contract in the event of a serious default by either party. They may also provide for either party to terminate if the works are seriously delayed or suspended by force majeure events that could not have been foreseen by the parties and were outside their control. Sometimes, they also provide for a party (usually only the employer) to terminate the contract for its convenience. They will then set out the rights and obligations of the parties following such termination and especially how the contractor's final account is to be handled and what is to be taken into consideration, including the value of work done before the termination and the parties' losses and damages resulting from the termination.

Terminations have become more common over recent years, especially in periods of significant economic turmoil, and this is reflected by the fact that this is a new chapter for the Third Edition of this book. This has also meant that there is a lack of wider commentary on the subject and a lack of precedent in relation to termination clauses under standard forms and the evaluation of the claims that result. Such termination largely involves heads of claim considered in other chapters of this book, but can also involve its own unique considerations. In practice, termination also often leads to claims and counterclaims calculated on opposing contractual bases, between the parties, because they disagree as to which of them is in default and whether a purported termination was valid under the contract or was wrongful. The complexities of the different heads of claim will then also be exacerbated by the need for alternative claims and counterclaims.

Where a party is contemplating termination it is essential that it takes good local legal advice on issues such as the basis of the termination and the drafting of a contractual notice. Those reacting to a termination or notice should similarly take advice. The complexities and costs arising out of wrongful termination should not be underestimated.

8

Other Sources of Claims

Previous chapters have considered in some detail the quantification of claims arising from the most common causes of claims for additional payment under construction and engineering contracts, i.e. changes to the works affecting the unit rates and prices, changes to the programme of works and termination. This chapter briefly considers some other less common causes of such claims. It is not practical to cover in any one text the full range of claims for additional payment that may arise, but here we consider some where these causes particularly tend to create some controversy or difficulty in terms of financial evaluation.

8.1   Letters of Intent

Claims arising from letters of intent are strictly not claims for additional payment under the contract as they arise from instructions that precede the formation of the contract. In many cases there may never be a subsequent construction contract where, for instance, the project is aborted after the issue of a letter of intent but before the contract is formed or where negotiations between the parties as to the terms of a full contract fail.

A project client generally issues letters of intent to the intended contractor in order that the contractor can then commence preparatory activities for the contract without awaiting the completion of all of the contractual formalities. A similar position can occur between a contractor and a subcontractor. Such preparatory activities might typically include procurement of materials that are on long delivery periods or the commencement of preliminary design work. The intention is that of saving time and allowing the works, or preparation for them, to proceed without the delay that might ensue if no activities were to be undertaken until all contract documentation was complete and signed by the parties.

However, a contractor undertaking such preparatory activities under a letter of intent that was strictly just that, i.e. a letter stating only that it was the employer's intention to place the contract for the works with the contractor, would be running a real risk that it might not be entitled to any payment in the event that a contract was not subsequently agreed. Strictly speaking, a letter of intent is exactly that, merely a letter expressing an intention to do something, i.e. to enter into a contract, in the future. If it has no more commitment than that, then there may be real difficulty in recovering payment

Evaluating Contract Claims, Third Edition. John Mullen and R. Peter Davison.
© 2020 John Wiley & Sons Ltd. Published 2020 by John Wiley & Sons Ltd.

for any actions (such as materials orders or design work) undertaken in anticipation of the intention (i.e. the construction or engineering contract) becoming a reality. Contractors, subcontractors and materials suppliers all too often make the mistake of believing that receipt of a letter of intent amounts to receipt of a contract. In fact, most letters of intent amount to a confirmation of the opposite – i.e. that there is, as yet, no contract. As the position was put by Judge Fay QC in *Turriff Construction Limited v Regalia Knitting Mills Limited* [1971] 9 BLR 20:

> A letter of intent is no more than an expression in writing of a party's present intention to enter into a contract at a future date. Save in exceptional circumstances it can have no binding effect.... A letter of intent would ordinarily have two characteristics, one, that it will express an intention to enter into a contract in future and, two, that it will itself create no liability in regard to that future contract.

The potential liabilities created by a letter of intent will depend upon its content. In some circumstances, where the sender states that if certain actions are taken by the recipient then payment for those actions will be made, a contract may be inferred on the letter itself. This is sometimes referred to as an 'if' contract. However, for a letter of intent to become an 'if' contract, the terms on which payment is to be made must be stated in the letter. Letters of intent must therefore be drafted with some thought and care, and the case of *British Steel Corporation v Cleveland Bridge & Engineering Co Ltd* (1981) 24 BLR 94 illustrates very well the type of problems that can arise from such letters if badly worded and ill considered.

In that case Cleveland Bridge & Engineering Ltd. (CBE) were engaged in a project in the Middle East for which they required cast steel nodes. They negotiated with British Steel Corporation (BSC) for the supply of the nodes and on 21 February 1979 sent the following letter of intent to BSC:

> We are pleased to advise you that it is [our] intention to enter into a subcontract with your company for the supply and delivery of the steel castings which form the roof nodes on this project .... We understand that you are already in possession of a complete set of our node detail drawings and we request that you proceed immediately with the works pending the preparation and issuing to you of the official form of subcontract.

BSC commenced work on the authority of the letter of intent and on 27 February 1979 CBE sent a telex to BSC setting out a sequence in which they required the nodes to be manufactured and delivered, this being the first intimation that BSC had received of any specific requirements in this respect.

No form of subcontract was completed, and price and delivery terms were never concluded between the two companies. BSC claimed £229,832.70 from CBE in respect of the manufacture and delivery of the nodes and CBE counterclaimed for £867,735.68 for late and out-of-sequence deliveries. There was no dispute that BSC had sold and delivered the nodes and CBE contended that the agreement between the two companies was contained in the letter of 21 February, the telex of 27 February setting out delivery sequence, and BSC's conduct.

The court held that no contract had come into existence on the basis of the letter of intent and BSC were entitled to be paid on a quantum meruit basis. BSC having based its claim in contract or alternatively on a quantum meruit. The Judge said in his judgment:

> In my judgement, the true analysis of the situation is simply this. Both parties confidently expected a formal contract to eventuate. In these circumstances, to expedite performance under that anticipated contract, one requested the other to commence the contract work, and the other complied with that request. If thereafter – as anticipated – a contract was entered into, the work done as requested will be treated as having been performed under that contract; if, contrary to their expectation, no contract was entered into, then the performance of the work is not referable to any contract of which the terms can be ascertained, and the law simply imposes an obligation on the party who made the request to pay a reasonable sum for such work as has been done pursuant to that request, such an obligation sounding in quasi-contract or, as we now say, in restitution. Consistently with that solution, the party making the request may find himself liable to pay for work which he would not have had to pay for as such if the anticipated contract had come into existence, e.g. preparatory work which will, if the contract is made, be allowed for in the price of the finished work.

The dangers for the sender of a letter of intent, and the uncertainties for the recipient, are obvious from this explanation of how the entitlement to payment will be assessed in the event that no contract is concluded. It therefore highlights the need for careful drafting and reading of a letter of intent and consideration of what its terms actually amount to. Key considerations are whether it amounts to a contract, what scope and detail of work it covers, to what extent work done under it will be paid for, when it will be paid and at what rates and prices. A further point, that is often overlooked, is the relationship between amounts stated to be paid for work under the letter of intent, with those contained in the subsequent full contract. Are these amounts separate, or do amounts paid under the letter of intent become subsumed by the later contract price?

In terms of the nature of a document that may or may not be regarded as a letter of intent, and interpreted by the courts to establish what it actually amounted to in contractual terms, the following cases are worth practitioners reading:

- In *OTM Limited v Hydranautics* [1981] 2 Lloyd's Rep. 211, OTM had sent a telex to the defendant stating 'It is our intention to place an order for one chain tensioner …. A purchase order will be prepared in the near future but you are directed to proceed with the tensioner fabrication on the basis of this telex'. The court held that this did not amount to acceptance of the defendant's offer and did not amount to a binding contract between the parties. It was only a letter of intent. The contract between the parties was later formed by subsequent exchanges between the parties.
- In contrast, in *Wilson Smithett & Cape (Sugar) v Bangladesh Sugar and Food Industries Corp* [1986] 1 Lloyd's Rep. 378, the defendants had sent a letter of intent to Wilson Smithett which stated 'We are pleased to issue this letter of intent to you for the supply of the following materials ...'. That letter then referred to the terms and conditions set out in an offer from Wilson Smithett. The court held that this did amount to a binding contract between the parties. The defendant's subsequent failure to proceed with their purchase amounted to a breach of that contract.

- In *Harris Calnan Construction Co Limited v Ridgewood (Kensington) Limited* [2007] EWHC 2738 (TCC), a case arising out of an adjudicator's decision and considering whether there was a construction contract in place giving that adjudicator jurisdiction, the contractor had proceeded on the basis of a 'letter of intent' that the Judge found contained:

  > ... complete agreement as to the parties to the contract; as to the contract workscope (because it was contained in what was described as 'Tender Documents dated 2nd November, 2005'); as to an agreed lump sum of £200,787.75; as to an agreed set of contract terms (namely the JCT 2005 Standard Form, Private with Quantities), with 5 percent retention and £5,000 per week liquidated damages; and as to a contract period of sixteen working weeks.

  On this basis the Judge agreed with the adjudicator that there was a binding contract between the parties in the form of the letter, that 'there appears to be nothing left for the parties to agree' and that all that was missing was a set of documents which made that agreement more formal.

- *Diamond Build Limited v Clapham Park Homes Limited* [2008] EWHC 1439 (TCC) also arose out of the decision of an adjudicator under the Housing (Grants) Construction Regeneration Act 1996. Mr Justice Akenhead considered the terms of a letter of intent and noted that:

  (a) Whilst the first paragraph merely confirms an intention to enter into a contract, the second paragraph effectively asks DB to proceed with the work.
  (b) There is an undertaking in effect pending the execution of a formal contract to pay for DB's reasonable costs, albeit up to a specific sum.
  (c) The fact in the penultimate paragraph that the undertakings given in the letter are to be 'wholly extinguished' upon the execution of the formal contract point very strongly to those undertakings having legal and enforceable effect until the execution of the formal contract.
  (d) The fact that the Specification referred to in the Letter required a contract under seal demonstrating that the parties were operating with that in mind.
  (e) The very fact that DB was asked to (and did) sign in effect by way of acceptance the Letter of Intent points clearly to the creation of a contract based on the terms of the Letter of Intent itself.

  > Although this is a simple contractual arrangement, it has sufficient certainty: there is a commencement date, requirement to proceed regularly and diligently, a completion date, an overall contract sum and an undertaking to pay reasonable costs in the interim.

  He concluded that there was a contract in writing between the parties for the purposes of the Housing (Grants) Construction Regeneration Act 1996.

- On the other hand, in *Hart Investments Ltd v Fidler & Another* [2006] EWHC 2857 (TCC), His Honour Judge Coulson found that the work scope was not discernible from the letter of intent and concluded that 'If the contract document does not even begin to define the contract work scope, it seems to me impossible to say that all the terms, or even all the material terms, are set out in writing'.

- In *Bryen and Langley Limited v Martin Rodney Boston* [2004] EWHC 2450 (TCC), His Honour Judge Richard Seymour considered whether a letter of intent that looked forward to the making of a contract under the JCT Form incorporated those terms. He found that, whilst the letter envisaged the formation of a contract in due course, it did amount to a simple contract in its own right but did not incorporate the terms of the JCT Form.

Some cases in relation to payment under letters of intent were considered in Chapter 5 of this book where we considered payment where there is no contract. The reader is referred to that earlier chapter, but those cases are summarised here for ease of reference:

- In *ERDC Group Limited v Brunel University* [2006] EWHC (TCC), the contractor started work on the basis of letters of intent, the last of which had an expiry date of 1st September 2002. It continued work after that date, but subsequent negotiations between the parties as to terms of a proper contract failed, and the contractor walked away from the part completed work. It was held that for the work done up to expiry of the final letter of intent, the value was to be based on the contract contemplated by the letter of intent. For works done after that date, this was to be valued on a quantum meruit basis, but on the basis of the same rates as the earlier work. This was because 'The conditions in which the remaining work was carried out did not differ materially from those which (it must be assumed) were originally contemplated'.
- In *Trollope and Colls Ltd v Atomic Power Construction Ltd* [1962] 3 All ER 1035, a contract was subsequently successfully concluded, setting out machinery for valuation of the work, which was held to act retrospectively to cover the valuation of work previously done under a letter of intent (in the absence of the letter of intent or the contract saying otherwise).
- In *Mitsui Babcock Energy Ltd v John Brown Engineering* (1996) 51 Con LR 129, John Brown proceeded on the basis of a letter of intent. When disputes arose regarding tolerances, Babcock argued that there was no concluded contract. The court disagreed, noting that the parties had agreed to enter into a contract and only subsequently disagreed over terms without which the contract could still be enforced. Accordingly, payment for the work done was to be based on the terms of the contract and not a quantum meruit basis.

These cases were all under English law and parties should take good legal advice whenever they are considering a letter of intent, either in drafting one to send to a contractor or subcontractor or commencing preparatory works in response to one.

The pitfalls for those relying on letters of intent, and their advisors, is illustrated by the judgment in *The Trustees of Ampleforth Abbey v Turner and Townsend Project Management Limited* [2012] EWHC 2137 (TCC). The Trust engaged the defendant to provide project management services including contractual management in relation to building works to Ampleforth College. Those services included advice in relation to eight letters of intent to the building contractor, Kier Northern. In the event, a subsequent building contract was never signed. Kier completed the works four months late and its final account was settled on the basis of the terms of the letters of intent, which failed to provide for liquidated damages for late completion. The Trust argued that it was hamstrung in those negotiations with Kier by Turner and Townsend's advice in relation to the contract. The Judge found that advice to be negligent and awarded the Trust damages on the basis that '.... the Trust would have taken advantage of its improved position when it came into dispute with Kier' and awarded the Trust damages against that practice.

8.1.1 Instructions to Proceed

Where a party intends, by way of what it labels a 'letter of intent', to do more than confirm an intention to create a contractual relationship in the future, the practical approach to avoiding the sort of problems that have been illustrated is not to think of letters of intent at all, but some other term that gets away from the impression that there is no more than an intention to create a legal relationship. A commonly applied alternative is to think of such letters as 'instructions to proceed'. This renaming focuses the sender's attention on the fact that it is issuing an instruction, and generally that there will be consequences flowing from that issuing, including the incurring of financial liabilities. In considering the implications of the instruction to proceed, the sender will need to take into account the contract it is anticipated will ultimately be put in place for the works and ensure that the terms of the instruction to proceed are compatible with the anticipated terms of the contract. A document that is an instruction to proceed, rather than a letter of intent, might also provide more reassurance for the contractor or subcontractor commencing works against it.

The precise content of an instruction to proceed will vary depending upon the type of works, the anticipated contract terms, the stage at which the instruction is being prepared and the actions required to be undertaken. It will generally entail the consideration of some or all of the following:

- Definition of what work is authorised by the instruction to proceed, whether it be just preliminary or preparatory activities, such as design work, ordering of materials or subcontracted works, mobilisation or the commencement of work on site.
- Precise definition of the terms of payment for work under the instruction and exactly how such payments will be calculated.
- The placing of a limit on the expenditure that can be undertaken under the instruction. See in this regard *Mowlem Plc (t/a Mowlem Marine) v Stena Line Ports Ltd* [2004] EWHC 2206 (TCC), where the court rejected Mowlem's claim for payment of a reasonable amount in respect of work done beyond the limit on its right to payment stated in a letter of intent at £10 million.
- A date beyond which the instruction is not valid without written confirmation of the extension to its validity.
- Confirmation that, in the event that the contract is completed and signed, the works executed and terms of the instruction to proceed (particularly amounts paid for work done) will be subsumed into the contract for the works and its contract price.
- Explanation of what is to happen in the event that no contract is ultimately concluded.

If such matters are considered and set out in the instruction to proceed then there is some prospect that the aims of the sender will be achieved, the preparatory works will not be delayed pending conclusion of contract negotiations and formalities, and the recipient will know exactly what it is required to do, how it is going to be paid and what is to happen if no contract is concluded.

Where work on-site is to commence under an instruction to proceed it may be necessary to consider not only how the subject works will be paid for but also how variations and changes required to works carried out under the instruction to proceed are to be measured and valued. Generally the intention will be that they should be treated exactly as they would be under the anticipated contract, but if the terms of that contract in relation to change and variation evaluation are not certain at the time of the instruction to proceed then the instruction will need to consider the matter.

The general rule if a contract is concluded is that work done prior to the formation of the contract, but in anticipation of that contract, will be subject to the terms and conditions of the contract. In *Trollope & Colls v Atomic Power*, mentioned elsewhere, the plaintiff company undertook civil engineering work under a letter of intent from June 1959. The contract for the works was concluded on 11 April 1960 but a dispute arose as to whether or not the terms of the contract applied to work carried out prior to the formation of the contract in April 1960, particularly in respect of the valuation of variations. It was held that the contract concluded on 11 April 1960 applied to all the works, including those executed under the letter of intent.

The general rule has therefore to be to consider each instruction to proceed as a 'mini contract' in its own right, as that is exactly what it will be. If it is considered and structured in that way there will be considerably less chance that there will be dispute as to the payment for any work executed under such an instruction.

8.2 Bonds

Construction and engineering contracts may include provision for the contractor to provide for the employer's benefit various types of bond, sometimes referred to as guarantees. These are intended to protect the employer against particular types of failure by the contractor. They usually require a third-party bondsman or surety to pay to the employer an amount up to the amount of the bond in the event of failure by the contractor as defined in the bond. Thus, the employer 'calls the bond' in the event of certain defined failure on the part of the contractor. If the bond is called, then the surety in turn will make a claim against the contractor and in this regard will have often required a counter-indemnity from the contractor or its parent company under which it will pursue that claim.

The benefit to an employer of obtaining bonds is to protect itself against default by the contractor, and this makes them particularly popular in periods of economic uncertainty and in countries where employers perceive that there is a particular risk of contractor default. However, they also have benefits for the contractor: firstly, in order to get on to a tender list where provision of bonds is a pre-condition and, secondly, some bonds help the contractor's cashflow. For example, the provision of a retention bond in lieu of the deduction of retention amounts from interim payments, and the provision of an advance payment bond in return for such an advance payment before work is commenced.

Such bonds are usually provided by a bank, insurance company or other surety such as a parent company, subject to the approval by the employer of that surety company as acceptable. The costs of such bonds are usually borne by the contractor and comprise charges made by the surety to the contractor for their provision, which may include initial setting-up costs and on-going financing costs. Where they are requirements of the contract, those costs should be allowed for in the contractor's tender pricing. Where bills of quantities are part of the contract documentation, an item or items will be included in these against which tenderers can set out such costs. Standard methods of measurement such as the NRM Rules (at 1.2.14) and CESMM4 (at class A division 1) require such items to be included in the preliminaries and general items section of bills of quantities that are measured applying those rules. Given how some of the types of bond described in this section are particularly common on international projects, it is perhaps surprising that the POMI Rules have no express item for such bonds, although in practice they are usually set out as bill items under the heading of rule A.1 'conditions of contract'.

Bonds can be conditional or on-demand. If conditional, then the employer has to supply the surety with evidence of actual default by the contractor under the terms of the bond and the damages incurred by the employer as a result of that breach. If a bond is on-demand, then the employer can make a call on such a bond without proof of default and without proof of actual damages incurred. On-demand bonds are rarer in the UK than many other parts of the world, although they are more regularly seen in projects related to petrochemical and power projects. The category into which a particular bond falls will depend on its detailed terms and it is surprising how often this is not clearly set out. Disagreement can result as to whether the bond is on-demand or conditional, what the nature of the breach is that allows a call and what information the surety should expect to evidence both the breach and the resulting damages. Given the risks for contractors that are inherent in an on-demand bond it is surprising how often bonds of that nature appear on international construction and engineering projects. It often seems to be simply because the contractor had no option but to accept such terms if it was to secure required turnover, and was willing to do so no matter what the risk inherent in that type of bond.

8.2.1 Types of Bond

Common claims arising out of bonds usually relate to allegations that they have been wrongly called by the employer or extended as a result of failures by the employer. That extension might be because of failure to release the bond when it should have been released or because of compensable delay to the project. The issues resulting may be limited to legal and factual ones, the resulting quantum being limited to the amount called against the bond. In such cases, the financial evaluation of such claims was therefore relatively simple. However, evaluation can also include some more complex issues as to extension costs, interest and even the amount called itself. These are explained in this section, after an outline of the types of typical bonds and guarantees that might be required under construction and engineering contracts. The types of potential costs claimable in relation to such bonds and guarantees under an engineering or construction claim become apparent from an understanding of their natures and workings. The following is not comprehensive of all types of construction and engineering contract bond but covers the most common types.

8.2.1.1 Bid Bonds

Sometimes called tender bonds or tender securities, these protect the employer against default by a successful contractor in honouring its bid. They are rare in the UK but more common on international projects. They are submitted along with each bidder's tender to guarantee that they will leave their bids open for the period stated in the enquiry and that if a contractor is successful then it will commence the works tendered for under the terms against which the bid was made. The intention is to provide the employer some protection against a frivolous bid, which is accepted by the employer but with the costs of delay and re-tendering because that contractor fails to honour its tender by commencing the work. Similarly, it will protect the employer where the successful contractor refuses to commence work on the terms against which it bid and attempts to re-negotiate those terms pre-contract. Also, enquiry documents should state the period over which tenders are to remain open for acceptance and a bid bond may guarantee that tenderers will leave their bids open for that full period and not withdraw them before acceptance.

To cover such situations a tender bond will usually state an amount, or the 'not exceeding' limit on that amount, that the surety will pay to the employer in the event of such failures by the contractor. Often a bid bond amount is expressed as a percentage of the bid amount, perhaps 3% or 5%, but sometimes as high as 10%. The costs covered by such a bond may include the extra-over costs to the employer of going to a second lowest bidder and its fees and costs arising from any delay that results to the works.

Bid bonds are by nature of short duration but it is important to ensure that their expiry date is clearly stated. This may be the date on which the contractor commences work or the date of coming into effect of a performance bond if the contract requires one of these as well.

The FIDIC Red Book provides a sample form of bid bond at Annex B Example of Tender Security. In that example, the failures that give rise to an employer's claim thereunder are: withdrawal of the contractor's tender before expiry of the tender validity period; refusal by the contractor to correct errors in its tender in accordance with the enquiry terms; failure by the contractor to enter into a contract if it is awarded to it as required by clause 1.6; and failure to provide a performance security under clause 4.2 of the FIDIC Red Book general conditions. The example requires an agreed 'not exceeding' amount to be entered. A demand under this form must be authenticated by the employer's bankers and/or a notary public. That form states an expiry date to be stated therein, of 35 days after the expiry of the validity of the contractor's Letter of Tender.

The most common claims in relation to bid bonds are on the basis that they have been wrongly called and as damages for breach of contract. They rarely if ever appear in the context of a prolongation claim for obvious reasons.

8.2.1.2 Advance Payment Bonds

Again, these are less common in the UK than internationally.

Particularly where a project involves significant start-up or procurement costs, it may be considered that the employer's interests are best served by its making an advance payment to the contractor at that stage. This is effectively an interest-free loan to the contractor to help with its initial costs. These can be great, given how the payment provisions of construction and engineering contracts usually provide for payments based on valuation and certification at each month's end and payment not due for perhaps some weeks after certification. This can place a heavy burden on the contractor's early cash flow. An advance payment may be made as one single payment or a number at different stages of the contractor's mobilisation and/or procurement periods. Potential benefits to the employer include widening the competitive field in terms of contractors willing to bid for such projects and also reducing the potential for default by the contractor if it might otherwise run into early cashflow difficulties. Understandably, the provision by the employer of such an advance payment is usually conditional on the contractor first providing the employer with such a bond.

The advance payment is usually progressively clawed back by the employer by deductions from interim payments to the contractor for the works as they progress, by the application of a deduction percentage stated in the contract. Thus, if the advance payment is expressed as 10% of the contract price, interim payments might be subject to a reduction by that same 10%, such that once the contracted works are completed the advance payment has all been returned. Under such an advance payment bond, the surety guarantees to pay the employer such an amount of the advance payment as the employer has failed to claw back from the contractor through interim payments to date. As a result, the potential liability of the surety will reduce with time, such that the charges it makes to the contractor for provision of the bond ought to similarly reduce over time. This is a matter for the contractor to administer under its agreement with the surety.

A typical provision is in FIDIC Red Book clause 14.2. A sample form is provided at Annex E, Example of Form of Advance Payment Guarantee. The failure giving rise to a claim under this example is non-payment of the advance payment as required by the contractor's contract. It requires an agreed 'not exceeding' guaranteed amount to be entered, which then reduces as the advance payment is clawed back by the employer through interim payments. A demand under this form must be authenticated by the employer's bankers and/or a notary public. It becomes effective on receipt of the advance payment, or first instalment thereof, and expires 70 days after the expected expiry of the Time for Completion.

SBC/Q provides the option of requiring an advance payment bond where an advance payment is to be made, in its clause 4.8. Where the contract particulars require such a bond, the employer is only required to make an advance payment if the contractor has provided a bond in the form set out in Part 1 of Schedule 6 to SBC/Q, which also has attached a standard 'Notice of Demand'. The amount of this bond is stated in its clause 3.1 as being up to the amount of the advance payment, with the actual amount recorded in its clause 6. The effective date of that form is that of the advance payment by the employer. Expiry occurs when the amount of the advance payment has reduced to nil through interim payments or the balance has been paid off by the contractor. This therefore gives the contractor the option to pay back the advance payment early, rather than progressively through deductions from its interim payments, if that better suits its finances, with the benefit to the contractor of releasing the bond early.

The relationship between bonds and claims for delay is recognised in clause 5.3 of the SBC/Q form of advance payment bond in its Schedule 6. This clause recognises how the granting to the contractor of extensions of time under the contract might affect the date on which such a bond expires, by delaying the rate at which work progresses and hence at which the amount of the advance payment reduces to nil through deductions from interim payments. This reflects a common basis of contractors' claims in relation to advance payment bonds – that an effect of compensable delay by the employer can be to extend the time over which they are required to maintain and fund an advance payment bond. As a result, claims for extending the financing periods of advance payment bonds are a common item in claims for prolongation as considered in Chapter 6 of this book.

8.2.1.3   Retention Bonds

The provision of a retention bond is an alternative to the deduction of a retention percentage from interim payments to the contractor. Either way, the intention is to protect the employer against any element of the works as certified and paid for that is in fact not in accordance with the requirements of the contract, is defective and will require remedial work. The use of a bond for this end may be a requirement of the contract, in lieu of the deduction of a retention percentage from interim payments. Alternatively, parties sometimes agree the early release of retained monies against the provision of a retention bond instead. This is particularly common where taking over or practical completion of the works is a condition to the release of a first half of the retention fund, but the contractor requests full release in exchange for the bond for the second half to help with its cashflow.

For the contractor, the provision of a retention bond will also aid its cashflow by replacing the deduction of perhaps 5% of its work value in interim payments. They are also popular with contractors where they are concerned that the employer will fail to release retained amounts, either on some spurious contractual pretext even when the precondition for its release has been satisfied or because of financial failure of the employer.

Under the FIDIC Red Book a sample form of retention bond is provided at Annex F, Example of Form of Retention Money Guarantee. The failure giving rise to a claim under this example is the contractor not rectifying defects for which it is responsible under its contract. It requires an agreed 'not exceeding' guaranteed amount to be entered, and the surety's liability is limited to the amount of retention at the time. A demand under this form must be authenticated by the employer's bankers and/or a notary public. It expires 70 days after the expected expiry of the Defects Notification Period.

SBC/Q provides the option of the contractor providing a retention bond in its clause 4.19. A standard form, agreed between the JCT and the British Bankers' Association, is attached at Schedule 6 Part 3 of SBC/Q. Under its clause 6, the surety is released from its liability under this bond on the earliest occurrence of either: the date of issue of a Certificate of Making Good Defects; or satisfaction of a demand up to the maximum aggregate amounts under the bond; or an agreed date inserted into the bond when its terms are agreed.

Retention bonds can particularly often feature in contractors' claims for prolongation, on the basis that delay to completion of the works has led to delay in the event that triggers their release. Because retention bonds are not released until the end of a project, they can also be more prone to claims for wrongful calling where some alleged failure by the contractor leads the employer to call whatever bonds are available to it without due regard for whether the terms of the bond allow the calling of that bond for that failure.

8.2.1.4   Parent Company Guarantees

An all too common feature of construction industries around the world is how contracting groups are formed of a large number of companies, which though they form a large, stable and secure overall entity, are in themselves much smaller and less secure. In such circumstances there may be a significant increase in confidence for an employer if it can obtain a guarantee from the parent company of one legal entity in such a group. From the contractor's viewpoint, offering such a guarantee may give it financial credibility that it would otherwise lack, and which might have prevented it from being appointed.

Under such guarantees the parent company receives the benefit of the employer awarding the contract to its subsidiary in return for guaranteeing that subsidiary's performance under its contract and indemnifying the employer against its losses should the subsidiary fail to perform its contract.

FIDIC Red Book Annex A contains an Example Form of Parent Company Guarantee. The parent company indemnifies the employer against all damages, losses and expenses arising from the contractor's failure to perform its obligations under its contract. The effective date is when the construction contract comes into effect. The guarantee expires on the discharge by the contractor of its obligations under its contract, which will effectively be when the engineer has issued a Performance Certificate under clause 11.9.

Parent company guarantees can particularly often feature in contractors' claims for prolongation, on the basis that delay to completion of the works has led to delay in the event that triggers their release. Because they are not released until the end of a project, they can also be more prone to claims for wrongful calling.

8.2.1.5 Performance Bonds

Performance bonds, sometimes referred to as a 'performance guarantee bond' or 'performance security', guarantee that the contractor will perform and complete the project as contracted. The amount of such a bond is usually expressed as a percentage of the contract sum. This is commonly 10%, but may be as high as 15% or more. The respective benefits to the parties are the same as those of parent company guarantees described elsewhere.

Under FIDIC Red Book clause 4.2, a performance bond is referred to as a performance security, and is valid and enforceable until the contractor has completed the works and remedied any defects, such that the engineer has issued a Performance Certificate under clause 11.9. Within 21 days thereafter, the employer returns the performance security to the contractor. The FIDIC Red Book provides two forms of performance security at Annexes C and D as follows:

- At Annex C is an Example Form of Performance Security – Demand Guarantee. This states that a total guaranteed amount is to be paid in the event that the contractor breaches its obligations under its contract. That amount is reduced by a stated percentage on taking over of the whole of the works under clause 10 of the FIDIC Red Book general conditions. A claim under the Annex C form must be authenticated by the employer's bankers and/or a notary public. This guarantee expires 70 days after the expected date of the expiry period of the Defects Notification Period, but if the Performance Certificate has not been issued by 28 days prior to that expiry date, then the employer can require it to be extended.
- At Annex D is an Example of Form of Performance Security – Surety Bond. Under this form the contractor and guarantor are held to a stated bond amount for the due performance of the contractor under its contract. On default by the contractor or the occurrence of an event under clause 15.2 of the FIDIC Red Book conditions

(Termination) the guarantor pays the resulting damages sustained by the employer, up to the bond amount. That bond amount reduces by an agreed percentage on the issuing of a taking over certificate for the whole of the works under clause 10 of the FIDIC Red Book conditions. The effective date is the Commencement Date defined in the construction contract. This bond expires six months after the expected date of the expiry period of the Defects Notification Period.

In the Infrastructure Conditions, clause 10(1) gives the employer the option to require a performance security, in an amount not to exceed 10% of the tender total within 28 days of award of the contract. The option to require or not require such a bond is set out in the Form of Tender, where the amount of the bond is inserted as a percentage of the tender total. The form of such a bond is annexed to the Infrastructure Conditions. Expiry under that standard form is 'the later of the date stated in the certificate of substantial completion issued by the Engineer and the Final Expiry Date'. This later date is to be inserted in a schedule to that standard form.

It is notable in the SBC/Q contract documents that, whereas Schedule 6 contains some other forms of bond agreed in standard terms, these do not include a standard form of performance bond. This perhaps reflects that SBC/Q is mostly used in the UK domestic market where performance bonds (and parent company guarantees already considered) are relatively rarely required by employers.

Claims relating to performance bonds usually assert that they have been extended as a result of compensable delays, released late or wrongly called. Often all three of these claims are made in a claim that can add complexity to quantification, as a claim for extended financing has to be split and allocated to its different causes.

The costs of extending performance bonds are a regular component of claims for prolongation as considered in Chapter 6. Where the expiry of such a security is related to the completion of the works, then it can be understood that the costs of delay to that completion will include any additional costs of extending the security. The nature of such resulting extension costs are also considered in this section.

Late release, independent of claims for compensable delays, will relate to assertions that the condition for the expiry of a performance bond were satisfied but not properly certified. For example, under the Infrastructure Conditions, the contractor might assert that the engineer should have issued a certificate of substantial completion but failed to do so. This might involve dispute as to whether the works were properly in accordance with the specifications and actually substantially completed or not.

8.2.1.6 Payment Bonds

Payment bonds might be to the interest of the contractor itself or its subcontractors and suppliers.

The FIDIC Red Book attaches an Example Form of Payment Guarantee by Employer at its Annex G. In this the surety guarantees to pay the contractor amounts up to a maximum stated therein in the event that the employer fails to pay the contractor amounts due under their contract. A demand under this form must be accompanied by documentary evidence as set out in the form and be authenticated by the contractor's bankers and/or a notary public. FIDIC is unusual in providing a standard form of this type of bond and this reflects how they are more popular on international construction and engineering projects where there may be significant doubt as to the employer's ability or willingness to pay the contractor.

A further example of types of payment bonds are guarantees that due payments will be made to subcontractors and suppliers providing labour, materials, work or supplies to the project. They are rare in many jurisdictions, but common in the United States, where the Miller Act requires the main contractor on some government contracts to provide such payment bonds and a performance bond. The purpose of requiring contractors to provide payment bonds in relation to its suppliers and subcontractors is to provide protection to those parties. The benefit intended is to encourage that end of the supply chain to bid for work that they might otherwise be reluctant to bid for if there was some doubt as to the contractor's willingness to pay for work done or supplies provided. This in turn is said to increase competition.

The most obvious area of contention in relation to such bonds is whether amounts unpaid by the contractor were properly due to the supplier or subcontractor such that the contractor is truly in default in not paying them and the surety is thus liable for their non-payment. This is particularly true in relation to contentious claims such as those for delay and disruption considered in Chapter 6. Furthermore, what of the situation where a valid claim for delay or disruption was the result of failure by the employer and nothing to do with the actions or liabilities of the contractor? Such issues will be resolved based on the terms of the payment bond and the effects of the applicable law, particularly precedents where such issues have been previously decided.

8.2.1.7 Materials Off-Site Bonds

Many forms of construction and engineering contract allow for payment to the contractor for materials that have not yet been delivered to site but are held off-site for the benefit of the project. Given the costs of large items of permanent plant and equipment, such costs can be significant. A provision that allows the contractor to be paid for such items before delivery is in the clear interests of the contractor's cash flow. From the employer's perspective the benefits of paying early for such items may include the protection of its programme, where critical items are on long delivery or fluctuating availability, and the indirect benefits of employing a properly funded contractor. The provision of an off-site materials bond is intended to protect those employer interests, where it has paid for materials off-site, but they are not subsequently delivered to the site.

The amount of an off-site materials bond will usually be set at the amount to be paid by the employer for those materials and the bond will provide for the amount to reduce as the materials are delivered to the site and the employer's risk in their being held off-site reduces.

SBC/Q provides the option of requiring a bond in relation to materials off-site in its clause 4.17. That form provides for what it refers to as 'Listed Items' and defines as 'materials, goods and/or items prefabricated for inclusion in the Works which are listed by the Employer in a list supplied to the Contractor and annexed to the Contract Bills'. Such items can be paid for as what are commonly referred to as 'materials off-site', provided several conditions as set out in clause 4.17 are met, including that the contractor has provided a bond in the terms as set out in a form at Part 2 of Schedule 6 to that standard form. It also attaches a standard form of 'Notice of Demand'.

In the SBC/Q form, the employer's maximum liability is the value stated in the contract which it considered would be sufficient to cover payments made for the Listed Items that had not been delivered at any one time. Expiry occurs when all the Listed Items have been delivered to or are adjacent to the site or a 'longstop date' if stated in the contract.

As with SBC/Q's standard form of advance payment bond, its bond for materials and goods off-site recognises the potential effects of delay to the works, which was considered elsewhere with reference to SBC/Q's clause 5.3, and is not repeated here.

8.2.2   Claims in Relation to Bonds

The surety providing such bonds as those outlined in this section will charge the contractor for their provision until their expiry. If that period is extended for any reason, then such charges can be expected to be extended accordingly. In addition to such charges, if the bond is called, the surety can be expected to recover the amount called from the contractor. This in turn can involve the contactor in costs of financing the amounts taken from it by the surety. For practitioners this can give rise to a number of claim evaluation issues.

Where costs are incurred by the contractor as a result of some failure by the employer, then it seems likely that most such failures will amount to breaches of contract by the employer such that those costs are claimable by the contractor. In the FIDIC Red Book, such situations are expressly provided for in relation to a performance security in clause 4.2 as follows:

> The Employer shall indemnify and hold the Contractor harmless against and from all damages, losses and expenses (including legal fees and expenses) resulting from a claim under the Performance Security to the extent to which the Employer was not entitled to make the claim.

This would cover a wrongful call on a bond under the FIDIC Red Book, but not the extension of the duration of a bond due to the employer's failures, whether in not releasing a bond when they should have or in causing compensable delays to the event that should have triggered their release. Costs incurred in both of these circumstances are considered as follows.

8.2.3   Costs Incurred When Bonds Are Called

When an employer makes a call on amounts under any of the types of bonds considered in this section, the surety can be expected to turn to the contractor to recover that amount called, under the agreement between them. The contractor may then have a choice as to how to fund the financial hole this creates. If it has the funds to enable it to pay the surety and considers that is the best use of those other funds, then it might pay off the amount called. However, if the contractor does not have sufficient other funds, or considers that they are better employed elsewhere where they can earn a higher return, then it might take out a new financing arrangement with the surety or even a second source, such as a loan. This choice may also be informed by the contractor's view of the legitimacy of the bond call. Clearly, for the call and financing costs resulting from it to be subject to a contractual claim, the contractor must consider that the call was wrongful, that it did not commit the failure or breach triggering the call, or perhaps that it was in turn the result of some failure or breach of the employer and its agents. The classic example of this is a call on a performance bond on the basis of failure to complete the works on time, where the contractor has a claim for delays caused by such as denial of access by the employer or late design from its engineer that it considers justifies its apparent lateness, but its claim for extension of time is being denied by the contract administrator.

Such a new financing arrangement will usually involve a capital amount, an interest rate (or source of varying interest rates) and a term. They may also include set-up charges, particularly if the source of funds is not the original surety, and government duties, in countries where these are levied on such financial transactions. A common solution is the provision of a promissory note, perhaps at a fixed rate and of 1 year's duration, to be paid off, replaced or renegotiated at the end of its term.

The term of such new facilities can often be short. It may be that lenders will only be willing to offer a short-term loan, in order to minimise their risk. It may also suit the contractor to take out a facility on this basis because it gives it greater flexibility in relation to future funding. It may also be on the contractor's expectation that it will resolve its issues with the employer in good time, thus securing the return of the amounts called, and at lower interest rates than would be applied to a longer-term arrangement.

The capital amount under a new financing arrangement is sometimes complicated in itself, particularly where more than one currency is involved. It may be that the contract price itself is split between into a number of currencies. However, even where the contract price is expressed in a single currency, it is common in some countries to provide for payments to be made in more than one currency, perhaps the local currency and United States dollars in a South American country or the local currency and euros in a European country that is not part of the eurozone. In such circumstances, amounts called on such as an advance payment bond or retention bond might be in more than one currency. If the amount thus called is then the subject of a loan taken out by the contractor, that loan may usually be able to be exchanged into a single currency at the then applicable exchange rate or rates.

In terms of the costs of financing a resulting loan, as noted elsewhere, this can start with an establishment or set-up charge from the lender and government tax where this is applicable. Such 'stamp duty' is usually charged as a percentage of the amount of the loan and has been levied in countries as diverse as Chile, Austria and India. An issue with it is how it is applied when a loan agreement comes into effect, a particularly relevant issue where the effects of the calling of a contractor's bond is financed by a series of consecutive facilities. It may also be that there is an annual limit on the amount of duty due to the government for such duty, meaning that some further facilities or extensions of existing facilities are not subject to it even in countries where the government generally charges it.

As noted elsewhere, for a call on a bond or bonds to be the subject of a contractor's claim, it can be assumed that it considers the call wrongful and expects that it will succeed in securing the release of the amounts called. Often the initial hope is that the presentation of a claim will soon persuade the employer and its advisors of the merits of the contractor's case. However, where the resulting claims and counterclaims proceed to formal dispute resolution, it is not uncommon that the funding of a bond call can run for several years. The effect of this can be that the contractor takes out a sequential series of facilities of different durations and interest rates. Durations can particularly depend on the contractor's view of how long it will need that financing to be in place, either because it expects the bonds to be repaid or because it has other funding coming available that would enable it to repay them. The interest rate can vary depending on the movement of local base rates and the lender's relationship and attitude to the contractor at the time of each facility.

The available interest rate can also depend on the currency in which the loan is taken out and this can lead to the funder and borrower also entering into swap agreements to convert the amount being loaned into one or more currencies with different interest rates for each tranche. Furthermore, the contractor might take the view that it wants to diversify its currency risk by converting a loan or part of it from local currency to one that is more stable, such as the United States Dollar or Euro, to avoid the effects of a declining local currency over the duration of the bond financing arrangement. This would not apply of course to the many countries whose currencies are 'pegged' to the US dollar. In addition to such more stable international currencies, some countries also have notional inflation linked currencies for banking applications, which run in parallel to their main currency. Loans based on repayment in such notional currencies are also likely to be the subject of lower interest rates. Such currency swap arrangements also tend to be used in parallel where the facility is of a short term.

A further consideration that is often missed in relation to such funding is the habit of banks in some parts of the world of applying the '365/360 method' in relation to their interest rates. This results from the fact of calendar months being variously of 28/29, 30 or 31 days duration. The use of a standard 30 days month in order to establish a standardised monthly interest rate, results in the duration of a one-year interest rate period being 360 days rather than 365 (or 366) days. The effect of this is that a quoted 5% annual rate on such a basis is a 5.07% (5% × 365/360) actual annual rate of interest. This might seem a minor difference to some, but the fact that the 360-day year is often hidden in the details of loan agreements has led to it being considered sharp practice in some parts, such that in several US states it is actually illegal. Others would say that it is a minor adjustment that avoids unnecessary complexity in the interest calculation.

The effect of these various issues is that calculations of claims for financing costs arising from the wrongful calling of bonds can be very complicated and needs careful consideration. To take one example of a project in South America, where the call on two of the contractor's bonds became part of claims and counterclaims considered in an International Chamber of Commerce arbitration. There the financing of the call ran for six years until the hearing of the matter, comprised five currency swaps and seven facilities of up to five tranches and with a myriad of interest rates and stamp duty on most of the facilities. This complexity was the result of careful financial predictions by the contractor's finance team at the end of each facility attempting to ensure the best interest rates and least currency risk in the next facility. However, in the event, had it simply renewed the initial facility on similar terms in relation to the central bank's base rate, its final financing total would have been less than was actually incurred under the complex arrangements it put in place!

8.2.3.1 Example of a Hypothetical Currency Swap

To illustrate the issue, take the example of where a contractor has a 12-month loan in a principal sum of 10 million in its local currency. The bank charges a high interest rate on this, of 5% per annum, because the loan and repayment obligation are in the local currency.

Under a currency swap arrangement, the bank promises to pay that same sum of 10 million and the same 5% per annum interest rate over the same term. At the same time the contractor now promises to pay the same principal sum but converted into United States dollars at the then current exchange rate of 500 local units to one United States dollar, that is US$ 20,000. The interest rate applicable to that promise is a lower 2% per annum because the repayment obligation is in the more stable United States dollar.

One year later, at the date of maturity, the local currency has slipped to a new exchange rate of 510 local units to one United States dollar. As a result, the respective obligations of the parties to the funding arrangement to pay principal sums and interest under the swap agreement are as follows:

Contractor owes on maturity:	Local currency
Principal amount:	US$ 20,000
Interest:	US$ 20,000 × 2% = US$ 400
Total:	US$ 20,400 @ 510 = 10,404,000
Bank owes on maturity:	
Principal amount in local currency:	10,000,000
Interest:	10,000,000 × 5% = 500,000
Total:	10,500,000 = 10,500,000
Contractor's net receipt on exchange of parties' debts:	= 96,000

The result in this example is that while the contractor has substantially benefitted from the lower interest rate it has suffered from the exchange rate movement. Much, but not all, of interest rate benefit has been lost as a result of the decline of the local currency against the dollar. It was of course that volatility in the local currency that led the bank to offer the much lower interest rate on a dollar loan in the first place. However, the decline in the currency was not apparently as great as the bank feared when it offered the significantly lower interest rate on the foreign currency. For its part, the contractor's gamble that its local currency would not decline sufficiently to wipe out the benefit of converting its loan unto US dollars has paid off in this example, although in practice the result could be quite different.

Many might suspect that in such a situation the bank is more likely to accurately predict the effect of currency exchange fluctuations than the borrower, and therefore offer an interest rate on a more stable international currency that will prove not lower than would be necessary to make up for exchange rate changes. For example, in one case, a contractor took out a range of swaps and facilities to finance the amount called against two bonds, but the actual currency fluctuations left it almost US$ 1 million worse off than it would have been had it taken out no currency swaps.

8.2.4 Costs Incurred When Bonds Are Extended

Bonds provided by contractors under construction and engineering contracts tend to be extended for one of two reasons: firstly, because of delays to the works which result in the event that triggers the release of the bond being delayed and, secondly, because although the event that should have triggered the release of the bond has occurred, the employer either denies that it has occurred or fails to release the bond anyway. Sometimes both of these circumstances arise on the same bond.

The costs that arise for a contractor in either of these circumstances depend in part on the nature of the business providing the bond. These tend generally to be either a bank or insurance company.

The basis of the claim for associated interest costs will depend on a number of issues including the factual background. If the release is delayed because of project completion delays to the event that triggers the release, then the contractor will make this interest claim part of its prolongation claim of the type considered in Chapter 6. This might be as a heading of 'loss and/or expense' under such as SBC/Q terms or 'costs' under the FIDIC Red Book or the Infrastructure Conditions terms or as damages where the delay claim is made on that basis. Where the late release is the result of some other failure by the employer to release the bond when the contractual trigger for its release has occurred, the claim is likely to be one for damages for breach of contract. The result is that different parts of the bond extension costs claim may need to be separated out and allocated to the causes as either the consequences of compensable delays or breaches of contract. Particularly in relation to that part associated with compensable delay, quantum may require adjustment if part of that delay is found to be the contractor's culpability or the result of a cost-neutral event.

It might be expected that establishing the costs of extending a bond would be a simple matter of presenting the initial bond agreement, notes extending it, the surety's charges for the extension and proof of payment of those charges. It might also be expected that whenever a bond is extended there will always be additional charges incurred by the contractor from the surety. However, there can be a number of complications in these respects.

It may be that the contractor has a company-wide arrangement, particularly with its bank, for the provision of bonds on all of its projects as and if they are required. In such cases it may be more difficult for the contractor to show the additional cost that it has incurred on the extension of one particular bond on a particular project. However, it should usually be possible to establish this, and it should be a requirement of any claim against the employer for such additional costs that such proof is provided. Failure to do so might mask the reality that the bank has made no additional charge for extending a particular bond at all. This may particularly occur where the contractor is a substantial and high value client for the bank and the particular bond was one of many under the company-wide bonding arrangement. The size of the particular bond relative to the overall total of bonds under the arrangement might also be relevant, together with the contractor's good track record of few bond calls in the past.

There is a further potential area of complication where the surety is the contractor's existing bank. It may be that, rather than making discrete charges against the contractor for the provision of bonds, the bank just off-sets the amount of the bond against the contractor's overdraft facility. In such circumstances, the bank may also charge a fee to set up the bond and an exit fee when it is released, but in the meantime, any extension of the bond attracts no additional bank charges. The additional cost to the contractor of extending the bond is then no more than being denied access to overdraft amounts. This may involve no cost at all, or if it does involve the contractor in costs then they may be very difficult to identify and prove.

The alternative to discrete charges or reduction of an overdraft facility will be for the surety to require the contractor to lodge a sum equivalent to the amount of the bond with the surety. In such a case the cost to the contractor of any extension of that bond may be the cost to it of having the equivalent sum tied up for longer. This cost may be the loss of return that the contractor would have obtained on that sum had it been available for investment elsewhere, usually in an interest earning account or, more rarely, in its other projects.

8.3   Fluctuations in Costs

When a contractor or subcontractor prices works of relatively small size and duration it can reasonably be expected that it will take the limited risk of any inflationary increases in its costs of labour, materials, plant and equipment over the duration of that contract (or the benefit of decreases in such costs during periods of deflation). Thus, when a homeowner obtains prices for the construction of an extension, or the owner of an office block obtains prices for replacement of its roof, they will expect that the contractor's prices will be fixed. However, a different approach may benefit both parties where works are of significant size and duration, for example the construction of a completely new office block.

From an employer's perspective, it may appear to always be to its benefit to procure works on a fixed price basis, and historically almost all construction engineering projects were procured that way. However, this will clearly not be to the employer's benefit in periods of falling prices, such as the Great Depression of the 1930s. Furthermore, in periods of rapidly rising prices, perhaps due to overheating of the market, such as occurred in Europe after the Second World War, it may not be to the employer's benefit that contractors price-in excessive amounts of inflationary risk or fail to make sufficient provision. In the former case, the employer may pay through the fixed price more that it should otherwise have done through the contractor's allowances for risk. In the latter case, the contractor's performance may be affected in ways that damage the employer. For example, the contractor may feel forced to make spurious claims in order to recover its losses, or choose to focus resources on other more profitable projects, or may even financially fail.

Internationally, contractual provision for the contractor to recover increases in its costs are common, especially in locations where cost changes are uncertain and can be significant. This particularly recognises how this approach can take some of the uncertainty out of tendering for potential contractors. It may widen the field of willing tenderers, thus increasing competition and making prices more competitive for the employer. It may also make prices keener to the extent that bidders do not have to allow for that element of pricing risk if they make over-conservative allowances that will involve unnecessary costs for the employer.

Lower down the supply chain, some suppliers of specialist products, such as lifts and permanent plant or machinery, may only contract on the basis that their contract supply price is subject to an escalation clause related to actual time of delivery. Often, they will only quote on the basis of the application of indices specifically published in relation to their supplies.

Rather like the risk of unforeseen ground conditions or exceptionally bad weather conditions that have been considered in Chapters 3 and 6 of this book, philosophical debate is common as who is best placed to assess, price, control and shoulder the risk of inflationary rises in the contactor's costs, and how both parties can be best served by the allocation between them of that risk.

The last few decades have seen construction and engineering contracts increasingly allow the contractor some additional payment for fluctuations in its costs. The titles of such clauses vary, but include such as 'price adjustment', 'fluctuations', 'inflation', 'cost escalation', 'price indexing', 'changes in legislation' and 'changes in costs'. Broadly, they can adopt one of three alternative approaches, as follows:

- Allowing only for the effects of government legislation such as changes in taxes, levies and statutory contributions.
- Allowing for changes in costs of specific items set out in the contract.
- Applying published cost indices to all of the contractor's rates and prices.

Some standard forms of contracts can set out alternatives provisions applying all three of these approaches as options, the parties then being able to apply whichever they wish to their particular project.

8.3.1 Taxes, Levies and Statutory Contributions

Construction and engineering contract policy as to where to place the 'tipping point' after which the contractor takes the risk of unforeseen events varies widely, particularly in relation to increases in its costs. However, it seems that one area in which it is acknowledged that contractors have no control, could not reasonably have provided for when pricing tenders and cannot reasonably avoid or overcome, is changes in government legislation affecting their costs. Thus, it is usual in contracts for major works of long duration that the contractor can recover additional costs arising from changes in government taxes, levies and other statutory contributions. There is, however, some inconsistency as to which taxes, levies and contributions are compensable.

In the UK, the detailed terms of the Infrastructure Conditions limit the contractor's recovery of fluctuations in its costs to those related to tax. They contain provision at clause 69 for recovery in relation to changes in labour taxes and landfill tax compared to their levels at the date for return of tenders. In subclause 69(a) this includes taxes, levies, contributions and refunds in relation to workpeople (but not changes in income tax and training levies). In subclause 69(b) it includes changes in landfill tax pursuant to the Finance Act 1996 and Landfill Tax Regulations 1996, a tax imposed by the UK government based on the weight of material disposed and whether it is active or inactive. This can be particularly relevant to UK civil engineering projects involving significant earthworks, and especially on 'brown field' sites. Clause 69(c) also includes changes to the Aggregates Levy pursuant to the Finance Act 2001, a tax imposed by the UK government based on the weight of extracted sand, gravel and crushed rock. The Infrastructure Conditions provide for payment of fluctuations in the contractor's costs but do not extend to changes in any taxes or levies, etc., that cause a change in material prices, except value added tax, which is dealt with under clause 70. Under that clause the contractor is deemed not to have allowed in its tender for value added tax payable to the Commissioners for Customs and Exercise, such that any changes in the rate of that tax are not at the contractor's risk (or benefit should its rate fall).

In contrast to the Infrastructure Conditions, but also in the UK, SBC/Q provides for all three of the alternative approaches to fluctuations in the contractor's costs as options. These are in clause 4.21, under the heading 'Choice of fluctuation provision'. Option A is for 'contribution, levy and tax fluctuations'. The details are in Schedule 7 to SBC/Q. This provides those for labour (part A.1) and materials (part A.2). The Contract Sum is deemed to have been based on the types and rates of contribution, tax and levy payable at the 'Base Date' as stated in the Contract Particulars. If any of these change, then the contractor is entitled to, or shall allow to the employer, the difference in costs.

Internationally, FIDIC Red Book clause 13.7 proves that:

> The Contract Price shall be adjusted to take account of any increase or decrease in Cost resulting from a change in the Laws of the Country (including the introduction of new Laws and the repeal or modification of existing Laws) or in the judicial or official governmental interpretation of such Laws, made after the Base Date, which affect the Contractor in the performance of obligations under the Contract.

The Base Date is defined in clause 1.1.3.1 as the date 28 days prior to the last date for submission of the tender.

FIDIC Red Book clause 13.7 also requires the contractor to give notice if it suffers (or will suffer) delay and/or incurs (or will incur) additional costs as a result of such changes. The contractor is then entitled to both an extension of time and payment of such additional costs.

Changes in the contractor's costs beyond those resulting from changes in legislation are provided for in the optional clause 13.8 of the FIDIC Red Book. This is considered further in this section.

It is always important to read carefully the definitions of the categories of costs and causes of increases that are recoverable; for example, in relation to labour, and payments made to them, in order to establish accurately the extent of persons employed on the works and payments made that will be subject to recovery of costs under such clauses.

Not all decisions of external bodies are automatically recoverable, even where the contract has a provision that appears to allow recovery of such costs. In the case of *William Sindall Ltd v North West Thames Regional Health Authority* (1997) 4 BLR 151, the House of Lords decided that increases in payments made in accordance with a bonus scheme established under the rules of the National Joint Council for the Building Industry (NJCBI) were not recoverable under clause 31A of the 1963 version of the JCT Standard Form.

The House of Lords came to their unanimous conclusion on the grounds that the amount of bonus payable was not prescribed by the NJCBI and the instigation of a bonus scheme was for the decision of the contractor; it was not compulsory. It could also be added that, presumably, such schemes are implemented to reward increased productivity and therefore, if implemented properly and successfully, they do not in theory increase the contractor's costs as the increased payments are compensated by a corresponding increase in productivity. The difference between theory and practice in this respect may, however, be marked.

It is therefore necessary to consider carefully the circumstances of any change in costs when considering claims for additional payment arising from changes in labour, or other costs by reference to an express term of a contract.

It is also necessary to consider carefully the scope of any contract provision in respect of the persons who may be included in such claims. In the same context of considering claims for additional costs arising from increases in employment costs under clause 31A of the 1963 edition of the JCT Standard Form, it was held in the case of *J. Murphy & Sons Ltd v London Borough of Southwark* (1982) 22 BLR 41 that costs incurred in respect of self-employed labour were not recoverable as such persons did not come within the meaning of 'workpeople' as used in the contract. Clause 31A provided that the contractor could recover increases in rates of pay, etc., 'in respect of workpeople engaged upon or in connection with the Works in accordance with the rules or decisions of the National Joint Council for the Building Industry'. Clause 31D(6)(c) defined the expression 'workpeople' as persons whose rate of payment, etc., was determined by the decisions or agreements of the NJCBI.

The court declined to include the self-employed labour within the NJCBI definition. However, if the self-employed labour had been part of approved subcontracting under the contract, then under clause 31C the contractor could have passed on similar provisions in the subcontract to those in clause 31A and any resulting increases as a result of NJCBI decisions would have been recoverable.

These cases are useful in illustrating the difficulties that can arise in determining the distinction between additional payments for which the contractor remains responsible and those that will give rise to the possibility of recovery under the contract, if they arise. They are equally important in underlining to a contractor the matters that should be considered when compiling the unit rates for use in tender calculations. Failure to make proper provision in the tender rates for all the contractors' obligations that are its risks cannot be corrected later when making claims for additional payment under the contract provisions.

One of the difficult problems that can arise at the tender stage is that of the notification of a change in a tax or levy, or the introduction of a new tax or levy, without any indication by the relevant statutory or other body responsible for the intended level of the change or new charge. The normal course of affairs is that the contractor is deemed to have allowed in its tender rates and prices for all known charges, taxes, levies, etc., as at a specified base date. However, a real difficulty can arise if it is known at that date that a tax or levy is to be introduced, or is to be charged at a different rate, at a date in the future but the rate of charge or increase is not known.

Such a situation arose in connection with the introduction of landfill taxation in the UK. The government had signalled its intention to introduce the tax well in advance of the date nominated for its introduction, but had given no indication of the rate of taxation that would apply, or how the rates would vary according to the type of waste being taken to landfill sites. In such a situation it was impossible for contractors to accurately include the effect of the taxation in their rates and prices, although they were aware that the tax would be implemented from a given date, until the rates and charging regime were published. This resulted in many contracts not having landfill tax included in the rates and prices although its introduction was known in principle at the relevant base date.

For the contractor's protection, the obvious course of action in such circumstances is to qualify the tender for the works and ensure that it is clearly stated in the contract documents that the relevant tax is to be recovered under the fluctuations in prices provisions once the rates and scheme are known and the costs incurred. In the absence of such a qualification or amendment the contractor is left to argue that if the rate of charging is not known then it could not be included in the rates and prices and falls to be recovered under the fluctuations provisions. The contractor will argue that it should not be expected to guess the possible rates of tax. This would seem to be a very reasonable position, but may not be what the contract says in terms of risk allocation, and the prevention of any doubt by flagging the potential problem in the tender and contract amendments will reduce the chances of later disagreement.

The situation can of course become even more fraught in respect of new, or changed, taxes or charges for which, even if the rate is known, there is no accurate means of forecasting their effect on a project. Such an instance would be the congestion charge introduced in London at the initial rate of £5 per day for each non-exempt vehicle entering the charge area in the city centre. Worldwide, a number of cities have followed suit or are planning to do so, for example the Salik system in Dubai. The first consideration will be whether such a charge falls within the definition of the contractor's entitlements to reimbursement. Even if the date of introduction and rate of the charge is known, it is difficult to know how many individual vehicle visits will be generated to a contract within the charge zone and therefore what the total amount of charges incurred by the contractor will be. For new contracts in the zone the contractor will have to make an estimate and, for existing contracts, record all chargeable visits if there is to be a claim for reimbursement. There could still be disagreement as to the *necessary* vehicle visits as opposed to the *actual* number of such visits, a problem that would presumably only be overcome by detailed records of the visitors and the purpose of the visits.

8.3.2   Labour, Materials and Tax Fluctuations

It was explained elsewhere how most construction and engineering contracts of anything other than minor size and duration will, as a minimum, relieve the contractor of the uncontrollable and unpredictable effects on its prices of changes in taxes, levies and statutory contributions. However, the next stage might be to consider whether there are particularly significant items of costs on a particular project that are unpredictable and uncontrollable, and which are best not left as contractor risk items in terms of resulting cost inflation.

For example, on a civil engineering project that involves large amounts of plant and equipment such as for earthmoving, carried out in a period of high volatility in oil prices, it might be considered that the risk of increases in fuel costs should not be left on the contractor's account. Similarly, on a project involving substantial amounts of imported steelwork, it might be considered that steelwork should be subject to a price adjustment clause. In either of these instances, it might be that the fuel and steelwork are respectively made exceptions to the fixed price nature of the contract price, with a base cost for a litre of fuel and tonne of steelwork agreed in the contract and any fluctuations between that base cost and actual costs being subject of adjustment of the contract price. As noted previously, making such exceptions might open the tendering process up to a larger number of bidding contractors, to the employer's benefit in terms of price competition and the avoidance of unnecessarily conservative risk allowances that it pays for through the contract price.

SBC/Q Schedule 7 Option B is for 'Labour and materials costs and tax fluctuations'. The details are in Schedule 7 to SBC/Q. This provides that for labour rates (part B.1), labour levies and taxes (part B.2) and materials, goods, electricity and fuel (part B.3) the Contract Sum is deemed to have been based on the rates of wages, emoluments expenses, levy, contribution and tax (for labour) and market prices (for materials, etc.) current at the 'Base Date', as stated in the Contract Particulars. If any of these change, then the contractor is entitled to be paid or allow the net amount of the difference.

The problem with the sort provision contained in SBC/Q Schedule 7 Option B is that it required detailed forensic analysis of both the contractor's actual costs and those that were current at the Base Date. Its clause B.8 understandably places the onus on the contractor to provide 'such evidence and computations as the Architect/Contract Administrator or the Quantity Surveyor may reasonably require'. This gives rise to two areas of particular difficulty in practice:

- The provision and checking of evidence of both actual and Base Date costs can be a lengthy and laborious task for all concerned. This includes both the contractor's commercial and accountancy staff in finding and presenting records such as comparative invoices, as well as the employer's quantity surveyors in checking them. In practice it often seems that the time spent in forensically analysing increases in costs for a particular item of materials was outweighed by the time and effort of proving that increase.
- Even with the resources and time available to provide and check relevant cost data, the process can suffer difficulty in proving the actual and base cost. A common complaint of those checking such claims is that contractors only exhibit evidence for a limited quantity of a particular material that may be higher than other records for the same materials, or that they have hidden supplier discounts. In relation to the base cost, these can sometimes only be established hypothetically given that they were never actually incurred. One way around this latter problem is to require the contract documents to include a schedule of the agreed base costs of key materials for this purpose. More broadly, the contract documents could include a document breaking down each of the contractor's bill of quantities rates into their components of labour, materials, plant/equipment, subcontract, overheads and profit as well as any other costs.

It is issues such as these that led to the publication and use of cost indices for the construction and engineering industries, as a short-cut to establishing the increased or decreased costs on a particular project.

8.3.3 Price Adjustment Formula

In the UK there have long been a number of price adjustment indices published specifically in relation to construction and engineering projects. Historically the building and civil engineering industries referred in their contracts to NEDO (National Economic Development Office), Baxter or Osborne Indices. Currently the BCIS (the 'Building Costs Information Service' of the RICS) publishes over 200 price adjustment formula indices including the following:

- For 'Civil engineering' contracts (formerly known as the Baxter Indices).
- For 'Building' (formerly known as NEDO Indices, having previously been published by the National Economic Development Office).

- For 'Specialist engineering', such as mechanical engineering, electrical installations, heating and ventilating and steelwork.
- For 'Highways Maintenance'.
- For 'Civil engineering and related specialist engineering'.

Before the advent of computers and spreadsheets, much of a contractor's project surveyor's month-end tasks could be taken up preparing by hand schedules of calculations of increased costs based on such indices. On receipt of a monthly application, the employer's quantity surveyor might then spend an equally laborious time correcting the calculations and amending them for matters such as any differences in the progressed value of works to which the indices were applied. Thankfully, the appearance of packages such as Lotus 1-2-3 and Excel enabled the more IT-savvy quantity surveyors to develop electronic spreadsheets that allowed the entry of each month's work values and indices into a spreadsheet that did all the maths automatically. Today, the BCIS provides such necessary electronic tools as part of its price adjustment formula services.

In jurisdictions where there are no indices published specifically for construction and engineering works, contracts may refer to locally published indices for the wider economy, such as a producer price index or a consumer price index. However, such wider economy indices are unlikely to accurately reflect actual changes in the costs of construction labour, plant and materials on construction and engineering projects and may at times significantly over- or under-compensate the contractor. However, where there are no specific and accurate alternatives, available indices such as those for producer and consumer prices might provide some mechanism for compensation. If that provides an agreeable pragmatic solution, then that is the parties' prerogative. They should, however, beware of their shortcomings where they are used for construction or engineering works.

In the UK, SBC/Q Schedule 7 Option C is for 'Formula adjustment'. The details are in Schedule 7 to SBC/Q. This provides for adjustment of the Contract Sum by application of the Formula Rules current at the Base Date as stated in the Contract Particulars. Those Formula Rules are published by the JCT and the current version is the 2016 revision, 'FR 2016'.

As noted elsewhere, the detailed terms of the Infrastructure Conditions clauses 69 and 70 limit the contractor's recovery of fluctuations in its costs to those related to tax. However, clause 72, 'Special Conditions' provides for provisions to be added to the general conditions in its clauses 1 to 71 and expressly refers to 'Contract Price Fluctuation' as such a potential incorporation. Furthermore, the Infrastructure Conditions provide a set of contract price fluctuation provisions which it describes as 'for use in appropriate cases as special conditions'. This provides for the application of the BCIS indices for 'Labour and Supervision', 'Plant and Road Vehicles' and for materials listed in its subclause 72(4), which also enables proportions to be entered against each listed type of material. The base indices are stated in subclause 72(2)(b) to be those applicable to the date 42 days before the date of return of tenders. The current indices are stated in subclause 72(2)(c) to be those applicable to the date 42 days before the earliest of: the contractually due date for completion (thus preventing recovery in relation to increased costs resulting from the contractor's own failings); the date certified pursuant to clause 48 as completion of the whole of the Works; or the last day of the period to which the certificate relates.

The FIDIC Red Book contains an optional clause 13.8 for adjustment of the Contract Sum for rises and falls in the contractor's costs. It requires the Appendix to Tender to include a 'table of adjustment data' for each of the currencies in which the Contract Price is payable. If there is no such table, this optional clause 13.8 does not apply. The clause requires the application of formulae based on indices stated in the 'table of adjustment data'. The base cost indices are those at the Base Date, being defined in FIDIC Red Book clause 1.1.3.1 as the date 28 days prior to the last date for submission of the tender. The current cost indices are those applicable on the date 49 days before the last day of the period to which the particular Payment Certificate relates.

8.3.4   Application to Other Claims

A common cause of disputes in relation to fluctuations in the contractor's costs is whether, and to what extent, a fluctuations clause, particularly one applying published indices, also applies to the valuation of other claims such as those for variations.

SBC/Q clause 4.22 seeks to clarify the position in relation to variations by providing that whichever of its Options A, B or C applies, it will only do so in relation to a variation that specifies the base date for the calculation of that variation.

Similarly, FIDIC Red Book clause 13.8 states that its optional provision for adjustments for changes in costs does not apply to works valued on the basis of 'Cost' (i.e. based on actual expenditure) or current prices. A perhaps obvious point, but which does not prevent the potential factual and legal issue, is identified as follows.

Many bespoke contracts do not make the position clear and the position can be clouded further where variations are agreed in terms that do not detail the basis on which they have been priced, for example because they were agreed on a broad lump sum basis. Whether the contractor is correct to later apply the fluctuations provision to the agreed values of such variations will be a matter of fact and law for each instance.

The inapplicability of price adjustment formula to claims for delay costs should be obvious, where such delay claims are based on actual costs. However, as explained further in Chapter 6, delay claims are sometimes agreed, for convenience, on the basis of rates in the contract, particularly for preliminaries and general items such as equipment, accommodation and staff. Furthermore, as also explained in Chapter 6, some contracts set out agreed rates for the remuneration of the contractor's delay costs in the event of some or all types of compensable delay event. Here again, it is important that the contract states the base date for those agreed contract rates and whether the price adjustment provision applies to them. For example, in a contract for the construction of a gas terminal, provision was made for the 'adjustment of the Contract Price' by application of the local national producer price index, without making it clear whether those indices should also apply to a delay claim also based on rates and prices set out in the contract. The contractor argued that as a matter of fact it could show that those contract delay rates were based on its allowances in the contract price. The employer argued that the schedule of delay rates attached to the contract did not form part of the Contract Price, such that it fell outside the adjustment provision as a matter of contract interpretation and that the question of how the delay rates were calculated was irrelevant. Whoever was right in these arguments, the important point is that they could have avoided referring it to arbitration, with the resulting costs, had the contract made the position clear.

8.3.5   Effects of Delays

Construction claims often feature some element of overlap between the reimbursement (or non-reimbursement where at the contractor's risk) of fluctuations under express conditions of the contract and contractor's claims for the costs of delay as considered in Chapter 6 of this book. This is another common cause of dispute in relation to fluctuations in the contractor's costs. There are two key areas to consider in this regard:

- Where a contract includes provision for reimbursement of increases in the contractor's prices, does that include those inflationary increases that were only encountered as a result of the contractor's culpable delays?
- Where a contract does not include provision for reimbursement of increases in the contractor's prices, does that preclude its recovery of those inflationary increases that were only encountered as a result of the delays for which the employer is responsible?

Stopping the contractor's recovery of fluctuations in a period of culpable contractor delay is provided for within all of the Options in SBC/Q Schedule 7. For Option A, clause A.9.1 provides that no amount shall be added (or deducted) 'in respect of which the payment or allowance would be made occurs after the Completion Date'. For Option B, clause B.10.1 is in similar terms. For Option C ('Formula Rules'), under clause C.6 the indices for calculation of fluctuations after the Completion date are frozen at those applicable to the valuation period in which the Completion Date falls.

What is notable about the SBC/C provisions, particularly those for Option C, is how they do not address the fact that if the contractor is in delay prior to the Completion Date, then some, and perhaps much, of its work to that date will have been carried out at times later than they would have been carried out but for the culpable delay (albeit at times before the contractually due date for completion). Thus, it may be that in each month's valuation of fluctuations, the amounts calculated based on work done are higher than they would have been had the contractor been progressing at a faster rate without culpable delay. This criticism can be made of many other provisions for the stopping of fluctuations at the contractually due date for completion.

As has been explained, the Infrastructure Conditions also limit the recovery of cost fluctuations, where the Special Conditions provision applies, by making the current indices for the formula calculation those not later than at the due date for completion.

Where a provision for payment of inflationary increases in costs does not stop those in periods of culpable contractor delay, then that part only encountered as a result of the contractor's culpable delay might be said to be an item of damages arising out of the contractor's breach. The extent to which that part was recoverable from the contractor and also not covered by rates of the employer's delay damages set in the contract may become an issue, and will particularly involve questions of contract and law.

Turning to contracts that do not provide for reimbursement of the contractor's inflationary increased costs, where compensatory delay occurs for which the employer is contractually liable, it is not uncommon for the contractors delay claim to include a heading for its increased costs of such as labour and materials and for this claim to be denied on the argument that the contract was on a fixed price basis, with no provision for reimbursement of inflationary cost increases. Contractors may respond that this misses the point that (unless the contract contains some particularly unusual wording) contractors will only allow in the contract price, and are only liable for, increases in costs on the basis of the contract period. If compensable delay occurs, then a proper head of the associated compensation will usually be increases in the effects of cost inflation caused by work being done later as a result of the delay. Alternatively to a claim under the express contract provision, the recovery of the additional increased costs could form an item of damages for breach of contract by the employer.

It was explained elsewhere how, in jurisdictions where there are no indices published specifically for construction and engineering works, and wider indices such as those for consumer or producer prices are applied, this may significantly over- or under-compensate the contractor. Where there is under-compensation and the contactor asserts that this in part results from compensatory delays, then part of the contractor's claim for that compensation may include a calculation of that part of actual price increases not covered by the contractual adjustment formula that arose out of the employer's culpable delays. In such a claim the contractor will assert that it took the risk that its actual cost increases would outstrip those of the contract indices, but not to the extent that this occurred as a result of its works being carried out later, and hence at higher costs, because of delays caused by the employer. Conversely, the employer may complain that the contractor took not only that risk but also the potential opportunity to the extent that actual cost increases were less than the contractor's tender allowances.

8.4 Suspension of Work

8.4.1 The Right to Suspend

There are a variety of circumstances in which it might be considered right to allow within the express terms of construction and engineering contract for either of the parties to suspend the works, part thereof or some of their activities under the contract. Those circumstances might include some unforeseen eventuality arising on the project, against which one or both parties need time to consider what actions to take as a consequence. Alternatively, they might include a default by a party, against which the other party needs time to consider its position, to protect itself against such a failure or as a means to apply pressure in relation to the failure. Standard forms of contract therefore set out various provisions for either party to suspend the works, and how the effects are to be dealt with, including the addressing of time and cost consequences that arise. They also tend to place a limit on the period over which such suspension of performance can last in some circumstances.

In the Infrastructure Conditions, adverse physical conditions or artificial obstructions, which could not have been foreseen by an experienced contractor, are covered by clause 12. If the contractor gives notice that it has encountered such conditions or obstructions, or informs the engineer that it intends to make a related claim for additional payment or extension of time, then subclause 12(4) empowers the engineer, among other things, to order a suspension under clause 40. It is a notable aspect of the Infrastructure Conditions that clause 12 should pick out adverse physical conditions or artificial obstructions as a particular issue that might lead to suspension of the work. It perhaps reflects that those conditions are most commonly used on civil engineering works that are likely to be potentially susceptible to such issues, and they therefore expressly allow the employer and engineer to gain time to consider what action to take in the face of such an occurrence.

In clause 40 of the Infrastructure Conditions, subclause 40(1) allows the engineer to order the contractor to suspend the progress of the works or any part thereof. The clause requires the contractor to comply and to 'properly protect and secure the work so far as is considered necessary by the Engineer'. Except where the suspension is otherwise provided for in the contract, or is due to weather conditions or the contractor's default, or is necessary for the proper construction and completion or safety of the works (where not due to default or engineer or employer or an 'Excepted Risk'), the contractor is entitled to an extension of time and 'such extra cost as may be reasonable' plus profit.

The limit on a period of suspension under the Infrastructure Conditions is at subclause 40(2) and is three months. Thereafter, the contractor may give notice of 28 days requiring permission to recommence on the suspended work. Failing that permission, the contractor may, by notice, elect to treat the suspended work as omitted per clause 51 or, where it affects the whole works, to treat the suspension as an abandonment of the contract by the employer.

It is notable that the Infrastructure Conditions contain no provision for the contractor to suspend the work, in the event of such employer failures as not making payments in accordance with the contract. This is in contrast to both the FIDIC Red Book and SBC/Q provisions as explained as follows.

In the FIDIC Red Book, clause 8.8 contains provisions for suspension to be instructed by the engineer that are not dissimilar to those of the Infrastructure Conditions, clause 40. For example, under clause 8.8, the contractor is required to 'protect, store and secure such part or the works against any deterioration, loss or damage'. Under clause 8.9, the contractor is entitled to claims in relation to delays and costs resulting from the suspension and/or the resumption of the work, other than to the extent that it results from the contractor's faulty design, workmanship or materials or its failure to 'protect, store or secure' as required by clause 8.9.

FIDIC Red Book clause 8.10 addresses any consequences of the suspension on 'Plant' or 'Materials' (both of which are defined terms under FIDIC as items intended to be incorporated into the permanent works) not delivered to the site at the date of suspension. If work to the plant or materials is suspended for more than 28 days and the contractor has marked it as the employer's property, the contractor is entitled to payment for those items.

FIDIC Red Book clause 8.11 places the limit on the period of suspension under clause 8 at 84 days. Thereafter, this clause is similar to Infrastructure Conditions clause 40(2). The contractor may request with 28 notice permission to recommence on the suspended work. Failing that permission, the contractor may by notice elect to treat the work as omitted per clause 13 or, where it affects the whole works, to give a notice of termination under clause 16.2.

Under FIDIC Red Book clause 8.12, after permission or instruction to resume work, the parties are required to jointly examine the works, plant and materials affected by the suspension and the contractor is required to 'make good any deterioration or defect in or loss of the Works or Plant or Materials, which has occurred during the suspension'.

FIDIC Red Book clause 16 addresses suspension (and termination) by the contractor in the event of some default by the employer or engineer. Clause 16.1 gives the contractor the right to suspend or reduce the rate of work in the event of: the engineer's failure to certify an interim payment in accordance with clause 14.6; or the employer's failure to pay in accordance with clause 14.7; or failure by the employer under clause 2.3 regarding its financial arrangements to enable it to pay the contract price. Such action follows 21 days' notice from the contractor. The clause entitles the contractor to an extension of time plus its 'Cost plus reasonable profit' suffered as a result of its suspending or reducing the rate of work under the clause. This right to reduce the rate of work is an unusual provision of the FIDIC Red Book, which often proves a very effective measure in terms of addressing the employer's failure and applying pressure, without some of the less desirable consequences of full suspension.

SBC/Q takes the opposite approach to the Infrastructure Conditions in terms of who can and cannot suspend works. As noted elsewhere, the SBC/Q conditions do provide for the contractor to suspend work. However, and in further contrast to both the Infrastructure Conditions and the FIDIC Red Book, SBC/Q contains no power for the employer to suspend the works.

The contractor's right to suspend is in SBC/Q clause 4.14 and is triggered by failure by the employer to pay a sum payable to the contractor under clause 4.12, 'Interim payments – final date and amount'. The contractor has to give seven days' notice to the employer and the architect/contract administrator of its intention to suspend and the grounds. Under clause 4.14.2, the contractor is entitled to its 'costs and expenses reasonably incurred' as a result of such suspension. In addition, clause 2.29 makes such suspension a 'Relevant Event' for which the contractor is entitled to extensions of time under clause 2.28.

8.4.2   Typical Financial Heads of Suspension Claims

A flavour of the type of financial claims that arise out of the suspension of construction and engineering works can be found in the summary of related terms of standard forms of contract set out in Section 8.4.1. They include such as: delays; the costs of security, storage and protection; claims in relation to omitted works (such as loss of margin); making good of deterioration, defects or loss occurring during the period of suspension; and payment for plant and materials not yet delivered to site. These heads of claim include a number of further potential subheadings, particularly the heading of delays and prolongation costs. In addition, there are other potential heads that can arise. The evaluation of most of these heads of claim are covered elsewhere in this book, but a few particular comments are necessary here.

Some of the cost consequences of suspension are obvious, but others are less so, particularly to some employers and their advisors. It is good practice for a contractor that has been instructed to suspend work to record to the employer and its advisors the type of costs that it is incurring and to keep that notification up to date as the costs develop.

8.4.2.1   Prolongation Costs

The contractor incurring delay and prolongation costs is the most obvious and inevitable consequence of most suspensions of its works, or part thereof. The evaluation of delay claims is the subject of Chapter 6 of this book. The subheadings and method of evaluation are likely to be very similar, although the costs might fall under different headings of a contractor's claims, with one under 'delays to completion' and the other under 'suspension costs'. Where the suspension is the cause of delay to completion, then the costs might go together. However, particularly where the suspension is only part of the works, and that part is not critical, then the suspension delay claim may need to stand on its own. For example, on a roads project in the Middle East, the wearing coarse part of a new highway was suspended whilst issues as to its specification were resolved by the engineer. However, that work was not then on the critical path of the project, such that the cost of idle times of the plant, drivers and labour affected were part of a specific claim for suspension and quite separate to the contractor's other claims for delay to completion caused by critical delay events elsewhere.

A further feature of delays caused by suspension that might distinguish it from other delays to completion is how the effects on plant, equipment, operatives, and supervision can be particularly 'clean' in terms of establishing that effect. This will, however, depend on whether the suspension is of the works as a whole or only part. If the whole of the works are suspended, the effect on the critical path ought to be inevitable and obvious (subject to arguments about float in the programme). This can be distinguished from situations where there is delay to a critical path activity, meriting an extension of time, but resources can be put on to other, non-critical, works that are continuing and not suspended, so that their time is not completely lost. As noted in Section 8.4.1, both the Infrastructure Conditions and the FIDIC Red Book provide for only part of the works to be suspended. The effect of this can be that resources affected by that suspension can be re-allocated to other works to some extent, thus partly reducing the contractor's resulting costs and claim.

The effect of such relocation is that records of the effects of suspension on resources are particularly important. If a critical delay to a project occurs, then the contractor may reasonably argue that its time-related resources such as plant and non-working people will automatically be prolonged in their involvement and costs. However, in a suspension of work, this does not follow if other works are continuing. Too often contractors make a claim for the loss of all resources that were working on the particular suspended works prior to their suspension, when what is actually required are records of the actual effects on those resources in terms of time actually lost as a consequence.

A prudent contractor ought to compile daily schedules of its resources and their idle time during a period of suspension and submit these to the engineer, architect or contract administrator for agreement This might seem an onerous task, but a cynic might sometimes ask what else are its staff doing in such a period of suspension! It is particularly useful where only part of the works are suspended. Even if a counter signature cannot be obtained, they might be a very helpful record in future in relation to a claim that is all too often very hard to quantify due to the lack of records.

8.4.2.2 Security, Storage and Protection

As noted in Section 8.4.1, this is a particular issue mentioned in Infrastructure Conditions clause 40(1) and FIDIC Red Book clause 8 in relation to suspension. However, it could in any event be covered by the wider terms such as SBC/Q clause 4.14.1, 'costs and expenses reasonably incurred'.

Security, storage and protection will already be part of the further such contractor's activities, but the issue in a suspension of works is the addition of measures. This in itself sometimes makes it difficult to establish what is additional and resulted from the suspension rather than being an existing activity. In terms of security against such issues as vandalism and theft, a suspended site or area of a site is likely to require additional measures, particularly during what would have been dayshifts but now see those areas unoccupied and requiring daytime patrols.

Even where no additional security measures are required, perhaps because it is a building site of limited size with a sound security fence and staff in place at the entrance, their ongoing costs during the suspension will be an obvious time-related effect of that event that will increase in proportion to the period of suspension and should be readily easy to establish.

Storage may be a matter of security (from vandalism or theft) or protection from deterioration by the elements. It may be that items of plant are best moved into secured storage areas, a particular issue in relation to earthmoving or road-laying plant on long highways projects. Similarly, stored materials that have been moved out of secure storage to place them near to their final permanent location, may need to be returned to storage.

Protection measures can include laying coverings over sensitive works that are left exposed by the suspension of following activities. Earthworks may be susceptible to water damage and require the continued maintenance of dewatering equipment and even a cofferdam.

One practical issue in relation to security, storage and protection is that of how long the suspension is expected to last when it first occurs. This may be completely unknown to the contractor or it might have a good idea that the issue (for example, what to do about an unforeseen artificial obstruction) will shortly be resolved. This may influence the extent of measures that the contractor takes and could be expected to take under clauses such as FIDIC Red Book clause 8.9, which limits its right to costs where it fails to take such measures. Problems can occur where the suspension is initially expected to be of short duration, but continues at length, but with the ongoing expectation that it will soon end. This is all too common in practice and it is easy to lose sight of this when considering with the benefit of hindsight whether a lack of security, storage and protection measures by the contractor was reasonable.

In addition, there may be a balance between the extent of reasonable costs to be expended in securing, storing and protecting works and items with the costs of deterioration and defects that are likely to result if no such measures are taken. If a length of new highway is suspended just after its base course is completed, and pending delivery of asphalt, it might be said that deterioration that requires remedial work to the base course post-suspension might have been avoided by suitable protection, but if the length is over a hundred metres of exposed surface, that might not be a realistic assertion, given the practicalities and costs of such protection compared to the likely costs of remedying deterioration and defects later.

8.4.2.3 Deterioration and Defects

As noted in Section 8.4.1, this is a particular issue mentioned in FIDIC Red Book clause 8 in relation to suspension. However, it could in any event be covered by the wider terms such as SBC/Q clause 4.14.1, 'costs and expenses reasonably incurred'.

As already noted, the entitlement to recover costs in relation to deterioration and defects arising from suspension may be linked to requirements to secure, store and protect. FIDIC Red Book clause 8.9 states in relation to the consequences of suspension that 'The Contractor shall not be entitled to an extension of time for, or to payment of the Cost incurred in, making good the consequences of ... the Contractor's failure to protect, store or secure in accordance with Sub-Clause 8.8'. Where an employer denies the costs of deterioration and defects on this basis, the practicalities of the particular situation must be considered. It is easy to say, for example, that rusting of exposed reinforcing bars could have been prevented by protection, but if the costs of covering them all would have been far more than the costs of wire brushing those few that require this to remove rust, then the contractor has taken the most efficient approach in the circumstances and should not be penalised in its claim for those remedial activities.

It is suggested that the requirement in FIDIC Red Book clause 8.12 for a joint inspection after permission or instruction to resume work should be a required practice under any form of contract where works are suspended. Recording the results, and getting them jointly signed and exchanged, is a particular risk for the contractor. It is not unknown for a joint survey to be carried out and agreed but for the contractor's site staff to then discover further problems as the works continue, which the employer then says were not the result of the suspension but of subsequent events at the contractor's risk. The need is therefore for as thorough a joint survey as possible. Supplementing its written schedules with photographic and video recordings is recommended although these must also have some narrative explanation of what is being recorded or filmed, where it is and what the exact issue of deterioration or defect is.

8.4.2.4   Payment for Items Not Yet on Site

As noted in Section 8.4.1, this is a particular head of cost mentioned in FIDIC Red Book clause 8.10 in relation to suspension, where this entitles the contractor to payment for defined 'Plant' or 'Materials' intended for incorporation into the permanent works, but which are still off-site. However, it could in any event be covered by the wider terms such as SBC/Q clause 4.14.1, 'costs and expenses reasonably incurred', where the particular circumstances arise.

Under FIDIC Red Book clause 8.10, it is necessary that the work on the Plant or delivery of the Plant or Materials has been suspended by more than 28 days and the items are marked as the employer's property in accordance with the engineer's instructions. Clause 7.7, 'Ownership of Plant and Materials', provides that when the contractor is entitled to payment for them under clause 8.10, such items of Plant and Materials become the property of the employer. This seems a very reasonable approach, where the contractor's liability for such items is extended by a suspension, including its incurrence of costs to its supplier or subcontractor. In practice, a further benefit of this approach for the employer is where suspension is followed by termination, such items will be retained by it, which could turn out to be either a potentially good point in terms of ownership or a bad point in terms of liability.

8.4.2.5   Resumption Costs

As noted elsewhere, FIDIC Red Book clause 8.9 is the only standard form considered here that specifically refers to such costs in relation to suspension. It is suggested that it should in any event be covered by the wider terms such as SBC/Q clause 4.14.1, 'costs and expenses reasonably incurred', or the Infrastructure Conditions clause 40(1), 'such extra costs as may be reasonable'. However, FIDIC's specific reference to it is helpful because such costs are often overlooked and usually underestimated.

A further problem with resumption costs after suspension is that they are often very difficult to isolate and prove. Such resumption costs may include such costs as:

- Relocation of plant, equipment and other resources to working locations. Requirements of forms such as the Infrastructure Conditions and the FIDIC Red Book to secure, store and protect were set out elsewhere. For example, where plant on a large linear site such as a railway is all retuned to a central plant yard during suspension, significant costs and time might be incurred in getting it back in place. If the plant has been off-hired, being not owned by the contractor, then it may also take time to get such items returned by the supplier. This may also include redelivery charges from the hire company to add to its collection charges that might have already been part of the contractor's claim for suspension.
- Similar considerations arise in relation to labour, particularly where the labour is from subcontractors or labour agencies. Even payroll employees might have been relocated to another project. In addition to remobilisation costs, it may be that the labour previously working on activities before they were suspended are no longer available to return after resumption of the work. This can have a number of cost consequences. Replacement labour might be only available at greater cost. Perhaps payroll labour now has to be obtained from agencies at a premium cost.
- 'Learning curves' for operatives are a commonly overlooked issue in construction claims, particularly where the works are very specialised, or the site is unusual. New operatives may also need to be subject to a site induction process to ensure they are fully familiar with the organisation and operation of that particular site. Labour may also require lengthy security clearances on, for example, a sensitive project such as one related to the defence or nuclear industries.
- Additional management and supervision costs can be incurred similar to those in relation to labour.

8.4.2.6 Reducing the Rate of Work Progress

As set out in Section 8.4.1, FIDIC Red Book clause 16.1 is unusual among the standard forms considered in this book in addressing this measure as a part of its suspension clause and as an alternative to suspension. In the event of the employer failures set out in that clause, the contractor has the option of reducing its rate of progress of work and is entitled to 'Cost plus reasonable profit' suffered as a result.

The advantages of applying this rather than suspension may include avoiding some of the more problematical consequences of suspension (such as loss of equipment, labour and staff and incurring controversial heads of costs that will then need to be claimed from the employer) whilst still applying sufficient pressure on the employer to cause it to put right its default. The practical 'downside' is often that claims for the resulting effects and costs from a slowdown in the rate of work can be particularly difficult to prove and quantify. They will much less 'clean' compared to the effects and additional costs of suspension.

The most obvious illustration of this difficulty is in quantifying the delay effects and their costs. If work is suspended, then the effect on the suspended programme activities, and potentially completion as a whole, is often readily established based on the duration of the suspension plus the effects of the resumption activities. However, the picture will be much less clear where works are just slowed down. As ever, the onus is on the contractor to prove its claim for the effects of such a reduced rate of working, and it may serve it to do the following:

- When giving its 21 days' notice to reduce its rate of work, under clause 16.1 of the FIDIC Red Book, the contractor might set out both the measures it intends to take and their likely effects. The measures might set out which works are being slowed and how, in terms such as which resources are being removed and what proportion of the total resources currently allocated to that work will be affected. The effects might assess the degree to which the measures will slow the work and the effect of that on progress and completion of both the activity and contractual sections of the work or the works as a whole.
- When the measures take effect, the contractor might re-confirm the measures it is taking and their likely effects, including any changes from those set out in its notice. At the same time, progress on the affected activities should be recorded in detail, for later comparison with progress when the normal rate of progress is later resumed.
- During the period over which progress is reduced, the contractor should give the employer updates on the measures it is taking and their effects in the detail suggested in relation to its notice.
- When the normal rate of progress is resumed, the contractor might record to the employer progress on the affected activities, for comparison with the record given when the measures took effect. This should be particularly useful in establishing the effect of the reduced rate.

If such notices and records are kept and submitted by the contractor, then the employer and its advisors should have no excuse for not being aware of the effects of the suspension, and those effects should be more readily capable of quantification and agreement.

8.4.2.7 Profit on the Contractor's Costs

As set out elsewhere, FIDIC Red Book clause 16 and Infrastructure Conditions clause 40(1) contain provision for the addition of profit to the contractor's suspension claims, whereas SBC/Q clause 4.14 only entitles it to 'costs and expenses' with no express mention of profit.

The FIDIC Red Book and the Infrastructure Conditions contain a subtle difference as to the extent of profit allowed. Under Infrastructure Conditions clause 40(1) the contractor is entitled to 'such extra costs as may be reasonable' and the clause then continues 'Profit shall be added thereto'. On the other hand, FIDIC Red Book clause 16.1(b) says that the contractor is entitled to 'any such Cost plus reasonable profit'. It is suggested that, in any event, profit under such a clause would need to be at a reasonable level, but FIDIC's addition of the qualification 'reasonable' leads directly to the question of what constitutes reasonable profit in these circumstances?

It is difficult to see how profit on such a claim can be addressed other than by the addition of a percentage to the other costs in the contractor's claim, but it is also suggested that there are the following available potential alternative approaches to the percentage in such claims:

- The percentage allowed by the contactor in its pricing of the contract sum, as evidenced by its build-up to that contract sum. This approach is advocated on the basis that this is the rate at which the contractor could be expected to have a priced profit on the costs of the suspension had it known of it, and allowed for it, in its tender.

- The percentage that the contractor was actually achieving on the works that were not suspended, as evidenced by its project accounts. This approach is advocated on the basis that this is what it could have expected to achieve on the suspended works had they not been suspended.
- The percentage that the contractor was actually achieving as a wider business, as evidenced by its company accounts. This approach is advocated on the basis that this is the rate at which resources expended on the suspended works would have earned profit had they been free to work on other projects within the contractor's portfolio.
- Some other measure, such as the rate at which contractors generally would be expected to earn profit, for example a market rate. This might be an approach proposed where the above alternatives give rise to what the employer considers an 'unreasonable' level, particularly where a qualification such as that in FIDIC Red Book clause 16.1(b) applies.

As illustrated in this last approach, the addition of the subjective qualification 'reasonable', such as in the FIDIC Red Book, might be considered unhelpful by a contractor who has won work on a very healthy profit allowance, claims to add that rate to a claim for suspension and is met with reference to that word as only entitling it to some lower percentage addition. Interestingly, the 2000 edition of the FIDIC Contracts Guide suggests that the following sentence may be included in the Particular Conditions:

> In these Conditions, provisions including the expression 'Cost plus reasonable profit' require this profit to be one-twentieth (5%) of this Cost.

In practice, and presumably based on this guidance, many FIDIC-based contracts include an agreement that 'reasonable profit' under their terms is at 5%.

8.5   Incomplete and/or Defective Work

8.5.1   The Requirement to Complete and a Defects Liability Period

It may seem essential to any commercial contract under which a first party provides goods or services and a second party pays the contract price for them, that the first party will provide those goods and services complete and defect free as specified in the contract for that price. In the Infrastructure Conditions the contractor's general obligations in clause 8 include, at subclause 8(1)(a), to 'construct and complete the Works'. Similarly, in SBC/Q clause 2.1, those general obligations include to '… carry out and complete the Works in a good and workmanlike manner and in compliance with the Contract Documents …'. More pointedly, in the FIDIC Red Book the contractor's obligation is put in clause 4.1 as being to '… execute and complete the Works in accordance with the Contract and the Engineer's instructions and shall remedy any defects'. FIDIC's separate emphasis on completing defects reflects how construction and engineering contracts usually contain detailed provisions for the completion of defective and incomplete works in a defects liability, rectification, maintenance, correction or notification period, sometimes referred to as a 'defects period'. Generally, that period runs after the works have been completed such that they have been taken over by the employer and used for their intended purpose, notwithstanding the presence of such defects and outstanding work items. That period may be as little as 3 months, is commonly 12 months but can be up to 5 years for some projects. Such an express provision for a defects period still leaves the contractor liable for defects under the law. For example, under the Limitation Act 1980 in England and Wales the contractor is liable for 6 years under a simple contract and 12 years where it is a deed.

The standard forms considered in this book address completion and the contractor's duties in relation to the defects period thereafter as follows:

- The Infrastructure Conditions clause 48 sets out machinery for the engineer to issue a Certificate of Substantial Completion. Clause 49 then provides for a Defects Correction Period during which any outstanding works and defects are to be completed by the contractor. Subclause 49(3) provides that such works are to be carried out at the contractor's expense. Should the contractor fail to do so, subclause 49(4) allows the employer to carry out the work at its own expense and to recover those costs from the contractor.
- SBC/Q clause 2.38 provides for a Rectification Period following the date of practical completion under clause 2.30. Subclauses 2.38.1 and 2.38.2 provide for the architect/contract administrator to issue a schedule of defects, shrinkages or other faults not later than 14 days after the expiry of that period, and, before that, to instruct that any such defects, shrinkages or other faults be made good by the contractor. Clause 2.38 continues by requiring the contractor to make them good within a reasonable time, unless the architect/contract administrator otherwise instructs, in which case 'an appropriate deduction shall be made from the Contract Sum'. No guidance is given in SBC/Q as to what an 'appropriate deduction' is and how it should be calculated. As explained elsewhere in this chapter, this can prove to be a problematical area of evaluation.
- FIDIC Red Book clause 11.1 requires the contractor to complete any work outstanding at the date of the Taking Over Certificate under clause 10 and to remedy 'defects or damage' notified by, or for, the employer during the Defects Notification Period. The contractor is required to complete and make these good by the end of that period or as soon as practicable thereafter. Under clause 11.2 the remedying of defects or damage is at the risk and cost of the contractor, if due to its design, plant, materials, workmanship or failures; otherwise it is treated as a variation under clause 13.3.

Such provisions operate to recognise the impracticality of requiring all works or the whole of sections thereof to be all completed and defect-free before taking over of a project, whilst also protecting the employer's interests in getting the works or such sections completed properly and fully in accordance with the specifications, drawings, etc., thereafter. However, they also act to protect the contractor from excessive claims for the costs of rectification of defects by the employer. Therefore, provisions for defects periods after handover of the works to the employer involve benefits for both parties, for example:

- For the employer benefits may include:
  - The works or sections can be taken over and the benefits of their use can be obtained even when it is known that there are incomplete or defective items, in

the knowledge that the contractor will still be obliged to complete or remedy them in the defects period.
- The employer and its consultants can have the time and opportunity provided by the period to identify any quality issues that were not apparent at taking over, rather than delay taking over whilst doing completely comprehensive surveys.
- Incomplete or defective works are addressed by the contractor at its cost, without the need for the employer to appoint another contractor, which would be likely to involve the employer in additional costs and require a claim against the contractor for their recovery.
- The requirement for the contractor to return also avoids an undesirable division of responsibility for the work between two contractors, with the potential for doubt that this might create.

These benefits for the employer will usually be supported by the incentive on the contractor of the employer withholding retention monies (or a part of them, usually the final half), pending the completion of all notified defects.

- For the contractor benefits may include:
  - Relief can be obtained from delay damages under the contract by having the completion of the works or a section thereof certified even when it is known that there are incomplete or defective items.
  - As the contractor is no longer in possession of the site, a requirement for the employer to notify the contractor of any defects will mean that the contractor is made aware of such issues.
  - Any incomplete or defective works are addressed by the contractor at its cost rather than at a probable premium by the employer using another contractor, which would then be recoverable from the contractor.
  - The contractor can maintain the quality of the works as a whole, rather than potentially sharing responsibility of parts of it with another contractor.

8.5.2   Potential Methods and Problems of Quantification

In terms of financial evaluation, the standard form provisions quoted in this section point to some of the potential bases. For example, the Infrastructure Conditions clause 49(4) reference to the employer's 'cost'. In Chapter 5 the use of percentage adjustments to the contract price for defective or incomplete work was discussed, but many construction and engineering contracts do not contain such provisions and therefore there has to be a means of addressing financial quantification in relation to any work that is defective or not completed in a proper manner. Of course, if the contractor corrects the defect or completes work itself the employer is likely to incur no costs in relation to that correction. Where the contractor does not correct a defect or complete work, it is suggested that the potential alternatives include:

- An appropriate amount for the work based on allowance for it in the contract price, particularly where the work is not done rather than being defective. This may be through the instruction of an omission, the valuation of which is covered in detail in Chapter 5 and Section 8.6 of this book.
- The costs the contractor would have incurred had it completed or corrected the work.

- An amount to reflect the reduced value of the work, particularly where it has been carried out but is not in accordance with the specification requirements of the contract in some way.
- The costs to the employer of completing or remedying the work either itself or through the employment of others.

Which of these approaches applies may depend on such variables as the nature of the work, its state of non-completion or defect, the terms of the contract, how the parties applied those terms and the applicable law.

A further consideration in relation to quantification may be whether the contractor has actually been paid for the work in question. In practice, this is often not as easy to establish as it sounds, particularly on large projects with voluminous bills of quantities and many variations. Particularly where a contract has been terminated and the work completed by the contractor at the date of termination has to be valued, including adjustment for incomplete or defective work, finding where the contractor has been paid, and in what amount, for that work can be even more difficult. The approximate and broad-brush approach to interim valuations on some major construction projects may not help with these problems. If the contractor has not been paid, particularly for incomplete work, then the employer's claim is likely to be only for its additional costs of having it carried out by itself or others compared to what it would have paid the contractor.

The approach in relation to incomplete work may differ depending upon the type of contract, i.e. whether it is a lump sum contract with adjustments for specified events, such as work not being carried out, or a remeasurement contract where the actual quantity of work properly executed is measured and valued.

The significance of which method of evaluation is adopted can be illustrated by a recent example from a project for the structures to the marine jetty for a liquid natural gas terminal. The parties agreed that a subcontract scope included asphaltic pavement for the jetty. They also agreed that the subcontractor did not carry out that work and that the contractor had therefore omitted it from that subcontractor's scope and engaged another to carry it out. However, whilst the subcontractor valued the omission as a negative variation to its subcontract based on its allowance in the subcontract price, the contractor contended that it had only omitted the work because of failures by the subcontractor and valued the omission as its actual costs in relation to the replacement subcontractor and management of it. As an omission based on the allowance in its subcontract price, the subcontractor valued the item at around US\$ 150,000. In contrast, the contractor's claim based on its costs of engaging the replacement subcontractor was for around US\$ 350,000, two-and-a-third times the subcontractor's amount.

A further problem in relation to the costs of work done by a replacement contractor may be in relation to betterment, where the scope or specification of works done by the replacement contractor is higher than that which the contractor should have followed. On high technology buildings of long construction duration, where that technology is changing fast, it can be particularly tempting for employers and their designers to have the replacement contractor complete or correct the works left by the contractor using a more up-to-date specification of products. This may be fine, providing they also ensure that the effect of that betterment on the costs of completing or correcting the work can be separated. In practice this is often not done, and the result is contention and complex calculation to isolate just the employer's costs in relation to what the contractor should have done but did not.

Obviously, the first, and key, question in relation to any claim for defective or incomplete work will be as to whether the work is actually incomplete or defective. This will largely centre on issues of interpretation of the scope of works for which the contractor has contracted and the specification of that work. There may be issues of law, design, engineering and fact here. In relation to facts, this may involve not just whether the work is incomplete or defective but also why, perhaps with the contractor asserting that it was the result of some act for which the employer was responsible. Such acts might include: the premature taking over of the works; that the cited works were actually the subject of an omission or variation of a specification instruction; or that they were the result of the failure of free-issue materials or plant supplied by the employer. On large complex projects where the specification documents are voluminous and the variations numerous, these issues may not be as easy to resolve as might be hoped.

8.5.3   The Employer Choses to Instruct a Covering Variation

Having established that the works are incomplete or defective, the next issue for the employer may be what it opts to do about this. It may be that it is best served by accepting that work is missing or not to specification, but in return for an adjustment of the contract price to reflect this. This may involve complex considerations for the employer and its advisors as to the effects of the defect or incomplete work on such as the sale or rental value of a building or the production of a process plant.

Under remeasurable contracts such as the FIDIC Red Book and the Infrastructure Conditions, incomplete work should not form part of the remeasured quantities, such that adjustment for incomplete work should be effected automatically. However, under a lump sum contract such as SBC/Q, some means may be required to effect an adjustment of the contract sum. One option is for the employer to have an instruction issued under the contract to omit the incomplete work. This would, for example, lead to an adjustment to the contract sum under clause 4.3.1.1 of SBC/Q. However, it sometimes happens that a contractor's response to this is that the work was already omitted by an instruction received by it. It may seem that the scope and nature of omission instructions should be readily apparent and easy to establish from the records. However, as suggested, in practice the nature of some larger engineering and construction projects, and the extent of instructions, drawings, sketches, queries, clarifications, etc., that are issued on them, can be such that this is not easy to establish.

As more fully explained in Chapter 5 of this book, the valuation of scope omissions is not always straightforward. Under the Infrastructure Conditions, for example, the engineer can order the omission of work from the contract under clause 51, and such an instruction then falls to be valued under clause 52. In most instances, variations of omission will be adjusted by the remeasurement of the works under clause 56 and the provisions of 56(2) will enable any adjustment of the contractor's rate or prices justified by the omission to be made. For instance, if it were decided to omit 80% of the works in laying topsoil, and the contract rate was based on a quantity of 2000 m^3, it might be that a re-rate would be justified for the 400 m^3 remaining after the instruction, due to the price for spreading and grading much smaller quantities and not allowing recovery of fixed costs for the operation in transportation of plant and equipment, etc. The provisions of Infrastructure Conditions clause 52(4) would also allow any required adjustment to preliminary and general items related to the omitted work. Such possibilities may further influence the employer's decision as to whether incomplete works should best be treated as an omission under the contract.

On this subject of accepting works that are not to specification in exchange for a reduction in the contract price, in some parts of the world government contracts for construction and engineering works have recently, and increasingly, tended to include provisions that limit the contractor's choices in relation to materials sources, where the specification does not prescribe a particular product. Such a provision might be worded to exclude some listed geographical locations or limit supply to some others. A typical clause to this end, from the contracts of a Middle East government agency, was quoted in Chapter 5 of this book, but is repeated here for ease of reference:

> Manufactured items, including all parts and components thereof, shall be manufactured and assembled in any of the following geographical locations:
>
> - Western Europe;
> - North America;
> - Australia; or
> - Japan.
>
> Manufactured items, including all parts and components thereof, which are manufactured and/or assembled, either in whole or in part, in geographical locations other than those listed above, shall not be acceptable.

Proponents of such a clause say that such a provision addresses the increasing internationalisation of the construction and engineering markets and concerns regarding the quality of products from some countries. Cynics allege that they are really intended to deny bidding contractors from countries excluded from such a list their competitive edge in sourcing materials from their home country where their governments give export incentives in relation to tax, creating the 'unlevel playing field' in international trade, particularly complained of by US President Trump.

Where contractors source manufactured items from geographical locations outside such a permitted list, the employer and its advisors are likely to have two choices:

- Firstly, they might condemn those works as not in accordance with the contract and require their removal and replacement with materials from a listed location. However, such an approach suffers from practical problems unless the scope of unacceptable manufactured items is limited, or the issue is spotted early. Otherwise, the programme effect of removing existing items and procuring replacements might be significant. Where employers have adopted this approach, contractors have been known to argue that it was a disproportionate reaction to a minor issue and was actually a ruse to create delays for which it could be said to be responsible to hide delays elsewhere to the works as a whole for which the employer was liable.
- Alternatively, the employer and its advisors might accept the 'inferior' products, but with a suitable adjustment to the contract price. The difficult issue here is likely to be establishing what that suitable adjustment is. For example, on an international airport the contractor sourced the following materials from countries outside those covered by the permitted list in the contract: pipe and duct cladding; chilled water pipes; and sanitary ware. There was no allegation that these items did not comply with the requirements of the specification other than to the extent that

they were manufactured in prohibited geographical locations. The issues that arose between the parties in relation to the employer's counterclaim for these works particularly included how the contract price should be adjusted on the basis of a reduction in their specification. The parties respectively argued for two contrasting approaches:

1. The employer claimed its estimate of the saving in cost between the materials as actually purchased by the contractor and those the contractor would have incurred had it sourced them from the permitted locations. This asserted saving was variously between 20% and 30% for the different materials. The problem in proving quantum on this basis was obtaining evidence of both what the contractor actually paid (information only in the contractor's possession) and what it would have paid (which could only ever be hypothetical), particularly some years after the event.

2. The contractor argued that the employer needed to show that the materials as actually purchased by it were of a lesser quality, functionality and value than those it could have obtained from the permitted locations, and to value the reduction in value on the basis of that different quality, functionality and value. The contractor asserted that, just because it had sourced materials from outside of the permitted locations, this did not mean the materials were inferior (and in this regard, even if they were cheaper, this did not show they were of a lower value and by how much). The contractor said that the employer had failed to prove its case in relation to quality, functionality and value and therefore no adjustment should be made to the contract price.

Perhaps employers wishing to incorporate such a list of permitted geographical locations for sourcing manufactured items in their contracts should consider also including a prescribed method of valuing an adjustment to the contract price for items manufactured in excluded locations.

8.5.4   The Employer Requires the Contractor to Complete or Remedy the Work

The avoidance of issues such as those identified in this section in valuing a covering variation instruction or otherwise adjusting for works that are incomplete or defective will be avoided where the employer requires the contractor to correct the defect or complete the incomplete works. If the contractor complies, then the costs involved will be to its account and of no concern to the employer.

However, what if the contractor fails or refuses to address the issue and the employer has to employ others to do so? What is the measure of the employer's financial claim against the contractor? This may principally depend on two issues. Firstly, and obviously, are the terms of the contract which may prescribe what the entitlement is. Secondly, is whether the employer gave the contractor the opportunity to address the issue, as part of any duty to mitigate its loss either expressly under the clause of the contract or as implied by the law. This second point can be particularly significant where the incomplete or defective works were carried out by a subcontractor, because it may be that the contractor can get them corrected at no or minimal cost to itself other than organising and managing the subcontractor. Similarly, if the defect is the result of the failure of a supplier of defective materials or permanent plant, the contractor is likely to have a claim for their replacement and recovery of the contractor's costs (such as that of labour to remove and install) against the supplier.

If the defects or incomplete works are extensive, or comprise an expensive element of the works, then a remedial works contract entered into by the employer with others should be well documented and recorded to allow the pursuit of the costs incurred against the defaulting contractor. Where remedial works are less extensive or are executed on a piecemeal basis, particularly being done by the employer's own staff, such as a maintenance team for the handed-over project, it is important to consider how the costs are to be recorded and how the reasonableness of the costs is to be demonstrated. From a contractor's perspective, for all but the most minor or urgent of works it might generally be expected that there will be evidence of competitive tendering for the remedial works or negotiation of the costs against independent criteria in order to ensure that such costs can be shown to have been subject to reasonable control. From an employer's perspective, it may be that the circumstances are such that the time pressures of getting the work put right were such that short-cuts were taken. This is a particular issue in relation to subcontractors who often complain that costs contracharged or backcharged to them for claims of remedial works have been treated by the contractor as a 'blank cheque' operation. The recording of such costs can be particularly problematical in relation to the work of such subcontractors, where the contractor uses its own equipment, operatives and other resources or those of another subcontractor to correct defects, rather than having the transparency of a discrete price for the remedial work.

As an extreme example of the need for the actions taken by a party in respect of defects to be reasonable and for the costs incurred to reflect such a course of actions, see the judgment in *Ian McGlinn v Waltham Contractors Ltd and Others* [2007] EWHC 149 (TCC). A house that had been built for the claimant had subsequently been demolished and rebuilt in order to achieve the rectification of numerous alleged defects. However, the court held that the alleged defects, which could not be examined as a result of the demolition, could be described as 'ordinary building defects', many of which were explicable on the grounds that the building works were not completed. There were no concerns about the structural integrity that could justify the course of action adopted in demolishing the whole house. The demolition and rebuilding had been an extreme course of action, particularly as many of the defects were aesthetic in nature, and should have been an option of last resort. It involved much more physical work than pure rectification of the alleged defects together with the inevitable danger of increased cost. The claimant's claim for the cost of demolition and rebuilding was contrary to common sense. The reasonable measure of loss in all the circumstances was the costs that would have been incurred on remedial work to the alleged defects.

As illustrated elsewhere by reference to clause 49(4) of the Infrastructure Conditions, many contracts include provisions for the rectification of defects by the employer or client in the event that the contractor does not, or is not given the opportunity to, rectify the defects itself. Such provisions often contemplate that the employer can recover the 'actual cost' of remedial works. However, in *Yorkshire Water Services Ltd v Taylor Woodrow Construction Northern Ltd and Others* [2004] BLR 409, it was held that the term 'actual costs incurred' in the context of rectification of defects under the IChemE Red Book form meant that it was necessary for the employer to have expended those costs, i.e. the costs must be factual and must have been incurred by it. It was further held that the employer could not recover in respect of defects that it had not rectified as the contract used the expression 'has made good' when addressing its entitlement to recover for the cost of rectifying defects. The case particularly emphasises the point made elsewhere regarding the need to consider the express terms of the contract in relation to remedies where there are defective or incomplete works.

A further point made was that the relevance of the conduct of the parties in terms of how they apply the provisions of the contract. In *William Tomkinson & Sons Ltd v The Parochial Church Council of St Michael & Others* [1990] CLJ 319, His Honour Judge Stannard addressed the meaning of Clause 2.5 of the JCT Standard Form of Agreement for Minor Works (1980 Edition):

> Any defects, excessive shrinkages or other faults which appear within three months of the date of the practical completion and are due to materials or workmanship not in accordance with the contract or frost occurring before practical completion shall be made good by the Contractor entirely at his own cost unless the Architect/Supervising Officer shall otherwise instruct.

A preliminary issue was whether this clause afforded the contractor a defence to the church's claim for the cost of remedying defects having regard to the circumstances that those defects were remedied by other contractors on the instructions of the employer prior to the date of practical completion. The case was therefore somewhat unusual in that it was not related to the rights and duties of the parties in relation to defects after completion, but those addressed before that date. The Judge held that:

> ... it does not follow that where workmanship falls short of the standard required by the contract and the employer remedies it prior to practical completion, there is no breach of contract, or the employer is not entitled to recover as damages his outlay in remedying the defective works .... Where [the defects] are not remedied by the contractor within the construction period, there is nothing in the wording of clause 2.5 to suggest that it is intended to exclude the employer's ordinary rights to damages for breach of contract, including the right to recover the cost of remedying defective workmanship.

Regarding the effect of clause 2.5, he considered that:

> Such a provision is generally to be regarded as providing an additional remedy for the employer, and not as releasing the contractor from his ordinary liability to pay damages for defective works.

However, he also held that:

> Clause 2.5 is concerned with the mitigation of loss in that it confers upon the contractor a right to reduce the costs of remedial works by undertaking them himself. Effect is given to this last aspect of the Clause if the damages recoverable by the employer in his outlay of cost in correcting defects in the works are limited to such a sum as represents the costs which the contractor would have incurred if he had been called on to remedy the defects.

Thus, where the contract gives the contractor a right to remedy defects, but the employer does not allow it that opportunity, the employer may be limited to recovery of the costs that the contractor would have incurred in remedying the defects had it been given the opportunity to.

On a similar theme, *Pearce and High v Baxter and Baxter* (1999) EWCA Civ. 789 was a case under the JCT Form for Minor Building Works 1995. Typically of JCT contracts, this gave the contractor the obligation, and the right, to remedy notified defects during a defects liability period. The employer was aware of defects during the defects liability period, but did not inform the contractor of them before pleading them as part of its claim for damages for breach of contract. In the Court of Appeal, Lord Justice Evans put the position as follows:

> The cost of employing a third-party repairer is likely to be higher than the cost to the contractor of doing the work himself would have been. So the right to return in order to repair the defect is valuable to him. The question arises whether, if he is denied that right the employer is entitled to employ another party and to recover the full cost of doing so as damages for the contractor's original breach. In my judgment, the contractor is not liable for the full cost of repairs in those circumstances. The employer cannot recover more than the amount which it would have cost the contractor himself to remedy the defects.

The decision that the employer could not recover its full costs, but only no more than what it would have cost the contractor to remedy the defects itself, was therefore based on the contractor having a contractual right under the terms of the JCT Minor Works Form to return to remedy the defects.

However, what if the contract does not expressly provide the contractor with the right to remedy its defects? Should the contractor still be given the opportunity to remedy the issues itself, as part of the employer's duty to mitigate its loss? In *Maersk Oil UK Limited formerly Kerr-McGee (UK) plc v Dresser-Rand (UK) Ltd* [2007] EWHC 752 (TCC) the contract provided for the position where the contractor did not comply with its obligations to make good defects or work as follows:

> If the contractor fails to do any of the Work or states, or by its actions indicates that it is unable or unwilling to proceed with corrective action in a reasonable time as aforesaid as required by the Company, the Company shall be entitled to have such work carried out by its own personnel or other contractors, without giving prior notice to the Contractor. If such work would have been carried out at the Contractor's own cost the Company shall be entitled to recover from the Contractor the total cost to the company, or may deduct the same from any monies due or which might become due to the Contractor.

His Honour Judge David Wilcox noted how the decision in *Pearce v Baxter* was based on the contractor having an express right to return to remedy the defects, but the contract between Maersk and Dresser-Rand contained no such right. He held that Maersk's failure to notify Dresser-Rand might be relevant to mitigation, but the defendant had led no evidence as to what it would have cost it to remedy the defects or that there would have been a cost saving.

Therefore, even where the contract contains no express right for the contractor to return to correct its defects, the employer may, by not offering such opportunity, fail in its duty to mitigate its loss.

In *Woodlands Oak Limited v Conwell and Another* [2011] EWCA Civ 254, the parties had entered into a very simple contract for works to the respondents' home under which they would pay Woodlands Oak their costs plus 5% margin. Among other issues referred to the County Court had been a claim by the respondents for the costs of rectifying snagging items that they had not given Woodlands Oak the opportunity to rectify and which could have been put right by Woodlands Oak's subcontractors at no cost to either of the parties. The court had held that the respondents had failed to mitigate their loss and awarded them nothing for their costs of engaging others to rectify those snagging items.

The Court of Appeal dismissed the appeal of this decision. It considered that 'where the employer fails to give the contractor an opportunity to rectify defects in the work that may amount to a failure to mitigate the losses'. It emphasised that the mere fact that an employer does not give the contractor an opportunity to rectify defects in the works will not always amount to a failure to mitigate the losses, but held that on the facts of this case the respondents had failed to mitigate their loss. Accordingly, the County Court had been correct to award them none of their costs of rectifying the snagging items.

Returning to a standard form of contract, rather than the very simple form in *Woodlands Oak*, in *Oksana Mul v Hutton Construction Ltd* [2014] EWHC 1979 (TCC), the employer claimant had engaged Hutton to carry out works to a large country house under the terms of the JCT Intermediate Form of Contract (2005). Clause 2.30 of that contract provided as follows:

> Any defects, shrinkages or other faults in the Works or a Section which appear and are notified by the Architect/Contract Administrator to the Contractor not later than 14 days after the expiry of the Rectification Period, and which are due to materials, goods or workmanship not in accordance with this Contract, shall at no cost to the Employer be made good by the Contractor unless the Architect/Contract Administrator with the consent of the Employer shall otherwise instruct. If he does so otherwise instruct, an appropriate deduction shall be made from the Contract Sum in respect of the defects, shrinkages or other faults not made good.

Practical completion was certified but with a substantial list of defects and incomplete works attached. Eleven months later the contract administrator wrote to the contractor listing those items that were still outstanding. The employer sued Hutton for over £1 million in relation to the outstanding items. The preliminary issue before the court was as to what was 'an appropriate deduction' under clause 2.30. Mr Justice Akenhead found that:

> It follows from the above that an 'appropriate deduction' under Clause 2.30 of the Contract means a deduction which is reasonable in all the circumstances and can be calculated by reference to one or more of the following, amongst possibly other factors:
>
> a. The Contract rates/priced schedule of works/Specification; or

      b. The cost to the Contractor of remedying the defect (including the sums to be paid to third party sub-contractors engaged by the Contractor); or

      c. The reasonable cost to the Employer of engaging another contractor to remedy the defect; or

      d. The particular factual circumstances and/or expert evidence relating to each defect and/or the proposed remedial works.

However, the court also suggested that whether the contractor had been given the opportunity to rectify any defects would have a bearing on the amount of the 'appropriate deduction' as follows:

> Of course, there can be no doubt here that, if the Employer acted unreasonably in not giving the Contractor a fair opportunity to put right the defects for which it was culpably responsible, she will probably have failed to mitigate her loss. If there has here been an 'otherwise' instruction under Clause 2.30 at all (and this is clearly contested), I do not see a peculiar difficulty in the CA doing the independent exercise of valuing the 'appropriate deduction' under Clause 2.30 by reference to fairly well known rules about mitigation of damage. The *William Tomkinson, Pearce & High* and *Woodlands Oak* cases point to the Employer being limited to what it would have cost the Contractor to effect the requisite remedial works for defects which it was, unreasonably on the Employer's part, not given the opportunity to put right.

Accordingly, whether the contract expressly requires the contractor to be given the opportunity to remedy defects or incomplete itself, or the law implies a duty on the employer to give it that opportunity, the fact of whether it was or was not may go to the quantification of the employer's damages where it remedies or completes work itself. However, in practice a further factual issue in some cases is whether the contractor was actually given such an opportunity, either at all or in accordance with the requirements of the contract or a test of reasonableness. This is particularly often an issue on highly secure projects that have been taken over, the contract administrator sends lists of defects to the contractor as required by the contract but those occupying the building or facility do not allow the necessary access.

    Finally, mention must be made of the NEC approach to defective work. This directly addresses the different positions in terms of quantifying the employer's claim where the contractor is given the opportunity to rectify defects or is not given such opportunity. In clause 46 of NEC4-ECC this is put as follows:

> 46.1 If the *Contractor* is given access in order to correct a notified Defect but the Defect is not corrected within its *defects correction period*, the *Project Manager* assesses the cost to the *Client* of having the Defect corrected by other people and the Contractor pays this amount. The Scope is treated as having been changed to accept the Defect.
>
> 46.1 If the *Contractor* is not given access in order to correct a notified Defect before the *defects date*, the *Project Manager* assesses the cost to the *Contractor* of correcting the Defect and the *Contractor* pays this amount. The Scope is treated as having been changed to accept the Defect.

8.6   Omitted Work

8.6.1   The Power to Omit Work

A fundamental component of construction and engineering contracts will be the scope of work agreed to be carried out by the contractor for the employer. However, there may be occasions where the employer wishes to omit items from that agreed scope of work. In the absence of a contract provision allowing the omission of parts of the works, the parties would be under an obligation to perform the full scope of works as contracted, with the contractor liable to carry it out and the employer under a duty to allow the contractor to do so and pay for it. As this was put by His Honour Judge Humphrey Lloyd in *Abbey Development Ltd v PP Brickwork Ltd* [2003] Adj.L.R. 07/14:

> A contract for the execution of work confers on the contractor not only the duty to carry out the work but the corresponding right to be able to complete the work which it contracted to carry out. To take away or to vary the work is an intrusion into and an infringement of that right and is a breach of contract.

Accordingly, any change in the scope of work would otherwise have to be subject to a mutually agreed amendment to the contract and therefore construction and engineering contracts will usually expressly empower the employer, through its representative, to omit work, just as they contain powers for it to add or change work by instructing a variation.

There are legitimate reasons for having the power to omit parts of the works, which is sometimes referred to as 'descoping'. From a design perspective, it may be that other work is to be substituted, or further design work has rendered some elements unnecessary, or the employer's requirements have changed since the contract was commenced. This latter cause of omissions is particularly relevant on projects of long duration as technology, legislation or other requirements change and give rise to the type of add and omit variation considered in Chapter 5 of this book. From a financial perspective, the need to omit work might arise where a client finds that it is having difficulty funding the project. This may be because it has run into wider financial difficulties unrelated to the project. However, it might also be because the cost of the project has increased significantly since the commencement of the contract or its value significantly reduces by reason of changes in the general economic climate. For example, following the global economic downturn of 2008 and 2009 many construction projects saw their employers pare down the scope of works in order to save costs as the value of those projects was seen to be diminishing, along with the return that the employer stood to achieve. Alternatively, where projects are in delay, it may be that the employer hopes to offset the effects of such delay on the completion date by reducing the scope of work to enable earlier completion. This is particularly the case where the employer has caused compensatory delays. On a less satisfactory note, on some occasions it may be that the employer's relations with the contractor have become strained or even acrimonious due to disagreements about matters relating to the contract. This might be because the contractor is making spurious claims, or the employer doubts the contractor's financial or technical ability to complete the works to time or quality, or because of the contractor's poor performance. In such cases descoping is sometimes applied rather than the more difficult and contentious recourse of its terminating the contract.

As a result, most standard and bespoke forms of engineering and construction contract allow the architect/contract administrator or engineer to issue variations that include the omission of work. In the Infrastructure Conditions this is in clause 51, which states that '... variations may include additions, omissions, substitutions, alternations .... Similarly, in SBC/Q clause 5.1 the definition of variation includes at subclause 5.1.1 'the addition, omission or substitution of any work'. FIDIC Red Book clause 13.1(d) contains an important qualification as to what work can be omitted, where it states that a variation may include 'omission of any work unless it is to be carried out by others'. This qualification in relation to omitting work so that it can be carried out by others is considered in more detail elsewhere in this chapter.

Otherwise, the Infrastructure Conditions, SBC/Q and the FIDIC Red Book do not expressly limit the power to instruct variations omitting works. This can be contrasted with the Government of Hong Kong General Conditions of Contract for Civil Engineering Works, 1999 Edition. Clause 60(1) of those conditions limits the power of the engineer to order variations to those that it considers '... Desirable for or to achieve the satisfactory completion and functioning of the Works'. It is suggested that this would preclude instructing omissions in order to save the employer money or time against the completion date.

There may also be limits on the extent of works that can be omitted from a contract without financial penalty for the employer, either through the express provisions of the contract or local governing law of the contract. In this regard, His Honour Judge Humphrey Lloyd's judgment in *Abbey Development v PP Brickwork* also included this:

> Provisions entitling an owner to vary the works have therefore to be construed carefully so as not to deprive the contractor of its contractual right to the opportunity to complete the works and realise such profit as may then be made.

That judgment was extensively quoted in the Hong Kong case of *Ipson Renovation Limited v The Incorporated Owners of Connie Towers* [2016] HKCFI 2117. That case arose out of a contract for repair and maintenance work to a building known as 'Connie Towers'. The employer omitted substantial parts of the contractor's scope of works, without any compensation for these substantial omissions. Whilst the contract allowed the employer to instruct the omission of works, the court held that clear and express language would be required to allow the omission of substantial parts of the works, and that this contract was not in such terms. The court held that, combined with the employer's suspension of the works, its actions of omitting work to such an extent amounted to a repudiation of the contract, which the contractor had rightly accepted.

8.6.2 How to Value Omissions

When work is omitted under a variation instruction, then the omission is generally valued, along with other forms of variations, at the rates and prices in the contract, and that part of the contract sum represented by the omitted work is consequently omitted from the sum to be paid to the contractor. In the FIDIC Red Book this is apparent in clause 12.3, 'Evaluation', which provides that 'for each item of work, the appropriate rate or price for the item shall be the rate or price specified for such item in the Contract...'. In SBC/Q clause 5.6.2 this is put as 'To the extent that a Variation relates to the omission of work set out in the Contract Bills … the rates and prices for such work therein set out shall determine the valuation of the work omitted'.

It seems obvious that omission of work must be valued on the basis of the amount included in the contract price for that work, if one considers that if this approach is not applied then the omission of the whole scope of work under the contract would not be valued by omission of the whole of the contract price. In other words, omission of all of the works would not lead to an adjusted contract price of nil. Whilst this may seem trite to traditionalists (though see regarding the NEC approach elsewhere), reference is made to the case *MT Hojgaard A/S v E.ON Climate Renewables UK Robin Rigg East Ltd* [2104] EWCA Civ. 710, considered in detail in Chapter 5 of this book. There, E.ON submitted that a variation omitting work should be valued on the basis of the amount of time, and hence costs, that Hojgaard would have spent had it continued with those omitted works. The Court of Appeal disagreed with E.ON and agreed with the decision of the Judge at first instance, that the omission should be based on the amount allowed for that work in the contact.

However, many omissions of work are not instructed in such a manner and timing that it is equitable that the whole of the amount allowed in the contract for that work should be omitted. This is often a question of timing of the instruction. It may be that part of the physical work was completed before the instruction. This should be relatively easy to measure and the omission adjusted to only omit those quantities not carried out. However, this can sometimes cause problems where the employer considers that the contractor carried out more work than it should have done, in the knowledge that the work was to be omitted. At a finer level, there may be preparatory work, at the least, such as design or ordering of materials, that has been carried out on the omitted work before the instruction to omit it was given. This is recognised by clause 12.4 of the FIDIC Red Book, which provides:

> Whenever the omission of any work forms part (or all) of a Variation, the value of which has not been agreed, if:
>
> (a) the Contractor will incur (or has incurred) cost which, if the work had not been omitted, would have been deemed to be covered by a sum forming part of the Accepted Contract Amount;
> (b) the omission of the work will result (or has resulted) in this sum not forming part of the Contract Price; and
> (c) this cost is not deemed to be included in the evaluation of any substituted work;
>
> then the Contractor shall give notice to the Engineer accordingly, with supporting particulars. Upon receiving this notice, the Engineer shall proceed in accordance with Sub-Clause 3.5 [Determinations] to agree or determine this cost, which shall be included in the Contract Price.

On this basis, the costs of activities, such as design and procurement of materials and subcontracts carried out before the instruction, should form part of the valuation of the omission offset against the omission of the amount included in the contract. The materials component of this may involve re-stocking charges or even the complete write-off of specialist items that cannot be returned to a supplier or fabricator or are of scrap value only. The fact that such items might be of long delivery will increase the chances of them being incurred on omitted work, again depending on timing. The subcontractor element may include claims from those subcontractors for their loss of contribution on works for which they had a subcontract but are now being denied. Abortive design costs are a perennial evaluation problem, where the design work is carried out by the contractor's own design staff, rather than an external consultancy, and those staff do not keep daily diaries of their activities and time spent on each of their activities in sufficient detail.

Omissions of significant parts of the works may also have implications for other work valuation provisions of the contract. A common complaint of contractors, particularly when their preliminaries and general items are not priced in their own section of the bills of quantities but in the rates for measured works, is that substantial omissions from the scope of works have led to an under-recovery of such preliminaries and general items. In this regard, FIDIC contracts have traditionally placed a limit on the scope of variations that can be instructed before the variation valuation rules require some adjustment to amounts due to the contractor. In the old 1987 edition of the FIDIC Red Book this was at clause 52.3 and was triggered where the value of all varied works taken together involved an increase or decrease of the contract price exceeding 15%. At that point the contract price was to be further adjusted '... having regard to the Contractor's Site and general overhead costs of the Contract'. Whilst this was in the old version of FIDIC, many contracts around the world, particularly government contracts, are still based on the 1987 edition and contain this same provision and threshold percentage. In the current FIDIC Red Book the trigger threshold is arguably more obscurely stated in clause 12.3(a), although the result is a rather broader requirement for a 'new rate' for an item where:

(i) the measured quantity of the item is changed by more than 10% from the quantity of this item in the Bill of Quantities or other Schedule,
(ii) this change in quantity multiplied by such specified rate for this item exceeds 0.01% of the Accepted Contract Amount,
(iii) this change in quantity directly changes the Cost per unit quantity of this item by more than 1%, and
(iv) this item is not specified in the Contract as a 'fixed rate item'.

In SBC/Q the contractor may have a claim for adjustment of its preliminaries and general items by the operation of clause 5.6.3.3, which requires that in any valuation of a variation 'allowance, where appropriate, shall be made for any addition or reduction of preliminary items of the type referred to in the Standard Method of Measurement ...'.

Of course, the converse of this is that from an employer's perspective, especially where preliminaries and general items have their own section in the bills of quantities, consideration must be given to omitting parts of those allowances as well as the measured work items directly relating to the omitted work.

As explained in Chapter 5, the approach of the NEC contracts to valuing omissions is very different. Under NEC4-ECC, clause 63.1 does not distinguish between compensation events that involve omissions from those that involve additions. The valuation is based on the effect on the actual of forecast Defined Cost. Since omitted works will not have been carried out, their valuation can only be based on an estimate of what the cost would have been.

8.6.3 Giving Omitted Work to Others

Whilst engineering and construction contracts do generally provide for work to be omitted, and for the valuation of that omission as a variation, the general position is that employer cannot, in the absence of some express power agreed in the contract, decide to omit work from the contract in order to give that work to others. The employer might be tempted to do this because another contractor can carry out those works cheaper or at a faster rate. If the employer does this, then it will be in breach of contract and liable to a claim for damages from the contractor.

There is also a practical implication here, given how contractors are usually entitled to undisturbed possession of the site, unless the contract expressly provides otherwise. It is for this reason that contracts such as the FIDIC Red Book state at clause 2.1 that the right of access and possession of the site 'shall not be exclusive to the Contractor' and clause 2.3 refers to 'the Employer's other contractors on the site'. Similarly, SBC/Q clause 2.7 refers to '… work not forming part of this Contract which the Employer requires to be carried out by the Employer himself or by any Employer's Persons'. It is on the basis of such provisions and the definition of variations as including omissions that employers sometimes, disingenuously, argue that they can omit works to give to other contractors. The employer's argument goes that 'the contract allows me to omit works from the contractor's scope and to employ others on the site at the same time as the contractor, therefore I can omit its work and give it to another'. The driver for this is sometimes that another contractor can do the work for a lower price. It more commonly occurs in relation to subcontractors, where there are already a large number working on the project at the same time. Alternatively, it may occur on projects let on a works package basis to a number of contractors under separate contracts with the employer. Particular examples of this are works at major international airports such as those in Hong Kong, Dubai and Heathrow.

In practice, it does occur occasionally that works packages or subcontracts are inadvertently let with overlapping work scopes such that two package contractors or two subcontractors have contracted to do some of the same work. It is suggested that this error would not excuse the employer or contractor for omitting work from one package contractor or subcontractor respectively, without being in breach of contract in allowing it to another package contractor or subcontractor, unless an unusual term of the contract allowed this.

The provision of the FIDIC Red Book's clause 13.1(d), that a variation may include 'omission of any work *unless it is to be carried out by others*' has been identified elsewhere. Even without such an express limit, if an employer omits works contracted for in order to give it to others, then it is likely to be in breach of contract and may be pursued by the contractor for damages, including matters such as reimbursement of profit and overhead contribution that would have been earned from the omitted work. The contractor would of course have to prove its case, both that the work was omitted for that motive and the amount of its loss, and neither of these are always easy to establish, as is discussed further elsewhere in this chapter.

The principle that the power to omit work under a contract is limited in the way suggested was adopted by the High Court in London in the case of *Amec Building Contracts Ltd v Cadmus Investments Co Ltd* (1996) 51 Con LR 105, which followed an earlier decision of the Australian High Court in the case of *Carr v J.A. Berriman Pty Ltd* (1953) 27 ALJR 273 in which the Judge said:

> The clause is a common and useful clause, the obvious power of which – so far as it is relevant to the present case – is to enable the architect to direct additions to, or substitutions in, or omissions from, the building as planned, which may turn out, in his opinion, to be desirable in the course of the performance of the contract. The words quoted from it would authorise the architect … to direct that particular items of work included in the plans and specifications shall not be carried out. But they do not, in my opinion, authorise him to say that particular items so included shall be carried out not by the builder with whom the contract is made but by some other builder or contractor. The words used do not, in their natural meaning, extend so far, and a power in the architect to handover at will any part of the contract to another contractor will be a most unreasonable power, which very clear words would be required to confer.

Establishing that work has been omitted in order to give it to others may involve issues of fact and law that might be difficult to establish. It has been suggested elsewhere how, on a traditionally procured project, the contractor has possession of the site, such that it would not be easy for the employer to bring in another contractor whilst the contractor still occupied the site. In any event, the most usual motives for omitting work from a contractor would dictate that it will be deferred to be carried out later. Take, for example, an employer's omission of a multistory car park to be attached to a high-rise residential tower because of an economic downturn. The employer might envisage adding the car park later when the economy picks up, or plan to add it after completion of the tower, when funds are available from the first sale of residences in the tower. If this occurs after the Final Certificate (to use the SBC/Q term) does the contractor still have a claim against the employer for breach when it discovers that another contractor has been appointed to construct the car park some time later?

Establishing the damages arising from the omission will usually involve loss of contribution to overheads (potentially both site and head office) and profit. If the test is what contribution the contractor would have earned on the omitted works, this can prove difficult to establish in practice. It is suggested that this is not the contractor's allowances for these items within its pricing of the omitted works in the contract price, unless they can be shown to have been achievable. Those allowances, made at tender stage, may or may not have been realistic and attainable. It may be that the contractor has some available record of the costs it would have incurred on the omitted works, such as materials orders and subcontracts let before the omission. However, this all depends on timing and is most unlikely to give a complete picture, especially in relation to such items as labour, plant and equipment costs. The obvious available measure is the actual contributions that the contractor achieved on other works that were not omitted. However, there are two issues with this, as follows:

- Whether the other works were sufficiently similar to those omitted? It may be that the contractor's cost/value records are such that only its financial performance on the works as a whole can be established. However, can it be assumed that the same contributions would have been achieved on the omitted works?
- It is the nature of many problem construction projects that issues of wrongful omission of works are often accompanied by other issues on the project (such as delay

and disruption). This will mean that the contractor will claim that its actual contributions on work that was not omitted are not representative of what it should have achieved on the omitted work. The result of this inter-relationship of claims by the contractor may be that the evaluation of the contractor's damages for breach in relation to the omitted work has to also be related to its claims for delay and disruption, ensuring that they do not overlap, but also ensuring that no claimable damages fall as unclaimed between the different claims.

A further complication in relation to calculating the loss of overhead contributions on works that have been omitted in breach of contract, is to establish what if any costs of such overheads the contractor would have incurred. This can be very difficult to establish with much certainty, but such saving would have to be offset against the loss of contribution.

8.7 Post-Handover Costs

This is a particularly idiosyncratic head of claim which is very rare in the UK but appears quite regularly in parts of the world where employers and their advisors take a more 'flexible' approach to what constitutes completion of the works and its certification and what can be required of the contractor by way of variations and additional works thereafter. This is particularly where there has been critical compensable delay to completion. To understand the basis of such a claim, it is necessary to understand the typical provisions of forms of construction and engineering contract in relation to what constitutes completion and what power the employer has to require varied works thereafter. Because this head occurs far more internationally that in the UK, the following discussion is based only on a consideration of the FIDIC Red Book as an example, although the same issues could arise under any form that is in materially similar terms and they are abused in the manner explained as follows.

Under FIDIC Red Book clause 10.1, the employer shall take over the works (or sections thereof) when they have been completed in accordance with the contract, except for any minor outstanding works and defects that would not substantially affect their intended use, and the engineer has issued a taking-over certificate (subject also to a deeming provision in that clause). Under clause 13.1 the engineer may instruct variations at any time up to the issuing of such a taking-over certificate. The result of these provisions should be that once the employer has taken over the works or a section for its occupation and use, the contractor should be left only to carry out minor outstanding works and defects therein and should not be required to carry out any further variations thereto. However, in practice actions adopted by some employers and engineers can prevent that result. These may include the following or a combination of the following:

- Premature taking-over of the works or parts or sections thereof.
- The imposition of unreasonable restrictions on the contractor's access.
- Back-dating of a taking-over certificate.
- Issuing instructions for employer-required additional works of changes after its taking over.

In terms of the timing of a taking-over certificate, FIDIC Red Book clause 10.1 should in theory provide the contractor with the machinery to ensure that it obtains this in a timely manner. It provides for the contractor to issue a notice within 14 days of a date when it considers that the works or a section will be completed and ready. Within 28 days of that notice, the engineer is obliged to either issue the taking-over certificate or reject the application, giving its reasons.

Particularly where an employer is responsible for compensatory delays, it is not unknown for a taking-over certificate to be issued at a date that is premature when an objective consideration is made of the actual stage of progress of the works. One apparent motive for this seems to be where the engineer believes that setting an early actual completion date will shorten the contractor's period of compensable delay. Where it is also the case that its justifiable claims for extension of time are being denied, it may be of some relief to the contractor that its apparent liability for delay damages is being stopped. However, the negative effect for the contractor may be that extensive lists of so-called 'snagging' items will have to be addressed in a project that is now in the possession of the employer. If, for example, this is a highway that has now been opened to traffic or a building that has now been occupied by a tenant, the cost implications may be significant.

A further problem with this occurs where the employer takes possession of the works or a section (and again the contactor is likely to be more than happy if that relieves it of the risk of delay damages) but a taking-over certificate is not issued. In practice, the engineer and contractor might then run into a lengthy disagreement as to whether the works or section were completed as required by clause 10.1. This can continue for months, or even become the subject of a dispute. However, the claim being considered here will arise if, as sometimes happens, the engineer eventually issues a back-dated certificate that taking-over was achieved a significant period earlier, often on the date on which the employer took possession.

This scenario might not be how the FIDIC Red Book and similar provisions expect completion and its certification to be handled, but it does happen. In some such cases the contractor has suspected that the engineer's motives were twofold. Firstly, not issuing the taking-over certificate in a timely manner would maintain pressure on the contractor to complete any minor outstanding works and defects rapidly under the pressure of delay damages. Secondly, the employer believed that back-dating that certificate would shorten any claim from the contractor for the financial consequences of delays to completion.

It is in such circumstances that contractors occasionally make claims for prolongation costs for a period after completion of the works, in the form of a post-handover delay claim. The contents of such a claim, and how it arises particularly from a back-dated taking-over certificate, can be considered by reference to those of the contractor's obligation that cease under construction and engineering contracts when the employer takes over the works or a section. An example of this is given in FIDIC Red Book clause 4.8 in relation to providing fencing, lighting, guarding and watching and in clause 17.2 by taking full responsibility for the care of the works and goods. Under such circumstances, the contractor may argue that because of the lack of a taking-over certificate, it maintained its site preliminaries and general requirements at levels commensurate with a construction period, not a post-completion defects period, and it will then claim for those greater resources.

As noted elsewhere, the shortening of a contractor's potential delay claim may be attractive where the employer has caused critical delay and knows that the contractor is entitled to an extension of time and associated costs. It is explained elsewhere in this book that these are not inextricably linked, but contracts such as given in the FIDIC Red Book address them together and in practice they do tend to be linked by both parties. If those delays were the results of failures by the engineer in executing its' functions under the contract, then it may also be motivated to reduce any potential consequential relationship damage or even a claim from the employer against it. An adverse practical effect for the contractor will be that works that should have been carried out while it had sole possession and control of the site will now be carried out under the employer's occupation and control. They may then be subject to all sorts of restrictions on the contractor's access to areas of the site to deliver and store materials and for its people to carry out, supervise and manage the work.

Difficulties in the evaluation of such post-handover claims will particularly need to show which of the contractor's resources were additional. In a defects liability period a contractor will normally be required to maintain on-site its staff to address such issues as those defects plus commercial staff to process its final account and those of its' subcontractors and suppliers. The evaluation question will be which of the contractor's actual resources in the period were extra to those normal requirements and were required as a result of the failures upon which the claim is based? For staff, this usually involves comparison of the schedule and tasks of actual staff with a similar list of those that would have been required in any event. However, the second part of this equation can be difficult to establish. Regarding the costs of items such as site establishment, offices and related facilities, the evaluation of what is additional can be even more difficult.

Such claims also often involve a delay analysis to justify the actual period over which the contractor maintained its additional resources after handover. The end date of that period may be the date on which the engineer issued its back-dated taking-over certificate or some other date on which the contractor was able to reduce resources to normal post-handover levels. It may also involve consideration of the impact on that period of additional works that the contractor has been required to carry out in the period.

In addition to the back-dating of the taking-over certificate, contractors also sometimes make claims in relation to the number of variations instructed in a post-taking-over period. If instructed to carry out variations of a post-handover, a contractor should consider whether it is contractually obliged to comply with that instruction. If, as is usually the case, it is not, then there should be a discussion with the employer regarding reaching a separate agreement in relation to those works and a proper price for them that fully takes into account their circumstances. As explained already, provisions such as FIDIC Red Book clause 13.1 allow the engineer to instruct variations at any time up to the issuing of the taking-over certificate. However, what if that certificate is not issued or is issued post-dated? In its absence the contractor is obliged to carry out instructed variations. If the employer is in possession of the works, then it may be that its users and occupiers come up with all kinds of amendments, modifications and additions and get the engineer to instruct these under the contract. The effect of this can be that the contractor finds itself still bound as responsible for the works and addressing a continuing and developing scope of works, when it should be just addressing defects and minor works that were incomplete at taking-over. This may mean that it maintains its preliminaries and general resources at high levels, certainly more than would have been required during a normal defects period.

The valuation of variations carried out post-handover is considered in Chapter 5 of this book. The premium in costs for contractors carrying out additional works in such a period should not be underestimated. For example, in the defects period of a project for construction of a new airport terminal a subcontractor was appointed to carry out works to installed low voltage switchboards and priced its preliminaries and general costs at 191% of the direct costs of the required work because the work value was spread across many rooms of an occupied airport with very difficult logistics in terms of notice, clearance and access for people, equipment and materials. An approach that is sometimes adopted is that contract documents include two sections for the pricing of Dayworks, the second to be the basis of valuing additional works in a defects period, where the contract does allow them to be required. This is more common in relation to subcontracts, given that some trades will finish well before the employer takes possession of the project. In the absence of this, proper application of, for example, FIDIC Red Book clause 12, with reference to its subclause 12.3(b)(iii), 'the item of work is not of similar character or is not executed under similar conditions, as any item in the Contract', should give a proper valuation, provided the engineer properly applies it and recognises the differences and how significant they can be to valuation.

The legal basis of claims for post-handover costs can be difficult, especially where the contractor has not taken a stance that fully exercised its rights under the contract. For example, by refusing to allow the employer to take possession in the absence of a taking-over certificate or by refusing to carry out additional works in a period when such a certificate had been issued, or ought to have been. However, practice shows that many contractors find it harder to take such stands than hindsight would require. They will say that they should not be penalised for taking a cooperative approach and did not stand on their full contractual rights, thereby benefiting the employer in giving it early possession and desired modifications and additions to the works, against a background where its engineer had failed to properly administer the contract.

8.8   The Costs of Preparing a Claim

It is not uncommon for financial claims for delay and disruption under construction and engineering contracts to include a heading for the costs of preparing that claim and/or other related submissions that preceded it. This might extend to preparation of notices and all related substantiation and submissions that followed, or just the preparation of a particular claim document. In extreme cases it might involve several iterations of claims on the same issues over several years. A claim for their preparation costs will often feature at the end of the contractor's, subcontractor's or supplier's submission, just before the addition of interest or finance charges. It may set out amounts paid to external consultants for the preparation of the claim(s) and also the time costs and expenses of directly employed staff to the extent that they have been involved and are not covered by other heads of claim such as prolongation of preliminaries or staff thickening. The staff component of such a claim will therefore usually be for non-project staff such as head office personnel diverted from their duties elsewhere in the business.

Whilst such a head of claim is not rare, it is rare that its contractual basis and the entitlement to it are considered at any more than a superficial level. Many claimants seem to include it either on an assumption of entitlement or as a 'give away' item for negotiation. Employers and contract administrators often dismiss it out of hand on the assumption that there can be no legal entitlement to it.

The conventional English law view as to whether a contractor is entitled to reimbursement for any costs incurred in preparing its claim was that they are not, except in specific circumstances. This was primarily based on the view that, when preparing such submissions, the contractor is merely complying with its obligations under the contract. The thinking here is that since clauses such as FIDIC Red Book clause 20.1 or the Infrastructure Conditions clause 53 or JCT clause 26 required the contractor to give notice and particulars of claims for such as extensions of time or 'loss and/or expense' or additional costs, they are required to maintain site staff to carry out such obligations. This approach is reflected by the SCL Protocol at paragraph 3.1 of its Guidance Part C.

Alternatively, many contracts do not envisage the contractor preparing claim submissions, as they simply require the contractor to give notice or make an application and then keep the information requested under the contract and send that to the contract administrator to make an ascertainment. Claim preparation costs, including external consultant costs, were therefore conventionally considered as not recoverable.

In essence, the response to such a claim was either that the contractor was only complying with its obligation to prepare the claim, or if no such obligation existed, then there was no need for the contractor to have incurred the claimed costs incurred in doing so.

However, notwithstanding this conventional wisdom, it seems that there are three alternative bases on which claims for the costs of preparing a claim are made. These are: as a head of a contractual claim; as damages for breach of contract; or as recoverable costs of formal dispute resolution proceedings.

8.8.1   A Claim Under the Contract

In *Walter Lilly & Company Limited v Giles Patrick Cyril Mackay and DMW Developments Limited* [2012] EWHC 1773 (TCC), the court considered a claim for 'Claim Preparation Costs', under clause 26 of the JCT Standard Form of Building Contract 1998 Edition Private Without Quantities. At paragraph 590 Mr Justice Akenhead stated the following in relation to the fees of a consultant employed by the contractor:

> The claim is for some £43,000 for Mr Parnham's [who elsewhere was claimed under the heading of 'professional fees'] time in preparing claims from time to time albeit that Mr Hunter only supports some £40,000 of this. Whilst in principle I do consider this could be a valid head of a loss and expense claim under Clause 26 [of the JCT form of contract] ....

The Judge therefore considered that the costs of preparing a claim might be admissible under clause 26 of the JCT form of contract. Furthermore, it is suggested that there does not appear to be any particular wording in that clause that would restrict the principle established in the judgment from applying to similar standard forms.

However, the Judge also observed that 'it is very difficult to unravel precisely what Mr Parnham actually did' and concluded that 'I am not satisfied that any additional sum has been proved over and above the (albeit conservative) allowance which I have already made in relation to Mr Parnham'. Those allowances were both under the heading of 'Thickening', where amounts were allowed against Mr Parnham's activities as follows:

> … bringing in Mr Parnham to assist in the commercial management and extension of time applications ….

> … the continued use of Mr Parnham to assist in the commercial management and extension of time applications ….

> … and some time and resources were applied by him in preparing extension of time claims in respect of matters upon which as a matter of liability WLC has succeeded.

Therefore Mr Justice Akenhead was also prepared to allow the costs of a consultant engaged 'to assist in the commercial management and extension of time applications' where the Claimant had succeeded in their liability argument in the circumstances of this case. It would therefore appear, applying this England and Wales judgment, that there is a route to frame a claim for preparation costs in this manner where there are good grounds for preparing the claims.

The activities of Mr Parnham against which amounts were allowed are sometimes described under the heading of 'preliminaries thickening', a head of claim considered in detail in Chapter 6 of this book. The focus of such claims is often on the addition of technical management, supervision and engineering staff to deal with works that are the subject of disruption or acceleration. However, there seems no reason why such claims should not equally include additional staff brought in to address the contractual and financial effects of such events, including to prepare necessary claims for recovery of time and/or money. As set out in relation to preliminaries thickening claims in Chapter 6, the key to such a claim includes establishing as a matter of factual evidence that such staff are additional, that they resulted from the events subject of the claim and would not have been required anyway, either to administer the project as originally planned or to address the results of the contractor's own failures. Such evidence would address the traditional response to such a claim that the staff preparing contractual claims are only carrying out what the claimant had to do under the contract. If doing those things (such as preparing notices or claims) involved the addition of staff specifically because of compensable failures of the other contracting party and their employment would not have been otherwise necessary, then they may be a claimable head of loss arising from the primary causes of the claim.

8.8.2 Breach of Contract

As has been explained, the conventional English law view is that in preparing claims submissions the contractor is just complying with existing obligations or doing something it was not obliged to do. However, one consideration may be whether this ought to depend on the employer or its advisers doing what they are required to do under the contract. This is ensuring that the contractor's entitlements were properly considered and recognised. In other words, the contractor is only obliged to do that which is necessary if the employer and its team similarly do what is properly required of them.

Otherwise, the contractor may claim that its claims preparation costs were incurred not in doing what the contract envisaged, but because it was forced to do more that comply with the contract, preparing something that should never properly have been required had the employer properly conducted itself in accordance with its obligations under the contract.

The potential for a claim on this basis is apparent from the Building Law Reports commentary in *James Longley & Co. Ltd v South West Regional Health Authority* 1983 25 BLR 56. In that case Mr Justice Lloyd held that a claims consultant's fees in preparing a final account were not allowable. However, at page 57 the editors said this: 'The costs of preparing a final account may be recovered as damages in a suitable case, e.g., for breach of an obligation on the part of the employer to provide a final account'. In addition to a final account, it is easy to identify a number of other actions that are typically required by employers or their contract administrators under engineering and construction contracts where failure in relation to those actions is likely to put the contractor to the costs of activities that should not otherwise have been required of it. Obvious examples are the preparation of interim payment certificates, formal confirmation of variations, granting of extensions of time and the ascertainment of financial claims. Against each of these the contractor will usually be required to prepare and submit something, for example an application for payment, confirmation of a verbal instruction, notice of delay or substantiation of its costs. The traditional approach has been that the contractor's activities in this regard are part of its contractual obligations and hence their costs are not recoverable. However, what if the contractor complies with those requirements, but the employer does not respond properly such that it is in breach of contract and the contractor has to incur additional costs making further and unnecessary submissions?

As described in the SCL Protocol at paragraph 3.1 of its Guidance Part C:

> Most construction contracts provide that the Contractor may only recover the cost, loss and/or expense it has actually incurred and that this be demonstrated or proved by documentary evidence. The Contractor should not be entitled to additional costs for the preparation of that information, unless it can show that it has been put to additional cost as a result of the unreasonable actions or inactions of the CA in dealing with the Contractor's claim.

This gives rise to the prospect of an alternative basis for claiming the costs of preparing claim submissions, where those entitlements have not been addressed properly. This approach relies on a secondary breach by the employer. The primary breach would be the events that gave rise to the claim for such as delay or costs in the first place. Thereafter, it may be that the employer's response, or that of the architect, engineer or contract administrator on its behalf, is such as to place the employer in a secondary breach of its contractual obligations in relation to the related claim. The contractor's damage resulting from that secondary breach could be the costs of preparing claims or revised or additional submissions that it should not otherwise have had to prepare, including the engagement of specialist staff or external consultants, whose engagement should equally not otherwise have been necessary.

It is a feature of standard forms of construction and engineering contracts that, whilst the parties are the employer and contractor, a separate administrator is appointed to carry out certain functions under the contract. Under FIDIC and ICE contracts this is 'the Engineer'. Under JCT contracts it is 'the Architect/Contract Administrator' and 'the Quantity Surveyor'. Under NEC contracts it is 'the Project Manager'. These may administer claims for such as extensions of time, 'loss and/or expense' or 'compensation events' under those contracts. Important principles in relation to such activities are whether they are doing so as an independent certifier or as the agent of the employer under the contract and the degree to which such a certifier is required to act impartially in its function. If the latter, then it may be that their failure to act impartially may place the employer in breach of contract.

These principles will vary between jurisdictions and the applicable law may have something to say about the duties of an engineer, for example, under an FIDIC-based contract. However, express clauses such as FIDIC Red Book 3.1(a) may expressly deem that in carrying out its duties or exercising authority under the contract, an engineer under such contracts is acting for the employer. On the other hand, under NEC-type terms, Mr Justice Jackson held in *Costain Ltd & Others v Bechtel Ltd & Another* [2005] EWHC 1018 (TCC) that, when assessing sums properly due to the contractor, a project manager under those terms, whilst required to act impartially under those terms, was not doing so as an agent of the employer.

Under JCT terms, the case of *Croudace v London Borough of Lambeth* (1986) 33 BLR 20 confirmed that the Architect's failure to ascertain, or to instruct the quantity surveyor to ascertain, that the loss and expense was a breach of contract for which the Employer was liable in damages. The Judge said:

> In my judgment Lambeth's acts and omissions after 2 February 1983, including but not limited to their failure to appoint a successor to Mr Jacoby, amounted to a failure by them to take such steps as were necessary to enable Croudace's claim for loss and expense to be ascertained, and as such amounted to a breach of contract on their part: cf Smith v Howden Union & Anor (1890) 2 Hudson's BC (4th Ed) 156.
>
> The real issues in this case are whether Croudace can establish that it has suffered any damage resulting from that breach of contract and, if so, whether the amount of that damage should be determined by the court or by arbitration.
>
> Unless it can be successfully maintained by Lambeth that there are no matters in respect of which Croudace are entitled to claim for loss and expense under Conditions 11(6) and 24(1)(e), it necessarily follows that Croudace must have suffered some damage as a result of there being no one to ascertain the amount of their claim.

A further defence to the recovery of the costs of preparing a claim is sometimes that, although the contractor did have a contractual obligation to prepare a claim and prepared and submitted one, the submission was not adequate. Thus, whereas the contractor asserts that the submission subject of the claim for preparation costs was only necessary because of failure by the employer or its contract administrator to properly carry out its function in relation to earlier submissions, the response is that the contractor's own earlier submissions were deficient and that this prevented them properly considering those claims. The habit of some contractor of submitting exaggerated claims for time and/or money in an apparent attempt to go in with a high claim and a view to settling at a proportion thereof, does not help with this in many cases.

There is no doubt that the quality of claims submissions from contractors, and particularly subcontractors and suppliers, can vary widely and enormously around the world, as can the quality of the consultants advising them. Furthermore, it is not uncommon for the same claim to be subject to numerous submissions. Hopefully each of these should be in greater detail than its predecessor, but why is more detail required? Was the preceding submission deficient or did the contract administrator respond unfairly to it? In practice, it is common that claims in relation to complex issues such as the method of delay analysis or quantification of disruption costs change significantly between submissions. For example, early delay claims might have been based on an 'impacted-as-planned' approach, notwithstanding that the planned programme was never approved or achievable given the contractor's own culpable delays. Alternatively, early disruption claims might have been based on a 'global total loss' approach with no linking of cause and effect, notwithstanding that the contractor is known to have underpriced its tender and/or to have been culpably inefficient. If these claims are rejected and the contractor has to employ additional or more senior staff or external consultants to submit further claims with proper particulars, it may have little basis for claiming the costs of those earlier bad submissions or for claiming to have been unfairly forced to engage the new resources on the new submissions. In extreme circumstances, employers have been known to consider counterclaims for resulting wasted time of their staff and their consultants' costs in dealing with poor and/or wholly unmerited claims from their contractors that prevented the contract administrator reaching a fair determination under the contract and required additional work for those consultants in attempting to some assessment of the contractor's reasonable entitlement or dealing with a new claim. As SCL Protocol continues in paragraph 3.1 of its Guidance Part C:

> Similarly, unreasonable actions or inactions by the Contractor in prosecuting its claim should entitle the Employer to recover its costs. The Protocol may be used as a guide as to what is reasonable or unreasonable

On the other hand, contractors, and particularly subcontractors and suppliers, may justifiably assert that they did not have the requisite skills in-house to prepare a 'time-impact-analysis' of delays or a 'measured mile' disruption analysis, such that they had to engage external consultants at additional costs.

In such circumstances the following questions might become relevant:

- What do the contract terms require the contractor to submit in terms of notice, particulars and substantiation of its claim?
- Did the contractor comply with those contractual requirements?
- What do the contract terms oblige the employer or contract administrator to do on receipt of such notice, particulars and substantiation?
- Did the contractor put the employer, or its contract administrator, in a position to comply with their obligations?
- If so, did the employer, or its contract administrator, comply with their obligations under the contract?

In this regard, it may help to consider, as an example, the provisions of FIDIC Red Book clause 20.1, 'Contractor's Claims'. Rather than set this clause out in full, the following passages illustrate the point:

… the Contractor shall give notice to the Engineer, describing the event or circumstance giving rise to the claim.

The notice shall be given as soon as practicable, and not later than 28 days after the Contractor became aware, or should have become aware, of the event or circumstance.

If the Contractor fails to give notice of a claim within such period of 28 days, the Time for Completion shall not be extended, the Contractor shall not be entitled to additional payment, and the Employer shall be discharged from all liability in connection with the claim.

The Contractor shall keep such contemporary records as may be necessary to substantiate any claim ….

… the Engineer may, after receiving any notice under this Sub-Clause, monitor the record-keeping and/or instruct the Contractor to keep further contemporary records.

The Contractor shall permit the Engineer to inspect all these records, and shall (if instructed) submit copies to the Engineer.

Within 42 days … the Contractor shall send to the Engineer a fully detailed claim which includes full supporting particulars of the basis of the claim and of the extension of time and/or additional payment claimed.

Within 42 days after receiving a claim or any further particulars supporting a previous claim, or within such other period as may be proposed by the Engineer and approved by the Contractor, the Engineer shall respond with approval, or with disapproval and detailed comments. He may also request any necessary further particulars, but shall nevertheless give his response on the principles of the claim within such time.

Each Payment Certificate shall include such amounts for any claim as have been reasonably substantiated as due under the relevant provision of the Contract.

This clause sets out detailed procedural requirements of both the contractor and the engineer. If they both follow these procedures then the contractor ought to see amounts for a meritorious claim included in interim payments within a reasonably short period of the events. However, it is easy to see in these procedures the potential for ether party to fail or abuse the process. For example:

- Given how the lack of a timely contractual notice may discharge the employer from any liability in connection with the claim, issues such as the timing and adequacy of notices may be crucial. Debate can ensue as to when the contractor 'should have become aware' of the event and, similarly, whether a purported notice properly described the event and its circumstances.

- Further subjective argument can arise on the application of terms such as 'fully detailed claim' and 'full supporting particulars'.
- Those details and particulars may depend on the records maintained by the contractor. However, again, terms like 'as may be necessary to substantiate any claim' may again play into debate as to what is reasonably necessary. To what extent is the engineer's existing knowledge of the project and issues in the claim relevant to what should have been 'necessary'?

The poor quality of some claim submissions from contractors, subcontractors and suppliers and their consultants has been mentioned elsewhere. However, they often argue that the submissions were adequate to secure an extension of time, assessment of financial recompense or even just a payment on account, but that those receiving them did not give them a fair hearing. This often includes assertions as to an FIDIC Engineer's existing knowledge of the project and its issues and that it is 'turning a blind eye' to what should be perfectly clear to it in that role under the contract. The assumption of provisions such as FIDIC Red Book clause 20.1 must be that claims should be capable of resolution between the contractor and engineer's existing project teams, provided they act reasonably. However, claims for the costs of preparing claim submissions often include the whole of an external consultant's fees, on the contention that its engagement was only necessary because of the refusal to give a fair hearing to submissions prepared by in-house personnel. In practice, it sometimes appears that those receiving claims prepared by the in-house project staff of a contractor, subcontractor or supplier do not take them seriously or seek to bluster the claim away. On the other hand, it may appear that the employment of an external consultant has only inflamed the position and moved the parties further apart. Among Mr Justice Akenhead's many critical comments on a claims consultant employed by the employer in *Walter Lilly v Mackay* was his statement that 'The introduction of [the consultant] was certainly to raise the temperature and did little to engender any feelings of trust between the employer and the contractor'. He also described that consultant's involvement as 'doubtless aggravating'.

The potential commercial motives for not giving a claim a fair hearing are obvious in relation to claims from subcontractors and suppliers, where a contractor receives what is a claim for 'domestic' issues that will fall due to its cost because it cannot be passed up to the employer. This may be exacerbated by the poor quality of many subcontractor and supplier claims mentioned previously. The fact is that many lack both the experience of such situations and the staff with the required knowledge and skills to prepare such claims. However, for contractors expecting a reasonable and professional hearing of their claims by an experienced engineer under FIDIC or ICE terms or an architect/contract administrator under JCT terms, it may be that there are malign influences that militate against that outcome. A common complaint is that architects considering, for example, 'discrepancy or divergence' in bills of quantities as a claimed 'Relevant Matter' under SBC/Q clause 4.24.3 or engineers considering errors in relation to setting out a delay event under FIDIC Red Book clause 4.7(a), are in fact being asked to admit their own failures. In practice there may be direct or indirect influences bearing on them as certifier in determining such issues. Internationally, this particularly seems to be a complaint on projects for public sector employers where the strictures of public financing and audit may mean that certifying engineers under the FIDIC Red Book fear that any certification of time or money arising from their own failures will have an effect on their own fee payments or even see a claim against their professional indemnity insurances.

It is in such circumstances that a contractor may have a legitimate complaint that it has been put to unnecessary expense in relation to claims submissions. As has been noted, the extent to which such a failure will put the employer in an actionable breach of contract depends on the law and terms of the contract as regards the duties of it and/or its certifier.

Where contractors consider that they have such a complaint, it would be prudent to notify both the certifier and employer directly of the alleged secondary failure, the actions that the contractor is taking as a result, the additional costs that will arise and that it intends to claim those additional costs from the employer as damages for breach of contract. It may be that such an open approach will encourage a more constructive response to the contractor's primary claims. Alternatively, it might later influence a tribunal if such a head of claim is later part of claims that are referred to a formal dispute resolution.

8.8.3   Costs Incurred in the Contemplation of Legal Proceedings

Where a claim is made for the costs of preparing a claim that proceeds to formal dispute resolution, a further consideration is its relationship to the claimant's costs that are recoverable in the formal process. Where the contract binds the parties' disputes to be determined by arbitration, it should set out the applicable rules, including those as to costs. For example, if the AFSA (Arbitration Foundation of Southern Africa) Rules for Commercial Arbitrations apply, its Article 13 states:

> 13.1 Unless the parties have in writing otherwise agreed, the arbitrator shall in his award deal also with 'the costs of the arbitration' and decide which parties shall bear the costs of the arbitration or in what proportions the parties shall bear such costs.
>
> 13.2.1 The 'costs of the arbitration' referred to in 13.1 include … the fees and expenses of expert witnesses specifically declared by the arbitrator to be recoverable costs and the normal legal costs incurred by the parties.

Ordinarily, such 'cost of the arbitration' would be expected to start from the date when a party seeks to get the other party to concur to the appointment of an arbitrator, normally in the form of a notice of arbitration. Such costs would therefore cover legal and expert fees during the arbitration, but not fees and costs expended in the previous preparation of the claims that are the subject of that arbitration.

The case of *James Longley & Co. Ltd v South West Regional Health Authority* 1983 25 BLR 56 established that, ordinarily, in the absence of a breach by the employer (as suggested elsewhere), fees related to work done in the preparation of the claimant's final account, and as a general advisor, were not recoverable. On the other hand, those costs in respect of work done in preparing the claimant's case for arbitration were. On that basis, it may be that costs of preparing a claim document that forms part of the pleading, notwithstanding it is prepared before the arbitration notice is issued, may be recoverable as costs incurred in the contemplation of arbitration.

The position in this regard will vary according to the applicable procedural rules and the substantive law. However, the potential for inclusion of such costs as part of the costs of formal proceedings is often completely overlooked by parties. A particular consideration is the extent to which a claim or submission will actually form part of the submissions in an eventual arbitration or litigation. There are two schools of thought as to the extent to which claims should be prepared in a form and detail such that they can be adopted as part of a claimant's pleadings should the matter proceed further. This is particularly pertinent where a contractor considers that its initial submissions under the contract have been unfairly rejected such that it has had to prepare a further claim, the costs of which it considers should be reimbursed, perhaps as damages for breach in relation to the earlier submissions. One approach suggests that all such submissions should be in a form and detail that can be used in formal proceedings if necessary. The alternative, and more common outcome, is that when claims are subsequently pleaded they are in a new, more formal format, not least because this is dictated by advisors specifically appointed for the proceedings. In addition, even in making new submissions having had earlier claims unfairly rejected, contractors are still hoping that the claim will be dealt with on a less formal basis. That is not to say that it does not often occur that parts of earlier claims form part of the pleading, as in the case of a set of schedules prepared by a claims consultant and which were eventually annexed to the Points of Claim in the *James Longley* case.

A contrary point that must be avoided is a claim submission as part of the pleadings in formal proceedings that is significantly different to the contractor's previous submissions whose rejection led to the referral. This is an occasional feature particularly in jurisdictions where contractors and their staff are typically inexperienced and prepare claims that would never have withstood the rigours of international arbitration. If they are then completely rewritten including in relation to such as the legal basis of claim headings or the methods of quantification such as extensions of time or difficult issues such as disruption, then the defendant might argue that there is no dispute on those referred claims. This is an issue that is outside the ambit of this book, but practitioners need to be aware of it.

8.8.4 In-House Management Time

Claims for the costs of preparing a contractor's claims sometimes also include the time and expenses of its own in-house staff and management. A popular additional defence to such a claim is that, since those employees were on the contractor's payroll, their costs would have been incurred anyway, or that it is not established that any loss of profit or revenue resulted from their being diverted from other activities for the business. It is therefore said that there is no additional cost or loss to the claimant in their work in preparing the claims.

Under English law, the precedent for the recovery of management time expended in the remedying of an actionable wrong done to a trading company was the case of *Tate & Lyle Food and Distribution Ltd and Another v Greater London Council and Another* 1.W.L.R. In that case it was held:

> ... (1) that, although the plaintiffs could properly recover damages for the managerial and supervisory expenses directly attributable to the defendants' failure to dredge the silt once it had become deposited, the plaintiffs had kept no records

of the time expended and, in those circumstances, the plaintiffs' loss could not be quantified either in cash or as a percentage of the damages awarded; and that, accordingly, the plaintiffs had failed to prove their loss ....

Therefore the failure of the claimant to recover its management time in the *Tate & Lyle* case was due to the lack of detailed records to prove its time spent rather than because such costs are not recoverable. The case therefore emphasised that if the costs of in-house staff preparing a claim are to be recovered, they should be instructed to record their time working on that claim.

The approach in *Tate & Lyle* was adopted by His Honour Judge Humphrey Lloyd in *Babcock Energy Ltd v Lodge Sturtevant Limited* [1994] CILL 981. He allowed Babcock to recover as a head of damage the time and cost incurred by certain members of its staff in dealing with the problems created by the defendant as a sum calculated on the basis of constructive charging rates applied to the total number of hours recorded for those staff at the material time.

Several more recent England and Wales judgments have further developed a claimant's entitlement to recovery of the costs of its in-house management, who are diverted on to matters related to a breach by the other party. They are summarised in *The Board of Trustees of National Museums and Galleries on Merseyside v AEW Architects and Designers Limited and PIHL UK Limited and Galliford Try Construction Limited (trading together in partnership as a Joint Venture 'PIHL Galliford Try JV')* [2013] EWHC 2403 (TCC).

The *Liverpool Museums* case concerned design problems related to the steps, seats and terraces to the new Museum of Liverpool constructed between 2007 and 2011. The problems were identified during the course of construction and resulted in extensive consultation between the various parties as to possible solutions, aborted remedial schemes and designs to finally overcome the problems. The client for the Museums was involved throughout the process through its Chief Executive and Premises Director. As the Judge put it: '... *it is clear that a substantial amount of time, energy and resource was applied by the Museum to seek ways to see what could be done and possibly to live with the problem*'.

From the previous authorities it might be thought that the time of the Museum resources could be claimed, without demonstrating loss of income, etc., that would have been earned elsewhere by those staff, but that records of the time expended by the Museum's staff would be required. However, the Museum had not kept such records and the Museum relied extensively on the witness evidence of Sharon Granville, its Executive Director. That evidence included how much time was spent by her and other members of staff included in the claim, and their pay grades and salary costs. The judgment helpfully summarises recent authorities and moves away from the *Tate and Lyle* requirement for detailed records, in certain circumstances, as set out in the following from paragraphs 135 to 143 of the judgment:

19. The Museum's Internal Management and Staff Costs to Practical Completion

135 ... I find, and indeed, there can be no doubt, that the time which these individuals spent in addressing the problems associated with the steps and seats would otherwise have been spent on other matters which would doubtless have inured for the benefit of the Museum.

136 ... The first is that Ms Granville was an impressive, believable and honest witness whose evidence I have no real difficulty in accepting, backed up as it largely was, at least in part, by other witnesses. In any event, she was not cross-examined as to credit and there was no suggestion that she was anything other than honest and competent. Secondly, whilst there may be cases in which the absence of underlying accounting documentation might undermine a party's quantum case (particularly where it was otherwise suspect), the supposedly missing documents such as payroll records were never sought on disclosure by AEW.

137 ... In *Aerospace Publishing Ltd v Thames Water Utilities Ltd* [2007] EWCA Civ. 3. One of the damages claims related to 'payments made to staff for work necessarily done by them in relation to, and consequent upon, the flood' (Paragraph 73). Wilson LJ who gave the judgement of the court on this topic reviewed the authorities and said at Paragraph 86:

(c) ... it is reasonable for the court to infer from the disruption that, had their time not been thus diverted, staff would have applied it to activities which would, directly or indirectly, have generated revenue for the claimant in an amount at least equal to the costs of employing them during that time.

138 Lord Justice Wilson also referred to the judgment of Mrs Justice Gloster (as she then was) in *R + V Versicherung AG v Risk Insurance and Reinsurance Solutions SA* [2006] EWHC 42 (Comm) in which she said:

'77. In my judgment, as a matter of principle, such head of loss (i.e. the cost of wasted staff time spent on the investigation and/or mitigation of the tort) is recoverable, notwithstanding that no additional expenditure "loss", or loss of revenue or profit can be shown. However, this is subject to the proviso that it has to be demonstrated with sufficient certainty that the wasted time was indeed spent on investigating and/or mitigating the relevant tort; namely that to be able to recover one has to show some significant disruption to the business; in other words that staff have been significantly diverted from their usual activities. Otherwise the alleged wasted expenditure on wages cannot be said to be "directly attributable" to the tort.'

139 A similar view was formed by Mr Justice Ramsey in *Bridge UK Com Ltd v Abbey Pynford Plc* [2007] EWHC 728 (TCC) in which there was a claim for management time incurred by the claimant's staff in dealing with the problems caused by the defendant. He said:

'123. Such a method of retrospective assessment is, I consider, a valid method of calculation. I have been referred to the judgment of His Honour Judge Peter Bowsher QC in *Holman Group v Sherwood* (Unreported, 7 November 2001) where he indicated that in the absence of records, evidence in the form of a reconstruction from memory was acceptable. I respectfully agree ....

125. I accept that the appropriate approach to the question of recovery of such management time is that set out by Gloster J in R + V Versicherung ....'

140 ... There is nothing intrinsically wrong with the retrospective assessment which Ms Granville has done; she was closely involved in this period and was

clearly taking a particular interest in these serious problems which were materially delaying the effective opening of the Museum. She was well placed to know how much time was spent at least in a general and broad sense.

141 Whilst I am by no means convinced that it is necessary as a matter of law to find that 'staff would have applied it [their time] to activities which would, directly or indirectly, have generated revenue for the claimant in an amount at least equal to the costs of employing them during that time' (because Lord Justice Wilson was not limiting the loss, waste or application of management time claims in this way), I can and do make this inference. The reason is that the staff actually had other jobs to do other than cope with the fallout from the steps and seats problems. For instance, Ms Granville was and is in charge of eight buildings for the Claimant; they all generate income albeit that it is clear also that some income is generated from grants or charitable giving. Her time, for instance, could and would have been better spent doing things which undoubtedly would have assisted in the generation of revenue. If one also brings in her evidence that there were what she called 'backfills', the justification for a recovery of management time becomes even more compelling. She said that, by reason of her involvement with the problems, other people took over parts of her other roles and there was a ripple effect with yet more people having to do the work which those other people did not have time for; she said that additional roles were actually created to fill the gaps that were left. She said that each of the people who took on additional duties were paid either a responsibility allowance or were temporarily re-graded to cover for her substantive role.

142 In my judgment, an innocent claimant which has established its cause of action can recover its management time reasonably spent dealing with the consequences of the negligence or breach of duty in question. Although it could be said that it would have to pay the salaries in any event to its staff and has therefore incurred no loss, the time of the staff is being deployed to remedy or otherwise address the otherwise recoverable loss and as a matter of causation it is equally being incurred for two causes, one the employment and the other the cause of action itself. This view is supported by Gloster J in the *R + V Versicherung AG* case.

143 ... Turning to the quantification, it is fair and reasonable given the relatively general retrospective assessment done by Ms Granville to adopt a reasonably cautious approach. I broadly accept the assessment which she makes in respect of herself which is 36.5 days' worth of her time over the period but I reduce it to 30 days to reflect the fact that she said that she had discussed the percentages with Mr Williams, they had both looked at their diaries and that it was difficult to estimate how much time they had spent discussing the particular issue .... The same can be said for Mr Williams and Mrs Green for whom similarly 36.5 days were claimed; I allow 30 days .... In relation to Mr Hemmings for whom 122 days are claimed, there is on the evidence a mathematical error which suggests that at most his time would have been 97.5 days. I round this down to 90 days because over the 18 months he clearly had other work to do and I can be confident that at least 90 days would have been applied by him to dealing with fallout from the steps and seats problems.

Similar adjustments were made to the cost claimed for prospective remedial works.

From the forgoing it is suggested that, in relation to claiming for in-house management time spent preparing claims under construction contracts where the legal basis for claiming that time is established:

- The fact that the staff would have been employed by the business in any event does not debar recovery of their costs.
- It must be established that the management time was diverted from their usual activities by the claims.
- It will also be necessary to show that the diversion caused significant disruption to the claimant's wider business.
- As to evidence of the time spent, in the absence of contemporaneous records, retrospective analysis or an impressive witness may suffice to demonstrate the costs incurred.
- However, having the involved staff record their activities and time spent on the claim would remove some uncertainty of outcome and avoid a cautious assessment of the loss.
- In the absence of proof that the diverted staff could have applied their time to revenue generating activities elsewhere, at least equal to their employment costs, it may be enough to infer that from the evidence.
- This may be established by those staff setting out the extent of disruption to the business, including what activities they were carrying out elsewhere from which they were diverted by the claim; and how those activities would have otherwise benefited the business.

Of course, the 'no additional cost' defence to a claim for staff time should not apply to staff who are rented from agencies or seconded from consultancies. This is a common occurrence in relation to some grades of staff, in particular on international construction and engineering projects. In this case they ought to be submitting timesheets on, for example, a monthly basis and the contractor should ensure that those timesheets contain sufficient detail of the activities being undertaken where there is a possibility that they may be the basis of a claim. A further situation is where salaried staff are employed on a particular claim and their diversion from other activities, such as another project, is replaced by hiring in of external staff. In such a case the contractor might base its claim on the external cost of that other staff.

8.9   Errors, Omissions and Contradictions

Chapter 5 of this book considered the position where the contractor has made an error in the insertion of its rates and prices into a contract document, such as a pricing schedule or bills of quantities intended to be used in the valuation of variations or remeasured valuation of the works. It was explained how, understandably, standard forms of construction and engineering contracts make such mistakes the risk of the contractor, but that they may give rise to argument between the parties, especially as to how such erroneous rates and prices are to be used to value variations or additional work. The corresponding position is that mistakes in the preparation of such documents by the employer and its advisors are generally at the employer's risk. Such errors can all too often lead to disagreement between the parties.

Unfortunately, it is not at all unusual to find that there are errors, omissions or contradictions in the contract documentation for an engineering or construction project. The larger the project, it seems the more common such problems are, and the greater the potential financial implications and greater likelihood of then being subject of a dispute. Given the extensive nature of the information provided in a typical construction and engineering contract, and the fact that it is generally supplied in different formats, i.e. contract documents using legal terms and phraseology prepared by legal advisors, alongside pricing documents prepared by quantity surveyors and specifications and drawn information using technical terms and phraseology prepared by technical advisors, the scope for error can be understandable to some extent. Add to this the effect of increasing globalisation of the industry and that documents, specifications, product lists, etc., might be in more than one language and the potential becomes even greater. However, these complications should not excuse documents that contain errors, omissions and contradictions as a result of poor or rushed preparation or that have been 'cut and pasted' together from documentation garnered from previous projects, such as the following examples:

- The contractual terms and procedures of over 8000 pages of contract documents for a rail project in Africa, which had largely been copied from the contract documents for a vaguely similar rail project in the UK.
- A contract for the construction of 11 residential towers in a Middle East city, which gave no more detail of specification, design and price than cross-referencing to a similar scheme elsewhere in that city. The parties appear to have assumed that the first scheme went so well and amicably between them that the second scheme could be contracted on a superficial 'same again please' basis with a negotiated contract price. In the event, the very different circumstances of the second project (such as its location, economic climate and individual management personnel) meant that the parties ended up in expensive disagreements over the design and detail of the second scheme and what did, and did not, constitute variations to it, and how they should be priced.

However, one must acknowledge the simple truth that some errors may creep into even a painstakingly well-prepared set of documents.

Some contract documents, particularly ad hoc contracts drawn up for a particular project, contain a hierarchy of documentation that prescribes the order of precedence of documents in the event that there is a problem between them. The hierarchy is usually in the following descending order of priority:

(1) Contract terms.
(2) Specification.
(3) Drawings.
(4) Bills of quantities or schedules of rates, etc.

Some standard forms set out such a hierarchy, others do not, although such a hierarchy is often added as a particular provision. The alternative is to provide that the contract documents are to be read 'as a whole', without stating which is to apply where two of them contradict each other. A key consideration in this regard is where the local substantive law provides a rule of contract construction that a term specifically drafted for a particular contract will override a general term in an adopted standard form.

In SBC/Q, clause 1.3 provides a limited hierarchy under which 'the Agreement and these Conditions are to be read as a whole, but nothing contained in the Contract Bills of the CDP Documents, nor anything in the Framework Agreement, shall override or modify the Agreement or these Conditions'.

The Infrastructure Conditions provisions merely provide in clause 5 that the contract documents shall be 'taken as mutually explanatory of one another'. Any ambiguities or discrepancies are to be explained and adjusted by the engineer who is required by clause 5 to issue appropriate instructions to the contractor in accordance with clause 13. The contractor can then recover costs, with profit, in relation to additions to the permanent or temporary works, in respect of such instructions. The engineer is also required by clause 13 to take such matters into account in assessing any extension to the completion date that might be required. Therefore, the Infrastructure Conditions allow the engineer greater freedom of operation in deciding how to resolve any discrepancies or ambiguities as the contract terms and conditions do not necessarily take precedence over information in other documents. However, it is suggested that the implications for evaluation are little different from the SBC/Q regime.

The FIDIC Red Book has, at clause 1.5, a similar term to that of the Infrastructure Conditions, 'taken as mutually explanatory of one another', but then continues by setting out that, for the purposes of interpretation of those documents, they shall be taken in accordance with the following sequence:

    (a) the Contract Agreement (if any),
    (b) the Letter of Acceptance,
    (c) the Letter of Tender,
    (d) the Particular Conditions,
    (e) these General Conditions,
    (f) the Specification,
    (g) the Drawings, and
    (h) the Schedules and any other documents forming part of the Contract.

The Schedules referred to in this final in subclause (h) are defined in the FIDIC Red Book as those completed by the contractor as part of its tender and included in the contract and that they may include bills of quantities. Clause 1.5 requires the Engineer to issue necessary clarification or instruction in relation to ambiguity or discrepancy in these documents, which would potentially bring it into the ambit of clause 13, 'Variations and Adjustments'.

Under such document hierarchies the intention is that the terms and conditions of the signed contract should incorporate the intention of the parties and in the event of conflict with other documents they are to take precedence. The specification is generally the technical document that sets out the details of the construction works and that is to take precedence over the drawn information showing the layout and arrangement of the works, including any required design detailing. The specification is secondary to the contract terms and conditions in the event of a conflict, although in practice because of the different nature of these documents such conflicts ought to be relatively rare. However, the specification takes precedence over the drawings which are intended to show the spatial requirements and detailing of the construction.

The bill of quantities, or schedule of rates, is therefore usually the lowest document in a typical hierarchy of contract documents, it being primarily a commercial document setting out the detail of the contract price. There is, however, often a contradiction in the use of the bill of quantities to also set out, in preambles to the various pricing sections for the works or in the preliminaries and general items section, certain matters that can seemingly modify or contradict documents higher in the hierarchy, particularly the contract terms and conditions. This often occurs in relation to the rules for pricing variations and interim payments. However, common (and arguably the most perverse) examples are the use of the preliminaries and general items section in the bills of quantities to set out timing requirements for the works including requirements for completion of the works in sections. Such matters should be addressed in the contract terms if problems are to be avoided and most standard form documents contain provisions for sectional completion of the works if that is required. Furthermore, it must also be remembered that if sectional completion is required then any provision for the employer's delay damages in the contract documents for non-completion of a section on time should be adapted to provide clear damages provisions for each section. Such matters should not be left to subordinate documents such as bills of quantities, to avoid the risk that the terms of the superior document will conflict and subvert the parties' wishes, although this still occurs.

Just as the JCT standard forms have traditionally provided for the contract terms to take precedence over the contract bills, they have also provided for any errors or omissions in the preparation of those bills to be corrected. In the SBC/Q form, this is in clauses 2.13 and 2.14. Clause 2.13 states that the bills of quantities have been prepared in accordance with 'the Standard Method of Measurement' except where specifically stated in the bills in respect of any specified items. That standard method was defined in the 2011 edition as 'the Standard Method of Measurement for Building Works, 7th Edition, produced by the Royal Institution of Chartered Surveyors and the Construction Confederation, current, unless otherwise stated in the Contract Bills, at the Base Date'. Clause 2.14 then provides that in the event that there are:

- departures from the method of measurement specified, or
- any errors in the description of work, or
- any errors in quantity of work, or
- omissions of items required by the method of measurement, or
- errors or omissions of information for any item which is the subject of a provisional sum for defined work, then

such errors or omissions shall be corrected, and any correction treated as if it were a variation ordered by the architect. Clause 2.14.3 then makes the correction of such errors or omissions a 'Variation' and hence subject to the valuation rules of clause 5.6 and the implications of such corrections can be evaluated using those rules.

For corrections to the descriptions of measured items this should not pose too many difficulties as the consequences will generally require the unit rate to be adjusted to reflect the corrected description or quantity. This is an area discussed at length in Chapter 5 of this book. If one bill item, or a series thereof, is set out erroneously, then it is likely that related items are also wrong and need similar attention. However, in some instances this could also require the re-evaluation of other items not themselves the subject of errors on their face, but whose pricing is affected by the correction to a related item. The most common example of this is items in the preliminaries and general items section of the bills that include allowances for costs such as plant, equipment, management and supervision related to erroneously described bill items for measured work.

Another particularly common area of difficulty in evaluation is that of the correction of any error or omission in information for any item that is the subject of a provisional sum for 'defined work' under such as rule 10 of SMM7. Provisional sums and the possible distinction between those that are 'defined' or 'undefined' was explained in Chapter 3. Under that rule, the contractor was deemed to have included in its pricing of preliminaries and general items and in its programme of works for the content of defined provisional sums and any correction of such deficiencies would therefore raise the prospect not only of corrections to the pricing of preliminaries and general items, but also corrections to the contract programme. If the correction were substantial enough, and if the work that was the subject of the allegedly 'defined' provisional sum were both substantial and on the critical path, it is possible that the contract completion date might be affected. In such circumstances the consequences of such a correction could be substantial both in terms of the impact on the employer's plans for the timing of the project and its delivery and the financial consequences of the correction.

Whilst errors, omissions and contradictions in construction and engineering contract documents are relatively uncommon in markets such as the UK and United States, there are some jurisdictions in which they are all too common occurrences. At times, the extent of these seems directly inversely proportional to the willingness of employers (particularly in the public sector) to properly spend money on good pre-contract advice and the thorough preparation of sound contract documentation. When it comes to the preparation of contract documents the proverb 'for want of a nail, the shoe was lost … etc.' seems particularly apposite. This said, it may seem a lame excuse for a consultant who prepared bad contract documents should blame the employer for being unwilling to pay for a proper pre-contract period and input. The practical problem for the contractor is that under such as FIDIC contracts, the engineer who prepared the contract documents, or at least the technical parts of them, is often the same engineer charged under the contract with responsibility for addressing (including admitting to) errors, omissions or contradictions in them.

8.10 Summary

In addition to those for the direct and indirect consequences of change discussed in Chapters 5 and 6, this is not a comprehensive discussion of all the financial claims that might be generated by a construction or engineering contract. However, it should cover most of the more common sources and types of claim. A common feature of the claims discussed in this book, and particularly those that are the subject of this chapter, is the extent to which good administration of the contract will limit, or even avoid, the consequences of such matters. If there is a general rule to be drawn from this discussion it is that a well-prepared and administered contract is the most likely to avoid difficulties, an obvious and simple rule but often more honoured in the breach than the observance! How this might be made more likely is the subject of the following chapter.

9

Minimising the Consequences of Change

9.1   Introduction

It is in the nature of construction projects, their length, complexity and often unique nature, and the inherent risks apportioned by typical contracts for a project, that changes and consequent claims for additional payment will often occur. Those changes may include such things as variations to details of the original works, additions or omissions from the scope of the works, or delays or disruption to the programme of works. The challenge facing the management teams for any construction project is to minimise the impact of such changes, should they occur, on the programme and cost of the works. In this regard the management teams should be understood to be: the employer's or client's project staff and any retained consultants, be they designers, quantity survey-ors or others; the main or prime contractor's project staff; and the project staff of any subcontractors for the project. It is only with genuine cooperation between all of those parties and teams that the consequences of change can be successfully managed and the potential negative impacts minimised.

Those negative impacts can be of two types. Firstly, there is the primary effect on the cost and programme of the project in terms of additional payments and overrun against the contract completion date(s). In addition, there may be secondary effects arising out of attempts to deal with and resolve those primary effects. This is in terms of manage-ment time and other costs, perhaps including the engagement of consultants, and the breakdown of project relationships. Potential indirect effects may even stretch to the costs of dispute resolution procedures, including management time and legal fees, that could have been avoided had the consequences of change been dealt with differently.

If the impact of changes can be minimised, then that is a significant contribution to reducing unnecessary disputes arising from differences as to the responsibility and lia-bility for, and consequences of, changes that have occurred on a project. The further significant contribution that can be made in this regard is obtaining the benefits that are available from good quality contract preparation. This includes choosing the right form of contract from extensive suites such as those published by such industry organisations as JCT, FIDIC and NEC.

Whichever form of contract is adopted, part of genuine cooperation by the parties should be a proper allocation of risk, responsibility and reward between them. Standard forms should be used in the unamended terms in which the industry bodies have agreed them, without attempts to shift risks on to a party that is not best positioned to manage such risks.

Evaluating Contract Claims, Third Edition. John Mullen and R. Peter Davison.
© 2020 John Wiley & Sons Ltd. Published 2020 by John Wiley & Sons Ltd.

Thereafter, the best chosen standard form may not help if it is not properly administered, both pre-contract and during its currency. This starts with the avoidance of attempts to make 'clever' amendments to those standard forms with the short-sighted aim of bending them to one party's advantage. During the currency of the works, it is essential that all involved are aware of the provisions and implement them properly. The example of the judgment in *Northern Ireland Housing v Healthy Buildings*, discussed in Chapter 5, is a good example of the consequences of failure to properly implement the carefully developed procedures of a standard form that, had they been operated fully and properly, might have prevented the subsequent adjudication and litigation. That example was under NEC terms, but a myriad of other examples could be referred to where failures to properly apply the procedures of other standard forms have led to expensive recourse to the courts to resolve disputes between the parties.

Good management requires the logical and detailed analysis of changes and their consequences when they occur. Too often the documentation of projects, including major contracts running into tens of millions of pounds or more, is rushed, ill-considered or inappropriate. This in itself sows the seed for problems in the management of the contract, which often surface when change consequences have to be identified and agreed. If the standard of preparation of change identification and the consequences are also substandard then there is every chance that, instead of a problem being analysed and agreed by the parties, a dispute will ensue between the parties to the contract for resolution by a third party.

This may seem trite, but there is no doubt that the UK construction industry is not alone in having produced many expensive and protracted disputes that should never have become disputes at all but should have been resolved by the project management teams. In the course of this failing the industry has spawned a whole 'subindustry' dealing with disputes, litigation, arbitration and 'alternative dispute resolution' in its various forms, including adjudication and busy times for local courts and arbitration centres.

It is arguable that the greatest step towards reducing this problem that the industry and its clients can make is to insist on greater attention to the proper analysis and documentation of change in its many forms.

A related subheading under this topic may be considered to be that of value engineering. Where change occurs its impacts can particularly be ameliorated if the contractor can be encouraged to be involved in obtaining the most cost-effective solution to those effects. All too often construction contracts are drafted in a form that actually serves only to encourage the contractor to make the most of any such changes that will result in additional payments and/or an extension of time. As explained in this chapter, approaches such as alliancing and partnering can also perhaps encourage a more helpful approach. Value engineering is considered in Chapter 2. Similarly, omission instructions are often used to minimise the effects of change, perhaps by making costs savings elsewhere or in direct relation to changed work itself, by omitting it in whole or in part. Such approaches are of course reliant on very early notification or warning of the event and its effects and these are topics considered elsewhere in this chapter. Again, omission instructions have already been considered elsewhere in this book, in Chapter 5 in particular.

In the UK there have been, in the recent past, a number of attempts to address such perceived problems with the construction and engineering industries and it is worth briefly considering at least two of those initiatives and their recommendations in some further detail. There is no doubt that similar initiatives could benefit the industries in many other parts of the world. Some of their initiatives have in fact started to provide benefit overseas. For example, in the increasing export of the NEC suite of contracts and the adoption by other countries of security of payment legislation similar to that in the UK's Housing Grants, Construction and Regeneration Act 1996.

9.1.1 Constructing the Team

Constructing the Team was the final report produced by Sir Michael Latham in July 1994 and the recommendations in the Executive Summary make interesting reading. To pick some of the principal comments from the summary:

- Implementation of the report begins with clients.
- The preparation of the project, its contract strategies and brief require patience and practical advice.
- The endless refinement of existing standard forms of contract will not solve the perceived problem of the adversarial nature of the industry.
- The role and duties of project managers in the industry need clarification.

There are many other such items in the summary, but this list already illustrates the problems that were perceived at the time of Sir Michael's report. These included that: clients needed to be more involved; greater care was needed with the preparation of contracts; tinkering with standard forms of contract would not reduce conflict; and that some persons in the industry with high profiles needed to have their roles better defined!

The report also made a number of recommendations, including the use of adjudication as the normal method of dispute resolution for construction projects, a recommendation that was subsequently introduced for the UK construction industry, with the exception of notable sectors such as the process industries, through Part II of the Housing Grants, Construction and Regeneration Act 1996. This is almost without doubt the greatest impact that Sir Michael's report had on the UK construction industry and adjudication is now widely used and, some would say, sometimes abused, on the vast majority of construction disputes in the UK. Adjudication is also being increasingly exported to construction markets elsewhere, along with other modern forms of 'alternative dispute resolution' such as mediation and dispute review boards. Furthermore, 14 other countries have since passed security of payment legislation similar to that of the UK.

The subject of adjudication has spawned a large number of textbooks devoted to the subject. Suffice to say here that it is ideally suited to resolving some problems by providing a speedy and economical decision where the parties are unable to negotiate their own solution. This is particularly so in relation to cash flow. It can provide an independent view and remove a source of conflict from the parties and allow them to proceed with their relationship and the project without the hindrance of an ongoing dispute and money being wrongly withheld by the employer from the contractor or the contractor from the subcontractor. It is, however, by virtue of its restricted timetable, often a 'quick and dirty' process that is not always sure to produce a reasoned and logical solution, especially if complex or very large matters are submitted to the process.

Unless the parties so agree, adjudication is not final and binding on the parties. Where adjudication addresses isolated disputes or issues that can be succinctly documented and presented there is always a good prospect that the outcome will be one that both parties are prepared to accept as a permanent solution. Where the dispute comprises a series of differences or is centred on complex arguments of delay and disruption the outcome is predictably less sure. In practice, since the introduction of adjudication under the 1996 Act, there has been a tendency for disputes of increasing complexity to be referred to adjudication, sometimes substantial final account disputes involving very many discrete items and sometimes complex delay claims involving analysis of programming, factual and contractual issues. While adjudication is still successful in providing the parties in the substantial majority of referrals with a decision which they are prepared to accept as final, or at least are not sufficiently disgruntled with it that they wish to pursue the matter further, it is argued by some that far from reducing the amount of conflict in the industry, the adjudication process has increased the number of disputes and made the negotiation of differences more difficult.

On the other hand, the accessibility and much lower cost of adjudication acts as an encouragement to parties to use it rather than try to resolve differences themselves, particularly if they believe their case may be less than totally sound but believe they may achieve some measure of success in a 'quick and dirty' adjudication.

The report produced by Sir Michael Latham was not the result of a systematic and logical research programme but was the outcome of an extended series of consultations with selected figures and institutions in the industry. While there is undoubtedly much of relevance in the final report, the introduction of adjudication is probably the most significant single innovation produced. While other bodies have taken up some of the other ideas and suggestions in the report, and others have been quietly dropped, it did not lead to any noticeable drive to educate all sides of the industry to the benefit of improving the standards of contract documentation and administration, although the issues are present in the report.

One of the areas in which significant benefits could be obtained, while reducing wasted time and costs, is in the better analysis and documentation of claims.

9.1.2 Rethinking Construction

The report *Rethinking Construction* was published in July 1998, just four years after Sir Michael Latham's report and two years after the Housing Grants, Construction and Regeneration Act. It was the report of a 'Construction Task Force' delivered to the Deputy Prime Minister with the objective of improving the quality and efficiency of the UK construction industry.

Chapter 3 of that report was entitled 'Improving the Project Process' and included many interesting and challenging ideas relating to the obtaining of a more integrated approach to the construction process, a reduction in competitive tendering and the reduction in the formation of new teams for the majority of contracts. The message from the report, often referred to by the name of its chairman as the Egan Report, was very much that using different ways of working such as partnering and alliance contracting would allow great gains in quality, efficiency and costs.

The report did not specifically address the problems arising from the nature of construction contracts, as the task force did not agree that construction was fundamentally different from manufacturing in that each project is unique. In the opinion of the task force, 80% of the input to a building is repeated, i.e. it is the same as used previously. This figure could be argued endlessly between designers and project managers with the likely outcome that it may be relevant for some types of construction such as housing and office developments where repeat use of components is possible, but may be optimistic for other types of construction, particularly civil and mechanical engineering where the materials and construction processes may be the same, but the design differences can be considerable. It might also be said that although 80% of the input to a building is repeated, its circumstances taking into account such items as project location and the parties involved, etc., are always unique.

If the Egan approach was to deal successfully with the potential changes that will still be inherent in any substantial project then it was to be left to the mechanism of partnering and alliance contracting to provide the appropriate process.

Disappointingly perhaps, the Egan Report failed to take the opportunity to make the point in the strongest possible terms that the biggest single improvement that could be made would be for the industry's clients to allow sufficient time and resources for their projects to be properly prepared before commencing construction work. This starts with preparation of the contract documents. This is a problem that particularly infects projects internationally. The focus of many clients seems to be solely on saving time and cost during the pre-contract stage, without consideration of how taking greater care might save time and cost during construction.

9.2   Contract Preparation

One of the oldest sayings in business is that 'time is money'. If this is combined with the prospect that practically none of the construction industry's clients are interested in construction per se, but are only interested in the end use of their one-off project, it is not difficult to imagine the pressures on the construction process to occupy as little time as possible in the lifetime of the final asset. This is often particularly true of the pre-contract phase of the construction process.

Whether a chemical company needs a new process plant, a developer needs a new office building or the national health service needs a new hospital, the construction process is merely a means to an end. The chemical company wants to sell its products, the developer wants an office to sell or let and the national health service wants to treat patients. Very often they want these as soon as possible. In such circumstances there is very real pressure to condense the whole process into as short a time period as possible. The first casualty of this pressure on pre-construction phase time is usually the time allowed for design, specification and project documentation. In some cases the pressure will be so intense as to lead to the consideration of means to obtain a start on the construction works before the design and documentation are complete – perhaps by the use of letters of intent, the subject of Section 8.1.

The construction management route for project procurement, where a prime contractor manages the works let in parcels to package contractors, is one often adopted solution in such circumstances, as is the use of enabling, or preliminary, works contracts to allow some early works to proceed in advance of the main works. Properly managed and monitored, there is no intrinsic reason why such procurement routes need to be more risk prone than other means of construction procurement. Indeed in some sectors, particularly the oil and process industries, the commencement of works with an incomplete design using such procurement methods is commonplace due to the high financial pressures to obtain output from the new facility.

However, all too often the pressure on time results in a condensing of the time available for pre-construction activity without proper consideration of the potential hazards and risks to which the project, and the parties, will thereby be exposed during the subsequent construction phase. Such condensing of the pre-contract period will often result in deficiencies, such as:

- Inadequate site investigation, or reliance on outdated or incomplete data.
- The excessive use of provisional sums, or other such sums, to allow in tender documentation for works yet to be designed or defined. See in this regard the judgment in *Lilly v Mackay* considered in Chapter 6.
- The over-reliance on provisional sums being compounded by combination with forms of contract that assume the works are fully, or practically fully, designed and defined at the time of tender (ditto).
- Tender documents produced using sketch designs, or 'indicative' drawings, as the basis. It is not unusual in such instances for drawings or specifications from a previous similar contract to be used. Again this can often be compounded by the adoption of contract terms that do not acknowledge that the project is not properly defined at the time of tender.
- Contract documents that are a 'thrown together' compendium of all documents and communications exchanged by the parties in the pre-contract phase with little thought as to how they should fit together and ignoring inconsistancies and contradictions between them.
- Failure to ensure that preconditions to the achievement of construction of the works are achievable or obtainable in the times anticipated or at all. A particular example of this is the procurement of land from many owners on a long infrastructure project. Alternatively, the procurement and supply by the client of permanent plant and equipment to be incorporated into an industrial project.
- Commencement of works under letters of intent that have been hurriedly, and hence poorly, drafted.

This is a sample of the problems that can be generated if condensed pre-construction periods are adopted without proper consideration. The result is often a difficult and fraught construction phase with differences and disputes between the parties as they try to cope with the problems caused by the inadequate pre-construction phase.

Whatever the pressures, the procurement route has to be thought through to produce a viable construction project with a sensible risk apportionment reflected in the documentation. Two fundamental steps that will go a long way to reducing differences and disputes, and enable claims arising from changes to be properly and logically documented, are:

1. The adoption of a procurement route and contract terms that properly reflect the status of the project at the time the construction contract is procured, be that in terms of design, site information or time, and that identify the risk apportionment and anticipate how claims for additional payment are to be calculated in detail.

2. The adoption of a thorough and reasoned means of defining the scope of the works at the outset. In contracts where bills of quantities are to be provided this means either the adoption of a Standard Method of Measurement or the provision of a comprehensive set of measurement rules so that the means of scope definition is understood by all parties. Where there are no bills of quantities there must be the means of defining the project scope and providing some mechanism for analysis of the financial content of the contract, be that in the form of a schedule of rates or tender sum analysis, on a defined and agreed basis.

If the basis of the documentation is reasonable and logical at the outset the prospect of any claims for additional payment as a consequence of change being similarly defined and costed may be immeasurably increased.

9.3   Alliance, Partnering and Framework Contracts

The perceived image, for some commentators, of the construction industry in past years has been one of companies with a willingness, or even eagerness, to engage in the mounting of extravagant claims followed by subsequent litigation or arbitration or alternative dispute resolution proceedings in order to resolve the disputes generated as a result. Whether such an image is justified, if the risks inherent in many construction projects are taken into consideration, is arguable. The image becomes even more questionable if the amount spent on construction litigation in a year is compared with the value of the construction industry's annual output, when it could be seen that such expenditure is in fact a relatively small proportion of the annual value. This suggests that the industry may not be as litigious as some perceive it to be. However, there is little doubt that any measures that can be taken to minimise the incidence of disputes will be to the benefit of all parties, be they constructors or clients.

This book is concerned with the evaluation of claims when they do arise, in order to minimise differences and disputes over that evaluation, rather than an examination of alternative methods of procuring construction contracts. However, if the development or promotion of other means of procuring construction work, other than the traditional competitive tender for a lump sum contract, has any relevance to the subject, then that should obviously be considered. In this respect, partnering, alliancing and framework agreements are worthy of particular consideration.

9.3.1   Partnering

One of the main thrusts of the Egan Report discussed elsewhere in this chapter was its emphasis on partnering as a route to increased efficiency and improved quality in construction. The concept of partnering itself, as a co-operative relationship between client and contractor where a joint approach to problem solving is adopted, was not new to the construction industry, but was previously not widely adopted. Some impetus was put behind it by the Egan Report's recommendations such as that competitive tendering be replaced by more long-term relationships.

Partnering is a broad term that can be applied to a number of approaches. Whilst it can be used on a one-off basis, it is usually applied to a long-term co-operation between contractor and employer on a number of projects. The aim of this is to improve understanding and lesson learning and develop attitudes of mutual respect and trust between the parties.

The opportunities for partnering on a regular basis are not of course available to every client, as many construction industry clients are one-off clients, using the industry on an infrequent rather than a regular basis. Even among clients with a continuous and ongoing programme of construction activity the benefits of partnering with a limited number of contractors are not universally acknowledged, with the construction director of one major client reportedly describing the prospect as 'synchronised swimming with sharks'! Perhaps the attitude suggested by that reported remark is the result of long exposure to the vagaries of badly procured and poorly documented projects, and the disputes they can produce. Perhaps long exposure to partnering under properly drafted contractual terms should change this individual's perception.

We are not concerned here with the undertaking of a full discourse on the mechanics of the partnering process for individual or series contracts, but it is relevant to consider the implications of such a contract route for the documentation and resolution of contractual claims. There is sometimes an expectation that claims for additional payments will not arise under such arrangements, but the risk apportionment in any contract will usually mean that it is unlikely. There is also evidence that the choice of the form of partnering arrangement is very much in the hands of the client with the contractor negotiating changes and amendments where relevant and possible. It seems therefore that partnering starts only when the partnering arrangement itself is in place.

The benefits of partnering lie not so much in the contract terms and conditions, or in the risk apportionment, as in the gains to be obtained from a collaborative approach to the project rather than an adversarial one. Surprisingly, some contractors have found that the adoption of a partnering route has led to significant gains in the management of their own supply chain, and even in the reduction of divisions within their own organisations. However, perhaps one of the key benefits of partnering from an operational point of view is the intent to minimise disputes and to resolve them when they do arise without undue conflict. Experience has shown that in partnering arrangements recourse to litigation, arbitration or other dispute resolution procedures is less frequent than with traditionally procured projects, and where disputes do arise they tend to be between the main contractor and his supply chain rather than between the main contractor and the project client.

The reason for the gains to be derived from partnering, and many contractors with a substantial partnering workload believe there are gains that feed directly through to their bottom line, is not the adoption of a particular contract or documentation for the project but the development of an attitude to the project that focuses on resolving operational problems and dealing with the financial consequences later. For this attitude to be applied throughout the project team some key factors need to be present:

- *Trust.* There must be a feeling that all elements in the team (contractor, client and their supply chains of subcontractor suppliers and consultants) can be trusted.
- *Integrity.* All the parties must act with integrity and be committed to the relationship.
- *Honesty.* If any party is perceived to be acting dishonestly the partnering arrangement will founder.
- *Mutual respect and understanding.* This should speak for itself.

There are other factors that will be needed to make a successful partnership but those in this list have to be present for the system to work. However, the question that should be asked is why these factors need to be present in the context of partnering specifically? Are these not attitudes that should be present in any contractual or working relationship?

Such questions may seem to some to be more than a little naïve. The objective of a commercial organisation is to make profit and it will therefore try to maximise its position at all times, be it client, contractor, subcontractor or supplier. However, does the adoption of bogus, ill-considered, poorly researched and often exaggerated claims for additional payments serve to maximise a position, or does such an attitude merely generate its own subculture and an industry engaged in resolving unnecessary differences and disputes?

There is no doubt that many claims for additional payment are badly recorded, researched and presented, sometimes deliberately so in the hope of camouflaging a poor claim or their own culpable failures. If such claims were presented with honesty and integrity, demonstrated by proper analysis and substantiation, there may be a chance of building the trust and mutual respect and understanding that characterises successful partnering arrangements whatever the contractual arrangement. If those presenting a claim took a reasonable approach to its preparation, then those receiving it might be similarly constructive in their response to it. Conversely, if those receiving claims took a more honest and open approach to their merits, then there should be no need for the claimant to exaggerate its claim in the first place.

These may be pious hopes, but experience suggests that the properly researched and substantiated claim, presented and received with trust, respect and understanding, generates fewer disputes and enables working relationships to continue notwithstanding the need to address claims for change.

In the UK, the best known standard form for a partnering approach is the Standard Form for Project Partnering Contracts, PPC 2000, which was launched by Sir John Egan in that year.

9.3.2   Alliance Contracts

Alliance contracts have many of the same aspects as partnering but are usually on a contract by contract basis with the added ingredient that they usually contain provisions for the sharing of financial gains, or pain, on an agreed basis. Such contracts have been particularly successful in the oil industry and other process type industries where the contractor often undertakes considerable design and management responsibilities on behalf of the client as well as construction of the project. In the UK they have particularly been adopted by Network Rail on its procurement of major infrastructure projects.

Like partnering arrangements, alliance contracts require qualities of openness, trust, integrity, honesty and risk sharing to develop the understanding and respect that are central to a successful project. While the contractual arrangements for such projects are generally a little different to those for projects of a more traditional building and civil engineering nature, the same comments about the benefits to be gained from developing claims for additional payment with these same qualities are as relevant here as with partnering.

In relation to change, a key component of the alliancing approach is how it allows flexibility in allowing the parties to address unplanned changes as they arise during the project. Attitudes of flexibility and innovation in dealing with such events include a flexible approach to addressing their effects. However, a potential disadvantage of this is that it means less certainty as to outcome because that outcome is not the result of a rigid application of contractual terms, rights and valuation methods.

The potentially adversarial allocation of risk between the parties under a traditional form of contract such as SBC/Q, the Infrastructure Conditions or the FIDIC Red Book is replaced in alliance contracts by the sharing of responsibility for dealing with changes and the outcome in terms of risk and benefits that result. It is therefore the performance of the project team as a whole that is in the interests of all members of that team. This prevents each party taking a narrow view of their own self-interests, by reference to the terms of its contract alone. The allocation of risk, and with it fault, under traditional approaches is replaced with a no-blame approach. A result of this is that, rather than such contracts providing for how each party's interests are catered for if a risk event or change occurs, including reference to formal dispute resolution if required, under an alliance agreement the aim is a flexible and amicable resolution. This in turn is intended to promote pro-active management in the equal interests of all parties, rather than each party looking to protect its own self-interest. This is often supported by a 'pain share/gain share' provision in the contract.

Perhaps reflecting an increase in interest in alliancing, the NEC suite of 2017 included an NEC4: Alliance Contract in draft form with a consultation period ending in late 2017. It was published in 2018.

9.3.3 Framework Agreements

In 2005 both the JCT and the NEC added 'Framework Agreements' to their suite of documents to reflect the growing incidence of such agreements, particularly between public sector clients and contractors. The idea of a framework contract is to foster a long-term relationship and understanding leading to more co-operative working. The approach is equally applicable to the client's contracts with both the contractor and its consultants. Historically, consultants have perhaps tended to operate under framework agreements more often than contractors. They are also more popular with subcontractors operating under framework agreements with contractors, particularly in sectors such as housing.

The JCT issued further editions of its framework agreement in 2011 and 2016. The NEC's 2017 suite includes a new framework agreement, the NEC4: Framework Contract.

A criticism of the 2005 edition of the JCT version of the framework agreement was that it did not contain any means for the creation of what it referred to as 'underlying contracts', i.e. the contracts for particular projects, and nor did it include a description of any particular type or programme of work to be awarded to the 'service provider' by the client body. It was therefore not a framework agreement in the sense envisaged by the EU Consolidated Directive that dealt with such agreements from January 2006; rather the JCT agreement had more of the features of a partnering agreement as it covered the working practices to be adopted by the parties to the agreement.

The JCT framework agreement can be binding or non-binding but even if it is binding it is stated to be subsidiary to the underlying contracts, which are to prevail in the event of conflicting or discrepancy provisions. It contains provisions for the dissemination of information, team working, and early warning and risk assessments before entering into the underlying contracts.

Any procedure that can encourage co-operative working is to be welcomed, but the possibility of creating two competing sets of rules governing a contract must be avoided in order to prevent the basic objective being compromised. There is the potential for conflict between, for instance, early warning/risk assessment under the framework agreement (or lack of it) and notice provisions in the underlying contracts. While these might be resolved legally, any confusion could damage the very working relationship the framework agreement is intended to encourage.

9.4   Early Warning Systems

One of the surest ways for trust between parties to be destroyed, or at least to be strained severely, is for one party to be taken by surprise when something unpleasant occurs, particularly where it considers that its contracting party could have warned it of the occurrence of that event. Most construction clients generally view substantial claims for additional payment as something decidedly unpleasant, for obvious reasons. Just as obviously there are occasions when the risk apportionment or events on the project make such claims unavoidable. However, the sooner they are advised, the easier it is likely to prove to resolve or reduce their effects and the less likely they are to result in dispute.

It is, however, generally the case that the cause of substantial claims does not occur without warning. Even problems such as unforeseen ground conditions are often discovered over a period rather than revealed instantaneously. The benefits of bringing such problems to the attention of all members of the project team are obvious. They include that any necessary records can be taken and monitored such that the effects are not the subject of later disagreement. Measures to overcome the problem can be discussed and if not agreed at least implemented with the full knowledge of all concerned. The party bearing the risk of the problem can also start to make appropriate provision in its budgets and internal reporting. Early advice can even mean that steps can be taken to avoid the event entirely, thus involving no additional costs to the contractor or need for a claim against the employer. In such circumstances the chances of being able to agree the financial consequences are much greater and the prospect of avoiding unnecessary dispute is much increased.

The traditional approach to encouraging the early notification of claim events, or potential claim events, has been by way of requirements for notices. These may be accompanied by sanctions, even including the loss of entitlement to claim for that event. Such 'condition precedent' notice requirements are discussed in Section 5.14 of this book. A more immediate approach than a notice may be a requirement for an 'early warning', again perhaps with a sanction for failure to give one. This is the approach adopted by the NEC suite of contracts. Whilst requirements for contractual notice were considered in Section 5.14, some typical notice requirements are outlined here before turning to early warning requirements themselves.

The Infrastructure Conditions provide at clause 53 for the contractor to give notice of any intention to claim additional payments. The provisions of clause 53(2) require the contractor to keep such contemporary records as may reasonably be necessary in order to support any subsequent claim. Clause 53(3) enables the engineer to require any further contemporary records it considers necessary, and provisions for the submission of interim accounts, etc., follow. The provision of such records is intended to make the agreement of the primary effects of an event easier, thus saving the potential secondary effects of time and resources spent on trying to agree them without such records. Other requirements of the contractor in making a claim under clause 53 include in clause 53(4) provision of detailed particulars of the amount claimed and the grounds on which the claim is based. Interestingly, clause 53(5) addresses 'If the Contractor fails to comply with any provision of this Clause …' but it does not make the notice a condition precedent to entitlement. The clause continues:

> … The Contractor shall only be entitled to payment in respect thereof to the extent that the Engineer has not been prevented from or substantially prejudiced by such failure in investigating the said claim.

Thus the penalty in clause 53 is in relation to a failure to provide particulars of the claim and that penalty is in the resulting quantum of the claim, rather than this clause stating a condition precedent to entitlement.

The keeping of contemporary records is no more than a prudent contractor should undertake in any event but the duty to bring the potential claim to the attention of the engineer is stated in the Infrastructure Conditions clause 53(2) to be only as soon as is reasonable and no later than 28 days after the events that give rise to the claim. Such a provision still leaves room for argument as to the trigger date for informing the engineer where the date of happening of the event is not clear or it happens over a passage of time. In such circumstances the point at which the engineer should be notified may not be possible to determine easily or precisely.

The SBC/Q Standard Form contains particular requirements for notices in relation to the consequences of change in clauses 2.27 (in relation to delay to progress) and 4.23 (in relation to loss and expense).

SBC/Q clause 2.27 relates to notice by the contractor of delay to progress. Clause 2.27.1 requires that:

> If and when it becomes reasonably apparent that the progress of the Works or any Section thereof is being or is likely to be delayed the Contractor shall forthwith give notice ….

This requirement therefore contains a prospective requirement to the extent that delay 'is likely', although in practice showing that future delay was 'reasonably apparent' to the contractor in advance, to establish when it should have given notice, may be difficult. The subjectivity of the terms 'reasonably' may also cause some difficulty. Clauses 2.27.2 and 2.27.3 then set out requirements on the contractor to provide particulars of the effect of such notified delays.

The sanction for the contractor's failure to give notice and particulars under these terms lies within clause 2.28 of SBC/Q, where the architect or contract administrator is required to fix new completion dates, as appropriate, on their receipt. Where events have not been notified by the contractor as required by clause 27, they only have to consider such events for an extension of time under clause 2.28.5.1 and this is after completion of the works or the relevant contractual section. The sanction for failure to give notice may therefore be the lack of an otherwise due extension of time until retrospective granting after completion of the works or relevant section.

SBC/Q clause 4.23 relates to notice of claims for loss and expense. It requires that the contractor makes an application for payment of loss and expense arising from the 'Relevant Matters' that are set out in clause 4.24.1. Such notice is required by clause 4.23.1 to be made:

> As soon as it has become, or should reasonably have become, apparent to him that the regular progress has been or is likely to be affected.

Again, the requirement for notice is triggered not just when progress has already been affected, but when it is 'likely to be affected', thus potentially in relation to a matter and/or its effect that has not yet occurred. The subjective requirement that it 'should reasonably have become apparent' to the contractor is a particular area for disagreement. Otherwise, no time limit for the notification or applications for payment under this provision are given and there is again room for argument in any particular circumstances as to when exactly notice should have been given. As to sanction for failure to provide notice, the architect or contract administrator's duty to ascertain, or instruct the quantity surveyor to ascertain, is 'provided always that' the requirements of clauses 4.23.1 to 4.23.3 have been met by the contractor, including the requirement for application under clause 42.3.1. On this basis, if the lack of an application can be said to have been given outside the requirements of clause 4.23.1, then this could arguably be something that can be taken into account in the ascertainment to the extent that part of the loss and expense could have been avoided had application been made as required by the contract.

SBC/Q's requirements for records follow in clauses 4.23.2 and 4.23.3. These are in less specific terms than those under the Infrastructure Conditions and merely require the contractor to submit such information and details of the loss and expense as should reasonably enable a decision on the claim to be made. There are no provisions for the client team to require any contemporary records, or indeed for the contractor to keep such in so many words, although in practice a contractor not keeping such records may well prejudice its claim, given where the burden of proof usually lies in relation to such issues.

These provisions of SBC/Q and the IInfrastructure Conditions relate to notice. Although aspects of them might be hoped to create an early warning, perhaps the most comprehensive provisions specifically for the early warning of potential problems are to be found in the core provisions of the NEC contracts. These are in addition to NEC's requirements for notice as discussed in Section 5.14 of this book. NEC4-ECC clause 15.1 'Early warning' reads as follows:

> The *Contractor* and the *Project Manager* give an early warning by notifying the other as soon as either becomes aware of any matter which could:
>
> - increase the total of the Prices,
> - delay Completion,
> - delay meeting a Key Date or

- impair the performance of the *works* in use.

> The *Project Manager* or the *Contractor* give an early warning to notify the other of any other matter which could increase the *Contractor's* total cost. The *Project Manager* enters the early warning matters in the Early Warning Register. Early warning of a matter for which a compensation event has previously been notified is not required.

This is then followed in clause 15 by requirements for an entry of the warning into the Early Warning Register and early warning meetings to be attended by the Project Manager and the Contractor, together with other people if these two agree and also subcontractors if it will assist. The objectives of such meetings are stated in clause 15 to include the making and consideration of proposals for the avoiding of the potential problem, or reducing its effect, seeking solutions that will bring advantage to all those affected for all concerned and deciding on the actions to be taken.

The proactive nature of these NEC provisions is clear, particularly in how they encourage the early flagging of issues and require the parties to attempt to co-operate to decide on actions to address any matters warned of. The clause culminates in sub-clause 15.4 with the Project Manager instructing any change to the scope of works if needed. These provisions can be particularly seen to serve one of the stated objectives of the NEC contracts, as this was stated in the NEC3 Guidance Notes:

> … the application of collaborative foresight and risk reduction procedures.

A danger with the NEC procedure is that every event that happens or could happen ends up being the subject of early warning notices, meetings and discussions, notwithstanding their perhaps insignificant nature. On the other hand, some NEC projects suffer as a result of failure to apply these provisions and reap the benefits of early intervention. This failure is particularly common where parties are unfamiliar with the procedures and their potential benefits. It is also the case that some parties regard early warning notices as being overused as a way for the other party to protect itself in relation to all events and potential events no matter how trivial, with the result of unnecessary over-burdening of all involved. Clearly, the drafters of these procedures anticipated that a degree of common sense would be applied by all.

Such a system may seem idealistic but it has been operated in practice on numerous contracts to the benefit of the parties. To be successful it needs the same qualities as make partnering and alliance contracting viable and successful, but if the benefits are to be accrued by all concerned there does not seem to be any reason why such an early warning system should not be successfully adopted as a matter of good practice on any project, rather than because the contract is under NEC terms. The gains in the definition and substantiation of claims for additional payment would be to the benefit of all parties.

In terms of sanctions where an early warning is not given, this is expressly provided in clause 63.7 as follows:

> If the *Project Manager* has stated in the instruction to submit quotations that the *Contractor* did not give an early warning of the event which an experienced contractor could have given, the compensation event is assessed as if the *Contractor* had given the early warning.

This is repeated in all the NEC Options. In addition, the definition of Disallowed Cost in NEC Options C, D, E and F, as explained in Chapter 5, includes cost which:

> was incurred only because the *Contractor* did not
>
> …
>
> give an early warning which this contract required it to give.

A potentially unwanted consequence of these sanctions might be that they serve to encourage the contractor to notify all manner of trivial issues through the early warning procedures, just to protect itself against such potential sanction, with the contractor adopting a 'better to be safe than sorry' approach.

9.4.1   Trend Analysis

There will inevitably be matters affecting projects that sometimes cannot be foreseen, particularly, for example, such as unforeseen physical obstructions in groundworks or exceptionally inclement weather, etc. Furthermore, many of the types of events that give rise to claims for additional payment develop progressively. For example, in the construction of new offices on the bank of the Thames in the 1980's early substructure excavation hit some unexpected brick structures and timber piles. These were initially believed to be isolated and the contractor was instructed to break out and remove them. Several weeks later it was realised that the site was on the location of one of the first landing points of early London and the whole foundations had to be redesigned. Similarly, gradual unfolding of the extent of a problem is especially true of many disruption claims where the progress of works is affected by extraneous factors such as information problems, the incidence of substantial numbers of small instructed variations to the works or the impact of works being carried out by other contractors on the site.

In such circumstances the point at which the contractor should give notice of the problem and its effects on the programme, and ensure that it is keeping sufficient records to enable a properly supported claim to be made, may be difficult to assess. Typically the contractor's site management will be fully occupied with the management of the project and a particular problem, although manifest, may be allowed to escalate before it is subject to the attention it requires in terms of giving early warning, contractual notice and keeping records of resources expended. Even where the formal requirements are met, the use and potential benefits of early warning meetings and actions as required by the formal requirements of the contract may be somewhat restricted.

In its purest sense, trend analysis is a project management technique that uses historical results in a mathematical model that can then be used to predict future outcomes based on those historical results. Given the one-off nature of many major projects that might particularly benefit from such techniques, in that purest form trend analysis is rarely applied to engineering and construction projects. However, to provide a safeguard against the escalation of problems before they are acknowledged and notified, the institution of simple trend analysis systems can be of tremendous benefit. Such systems were developed in the mechanical engineering industry to monitor production on site, often by use of crude indicators such as welds per man per day, pipe erection per gang per day or metres of insulation applied per team per day. The data would be available for different sections of the works and would be monitored against the expected outputs with any significant variances investigated to establish the cause. The monitoring of such trends might then serve to give the basis for an early warning of a claim event and its consequences.

It is generally unusual to find trend analysis routinely undertaken in such terms on building and civil engineering sites, but the adoption of such simple analysis, coupled with the type of early warning system envisaged by such as the NEC contracts, could be very beneficial. In addition to highlighting problems so that they can be warned of, notified, addressed and recorded at an early stage, it may also provide a powerful tool for the simplification and resolution of subsequent claims for additional payment. In particular, benefits might be realised in relation to claims for disruption, where Chapter 6 of this book highlighted the use of a measured mile analysis as a particularly useful approach to quantification.

9.5   The Claims Industry

In the last edition of this book, we said this:

> Mention was made earlier in this chapter of the sometimes common perception of the construction industry as an industry populated by litigious companies bent on pursuing exaggerated and often ill-founded claims for additional payment. The validity of that image is questionable but the fact that it exists at all is regrettable and a matter of concern for everyone connected with the industry.

That was written in 2008. Ten years on, it is suggested that the UK construction market has since improved in this regard. Contracts such as the NEC and the increased use of partnering and alliancing approaches seem to have been beneficial influences in this regard, as has a recognition of the high costs of formal dispute resolution. However, internationally these words still appear valid in many parts of the world.

9.5.1   The Numbers Game

One of the reasons why many seemingly exaggerated claims are submitted is the popular perception among many personnel in the construction and engineering industries that the amount paid for a claim is usually a proportion of the amount claimed. There has long existed a rule of thumb among some practitioners that said 'good claim two-thirds, bad claim one-third'. On this basis, it was expected that the more that was claimed the more that would be paid. At its crudest it would mean that the presenter of a bad claim would set its amount at three times its actual loss, on the basis that it might end up settling at that actual amount.

Such attitudes may seem naïve and simplistic but, like many such 'urban myths', it has its foundation in an element of truth, in that statistically many bad claims do seem to settle at around one-third of the claimed value. The perception of this level of settlement, however, seems often to be based on the experience of the settlement of claims that have been the subject of some controversy and dispute. It is always the problem children that are remembered. In fact there is much experience of higher levels of settlement, often without any substantial dispute, where claims for additional payment are fully and properly notified and supported.

It may be that the exaggeration in the submission of many claim quantifications is itself the cause of the problem. After all, if you multiply something by three times what it is worth you would only expect to receive one-third of the submitted amount. This may not mean that had you submitted a proper claim for what it was really worth in the first place, you would still have ended up at the same amount.

On the other hand, those receiving claims may also have come to assume that any claim received from a contractor will have been exaggerated to the same proportion. Furthermore, it may be perceived that a certifier approving a substantial proportion of a claim as presented is not doing his or her job properly. In countries where public sector clients need close scrutiny on their employees to avoid the perception that there may have been some act of corruption, the pressure on a certifier to greatly reduce a claim can be immense, no matter what the quality of the claim and its substantiation. Accordingly, it may be a brave contractor indeed that presents a claim valued at the amount it reasonably considers it is entitled to, with no allowance for 'fat' to be trimmed off it.

9.5.2   The First Number

A further cause of difficulty is often the unrealistic nature of the first amount that is claimed as being the value of a claim for additional payment. The effect can be the same for both the claimant and the recipient, as the figure may be entered into financial forecasts by both, either as an anticipated income or an anticipated liability. If the figure is unduly optimistic in either case, too high for the claimant or too low for the recipient, the result will be a figure in the financial forecasts that can become difficult to alter without causing significant problems.

This can be a serious problem where the amounts assessed are significant in terms of the trading position of the company, and subsequent alteration may result in more than just a red face in admitting the initial assessment was incorrect. For contractors in such circumstances, it is not uncommon to find that the figure incorporated into the financial accounts, and an unwillingness or inability to 'write it down' is the driving force behind the claim being pursued, and not its merits and proper worth.

On the other hand, it is not uncommon to see claims that should have been settled run on for no better reason than that the party advising on that claim gave an unrealistically low report or recommendation in relation to it. If the party bearing the risk of that claim then makes a low provision for it in its books or budget, it may become harder to shift that position the longer the matter drags on. Where the advisor has a vested interest in that claim, such problems can be particularly acute. This is a common criticism of contracts such as those of FIDIC, where the Engineer has the role of certifying the value of claims that it caused, such as through its issuing of late or incorrect approvals, instructions or design information.

9.5.3   The Claims Industry and Experts

There is often a further perception that there is a 'claims industry' within the construction industry that serves to perpetuate the generation and negotiation of claims and litigation and that the UK and USA have been particularly good at exporting that part of their service industries. This perception is also possibly overplayed and given more credence at times than is justified. There is no doubt that the incidence of poorly researched and badly supported claims is not uncommon and serves to perpetuate this perception.

However, the UK has actually achieved much in exporting good practice in the preparation of construction claims. The most obvious example of this is the SCL Protocol, which has been highlighted several time in this book as being applied and referred to extensively around the world. The reality is that whilst many other jurisdictions may not have a 'claims industry', they have long had a mixed quality in the preparation of construction claims and in many cases the importation or development of a 'claims industry' has only improved their practices.

Happily for the UK industry and its image, recent years have seen positive steps to control and reduce some of the worst aspects of the abuse of the contract system. Perhaps the antidote to the claims industry is the expert witness industry. The book *The Expert Witness in Construction*[1] contains a detailed analysis of the role and duties of expert witnesses in relation to construction disputes. Whilst 15 years ago many experts were seen by some as no more than 'hired guns' in the UK, and that is still the case in some countries, in the UK the quality of experts and their evidence has improved greatly.

Professional institutions have introduced codes of practice, to be followed by members providing services as experts in connection with disputes, aimed at ensuring the independence and veracity of the expert opinions they provide, and thereby combating the 'hired gun' syndrome where an 'expert' is used merely to support his or her client's case regardless of its content. In this respect the Royal Institution of Chartered Surveyors' practice statement, entitled Surveyors Acting as Expert Witnesses, is a particularly detailed and rigorous code for members of that institution.

In the UK courts, these moves have been reinforced by the requirements for experts and assessors introduced by Part 35 of the Civil Procedure Rules in 1998. These rules apply to litigation and introduce the fundamental concept that the expert's duty is to the court rather than his client or instructing solicitors. Outside litigation and the decision of the England and Wales commercial court in the case of *National Justice Compania Naviera SA v Prudential Assurance Co Ltd* [1993] 2 Lloyds Rep 68 ('*The Ikarian Reefer*') provided similar guidance for experts in arbitration.

The consequences of failing to adhere to such requirements can be painful, as demonstrated in the case of *Gareth Pearce v Ove Arup Partnership Ltd and Others* (2002) IPD 25011, where the Judge considered that the evidence given by the claimant's expert fell far short of the standards of objectivity required by the court. The expert had reportedly adopted the role of supporting his client's case regardless and 'any point which might support that case he took'. The expert in this particular case had apparently been central to the issue proceeding to court and the Judge considered that, in cases where the expert had seriously breached his duties to the court, he should be referred to his professional body. It is understood that in this case the relevant professional body decided, on investigation, that there was no case to answer. However, the warning is clear that such action is not only possible but will be taken in appropriate situations. Given how expensive formal dispute resolution proceedings can be, where they result because of the

---

1 Horne R and Mullen J (2013). *The Expert Witness in Construction*. Wiley, Oxford, UK.

over-zealous activities of an advisor, it is easy to see how the secondary consequences of change could have been minimised by taking good, rather than bad, advice.

Such actions in the courts, and the practice requirements of professional institutions, should serve as a lever towards raising the standards of supporting analysis and objectivity in litigation and other dispute resolution forums. There is a constant pressure now for experts giving evidence in court to have 'qualified' as experts in addition to being competent in their own field, and professional bodies for expert witnesses are now well established. In one way, such developments are to be welcomed as providing ever more realistic prospects that experts will act independently and objectively, while at the same time there is a very real danger that the role of the expert becomes that of a profession in itself rather than something that a well-qualified and experienced member of a primary profession can undertake after properly considering the duties required of him or her.

However, the reasons why experts should apply such standards can be argued to apply equally to the preparation and analysis of claims generally. As demonstrated in the *Pearce v Ove Arup* judgment, the consequence of an expert's lacking objectivity can be the expenditure of large amounts of wasted costs, and probably some not inconsiderable inconvenience, in undertaking a futile court action. To the costs of such action can be added the detrimental effects on relationships. The same consequences will often result from poorly supported and substantiated claims that will fail, partially or entirely, only after the expenditure of time, cost and effort in their pursuit.

The poorly supported and substantiated claim does not serve anyone well. One approach to making the best use of an expert witness in this regard is early appointment in the role of 'expert advisor'. In this role the expert provides early advice on the preparation of a claim, or the defence to a claim, before it is submitted. This allows others to prepare the position and submission under advice as to what the expert would be willing to support under oath or affirmation in a hearing in terms of for example, how disruption should be quantified, or an extension of time calculated and what that tribunal is likely to accept.

9.5.4 Resolving Disputed Claims

In some jurisdictions, recent years have seen a dramatic change in the means by which the construction industry tries to resolve disputed claims. Both the introduction of adjudication as a result of the Latham initiatives and the move away from arbitration, at least for UK domestic contracts, have resulted in marked changes in approach there. As noted before, 14 other jurisdictions have followed the UK's lead in passing security of payment legislation including statutory adjudication. Elsewhere that method of dispute resolution is also gaining currency.

There is now a broad 'alternative dispute resolution sub-industry' and a not much narrower 'adjudication sub-industry' within the construction industry worldwide, with the stated aim of providing fast and economical resolution of disputes, on an interim if not permanent basis. The discussion of adjudication as adopted by the UK construction industry is outside the remit of this text, but it can be said that poorly supported and substantiated claims are no more appropriate to adjudication than to other forms of dispute resolution. On the other hand, there is a perception that some claimants believe that this forum might give it the chance to 'get away with' a poorly presented and substantiated claim where the procedure is of a 'quick and dirty' nature.

There is no doubt that some within the construction industry view adjudication as a means of obtaining a result from a poorly substantiated claim that could not otherwise be obtained in other forums, due to the generally short timetables applied to adjudication. This should not be so if adjudicators have the courage to reject claims submitted without a reasonable level of support and substantiation, and while the timetable for adjudication might be restricted much preparation work can be undertaken before any adjudication notice is given.

Adjudication should not be allowed to encourage the submission of poorly presented and substantiated claims as it will otherwise encourage disputes rather than offer the faster and more economical determination it was intended to provide.

The move away from arbitration in UK dispute resolution is, in the authors' opinion, at least partly due to the perception of arbitration as an overlong and expensive process. 'Old school' practitioners often complain that arbitration has changed in the last 30 years to become no different to (and perhaps worse than) litigation in its formality, time scales and costs, without the types of controls on timetable and costs that the UK courts in particular now impose. Certainly many arbitrations of disputes on large projects had extremely long time scales and were no less expensive than the same proceedings would have been in court and in fact were probably much more expensive. Whatever the cause, there is no doubt that the move to litigation as the ultimate dispute resolution process in contracts has reinforced the need for quantum assessments to be thorough and well supported.

9.6 Summary

Claims for additional payment are almost always going to arise on anything other than very small construction and engineering projects in one form or another as a result of their nature and the contractual regime for risk apportionment and management. This chapter has considered some aspects of the management of projects and claims that might minimise the consequences of change. These consequences have been categorised as either primary effects, such as additional costs or delays on the works, or indirect effects, such as claims becoming contentious or ending in formal dispute resolution procedures with the related time and costs.

Initially, it is contract drafting that can work to minimise the consequences of change. Choosing the right standard form of contract for the project, without prejudiced and one-side amendments, preparing the documentation well, and ensuring that project personnel all understand the relevant provisions and properly apply them are key issues. If those contract provisions include requirements for notices and early warnings, then these can be especially beneficial if applied in such a way as to achieve their true potential.

Moves to more constructively manage and minimise the effects of change have been led in the UK by, for example, the Latham and Egan Reports. The advocacy of partnering and alliancing approaches, in place of traditional and more adversarial contractual relationships, has seen these increasingly adopted and standard forms published for partnering and alliancing. The adoption of security of payment legislation for the industry can particularly help.

Perhaps the simplest management strategy, for contractor and employer, is that a thorough, informed and realistic appraisal of claims at the earliest possible time is the surest way of preventing the consequences of change from becoming distorted and a source of aggravation and costs at a later stage. This applies to both those presenting claims and those responding to them. It was stated at the start of this book that the evaluation of claims under engineering and construction contracts is not an exact science. There is no single approach that will apply to the circumstances of every type of claim. The variables include items such as: the relationship between the parties; the particular facts of each claim; the available records; the size of the claim; the contract terms; and the law of the relevant jurisdiction. However, it is hoped that this book will have added to its readers' knowledge in terms of alternative approaches and best practice.

Appendix A

Example of Financial Accounts

The figures below are an example of those typically found in the published company accounts of a construction contracting company in the UK, and demonstrate the brevity and high level of such financial information. The figures are fictitious although the format is based on actual published accounts from a number of companies.

Consolidated profit and loss account summary

	Year ended 30 September 2003 £ ,000	Year ended 30 September 2002 £ ,000
Turnover	55,876	56,176
Gross profit	1,097	(2,897)
Administration expenses	2,435	2,237
Operating profit	(1,338)	(5,224)
Investment income	2,099	276
Interest payable, etc.	(217)	(453)
Profit before taxation	544	(5,401)

The object of this account is to give a summary of the company's trading results over the financial period, with figures from the previous period for comparison. Elsewhere in the published accounts will be a consolidated balance sheet showing, again at high level, the source of funds in the business at the reporting date and the manner in which they are used.

There will also be notes explaining exceptional items and the trading experienced in that period. The figures used as an example above suggest a very poor year ending 30 September 2002, followed by a recovery in the year to 30 September 2003, by which time the annual loss has been converted to a profit of £544,000 against a loss of almost ten times that figure in the previous year. A closer examination of the figures shows a significant drop in interest paid against a sudden increase in investment income, while turnover remains static.

The accounts themselves do not explain why this pattern has emerged, but the Director's Report contained with the accounts should do so. Perhaps the loss in the year to 30 September 2002 was generated because the firm was engaged in a project with no revenues until completion, and the sudden increase in investment revenues in the year

Evaluating Contract Claims, Third Edition. John Mullen and R. Peter Davison.
© 2020 John Wiley & Sons Ltd. Published 2020 by John Wiley & Sons Ltd.

to 30 September 2003 marks the completion of that project and commencement of the revenue stream.

Only further explanation will allow the figures to be understood in the proper context but they represent a high level view of the firm's activities and results. If the company has a property development and investment division, a small projects division and a plant hire company as well as the main contracting business, all trading under the one set of accounts, then it is not possible to identify the relevant figures for each division or the profitability and overhead cost of each division.

Appendix B

Example of Management Accounts

If the financial accounts contained in Appendix A incorporate the activities of the main contracting division of the company, in addition to the property development and investment, small projects and plant hire divisions, then there will be a set of management accounts that detail the profit and loss account for the main contracting activity in some detail. Typically the account would contain details such as the following:

		Year ending 30 September 2003 £ ,000
Sales		43 876
Cost of sales		
Salaries and wages	5,653	
Suppliers and subcontractors	28,764	
Plant and equipment	4,343	38,760
Gross profit		5,116
Other income		
Rent receivable		65
Gross profit and other income		5,181
Salary, pension and benefits cost		
Directors		465
Administration staff		751
Other employee costs		77
Administration costs		
Rent and rates		265
Light and heat		92
Repairs and renewals		60
Telephone		105
Insurances		642
Printing, stationery and advertising		51
Computer expenses		88

Evaluating Contract Claims, Third Edition. John Mullen and R. Peter Davison.
© 2020 John Wiley & Sons Ltd. Published 2020 by John Wiley & Sons Ltd.

	Year ending 30 September 2003 £ ,000
Travel and entertaining	42
Motor expenses	168
Legal and professional	354
Audit and accountancy	114
Subscriptions and donations	17
Marketing and promotions	81
Depreciation	46
Bad debts	143
Bank charges and interest	207
Loan interest	<u>116</u>
Net profit/(loss) before management charge	1,297
Management charge	<u>465</u>
Net profit/(loss) before dividend and tax	<u>832</u>

Similar sets of profit and loss accounts will exist for the other company activities as part of the information required for effective management. In total they will comprise the consolidated accounts for each period presented in the financial accounts.

The above accounts, if related to the contracting activity alone, may be useful for the analysis of the overhead charges related to that activity. It should, however, be noted that, as with most accounts, it will be necessary to understand the trading activity that lies behind the figures.

Table of Cases

The following abbreviations of reports are used:

AC	Law Reports Appeal Cases
AdjLR	Adjudication Law Reports
ALJR	Australian Law Journal
All ER	All England Law Reports
BLR	Building Law Reports
CILL	Construction Industry Law Letter
CLJ	Cambridge Law Journal
Com LR	Commercial Law Reports
Con LR	Construction Law Reports
Const LJ	Construction Law Journal
CSOH	Court of Session, Outer House
EWHC	England and Wales High Court
Ex	Law Reports Exchequer Cases
IPD	Intellectual Property Decision Reports
KB	Law Reports Kings Bench Division
LGR	Local Government Reports
Lloyds Rep	Lloyds List Law Reports
LT	Law Times Reports
PIQR	Personal Injuries Quantum Reports
UKHL	United Kingdom House of Lords
WLR	Weekly Law Reports

Evaluating Contract Claims, Third Edition. John Mullen and R. Peter Davison.
© 2020 John Wiley & Sons Ltd. Published 2020 by John Wiley & Sons Ltd.

Index

Evaluating Contract Claims, Third Edition. John Mullen and R. Peter Davison.
© 2020 John Wiley & Sons Ltd. Published 2020 by John Wiley & Sons Ltd.

Printed and bound by CPI Group (UK) Ltd, Croydon, CR0 4YY

27/10/2024

14580361-0004